AF441842

Advances in Experimental
Medicine and Biology

970

For further volumes:
http://www.springer.com/series/5584

Preface

Synapses are sites of a specialized cell-cell contact between neuronal cells and represent the major structure involved in chemical neurotransmission in the nervous system. It is widely believed that glutamatergic synapses are important loci for modifying the functional properties of CNS networks, possibly providing the basis for phenomena collectively referred to as "learning and memory". Given their importance, it is not surprising that enormous efforts are being made to understand the formation, structure, function and regulation of glutamatergic synapses. To date, significant progress has been made in our understanding of their ultrastructure, molecular composition, and physiological properties, as well as the principles of how these synapses are initially assembled and "plastically" modified.

The term synaptic plasticity covers many different aspects of use-dependent synaptic modifications and is commonly used in a broader sense describing aspects of synaptic signal transmission as well as structural alterations in the molecular make-up of the synapse related to synaptic signaling events. The capacity of the nervous system to modify its own organization is remarkable; plastic changes can occur as a consequence of many events, including the normal development and maturation of the organism, the acquisition of new skills ('learning') and after brain damage. This response specificity is not always intrinsic to neurons; rather, it can develop as a consequence of interaction with the environment and thus involves information processing and memory storage.

Disturbances of synaptic and neuronal plasticity ultimately result in diseases affecting cognitive functions and it has become increasingly clear during recent years that synaptopathies – disruptions in synaptic structure and function – in many cases are the major determinant of many brain disorders. These diseases and related animal models include Alzheimer's, prion diseases, Down's syndrome, Huntington's or Parkinson's diseases as well as schizophrenia and autism spectrum disorders that almost ultimately result in disturbed plasticity processes. In accord, it is becoming increasingly clear that studies of synaptic plasticity and memory formation are critical for understanding the initial stages of these disorders. At an early stage in most of these diseases no structural damage exists but rather an impairment of synaptic function. Interventions aimed to preserve or even restore synaptic function will probably delay the onset or might even provide a cure for such disorders. A general strategy to treat these types of diseases can therefore be plausibly based on knowledge, how to maintain the plastic properties of neurons in the adult and aging brain.

Crucial technological advancements have recently accelerated progress in our understanding of synaptic processes, five of them are listed here: (1) Live-cell imaging has provided essential constraints regarding the kinetics (rate constants) and spatial organization of signaling pathways, (2) the development of next generation sequencing allows individual transcriptome profiling, (3) quantitative synaptic proteome profiling of normal and disease brain has established protein interaction networks databases for signaling pathway analysis, (4) optogenetic tools are available to study neuronal function in vivo, (5) and finally, progress in computer simulation and neuroinformatics has been crucial for improving the temporal and spatial scale of synaptic plasticity models, because simulating large spatial structures for long durations with high resolution requires trillions of calculations.

In higher mammals the majority of brain glutamatergic excitatory synapses is found on the principal neuron of the cortex and hippocampus, the pyramidal cell. Pyramidal cells are characterized by a complex dendritic cytoarchitecture harboring approximately 10^4–10^5 synaptic contact sites with other neurons. It is estimated less than 1% of all synaptic contacts of cortical pyramids is concerned with the wiring to subcortical areas, implying that the predominant synapse of the mammalian telencephalon is concerned with input from a closely related neuron in terms of cell lineage, morphology and functional characteristics. This fact is mainly emphasized because our knowledge about synaptic plasticity of this type of synaptic input is still very sparse.

We have focused our attention in this book mostly on postsynaptic molecular mechanisms because a lot more knowledge exists with respect to this side of the synapse especially with respect mechanisms underlining synaptic dysfunction and synaptopathies. The purpose of this book is to summarize this knowledge and to provide insights into the most recent developments in the field including the major technological advancements in recent years. The first part of the book concerns structural plasticity at the pre- and postsynaptic scaffold, the molecular dynamics of the synapse and synapto-dendritic plasticity in development and learning. In the second part the book includes chapters on synapse-to-nucleus communication and synaptic dysfunction and synaptopathies. Finally, we want to particularly thank the authors for their contribution. We are very happy that we could convince leading experts in this field to provide a detailed account of the most exciting recent developments.

Michael R. Kreutz
Carlo Sala

Contents

Part IV   Synaptic Dysfunction and Synaptopathies

Part I
Molecular Organization of the Pre- and Postsynaptic Scaffold

Chapter 1
Glutamate Receptors in Synaptic Assembly and Plasticity: Case Studies on Fly NMJs

Ulrich Thomas and Stephan J. Sigrist

Abstract The molecular and cellular mechanisms that control the composition and functionality of ionotropic glutamate receptors may be considered as most important "set screws" for adjusting excitatory transmission in the course of developmental and experience-dependent changes within neural networks. The *Drosophila* larval neuromuscular junction has emerged as one important invertebrate model system to study the formation, maintenance, and plasticity-related remodeling of glutamatergic synapses in vivo. By exploiting the unique genetic accessibility of this organism combined with diverse tools for manipulation and analysis including electrophysiology and state of the art imaging, considerable progress has been made to characterize the role of glutamate receptors during the orchestration of junctional development, synaptic activity, and synaptogenesis. Following an introduction to basic features of this model system, we will mainly focus on conceptually important findings such as the selective impact of glutamate receptor subtypes on the formation of new synapses, the coordination of presynaptic maturation and receptor subtype composition, the role of nonvesicularly released glutamate on the synaptic localization of receptors, or the homeostatic feedback of receptor functionality on presynaptic transmitter release.

Keywords BMP signaling • Development • Glutamate receptors • Neuromuscular junction • Wnt signaling

U. Thomas (✉)
Leibniz Institute for Neurobiology, Brenneckestr. 6, 39118 Magdeburg, Germany
e-mail: thomas@ifn-magdeburg.de

S.J. Sigrist
Genetics Institute of Biology, Freie Universität Berlin, Takustr. 6, 14195 Berlin, Germany
e-mail: stephan.sigrist@fu-berlin.de

M.R. Kreutz and C. Sala (eds.), *Synaptic Plasticity*,
Advances in Experimental Medicine and Biology 970,
DOI 10.1007/978-3-7091-0932-8_1, © Springer-Verlag/Wien 2012

1.1   Introduction

Excitatory synapses categorized as "glutamatergic" are no less heterogeneous than the many types of neurons that use glutamate in synaptic transmission. In-depth studies on a variety of glutamatergic synapses from different species are thus required to elucidate both general as well as subcategory-defining principles of synapse assembly, function, and plasticity. L-glutamate is the primary transmitter not only at the vast majority of excitatory synapses in the vertebrate CNS but also at arthropod neuromuscular junctions (NMJs). Concerning the latter, illuminative physiological, pharmacological, and ultrastructural analyses have been performed on NMJs from various crustacean and insect model organisms. Undeniably, *Drosophila* has reached a pole position in this respect, mainly because of its unbeatable genetic accessibility. Preceded by pilot studies from Jan and Jan (1976a, b), thorough electrophysiological and morphological studies on NMJs from both wild-type and excitability mutants paved the road for intensive research on this model system. Various aspects of synapse biology such as pre- and postsynaptic assembly, activity-dependent and homeostatic plasticity, or disease-related synaptic dysfunctions are addressed at the larval NMJ. It in fact turns out that many of the underlying molecular mechanisms can be assorted to well-conserved proteins.

In its main part, this chapter is centered around ionotropic glutamate receptors (GluRs) at NMJs. As for studies on neurotransmitter receptors in other systems, findings on junctional GluRs are multifaceted and therefore allow us to bring up a number of current issues in cellular neurobiology. We start out introducing the system, including a brief overview on various forms of plasticity that regulate its structural and functional properties.

1.2   Basic Morphological and Functional Features of Glutamatergic Nerve Terminals at Larval NMJs

The pattern of motoneuron innervation on *Drosophila* abdominal body wall muscles is well-defined and rather stereotypic across segments (Keshishian et al. 1996). A surprising level of complexity is added by the fact that each muscle is innervated by up to four neurons (Fig. 1.1a), which differ by their nerve terminal morphology and transmissive properties. While a few motoneurons are primarily specializing in the release of neuropeptides or catecholamines on subsets of muscles, all muscles receive input from axon terminals of glutamatergic motoneurons (Jia et al. 1993). Such terminals form arrays of boutons, and each bouton is equipped with several, quite evenly spaced active zones (AZs; Fig. 1.1b–c) (Meinertzhagen et al. 1998; Reiff et al. 2002; Dickman et al. 2006), i.e., presynaptic membrane specializations designated for synaptic vesicle release. AZs are commonly associated with protein-rich, electron-dense cytomatrices, which at *Drosophila* synapses typically occur as T-shaped dense bodies (T-Bars;

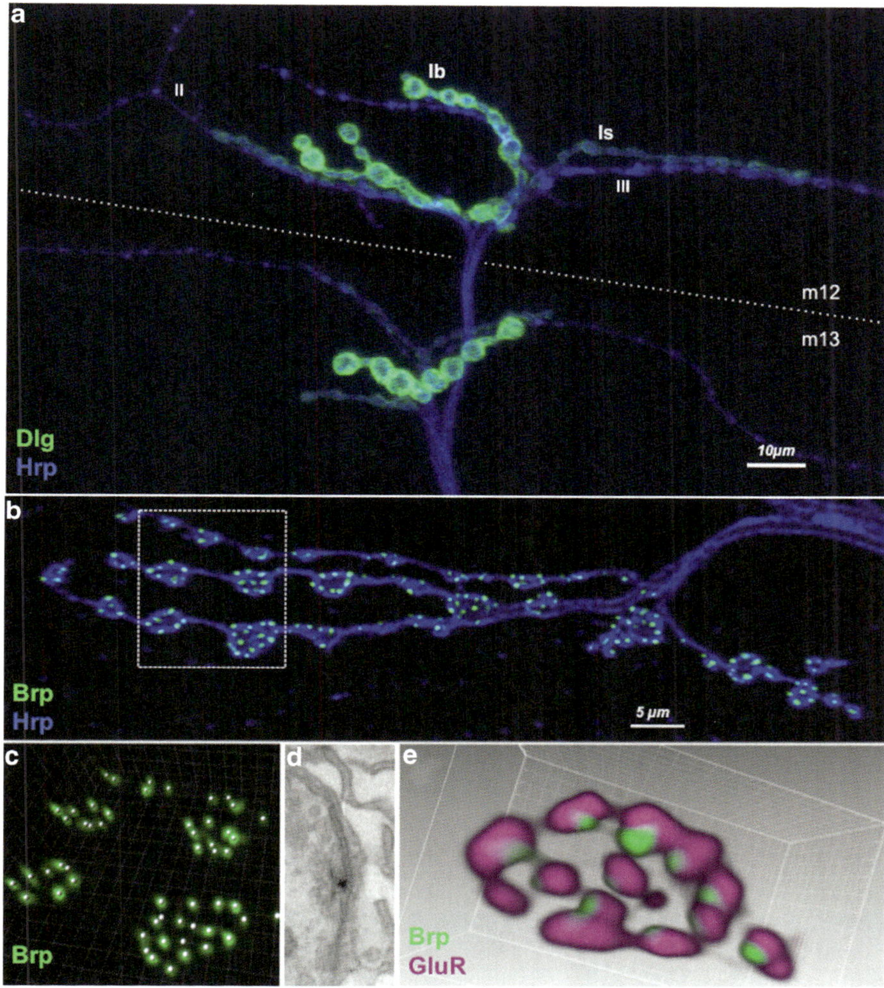

Fig. 1.1 Larval NMJs at various levels of organization. (**a**) Motor nerve terminals on muscle fibers 12 and 13 (*above and below the dotted line*). Type Ib and Is nerve terminals are primarily glutamatergic, whereas type II boutons release octopamine (and glutamate). Peptidergic type III terminals are restricted to muscle 12. A neuronal surface marker (Hrp) was used to visualize all nerve terminals. The scaffold protein Dlg is abundant around type Ib boutons and is also present at type Is boutons, yet at much lower levels. (**b**) Branches of type I nerve terminals on muscle 6, stained for the Hrp-epitope and for the presynaptic marker protein Brp. (**c**) 3D representation of Brp-positive sites corresponding to the delineated region in B. Most Brp-spots correspond to T-bar-shaped dense bodies. (**d**) Electron micrograph displaying a single synaptic contact including an AZ with a T-Bar. Note the clustering of SVs around the T-Bar. (**e**) 3D reconstruction from a confocal stack of a type Ib bouton containing several synaptic contacts that were stained for Brp and the GluR subunit IID

Fig. 1.1d) (Zhai and Bellen 2004; Owald and Sigrist 2009). The presence of a T-Bar indeed indicates a ready for use release machinery and most, though not all AZs of larval NMJs, are decorated by a T-Bar (Atwood et al. 1993; Wichmann and Sigrist 2010).

Each AZ, together with an opposing postsynaptic receptor field (Fig. 1.1e), is referred to as a synaptic contact or simply as a synapse, and a few hundred of these are present at full-grown terminals (Please note, however, that in the literature, the term "synapse" is sometimes used to assign entire NMJs.) Mature wild-type boutons of type 1s motoneurons range in diameter from 1 to 3 μm, thereby harboring 7 AZs on average, whereas boutons of type 1b motoneurons typically are 1–7 μm in diameter and can harbor more than 40 AZs (Johansen et al. 1989; Atwood et al. 1993). Next to this morphological diversification, type 1s and 1b terminals, which are found side by side on many muscles, show meaningful differences regarding synaptic output. The threshold to elicit spiking activity appears higher in type 1s than in type 1b motoneurons, and the former also display a marked delay between stimulation and the appearance of first spikes (Kurdyak et al. 1994; Choi et al. 2004; Schaefer et al. 2010). At the NMJ, however, type 1s terminals evoke stronger postsynaptic responses than type 1b terminals, in agreement with the findings that (1) their AZs display a higher probability to release synaptic vesicles (SVs) upon arrival of an action potential (Atwood et al. 1997) and (2) their SVs are of much greater volume, giving rise to larger quantal size, i.e., postsynaptic current per fusion event (Karunanithi et al. 2002). Type 1b terminals in turn easily facilitate upon repeated stimulation (Kurdyak et al. 1994; Lnenicka and Keshishian 2000), reflecting short-term plasticity in a way typical for synapses with low release probability (Thomson 2000). By analogy to the different types of motoneurons innervating crayfish muscles, type 1b and 1s motor nerve terminals have been referred to as tonic or phasic inputs, respectively (Kurdyak et al. 1994). This dichotomy likely contributes to the versatility of locomotor behaviors (Schaefer et al. 2010). Most remarkably, by using a truly physiological behavioral paradigm, namely, hunger-induced increase in larval locomotion as a means to search of food, Koon et al. (2011) could recently demonstrate cross talk between different motoneuron populations concerning their structural and functional plasticity. Specifically, they showed that starvation-triggered locomotion depends on octopamine release from type II motoneurons, which in turn promotes outgrowth of type I glutamatergic motor nerve terminals, thus revealing a case of metaplasticity at larval NMJs (Sigrist and Andlauer 2011).

1.2.1 Morphogen-Dependent Control of NMJ Development and Plasticity

The formation of NMJs first becomes manifest when filopodia from a motoneuronal growth cone intermingle with filopodia (myopodia) protruding from the respective target muscle (Ritzenthaler et al. 2000; Ritzenthaler and Chiba 2003). This initial event is followed by the transition of the muscle-attached, flattened axonal growth cone into the bouton-forming motor nerve terminal (Yoshihara et al. 1997). Size matching of the terminals according to the size of their respective target muscles becomes evident shortly after NMJ formation (Nakayama et al. 2006). Challenged by rapid muscle growth during the larval phase, NMJs expand by both elongation and branching of their motor nerve terminals, thereby continuously forming new boutons (Gorczyca et al. 1993; Schuster et al. 1996a; Zito et al. 1999). Concurrently, new synaptic contacts are implemented on both newborn as well as preexisting boutons (Rasse et al. 2005), thus resulting in synaptic strengthening according to muscle size. This process involves retrograde as well as anterograde and autocrine signaling by morphogens, and most notably, it is subject to plasticity. In the following, we briefly elaborate on these aspects.

1.2.1.1 BMP Signaling

A canonical bone morphogenetic protein (BMP) pathway is employed for retrograde control of NMJ growth and function (Marqués and Zhang, 2006). The BMP-7 homolog Glass bottom boat (Gbb) is secreted by muscles to activate presynaptic BMP/TGF-β receptors which then phosphorylate the R-SMAD protein Mothers against decapentaplegic (MAD) (McCabe et al. 2003). Phospho-MAD, upon association with its co-SMAD Medea, shuttles to the nucleus to exert its role as a transcription factor. Recently, expression of the Rac GTPase exchange factor Trio, a regulator of the nerve terminal cytoskeleton, was found to be strongly upregulated by phospho-MAD (Ball et al. 2010). This finding thus provides a mechanistic link between Gbb-triggered synapse-to-nucleus signaling and its growth promoting effect on NMJs (Fuentes-Medel and Budnik 2010). In fact, mutations that block the Gbb pathway cause a substantial reduction in the number of boutons (Aberle et al. 2002; Marqués et al. 2002; McCabe et al. 2003, 2004). Consequently, several factors that attenuate Gbb signaling were identified based on mutant NMJ overgrowth phenotypes. Consistent with the widely observed endocytic control of TGF-β signaling (Chen 2009), most of them have been implicated in the local trafficking of the TGF-β receptors, including their passage through signaling endosomes (Sweeney and Davis 2002; Wang et al. 2007; O'Connor-Giles et al. 2008; Rodal et al. 2008; Kim et al. 2010; see also Bayat et al. 2011 for a review on disease-related aspects of BMP signaling at NMJs). Postsynaptic downregulation of the Gbb signaling pathway involves CIP4, a Cdc42-dependent regulator of membrane-associated actin polymerization, that restricts the release of Gbb, a function that is in turn counteracted by the Cdc42GAP Rich (Nahm et al. 2010a, b). Interestingly, postsynaptic Ca^{2+}/calmodulin-dependent kinase II (CaMKII)-activity, triggered by Ca^{2+}-influx through GluRs, may also act as a negative regulator of retrograde Gbb signaling (Haghighi et al. 2003), and it is therefore tempting to speculate that CaMKII might exert this role by activating CIP4 (Nahm et al. 2010a).

The Gbb signaling pathway has also been implicated in synaptic homeostasis, which is expressed by increased presynaptic glutamate release in response to reduced postsynaptic activity (Haghighi et al. 2003; Goold and Davis 2007). Both decreased muscle membrane excitability (Paradis et al. 2001) and reduced postsynaptic sensitivity to glutamate (Petersen et al. 1997) were initially found to elicit this form of plasticity with little or no effect on junctional growth. Using the glutamate receptor antagonist Philanthotoxin-433 (PhTX) at subblocking concentrations on semi-intact larval body wall muscle preparations, Frank et al. (2006) could demonstrate that the compensatory upregulation of presynaptic release occurs gradually within a few minutes, i.e., on an unexpectedly fast time scale. Interestingly, evoked neuronal activity is not required here, suggesting that integration of relatively few spontaneous release events (measured as miniature excitatory postsynaptic potentials or currents, mEPSPs, or mEPSCs) is sufficient to elicit a retrograde signal that triggers a rapid compensatory upregulation of release. A screen for genetic modifiers has led to the identification of factors acting during the acute phase of synaptic homeostasis including the presynaptic calcium channel cacophony (Cac) and the schizophrenia susceptibility factor dysbindin (Frank et al. 2006, 2009; Dickman and Davis 2009). Gbb signaling, on the other hand, is required to confer a principal, transcription-dependent competence onto motoneurons to express rapid synaptic homeostasis (Goold and Davis 2007). Possibly related to this role of Gbb signaling, some discrete subcellular defects are commonly encountered in mutants that interfere with the pathway. These defects include T-bars that are not properly associated with the presynaptic membrane, local detachment of the pre- and postsynaptic membranes, and decreased stability of axonal and nerve terminal microtubules (Aberle et al. 2002; Marqués et al. 2002; McCabe et al. 2003; Eaton and Davis 2005; Wang et al. 2007). Moreover, the respective mutants display up to 90% reduction in baseline synaptic transmission (Aberle et al. 2002; McCabe et al. 2004). This pleiotropism, further complicated by a regulatory role of Gbb in the expression of the circulating neuromodulatory peptide FMRFamide (Keshishian and Kim 2004), has made it difficult, to distinguish between chronic developmental and more acute, plasticity-related functions of the BMP pathway at the NMJ. Likewise, early requirements for BMP/TGF-β signaling during dendrito- and axonogenesis of vertebrate neurons (Lee-Hoeflich et al. 2004; Podkowa et al. 2010; Yi et al. 2010) might conceal subsequent synaptic functions. Clearly, an identification and functional evaluation of transcriptional targets of the BMP/TGF-β signaling pathway will be relevant here.

1.2.1.2 Wnt Signaling

The prototypic *Drosophila* Wnt morphogen Wingless (Wg) is secreted from motor nerve terminals (Packard et al. 2002; Korkut et al. 2009) and governs NMJ growth and synapse assembly. Evidence for comparable roles of Wnts at mammalian synapses accumulates (Korkut and Budnik 2009; Budnik and Salinas 2011. Downstream of the Wg receptor Frizzled-2 (Fz2), different pathways are employed to execute Wg instructions on the pre- or postsynaptic side, respectively. Activation of presynaptic Fz2 triggers a cascade which diverts from the canonical Wnt pathway to locally regulate the microtubular cytoskeleton as a prerequisite for proper bouton formation (Miech et al. 2008). Binding of Wg to postsynaptic Fz2 elicits an unconventional synapse-to-nucleus signal that might control the expression of genes involved in postsynaptic differentiation (Mathew et al. 2005; Ataman et al. 2006a; Mosca and Schwarz 2010). Importantly, Wg secretion is upregulated upon acute spaced stimulation and induces profound junctional growth followed by differentiation as reflected by the assembly of new synaptic contacts (Ataman et al. 2008). As this process is dependent on both transcription and translation, it clearly constitutes a mechanism for activity-inducible long-term plasticity.

1.2.2 Activity-Dependent Plasticity

Both the morphological and physiological properties of the growing NMJs are subject to activity-dependent plasticity. Next to mutants with globally altered electrical activity, long-term memory mutants displaying elevated or reduced intracellular cAMP levels, respectively, provide classical examples for this notion (Budnik et al. 1990; Zhong and Wu 1991; Zhong et al. 1992). Chronically elevated cAMP levels due to mutations in the cAMP-phosphodiesterase gene *dunce (dnc)*, for instance, cause junctional overgrowth. In addition, *dnc*-mutant NMJs display increased synaptic strength at low or moderate Ca^{2+}-concentrations, an effect that is due to increased release probability (Zhong and Wu 1991; Zhong et al. 1992). At the same time, short-term plasticity as expressed by posttetanic potentiation and paired-pulse facilitation is impaired in these mutants, likely reflecting changes in the functional status of presynaptic release sites due to chronic changes in cAMP.

Various downstream effectors of cAMP may account for different aspects of the *dnc* phenotype, including ion channels and transcription factors. For instance, cAMP promotes downregulation of the perisynaptic cell adhesion molecule Fasciclin II (FasII) and in parallel activates the cAMP response element binding transcription factor CREB. Both events converge in the observed structural and functional strengthening of NMJs (Schuster et al. 1996b; Davis et al. 1996). The prevailing model for activity-dependent plasticity, in which CREB induces the expression of the immediate early gene products Fos and Jun, however, might not apply here. Instead, Sanyal et al. (2002) could demonstrate that the heterodimeric Fos-Jun transcription factor AP-1 promotes synaptic strengthening as a downstream effector of Jun N-terminal kinase (JNK), thereby in turn inducing CREB expression. Like in *dnc* mutants (Kuromi and Kidokoro 2000), long-term enhancement of synaptic strength by overexpression of AP-1 is likely achieved through persistent mobilization of synaptic vesicles from the reserve pool, possibly involving activation of myosin light chain kinase in the motor nerve terminals (Verstreken et al. 2005; Kim et al. 2009). However, whereas synaptic strengthening in *dnc* mutants involves an increase in the number of synaptic contacts (Renger et al. 2000),

AP-1-induced synaptic strengthening is accompanied by a moderate decrease in the number of synaptic contacts (Kim et al. 2009). In fact, while NMJ growth, the addition of synaptic contacts, and synaptic strengthening occur simultaneously during normal development, they are controlled by divergent pathways as revealed by phenotypes of numerous mutants (e.g., Stewart et al. 1996; Schuster et al. 1996b; Wan et al. 2000; Reiff et al. 2002; Merino et al. 2009).

1.3 Ionotropic Glutamate Receptors at Larval NMJs

The overall synaptic strength of a given NMJ is determined by the number of individual synaptic contacts and by their pre- and postsynaptic properties such as the pool of release-ready vesicles, their individual release probabilities, and post-synaptic responsiveness. As will be described in the following sections, all three determinants are strongly influenced by the composition and availability of synaptic glutamate receptors, which in turn are regulated at various levels.

1.3.1 Subtypes of Junctional GluRs

Approximately 30 putative ionotropic GluR subunits are encoded in the *Drosophila* genome (Littleton and Ganetzky 2000), but only five of them, GluRIIA, GluRIIB, GluRIIC (GluRIII), GluRIID, and GluRIIE, were found to assemble into functional, calcium-permeable receptors at larval NMJs (Schuster et al. 1991; Chang et al. 1994; Petersen et al. 1997; Marrus et al. 2004; Featherstone et al. 2005; Qin et al. 2005). All five are non-NMDA-type receptor subunits, which display significant sequence homologies to vertebrate AMPA- and kainate-type GluR subunits (30–40% identity, 50–60% similarity). Notably, however, fly GluRI, which is even more closely related to AMPA-type receptor subunits, as well as the two fly homologs of NMDA-type receptor subunits have not been detected at NMJs.

Attempts to approach the subunit composition of junctional GluRs by reconstituting them in in vitro systems have failed so far (S.J.S., unpublished). Their stoichiometry was rather inferred from genetic analyses, which point to the existence of two, most likely heterotetrameric receptor subtypes. Apparently, both subtypes contain subunits GluRIIC, IID, and IIE but differ by the incorporation of either GluRIIA or GluRIIB and hence are referred to as A- or B-type receptors, respectively (Fig. 1.2a). Lack of any single of the three default subunits (GluRIIC, IID, IIE) prevents the formation of ionotropic GluRs at body wall muscle NMJs, thus leading to paralysis and late embryonic lethality (DiAntonio et al. 1999; Marrus et al. 2004; Featherstone et al. 2005; Qin et al. 2005). In contrast, mutants lacking either GluRIIA or IIB are viable, and only concomitant loss of both subunits is embryonic lethal. The fact that the presence of just one GluR subtype is sufficient for viability reflects a certain level of redundancy among the two receptor subtypes

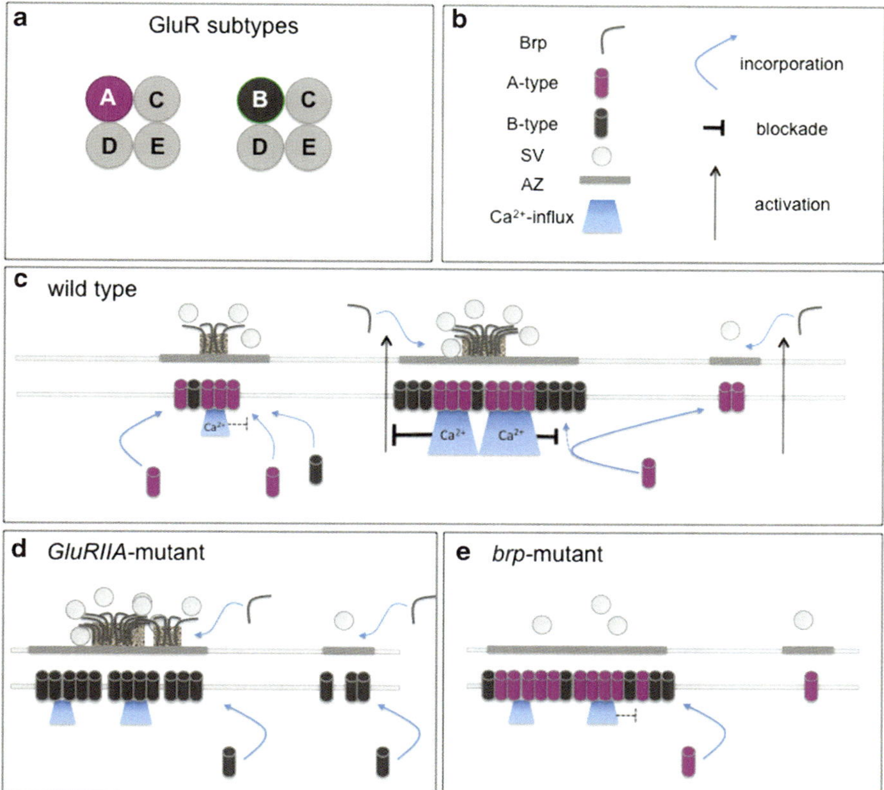

Fig. 1.2 Model for the control of synaptic structure and function by GluRIIA. (**a**) Proposed subunit composition of A- and B-type receptors at NMJs. (**b**) Symbols used in C to E. (**c**) Schematic representation of three synaptic contacts. Mature synapses (*middle*) harbor similar amounts of A- and B-type receptors. Local Ca^{2+}-influx, mainly through A-type receptors, counteracts further incorporation of A-type receptors, which may instead be consumed by premature (*left*) and newborn synapses (*right*). Ca^{2+} may further suppress a retrograde signal that otherwise promotes the strengthening of presynaptic T-Bars by incorporating additional Brp. (**d**) In the absence of GluRIIA, postsynaptic Ca^{2+}-influx is low, thus leading to continued Brp recruitment to AZs, sometimes resulting in AZ profiles with two or more T-Bars. Similar effects can be triggered by acute pharmacological inhibition of GluRs. (**e**) In the absence of Brp, evoked postsynaptic responses are largely diminished. Chronically reduced postsynaptic Ca^{2+}-influx then allows for persistent recruitment of A-type receptors to individual synapses

(Petersen et al. 1997; DiAntonio et al. 1999). In fact, both receptors (1) display virtually identical, high single-channel conductances of approximately 120–150 pS, i.e., approximately five- to tenfold higher than vertebrate AMPA- or kainate receptors (Heckmann and Dudel 1995; Nishikawa and Kidokoro 1995; DiAntonio et al. 1999) and (2) appear similarly abundant, when averaged across all synapses of a larval NMJ (Pawlu et al. 2004; Schmid et al. 2008). B-type receptors, however, desensitize much more rapidly than A-type receptors as revealed by single-channel analysis on extrajunctional muscle membrane outside-out patches from larvae expressing either receptor subtype alone (DiAntonio et al. 1999). Selective absence of A- or B-type GluRs was accomplished by complementing a deletion of the tandem-arranged *GluRIIA* and *GluRIIB* genes with genomic transgenes encoding either GluRIIA or GluRIIB. In this approach, which precludes unwanted over-expression effects, quantal size was found to be three- to fourfold larger upon complementation by GluRIIA as compared to complementation by GluRIIB (DiAntonio et al. 1999; Schmid et al. 2008). Thus, the postsynaptic sensitivity to glutamate is much higher at synapses enriched for A-type than at synapses enriched for B-type receptors. From this, it follows that the number of synaptic A-type GluRs is the primary determinant of postsynaptic strength and, importantly, for the entry of Ca^{2+} as a second messenger (see below).

1.3.2 Synaptic Clustering of GluRs

The picture of how GluRs get targeted to and clustered at NMJs is far from being complete, especially in terms of the physical interactions involved. Nonetheless, the process has been described in quite some detail, including studies on synaptogenesis during the initial phase of NMJ formation and long-term in vivo imaging of GluR incorporation into synapses, and these studies have revealed important regulatory principles underlying the control of synapse formation.

Transcripts encoding GluR subunits are clearly detectable in muscles before motoneurons and target muscles contact each other (Schuster et al. 1991; Currie et al. 1995; Petersen et al. 1997; Marrus et al. 2004; Qin et al. 2005), implying that innervation does not per se induce GluR expression. In turn, GluRs do not prefigure the postsynaptic area of the presumptive neuromuscular junction, nor are they required to induce presynaptic differentiation (Prokop et al. 1996; Qin et al. 2005; Schmid et al. 2006). Whole-cell patch clamp recordings during focal ionto-phoresis of glutamate and cell-attached patch clamp recordings at different positions on embryonic muscles revealed that shortly before innervation functional receptors are evenly distributed on the muscle surface (Broadie and Bate 1993a; Nishikawa and Kidokoro 1995). Innervation then induces (1) the recruitment of extrajunctional receptors to the developing receptive fields opposite to AZs (Broadie and Bate 1993b; Saitoe et al. 1997; Chen and Featherstone 2005a) and (2) an increase in receptor expression (Broadie and Bate 1993b). Both effects require at least a minimum of electrical activity within motoneurons (Broadie and Bate 1993c; Saitoe et al. 1997). The mechanism, however, by which neural activity translates into the synaptic clustering of GluRs still remains obscure. Vesicular release of glutamate was ruled out to play a role in this process, as null mutants for the single vesicular glutamate transporter in motoneurons display normal clustering of functional receptors (Daniels et al. 2006). This result is paralleled by findings in Munc13-1, -2 double knockout mice, where postsynaptic differentiation including the clustering of glutamate receptors appears unaffected in the complete absence of synaptic vesicle release (Varoqueaux et al. 2002). Moreover, a total block of both evoked and spontaneous release did not markedly interfere with synaptic GluR clustering (Schmid et al. 2006), arguing against the previously controversial idea, that the machinery for spontaneous neurotransmitter release might be employed to secrete factor(s) that induce junctional accumulation of GluRs (Saitoe et al. 2001, 2002; Featherstone et al. 2002a, Featherstone and Broadie 2002; Verstreken and Bellen 2002). Signaling pathways across the junctional cleft, that would initiate postsynaptic GluR clustering, thus remain to be identified.

1.3.3 Extracellular Matrix and Cell Adhesion Molecules in Synaptic GluR Clustering

Recently, the presynaptically secreted N-acetyl-glycosaminoglycan-binding glycoprotein termed Mind-the-Gap (Mtg) has been reported to play an important role early in postsynaptic differentiation (Rohrbough et al. 2007; Rushton et al. 2009). A *mtg* null allele was identified in a screen for mutants causing late embryonic lethality associated with severe paralysis. In fact, the junctional recruitment of GluRs and other postsynaptic marker proteins appears strongly diminished, though not completely abrogated in these mutants. Mtg might exert its role by organizing the synaptic cleft matrix as an environment for effective signaling and/or physical trapping of synaptic transmembrane proteins including integrins and the receptor tyrosine kinase Alk (Rushton et al. 2009; Rohrbough and Broadie 2010). While there is no obvious ortholog of Mtg in vertebrates, the crucial involvement of extracellular matrix proteins emerges as a common theme in the clustering of various neurotransmitter receptors (Kummer et al. 2006; Dityatev and Schachner 2006; Wu et al. 2010; see also accompanying article by Frischknecht and Gundelfinger).

By analogy to findings in vertebrates, it appears conceivable that transsynaptic adhesion molecules and synaptic scaffold molecules may play important roles in GluR recruitment and clustering (Gerrow and El-Husseini 2007; Han and Kim 2008). FasII, which is related to vertebrate neural cell adhesion molecule (NCAM), is present at both motor nerve growth cones and their target muscles prior to NMJ formation. Although various aspects of postsynaptic differentiation are impaired by complete loss of FasII, it causes only a moderate decrease in the recruitment of both GluR subtypes (Kohsaka et al. 2007).

The only *Drosophila* neurexin (DNrx1) and at least two out of four neuroligins (DNlg1, DNlg2) were recently shown to be present at NMJs from early onward (Li et al. 2007; Sun et al. 2009; Chen et al. 2010; Banovic et al. 2010; Sun et al. 2011). Loss of function alleles for either protein display discrete phenotypes at NMJs, including altered composition of GluR subtypes (Sun et al. 2011) and accumulation defects of GluRs (Banovic et al. 2010). In *dnlg1* mutants, boutons with differentiated AZs, yet without corresponding accumulation of postsynaptic

GluRs, are frequently observed, and the remaining receptor fields are often enlarged and misshapen (Banovic et al. 2010). Thus, the formation of GluR fields (and other postsynaptic structures) is severely affected by loss of DNlg1. Notably, the conventional concept of presynaptic neurexin primarily forming a functional unit with postsynaptic neuroligin is challenged by the facts that (1) *dnlg1* mutants display stronger physiological and structural phenotypes than *dnrx* mutants (Banovic et al. 2010), (2) *dnrx-dnlg2* double mutants show strongly enhanced phenotypes as compared to the respective single mutants (while *dnrx-* and *dnlg1* mutations are nonadditive), and (3) Dnrx might also be expressed in muscles (Chen et al. 2010).

1.3.4 Cytoskeletal and Scaffold Components

Among the few proteins hitherto shown to co-enrich with GluRs within the discrete, several hundred nanometers wide postsynaptic sites are the p21-activated kinase (PAK), the Rho-type GTPase exchange factor dPix (Parnas et al. 2001), and Dreadlocks (Dock), a functional homolog of human Nck, which can act as an adaptor protein to link receptor tyrosine kinases (or other transmembrane proteins) to effectors of the actin cytoskeleton, including activated Pak (Li et al. 2001). GluRs, in particular GluRIIA, are reduced by about 50% at synapses that lack Pak, dPix, or Dock, leading to a significant decrease in quantal size (Albin and Davis 2004). Genetic analyses further implied that Pak can be activated by both Rac and Cdc42 and acts with Dock in the same pathway to control GluR abundance. The existence and identity of a postsynaptic Dock-binding receptor, however, remains obscure. Moreover, although Pak localizes to postsynaptic sites opposite to embryonic AZs even in the complete absence of GluRs, its further accumulation is strongly impaired when GluRs are expressed at a minimum level ($\leq$5% of normal) required for survival (Schmid et al. 2006). This observation not only reflects interdependency between Pak and GluRs during synaptic maturation but also correlates with an important structural role of GluRIIA in synapse development that was unraveled in the course of that study.

A possible role for the actin cytoskeleton in GluR anchorage was assessed by applying via a patch pipette (1) latrunculin A, which precludes actin polymerization and (2) cofilin, which induces depolymerization of filamentous actin (Chen et al. 2005b). Both drugs were found to reduce synaptic GluRIIA, but not GluRIIB. This subunit-selective effect may be attributable to Coracle (Cor), a *Drosophila* homolog of the mammalian cytocortical protein 4.1, which was shown to physically interact with GluRIIA in vitro (but not with GluRIIB). In fact, synaptic GluRIIA was found severely reduced in the absence of Cor, whereas GluRIIB levels remained normal (Chen et al. 2005b).

A number of well-conserved scaffold molecules have been implicated in NMJ organization and function (Ataman et al. 2006b). In mammalian neurons, members of the family of Dlg-like membrane-associated guanylate kinases (MAGUKs) play a pivotal role in the trafficking, surface expression, and synaptic clustering of both

NMDA- and non-NMDA-type receptors (Funke et al. 2005; Elias and Nicoll 2007). Discs large (Dlg), the prototypic *Drosophila* MAGUK, is highly enriched at the postsynaptic site of larval NMJs; however, it is virtually excluded from the GluR fields and is therefore, like many other NMJ components, considered as a perisynaptic component (Thomas et al. 2010). In conjunction with two other scaffold proteins, Metro and DLin-7, Dlg limits the size of GluR fields (Karunanithi et al. 2002; Mendoza-Topaz et al. 2008; Bachmann et al. 2010), and this role may involve direct interactions with FasII (Stewart et al. 1996; Thomas et al. 1997; Zito et al. 1997). Pre- and postsynaptic spectrin is required for proper junctional recruitment of Dlg (Featherstone et al. 2001), and RNAi-mediated disruption of the postsynaptic spectrin lattice alone causes conspicuous disorganization of junctional Dlg along with a pronounced broadening of synaptic areas, suggesting that Dlg acts downstream of spectrin to execute its role in proper dimensioning of GluR fields (Pielage et al. 2006).

1.3.5   In Vivo Observation of GluRs During Synapse Addition at NMJs

Based on the relative transparency of the larval cuticle, NMJs can be assessed by confocal microscopy on intact animals provided that they express a fluorescently tagged protein that enriches at NMJs (Zito et al. 1999). This approach has been used to analyze the dynamics of GluRs at individual synapses of selected NMJs over a period corresponding to about 20% of larval development. Fluorescently tagged GluRIIA and GluRIIB expressed from "genomic" transgenes comprising all introns and regulatory upstream sequences were found to mimic the respective endogenous GluR subunits in terms of functionality, expression levels, and subcellular distribution (Rasse et al. 2005; Schmid et al. 2008). More than hundred newborn receptor fields per NMJ were detected during the observation period. Within 6–8 h (at 25°C), each of them reaches final size of 0.24 μm^2 on average. Fluorescence recovery after photobleaching (FRAP) and photoactivation analyses revealed that, once arrived at the synapse, GluRs, especially type A, are largely immobile, although local internalization and recycling at individual synapses cannot be ruled out. This is somewhat in contrast to the mobility of a large percentage of synaptic AMPA receptors in cultured mammalian neurons (Newpher and Ehlers 2008; see also accompanying article by Heine). In particular, there is little if any exchange of GluRs between synapses within a given NMJ. Instead, GluRs from all over the muscle contribute to the maturation of growing GluR fields.

Simultaneous quantification of differently fluorescently tagged GluRIIA and GluRIIB showed that the former is prevalent at small receptor fields (Fig. 1.2b–c). With subsequent growth, a clear shift from GluRIIA-dominated immature synapses to a balanced composition at mature synapses became evident. Thereby, a strong negative correlation between presynaptic maturation as indicated by increasing amounts of the major constituent of T-Bars, i.e., the protein Bruchpilot (Brp; related to vertebrate ELKS/ERC/CAST) (Fouquet et al. 2009), and the ongoing incorporation of GluRIIA was observed (Fig. 1.2c). Consistent with the idea that glutamate-triggered Ca^{2+}-influx through GluRIIA might itself constitute a negative feedback signal, Schmid et al. (2008) could show that blockade of evoked glutamate release leads to unrestricted synaptic recruitment of GluRIIA on the expense of GluRIIB incorporation (Fig. 1.2d). Likewise, in mutants for Brp and mutants for the Brp-binding partner DSyd-1, where the prevention of T-Bar formation causes a severe deficit in evoked release (Wagh et al. 2006; Kittel et al. 2006; Owald et al. 2010), increased accumulation of GluRIIA becomes apparent (Fig. 1.2e).

1.3.6 GluRIIA-Dependent Synapse Formation and Plasticity

In animals raised under standard conditions, loss of GluRIIA causes only a moderate reduction in the number of boutons and synaptic contacts (Reiff et al. 2002; Schmid et al. 2008), but a severe reduction in quantal size. Nonetheless, *GluRIIA* mutant NMJs show normal evoked response when stimulated with single action potentials (Petersen et al. 1997). This is brought about by an increase in presynaptic release (quantal content), which in turn is tightly correlated with ultrastructural adaptations toward more and bigger T-Bars per synapse, most likely due to recruitment of additional Brp protein (Reiff et al. 2002). Recently, Weyhersmüller et al. (2011) showed that paired-pulse ratios are similar in *GluRIIA* mutants and controls, implying that the release probability of individual vesicles remains unchanged in the mutants. At the same time, the number of release-ready vesicles was found to be almost doubled. Moreover, this presynaptic adaptation to loss of GluRIIA was accompanied by a moderate but significant increase in the size of Brp clusters at AZs. Interestingly, an increase in synaptic Brp content was even traceable on a short-term scale after blocking GluRs acutely with PhTX (see above) (Weyhersmüller et al. 2011). Increases in Brp at individual active zones triggered by loss of conductance through postsynaptic GluRIIA may therefore be part of the observed homeostatic response. This view is further substantiated by a study, in which the probability of SV release at individual AZs was determined based on real-time imaging and was found to correlate with the amount of Brp at these sites (Peled and Isacoff 2011). At this point, we refer the reader to recent reviews on plasticity phenomena concerning AZs, T-Bars, and related specializations in other species as substrates for presynaptic modes of plasticity (Owald and Sigrist 2009; Wichmann and Sigrist 2010; Sigrist and Schmitz 2011).

It is conceivable that Ca^{2+}-entry through GluRs, mainly GluRIIA itself, constitutes a crucial determinant for negative feedback onto presynaptic release. In fact, constitutive expression of a CaMKII-inhibiting peptide leads to increased quantal content (Haghighi et al. 2003; Morimoto et al. 2010), and expression of a constitutively activated CaMKII was found to interfere with homoeostatic compensation in one study (Haghighi et al. 2003). In a similar, though not identical approach, Morimoto et al. (2010) found that active CaMKII downregulates synaptic GluRIIA levels along with proper homeostatic upregulation of synaptic vesicle release. The way CaMKII joins in here thus remains to be specified, e.g., in terms of substrates, that may play a role in retrograde signaling (see also above notes on retrograde BMP signaling). The postsynaptic scaffold protein Dystrophin (Dys) is to be named here, as *dys* mutants display increased presynaptic release despite normal levels of GluRIIA and no change in the number of synaptic contacts, yet paralleled by a more prominent appearance of T-Bars (van der Plas et al. 2006). This phenotypic constellation is consistent with Dys acting downstream of GluRIIA in controlling presynaptic release. The identification of mammalian Dys as a target for phosphorylation by CaMKII is further suggestive, and it is tempting to speculate that Dys mediates negative feedback upon phosphorylation by CaMKII. NMJ localization of Dys depends on the transmembrane protein dystroglycan (Bogdanik et al. 2009), and a genetic modifier screen led to the finding that the RhoGAP crossveinless-c and its target Cdc42 act in concert with Dys (Pilgram et al. 2011).

A second Ca^{2+}-sensor that has been implicated in GluRIIA-dependent retrograde control at NMJs is synaptotagmin 4 (Syt4) (Yoshihara et al. 2005). In contrast to CaMKII, however, Syt4 is required for a positive feedback loop, activated by synapse-specific Ca^{2+}-influx. High-frequency stimulation induces Syt4-dependent postsynaptic vesicle fusion, which is required to induce facilitated presynaptic release (expressed as an increased mEPSP rate) and, presumably, cytoskeletal rearrangements, that in turn lead to NMJ expansion. These responses are dependent on presynaptic protein kinase A (PKA), a major target of cAMP (Yoshihara et al. 2005). Thus, the role of Syt4 may be well related to activity-induced synaptic strengthening, which is mimicked by elevated cAMP levels in *dnc* mutants and which is accompanied by increased numbers of boutons and synaptic contacts. Notably, a concomitant increase in GluRIIA levels was observed (Sigrist et al. 2000) and indeed shown to be limiting in this process (Sigrist et al. 2002). Muscle-specific overexpression of GluRIIA is indeed sufficient to induce junctional over-growth and a proportional increase in synapse numbers (Sigrist et al. 2002). The in vivo relevance of this regulation was highlighted by the finding that enhanced larval locomotor activity induces qualitatively equivalent changes in a GluRIIA-dependent manner (Sigrist et al. 2003). Strikingly, the amount of GluRIIA within individual synapses does not change during activity-induced synaptic strengthening, consistent with the aforementioned, maturation-related restriction of GluRIIA recruitment to individual synaptic contacts (Schmid et al. 2008). Therefore, limited consumption of GluRIIA at individual synapses warrants the availability of this receptor subtype for the formation of additional synapses. While GluRs get recruited to synapses from all over the muscle during NMJ expansion (Rasse et al. 2005), there is evidence that local synthesis of GluRIIA at NMJs might be employed to meet the requirement for activity-induced formation of synapses and that posttranscriptional control of the translational initiation factor eIF-4E is crucially involved in this regulation (Sigrist et al. 2000, 2002; Reiff et al. 2002;

Menon et al. 2004, 2009). The ultimate proof for such a translational control of GluRIIA, however, is still missing.

It appears conceivable that GluRIIA and GluRIIB compete for default subunits and/or by occupying slots within presumptive receptor field areas. Overexpression of GluRIIB was indeed found to interfere with functional A-type receptor expressivity, thus leading to a decrease in quantal size (DiAntonio et al. 1999; Sigrist et al. 2002). It remains questionable, however, whether competition is relevant within the physiological range of expression. A recent study by Karr et al. (2009) demonstrated that limited availability of default subunits accounts for only a moderate increase in the number of surface-expressed and synaptic GluRs when expression levels of both GluRIIA and GluRIIB were substantially elevated due to a lack in posttranscriptional suppression by microRNAs. Thus, competition at the level of receptor assembly may become effective as a mechanism to adjust synaptic receptor composition when expression levels of GluRIIA and GluRIIB change asymmetrically.

1.3.7 Control of Synaptic GluRs by Ambient Glutamate

Following initial observations which pointed to nonvesicular release of glutamate as a negative regulator of synaptic GluR accumulation (Featherstone et al. 2000, 2002), Augustin et al. (2007) investigated Genderblind (Gb), the xCT subunit of the cystine/glutamate transporter for its impact in this mode of regulation. Reminiscent to obvious expression of its mammalian homolog in border areas of the brain, Gb is expressed in surface glia of the CNS and within a particular glia cell (now termed Gb cell) associated with NMJs. Compared to ambient glutamate concentrations in the mammalian brain, which are normally in a low micromolar range (Featherstone and Shippy 2008), glutamate concentrations in the larval hemolymph are in a millimolar range, yet close to the level required for half-maximal activation of junctional receptors (Heckmann et al. 1996). Loss of Gb causes an ~50% reduction in hemolymph glutamate concentration. Under this condition, the junctional amount of both GluR subtypes was more than doubled, and this effect was further enhanced when glutamate was completely omitted experimentally. Importantly, by using a glutamate antagonist (γ-D-glutamylglycine), which prevents glutamate-dependent desensitization, the effect of reduced glutamate was mimicked, implying that ambient glutamate interferes with synaptic clustering of GluRs via desensitization of the receptors. As pointed out by Featherstone and Shippy (2008), a steady-state desensitization of GluRs offers additional options for regulating synaptic strength, either cell-autonomously by altering desensitization kinetics of GluRs or systemically by altering ambient glutamate levels. The role of xCT in regulating neural functions has recently been further corroborated by demonstrating that it is required for regulating synaptic strength in adult flies and thereby controls courtship behavior (Grosjean et al. 2008).

1.4   Concluding Remarks

The *Drosophila* larval NMJ has proven a highly versatile model to unravel structure–function relationships during synaptogenesis, synaptic transmission, synapse maintenance, and synapse remodeling. Time and again, proteins relevant for these processes emerge from unbiased forward genetic screens for mutants affecting synaptic function and/or structure, a trademark strategy so far largely confined to *Drosophila melanogaster* and the nematode *Caenorhabditis elegans*. Typically, the merit of such screens lies in the shortcut toward the functional characterization of the identified genes or proteins. For instance, fruitful screens were built onto the ability to generate flies in which all photoreceptor cells in the eyes are homozygous for a given mutation, whereas cells of all other tissues remain heterozygous (Stowers and Schwarz 1999). This way, mutations affecting synaptic transmission can be screened for by performing electroretinograms on the easily accessible compound eyes. Once identified, mutants are then often subjected to further analysis at the NMJ. In several instances, factors identified in these or other screens were found to be homologous to hitherto poorly characterized proteins implicated in human neurodegenerative or other neurological disorders (e.g., Zhai et al. 2006, 2008; Dickman and Davis 2009; Kim et al. 2010).

Complementary to forward genetic approaches, well-conserved synaptic proteins such as the GluRs are often directly assessed for their role at NMJs. The genetic toolbox for refined analyses is constantly upgraded, including comprehensive collections of transgenic fly stocks allowing for cell-type-specific RNAi-mediated knockdown of almost every gene (Dietzl et al. 2007; Ni et al. 2009) and the establishment of techniques for generating small deletions (Parks et al. 2004) or even predefined gene targeting by homologous recombination (Gong and Golic 2003). Moreover, by optimizing recombineering techniques, it has now become possible to generate genomic constructs of more than 100 kb, which, for instance, can be used to express fluorescence-tagged synaptic proteins at or near endogenous levels, an advantage that can hardly be overestimated (Venken et al. 2006, 2009; Ejsmont et al. 2009). In fact, the benefit of this approach has been well exemplified by the aforementioned expression of junctional GluRs from conventionally cloned genomic transgenes and their subsequent assessment by life imaging (Rasse et al. 2005; Schmid et al. 2008). Future studies may therefore be expected to include other synaptic proteins in this sort of analysis, thus leading to a more detailed view on the molecular dynamics of glutamatergic synapses.

References

Aberle, H., Haghighi, A. P., Fetter, R. D., McCabe, B. D., Magalhães, T. R., & Goodman, C. S. (2002). Wishful thinking encodes a BMP type II receptor that regulates synaptic growth in Drosophila. *Neuron, 33*, 545–558.
Albin, S. D., & Davis, G. W. (2004). Coordinating structural and functional synapse development: Postsynaptic p21-activated kinase independently specifies glutamate receptor abundance and postsynaptic morphology. *Journal of Neuroscience, 24*, 6871–6879.

Ataman, B., Ashley, J., Gorczyca, D., Gorczyca, M., Mathew, D., Wichmann, C., Sigrist, S. J., & Budnik, V. (2006a). Nuclear trafficking of Drosophila frizzled-2 during synapse development requires the PDZ protein dGRIP. *Proceedings of the National Academy of Sciences of the United States of America, 103*, 7841–7846.

Ataman, B., Ashley, J., Gorczyca, M., Ramachandran, P., Fouquet, W., Sigrist, S. J., & Budnik, V. (2008). Rapid activity-dependent modifications in synaptic structure and function require bidirectional Wnt signaling. *Neuron, 57*, 705–718.

Ataman, B., Budnik, V., & Thomas, U. (2006b). Scaffolding proteins at the Drosophila neuromuscular junction. *International Review of Neurobiology, 75*, 181–216.

Atwood, H. L., Govind, C. K., & Wu, C. F. (1993). Differential ultrastructure of synaptic terminals on ventral longitudinal abdominal muscles in Drosophila larvae. *Journal of Neurobiology, 24*, 1008–1024.

Atwood, H. L., Karunanithi, S., Georgiou, J., & Charlton, M. P. (1997). Strength of synaptic transmission at neuromuscular junctions of crustaceans and insects in relation to calcium entry. *Invertebrate Neuroscience, 3*, 81–87.

Augustin, H., Grosjean, Y., Chen, K., Sheng, Q., & Featherstone, D. E. (2007). Nonvesicular release of glutamate by glial xCT transporters suppresses glutamate receptor clustering in vivo. *Journal of Neuroscience, 27*, 111–123.

Bachmann, A., Kobler, O., Kittel, R. J., Wichmann, C., Sierralta, J., Sigrist, S. J., Gundelfinger, E. D., Knust, E., & Thomas, U. (2010). A perisynaptic ménage à trois between Dlg, DLin-7, and Metro controls proper organization of Drosophila synaptic junctions. *Journal of Neuroscience, 30*, 5811–5824.

Ball, R. W., Warren-Paquin, M., Tsurudome, K., Liao, E. H., Elazzouzi, F., Cavanagh, C., An, B. S., Wang, T. T., White, J. H., & Haghighi, A. P. (2010). Retrograde BMP signaling controls synaptic growth at the NMJ by regulating trio expression in motor neurons. *Neuron, 66*, 536–549.

Banovic, D., Khorramshahi, O., Owald, D., Wichmann, C., Riedt, T., Fouquet, W., Tian, R., Sigrist, S. J., & Aberle, H. (2010). Drosophila neuroligin 1 promotes growth and postsynaptic differentiation at glutamatergic neuromuscular junctions. *Neuron, 66*, 724–738.

Bayat, V., Jaiswal, M., & Bellen, H. J. (2011). The BMP signaling pathway at the Drosophila neuromuscular junction and its links to neurodegenerative diseases. *Current Opinion in Neurobiology, 21*, 182–188.

Bogdanik, L., Framery, B., Frölich, A., Franco, B., Mornet, D., Bockaert, J., Sigrist, S. J., Grau, Y., & Parmentier, M. L. (2009). Muscle dystroglycan organizes the postsynapse and regulates presynaptic neurotransmitter release at the Drosophila neuromuscular junction. *PloS One, 3*, e2084.

Broadie, K. S., & Bate, M. (1993a). Development of the embryonic neuromuscular synapse of *Drosophila melanogaster*. *Journal of Neuroscience, 13*, 144–166.

Broadie, K., & Bate, M. (1993b). Innervation directs receptor synthesis and localization in Drosophila embryo synaptogenesis. *Nature, 361*, 350–353.

Broadie, K., & Bate, M. (1993c). Activity-dependent development of the neuromuscular synapse during Drosophila embryogenesis. *Neuron, 11*, 607–619.

Budnik, V., & Salinas, P. C. (2011). Wnt signaling during synaptic development and plasticity. *Current Opinion in Neurobiology, 21*, 151–159.

Budnik, V., Zhong, Y., & Wu, C. F. (1990). Morphological plasticity of motor axons in Drosophila mutants with altered excitability. *Journal of Neuroscience, 10*, 3754–3768.

Chang, H., Ciani, S., & Kidokoro, Y. (1994). Ion permeation properties of the glutamate receptor channel in cultured embryonic Drosophila myotubes. *The Journal of Physiology, 476*, 1–16.

Chen, Y. G. (2009). Endocytic regulation of TGF-beta signaling. *Cell Research, 19*, 58–70.

Chen, K., & Featherstone, D. E. (2005a). Discs-large (DLG) is clustered by presynaptic innervation and regulates postsynaptic glutamate receptor subunit composition in Drosophila. *BMC Biology, 3*(1).

Chen, K., Gracheva, E. O., Yu, S. C., Sheng, Q., Richmond, J., & Featherstone, D. E. (2010). Neurexin in embryonic Drosophila neuromuscular junctions. *PloS One, 5*, e11115.

Chen, K., Merino, C., Sigrist, S. J., & Featherstone, D. E. (2005). The 4.1 protein coracle mediates subunit-selective anchoring of Drosophila glutamate receptors to the postsynaptic actin cytoskeleton. *Journal of Neuroscience, 25*, 6667–6675.

Choi, J. C., Park, D., & Griffith, L. C. (2004). Electrophysiological and morphological characterization of identified motor neurons in the Drosophila third instar larva central nervous system. *Journal of Neurophysiology, 91*, 2353–2365.

Currie, D. A., Truman, J. W., & Burden, S. J. (1995). Drosophila glutamate receptor RNA expression in embryonic and larval muscle fibers. *Developmental Dynamics, 203*, 311–316.

Daniels, R. W., Collins, C. A., Chen, K., Gelfand, M. V., Featherstone, D. E., & DiAntonio, A. (2006). A single vesicular glutamate transporter is sufficient to fill a synaptic vesicle. *Neuron, 49*, 11–16.

Davis, G. W., Schuster, C. M., & Goodman, C. S. (1996). Genetic dissection of structural and functional components of synaptic plasticity III. CREB is necessary for presynaptic functional plasticity. *Neuron, 17*, 669–679.

DiAntonio, A., Petersen, S. A., Heckmann, M., & Goodman, C. S. (1999). Glutamate receptor expression regulates quantal size and quantal content at the Drosophila neuromuscular junction. *Journal of Neuroscience, 19*, 3023–3032.

Dickman, D. K., & Davis, G. W. (2009). The schizophrenia susceptibility gene dysbindin controls synaptic homeostasis. *Science, 326*, 1127–1130.

Dickman, D. K., Lu, Z., Meinertzhagen, I. A., & Schwarz, T. L. (2006). Altered synaptic development and active zone spacing in endocytosis mutants. *Current Biology, 16*, 591–598.

Dietzl, G., Chen, D., Schnorrer, F., Su, K. C., Barinova, Y., Fellner, M., Gasser, B., Kinsey, K., Oppel, S., Scheiblauer, S., Couto, A., Marra, V., Keleman, K., & Dickson, B. J. (2007). A genome-wide transgenic RNAi library for conditional gene inactivation in Drosophila. *Nature, 448*, 151–158.

Dityatev, A., & Schachner, M. (2006). The extracellular matrix and synapses. *Cell and Tissue Research, 326*, 647–654.

Eaton, B. A., & Davis, G. W. (2005). LIM Kinase1 controls synaptic stability downstream of the type II BMP receptor. *Neuron, 47*, 695–708.

Ejsmont, R. K., Sarov, M., Winkler, S., Lipinski, K. A., & Tomancak, P. (2009). A toolkit for high-throughput, cross-species gene engineering in Drosophila. *Nature Methods, 6*, 435–437.

Elias, G. M., & Nicoll, R. A. (2007). Synaptic trafficking of glutamate receptors by MAGUK scaffolding proteins. *Trends in Neurosciences, 17*, 342–352.

Featherstone, D., & Broadie, K. (2002). Response: Meaningless minis? *Trends in Neurosciences, 25*, 386–387.

Featherstone, D. E., Davis, W. S., Dubreuil, R. R., & Broadie, K. (2001). Drosophila alpha- and beta-spectrin mutations disrupt presynaptic neurotransmitter release. *Journal of Neuroscience, 21*, 4215–4224.

Featherstone, D. E., Rushton, E., & Broadie, K. (2002). Developmental regulation of glutamate receptor field size by nonvesicular glutamate release. *Nature Neuroscience, 5*, 141–146.

Featherstone, D. E., Rushton, E., Rohrbough, J., Liebl, F., Karr, J., Sheng, Q., Rodesch, C. K., & Broadie, K. (2005). An essential Drosophila glutamate receptor subunit that functions in both central neuropil and neuromuscular junction. *Journal of Neuroscience, 25*, 3199–3208.

Featherstone, D. E., & Shippy, S. A. (2008). Regulation of synaptic transmission by ambient extracellular glutamate. *The Neuroscientist, 14*, 171–181.

Fouquet, W., Owald, D., Wichmann, C., Mertel, S., Depner, H., Dyba, M., Hallermann, S., Kittel, R. J., Eimer, S., & Sigrist, S. J. (2009). Maturation of active zone assembly by Drosophila bruchpilot. *The Journal of Cell Biology, 186*, 129–145.

Frank, C. A., Kennedy, M. J., Goold, C. P., Marek, K. W., & Davis, G. W. (2006). Mechanisms underlying the rapid induction and sustained expression of synaptic homeostasis. *Neuron, 52*, 663–677.

Frank, C. A., Pielage, J., & Davis, G. W. (2009). A presynaptic homeostatic signaling system composed of the Eph receptor, ephexin, Cdc42, and CaV2.1 calcium channels. *Neuron, 61*, 556–569.

Fuentes-Medel, Y., & Budnik, V. (2010). Ménage à Trio during BMP-mediated retrograde signaling at the NMJ. *Neuron, 66*, 473–475.

Funke, L., Dakoji, S., & Bredt, D. S. (2005). Membrane-associated guanylate kinases regulate adhesion and plasticity at cell junctions. *Annual Review of Biochemistry, 74*, 219–245.

Gerrow, K., & El-Husseini, A. (2007). Receptors look outward: Revealing signals that bring excitation to synapses. *Science's STKE, 2007*, 56.

Gong, W. J., & Golic, K. G. (2003). Ends-out, or replacement, gene targeting in Drosophila. *Proceedings of the National Academy of Sciences of the United States of America, 100*, 2556–2561.

Goold, C. P., & Davis, G. W. (2007). The BMP ligand Gbb gates the expression of synaptic homeostasis independent of synaptic growth control. *Neuron, 56*, 109–123.

Gorczyca, M., Augart, C., & Budnik, V. (1993). Insulin-like receptor and insulin-like peptide are localized at neuromuscular junctions in Drosophila. *Journal of Neuroscience, 13*, 3692–3704.

Grosjean, Y., Grillet, M., Augustin, H., Ferveur, J. F., & Featherstone, D. E. (2008). A glial amino-acid transporter controls synapse strength and courtship in Drosophila. *Nature Neuroscience, 11*, 54–61.

Han, K., & Kim, E. (2008). Synaptic adhesion molecules and PSD-95. *Progress in Neurobiology, 84*, 263–283.

Haghighi, A. P., McCabe, B. D., Fetter, R. D., Palmer, J. E., Hom, S., & Goodman, C. S. (2003). Retrograde control of synaptic transmission by postsynaptic CaMKII at the Drosophila neuro-muscular junction. *Neuron, 39*, 255–267.

Heckmann, M., & Dudel, J. (1995). Recordings of glutamate-gated ion channels in outside-out patches from Drosophila larval muscle. *Neuroscience Letters, 196*, 53–56.

Heckmann, M., Parzefall, F., & Dudel, J. (1996). Activation kinetics of glutamate receptor channels from wild-type Drosophila muscle. *Pflügers Archiv, 432*, 1023–1029.

Jan, L. Y., & Jan, Y. N. (1976a). Properties of the larval neuromuscular junction in *Drosophila melanogaster*. *The Journal of Physiology, 262*, 189–214.

Jan, L. Y., & Jan, Y. N. (1976b). L-glutamate as an excitatory transmitter at the Drosophila larval neuromuscular junction. *The Journal of Physiology, 262*, 215–236.

Jia, X. X., Gorczyca, M., & Budnik, V. (1993). Ultrastructure of neuromuscular junctions in Drosophila: Comparison of wild type and mutants with increased excitability. *Journal of Neurobiology, 24*, 1025–1044.

Johansen, J., Halpern, M. E., Johansen, K. M., & Keshishian, H. (1989). Stereotypic morphology of glutamatergic synapses on identified muscle cells of Drosophila larvae. *Journal of Neuro-science, 9*, 710–725.

Karr, J., Vagin, V., Chen, K., Ganesan, S., Olenkina, O., Gvozdev, V., & Featherstone, D. E. (2009). Regulation of glutamate receptor subunit availability by microRNAs. *The Journal of Cell Biology, 185*, 685–607.

Karunanithi, S., Marin, L., Wong, K., & Atwood, H. L. (2002). Quantal size and variation determined by vesicle size in normal and mutant Drosophila glutamatergic synapses. *Journal of Neuroscience, 22*, 10263–10276.

Keshishian, H., Broadie, K., Chiba, A., & Bate, M. (1996). The Drosophila neuromuscular junction: A model system for studying synaptic development and function. *Annual Review of Neuroscience, 19*, 545–575.

Keshishian, H., & Kim, Y. S. (2004). Orchestrating development and function: Retrograde BMP signaling in the Drosophila nervous system. *Trends in Neurosciences, 27*, 143–147.

Khuong, T. M., Habets, R. L., Slabbaert, J. R., & Verstreken, P. (2010). WASP is activated by phosphatidylinositol-4,5-bisphosphate to restrict synapse growth in a pathway parallel to bone morphogenetic protein signaling. *Proceedings of the National Academy of Sciences of the United States of America, 107*, 17379–17384.

Kim, S. M., Kumar, V., Lin, Y.-Q., Karunanithi, S., & Ramaswami, M. (2009). Fos and Jun potentiate individual release sites and mobilize the reserve synaptic-vesicle pool at the *Drosophila* larval motor synapse. *Proceedings of the National Academy of Sciences of the United States of America, 106*, 4000–4005.

Kim, S., Wairkar, Y. P., Daniels, R. W., & DiAntonio, A. (2010). The novel endosomal membrane protein Ema interacts with the class C Vps-HOPS complex to promote endosomal maturation. *Journal of Cell Biology, 188*, 717–734.

Kittel, R. J., Wichmann, C., Rasse, T. M., Fouquet, W., Schmidt, M., Schmid, A., Wagh, D. A., Pawlu, C., Kellner, R. R., Willig, K. I., Hell, S. W., Buchner, E., Heckmann, M., & Sigrist, S. J. (2006). Bruchpilot promotes active zone assembly, Ca^{2+} channel clustering, and vesicle release. *Science, 312*, 1051–1054.

Kohsaka, H., Takasu, E., & Nose, A. (2007). In vivo induction of postsynaptic molecular assembly by the cell adhesion molecule Fasciclin2. *The Journal of Cell Biology, 179*, 1289–1300.

Koon, A. C., Ashley, J., Barria, R., DasGupta, S., Brain, R., Waddell, S., Alkema, M. J., & Budnik, V. (2011). Autoregulatory and paracrine control of synaptic and behavioral plasticity by octopaminergic signaling. *Nature Neuroscience, 14*, 190–199.

Korkut, C., Ataman, B., Ramachandran, P., Ashley, J., Barria, R., Gherbesi, N., & Budnik, V. (2009). Trans-synaptic transmission of vesicular Wnt signals through Evi/Wntless. *Cell, 139*, 393–404.

Korkut, C., & Budnik, V. (2009). WNTs tune up the neuromuscular junction. *Nature Reviews Neuroscience, 10*, 627–634.

Kummer, T. T., Misgeld, T., & Sanes, J. R. (2006). Assembly of the postsynaptic membrane at the neuromuscular junction: Paradigm lost. *Current Opinion in Neurobiology, 16*, 74–82.

Kurdyak, P., Atwood, H. L., Stewart, B. A., & Wu, C. F. (1994). Differential physiology and morphology of motor axons to ventral longitudinal muscles in larval Drosophila. *The Journal of Comparative Neurology, 350*, 463–472.

Kuromi, H., & Kidokoro, Y. (2000). Tetanic stimulation recruits vesicles from reserve pool via a cAMP-mediated process in Drosophila synapses. *Neuron, 127*, 133–143.

Lee-Hoeflich, S. T., Causing, C. G., Podkowa, M., Zhao, X., Wrana, J. L., & Attisano, L. (2004). Activation of LIMK1 by binding to the BMP receptor, BMPRII, regulates BMP-dependent dendritogenesis. *EMBO Journal, 23*, 4792–4801.

Li, J., Ashley, J., Budnik, V., & Bhat, M. A. (2007). Crucial role of Drosophila neurexin in proper active zone apposition to postsynaptic densities, synaptic growth, and synaptic transmission. *Neuron, 55*, 741–755.

Li, W., Fan, J., & Woodley, D. T. (2001). Nck/Dock: An adapter between cell surface receptors and the actin cytoskeleton. *Oncogene, 20*, 6403–6417.

Littleton, J. T., & Ganetzky, B. (2000). Ion channels and synaptic organization: Analysis of the Drosophila genome. *Neuron, 26*, 35–43.

Lnenicka, G. A., & Keshishian, H. (2000). Identified motor terminals in Drosophila larvae show distinct differences in morphology and physiology. *Journal of Neurobiology, 43*, 186–197.

Marqués, G., Bao, H., Haerry, T. E., Shimell, M. J., Duchek, P., Zhang, B., & O'Connor, M. B. (2002). The Drosophila BMP type II receptor Wishful Thinking regulates neuromuscular synapse morphology and function. *Neuron, 33*, 529–543.

Marqués, G., & Zhang, B. (2006). Retrograde signaling that regulates synaptic development and function at the Drosophila neuromuscular junction. *Int. Rev Neurobiology, 75*, 267–285.

Marrus, S. B., Portman, S. L., Allen, M. J., Moffat, K. G., & DiAntonio, A. (2004). Differential localization of glutamate receptor subunits at the Drosophila neuromuscular junction. *Journal of Neuroscience, 24*, 1406–1415.

Mathew, D., Ataman, B., Chen, J., Zhang, Y., Cumberledge, S., & Budnik, V. (2005). Wingless signaling at synapses is through cleavage and nuclear import of receptor DFrizzled2. *Science, 310*, 1344–1347.

McCabe, B. D., Hom, S., Aberle, H., Fetter, R. D., Marques, G., Haerry, T. E., Wan, H., O'Connor, M. B., Goodman, C. S., & Haghighi, A. P. (2004). Highwire regulates presynaptic BMP signaling essential for synaptic growth. *Neuron, 41*, 891–905.

McCabe, B. D., Marqués, G., Haghighi, A. P., Fetter, R. D., Crotty, M. L., Haerry, T. E., Goodman, C. S., & O'Connor, M. B. (2003). The BMP homolog Gbb provides a retrograde signal that regulates synaptic growth at the Drosophila neuromuscular junction. *Neuron, 39*, 241–254.

Meinertzhagen, I. A., Govind, C. K., Stewart, B. A., Carter, J. M., & Atwood, H. L. (1998). Regulated spacing of synapses and presynaptic active zones at larval neuromuscular junctions in different of the flies *Drosophila* and *Sarcophaga*. *The Journal of Comparative Neurology, 393*, 482–492.

Mendoza-Topaz, C., Urra, F., Barría, R., Albornoz, V., Ugalde, D., Thomas, U., Gundelfinger, E. D., Delgado, R., Kukuljan, M., Sanxaridis, P. D., Tsunoda, S., Ceriani, M. F., Budnik, V., & Sierralta, J. (2008). DLGS97/SAP97 is developmentally upregulated and is required for complex adult behaviors and synapse morphology and function. *Journal of Neuroscience, 28*, 304–314.

Menon, K. P., Andrews, S., Murthy, M., Gavis, E. R., & Zinn, K. (2009). The translational repressors Nanos and Pumilio have divergent effects on presynaptic terminal growth and postsynaptic glutamate receptor subunit composition. *Journal of Neuroscience, 29*, 5558–5572.

Menon, K. P., Sanyal, S., Habara, Y., Sanchez, R., Wharton, R. P., Ramaswami, M., & Zinn, K. (2004). The translational repressor Pumilio regulates presynaptic morphology and controls postsynaptic accumulation of translation factor eIF-4E. *Neuron, 44*, 663–676.

Merino, C., Penney, J., González, M., Tsurudome, K., Moujahidine, M., O'Connor, M. B., Verheyen, E. M., & Haghighi, P. (2009). Nemo kinase interacts with Mad to coordinate synaptic growth at the Drosophila neuromuscular junction. *The Journal of Cell Biology, 185*, 713–725.

Miech, C., Pauer, H. U., He, X., & Schwarz, T. L. (2008). Presynaptic local signaling by a canonical wingless pathway regulates development of the Drosophila neuromuscular junction. *Journal of Neuroscience, 28*, 10875–10884.

Morimoto, T., Nobechi, M., Komatsu, A., Miyakawa, H., & Nose, A. (2010). Subunit-specific and homeostatic regulation of glutamate receptor localization by CaMKII in Drosophila neuromuscular junctions. *Neuroscience, 165*, 1284–1292.

Mosca, T. J., & Schwarz, T. L. (2010). The nuclear import of Frizzled2-C by Importins-beta11 and alpha2 promotes postsynaptic development. *Nature Neuroscience, 13*, 935–943.

Nahm, M., Kim, S., Paik, S. K., Lee, M., Lee, S., Lee, Z. H., Kim, J., Lee, D., Bae, Y. C., & Lee, S. (2010a). The Cdc42-selective GAP rich regulates postsynaptic development and retrograde BMP transsynaptic signaling. *Journal of Neuroscience, 30*, 8138–8150.

Nahm, M., Long, A. A., Paik, S. K., Kim, S., Bae, Y. C., Broadie, K., & Lee, S. (2010b). The Cdc42-selective GAP rich regulates postsynaptic development and retrograde BMP transsynaptic signaling. *The Journal of Cell Biology, 191*, 661–675.

Nakayama, H., Kazama, H., Nose, A., & Morimoto-Tanifuji, T. (2006). Activity-dependent regulation of synaptic size in Drosophila neuromuscular junctions. *Journal of Neurobiology, 66*, 929–939.

Newpher, T. M., & Ehlers, M. D. (2008). Glutamate receptor dynamics in dendritic microdomains. *Neuron, 58*, 472–497.

Ni, J. Q., Liu, L. P., Binari, R., Hardy, R., Shim, H. S., Cavallaro, A., Booker, M., Pfeiffer, B. D., Markstein, M., Wang, H., Villalta, C., Laverty, T. R., Perkins, L. A., & Perrimon, N. (2009). A Drosophila resource of transgenic RNAi lines for neurogenetics. *Genetics, 182*, 1089–1100.

Nishikawa, K., & Kidokoro, Y. (1995). Junctional and extrajunctional glutamate receptor channels in Drosophila embryos and larvae. *Journal of Neuroscience, 15*, 7905–7915.

O'Connor-Giles, K. M., Ho, L. L., & Ganetzky, B. (2008). Nervous wreck interacts with thick veins and the endocytic machinery to attenuate retrograde BMP signaling during synaptic growth. *Neuron, 58*, 507–518.

Owald, D., Fouquet, W., Schmidt, M., Wichmann, C., Mertel, S., Depner, H., Christiansen, F., Zube, C., Quentin, C., Körner, J., Urlaub, H., Mechtler, K., & Sigrist, S. J. (2010). A Syd-1 homologue regulates pre- and postsynaptic maturation in Drosophila. *The Journal of Cell Biology, 188*, 565–579.

Owald, D., & Sigrist, S. J. (2009). Assembling the presynaptic active zone. *Current Opinion in Neurobiology, 19*, 311–318.

Packard, M., Koo, E. S., Gorczyca, M., Sharpe, J., Cumberledge, S., & Budnik, V. (2002). The Drosophila Wnt, wingless, provides an essential signal for pre- and postsynaptic differentiation. *Cell, 111*, 319–330.

Paradis, S., Sweeney, S. T., & Davis, G. W. (2001). Homeostatic control of presynaptic release is triggered by postsynaptic membrane depolarization. *Neuron, 30*, 737–749.

Parks, A. L., Cook, K. R., Belvin, M., Dompe, N. A., Fawcett, R., Huppert, K., Tan, L. R., Winter, C. G., Bogart, K. P., Deal, J. E., Deal-Herr, M. E., Grant, D., Marcinko, M., Miyazaki, W. Y., Robertson, S., Shaw, K. J., Tabios, M., Vysotskaia, V., Zhao, L., Andrade, R. S., Edgar, K. A., Howie, E., Killpack, K., Milash, B., Norton, A., Thao, D., Whittaker, K., Winner, M. A., Friedman, L., Margolis, J., Singer, M. A., Kopczynski, C., Curtis, D., Kaufman, T. C., Plowman, G. D., Duyk, G., & Francis-Lang, H. L. (2004). Systematic generation of high-resolution deletion coverage of the *Drosophila melanogaster* genome. *Nature Genetics, 36*, 288–292.

Parnas, D., Haghighi, A. P., Fetter, R. D., Kim, S. W., & Goodman, C. S. (2001). Regulation of postsynaptic structure and protein localization by the Rho-type guanine nucleotide exchange factor dPix. *Neuron, 32*, 415–424.

Pawlu, C., DiAntonio, A., & Heckmann, M. (2004). Postfusional control of quantal current shape. *Neuron, 42*, 607–618.

Peled, E. S., & Isacoff, E. Y. (2011). Optical quantal analysis of synaptic transmission in wild-type and rab3-mutant Drosophila motor axons. *Nature Neuroscience, 14*, 519–526.

Petersen, S. A., Fetter, R. D., Noordermeer, J. N., Goodman, C. S., & DiAntonio, A. (1997). Genetic analysis of glutamate receptors in Drosophila reveals a retrograde signal regulating presynaptic transmitter release. *Neuron, 19*, 1237–1248.

Pielage, J., Fetter, R. D., & Davis, G. W. (2006). A postsynaptic spectrin scaffold defines active zone size, spacing, and efficacy at the Drosophila neuromuscular junction. *The Journal of Cell Biology, 175*, 491–503.

Pilgram, G. S., Potikanond, S., van der Plas, M. C., Fradkin, L. G., & Noordermeer, J. N. (2011). The RhoGAP crossveinless-c interacts with Dystrophin and is required for synaptic homeostasis at the Drosophila neuromuscular junction. *Journal of Neuroscience, 31*, 492–500.

Podkowa, M., Zhao, X., Chow, C. W., Coffey, E. T., Davis, R. J., & Attisano, L. (2010). Microtubule stabilization by bone morphogenetic protein receptor-mediated scaffolding of c-Jun N-terminal kinase promotes dendrite formation. *Molecular and Cellular Biology, 30*, 2241–2250.

Prokop, A., Landgraf, M., Rushton, E., Broadie, K., & Bate, M. (1996). Presynaptic development at the Drosophila neuromuscular junction: Assembly and localization of presynaptic active zones. *Neuron, 17*, 617–626.

Qin, G., Schwarz, T., Kittel, R. J., Schmid, A., Rasse, T. M., Kappei, D., Ponimaskin, E., Heckmann, M., & Sigrist, S. J. (2005). Four different subunits are essential for expressing the synaptic glutamate receptor at neuromuscular junctions of Drosophila. *Journal of Neuroscience, 25*, 3209–3218.

Rasse, T. M., Fouquet, W., Schmid, A., Kittel, R. J., Mertel, S., Sigrist, C. B., Schmidt, M., Guzman, A., Merino, C., Qin, G., Quentin, C., Madeo, F. F., Heckmann, M., & Sigrist, S. J. (2005). Glutamate receptor dynamics organizing synapse formation in vivo. *Nat. Neuroscience, 8*, 898–905.

Reiff, D. F., Thiel, P. R., & Schuster, C. M. (2002). Differential regulation of active zone density during long-term strengthening of Drosophila neuromuscular junctions. *Journal of Neuroscience, 22*, 9399–9409.

Renger, J. J., Ueda, A., Atwood, H. L., Govind, C. K., & Wu, C. F. (2000). Role of cAMP cascade in synaptic stability and plasticity: Ultrastructural and physiological analyses of individual synaptic boutons in Drosophila memory mutants. *Journal of Neuroscience, 20*, 3980–3992.

Ritzenthaler, S., & Chiba, A. (2003). Myopodia (postsynaptic filopodia) participate in synaptic target recognition. *Journal of Neurobiology, 55*, 31–40.

Ritzenthaler, S., Suzuki, E., & Chiba, A. (2000). Postsynaptic filopodia in muscle cells interact with innervating motoneuron axons. *Nature Neuroscience, 3*, 1012–1017.

Rodal, A. A., Motola-Barnes, R. N., & Littleton, J. T. (2008). Nervous wreck and Cdc42 cooperate to regulate endocytic actin assembly during synaptic growth. *Journal of Neuroscience, 28*, 8316–8325.

Rohrbough, J., & Broadie, K. (2010). Anterograde Jelly belly ligand to Alk receptor signaling at developing synapses is regulated by Mind the gap. *Development, 137*, 3523–3533.

Rohrbough, J., Rushton, E., Woodruff, E., 3rd, Fergestad, T., Vigneswaran, K., & Broadie, K. (2007). Presynaptic establishment of the synaptic cleft extracellular matrix is required for post-synaptic differentiation. *Genes & Development, 21*, 2607–2628.

Rushton, E., Rohrbough, J., & Broadie, K. (2009). Presynaptic secretion of mind-the-gap organizes the synaptic extracellular matrix-integrin interface and postsynaptic environments. *Developmental Dynamics, 238*, 554–571.

Saitoe, M., Schwarz, T. L., Umbach, J. A., Gundersen, C. B., & Kidokoro, Y. (2001). Absence of junctional glutamate receptor clusters in Drosophila mutants lacking spontaneous transmitter release. *Science, 293*, 514–517.

Saitoe, M., Schwarz, T. L., Umbach, J. A., Gundersen, C. B., & Kidokoro, Y. (2002). Response: Meaningless minis? *Trends in Neurosciences, 25*, 385–386.

Saitoe, M., Tanaka, S., Takata, K., & Kidokoro, Y. (1997). Neural activity affects distribution of glutamate receptors during neuromuscular junction formation in Drosophila embryos. *Developmental Biology, 184*, 48–60.

Sanyal, S., Sandstrom, D. J., Hoeffer, C. A., & Ramaswami, M. (2002). AP-1 functions upstream of CREB to control synaptic plasticity in Drosophila. *Nature, 416*, 870–874.

Schaefer, J. E., Worrell, J. W., & Levine, R. B. (2010). Role of intrinsic properties in Drosophila motoneuron recruitment during fictive crawling. *Journal of Neurophysiology, 104*, 1257–1266.

Schmid, A., Hallermann, S., Kittel, R. J., Khorramshahi, O., Frölich, A. M., Quentin, C., Rasse, T. M., Mertel, S., Heckmann, M., & Sigrist, S. J. (2008). Activity-dependent site-specific changes of glutamate receptor composition in vivo. *Nature Neuroscience, 11*, 659–666.

Schmid, A., Qin, G., Wichmann, C., Kittel, R. J., Mertel, S., Fouquet, W., Schmidt, M., Heckmann, M., & Sigrist, S. J. (2006). Non-NMDA-type glutamate receptors are essential for maturation but not for initial assembly of synapses at Drosophila neuromuscular junctions. *Journal of Neuroscience, 26*, 11267–11277.

Schuster, C. M., Davis, G. W., Fetter, R. D., & Goodman, C. S. (1996a). Genetic dissection of structural and functional components of synaptic plasticity I. Fasciclin II controls synaptic stabilization and growth. *Neuron, 17*, 567–570.

Schuster, C. M., Davis, G. W., Fetter, R. D., & Goodman, C. S. (1996b). Genetic dissection of structural and functional components of synaptic plasticity II. Fasciclin II controls presynaptic structural plasticity. *Neuron, 17*, 655–667.

Schuster, C. M., Ultsch, A., Schloss, P., Cox, J. A., Schmitt, B., & Betz, H. (1991). Molecular cloning of an invertebrate glutamate receptor subunit expressed in Drosophila muscle. *Science, 254*, 112–114.

Sigrist, S. J., & Andlauer, T. F. (2011). Fighting the famine with an amine: Synaptic strategies for smart search. *Nature Neuroscience, 14*, 124–126.

Sigrist, S. J., Reiff, D. F., Thiel, P. R., Steinert, J. R., & Schuster, C. M. (2003). Experience-dependent strengthening of Drosophila neuromuscular junctions. *Journal of Neuroscience, 23*, 6546–6556.

Sigrist, S. J., & Schmitz, D. (2011). Structural and functional plasticity of the cytoplasmic active zone. *Current Opinion in Neurobiology, 21*, 144–150.

Sigrist, S. J., Thiel, P. R., Reiff, D. F., Lachance, P. E., Lasko, P., & Schuster, C. M. (2000). Postsynaptic translation affects the efficacy and morphology of neuromuscular junctions. *Nature, 405*, 1062–1065.

Sigrist, S. J., Thiel, P. R., Reiff, D. F., & Schuster, C. M. (2002). The postsynaptic glutamate receptor subunit DGluR-IIA mediates long-term plasticity in Drosophila. *Journal of Neuroscience, 22*, 7362–7372.

Stewart, B. A., Schuster, C. M., Goodman, C. S., & Atwood, H. L. (1996). Homeostasis of synaptic transmission in Drosophila with genetically altered nerve terminal morphology. *Journal of Neuroscience, 16*, 3877–3886.

Stowers, R. S., & Schwarz, T. L. (1999). A genetic method for generating Drosophila eyes composed exclusively of mitotic clones of a single genotype. *Genetics, 152*, 1631–1639.

Sun, M., Liu, L., Zeng, X., Xu, M., Liu, L., Fang, M., & Xie, W. (2009). Genetic interaction between Neurexin and CAKI/CMG is important for synaptic function in Drosophila neuromuscular junction. *Neuroscience Research, 64*, 362–371.

Sun, M., Xing, G., Yuan, L., Gan, G., Knight, D., With, S. I., He, C., Han, J., Zeng, X., Fang, M., Boulianne, G. L., & Xie, W. (2011). Neuroligin 2 is required for synapse development and function at the Drosophila neuromuscular junction. *Journal of Neuroscience, 31*, 687–699.

Sweeney, S. T., & Davis, G. W. (2002). Unrestricted synaptic growth in spinster-a late endosomal protein implicated in TGF-beta-mediated synaptic growth regulation. *Neuron, 36*, 403–416.

Thomas, U., Kim, E., Kuhlendahl, S., Koh, Y. H., Gundelfinger, E. D., Sheng, M., Garner, C. C., & Budnik, V. (1997). Synaptic clustering of the cell adhesion molecule fasciclin II by discs-large and its role in the regulation of presynaptic structure. *Neuron, 19*, 787–799.

Thomas, U., Kobler, O., & Gundelfinger, E. D. (2010). The Drosophila larval neuromuscular junction as a model for scaffold complexes at glutamatergic synapses: Benefits and limitations. *Journal of Neurogenetics, 24*, 109–119.

Thomson, A. M. (2000). Facilitation, augmentation and potentiation at central synapses. *Trends in Neurosciences, 23*, 305–312.

van der Plas, M. C., Pilgram, G. S., Plomp, J. J., de Jong, A., Fradkin, L. G., & Noordermeer, J. N. (2006). Dystrophin is required for appropriate retrograde control of neurotransmitter release at the Drosophila neuromuscular junction. *Journal of Neuroscience, 26*, 333–344.

Varoqueaux, F., Sigler, A., Rhee, J. S., Brose, N., Enk, C., Reim, K., & Rosenmund, C. (2002). Total arrest of spontaneous and evoked synaptic transmission but normal synaptogenesis in the absence of Munc13-mediated vesicle priming. *Proceedings of the National Academy of Sciences of the United States of America, 99*, 9037–9042.

Venken, K. J., Carlson, J. W., Schulze, K. L., Pan, H., He, Y., Spokony, R., Wan, K. H., Koriabine, M., de Jong, P. J., White, K. P., Bellen, H. J., & Hoskins, R. A. (2009). Versatile P[acman] BAC libraries for transgenesis studies in *Drosophila melanogaster*. *Nature Methods, 6*, 431–434.

Venken, K. J., He, Y., Hoskins, R. A., & Bellen, H. J. (2006). P[acman]: A BAC transgenic platform for targeted insertion of large DNA fragments in *D. melanogaster*. *Science, 314*, 1747–1751.

Verstreken, P., & Bellen, H. J. (2002). Meaningless minis? Mechanisms of neurotransmitter-receptor clustering. *Trends in Neurosciences, 25*, 383–385.

Verstreken, P., Ly, C. V., Venken, K. J., Koh, T. W., Zhou, Y., & Bellen, H. J. (2005). Synaptic mitochondria are critical for mobilization of reserve pool vesicles at Drosophila neuromuscular junctions. *Neuron, 47*, 365–378.

Wagh, D. A., Rasse, T. M., Asan, E., Hofbauer, A., Schwenkert, I., Dürrbeck, H., Buchner, S., Dabauvalle, M. C., Schmidt, M., Qin, G., Wichmann, C., Kittel, R., Sigrist, S. J., & Buchner, E. (2006). Bruchpilot, a protein with homology to ELKS/CAST, is required for structural integrity and function of synaptic active zones in Drosophila. *Neuron, 49*, 833–849.

Wan, H. I., DiAntonio, A., Fetter, R. D., Bergstrom, K., Strauss, R., & Goodman, C. S. (2000). Highwire regulates synaptic growth in Drosophila. *Neuron, 26*, 313–329.

Wang, X., Shaw, W. R., Tsang, H. T., Reid, E., & O'Kane, C. J. (2007). Drosophila spichthyin inhibits BMP signaling and regulates synaptic growth and axonal microtubules. *Nature Neuroscience, 10*, 177–185.

Weyhersmüller, A., Hallermann, S., Wagner, N., & Eilers, J. (2011). Rapid active zone remodeling during synaptic plasticity. *Journal of Neuroscience, 31*, 6041–6052.

Wichmann, C., & Sigrist, S. J. (2010). The active zone T-bar – A plasticity module? *Journal of Neurogenetics, 24*, 133–145.

Wu, H., Xiong, W. C., & Mei, L. (2010). To build a synapse: Signaling pathways in neuromuscular junction assembly. *Development, 137*, 1017–1033.

Yi, J. J., Barnes, A. P., Hand, R., Polleux, F., & Ehlers, M. D. (2010). TGF-beta signaling specifies axons during brain development. *Cell, 142*, 144–157.

Yoshihara, M., Rheuben, M. B., & Kidokoro, Y. (1997). Transition from growth cone to functional motor nerve terminal in Drosophila embryos. *Journal of Neuroscience, 17*, 8408–8426.

Yoshihara, M., Adolfsen, B., Galle, K. T., & Littleton, J. T. (2005). Retrograde signaling by Syt 4 induces presynaptic release and synapse-specific growth. *Science, 310*, 858–863.

Zhai, R. G., & Bellen, H. J. (2004). The architecture of the active zone in the presynaptic nerve terminal. *Physiology, 19*, 262–270.

Zhai, R. G., Cao, Y., Hiesinger, P. R., Zhou, Y., Mehta, S. Q., Schulze, K. L., Verstreken, P., & Bellen, H. J. (2006). Drosophila NMNAT maintains neural integrity independent of its NAD synthesis activity. *PLoS Biology, 4*, e416.

Zhai, R. G., Zhang, F., Hiesinger, P. R., Cao, Y., Haueter, C. M., & Bellen, H. J. (2008). NAD synthase NMNAT acts as a chaperone to protect against neurodegeneration. *Nature, 452*, 887–891.

Zhong, Y., Budnik, V., & Wu, C. F. (1992). Synaptic plasticity in Drosophila memory and hyperexcitable mutants: Role of cAMP cascade. *Journal of Neuroscience, 12*, 644–651.

Zhong, Y., & Wu, C. F. (1991). Altered synaptic plasticity in Drosophila memory mutants with a defective cyclic AMP cascade. *Science, 251*, 198–201.

Zito, K., Fetter, R. D., Goodman, C. S., & Isacoff, E. Y. (1997). Synaptic clustering of Fascilin II and Shaker: Essential targeting sequences and role of Dlg. *Neuron, 19*, 1007–1016.

Zito, K., Parnas, D., Fetter, R. D., Isacoff, E. Y., & Goodman, C. S. (1999). Watching a synapse grow: Noninvasive confocal imaging of synaptic growth in Drosophila. *Neuron, 22*, 719–729.

Chapter 2
Scaffold Proteins at the Postsynaptic Density

Chiara Verpelli, Michael J. Schmeisser, Carlo Sala, and Tobias M. Boeckers

Abstract Scaffold proteins are abundant and essential components of the postsynaptic density (PSD). They play a major role in many synaptic functions including the trafficking, anchoring, and clustering of glutamate receptors and adhesion molecules. Moreover, they link postsynaptic receptors with their downstream signaling proteins and regulate the dynamics of cytoskeletal structures. By definition, PSD scaffold proteins do not have intrinsic enzymatic activities but are formed by modular and specific domains deputed to form large protein networks. Here, we will discuss the latest findings regarding the structure and functions of major PSD scaffold proteins.

Given that scaffold proteins are central components of PSD architecture, it is not surprising that deletion or mutations in their human genes cause severe neuropsychiatric disorders including autism, mental retardation, and schizophrenia. Thus, their dynamic organization and regulation are directly correlated with the essential structure of the PSD and the normal physiology of neuronal synapses.

Keywords BMP signaling • Development • Glutamate receptors • Neuromuscular junction • Wnt signaling

C. Verpelli • C. Sala (✉)
CNR Institute of Neuroscience, Department of Pharmacology, University of Milan and Neuromuscular Diseases and Neuroimmunology, Neurological Institute Foundation Carlo Besta, Via Vanvitelli 32, 20129 Milan, Italy
e-mail: c.sala@cnr.it

M.J. Schmeisser • T.M. Boeckers (✉)
Institute of Anatomy and Cell Biology, Ulm University Faculty of Medicine, Albert Einstein Allee 11, 89081 Ulm, Germany
e-mail: tobias.boeckers@uni-ulm.de

M.R. Kreutz and C. Sala (eds.), *Synaptic Plasticity*,
Advances in Experimental Medicine and Biology 970,
DOI 10.1007/978-3-7091-0932-8_2, © Springer-Verlag/Wien 2012

2.1   The MAGUK Family

2.1.1   Structural Organization of the MAGUK Proteins

Genetic and biochemical studies over the past 20 years have identified the membrane-associated guanylate kinases (MAGUKs) as ubiquitous scaffolding molecules concentrated at sites of cell-cell contact such as synapses (Craven and Bredt 1998; Kornau et al. 1997; Sheng and Sala 2001; Sheng and Kim 2002). MAGUK members include SAP90/PSD-95, SAP102, SAP97, Chapsyn 110/PSD93, and p55 (SAP = synapse-associated protein). They represent a superfamily of multidomain proteins that are related by the presence of a shared set of structural domains. The defining feature of MAGUKs is the presence of a region of approximately 300 amino acids at the C-terminus with homology to yeast guanylate kinase (GK), which catalyzes the ATP-dependent phosphorylation of GMP to GDP. Curiously, the GK domain in MAGUKs is catalytically inactive (Olsen and Bredt 2003), but it is always accompanied by either a preceding SH3 (Src homology 3) domain or followed closely by a WW (two conserved Trp residues) motif. Also, MAGUKs always contain PDZ (PSD-95/DLG/ZO-1) domains (in most cases three), and all these modular motifs in MAGUKs mediate protein-protein interactions.

PDZ domains typically bind specific C-terminal sequences in target proteins (Kim et al. 1995; Kornau et al. 1995). However, some PDZ domains can heterodimerize (Brenman et al. 1996a). Several structures of isolated PDZ domains by x-ray crystallography and three-dimensional nuclear magnetic resonance (NMR) spectroscopy reveal that PDZ domains are compact and modular. Interestingly, a number of recent studies have demonstrated that two or more PDZ domains connected in tandem often display target-binding properties that are distinct from those of each isolated domain or even the simple sum of the isolated PDZ domains (Long et al. 2003). For PSD-95, the linking sequence between the first two PDZ domains is formed by five residues, rigid and highly conserved, suggesting that it might reduce interdomain movement rather than simply function as a passive linker. The structure of the PDZ1 and 2 tandem showed that the two PDZ domains indeed contact each other in a side-by-side manner and that two of their target-binding grooves point in directions that are favorable for binding to the tails of multimeric transmembrane proteins extending from the membrane surface into the cytoplasm including NMDA receptors (NMDARs). Mutations that increase the length of the interdomain linker impaired the supramodular nature of PDZ1 and 2 of PSD-95 which then displayed weaker binding to dimeric targets and a decreased capacity in clustering (Long et al. 2003).

Like PDZ domains, SH3 domains are protein-protein interaction modules that commonly occur in proteins with widely divergent functions (Kuriyan and Cowburn 1997). SH3 domains typically bind to polyproline motifs (ProXXPro); however, MAGUK SH3 domains rarely bind to such ProXXPro-containing sequences. One of the few is the proline-rich C-terminus of α-secretase ADAM10 that binds to the SH3 domain of SAP97 (Marcello et al. 2007). Conversely, numerous ligands bind with high affinity to the GK domain of MAGUKs. These ligands include, for example, guanylate kinase-associated proteins (GKAPs) (Kim et al. 1997; Takeuchi et al. 1997). In addition to their interaction with downstream signaling proteins, GK domains in MAGUKs bind to their SH3 motifs, preferentially in an intramolecular fashion (Shin et al. 2000; McGee et al. 2001). As for the PDZ1 and 2, the crystal structures of the PSD-95-SH3-GK tandem revealed that the SH3 domain and the GK domain pack extensively with each other to form an integral structural unit as an integral supramodule required for the proper functioning of PSD-95. Indeed, disruption of the SH3-GK interaction compromised PSD-95-mediated clustering properties (Hsueh and Sheng 1999), and mutations that disrupt SH3-GK packing in the only PSD-95 family MAGUK in *Drosophila melanogaster* (DLG1) resulted in a tumorigenic phenotype of larval imaginal discs (Woods et al. 1996).

2.1.2 Interactions and Functional Properties of MAGUK Family Members at the PSD

Mammalian brain is the tissue expressing the greatest diversity of MAGUK proteins. In each synapse, a typical PSD is composed of a huge complex protein network consisting of several hundred different proteins whereas MAGUK family members are of crucial importance. They are localized at CNS glutamatergic synapses (Garner et al. 2000; Aoki et al. 2001) as well as at cholinergic synapses (Conroy et al. 2003). As modular proteins, it has often been hypothesized that the most likely function of PSD-MAGUKs is being central organizers of vertebrate CNS synapses. They are, in fact, key scaffold proteins determining the steady state as well as the activity-dependent changes of glutamate receptor numbers in excitatory synapses (Elias and Nicoll 2007) (Fig. 2.1).

PSD-95 (also named SAP90), a major complex of the PSD fraction, can be seen as the prototypical best characterized MAGUK protein of the PSD. It is now clear that the most important function of PSD-95 is to organize signaling complexes at the postsynaptic membrane. However, the amount of PSD-95 can also regulate the balance between the number of inhibitory and excitatory synapses (Levinson and El-Husseini 2005). PSD-95 interacts with a large variety of molecules and thus, by physically bringing together cytoplasmic signal transduction proteins and surface receptors, may facilitate the coupling of various signaling cascades within the PSD. More than 15 years ago, the first and second PDZ domains of PSD-95 (PDZ1 and 2) were described to bind the extreme C-termini of Kiv1.4 of the Shaker K^+ channels (Kim et al. 1995) and of NR2A/B subunits of the NMDARs (Cho et al. 1992; Kistner et al. 1993; Kornau et al. 1995; Niethammer et al. 1996). Since then, a whole variety of proteins associated to the distinct domains of PSD-95 has emerged. Here is a list of the most relevant and recently described ones.

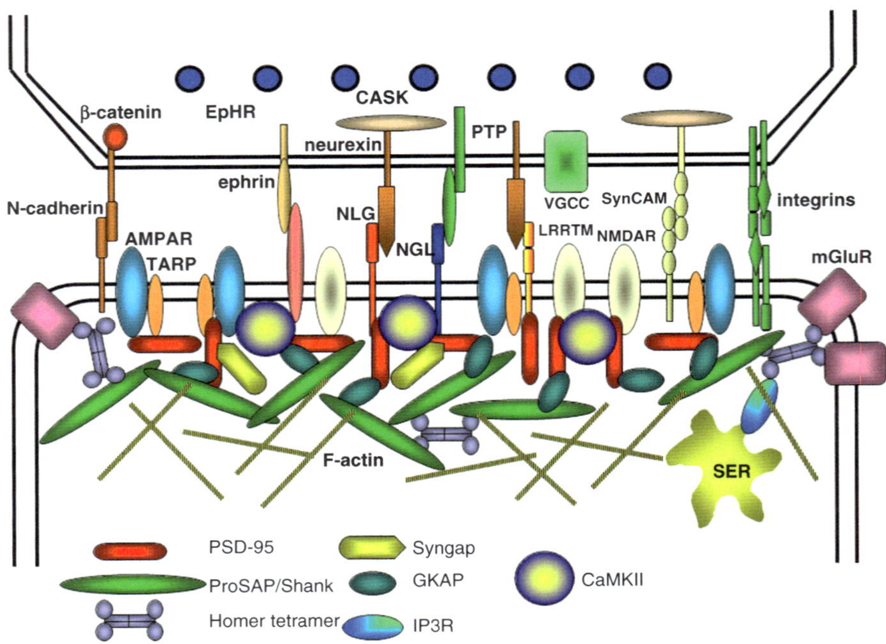

Fig. 2.1 The diagram shows a schematic organization of the protein network in the PSD. Only major families and certain classes of PSD proteins are shown. The interaction between them is schematically indicated

The other glutamate receptor that directly binds to PSD-95 is the kainate receptor whose KA2 subunit was shown to be linked at the SH3-GK domains of PSD-95 (Garcia et al. 1998). The second major class of transmembrane proteins that bind to PSD-95 are the synaptic cell adhesion molecules (SynCAMs). The neurexin transmembrane ligand, neuroligin, binds to the third PDZ domain of PSD-95 (PDZ3) (Irie et al. 1997; Song et al. 1999). Neuroligin (NLGN) is an adhesion molecule with the ability to induce synapse formation. Interestingly, a significant subset of SynCAMs including NGLNs, synaptic-adhesion-like molecules (SALMs), ADAM22, and leucine-rich repeat transmembrane proteins (LRRTMs) associated with PSD-95, suggesting that they may act in concert to couple trans-synaptic adhesion to the molecular organization of synaptic proteins. Thus, PSD-95 may be one of the central organizers that recruits diverse proteins to sites of synaptic adhesion, promotes trans-synaptic signaling, and couples neuronal activity with changes in synaptic adhesion (Han and Kim 2008; Margeta and Shen 2010).

Importantly, PSD-95 greatly influences synaptic transmission and plasticity mainly because it recruits the stargazin tetraspanning membrane protein to synapses via binding to its C-terminus with PDZ1 and 2. Stargazin and its relatives are associated with AMPA receptors (AMPARs) and are essential for their surface expression, surface diffusion, synaptic accumulation, and function (Chen et al. 2000;

Tomita et al. 2005). These data may explain why PSD-95 overexpression potentiates AMPAR-mediated excitatory postsynaptic currents (EPSCs), but not the currents of the directly linked NMDAR (Elias et al. 2006; Sumioka et al. 2010). The role of PSD-95 in regulating AMPAR number at the PSD can also be mediated by the interaction with other proteins modulating AMPAR internalization. Han et al. showed that a regulated interaction of the endocytic adaptor RalBP1 with the small GTPase RalA and PSD-95 controls NMDAR-dependent AMPAR endocytosis during LTD. NMDAR activation brings RalBP1 close to PSD-95 to promote the interaction of RalBP1-associated endocytic proteins with PSD-95-associated AMPARs (Han et al. 2009). Similarly, Bhattacharyya et al. suggest that interaction of PSD-95 and calcineurin with A kinase anchoring protein AKAP150 is critical for NMDAR-triggered AMPAR endocytosis and LTD (Bhattacharyya et al. 2009). How these two mechanisms are functionally connected remains to be determined.

The second PSD-95 PDZ domain (PDZ2) can also bind to the PDZ domain in neuronal nitric oxide synthase (nNOS) (Brenman et al. 1996a, b). nNOS is a Ca^{2+}-/ calmodulin-activated enzyme that produces the nitric oxide involved in neurotransmission and excitotoxicity. Interestingly, the ternary NMDAR-PSD-95-nNOS complex may functionally couple NMDAR gating to nNOS activation, as it is suggested by the observation that disrupting the NMDAR-PSD-95 interaction with a synthetic peptide that mimics the last nine residues of NR2B or a synthetic compound that blocks the interaction between PSD-95 and nNOS reduces NMDAR-induced excitotoxicity in vitro and in vivo without affecting NMDAR function (Aarts et al. 2002; Zhou et al. 2010). Recently, it has been shown that the interaction with nNOS is required for the ability of PSD-95 to regulate synaptogenesis and multi-innervated dendritic spines suggesting a physiological role for the NMDAR-PSD-95-nNOS complex at synapses (Nikonenko et al. 2008).

Several other binding partners of PSD-95 are scaffold proteins and regulators or effectors of small GTPases. The synaptic GTPase-activating protein for Rac, SynGAP, can interact with all three PDZ domains of PSD-95 via its C-terminus (Chen et al. 1998; Kim et al. 1998). The SH3-GK of PSD-95 also binds the spine-associated Rap-Gap SPAR (Pak et al. 2001), AKAP (Bhattacharyya et al. 2009), SPIN90/WISH (Kim et al. 2009), MAP1a (Reese et al. 2007), Preso, and other scaffold proteins such as the four members of the GKAP family (Kim et al. 1997; Takeuchi et al. 1997). Most of these interactions have been implicated in the regulation of both the size and the number of spines and synapses (Brenman et al. 1998; Kim et al. 1998; Colledge et al. 2000; Pak et al. 2001; Vazquez et al. 2004). Finally, a nuclear protein, AIDA-1d, has been identified to interact with PDZ1 and 2 of PSD-95 and to shuttle between the synapse and the nucleus. Synaptic activity induces a Ca^{2+}-independent translocation of AIDA-1d to the nucleus, where it couples to Cajal bodies and increases nucleolar numbers and protein synthesis thus linking synaptic activity and protein biosynthetic capacity (Jordan et al. 2007).

It is now important to underline individual localization and expression characteristics of PSD-MAGUK family members. Each of the MAGUK proteins shows a different distribution in respect to subcellular compartments of the brain.

PSD-95 and PSD-93 are highly enriched in the postsynaptic density (PSD), especially due to their high palmitoylation degree (El-Husseini et al. 2000). SAP102 and SAP97 are found in dendrites and axons and are abundant in the cytoplasm as well as at synapses. Further, PSD-MAGUK proteins exhibit a distinct developmental expression pattern: SAP102 is highly expressed and functionally dominates in early postnatal development, whereas PSD-95 and PSD-93 predominate at later stages (Sans et al. 2000; Elias et al. 2006). Regarding 3D-structure, there are further differences among the PSD-MAGUKs. Negative stain images of PSD-95 and SAP97 suggest that these two highly related proteins are in fact adopting different shapes. PSD-95 monomers are relatively compact whereas SAP97 monomers are relatively extended rod shapes that tend to dimerize (Nakagawa et al. 2004). However, both proteins contain alternative N-termini, expressing either an L27 domain (beta-isoform) or double cysteines that are normally palmitoylated (alpha-isoforms) (Schlüter et al. 2006).

In vivo, MAGUK family members apparently interact with different, but overlapping, sets of proteins with PSD-95 and PSD-93 being preferentially associated with the NR2A and SAP102 with the NR2B subunit of the NMDAR (Sans et al. 2000). This phenomenon suggests that the properties of the NR2B-SAP102 complex may be different from those of the NR2A-PSD-95/PSD-93 complex and that the functional properties of synaptic NMDARs may depend on the prevalence of one or the other (Kim et al. 2005). However, also PSD-93 and PSD-95 may have opposite roles in regulating LTP (Carlisle et al. 2008). SAP102 and SAP97 are involved in the trafficking of NMDARs and AMPARs, respectively. By interacting with the PDZ-binding domain of Sec8, SAP102 can associate with the exocyst complex and regulate the delivery of NMDARs to the surface of neuronal cells (Sans et al. 2003). SAP97 directly interacts with the AMPAR GluR1 subunit (Leonard et al. 1998), and the fact that the SAP97-GluR1 complex has been found early in the secretory pathway indicates that SAP97 can regulate the trafficking of GluR1 (Sans et al. 2001). CaMKII phosphorylation of SAP97 in the N-terminal L27 domain promotes the synaptic targeting of SAP97 and GluR1 (Mauceri et al. 2004). To some extent like PSD-95, the overexpression of SAP97 increases the number of synaptic AMPARs, induces spine enlargement, and increases the frequency of miniature EPSCs (mEPSCs) (Rumbaugh et al. 2003; Nakagawa et al. 2004; Howard et al. 2010). The abundance of PDZ scaffold proteins in synapses with overlapping targets for interaction raises questions regarding the specificity redundancy of the scaffolds. For example, PSD-95, PSD-93, SAP102, and SAP97 are expressed in excitatory synapses, and each of these MAGUKs can mediate the trafficking of glutamate receptors at different developmental stages (Sans et al. 2000; Elias et al. 2006). Knockout studies in mice have revealed that MAGUKs have in part the ability to functionally compensate for each other (Migaud et al. 1998), but only a correct highly interconnected MAGUK system is assuring appropriate glutamate receptor expression and localization at synapses (Elias and Nicoll 2007).

2.1.3 Synaptic Localization and Spatial Regulation of the MAGUK PSD-95

Spatially, PSD-95 is closely associated to membrane receptors and ion channels and seems to be arranged perpendicular to the PSD membrane (Chen et al. 2008). On the ultrastructural level, the PSD can anatomically be divided into three layers: the first layer mainly contains membrane receptors, ion channels, and CAMs, with NMDARs at the center and AMPARs at the periphery; the second layer is enriched with MAGUK proteins, in particular PSD-95, which are closely coupled to the membrane receptors and ion channels; the third layer is comprised of ProSAP/Shank and GKAPs (see the following paragraphs) (Petralia et al. 1994; Valtschanoff and Weinberg 2001). Synaptic localization of PSD-95 depends on the palmitoylation of two N-terminal cysteines (Cys3 and Cys5) (Craven et al. 1999), and synaptic activity induces the removal of PSD-95 by depalmitoylation of the two Cys residues (El-Husseini Ael et al. 2002). A set of enzymes capable of inducing PSD-95 palmitoylation has recently been identified, but some controversy remains as to which of them is specific for PSD-95, and only one of these, the palmitoyl transferase DHHC2, seems to be regulated by synaptic activity (Fukata et al. 2004; Huang et al. 2004; Noritake et al. 2009). PSD-95 can be degraded through the ubiquitin-proteasome pathway by means of direct ubiquitylation (Colledge et al. 2003) or indirectly, via the ubiquitylation and degradation of its interacting protein SPAR (Pak and Sheng 2003). Two different phosphorylation sites have been identified on PSD-95 with opposite effects. Phosphorylation of serine 295 by JNK-1 enhances the synaptic accumulation of PSD-95 (Pavlowsky et al. 2010), while trafficking of PSD-95 to synapses is inhibited by activity-dependent CaMKII phosphorylation at serine 73 (Steiner et al. 2008). In general, the activity-dependent accumulation, dispersal, or degradation of PSD-95 is often associated with an increase or loss of synaptic AMPARs, strengthened or weakened synapses, and changes in glutamate-receptor-induced intracellular signaling such as CREB and MAPK phosphorylation (Ehlers 2003).

2.2 The ProSAP/Shank Family

2.2.1 Molecular Composition and Expression Profile of the ProSAP/Shank Family

The ProSAP/Shank family of scaffold proteins consists of three members all highly enriched in the PSD and localized at the interface between membrane receptors and cytoskeletal elements: Shank1 (also named Shank1a, Synamon, or SSTRIP), ProSAP1/Shank2 (also named CortBP1), and ProSAP2/Shank3 (Boeckers et al. 1999a, b; Naisbitt et al. 1999; Yao et al. 1999). These molecules contain multiple domains which are essential for various protein-protein interactions within the PSD: N-terminal ankyrin repeats, an SH3 domain, a PDZ domain, a proline-rich domain, and a sterile alpha motif (SAM) domain. The name ProSAP (*pro*line-rich *s*ynapse-*a*ssociated *p*rotein) derives from proline-rich clusters that are conserved among all family members (Boeckers et al. 1999a, b), while the term Shank reflects the *SH*3 domain and multiple *ank*yrin repeats (Naisbitt et al. 1999). ProSAPs/ Shanks are large proteins with a molecular mass of more than 180 kDa (Boeckers et al. 1999a). All three share 63–87% amino acid identity while SH3, PDZ, and SAM domains are conserved at the highest level. Shank1 is only expressed in brain (Yao et al. 1999); ProSAP1/Shank2 also appears in non-neuronal tissue like pancreas, pituitary, lung, liver, kidney, and testis (Redecker et al. 2001, 2003; McWilliams et al. 2004, 2005; Dobrinskikh et al. 2010); and ProSAP2/Shank3 has been detected in almost every tissue examined (Lim et al. 1999). Interestingly, all three ProSAP/Shank family members show a distinct expression pattern in *Xenopus laevis* embryos indicating a functional role of this protein family not only in the adult organism but also in embryonic development (Gessert et al. 2011). Within the nervous system, ProSAP/Shank expression is not limited to cortical areas but has further been detected in glial cells (Redecker et al. 2001), at olfactory cilia membranes (Saavedra et al. 2008) and at postsynaptic specializations of retinal (Brandstätter et al. 2004), and various peripheral synapses (Raab et al. 2010). However, all three family members are highly expressed in the hippocampus and cortex. In the cerebellum, Shank1 and ProSAP1/Shank2 primarily appear in Purkinje cells, while ProSAP2/Shank3 is only found in the granular cell layer (Boeckers et al. 1999a, b, 2004). On the subcellular level, ProSAPs/Shanks are not localized directly underneath the postsynaptic membrane but extend up to 120 nm deep inside the PSD (Naisbitt et al. 1999; Tao-Cheng et al. 2010). Alternative splicing events regulate ProSAP/Shank domain composition (Boeckers et al. 1999b; Lim et al. 1999). Shank2E, for example, one of the two major alternative splice variants of ProSAP1/Shank2, is only expressed in epithelial cells and contains an SH3 domain and N-terminal ankyrin repeats like ProSAP2/ Shank3 and Shank1 (McWilliams et al. 2004). The other major alternative splice variant of ProSAP1/Shank2 called ProSAP1A misses the ankyrin repeats, but still includes the SH3 domain (Boeckers et al. 1999a, b). Further, there is knowledge of an alternatively spliced Shank1 lacking the SAM domain called Shank1b (Sala et al. 2001). Interestingly and contrary to Shank1 and ProSAP1/Shank2, tissue-specific expression of the ProSAP2/Shank3 gene is exclusively regulated by DNA methylation (Ching et al. 2005; Beri et al. 2007). A recent study further demonstrates that methylation of the ProSAP2/Shank3 gene predominantly happens at intragenic CpG island promoters. Thus, alternative transcripts are generated and expressed differentially not only in a tissue- and cell-type-specific manner but even within the same cell types from distinct brain regions (Maunakea et al. 2010) (Fig. 2.2).

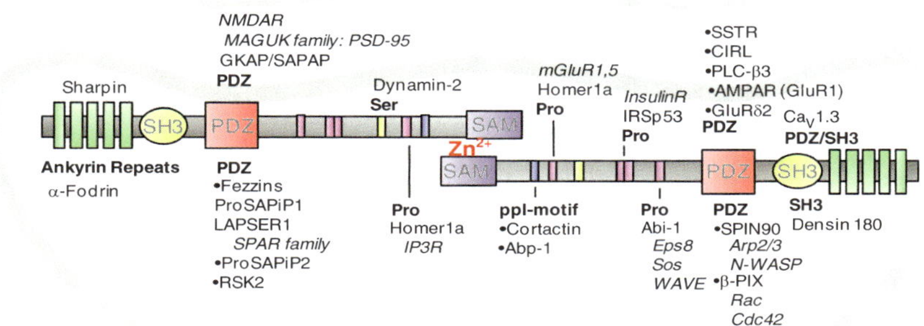

Fig. 2.2 The diagram shows the domain composition of ProSAP/Shank family members at the PSD (the three major domains, SH3, PDZ, and SAM, are clearly depicted). For each domain/interacting motif, direct binding partners as well as indirect ones (*italicized* letters) are listed

2.2.2 *Synaptic Recruitment and Assembly of ProSAP/Shank Family Members*

Proper formation and maturation of a synapse requires the specific localization of proteins at both sites of the contact. A noticeable feature of all three ProSAP/Shank mRNAs is their strong dendritic localization, which – in response to the appropriate stimuli – makes local translation directly at the site of spines and synapses probable (Bockers et al. 2004; Falley et al. 2009). The postsynaptic recruitment of ProSAP/ Shank family members depends on certain amino acid sequences within these molecules called synaptic targeting signals. Shank1, for example, only requires an intact PDZ domain for synaptic localization (Sala et al. 2001; Romorini et al. 2004). In contrast, C-terminal elements of ProSAP1/Shank2 and ProSAP2/Shank3 including the SAM domain, the ppI motif, and a serine-rich stretch of approximately 50 amino acids (Okamoto et al. 2001; Boeckers et al. 2005; Grabrucker et al. 2009) are responsible for synaptic targeting. Interestingly, high turnover rates of ProSAP/Shank family members have been observed at single PSDs in immature neuronal cultures (Bresler et al. 2004). Additional in vitro studies have revealed that ProSAP1/Shank2 and ProSAP2/Shank3 are core elements of newly formed PSDs at nascent synapses, while Shank1 is only recruited during the later process of synapse maturation (Sala et al. 2001; Boeckers et al. 2005; Grabrucker et al. 2011). These observations are supported by the fact that transient expression of ProSAP2/Shank3 is sufficient to induce functional dendritic spines in aspiny cultured cerebellar granule cells while Shank1 is not capable to induce the same effect (Roussignol et al. 2005). Recent studies show that the SAM domains of ProSAP1/Shank2 and

ProSAP2/Shank3 are not only crucial for postsynaptic targeting but, due to their oligomerization ability, further mediate assembly of large ProSAP/Shank sheets via zinc ions thus forming a postsynaptic platform (PSP) (Baron et al. 2006; Gundelfinger et al. 2006; Grabrucker et al. 2011). Zinc ions are located in and released from presynaptic vesicles of glutamatergic terminals, can bind glutamate receptors at hippocampal synapses, and enter the postsynaptic compartment (Assaf and Chung 1984); (Li et al. 2001). Whereas Zn^{2+} binding regulates the packaging density of ProSAP2/Shank3 (Baron et al. 2006), Shank1 seems to stabilize synapses in a Zn^{2+}-insensitive mechanism (Grabrucker et al. 2011). Sharpin, whose C-terminal part has been shown to interact with the ankyrin repeats of Shank1, can dimerize through its N-terminal half further cross-linking ProSAPs/Shanks (Lim et al. 2001). Shank1 is also able to multimerize via homomeric attachment of the ankyrin repeats and the SH3 domain (Romorini et al. 2004). All these mechanisms of gigantic ProSAP/Shank multimerization are key events in forming a polymeric network structure that – together with postsynaptic Homer – resides at the core of the PSD (Hayashi et al. 2009). Interestingly, the oligomerization state of certain ProSAP/Shank interaction domains such as the PDZ domain is known to regulate the binding affinity of partner ligands thus rendering ProSAP/Shank platforms even more diverse in respect to multiprotein complex formation at the PSD (Iskenderian-Epps and Imperiali 2010). Among PSDs of cultured neurons, continuous loss and redistribution has been shown for ProSAP2/Shank3, a phenomenon that was independent from protein synthesis or degradation and could be accelerated by electrophysiological stimulation (Tsuriel et al. 2006). This might result from activity-dependent changes of ProSAP/Shank levels and/or their subcellular distribution mediated by secondary modifications such as palmitoylation or phosphorylation. In fact, phosphorylation sites have already been found in ProSAP2/Shank3 (Jaffe et al. 2004), and the ribosomal S6 kinase (RSK) has been identified to interact with and phosphorylate Shank1 and ProSAP2/Shank3 (Thomas et al. 2005).

2.2.3 Clustering of Receptor Complexes at the PSD by ProSAPs/Shanks

Located underneath the postsynaptic membrane, ProSAP/Shank scaffolds are primarily involved in the recruitment, clustering, and functional coupling of transmembrane proteins like postsynaptic glutamate receptors while interaction can happen directly or indirectly via adaptor proteins. Among the latter are the GKAPs. These molecules are attached to the ProSAP/Shank PDZ domain via their C-termini and via their N-termini associated with MAGUKs such as PSD-95 (Boeckers et al. 1999a; Naisbitt et al. 1999). Interestingly, the GluR1 subunit of the AMPAR directly binds to the PDZ domain of all ProSAP/Shank family members via its C-terminal PDZ domain-binding motif (Uchino et al. 2006). Binding partner of the proline-rich stretch right next to the serine-rich region of Shank1 and ProSAP2/Shank3 is Homer1a, a protein which clusters metabotropic glutamate receptors (mGluRs) mGluR1a and mGluR5 at the PSD via their C-termini and further interacts with the inositol trisphosphate receptor (IP$_3$R) of the spine apparatus (Tu et al. 1999; Sala et al. 2001, 2005). In this context, it has been shown that a ternary complex composed of ProSAP1/Shank2, its PDZ domain interaction partner phospholipase β-3, and Homer1b contributes to mGluR-evoked calcium mobilization (Hwang et al. 2003). Moreover, the GluRδ2 subunit has been introduced as interactor of the Shank1 and ProSAP1/Shank2 PDZ domains selectively at the PSDs of parallel fiber-Purkinje cell synapses in the cerebellum (Uemura et al. 2004). A short serine-rich sequence of Shank1 and ProSAP1/Shank2 proteins further mediates direct association with the proline-rich region of dynamin-2, a PSD molecule participating in membrane turnover and glutamate receptor recycling (Okamoto et al. 2001; Boeckers et al. 2002). All those interactions define the crucial role of the ProSAP/Shank-based scaffolding network in cross-linking distinct glutamate receptor subtypes to each other and to intracellular calcium stores. However, not only glutamate receptors but also other G-protein-coupled receptors and voltage-gated ion channels depend on ProSAP/Shank presence at the PSD. Among the G-protein-coupled receptors, the neurotransmission-related somatostatin receptor 2 (SSTR2) has been identified as being clustered at the PSD by all three ProSAP/Shank family members via PDZ domain interaction as well as the calcium-independent receptor (CIRL) for α-latrotoxin that binds to the PDZ domains of Shank1 and ProSAP1/Shank2 and may take part in cell adhesion (Zitzer et al. 1999; Kreienkamp et al. 2000). Furthermore, two studies have shown that the voltage-gated L-type calcium channel Ca$_V$ 1.3 binds to the SH3 and/or the PDZ domains of Shank1 and ProSAP2/Shank3, interactions that tend to be important for linking calcium influx to pCREB signaling (Olson et al. 2005; Zhang et al. 2005).

2.2.4 Association of ProSAPs/Shanks with the Postsynaptic Cytoskeleton

The dynamical interplay of the postsynaptic ProSAP/Shank protein scaffold with the cytoskeleton of the dendritic spine is accomplished via proteins directly attached to or indirectly involved in the regulation of actin. In this context, it is important to mention three key in vitro studies. The first one demonstrates that single overexpression of a Shank1 isoform (Shank1b) in hippocampal neurons is capable to promote the maturation and growth of preestablished dendritic spines (Sala et al. 2001). The second one shows that the cortactin-binding site and the ankyrin repeat regions of ProSAP2/Shank3 are both indispensable for proper spine maturation thus clearly implicating concerted involvement of ProSAP/Shank-binding actin-associated proteins in the formation of plastic spines and functional synapses (Roussignol et al. 2005). Interestingly, the third study has implicated

ProSAP1/Shank2 as part of a transient postsynaptic-signaling complex whose further members include PSD-95 and GKAP and which regulates activity-dependent spine growth (Steiner et al. 2008).

The interaction of ProSAPs/Shanks with Densin-180 antagonizes dendritic branching in order to promote the development of functional spines and synapses (Quitsch et al. 2005). While physically linked to F-actin of dendritic spines, α-fodrin interacts with the N-terminal ankyrin repeats of Shank1 and ProSAP2/Shank3 via one of its spectrin motifs (Bockers et al. 2001) and is further processed in a calmodulin-dependent manner whenever intracellular calcium levels are elevated followed by the reorganization of cytoskeletal elements within spines. ProSAP-interacting protein 2 (ProSAPiP2), one of the most recently identified molecules binding to the PDZ domain of ProSAP2/Shank3 might also be involved in the attachment and modulation of cytoskeletal elements due to its actin binding properties (Liebau et al. 2009). The C-terminal ppI motif of ProSAP1/Shank2 and ProSAP2/Shank3 mediates interaction with the SH3 domain of two proteins that are tightly attached to the actin cytoskeleton, cortical-actin-binding protein (cortactin) and actin-binding protein1 (Abp1) (Du et al. 1998; Boeckers et al. 1999b; Qualmann et al. 2004). Abp1 has been shown to regulate spine morphology by controlling actin polymerization within spine heads (Haeckel et al. 2008), while cortactin has long been known as an effector of activity-dependent, actin-based spine morphology regulation (Hering and Sheng 2003). Moreover, SH3 protein interacting with Nck, 90 kDa (SPIN90), a well-known binding partner of F-actin and of actin regulators like the Arp2/3 complex and N-WASP, especially promotes Shank1b-mediated spine enlargement by interaction with the Shank1 C-terminus (Kim et al. 2009). In addition to all proteins that directly interact with actin, further ProSAP/Shank binding partners exist that are indirectly involved in actin-based cytoskeletal rearrangements within dendritic spines, mainly via small-GTPase-dominated signaling pathways. The signal transduction molecule βPIX is among them, interacting with the PDZ domains of all ProSAP/Shank family members and contributing to cytoskeletal reorganization within dendritic spines as being a guanine nucleotide exchange factor (GEF) for the Rac1 and Cdc42 small GTPases (Park et al. 2003). The latter two molecules further induce the binding of insulin receptor substrate IRSp53 to two N-terminally positioned, consecutive proline-rich clusters of Shank1 and ProSAP2/Shank3, respectively, thereby implying the involvement of ProSAP/Shank platforms in insulin-dependent remodeling of the postsynaptic cytoskeleton (Bockmann et al. 2002; Soltau et al. 2002, 2004). Another study has identified Abelson interacting protein 1 (Abi-1) as interaction partner of the ProSAP2/Shank3 proline-rich clusters and implies that it controls actin assembly within developing dendritic spines by regulating Rac-dependent pathways in a complex together with Eps8, Sos1, and WAVE1 (Proepper et al. 2007). Furthermore, proteins of the SPAR family (SPAR, SPAR2), which crucially regulate the actin cytoskeleton within dendritic spines by activating the small GTPases Rap1 and Rap2, are cross-linked to the PDZ domain of ProSAP2/Shank3 via the Fezzin family members ProSAP-interacting protein 1 (ProSAPiP1), postsynaptic density protein containing leucine-zippers, 70 kDa (PSD-Zip70), and LAPSER1 (Maruoka et al. 2005; Wendholt et al. 2006; Schmeisser et al. 2009).

To summarize, the ProSAP/Shank platform is a core element of the PSD which mainly clusters postsynaptic receptor complexes by cross-linking them to the actin cytoskeleton of dendritic spines. Because of their gigantic multimerization ability and various protein-protein interaction domains, ProSAP/Shank scaffolds serve as a meshwork for the integration of multiple other molecules into the PSD. By reorganizing cytoskeletal elements, ProSAPs/Shanks have further emerged to be essential modulators of activity-dependent remodeling of synaptic contacts in the mammalian nervous system.

2.3 Other Major Scaffold Proteins

A number of studies has provided quantitative information on the stoichiometry of proteins in the PSD using several approaches including EM combined with quantitative immunoblotting (Chen et al. 2005), quantitative mass spectroscopy (MS) (Cheng et al. 2006), and green fluorescent protein (GFP)-based quantitative fluorescence calibration (Sugiyama et al. 2005). The PSD can be biochemically isolated by extracting synaptosome preparations with nonionic detergents, such as Triton X-100, which does not solubilize the PSD.

An average PSD of 360 nm diameter might contain a total molecular mass of 1.10 ± 0.36 gigadaltons (GDa) (Chen et al. 2005), for example, there would be 10,000 proteins of approximately 100 kDa on average. These studies have also definitively demonstrated that scaffold proteins are major components of the PSD. However, the most abundant proteins are two enzymes: CaMKII and SynGAP together represent more than 8% of the PSD protein mass (Cheng et al. 2006). Although it remains a mystery, why two enzymes should be so plentiful in the PSD, some recent evidences suggest that CaMKII and SynGAP could play a structural as well as a regulatory role in synaptic homeostasis. For example, CaMKIIβ binds to F-actin and several other abundant PSD proteins (Colbran and Brown 2004) while autophosphorylated CaMKIIα acts as a scaffold to recruit proteasomes to dendritic spines (Bingol et al. 2010). Because SynGAP, in addition to its RasGAP activity, contains multiple protein-protein interaction motifs, one cannot exclude that it may also have a scaffolding function in the PSD (Rama et al. 2008). Among the more classical scaffold proteins, PSD-95 was found to be highly enriched and much more abundant than its closest relatives PSD-93 and SAP102 (see previous paragraph).

At the PSD, MAGUKs (PSD-95) bind to GKAPs which interact with ProSAPs/ Shanks that in turn bind the Homers. According to the quantitative MS studies described above, GKAP family proteins are approximately equimolar with ProSAP/ Shank family proteins, but only about 30–40% as abundant as PSD-95 family proteins, and twice as abundant as Homer family proteins (Cheng et al. 2006). If an average PSD contains 300 molecules of PSD-95, we can assume the following stoichiometry: 400 MAGUK family members, 150 GKAP family members, 150

ProSAP/Shank family members, and 60 Homer family members. Indeed, using a quantitative fluorescence imaging approach in cultured hippocampal neurons (Sugiyama et al. 2005), Okabe and colleagues have obtained similar quantitative results for PSD-95, but higher values for ProSAP/Shank family proteins (310 copies) and Homer family proteins (340). These differences might be explained by the distinct metrology used for quantification on the one hand and by the different synaptic protein expression in cultured neurons versus adult brain on the other hand. It is important to mention that the numbers are higher in large PSDs and lower in small PSDs, but their stoichiometry seems to be preserved among each PSD, suggesting the presence of a possible "master" organizing scaffold protein.

Despite those quantitative discrepancies, it is clear that the MAGUK(PSD-95)-GKAP-ProSAP/Shank-Homer platform accounts for a substantial proportion of the total protein mass within the PSD and represents the core scaffold structure of the PSD (Sugiyama et al. 2005).

2.3.1   The GKAP Family

The four members of the GKAP (also named GKAP/SAPAP) family of proteins were originally identified as proteins interacting with the GK domain of PSD-95 (Kim et al. 1997; Takeuchi et al. 1997). The N-terminal domain of GKAP binds to PSD-95 while the rest of the protein exhibits binding domains for synaptic scaffolding molecule (S-SCAM), nArgBP2, and dynein light chain, thus suggesting a function as a scaffold protein that links PSD protein complexes to motor proteins (Naisbitt et al. 2000). The very C-terminal part further interacts with the PDZ domain of ProSAPs/Shanks. It has recently been demonstrated that GKAP is a specific substrate of one E3 ubiquitin ligase, the RING finger-containing protein TRIM3. TRIM3 stimulates ubiquitylation and proteasome-dependent degradation of GKAP and the associated protein Shank1. The suppression of endogenous TRIM3 results in increased accumulation of GKAP and Shank1 at synapses and prevents the loss of GKAP induced by synaptic activity (Hung et al. 2010). Interestingly, degradation of GKAP and ProSAP/Shank occurs during memory consolidation and reconsolidation (Lee et al. 2008).

2.3.2   The Homer Family

The Homer proteins are encoded by three genes (Homer1-3) and structurally formed by an N-terminal Ena/VASP homology 1 (EVH1) domain followed by a coiled-coil domain that mediates dimerization with other Homer proteins. The Homers were originally discovered by cloning of Homer1a, a splice variant of the Homer1 gene, which is regulated as an immediate early gene (IEG). Homer1a is rapidly upregulated in neurons in response to synaptic activity induced by seizure or during induction of LTP and is selectively induced in cells of the hippocampus when rodents engage in exploratory behavior (Brakeman et al. 1997; Kato et al. 1998; Fagni et al. 2002). The Homer1 gene encodes for two additional and longer transcripts, Homer1b and Homer1c, which are more similar to the other Homer genes, Homer2 and Homer3, that have also been reported to encode for several transcripts, but none of them is induced by neuronal activity (Soloviev et al. 2000; Xiao et al. 2000). The EVH1 domain of Homer1 binds to a PPXXF or very similar sequence motif in ProSAP/Shank, mGluR1/5, the inositol-1,4,5-trisphosphate (IP$_3$) receptor, the ryanodine receptor, different members of the TRPC family of ion channels, PLCβ (Nakamura et al. 2004; Hwang et al. 2005), selective L-type Ca^{2+} channel isoforms (Yamamoto et al. 2005; Huang et al. 2007), and oligophrenin (Govek et al. 2004). Through their ability to self-associate, Homer isoforms containing the coiled-coil domain (called "CC-Homer" or Homer1b in the case of the Homer1 gene) can physically and functionally link the proteins and receptors that bind to the EVH1 domain. This scaffold-like activity has well been demonstrated for the ability of Homer to facilitate a physical association between type I mGluRs and the IP$_3$R or TRPC1 and the IP$_3$R. In each case, this association is required for the mGluRs and for the TRP channel to respond to signals (Yuan et al. 2003; Sala et al. 2005). Homer1a only contains the EVH1 domain, but lacks the coiled-coil domain; it functions as a natural dominant negative because it cannot dimerize. As on the mRNA level, Homer1a expression is induced by synaptic activity; it might function as a regulator of synaptic structure and activity (Sala et al. 2003). More recently, Homer1a has been implicated in the regulation of homeostatic scaling by regulating agonist-independent signaling of group I mGluRs, a process which scales down the expression of synaptic AMPARs (Hu et al. 2010). These data suggest that behind their function as scaffolds, Homer proteins exhibit a specialized signaling function at the synapse by regulating the activity of type I mGluR.

2.3.3 The GRIP Family

The four previously discussed PSD scaffold protein families (MAGUKs, GKAPs, ProSAPs/Shanks, Homers) are mostly directly or indirectly linked to the NMDAR complex, while the other major ionotropic glutamate receptors, the AMPARs, are classically linked to a different set of scaffold proteins including GRIP/ABP (encoded by the two distinct genes GRIP1 and ABP/GRIP2) and the protein interacting with C kinase 1 (PICK1). These interactions may account for the dynamic behavior of AMPARs at synapses. Several evidences suggest that GRIP proteins are involved in the synaptic trafficking and/or stabilization of AMPARs and other interacting proteins. However, GRIP, being formed by seven PDZ domains, interacts with many proteins, including Eph receptors and their ephrin ligands, RasGEF, liprin-α, PICK1, the transmembrane protein Fraser syndrome 1 (FRAS1), metabotropic and kainate-type glutamate receptors, the cadherin-associated protein neural-plakophilin-related arm protein (NPRAP), and the metalloproteinase membrane-type 5 MMP (MT5-MMP) (Dong et al. 1997; Bruckner et al. 1999; Dong et al. 1999; Ye et al. 2000; Hirbec et al. 2002; Setou et al. 2002; Wyszynski et al. 2002; Takamiya et al. 2004; Hoogenraad et al. 2005; Monea et al. 2006; Silverman et al. 2007). Thus, not surprisingly, GRIP proteins can participate in synaptic and neuronal functions not only by interacting with AMPARs but also by interacting with Eph receptors and their ephrin ligands, a signaling complex known to be involved in dendritic spine morphogenesis and hippocampal synaptic plasticity (Hoogenraad et al. 2005). GRIP interaction with motor proteins (directly with conventional kinesin KIF5 or indirectly with KIF1A via liprin-α) further suggests that this protein family may also contribute to the transport of AMPARs to the synapse (Shin et al. 2003; Hoogenraad et al. 2005). GRIP is widely expressed in different tissues and in neurons, and it is present in both axons and dendrites (Wyszynski et al. 2002). Therefore, the function of GRIP has to be considered beyond the regulation of AMPARs. This is supported by the fact that GRIP1 knockout mice show hemorrhagic blisters and embryonic lethality (Bladt et al. 2002; Takamiya et al. 2004). Only the GluR2/3 subunits of the AMPAR specifically bind to the PDZ5 domain of GRIP, although the PDZ4 domain is also required for strengthening this interaction. Several studies have demonstrated the role of GRIP in AMPAR trafficking to synapses using primary neuronal cultures. A more conclusive work by Takamiya et al. showed that the genetic ablation of both *GRIP* genes blocks LTD expression in cerebellar neurons while single deletion of either isoform allows LTD to occur, suggesting the ability of the two proteins to functional compensate each other at least in part (Takamiya et al. 2008). Finally, certain splice variants of GRIP can be palmitoylated thereby regulating its association with the plasma membrane and localization at the synapse (DeSouza et al. 2002). In contrast, nonpalmitoylated GRIP mostly associates with intracellular membranes (Fu et al. 2003). To conclude, these differentially modified subpopulations of GRIP might control synaptic and intracellular pools of AMPARs, respectively.

2.3.4 PICK1

The BAR (bin/amphiphysin/rvs) and PDZ-domain-containing protein PICK1 (protein interacting with C-kinase 1) directly binds to the GluR2/3 subunits of AMPARs (Dev et al. 1999; Xia et al. 1999). PICK1 is present at synaptic and nonsynaptic sites in neurons, and its PDZ domain shows relatively promiscuous binding. In addition to PKCα and GluR2/3, it has many other binding partners (both pre- and postsynaptic), including the netrin receptor UNC5H (Williams et al. 2003), various metabotropic glutamate receptor subtypes (Hirbec et al. 2002; Perroy et al. 2002), the dopamine plasma membrane transporter, and the erythroblastic leukemia viral oncogene homologue 2 (ErbB2) receptor tyrosine kinase (Jaulin-Bastard et al. 2001; Torres et al. 2001). PICK1 plays a clear and important role in AMPAR surface expression, trafficking, and synaptic targeting (Jin et al. 2006; Hanley 2008).

Recently, Anggono et al. demonstrated that PICK1 participates in homeostatic plasticity by regulating the subunit composition, abundance, and trafficking of GluR2-containing AMPARs (Anggono et al. 2011). However, PICK1 negatively regulates Arp2/3-mediated actin polymerization thus influencing both NMDA-induced AMPAR internalization and dendritic spine morphology (Rocca et al. 2008; Nakamura et al. 2011). The pleiotropic role of PICK1 can probably explain the finding that loss of PICK1 has no significant effect on synaptic plasticity in juvenile mice but impairs some forms of LTP and multiple distinct forms of LTD in adult mice, suggesting that PICK1 is selectively required for hippocampal synaptic plasticity and learning in adult rodents (Volk et al. 2010).

The role of GRIP and PICK1 in the stabilization of synaptic versus intracellular AMPAR pools has been studied extensively (Daw et al. 2000; Kim et al. 2001; Chung et al. 2003; Hirbec et al. 2003; Seidenman et al. 2003). However, not all the controversies have been resolved, also because of overlapping specificities of PDZ domain interactions. The peptides that are typically used to interfere with the PDZ interactions of GRIP and PICK1 are probably not highly specific for these proteins or for GluR2/3 interactions. Recent experiments in knockout mice are promising to elucidate the specific functions for GRIP and PICK1 (Takamiya et al. 2008; Anggono et al. 2011).

2.4 Postsynaptic Scaffolds and Their Relation to Neuropsychiatric Disorders

As several neuropsychiatric disorders have directly been linked to altered synaptic morphology and function throughout recent years, future therapeutic implications will require an in-depth understanding of the molecular mechanisms underlying the disruption of subcellular structures at the synapse like the active zone or the PSD. Referring to the latter, mislocalization and dysregulation of postsynaptic scaffold proteins are crucial events during the pathophysiological course of several so-called synaptopathies like distinct forms of autism, schizophrenia, or dementia.

2.4.1 The MAGUK Family and Other Scaffold Proteins

Considering that PSD-95 is the most abundant scaffold protein in the PSD, it is not surprising that its human gene DLG4 was extensively studied for the presence of specific polymorphisms and mutations to be associated with mental diseases. However, up to now, only one preliminary study suggests the association between DLG4 gene variation, autism spectrum disorders (ASDs), and Williams' syndrome (Feyder et al. 2010), and perhaps a polymorphism in the promoter gene is linked to schizophrenia (Cheng et al. 2010). Interestingly, alterations in PSD-95 expression have been found in patients with Alzheimer's disease (AD) and in a mouse model of Fragile X syndrome, a genetically linked autistic-like disorder (Zalfa et al. 2007; Leuba et al. 2008; Zhu et al. 2010). Moreover, mental retardation is clearly associated with the human DLG3 gene that encodes SAP102 (Tarpey et al. 2004; Zanni et al. 2010). The mutations identified by Tarpey et al. introduce premature stop codons within or before the third PDZ domain, and it is likely that these alterations impair the ability of SAP102 to interact with the NMDAR and/or other proteins involved in downstream NMDAR signaling pathways.

The first evidence for a major role of SAPAP3 in brain function was demonstrated by Feng's lab. Mice deficient for SAPAP3 developed an obsessive-compulsive disorder (OCD)-like phenotype, including compulsive grooming and increased anxiety (Welch et al. 2007), while multiple rare SAPAP3 missense variants in humans have been found associated with trichotillomania (TTM) and OCD (Züchner et al. 2009). Further genetic studies should provide more insights into SAPAP3 mutations in humans and TTM or OCD. Similarly, research should provide information whether there is an association between a Homer2 gene mutation and schizophrenia or development and maintenance of alcohol- and substance-use disorders which have been found in KO mice (Szumlinski et al. 2004) but not in humans (Szumlinski et al. 2006).

2.4.2 The ProSAP/Shank Family

Haploinsufficiency of the ProSAP2/Shank3 gene as underlying cause for the Phelan-McDermid syndrome (also named PMS, 22q13.3 deletion syndrome) provides the most direct link between the loss of a postsynaptic scaffold protein and a disorder whose major clinical features include neuropsychiatric symptoms, among them global developmental delay, absent or severely delayed speech, muscular hypotonia, and "autistic-like" behavior (Bonaglia et al. 2001; Phelan et al. 2001; Wilson et al. 2003; Manning et al. 2004). To exclude that the disruption of genes other than ProSAP2/Shank3, but also located on 22q13.3, is the genetically determined reason for the above-described neuropsychiatric symptoms, researchers have defined a minimal critical region on the chromosome including ProSAP2/Shank3 (Bonaglia et al. 2001; Wilson et al. 2003; Delahaye et al. 2009). A balanced translocation between chromosome 12 and 22 with the breakpoint in the ProSAP2/Shank3 gene (Bonaglia et al. 2001) and another, recurrent breakpoint within intron 8 of the ProSAP2/Shank3 gene exclusively affecting the latter, clearly supported its crucial role in the molecular pathology of PMS (Bonaglia et al. 2006). Further studies have identified de novo mutations in the ProSAP2/Shank3 (Durand et al. 2007; Moessner et al. 2007; Gauthier et al. 2009) and ProSAP1/Shank2 (Berkel et al. 2010) genes in individuals diagnosed with ASD. These mutations might all have a severe impact on the molecular setup of the whole postsynaptic protein platform in the affected patients by disrupting its physiological properties which are crucial for normal synaptic homeostasis and the balance between excitation and inhibition. Considering the fact that alterations in other synaptic proteins like presynaptic neurexins (Kim et al. 2008; Ching et al. 2010) or postsynaptic neuroligins (Jamain et al. 2003) have been identified in autistic individuals and which are directly attached to the ProSAP/Shank platform via protein-protein interactions, a synaptic NRXN-NLGN-ProSAP/Shank pathway has been proposed whose dysregulation might be one of the core pathophysiological causes in the development of ASDs (Bourgeron 2009). In this context, some in vivo data have already been collected by analysis of transgenic mouse models harboring a targeted disruption of this pathway on the level of the ProSAP/Shank platform, like the Shank1 knockout mouse which exhibits smaller dendritic spines, weaker synaptic transmission (Hung et al. 2008), and reduced motor functions (Silverman et al. 2011), or the ProSAP2/Shank3 haploinsufficiency mouse that shows delayed synaptic development, a decrease in synaptic transmission and reduced social sniffing and ultrasonic vocalizations (Bozdagi et al. 2010). Furthermore, mice with distinct deletions of the ProSAP2/Shank3 gene exhibit self-injurious repetitive grooming and deficits in social interaction most probably due to defects at striatal synapses and corticostriatal circuits (Peça et al. 2011).

Schizophrenia has emerged to be another neuropsychiatric disorder related to ProSAP/Shank malfunction as revealed by de novo mutations in the ProSAP2/Shank3 gene in patients ascertained for this neuropsychiatric disease (Gauthier et al. 2010). However, both individuals identified in this study exhibited impaired intellectual abilities even before the diagnosis of schizophrenia. Most interestingly, a very recent study from the same consortium (Hamdan et al. 2011) found truncating and/or splicing mutations in several synaptic scaffold proteins including ProSAP2/Shank3 in patients with nonsyndromic intellectual disability (NSID) thus again supporting the crucial role of scaffolds like ProSAP/Shank for the proper development and maintenance of higher brain function.

Recent investigations have additionally implicated ProSAP/Shank platform disassembly in neurodegeneration. Accumulation of soluble β-Amyloid oligomers in rat frontocortical cell culture, in the cortex of transgenic Alzheimer's Disease (AD) mouse models, and in the cortex of AD patients is accompanied by a reduction of Shank1 (Roselli et al. 2009; Pham et al. 2010) and ProSAP2/Shank3 (Gong et al. 2009; Pham et al. 2010) while ProSAP1/Shank2 levels seem to be upregulated (Gong et al. 2009). In this context, it is important to mention that ProSAP/Shank platforms are organized and stabilized via zinc at their C-terminal SAM domains (Baron et al. 2006; Gundelfinger et al. 2006; Grabrucker et al. 2011). Zinc is a metal ion that has the capacity to directly bind to β-Amyloid (Matsubara et al. 2003) and is known to promote its aggregation (Friedlich et al. 2004; Miller et al. 2010). As β-Amyloid accumulation within the synaptic cleft has previously been suggested to contribute to the development of cognitive impairment in AD by trapping synaptic zinc rather than through direct neuronal toxicity (Deshpande et al. 2009; Adlard et al. 2010), one could imagine a synaptopathic mechanism involving β-Amyloid accumulation entrapment of synaptic zinc and a disruption of the postsynaptic ProSAP/Shank platform.

References

Aarts, M., Liu, Y., Liu, L., Besshoh, S., Arundine, M., Gurd, J. W., Wang, Y. T., Salter, M. W., & Tymianski, M. (2002). Treatment of ischemic brain damage by perturbing NMDA receptor-PSD-95 protein interactions. *Science, 298*, 846–850.

Adlard, P. A., Parncutt, J. M., Finkelstein, D. I., & Bush, A. I. (2010). Cognitive loss in zinc transporter-3 knock-out mice: A phenocopy for the synaptic and memory deficits of Alzheimer's disease? *Journal of Neuroscience, 30*, 1631–1636.

Anggono, V., Clem, R. L., & Huganir, R. L. (2011). PICK1 loss of function occludes homeostatic synaptic scaling. *Journal of Neuroscience, 31*, 2188–2196.

Aoki, C., Miko, I., Oviedo, H., Mikeladze-Dvali, T., Alexandre, L., Sweeney, N., & Bredt, D. S. (2001). Electron microscopic immunocytochemical detection of PSD-95, PSD-93, SAP-102, and SAP-97 at postsynaptic, presynaptic, and nonsynaptic sites of adult and neonatal rat visual cortex. *Synapse, 40*, 239–257.

Assaf, S. Y., & Chung, S. H. (1984). Release of endogenous Zn^{2+} from brain tissue during activity. *Nature, 308*, 734–736.

Baron, M. K., Boeckers, T. M., Vaida, B., Faham, S., Gingery, M., Sawaya, M. R., Salyer, D., Gundelfinger, E. D., & Bowie, J. U. (2006). An architectural framework that may lie at the core of the postsynaptic density. *Science, 311*, 531–535.

Beri, S., Tonna, N., Menozzi, G., Bonaglia, M. C., Sala, C., & Giorda, R. (2007). DNA methylation regulates tissue-specific expression of Shank3. *Journal of Neurochemistry, 101*, 1380.

Berkel, S., Marshall, C. R., Weiss, B., Howe, J., Roeth, R., Moog, U., Endris, V., Roberts, W., Szatmari, P., Pinto, D., Bonin, M., Riess, A., Engels, H., Sprengel, R., Scherer, S. W., & Rappold, G. A. (2010). Mutations in the SHANK2 synaptic scaffolding gene in autism spectrum disorder and mental retardation. *Nature Genetics, 42*, 489–491.

Bhattacharyya, S., Biou, V., Xu, W., Schlüter, O., & Malenka, R. C. (2009). A critical role for PSD-95/AKAP interactions in endocytosis of synaptic AMPA receptors. *Nature Neuroscience, 12*, 172–181.

Bingol, B., Wang, C. F., Arnott, D., Cheng, D., Peng, J., & Sheng, M. (2010). Autophosphorylated CaMKIIalpha acts as a scaffold to recruit proteasomes to dendritic spines. *Cell, 140*, 567–578.

Bladt, F., Tafuri, A., Gelkop, S., Langille, L., & Pawson, T. (2002). Epidermolysis bullosa and embryonic lethality in mice lacking the multi-PDZ domain protein GRIP1. *Proceedings of the National Academy of Sciences of the United States of America, 99*, 6816–6821.

Bockers, T. M., Mameza, M. G., Kreutz, M. R., Bockmann, J., Weise, C., Buck, F., Richter, D., Gundelfinger, E. D., & Kreienkamp, H. J. (2001). Synaptic scaffolding proteins in rat brain. Ankyrin repeats of the multidomain Shank protein family interact with the cytoskeletal protein alpha-fodrin. *Journal of Biological Chemistry, 276*, 40104–40112.

Bockers, T. M., Segger-Junius, M., Iglauer, P., Bockmann, J., Gundelfinger, E. D., Kreutz, M. R., Richter, D., Kindler, S., & Kreienkamp, H. J. (2004). Differential expression and dendritic transcript localization of Shank family members: Identification of a dendritic targeting element in the 3' untranslated region of Shank1 mRNA. *Molecular and Cellular Neuroscience, 26*, 182–190.

Bockmann, J., Kreutz, M. R., Gundelfinger, E. D., & Bockers, T. M. (2002). ProSAP/Shank postsynaptic density proteins interact with insulin receptor tyrosine kinase substrate IRSp53. *Journal of Neurochemistry, 83*, 1013–1017.

Boeckers, T. M., Bockmann, J., Kreutz, M. R., & Gundelfinger, E. D. (2002). ProSAP/Shank proteins – A family of higher order organizing molecules of the postsynaptic density with an emerging role in human neurological disease. *Journal of Neurochemistry, 81*, 903–910.

Boeckers, T. M., Kreutz, M. R., Winter, C., Zuschratter, W., Smalla, K. H., Sanmarti-Vila, L., Wex, H., Langnaese, K., Bockmann, J., Garner, C. C., & Gundelfinger, E. D. (1999a). Proline-rich synapse-associated protein-1/cortactin binding protein 1 (ProSAP1/CortBP1) is a PDZ-domain protein highly enriched in the postsynaptic density. *Journal of Neuroscience, 19*, 6506–6518.

Boeckers, T. M., Liedtke, T., Spilker, C., Dresbach, T., Bockmann, J., Kreutz, M. R., & Gundelfinger, E. D. (2005). C-terminal synaptic targeting elements for postsynaptic density proteins ProSAP1/Shank2 and ProSAP2/Shank3. *Journal of Neurochemistry, 92*, 519–524.

Boeckers, T. M., Winter, C., Smalla, K. H., Kreutz, M. R., Bockmann, J., Seidenbecher, C., Garner, C. C., & Gundelfinger, E. D. (1999b). Proline-rich synapse-associated proteins ProSAP1 and ProSAP2 interact with synaptic proteins of the SAPAP/GKAP family. *Biochemical and Biophysical Research Communications, 264*, 247–252.

Bonaglia, M. C., Giorda, R., Borgatti, R., Felisari, G., Gagliardi, C., Selicorni, A., & Zuffardi, O. (2001). Disruption of the ProSAP2 gene in a t(12;22)(q24.1;q13.3) is associated with the 22q13.3 deletion syndrome. *American Journal of Human Genetics, 69*, 261–268.

Bonaglia, M. C., Giorda, R., Mani, E., Aceti, G., Anderlid, B. M., Baroncini, A., Pramparo, T., & Zuffardi, O. (2006). Identification of a recurrent breakpoint within the SHANK3 gene in the 22q13.3 deletion syndrome. *Journal of Medical Genetics, 43*, 822–828.

Bourgeron, T. (2009). A synaptic trek to autism. *Current Opinion in Neurobiology, 19*, 231–234.

Bozdagi, O., Sakurai, T., Papapetrou, D., Wang, X., Dickstein, D. L., Takahashi, N., Kajiwara, Y., Yang, M., Katz, A. M., Scattoni, M. L., Harris, M. J., Saxena, R., Silverman, J. L., Crawley, J. N., Zhou, Q., Hof, P. R., & Buxbaum, J. D. (2010). Haploinsufficiency of the autism-associated Shank3 gene leads to deficits in synaptic function, social interaction, and social communication. *Molecular Autism, 1*, 15.

Brakeman, P. R., Lanahan, A. A., O'Brien, R., Roche, K., Barnes, C. A., Huganir, R. L., & Worley, P. F. (1997). Homer: A protein that selectively binds metabotropic glutamate receptors. *Nature, 386*, 284–288.

Brandstätter, J. H., Dick, O., & Boeckers, T. M. (2004). The postsynaptic scaffold proteins ProSAP1/Shank2 and Homer1 are associated with glutamate receptor complexes at rat retinal synapses. *The Journal of Comparative Neurology, 475*, 551–563.

Brenman, J. E., Chao, D. S., Gee, S. H., McGee, A. W., Craven, S. E., Santillano, D. R., Wu, Z., Huang, F., Xia, H., Peters, M. F., Froehner, S. C., & Bredt, D. S. (1996a). Interaction of nitric oxide synthase with the postsynaptic density protein PSD-95 and a1-syntrophin mediated by PDZ domains. *Cell, 84*, 757–767.

Brenman, J. E., Christopherson, K. S., Craven, S. E., McGee, A. W., & Bredt, D. S. (1996b). Cloning and characterization of postsynaptic density 93, a nitric oxide synthase interacting protein. *Journal of Neuroscience, 16*, 7407–7415.

Brenman, J. E., Topinka, R. J., Cooper, E. C., McGee, A. W., Rosen, J., Milroy, T., Ralston, H. J., & Bredt, D. S. (1998). Localization of postsynaptic density-93 to dendritic microtubules and interaction with microtubule-associated protein 1A. *Journal of Neuroscience, 18*, 8805–8813.

Bresler, T., Shapira, M., Boeckers, T., Dresbach, T., Futter, M., Garner, C. C., Rosenblum, K., Gundelfinger, E. D., & Ziv, N. E. (2004). Postsynaptic density assembly is fundamentally different from presynaptic active zone assembly. *Journal of Neuroscience, 24*, 1507–1520.

Bruckner, K., Pablo Labrador, J., Scheiffele, P., Herb, A., Seeburg, P. H., & Klein, R. (1999). EphrinB ligands recruit GRIP family PDZ adaptor proteins into raft membrane microdomains. *Neuron, 22*, 511–524.

Carlisle, H. J., Fink, A. E., Grant, S. G., & O'Dell, T. J. (2008). Opposing effects of PSD-93 and PSD-95 on long-term potentiation and spike timing-dependent plasticity. *Journal de Physiologie, 586*, 5885–5900.

Chen, L., Chetkovich, D. M., Petralia, R. S., Sweeney, N. T., Kawasaki, Y., Wenthold, R. J., Bredt, D. S., & Nicoll, R. A. (2000). Stargazin regulates synaptic targeting of AMPA receptors by two distinct mechanisms. *Nature, 408*, 936–943.

Chen, H. J., Rojas-Soto, M., Oguni, A., & Kennedy, M. B. (1998). A synaptic Ras-GTPase activating protein (p135 SynGAP) inhibited by CaM kinase II. *Neuron, 20*, 895–904.

Chen, X., Vinade, L., Leapman, R. D., Petersen, J. D., Nakagawa, T., Phillips, T. M., Sheng, M., & Reese, T. S. (2005). Mass of the postsynaptic density and enumeration of three key molecules. *Proceedings of the National Academy of Sciences of the United States of America, 102*, 11551–11556.

Chen, X., Winters, C., Azzam, R., Li, X., Galbraith, J. A., Leapman, R. D., & Reese, T. S. (2008). Organization of the core structure of the postsynaptic density. *Proceedings of the National Academy of Sciences of the United States of America, 105*, 4453–4458.

Cheng, D., Hoogenraad, C. C., Rush, J., Ramm, E., Schlager, M. A., Duong, D. M., Xu, P., Rukshan, S., Hanfelt, J., Nakagawa, T., Sheng, M., & Peng, J. (2006). Relative and absolute quantification of postsynaptic density proteome isolated from rat forebrain and cerebellum. *Molecular & Cellular Proteomics, 5*, 1158.

Cheng, M. C., Lu, C. L., Luu, S. U., Tsai, H. M., Hsu, S. H., Chen, T. T., & Chen, C. H. (2010). Genetic and functional analysis of the DLG4 gene encoding the post-synaptic density protein 95 in schizophrenia. *PloS One, 5*, e15107.

Ching, T. T., Maunakea, A. K., Jun, P., Hong, C., Zardo, G., Pinkel, D., Albertson, D. G., Fridlyand, J., Mao, J. H., Shchors, K., Weiss, W. A., & Costello, J. F. (2005). Epigenome analyses using BAC microarrays identify evolutionary conservation of tissue-specific methylation of SHANK3. *Nature Genetics, 37*, 645–651.

Ching, M. S., et al. (2010). Deletions of NRXN1 (neurexin-1) predispose to a wide spectrum of developmental disorders. *American Journal of Medical Genetics. Part B, Neuropsychiatric Genetics, 153B*, 937–947.

Cho, K.-O., Hunt, C. A., & Kennedy, M. B. (1992). The rat brain postsynaptic density fraction contains a homolog of the Drosophila discs-large tumor suppressor protein. *Neuron, 9*, 929–942.

Chung, H. J., Steinberg, J. P., Huganir, R. L., & Linden, D. J. (2003). Requirement of AMPA receptor GluR2 phosphorylation for cerebellar long-term depression. *Science, 300*, 1751–1755.

Colbran, R. J., & Brown, A. M. (2004). Calcium/calmodulin-dependent protein kinase II and synaptic plasticity. *Current Opinion in Neurobiology, 14*, 318–327.

Colledge, M., Dean, R. A., Scott, G. K., Langeberg, L. K., Huganir, R. L., & Scott, J. D. (2000). Targeting of PKA to glutamate receptors through a MAGUK-AKAP complex. *Neuron, 27*, 107–119.

Colledge, M., Snyder, E. M., Crozier, R. A., Soderling, J. A., Jin, Y., Langeberg, L. K., Lu, H., Bear, M. F., & Scott, J. D. (2003). Ubiquitination regulates PSD-95 degradation and AMPA receptor surface expression. *Neuron, 40*, 595–607.

Conroy, W. G., Liu, Z., Nai, Q., Coggan, J. S., & Berg, D. K. (2003). PDZ-containing proteins provide a functional postsynaptic scaffold for nicotinic receptors in neurons. *Neuron, 38*, 759–771.

Craven, S. E., & Bredt, D. S. (1998). PDZ proteins organize synaptic signaling pathways. *Cell, 93*, 495–498.

Craven, S. E., El-Husseini, A. E., & Bredt, D. S. (1999). Synaptic targeting of the postsynaptic density protein PSD-95 mediated by lipid and protein motifs. *Neuron, 22*, 497–509.

Daw, M. I., Chittajallu, R., Bortolotto, Z. A., Dev, K. K., Duprat, F., Henley, J. M., Collingridge, G. L., & Isaac, J. T. (2000). PDZ proteins interacting with C-terminal GluR2/3 are involved in a PKC-dependent regulation of AMPA receptors at hippocampal synapses. *Neuron, 28*, 873–886.

Delahaye, A., Toutain, A., Aboura, A., Dupont, C., Tabet, A. C., Benzacken, B., Elion, J., Verloes, A., Pipiras, E., & Drunat, S. (2009). Chromosome 22q13.3 deletion syndrome with a de novo interstitial 22q13.3 cryptic deletion disrupting SHANK3. *European Journal of Medical Genetics, 52*, 328–332.

Deshpande, A., Kawai, H., Metherate, R., Glabe, C. G., & Busciglio, J. (2009). A role for synaptic zinc in activity-dependent Abeta oligomer formation and accumulation at excitatory synapses. *Journal of Neuroscience, 29*, 4004–4015.

DeSouza, S., Fu, J., States, B. A., & Ziff, E. B. (2002). Differential palmitoylation directs the AMPA receptor-binding protein ABP to spines or to intracellular clusters. *Journal of Neuroscience, 22*, 3493–3503.

Dev, K. K., Nishimune, A., Henley, J. M., & Nakanishi, S. (1999). The protein kinase C alpha binding protein PICK1 interacts with short but not long form alternative splice variants of AMPA receptor subunits. *Neuropharmacology, 38*, 635–644.

Dobrinskikh, E., Giral, H., Caldas, Y. A., Levi, M., & Doctor, R. B. (2010). Shank2 redistributes with NaPilla during regulated endocytosis. *American Journal of Physiology: Cell Physiology, 299*, C1324–C1334.

Dong, H., O'Brien, R. J., Fung, E. T., Lanahan, A. A., Worley, P. F., & Huganir, R. L. (1997). GRIP: A synaptic PDZ domain-containing protein that interacts with AMPA receptors. *Nature, 386*, 279–284.

Dong, H., Zhang, P., Song, I., Petralia, R. S., Liao, D., & Huganir, R. L. (1999). Characterization of the glutamate receptor-interacting proteins GRIP1 and GRIP2. *Journal of Neuroscience, 19*, 6930–6941.

Du, Y., Weed, S. A., Xiong, W.-C., Marshall, T. D., & Parsons, J. T. (1998). Identification of a novel cortactin SH3 domain-binding protein and its localization to growth cones of cultured neurons. *Molecular and Cellular Biology, 18*, 5838–5851.

Durand, C. M., et al. (2007). Mutations in the gene encoding the synaptic scaffolding protein SHANK3 are associated with autism spectrum disorders. *Nature Genetics, 39*, 25–27.

Ehlers, M. D. (2003). Activity level controls postsynaptic composition and signaling via the ubiquitin-proteasome system. *Nature Neuroscience, 6*, 231–242.

El-Husseini Ael, D., Schnell, E., Dakoji, S., Sweeney, N., Zhou, Q., Prange, O., Gauthier-Campbell, C., Aguilera-Moreno, A., Nicoll, R. A., & Bredt, D. S. (2002). Synaptic strength regulated by palmitate cycling on PSD-95. *Cell, 108*, 849–863.

El-Husseini, A. E., Craven, S. E., Chetkovich, D. M., Firestein, B. L., Schnell, E., Aoki, C., & Bredt, D. S. (2000). Dual palmitoylation of PSD-95 mediates its vesiculotubular sorting, postsynaptic targeting, and ion channel clustering. *The Journal of Cell Biology, 148*, 159–172.

Elias, G. M., Funke, L., Stein, V., Grant, S. G., Bredt, D. S., & Nicoll, R. A. (2006). Synapse-specific and developmentally regulated targeting of AMPA receptors by a family of MAGUK scaffolding proteins. *Neuron, 52*, 307–320.

Elias, G. M., & Nicoll, R. A. (2007). Synaptic trafficking of glutamate receptors by MAGUK scaffolding proteins. *Trends in Cell Biology, 17*, 343–352.

Fagni, L., Worley, P. F., & Ango, F. (2002). Homer as both a scaffold and transduction molecule. *Science's STKE, 2002*, RE8.

Falley, K., Schütt, J., Iglauer, P., Menke, K., Maas, C., Kneussel, M., Kindler, S., Wouters, F. S., Richter, D., & Kreienkamp, H. J. (2009). Shank1 mRNA: Dendritic transport by kinesin and translational control by the 5′untranslated region. *Traffic, 10*, 844–857.

Feyder, M., et al. (2010). Association of mouse Dlg4 (PSD-95) gene deletion and human DLG4 gene variation with phenotypes relevant to autism spectrum disorders and Williams' syndrome. *The American Journal of Psychiatry, 167*, 1508–1517.

Friedlich, A. L., Lee, J. Y., van Groen, T., Cherny, R. A., Volitakis, I., Cole, T. B., Palmiter, R. D., Koh, J. Y., & Bush, A. I. (2004). Neuronal zinc exchange with the blood vessel wall promotes cerebral amyloid angiopathy in an animal model of Alzheimer's disease. *Journal of Neuroscience, 24*, 3453–3459.

Fu, J., deSouza, S., & Ziff, E. B. (2003). Intracellular membrane targeting and suppression of Ser880 phosphorylation of glutamate receptor 2 by the linker I-set II domain of AMPA receptor-binding protein. *Journal of Neuroscience, 23*, 7592–7601.

Fukata, M., Fukata, Y., Adesnik, H., Nicoll, R. A., & Bredt, D. S. (2004). Identification of PSD-95 palmitoylating enzymes. *Neuron, 44*, 987–996.

Garcia, E. P., Mehta, S., Blair, L. A., Wells, D. G., Shang, J., Fukushima, T., Fallon, J. R., Garner, C. C., & Marshall, J. (1998). SAP90 binds and clusters kainate receptors causing incomplete desensitization. *Neuron, 21*, 727–739.

Garner, C. C., Nash, J., & Huganir, R. L. (2000). PDZ domains in synapse assembly and signalling. *Trends in Cell Biology, 10*, 274–280.

Gauthier, J., Spiegelman, D., Piton, A., Lafreniere, R. G., Laurent, S., St-Onge, J., Lapointe, L., Hamdan, F. F., Cossette, P., Mottron, L., Fombonne, E., Joober, R., Marineau, C., Drapeau, P., & Rouleau, G. A. (2009). Novel de novo SHANK3 mutation in autistic patients. *American Journal of Medical Genetics. Part B, Neuropsychiatric Genetics, 150B*, 421–424.

Gauthier, J., et al. (2010). De novo mutations in the gene encoding the synaptic scaffolding protein SHANK3 in patients ascertained for schizophrenia. *Proceedings of the National Academy of Sciences of the United States of America, 107*, 7863–7868.

Gessert, S., Schmeisser, M. J., Tao, S., Boeckers, T. M., & Kühl, M. (2011). The spatio-temporal expression of ProSAP/shank family members and their interaction partner LAPSER1 during Xenopus laevis development. *Developmental Dynamics, 240*, 1528.

Gong, Y., Lippa, C. F., Zhu, J., Lin, Q., & Rosso, A. L. (2009). Disruption of glutamate receptors at Shank-postsynaptic platform in Alzheimer's disease. *Brain Research, 1292*, 191–198.

Govek, E. E., Newey, S. E., Akerman, C. J., Cross, J. R., Van der Veken, L., & Van Aelst, L. (2004). The X-linked mental retardation protein oligophrenin-1 is required for dendritic spine morphogenesis. *Nature Neuroscience, 7*, 364–372.

Grabrucker, A. M., Knight, M. J., Proepper, C., Bockmann, J., Joubert, M., Rowan, M., Nienhaus, G. U., Garner, C. C., Bowie, J. U., Kreutz, M. R., Gundelfinger, E. D., & Boeckers, T. M. (2011). Concerted action of zinc and ProSAP/Shank in synaptogenesis and synapse maturation. *EMBO Journal, 30*, 569–581.

Grabrucker, A. M., Vaida, B., Bockmann, J., & Boeckers, T. M. (2009). Efficient targeting of proteins to post-synaptic densities of excitatory synapses using a novel pSDTarget vector system. *Journal of Neuroscience Methods, 181*, 227–234.

Gundelfinger, E. D., Boeckers, T. M., Baron, M. K., & Bowie, J. U. (2006). A role for zinc in postsynaptic density asSAMbly and plasticity? *Trends in Biochemical Sciences, 31*, 366–373.

Haeckel, A., Ahuja, R., Gundelfinger, E. D., Qualmann, B., & Kessels, M. M. (2008). The actin-binding protein Abp1 controls dendritic spine morphology and is important for spine head and synapse formation. *Journal of Neuroscience, 28*, 10031–10044.

Hamdan, F. F., et al. (2011). Excess of de novo deleterious mutations in genes associated with glutamatergic systems in nonsyndromic intellectual disability. *American Journal of Human Genetics, 88*, 306–316.

Han, K., & Kim, E. (2008). Synaptic adhesion molecules and PSD-95. *Progress in Neurobiology, 84*, 263–283.

Han, K., Kim, M. H., Seeburg, D., Seo, J., Verpelli, C., Han, S., Chung, H. S., Ko, J., Lee, H. W., Kim, K., Heo, W. D., Meyer, T., Kim, H., Sala, C., Choi, S. Y., Sheng, M., & Kim, E. (2009). Regulated RalBP1 binding to RalA and PSD-95 controls AMPA receptor endocytosis and LTD. *PLoS Biology, 7*, e1000187.

Hanley, J. G. (2008). PICK1: A multi-talented modulator of AMPA receptor trafficking. *Pharmacology and Therapeutics, 118*, 152–160.

Hayashi, M. K., Tang, C., Verpelli, C., Narayanan, R., Stearns, M. H., Xu, R. M., Li, H., Sala, C., & Hayashi, Y. (2009). The postsynaptic density proteins Homer and Shank form a polymeric network structure. *Cell, 137*, 159–171.

Hering, H., & Sheng, M. (2003). Activity-dependent redistribution and essential role of cortactin in dendritic spine morphogenesis. *Journal of Neuroscience, 23*, 11759–11769.

Hirbec, H., Francis, J. C., Lauri, S. E., Braithwaite, S. P., Coussen, F., Mulle, C., Dev, K. K., Coutinho, V., Meyer, G., Isaac, J. T., Collingridge, G. L., & Henley, J. M. (2003). Rapid and differential regulation of AMPA and kainate receptors at hippocampal mossy fibre synapses by PICK1 and GRIP. *Neuron, 37*, 625–638.

Hirbec, H., Perestenko, O., Nishimune, A., Meyer, G., Nakanishi, S., Henley, J. M., & Dev, K. K. (2002). The PDZ proteins PICK1, GRIP, and syntenin bind multiple glutamate receptor subtypes. Analysis of PDZ binding motifs. *Journal of Biological Chemistry, 277*, 15221–15224.

Hoogenraad, C. C., Milstein, A. D., Ethell, I. M., Henkemeyer, M., & Sheng, M. (2005). GRIP1 controls dendrite morphogenesis by regulating EphB receptor trafficking. *Nature Neuroscience, 8*, 906.

Howard, M. A., Elias, G. M., Elias, L. A., Swat, W., & Nicoll, R. A. (2010). The role of SAP97 in synaptic glutamate receptor dynamics. *Proceedings of the National Academy of Sciences of the United States of America, 107*, 3805–3810.

Hsueh, Y.-P., & Sheng, M. (1999). Requirement of N-terminal cysteines of PSD-95 for PSD-95 multimerization and ternary complex formation, but not for binding to potassium channel Kv1.4. *Journal of Biological Chemistry, 174*, 532–536.

Hu, J. H., Park, J. M., Park, S., Xiao, B., Dehoff, M. H., Kim, S., Hayashi, T., Schwarz, M. K., Huganir, R. L., Seeburg, P. H., Linden, D. J., & Worley, P. F. (2010). Homeostatic scaling requires group I mGluR activation mediated by Homer1a. *Neuron, 68*, 1128–1142.

Huang, G., Kim, J. Y., Dehoff, M., Mizuno, Y., Kamm, K. E., Worley, P. F., Muallem, S., & Zeng, W. (2007). Ca^{2+} signaling in microdomains: Homer1 mediates the interaction between RyR2 and Cav1.2 to regulate excitation-contraction coupling. *Journal of Biological Chemistry, 282*, 14283–14290.

Huang, K., Yanai, A., Kang, R., Arstikaitis, P., Singaraja, R. R., Metzler, M., Mullard, A., Haigh, B., Gauthier-Campbell, C., Gutekunst, C. A., Hayden, M. R., & El-Husseini, A. (2004). Huntingtin-interacting protein HIP14 is a palmitoyl transferase involved in palmitoylation and trafficking of multiple neuronal proteins. *Neuron, 44*, 977–986.

Hung, A. Y., Futai, K., Sala, C., Valtschanoff, J. G., Ryu, J., Woodworth, M. A., Kidd, F. L., Sung, C. C., Miyakawa, T., Bear, M. F., Weinberg, R. J., & Sheng, M. (2008). Smaller dendritic spines, weaker synaptic transmission, but enhanced spatial learning in mice lacking Shank1. *Journal of Neuroscience, 28*, 1697–1708.

Hung, A. Y., Sung, C. C., Brito, I. L., & Sheng, M. (2010). Degradation of postsynaptic scaffold GKAP and regulation of dendritic spine morphology by the TRIM3 ubiquitin ligase in rat hippocampal neurons. *PloS One, 5*, e9842.

Hwang, J. I., Kim, H. S., Lee, J. R., Kim, E., Ryu, S. H., & Suh, P. G. (2005). The interaction of phospholipase C-beta3 with Shank2 regulates mGluR-mediated calcium signal. *Journal of Biological Chemistry, 280*, 12467–12473.

Hwang, S. Y., Wei, J., Westhoff, J. H., Duncan, R. S., Ozawa, F., Volpe, P., Inokuchi, K., & Koulen, P. (2003). Differential functional interaction of two Vesl/Homer protein isoforms with ryanodine receptor type 1: A novel mechanism for control of intracellular calcium signaling. *Cell Calcium, 34*, 177–184.

Irie, M., Hata, Y., Takeuchi, M., Ichtchenko, K., Toyoda, A., Hirao, K., Takai, Y., Rosahl, T. W., & Südhof, T. C. (1997). Binding of neuroligins to PSD-95. *Science, 277*, 1511–1515.

Iskenderian-Epps, W. S., & Imperiali, B. (2010). Modulation of Shank3 PDZ domain ligand-binding affinity by dimerization. *ChemBioChem, 11*, 1979–1984.

Jaffe, H., Vinade, L., & Dosemeci, A. (2004). Identification of novel phosphorylation sites on postsynaptic density proteins. *Biochemical and Biophysical Research Communications, 321*, 210–218.

Jamain, S., Quach, H., Betancur, C., Rastam, M., Colineaux, C., Gillberg, I. C., Soderstrom, H., Giros, B., Leboyer, M., Gillberg, C., & Bourgeron, T. (2003). Mutations of the X-linked genes encoding neuroligins NLGN3 and NLGN4 are associated with autism. *Nature Genetics, 34*, 27–29.

Jaulin-Bastard, F., Saito, H., Le Bivic, A., Ollendorff, V., Marchetto, S., Birnbaum, D., & Borg, J. P. (2001). The ERBB2/HER2 receptor differentially interacts with ERBIN and PICK1 PSD-95/DLG/ZO-1 domain proteins. *Journal of Biological Chemistry, 276*, 15256–15263.

Jin, W., Ge, W. P., Xu, J., Cao, M., Peng, L., Yung, W., Liao, D., Duan, S., Zhang, M., & Xia, J. (2006). Lipid binding regulates synaptic targeting of PICK1, AMPA receptor trafficking, and synaptic plasticity. *Journal of Neuroscience, 26*, 2380–2390.

Jordan, B. A., Fernholz, B. D., Khatri, L., & Ziff, E. B. (2007). Activity-dependent AIDA-1 nuclear signaling regulates nucleolar numbers and protein synthesis in neurons. *Nature Neuroscience, 10*, 427–435.

Kato, A., Ozawa, F., Saitoh, Y., Fukazawa, Y., Sugiyama, H., & Inokuchi, K. (1998). Novel members of the Vesl/Homer family of PDZ proteins that bind metabotropic glutamate receptors. *Journal of Biological Chemistry, 273*, 23969–23975.

Kim, S. M., Choi, K. Y., Cho, I. H., Rhy, J. H., Kim, S. H., Park, C. S., Kim, E., & Song, W. K. (2009). Regulation of dendritic spine morphology by SPIN90, a novel Shank binding partner. *Journal of Neurochemistry, 109*, 1106–1117.

Kim, C. H., Chung, H. J., Lee, H. K., & Huganir, R. L. (2001). Interaction of the AMPA receptor subunit GluR2/3 with PDZ domains regulates hippocampal long-term depression. *Proceedings of the National Academy of Sciences of the United States of America, 98*, 11725–11730.

Kim, M. J., Dunah, A. W., Wang, Y. T., & Sheng, M. (2005). Differential roles of NR2A- and NR2B-containing NMDA receptors in Ras-ERK signaling and AMPA receptor trafficking. *Neuron, 46*, 745–760.

Kim, J. H., Liao, D., Lau, L. F., & Huganir, R. L. (1998). SynGAP: A synaptic RasGAP that associates with the PSD-95/SAP90 protein family. *Neuron, 20*, 683–691.

Kim, E., Naisbitt, S., Hsueh, Y.-P., Rao, A., Rothschild, A., Craig, A. M., & Sheng, M. (1997). GKAP, a novel synaptic protein that interacts with the guanylate kinase-like domain of the PSD-95/SAP90 family of channel clustering molecules. *The Journal of Cell Biology, 136*, 669–678.

Kim, E., Niethammer, M., Rothschild, A., Jan, Y. N., & Sheng, M. (1995). Clustering of shaker-type K + channels by interaction with a family of membrane-associated guanylate kinases. *Nature, 378*, 85–88.

Kim, H. G., et al. (2008). Disruption of neurexin 1 associated with autism spectrum disorder. *American Journal of Human Genetics, 82*, 199–207.

Kistner, U., Wenzel, B. M., Veh, R. W., Cases-Langhoff, C., Garner, A. M., Appeltauer, U., Voss, B., Gundelfinger, E. D., & Garner, C. C. (1993). SAP90, a rat presynaptic protein related to the product of the Drosophila tumor suppressor gene dlg-A. *Journal of Biological Chemistry, 268*, 4580–4583.

Kornau, H.-C., Schenker, L. T., Kennedy, M. B., & Seeburg, P. H. (1995). Domain interaction between NMDA receptor subunits and the postsynaptic density protein PSD-95. *Science, 269*, 1737–1740.

Kornau, H. C., Seeburg, P. H., & Kennedy, M. B. (1997). Interaction of ion channels and receptors with PDZ domain proteins. *Current Opinion in Neurobiology, 7*, 368–373.

Kreienkamp, H. J., Zitzer, H., Gundelfinger, E. D., Richter, D., & Bockers, T. M. (2000). The calcium-independent receptor for alpha-latrotoxin from human and rodent brains interacts with members of the ProSAP/SSTRIP/Shank family of multidomain proteins. *Journal of Biological Chemistry, 275*, 32387–32390.

Kuriyan, J., & Cowburn, D. (1997). Modular peptide recognition domains in eukaryotic signaling. *Annual Review of Biophysics and Biomolecular Structure, 26*, 259–288.

Lee, S. H., Choi, J. H., Lee, N., Lee, H. R., Kim, J. I., Yu, N. K., Choi, S. L., Kim, H., & Kaang, B. K. (2008). Synaptic protein degradation underlies destabilization of retrieved fear memory. *Science, 319*, 1253–1256.

Leonard, A. S., Davare, M. A., Horne, M. C., Garner, C. C., & Hell, J. W. (1998). SAP97 is associated with the alpha-amino-3-hydroxy-5-methylisoxazole-4-propionic acid receptor GluR1 subunit. *Journal of Biological Chemistry, 273*, 19518–19524.

Leuba, G., Savioz, A., Vernay, A., Carnal, B., Kraftsik, R., Tardif, E., Riederer, I., & Riederer, B. M. (2008). Differential changes in synaptic proteins in the Alzheimer frontal cortex with marked increase in PSD-95 postsynaptic protein. *Journal of Alzheimer's Disease, 15*, 139–151.

Levinson, J. N., & El-Husseini, A. (2005). Building excitatory and inhibitory synapses: Balancing neuroligin partnerships. *Neuron, 48*, 171–174.

Li, Y., Hough, C. J., Frederickson, C. J., & Sarvey, J. M. (2001). Induction of mossy fiber → Ca3 long-term potentiation requires translocation of synaptically released Zn^{2+}. *Journal of Neuroscience, 21,* 8015–8025.

Liebau, S., Proepper, C., Schmidt, T., Schoen, M., Bockmann, J., & Boeckers, T. M. (2009). ProSAPiP2, a novel postsynaptic density protein that interacts with ProSAP2/Shank3. *Biochemical and Biophysical Research Communications, 385,* 460–465.

Lim, S., Naisbitt, S., Yoon, J., Hwang, J. I., Suh, P. G., Sheng, M., & Kim, E. (1999). Characterization of the Shank family of synaptic proteins. Multiple genes, alternative splicing, and differential expression in brain and development. *Journal of Biological Chemistry, 274,* 29510–29518.

Lim, S., Sala, C., Yoon, J., Park, S., Kuroda, S., Sheng, M., & Kim, E. (2001). Sharpin, a novel postsynaptic density protein that directly interacts with the shank family of proteins. *Molecular and Cellular Neuroscience, 17,* 385–397.

Long, J. F., Tochio, H., Wang, P., Fan, J. S., Sala, C., Niethammer, M., Sheng, M., & Zhang, M. (2003). Supramodular structure and synergistic target binding of the N-terminal tandem PDZ domains of PSD-95. *Journal of Molecular Biology, 327,* 203–214.

Manning, M. A., Cassidy, S. B., Clericuzio, C., Cherry, A. M., Schwartz, S., Hudgins, L., Enns, G. M., & Hoyme, H. E. (2004). Terminal 22q deletion syndrome: A newly recognized cause of speech and language disability in the autism spectrum. *Pediatrics, 114,* 451–457.

Marcello, E., Gardoni, F., Mauceri, D., Romorini, S., Jeromin, A., Epis, R., Borroni, B., Cattabeni, F., Sala, C., Padovani, A., & Di Luca, M. (2007). Synapse-associated protein-97 mediates alpha-secretase ADAM10 trafficking and promotes its activity. *Journal of Neuroscience, 27,* 1682–1691.

Margeta, M. A., & Shen, K. (2010). Molecular mechanisms of synaptic specificity. *Molecular and Cellular Neuroscience, 43,* 261–267.

Maruoka, H., Konno, D., Hori, K., & Sobue, K. (2005). Collaboration of PSD-Zip70 with its binding partner, SPAR, in dendritic spine maturity. *Journal of Neuroscience, 25,* 1421–1430.

Matsubara, E., et al. (2003). Melatonin increases survival and inhibits oxidative and amyloid pathology in a transgenic model of Alzheimer's disease. *Journal of Neurochemistry, 85,* 1101–1108.

Mauceri, D., Cattabeni, F., Di Luca, M., & Gardoni, F. (2004). Calcium/calmodulin-dependent protein kinase II phosphorylation drives synapse-associated protein 97 into spines. *Journal of Biological Chemistry, 279,* 23813–23821.

Maunakea, A. K., et al. (2010). Conserved role of intragenic DNA methylation in regulating alternative promoters. *Nature, 466,* 253–257.

McGee, A. W., Dakoji, S. R., Olsen, O., Bredt, D. S., Lim, W. A., & Prehoda, K. E. (2001). Structure of the SH3-guanylate kinase module from PSD-95 suggests a mechanism for regulated assembly of MAGUK scaffolding proteins. *Molecular Cell, 8,* 1291–1301.

McWilliams, R. R., Breusegem, S. Y., Brodsky, K. F., Kim, E., Levi, M., & Doctor, R. B. (2005). Shank2E binds NaP(i) cotransporter at the apical membrane of proximal tubule cells. *American Journal of Physiology: Cell Physiology, 289,* C1042–C1051.

McWilliams, R. R., Gidey, E., Fouassier, L., Weed, S. A., & Doctor, R. B. (2004). Characterization of an ankyrin repeat-containing Shank2 isoform (Shank2E) in liver epithelial cells. *Biochemical Journal, 380,* 181–191.

Migaud, M., Charlesworth, P., Dempster, M., Webster, L. C., Watabe, A. M., Makhinson, M., He, Y., Ramsay, M. F., Morris, R. G., Morrison, J. H., O'Dell, T. J., & Grant, S. G. (1998). Enhanced long-term potentiation and impaired learning in mice with mutant postsynaptic density-95 protein. *Nature, 396,* 433–439.

Miller, Y., Ma, B., & Nussinov, R. (2010). Zinc ions promote Alzheimer Abeta aggregation via population shift of polymorphic states. *Proceedings of the National Academy of Sciences of the United States of America, 107,* 9490–9495.

Moessner, R., Marshall, C. R., Sutcliffe, J. S., Skaug, J., Pinto, D., Vincent, J., Zwaigenbaum, L., Fernandez, B., Roberts, W., Szatmari, P., & Scherer, S. W. (2007). Contribution of SHANK3 mutations to autism spectrum disorder. *American Journal of Human Genetics, 81*, 1289–1297.

Monea, S., Jordan, B. A., Srivastava, S., DeSouza, S., & Ziff, E. B. (2006). Membrane localization of membrane type 5 matrix metalloproteinase by AMPA receptor binding protein and cleavage of cadherins. *Journal of Neuroscience, 26*, 2300–2312.

Naisbitt, S., Kim, E., Tu, J. C., Xiao, B., Sala, C., Valtschanoff, J., Weinberg, R. J., Worley, P. F., & Sheng, M. (1999). Shank, a novel family of postsynaptic density proteins that binds to the NMDA receptor/PSD-95/GKAP complex and cortactin. *Neuron, 23*, 569–582.

Naisbitt, S., Valtschanoff, J., Allison, D., Sala, C., Kim, E., Craig, A., Weinberg, R., & Shen, M. (2000). Interaction of the postsynaptic density-95/guanylate kinase domain-associated protein complex with a light chain of myosin-V and dynein. *Journal of Neuroscience, 20*, 4524–4534.

Nakagawa, T., Futai, K., Lashuel, H. A., Lo, I., Okamoto, K., Walz, T., Hayashi, Y., & Sheng, M. (2004). Quaternary structure, protein dynamics, and synaptic function of SAP97 controlled by L27 domain interactions. *Neuron, 44*, 453–467.

Nakamura, M., Sato, K., Fukaya, M., Araishi, K., Aiba, A., Kano, M., & Watanabe, M. (2004). Signaling complex formation of phospholipase Cbeta4 with metabotropic glutamate receptor type 1alpha and 1,4,5-trisphosphate receptor at the perisynapse and endoplasmic reticulum in the mouse brain. *European Journal of Neuroscience, 20*, 2929–2944.

Nakamura, Y., Wood, C. L., Patton, A. P., Jaafari, N., Henley, J. M., Mellor, J. R., & Hanley, J. G. (2011). PICK1 inhibition of the Arp2/3 complex controls dendritic spine size and synaptic plasticity. *EMBO Journal, 30*, 719–730.

Niethammer, M., Kim, E., & Sheng, M. (1996). Interaction between the C terminus of NMDA receptor subunits and multiple members of the PSD-95 family of membrane-associated guanylate kinases. *Journal of Neuroscience, 16*, 2157–2163.

Nikonenko, I., Boda, B., Steen, S., Knott, G., Welker, E., & Muller, D. (2008). PSD-95 promotes synaptogenesis and multiinnervated spine formation through nitric oxide signaling. *The Journal of Cell Biology, 183*, 1115–1127.

Noritake, J., Fukata, Y., Iwanaga, T., Hosomi, N., Tsutsumi, R., Matsuda, N., Tani, H., Iwanari, H., Mochizuki, Y., Kodama, T., Matsuura, Y., Bredt, D. S., Hamakubo, T., & Fukata, M. (2009). Mobile DHHC palmitoylating enzyme mediates activity-sensitive synaptic targeting of PSD-95. *The Journal of Cell Biology, 186*, 147–160.

Okamoto, P. M., Gamby, C., Wells, D., Fallon, J., & Vallee, R. B. (2001). Dynamin isoform-specific interaction with the shank/ProSAP scaffolding proteins of the postsynaptic density and actin cytoskeleton. *Journal of Biological Chemistry, 276*, 48458–48465.

Olsen, O., & Bredt, D. S. (2003). Functional analysis of the nucleotide binding domain of membrane-associated guanylate kinases. *Journal of Biological Chemistry, 278*, 6873–6878.

Olson, P. A., Tkatch, T., Hernandez-Lopez, S., Ulrich, S., Ilijic, E., Mugnaini, E., Zhang, H., Bezprozvanny, I., & Surmeier, D. J. (2005). G-protein-coupled receptor modulation of striatal CaV1.3 L-type Ca^{2+} channels is dependent on a Shank-binding domain. *Journal of Neuroscience, 25*, 1050–1062.

Pak, D. T., & Sheng, M. (2003). Targeted protein degradation and synapse remodeling by an inducible protein kinase. *Science, 302*, 1368–1373.

Pak, D. T., Yang, S., Rudolph-Correia, S., Kim, E., & Sheng, M. (2001). Regulation of dendritic spine morphology by SPAR, a PSD-95-associated RapGAP. *Neuron, 31*, 289–303.

Park, E., Na, M., Choi, J., Kim, S., Lee, J. R., Yoon, J., Park, D., Sheng, M., & Kim, E. (2003). The Shank family of postsynaptic density proteins interacts with and promotes synaptic accumulation of the beta PIX guanine nucleotide exchange factor for Rac1 and Cdc42. *Journal of Biological Chemistry, 278*, 19220–19229.

Pavlowsky, A., Gianfelice, A., Pallotto, M., Zanchi, A., Vara, H., Khelfaoui, M., Valnegri, P., Rezai, X., Bassani, S., Brambilla, D., Kumpost, J., Blahos, J., Roux, M. J., Humeau, Y., Chelly, J., Passafaro, M., Giustetto, M., Billuart, P., & Sala, C. (2010). A postsynaptic signaling pathway that may account for the cognitive defect due to IL1RAPL1 mutation. *Current Biology, 20*, 103–115.

Peça, J., Feliciano, C., Ting, J. T., Wang, W., Wells, M. F., Venkatraman, T. N., Lascola, C. D., Fu, Z., & Feng, G. (2011). Shank3 mutant mice display autistic-like behaviours and striatal dysfunction. *Nature, 472*, 437.

Perroy, J., El Far, O., Bertaso, F., Pin, J. P., Betz, H., Bockaert, J., & Fagni, L. (2002). PICK1 is required for the control of synaptic transmission by the metabotropic glutamate receptor 7. *EMBO Journal, 21*, 2990–2999.

Petralia, R. S., Yokotani, N., & Wenthold, R. J. (1994). Light and electron microscope distribution of the NMDA receptor subunit NMDAR1 in the rat nervous system using a selective anti-peptide antibody. *Journal of Neuroscience, 14*, 667–696.

Pham, E., Crews, L., Ubhi, K., Hansen, L., Adame, A., Cartier, A., Salmon, D., Galasko, D., Michael, S., Savas, J. N., Yates, J. R., Glabe, C., & Masliah, E. (2010). Progressive accumulation of amyloid-beta oligomers in Alzheimer's disease and in amyloid precursor protein transgenic mice is accompanied by selective alterations in synaptic scaffold proteins. *FEBS Journal, 277*, 3051–3067.

Phelan, M. C., Rogers, R. C., Saul, R. A., Stapleton, G. A., Sweet, K., McDermid, H., Shaw, S. R., Claytor, J., Willis, J., & Kelly, D. P. (2001). 22q13 deletion syndrome. *American Journal of Medical Genetics, 101*, 91–99.

Proepper, C., Johannsen, S., Liebau, S., Dahl, J., Vaida, B., Bockmann, J., Kreutz, M. R., Gundelfinger, E. D., & Boeckers, T. M. (2007). Abelson interacting protein 1 (Abi-1) is essential for dendrite morphogenesis and synapse formation. *EMBO Journal, 26*, 1397–1409.

Qualmann, B., Boeckers, T. M., Jeromin, M., Gundelfinger, E. D., & Kessels, M. M. (2004). Linkage of the actin cytoskeleton to the postsynaptic density via direct interactions of Abp1 with the ProSAP/Shank family. *Journal of Neuroscience, 24*, 2481–2495.

Quitsch, A., Berhorster, K., Liew, C. W., Richter, D., & Kreienkamp, H. J. (2005). Postsynaptic shank antagonizes dendrite branching induced by the leucine-rich repeat protein Densin-180. *Journal of Neuroscience, 25*, 479–487.

Raab, M., Boeckers, T. M., & Neuhuber, W. L. (2010). Proline-rich synapse-associated protein-1 and 2 (ProSAP1/Shank2 and ProSAP2/Shank3)-scaffolding proteins are also present in postsynaptic specializations of the peripheral nervous system. *Neuroscience, 171*, 421–433.

Rama, S., Krapivinsky, G., Clapham, D. E., & Medina, I. (2008). The MUPP1-SynGAPalpha protein complex does not mediate activity-induced LTP. *Molecular and Cellular Neuroscience, 38*, 183–188.

Redecker, P., Gundelfinger, E. D., & Boeckers, T. M. (2001). The cortactin-binding postsynaptic density protein proSAP1 in non-neuronal cells. *Journal of Histochemistry and Cytochemistry, 49*, 639–648.

Redecker, P., Kreutz, M. R., Bockmann, J., Gundelfinger, E. D., & Boeckers, T. M. (2003). Brain synaptic junctional proteins at the acrosome of rat testicular germ cells. *Journal of Histochemistry and Cytochemistry, 51*, 809–819.

Reese, M. L., Dakoji, S., Bredt, D. S., & Dötsch, V. (2007). The guanylate kinase domain of the MAGUK PSD-95 binds dynamically to a conserved motif in MAP1a. *Nature Structural and Molecular Biology, 14*, 155–163.

Rocca, D. L., Martin, S., Jenkins, E. L., & Hanley, J. G. (2008). Inhibition of Arp2/3-mediated actin polymerization by PICK1 regulates neuronal morphology and AMPA receptor endocytosis. *Nature Cell Biology, 10*, 259–271.

Romorini, S., Piccoli, G., Jiang, M., Grossano, P., Tonna, N., Passafaro, M., Zhang, M., & Sala, C. (2004). A functional role of postsynaptic density-95-guanylate kinase-associated protein complex in regulating Shank assembly and stability to synapses. *Journal of Neuroscience, 24*, 9391–9404.

Roselli, F., Hutzler, P., Wegerich, Y., Livrea, P., & Almeida, O. F. (2009). Disassembly of shank and homer synaptic clusters is driven by soluble beta-amyloid(1–40) through divergent NMDAR-dependent signalling pathways. *PloS One, 4*, e6011.

Roussignol, G., Ango, F., Romorini, S., Tu, J. C., Sala, C., Worley, P. F., Bockaert, J., & Fagni, L. (2005). Shank expression is sufficient to induce functional dendritic spine synapses in aspiny neurons. *Journal of Neuroscience, 25*, 3560–3570.

Rumbaugh, G., Sia, G. M., Garner, C. C., & Huganir, R. L. (2003). Synapse-associated protein-97 isoform-specific regulation of surface AMPA receptors and synaptic function in cultured neurons. *Journal of Neuroscience, 23*, 4567–4576.

Saavedra, M. V., Smalla, K. H., Thomas, U., Sandoval, S., Olavarria, K., Castillo, K., Delgado, M. G., Delgado, R., Gundelfinger, E. D., Bacigalupo, J., & Wyneken, U. (2008). Scaffolding proteins in highly purified rat olfactory cilia membranes. *NeuroReport, 19*, 1123–1126.

Sala, C., Futai, K., Yamamoto, K., Worley, P. F., Hayashi, Y., & Sheng, M. (2003). Inhibition of dendritic spine morphogenesis and synaptic transmission by activity-inducible protein Homer1a. *Journal of Neuroscience, 23*, 6327–6337.

Sala, C., Piech, V., Wilson, N. R., Passafaro, M., Liu, G., & Sheng, M. (2001). Regulation of dendritic spine morphology and synaptic function by Shank and Homer. *Neuron, 31*, 115–130.

Sala, C., Roussignol, C., Meldolesi, J., & Fagni, L. (2005). Key role of the postsynaptic density scaffold proteins Shank and Homer in the functional architecture of Ca^{2+} homeostasis at dendritic spines in hippocampal neurons. *Journal of Neuroscience, 25*, 4587–4592.

Sans, N., Petralia, R. S., Wang, Y. X., Blahos, J., 2nd, Hell, J. W., & Wenthold, R. J. (2000). A developmental change in NMDA receptor-associated proteins at hippocampal synapses. *Journal of Neuroscience, 20*, 1260–1271.

Sans, N., Prybylowski, K., Petralia, R. S., Chang, K., Wang, Y. X., Racca, C., Vicini, S., & Wenthold, R. J. (2003). NMDA receptor trafficking through an interaction between PDZ proteins and the exocyst complex. *Nature Cell Biology, 5*, 520–530.

Sans, N., Racca, C., Petralia, R. S., Wang, Y. X., McCallum, J., & Wenthold, R. J. (2001). Synapse-associated protein 97 selectively associates with a subset of AMPA receptors early in their biosynthetic pathway. *Journal of Neuroscience, 21*, 7506–7516.

Schlüter, O. M., Xu, W., & Malenka, R. C. (2006). Alternative N-terminal domains of PSD-95 and SAP97 govern activity-dependent regulation of synaptic AMPA receptor function. *Neuron, 51*, 99–111.

Schmeisser, M. J., Grabrucker, A. M., Bockmann, J., & Boeckers, T. M. (2009). Synaptic crosstalk between N-methyl-D-aspartate receptors and LAPSER1-beta-catenin at excitatory synapses. *Journal of Biological Chemistry, 284*, 29146–29157.

Seidenman, K. J., Steinberg, J. P., Huganir, R., & Malinow, R. (2003). Glutamate receptor subunit 2 Serine 880 phosphorylation modulates synaptic transmission and mediates plasticity in CA1 pyramidal cells. *Journal of Neuroscience, 23*, 9220–9228.

Setou, M., Seog, D. H., Tanaka, Y., Kanai, Y., Takei, Y., Kawagishi, M., & Hirokawa, N. (2002). Glutamate-receptor-interacting protein GRIP1 directly steers kinesin to dendrites. *Nature, 417*, 83–87.

Sheng, M., & Kim, M. J. (2002). Postsynaptic signaling and plasticity mechanisms. *Science, 298*, 776–780.

Sheng, M., & Sala, C. (2001). PDZ domains and the organization of supramolecular complexes. *Annual Review of Neuroscience, 24*, 1–29.

Shin, H., Hsueh, Y. P., Yang, F. C., Kim, E., & Sheng, M. (2000). An intramolecular interaction between Src homology 3 domain and guanylate kinase-like domain required for channel clustering by postsynaptic density-95/SAP90. *Journal of Neuroscience, 20*, 3580–3587.

Shin, H., Wyszynski, M., Huh, K. H., Valtschanoff, J. G., Lee, J. R., Ko, J., Streuli, M., Weinberg, R. J., Sheng, M., & Kim, E. (2003). Association of the kinesin motor KIF1A with the multimodular protein liprin-alpha. *Journal of Biological Chemistry, 278*, 11393–11401.

Silverman, J. B., Restituito, S., Lu, W., Lee-Edwards, L., Khatri, L., & Ziff, E. B. (2007). Synaptic anchorage of AMPA receptors by cadherins through neural plakophilin-related arm protein AMPA receptor-binding protein complexes. *Journal of Neuroscience, 27*, 8505–8516.

Silverman, J. L., Turner, S. M., Barkan, C. L., Tolu, S. S., Saxena, R., Hung, A. Y., Sheng, M., & Crawley, J. N. (2011). Sociability and motor functions in Shank1 mutant mice. *Brain Research, 1380*, 120–137.

Soloviev, M. M., Ciruela, F., Chan, W. Y., & McIlhinney, R. A. (2000). Molecular characterisation of two structurally distinct groups of human homers, generated by extensive alternative splicing. *Journal of Molecular Biology, 295*, 1185–1200.

Soltau, M., Berhörster, K., Kindler, S., Buck, F., Richter, D., & Kreienkamp, H. J. (2004). Insulin receptor substrate of 53 kDa links postsynaptic shank to PSD-95. *Journal of Neurochemistry, 90*, 659–665.

Soltau, M., Richter, D., & Kreienkamp, H. J. (2002). The insulin receptor substrate IRSp53 links postsynaptic shank1 to the small G-protein cdc42. *Molecular and Cellular Neuroscience, 21*, 575–583.

Song, J. Y., Ichtchenko, K., Sudhof, T. C., & Brose, N. (1999). Neuroligin 1 is a postsynaptic cell-adhesion molecule of excitatory synapse. *Proceedings of the National Academy of Sciences of the United States of America, 96*, 1100–1105.

Steiner, P., Higley, M. J., Xu, W., Czervionke, B. L., Malenka, R. C., & Sabatini, B. L. (2008). Destabilization of the postsynaptic density by PSD-95 serine 73 phosphorylation inhibits spine growth and synaptic plasticity. *Neuron, 60*, 788–802.

Sugiyama, Y., Kawabata, I., Sobue, K., & Okabe, S. (2005). Determination of absolute protein numbers in single synapses by a GFP-based calibration technique. *Nature Methods, 2*, 677–684.

Sumioka, A., Yan, D., & Tomita, S. (2010). TARP phosphorylation regulates synaptic AMPA receptors through lipid bilayers. *Neuron, 66*, 755–767.

Szumlinski, K. K., Dehoff, M. H., Kang, S. H., Frys, K. A., Lominac, K. D., Klugmann, M., Rohrer, J., Griffin, W., Toda, S., Champtiaux, N. P., Berry, T., Tu, J. C., Shealy, S. E., During, M. J., Middaugh, L. D., Worley, P. F., & Kalivas, P. W. (2004). Homer proteins regulate sensitivity to cocaine. *Neuron, 43*, 401–413.

Szumlinski, K. K., Kalivas, P. W., & Worley, P. F. (2006). Homer proteins: Implications for neuropsychiatric disorders. *Current Opinion in Neurobiology, 16*, 251–257.

Takamiya, K., Kostourou, V., Adams, S., Jadeja, S., Chalepakis, G., Scambler, P. J., Huganir, R. L., & Adams, R. H. (2004). A direct functional link between the multi-PDZ domain protein GRIP1 and the Fraser syndrome protein Fras1. *Nature Genetics, 36*, 172–177.

Takamiya, K., Mao, L., Huganir, R. L., & Linden, D. J. (2008). The glutamate receptor-interacting protein family of GluR2-binding proteins is required for long-term synaptic depression expression in cerebellar Purkinje cells. *Journal of Neuroscience, 28*, 5752–5755.

Takeuchi, M., Hata, Y., Hirao, K., Toyoda, A., Irie, M., & Takai, Y. (1997). SAPAPs: A family of PSD-95/SAP90-associated proteins localized at postsynaptic density. *Journal of Biological Chemistry, 272*, 11943–11951.

Tao-Cheng, J. H., Dosemeci, A., Gallant, P. E., Smith, C., & Reese, T. (2010). Activity induced changes in the distribution of Shanks at hippocampal synapses. *Neuroscience, 168*, 11–17.

Tarpey, P., et al. (2004). Mutations in the DLG3 gene cause nonsyndromic X-linked mental retardation. *American Journal of Human Genetics, 75*, 318–324.

Thomas, G. M., Rumbaugh, G. R., Harrar, D. B., & Huganir, R. L. (2005). Ribosomal S6 kinase 2 interacts with and phosphorylates PDZ domain-containing proteins and regulates AMPA receptor transmission. *Proceedings of the National Academy of Sciences of the United States of America, 102*, 15006–15011.

Tomita, S., Adesnik, H., Sekiguchi, M., Zhang, W., Wada, K., Howe, J. R., Nicoll, R. A., & Bredt, D. S. (2005). Stargazin modulates AMPA receptor gating and trafficking by distinct domains. *Nature, 435*, 1052–1058.

Torres, G. E., Yao, W. D., Mohn, A. R., Quan, H., Kim, K. M., Levey, A. I., Staudinger, J., & Caron, M. G. (2001). Functional interaction between monoamine plasma membrane transporters and the synaptic PDZ domain-containing protein PICK1. *Neuron, 30*, 121–134.

Tsuriel, S., Geva, R., Zamorano, P., Dresbach, T., Boeckers, T., Gundelfinger, E. D., Garner, C. C., & Ziv, N. E. (2006). Local sharing as a predominant determinant of synaptic matrix molecular dynamics. *PLoS Biology, 4*, e271.

Tu, J. C., Xiao, B., Naisbitt, S., Yuan, J. P., Petralia, R. S., Brakeman, P., Doan, A., Aakalu, V. K., Lanahan, A. A., Sheng, M., & Worley, P. F. (1999). Coupling of mGluR/Homer and PSD-95 complexes by the Shank family of postsynaptic density proteins. *Neuron, 23*, 583–592.

Uchino, S., Wada, H., Honda, S., Nakamura, Y., Ondo, Y., Uchiyama, T., Tsutsumi, M., Suzuki, E., Hirasawa, T., & Kohsaka, S. (2006). Direct interaction of post-synaptic density-95/Dlg/ZO-1 domain-containing synaptic molecule Shank3 with GluR1 alpha-amino-3-hydroxy-5-methyl-4-isoxazole propionic acid receptor. *Journal of Neurochemistry, 97*, 1203–1214.

Uemura, T., Mori, H., & Mishina, M. (2004). Direct interaction of GluRdelta2 with Shank scaffold proteins in cerebellar Purkinje cells. *Molecular and Cellular Neuroscience, 26*, 330–341.

Valtschanoff, J. G., & Weinberg, R. J. (2001). Laminar organization of the NMDA receptor complex within the postsynaptic density. *Journal of Neuroscience, 21*, 1211–1217.

Vazquez, L. E., Chen, H. J., Sokolova, I., Knuesel, I., & Kennedy, M. B. (2004). SynGAP regulates spine formation. *Journal of Neuroscience, 24*, 8862–8872.

Volk, L., Kim, C. H., Takamiya, K., Yu, Y., & Huganir, R. L. (2010). Developmental regulation of protein interacting with C kinase 1 (PICK1) function in hippocampal synaptic plasticity and learning. *Proceedings of the National Academy of Sciences of the United States of America, 107*, 21784.

Welch, J. M., Lu, J., Rodriguiz, R. M., Trotta, N. C., Peca, J., Ding, J. D., Feliciano, C., Chen, M., Adams, J. P., Luo, J., Dudek, S. M., Weinberg, R. J., Calakos, N., Wetsel, W. C., & Feng, G. (2007). Cortico-striatal synaptic defects and OCD-like behaviours in Sapap3-mutant mice. *Nature, 448*, 894–900.

Wendholt, D., Spilker, C., Schmitt, A., Dolnik, A., Smalla, K. H., Proepper, C., Bockmann, J., Sobue, K., Gundelfinger, E. D., Kreutz, M. R., & Boeckers, T. M. (2006). ProSAP-interacting protein 1 (ProSAPiP1), a novel protein of the postsynaptic density that links the spine-associated Rap-Gap (SPAR) to the scaffolding protein ProSAP2/Shank3. *Journal of Biological Chemistry, 281*, 13805–13816.

Williams, M. E., Wu, S. C., McKenna, W. L., & Hinck, L. (2003). Surface expression of the netrin receptor UNC5H1 is regulated through a protein kinase C-interacting protein/protein kinase-dependent mechanism. *Journal of Neuroscience, 23*, 11279–11288.

Wilson, H. L., Wong, A. C., Shaw, S. R., Tse, W. Y., Stapleton, G. A., Phelan, M. C., Hu, S., Marshall, J., & McDermid, H. E. (2003). Molecular characterisation of the 22q13 deletion syndrome supports the role of haploinsufficiency of SHANK3/PROSAP2 in the major neurological symptoms. *Journal of Medical Genetics, 40*, 575–584.

Woods, D. F., Hough, C., Peel, D., Callaini, G., & Bryant, P. J. (1996). Dlg protein is required for junction structure, cell polarity, and proliferation control in Drosophila epithelia. *The Journal of Cell Biology, 134*, 1469–1482.

Wyszynski, M., Kim, E., Dunah, A. W., Passafaro, M., Valtschanoff, J. G., Serra-Pages, C., Streuli, M., Weinberg, R. J., & Sheng, M. (2002). Interaction between GRIP and liprin-alpha/SYD2 is required for AMPA receptor targeting. *Neuron, 34*, 39–52.

Xia, J., Zhang, X., Staudinger, J., & Huganir, R. L. (1999). Clustering of AMPA receptors by the synaptic PDZ domain-containing protein PICK1. *Neuron, 22*, 179–187.

Xiao, B., Tu, J. C., & Worley, P. F. (2000). Homer: A link between neural activity and glutamate receptor function. *Current Opinion in Neurobiology, 10*, 370–374.

Yamamoto, K., Sakagami, Y., Sugiura, S., Inokuchi, K., Shimohama, S., & Kato, N. (2005). Homer 1a enhances spike-induced calcium influx via L-type calcium channels in neocortex pyramidal cells. *European Journal of Neuroscience, 22*, 1338–1348.

Yao, I., Hata, Y., Hirao, K., Deguchi, M., Ide, N., Takeuchi, M., & Takai, Y. (1999). Synamon, a novel neuronal protein interacting with synapse-associated protein 90/postsynaptic density-95-associated protein. *Journal of Biological Chemistry, 274*, 27463–27466.

Ye, B., Liao, D., Zhang, X., Zhang, P., Dong, H., & Huganir, R. (2000). GRASP-1: A neuronal RasGEF associated with the AMPA receptor/GRIP complex. *Neuron, 26*, 603–617.

Yuan, J. P., Kiselyov, K., Shin, D. M., Chen, J., Shcheynikov, N., Kang, S. H., Dehoff, M. H., Schwarz, M. K., Seeburg, P. H., Muallem, S., & Worley, P. F. (2003). Homer binds TRPC family channels and is required for gating of TRPC1 by IP3 receptors. *Cell, 114*, 777–789.

Zalfa, F., Eleuteri, B., Dickson, K. S., Mercaldo, V., de Rubeis, S., di Penta, A., Tabolacci, E., Chiurazzi, P., Neri, G., Grant, S. G., & Bagni, C. (2007). A new function for the fragile X mental retardation protein in regulation of PSD-95 mRNA stability. *Nat Neurosci 10*, 578–587.

Zanni, G., van Esch, H., Bensalem, A., Saillour, Y., Poirier, K., Castelnau, L., Ropers, H. H., de Brouwer, A. P., Laumonnier, F., Fryns, J. P., & Chelly, J. (2010). A novel mutation in the DLG3 gene encoding the synapse-associated protein 102 (SAP102) causes non-syndromic mental retardation. *Neurogenetics, 11*, 251–255.

Zhang, H., Maximov, A., Fu, Y., Xu, F., Tang, T. S., Tkatch, T., Surmeier, D. J., & Bezprozvanny, I. (2005). Association of CaV1.3 L-type calcium channels with Shank. *Journal of Neuroscience, 25*, 1037–1049.

Zhou, L., Li, F., Xu, H. B., Luo, C. X., Wu, H. Y., Zhu, M. M., Lu, W., Ji, X., Zhou, Q. G., & Zhu, D. Y. (2010). Treatment of cerebral ischemia by disrupting ischemia-induced interaction of nNOS with PSD-95. *Nature Medicine, 16*, 1439–1443.

Zhu, Z. W., Xu, Q., Zhao, Z. Y., Gu, W. Z., & Wu, D. W. (2010). Spatiotemporal expression of PSD-95 in Fmr1 knockout mice brain. *Neuropathology, 31*, 223.

Zitzer, H., Honck, H. H., Bachner, D., Richter, D., & Kreienkamp, H. J. (1999). Somatostatin receptor interacting protein defines a novel family of multidomain proteins present in human and rodent brain. *Journal of Biological Chemistry, 274*, 32997–33001.

Züchner, S., Wendland, J. R., Ashley-Koch, A. E., Collins, A. L., Tran-Viet, K. N., Quinn, K., Timpano, K. C., Cuccaro, M. L., Pericak-Vance, M. A., Steffens, D. C., Krishnan, K. R., Feng, G., & Murphy, D. L. (2009). Multiple rare SAPAP3 missense variants in trichotillomania and OCD. *Molecular Psychiatry, 14*, 6–9.

Chapter 3
Diversity of Metabotropic Glutamate Receptor–Interacting Proteins and Pathophysiological Functions

Laurent Fagni

Abstract In the mammalian brain, the large majority of excitatory synapses express pre- and postsynaptic glutamate receptors. These are ion channels and G protein–coupled membrane proteins that are organized into functional signaling complexes. Here, we will review the nature and pathophysiological functions of the scaffolding proteins associated to these receptors, focusing on the G protein–coupled subtypes.

Keywords Brain disorders • Homer • Metabotropic glutamate receptors • Postsynaptic density • Scaffolding proteins

3.1 Introduction

Excitatory synaptic transmission in the mammalian brain is predominantly mediated by glutamate. The neurotransmitter binds to ionotropic (AMPA, kainate, and NMDA receptor-channels) and G protein–coupled/metabotropic (mGlu) receptors. The mGlu receptors are classified into three groups. Group I mGlu receptors (mGlu1 and mGlu5) are mainly localized postsynaptically, whereas group II (mGlu2 and mGlu3) and group III (mGlu4, mGlu7, and mGlu8) receptors are preferentially found on axon terminals. The group III mGlu6 receptor is solely expressed in the retina. Within group I, the mGlu1 receptor subtype is expressed as four splice variants: a long C-terminal form (mGlu1a, 350 residues) and three shorter C-terminal forms (mGlu1b, c, d), whereas the mGlu5 receptor is expressed as two long C-terminal splice variants (mGlu5a, b). The mGlu7 and mGlu8 receptors also display two splice variants (mGlu7a, b and mGlu8a, b) of roughly similar length.

L. Fagni (✉)

Institute of Functional Genomics, CNRS-UMR5203, INSERM-U661, Université Montpellier 1, Université Montpellier 2, 141 Rue de la Cardonille, 34000 Montpellier, France

e-mail: laurent.fagni@igf.cnrs.fr

M.R. Kreutz and C. Sala (eds.), *Synaptic Plasticity*,
Advances in Experimental Medicine and Biology 970,
DOI 10.1007/978-3-7091-0932-8_3, © Springer-Verlag/Wien 2012

It is well established that membrane neurotransmitter receptors, including mGlu receptors, are organized into functional networks by multiprotein complexes. The receptor-associated proteins can be other membrane neurotransmitter receptors or intracellular scaffolding/signaling proteins that physically link the membrane receptors to their intracellular effectors and cytoskeleton. During the past 10 years, the structure and functions of some of these receptor complexes (also called receptosomes) have been extensively studied. Here, we will review their implications in physiological and pathological functions, focusing on the mGlu receptosomes.

3.2 Membrane Proteins

Metabotropic glutamate receptors can interact with other membrane receptors and channels. These can be mGlu receptors of the same subtype, thus forming homodimers that are stabilized by an external N-terminal disulphur bound (Kunishima et al. 2000). The mGlu2 receptor subtype can also hetrodimerize with the serotonin 5-HT_{2A} receptor (Gonzalez-Maeso et al. 2008). The mGlu1 receptor can interact with the A_1 adenosine receptor (Ciruela et al. 2001), and the mGlu5 receptor can form a higher order trimeric receptor complex with the D_2 dopamine receptor and the A_{2A} adenosine receptor in striatal neurons (Cabello et al. 2009). The mGlu1a receptor variant can interact with the calcium-sensing receptor (CaR) (Gama et al. 2001), Caveolin1/2β (Burgueno et al. 2004), the GB1 subunit of $GABA_B$ receptor (Tabata et al. 2004), and the $Ca_{v2.1}$ pore-forming subunit of voltage-gated calcium channels (Kitano et al. 2003). Coassembly of mGlu5 receptor with N-methyl-D-aspartate (NMDA) (Perroy et al. 2008) and μ-opioid receptors (Schroder et al. 2009) has also been described (Table 3.1). Such a complexity of receptor associations gives rise to diversity of intracellular signalings.

Heteromeric mGlu receptors have been described in heterologous expression systems, using combined time-resolved fluorescence resonance energy transfer (trFRET) technology, and specific cell surface labeling of single nucleotide–amplified polymorphism (SNAP)- and click-enabled linker for interacting proteins (CLIP)-tagged mGlu receptor subunits. Different mGlu receptor subunits of a same group can dimerize, but mGlu subunits of group II and group III can also associate to generate heteromeric receptor complexes. In contrast, neither group II nor group III mGlu receptor subunits can dimerize with group I mGlu receptor subunits (Doumazane et al. 2010). Whether such heterodimerization exists in neurons remains to be investigated. If this is the case, such a specificity of heterodimerization would prevent formation of receptor complexes coupled to different G proteins. Indeed group I mGlu receptors activate Gq proteins, whereas group II and group III mGlu receptors are coupled to Gi/Go proteins (Table 3.1). Also such a specificity of heterodimerization is consistent with the described localization of mGlu receptor subtypes in neurons, group I mGlu receptors being mostly localized at postsynaptic sites, whereas group II and group III mGlu receptors are expressed at axon terminals.

Table 3.1 Metabotropic glutamate receptor–interacting proteins

mGlu receptors		Interacting proteins			
Group	Subunits	Membrane	Scaffolding	Cytoskeleton	Signaling
I (Gq)	1a	mGlu1/5, CaR, A_1R/A_2R, Caveolin1/2β, GB1, $Ca_v2.1$	Homer, CAL, Tamalin	α-Actinin 4, band 4.1, tubulin α/β	Norbin, optineurin, GRK2, CAIN, PKC, PP1γ, Siah-1A
	1b	mGlu1/5	–	α-Actinin 4, tubulin α/β, Vimentin	GRK2
	5a	mGlu1/5, A_1R/A_2R, NMDAR, μ-OpioidR	Homer, CAL, Tamalin, NHERF, Aβ	Filamin-A	Norbin, optineurin, CAIN, PKC, Siah-1A, P1γ, CaM, NECAB2
	5b	mGlu1/5, D_2R, μ-OpioidR	Homer, Tamalin, Aβ	α-Actinin 4, filamin-A	Norbin, CaM, NECAB2, PKC, PP1γ, Siah-1A
II (Gi/Go)	2	mGlu2/3/4/7/8, 5-HT$_{2A}$R	Tamalin	–	PKA, RanBPM
	3	mGlu2/3/4/7/8	Tamalin, GRIP1, PICK1	–	PKA, PP2α, RanBPM
III (Gi/Go)	4	mGlu2/3/4/7/8	GRIP-1, PICK1, Syntenin	Filamin-A, MAP1B	CaM, PIAS1, PKA, RanBPM
	6	mGlu2/3/4/7/8	GRIP1, Syntenin	MAP1B	PIAS1
	7a	mGlu2/3/4/7/8	PICK1, GRIP1, Syntenin	Tubulin α/β, MAP1B	CaM, G-$\beta\gamma$, MacMARCKS, PKA, PKC, PIAS1, RanBPM
	7b	mGlu2/3/4/7/8	PICK1, Syntenin	Filamin-A, MAP1B	CaM, G-$\beta\gamma$, MacMARCKS, PIAS1, PP1γ, RanBPM
	8a	mGlu2/3/4/7/8	PICK1	Filamin-A, MAP1B	CaM, PIAS1, PKA, RanBPM
	8b	mGlu2/3/4/7/8	PICK1	MAP1B	CaM, PIAS1, RanBPM

5-HT$_{2A}$R serotonin type 2A receptor, *A_1R/A_2R* type1/type2 adenosine receptors, *CAIN* calcineurin inhibitor protein, *CAL* CFTR-associated ligand, *CaM* calmodulin, *CaR* Ca^{2+}-sensing receptor, *$Ca_{v2.1}$* pore-forming unit of voltage-gated Ca^{2+} channel, *D_2R* dopamine receptor type2, *GB1* GB1 subunit of GABA$_B$ receptor, *GRIP-1* glutamate receptor–interacting protein 1, *GRK2* G Protein–coupled receptor kinase 2, *MAP1B* microtubule-associated protein 1B, *μ-OpioidR* m-opioid receptor, *NECAB2* neuronal Ca^{2+}-binding protein 2, *NHERF-2* Na^+/H^+-exchanger regulatory factor 2, *NMDAR* NMDA receptor, *PIAS1* protein inhibitor of activated STAT 1, *PICK1* protein interacting with C-kinase 1, *PP1γ* protein phosphatase 1γ, *PP2α* protein phosphatase 2α, *RanBPM* Ran-binding protein in the microtubule-organizing center, *RanBPM* Ran-binding protein in the microtubule-organizing center, *Siah-1A* seven in absentia homolog 1A

The mGlu5 and NMDA receptors have been reported to directly interact when coexpressed in heterologous expression systems. Formation of the complex results in reciprocal and agonist-independent inhibition of the two receptors (Perroy et al. 2008). Both receptors coexist in dendritic spines, but are physically distant one from each other. NMDA receptors are localized within the synaptic density and face the neurotransmitter (glutamate) release site, whereas mGlu5 receptors are localized at the periphery of the synapse. Although not definitively proved, the biological substrate responsible for this specific postsynaptic membrane localization might be the multiprotein (Homer-Shank-GKAP-PSD95) scaffolding complex that physically links these receptors together. It has been hypothesized that in conditions of elevated synaptic activity, the scaffold can be disrupted by the immediate early gene Homer1a, and the mGlu5 receptor may then move toward and interact with the NMDA receptor (Perroy et al. 2008). Thus, this association should give rise to reduced NMDA receptor–mediated currents and protect the neuron from excitotoxicity.

Both 5-HT_{2A} and mGlu2 receptors have been implicated in schizophrenia (Roth et al. 1998; Benneyworth et al. 2007, 2008). Activation of mGlu2 receptor increases affinity of 5-HT_{2A} receptors, whereas activation of 5-HT_{2A} receptors decreases mGlu2 receptor affinity. Moreover the two receptors colocalize and coimmunoprecipitate. These observations suggest that mGlu2 and 5-HT_{2A} receptors oligomerize (Gonzalez-Maeso et al. 2008). Formation of the complex would depend on the TM4 and TM5 transmembrane helices of mGlu2 receptor, as chimeric mGlu3 receptor containing the TM4 and TM5 domains of mGlu2 receptor does not associate with 5-HT_{2A} receptor. The mGlu2-5-HT_{2A} receptor heterodimer thus represents a promising target for the treatment of schizophrenia.

The A_{2A} adenosin receptor couples to Gs protein and mediates phosphorylation of 32 kDa dopamine- and cyclic AMP-regulated phosphoprotein (DARPP-32) via PKA activation, in striatal neurons. On the other hand, the D_2 dopamine receptor couples to Gi/Go proteins and counteracts the effect of A_{2A} receptor (Agnati et al. 2003). When coexpressed with mGlu5 receptor, A_{2A} receptors can trigger Ca^{2+} responses in HEK293 cells, similarly to mGlu5 receptor activation. Costimulation of A_{2A} and mGlu5 receptors synergistically increases DARPP-32 phosphorylation in striatal slices, making DARPP-32 a crossroad for the A_{2A}, mGlu5 and D_2 receptor signaling pathways. This functional interplay may be supported by oligomerization of the receptors, which have been evidenced using bimolecular fluorescence complementation, bioluminescence resonance energy transfer (BRET) and sequential resonance energy transfer (RET) approaches. Interactions depend on the third intracellular loop of the D_2 receptor and a C-terminal serine containing motif of the A_{2A} receptor (Ciruela et al. 2004; Ferre et al. 2007; Azdad et al. 2009), while interactions with mGlu5 receptor depends on the C-terminus of the receptor. Indeed not only mGlu5 receptor can interact with D_2 receptor, but all three A_{2A}, D_2, and mGlu5 receptors can form hetero-oligomers, as supported by high-resolution electron microscopy that shows the presence of these receptors at extrasynaptic membranes of the same striatal dendritic spines (see Cabello et al. 2009; Prezeau et al. 2010, for a review).

A synergistic action of A_1 receptor activation on mGlu5 receptor–mediated Ca^{2+} responses is observed in transfected HEK293 cells. Interestingly, the A_1 and mGlu1 receptor coimmunoprecipitate and colocalize in cortical and cerebellar neurons (Ciruela et al. 2001), further suggesting heterodimerization of the two receptors.

3.3 Scaffolding Proteins

Synaptic scaffolding proteins are present at both pre- and postsynaptic sites. They form multiprotein complexes that organize receptors and channels at the membrane, and link them to their intracellular signaling pathways, as well as to cytoskeleton. A number of these proteins also serve as signaling proteins in developing and adult neurons, thus playing an essential role in dendritic spine morphogenesis and synaptic transmission. Several scaffolding proteins have been shown to interact with mGlu receptors. Group I mGlu receptors can interact with Homer (Tu et al. 1998), tamalin (Kitano et al. 2002), Na^+/H^+-exchanger regulatory factor 2 (NHERF2) (Paquet et al. 2006), and CFTR-associated ligand (CAL) proteins (Zhang et al. 2008; Cheng et al. 2010). The group II mGlu3 and group III mGlu7 receptors have been reported to interact with glutamate receptor–interacting protein 1 (GRIP1) (Hirbec et al. 2002) and protein interacting with C-kinase 1 (PICK1, (El Far et al. 2000) (Table 3.1)).

Functional intercellular communications, including synaptic transmission, require adequate membrane and subcellular localization of ion channels and neurotransmitter receptors. Scaffolding proteins can provide such cellular functions, namely during early stages of development and in the adult cell communication systems. Best examples of these are given by the mGlu receptor–associated proteins Homer. Within group I, the three long C-terminus of mGlu1a and mGlu5 subunits contain a consensus sequence (-PPxxF-) that is a binding motif for the N-terminal EVH-like domain of Homer proteins (Tu et al. 1998). These proteins are the products of 3 genes (*Homer1*, *Homer2*, and *Homer3*) that give rise to several splice variants (Brakeman et al. 1997; Kato et al. 1997; Xiao et al. 1998). We have previously shown that constitutively expressed Homer3 controls the neuritic sorting of mGlu5 receptor in cultured neurons, thus excluding the receptor from axons and triggering its localization at postsynaptic sites (Ango et al. 2000). Notwithstanding, the receptor complex was predominantly found in the intracellular compartment. The short variant Homer1a was found to trigger the membrane targeting of the receptor (Ango et al. 2002). These studies suggest complementary roles of Homer proteins in functional expression of mGlu5 receptors at the synapse.

Although the mGlu1a receptor also interacts with Homer proteins, the same regulation did not apply to this receptor. The synaptic localization of mGlu1a seems to be rather regulated by the PDZ protein Tamalin (Kitano et al. 2002). Another PDZ domain–interacting protein, CAL, inhibits polyubiquitination and degradation of mGlu5a receptor, thus enhancing functional expression of the receptor in neurons (Zhang et al. 2008; Cheng et al. 2010).

Homer proteins also control efficacy of receptor signaling. The Homer ligand motif (-PPxxF-) is not only present at the C-terminus of mGlu1a and mGlu5 receptors, but also in the inositol trisphosphate (IP3) and ryanodine receptor-channels (Tu et al. 1998, Xiao et al. 1998), transient receptor-channels 1 and 4 (TRPC1, TRPC4; (Yuan et al. 2003) and $Ca_{v2.1}$ subunit of voltage-gated Ca^{2+} channels (Huang et al. 2007). Thus, Homer proteins can link mGlu1a/5 receptors to both membrane and intracellular Ca^{2+}-permeable ion channels. Formation of such a multiprotein association by Homer proteins constitutes an ideal machinery for controlling intracellular Ca^{2+} signaling (Fagni et al. 2004 for a review).

Macromolecular complexes can be responsible for restriction of intracellular signaling to specific subcellular microdomains. For instance, Homer proteins promote faster receptor Ca^{2+} responses by bringing mGlu1a/5 receptors in close proximity to Ca^{2+} stores and TRPCs (Tu et al. 1998). The specific localization of the receptor complex in the dendritic spine also contributes to confination of the mGluRa/5 receptor Ca^{2+} signaling at the synapse.

The C-terminus of presynaptic mGlu7 receptors binds to several proteins, including the PDZ protein PICK1, G protein $\beta\gamma$ subunits (G$\beta\gamma$), Ca^{2+}/calmodulin (Ca/CaM), and the PKC substrate macrophage myristolated alanine-rich C-kinase substrate (MacMARCKS) (Jess et al. 2002). PICK1 is an adaptor protein that dimerizes and physically links PKCα to mGluR7. The complex integrity is required for mGlu7a receptor intracellular signaling, which consists of self- and voltage-sensitive Ca^{2+} channel phosphorylation by PKCα (Dev et al. 2000). This effect is responsible for inhibition of the channel and reduced synaptic release of glutamate (Perroy et al. 2002). The mGlu7 receptor–mediated downregulation of synaptic transmission plays an essential role in condition of high-frequency synaptic activity. Indeed, mice deleted from the *GRM7* gene (Masugi et al. 1999) or expressing a mutated mGlu7 receptor that does not bind to PICK1 (Bertaso et al. 2008) display epileptic activities.

The following example illustrates the importance of receptor scaffolding proteins in agonist-independent/constitutive regulation of G-protein receptor function. The mGlu1a and mGlu5 receptors display marked agonist-independent/constitutive activity in heterologous expression systems, but not in neurons. We have shown that blockade of Homer3 synthesis by a specific antisense oligonucleotide or induction/transfection of Homer1a in cerebellar neurons results in marked constitutive activity of endogenous mGlu1a receptor. This suggests that Homer3 stabilizes mGlu1a/5 receptors in an inactive conformation, thus preventing their spontaneous activity in the absence of agonist. On the other hand, activity-dependent induction of Homer1a and disruption of Homer3-mGlu1a/5 receptor interaction results in spontaneous activity of the receptors. Thus, the multimeric Homer3-IP3R/RyR complex would create sufficient physical constraints to the C-terminus of mGlu1a receptor to prevent agonist-independent activity of the receptor. Disruption of the interaction between mGlu1a receptor and its C-terminal complex would decrease the physical constraints, thus allowing spontaneous activation of the G protein. These results provide the first evidence that an intracellular protein can activate a G protein–coupled receptor, independently of extracellular ligand receptor binding, in neurons (Ango et al. 2001). Homer proteins have later been found to exert a similar control over the spontaneous activity of TRPCs (Yuan et al. 2003). These studies support the notion that in addition to extracellular ligands, receptors and channels can be activated by intracellular partners. This noncanonical mode of activation of receptors and channels appears to have slower kinetics than conventional synaptic stimulation, membrane potential variation or even second messenger–mediated activation of receptors and channels, and may therefore represent a more stable signal of cell communication.

Mutations of genes encoding Homer proteins are candidates for neuropsychiatric phenotypes resulting from abnormal functioning of glutamate receptors, such as memory impairments, epilepsy, schizophrenia, affective syndromes, neuropathic disorders, and neurodegenerative diseases (Table 3.2). For example, studies performed in Homer1 deficient mice indicate a fundamental role of this protein in motivational, emotional, cognitive, and sensorimotor processes that were consistent with schizophrenia and altered cocaine-stimulated increase in circulating glutamate, in the prefrontal cortex (Szumlinski et al. 2005). Cognitive performance is closely correlated to the density and shape of dendritic spines. Thus, mental disorders are often associated with reduced number and abnormal spine morphology. The mGluR1a/5-associated Homer proteins, in combination with Shank

Table 3.2 List of mGlu receptors related brain diseases

Diseases	mGlu receptor subtypes (references)
Psychiatric disorders	
Schizophrenia	mGlu2 (Gonzalez-Maeso et al. 2008)
Mental retardation (Fragile X syndrome)	mGlu5 (Bear et al. 2004)
Neurological disorders	
Addiction to cocaine	mGlu1 (Mameli et al. 2007)
Stress, anxiety, depression	mGlu5 (Porter et al. 2005)
	mGlu2/3 (Swanson et al. 2005)
Temporal lobe epilepsy	mGlu1/5 (Blumcke et al. 2000)
	mGlu2/3 (Tang et al. 2004)
Absence epilepsy	mGlu7 (Bertaso et al. 2008)
Pain	mGlu1/5 (Karim et al. 2001)
Neurodegenerative diseases	
Ischemia	mGlu1/5 (Rao et al. 2000; Bao et al. 2001)
Alzheimer's	mGlu1 (Albasanz et al. 2005)
	mGlu5 (Renner et al. 2010)
	mGlu2 (Lee et al. 2004),
Parkinson's	mGlu1/5/2/3 (Pisani et al. 2003)
Sporadic amyotrophic sclerosis	mGlu1 (Valerio et al. 2002)
	mGlu2 (Poulopoulou et al. 2005)
Multiple sclerosis	mGlu4/8 (Geurts et al. 2005)
Retinitis pigmentosa	mGlu8 (Scherer et al. 1996, 1997)
Age-related hearing impairment	mGlu7 (Friedman et al. 2009)

proteins, play important roles in spine morphogenesis. Interestingly, in the Fragile X mental retardation syndrome, mGluR5-Homer association is reduced at the postsynaptic density (Giuffrida et al. 2005), and synaptic plasticity, as well as spine morphogenesis, are profoundly modified (see Bear et al. 2004 for a review).

Glutamatergic Ca^{2+} signaling is crucial for the development of nociceptive plasticity associated with chronic pain. In vivo studies showed that Homer1a is up-regulated in spinal cord neurons after peripheral inflammation and downregulates excitability of the pain pathway in an activity-dependent manner. Indeed, activity-dependent uncoupling of glutamate receptors from intracellular Ca^{2+} signaling pathway by Homer1a provides a mechanism to counteract sensitization of the first synapses in the pain pathway (Tappe et al. 2006). Homer proteins thus appear as promising targets for the treatment of hyperalgesia.

3.4 Proteins Associated with Cytoskeleton

Metabotropic glutamate receptors directly interact with proteins of the cytoskeleton. Thus, α-actinin-4 interacts with group I mGlu receptors (Cabello et al. 2007; Francesconi et al. 2009a); vimentin with mGlu1b (Francesconi et al. 2009b); α/β-tubulin with mGlu1a/b and mGlu7 (Francesconi et al. 2009b); band 4.1 proteins with mGlu1a and mGlu8 (Lu et al. 2004; Rose et al. 2008); filamin-A with mGlu5 (Enz 2002a), mGlu4, mGlu7, and mGlu8; and microtubule-associated protein 1B (MAP1B) with all group III mGlu subunits (Moritz et al. 2009).

Among these interactions, cytoskeleton and associated proteins are believed to control trafficking of the receptors to synapses. Vimentin is a phosphorylated intermediate filament protein involved in multiple cellular functions, including trafficking of protein complexes at the cell membrane. Interestingly vimentin is expressed in glia, but not in postmitotic neurons, except following injury. Then vimentin participates in retrograde axonal transport of proteins, which therefore may include mGlu1b receptor subunit.

Band 4.1 protein is part of the ezrin/radixin/moesin protein superfamily that functions as adaptor proteins, linking membrane proteins to actin cytoskeleton. This family of proteins not only serve as structural elements but also regulate signal transduction and neuronal cell development. However, the role of these proteins in mGlu1a and mGlu8 receptor signaling is not yet characterized.

3.5 Signaling Proteins

Although mGlu receptors are coupled to G proteins, they can also directly bind other signaling proteins, thus controlling additional intracellular signaling pathways. The mGlu receptors signaling protein partners are G$\beta\gamma$ subunits for mGlu7 receptors (Minakami et al. 1997; O'Connor et al. 1999; El Far et al.

2001), Ca^{2+} binding proteins such as calmodulin for mGlu4/5/7/8 (Minakami et al. 1997; O'Connor et al. 1999; El Far et al. 2001) and neuronal Ca^{2+}-binding protein 2 (NECAB2) for mGlu5 subunits (Canela et al. 2009). They can also bind protein kinsases such as G protein-coupled receptor kinase 2 (GRK2) for group I mGlu receptors (Dhami et al. 2005), PKA for group II and group III mGlu receptors (Cai et al. 2001), PKC for group I mGlu and mGlu7 receptors (Airas et al. 2001; Mao et al. 2008); protein phosphatases such as PP1γ, (Enz 2002b) and PP2α (Flajolet et al. 2003) for group I and mGlu7 receptors; associated modulatory protein such as calcineurin inhibitor (CAIN) for group I mGlu receptors (Ferreira et al. 2009), MacMARCKS for mGlu7 receptor (Bertaso et al. 2006), protein inhibitor of activated STAT1 (PIAS1) for group III and mGlu4 receptors (Tang et al. 2005). And G protein–coupled receptor kinase 2 (GRK2) for group I mGlu receptors (Dhami et al. 2005). Finally, group I mGlu receptors can also interact with non-enzymatically active proteins such as Norbin/Neurochodrin (Wang et al. 2009) and the huntingtin-binding protein, optineurin (Anborgh et al. 2005). Group I mGlu receptors can bind to the E3 ubiquitin ligase, seven in absentia homolog 1 (Siah-1A; (Ishikawa et al. 1999)). Group II and group III mGlu receptors are associated with Ran-binding protein in the microtubule-organizing center (RanBPM; (Seebahn et al. 2008)), and mGlu4 receptor binds to Munc18-1 (Nakajima et al. 2009) and RGS4 (Schwendt and McGinty 2007). It has recently been shown that mGlu5 receptor coimmunoprecipitates with amyloid peptide APP (Renner et al. 2010) (Table 3.1).

Interaction between Gβγ subunits and mGlu7a receptor is modulated by the binding of Ca^{2+}/CaM and MacMARCKS on the receptor C-terminus. The Gβγ binding site overlaps with that of Ca^{2+}/CaM, leading to mutually exclusive association of these molecules with the mGlu7a receptor. Thus Ca^{2+}/CaM binding leads to the release of Gβγ subunits, independently of mGlu7a receptor stimulation by an agonist, and Gβγ can then directly inhibit voltage-sensitive Ca^{2+} channels and synaptic transmission. Importantly, the concomitant binding of MacMARCKS to mGlu7a receptor antagonizes the binding of Ca^{2+}/CaM and impairs the agonist-independent inhibition of Ca^{2+} channels by Gβγ (Bertaso et al. 2006). This system illustrates how competitive binding between intracellular proteins can organize receptor signaling, independently of agonist stimulation.

Norbin/neurochondrin is a 75-kD neuronal protein without any known functional domain. It coimmunoprecipitates with mGlu5 receptor from rat brain lysates. The proximal C-terminus of the receptor is responsible for interaction with the protein. Coexpression of norbin increases surface expression of mGlu5 receptor in heterologous cells, and knock-down of norbin decreases the amount of functional mGlu5 receptors in neurons. This indicated an important role of this protein in regulating the receptor targeting to the plasma membrane. Indeed genetic deletion of norbin in mouse alters plasticity of excitatory synaptic transmission (both long-term potentiation and long-term depression) in the hippocampus, and induces a phenotype that is related to that of rodent schizophrenia models (Wang et al. 2009).

Optineurin belongs to the large group of proteins that bind to huntingtin, a protein that is responsible for Huntington's disease when displaying polyglutamine expansion in its amino-terminal region. Optineurin is a coiled-coil domain containing protein with no clear specific cellular function. The protein interacts with group I mGlu receptors and antagonizes the agonist-stimulated receptors signaling. Interestingly, mutated huntingtin facilitates optineurin-mediated attenuation of mGlu1a receptor signaling, thus indicating a possible role of the receptor complex in Huntington's disease. It has been proposed that in striatal tissue, where GRK2 is expressed at relatively low level, optineurin substitutes for GRK2 to mediate mGlu5 receptor desensitization. These observations provide additional biochemical link between mGlu receptors and Huntington's disease (Anborgh et al. 2005).

Oligomers of amyloid β (Aβo) peptide greatly contributes to memory loss and neurodegeneration in Alzheimer's disease. Using single particle tracking of Aβo labeled with quantum dots, it has recently been shown that membrane-bound Aβo diffuses laterally and accumulates at excitatory synapses in cultured hippocampal neurons. The newly generated pathogenic synaptic clusters significantly reduce the mobility of mGlu5 receptors, which normally diffuse readily in the plasma membrane. Consequently, mGlu5 receptors form ectopic aggregates, which impact on Ca^{2+} signaling at synapses. Consistent with these findings, cultured hippocampal neurons isolated from mGlu5 receptor KO mouse were protected from neuronal surface binding of Aβo (Renner et al. 2010). These results provide new pharmacological targets for therapeutic improvement of synaptic and cognitive functions in Alzheimer's disease.

3.6 Concluding Remarks

Receptor-interacting proteins have been initially thought to organize receptors, channels and transporters to plasma membrane solely. The studies described here show that these proteins can also play important roles in receptor targeting and signaling, at least for the mGlu receptor family. By interacting with the C-terminus of the receptors, they may also specify the function of receptor variants. It is remarkable that such a concept applies to both pre- and post-synaptic mGlu receptors (Fig. 3.1).

Metabotropic glutamate receptors have been implicated in several neurological and psychiatric disorders. These are neurodegenerative diseases such as Alzheimer's and Parkinson's diseases for group I mGlu receptors and mGlu2 receptor, sporadic amyotrophic lateral sclerosis (for mGlu1 receptor) and multiple sclerosis (for mGlu4/8 receptors), age-related hearing impairment (for mGlu7 receptor) and retinitis pigmentosa (for mGlu8 receptor). Group I and group II mGlu receptors are implicated in temporal lobe epilepsy, while the group III mGlu7 receptor may participate in absence epilepsy. Metabotropic glutamate receptors have also been associated with drug addiction (mGlu1 receptor), mental retardation (mGlu5 receptor, Down's and Fragile X syndromes), as well as anxiety

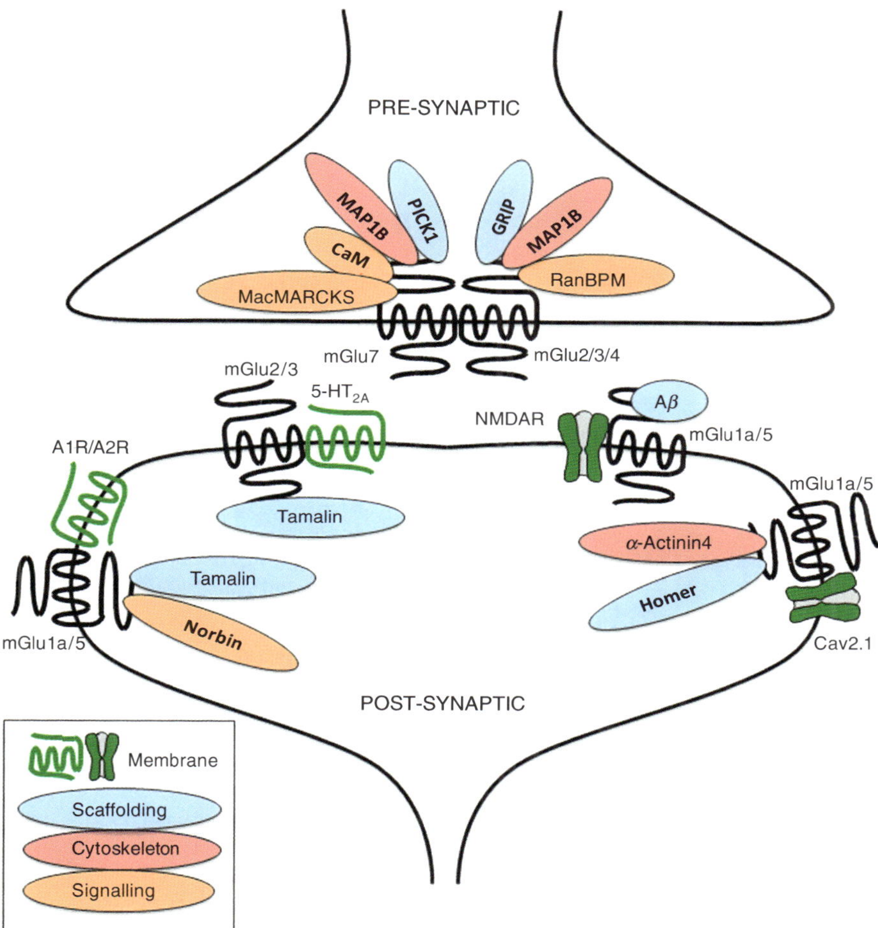

Fig. 3.1 Schematic view of mGlu receptor interactomes. MGlu receptors are represented in *black*. The mGlu receptor–interacting proteins are indicated with different colors, depending on their functional role. These colors are *blue* for the scaffolding proteins, *pink* for the cytoskeleton adaptor proteins, *donut* for the signaling proteins, and *green* for the membrane proteins. Only one representative protein of the different classes is represented for each mGlu receptor group

and related diseases (mGlu5 receptor and group II receptors; see Table 3.2 for references). Studies reported here show that mGlu receptor–interacting proteins play a role in several of these pathologies, thus pointing out that these receptor-associated proteins provide novel potential therapeutic targets for the treatment of neurological and psychiatric disorders.

References

Agnati, L. F., Ferre, S., Lluis, C., Franco, R., & Fuxe, K. (2003). Molecular mechanisms and therapeutical implications of intramembrane receptor/receptor interactions among heptahelical receptors with examples from the striatopallidal GABA neurons. *Pharmacological Reviews, 55*, 509–550.

Airas, J. M., Betz, H., & El Far, O. (2001). PKC phosphorylation of a conserved serine residue in the C-terminus of group III metabotropic glutamate receptors inhibits calmodulin binding. *FEBS Letters, 494*, 60–63.

Albasanz, J. L., Dalfo, E., Ferrer, I., & Martin, M. (2005). Impaired metabotropic glutamate receptor/phospholipase C signaling pathway in the cerebral cortex in Alzheimer's disease and dementia with Lewy bodies correlates with stage of Alzheimer's-disease-related changes. *Neurobiology of Disease, 20*, 685–693.

Anborgh, P. H., Godin, C., Pampillo, M., Dhami, G. K., Dale, L. B., Cregan, S. P., Truant, R., & Ferguson, S. S. (2005). Inhibition of metabotropic glutamate receptor signaling by the huntingtin-binding protein optineurin. *The Journal of Biological Chemistry, 280*, 34840–34848.

Ango, F., Pin, J. P., Tu, J. C., Xiao, B., Worley, P. F., Bockaert, J., & Fagni, L. (2000). Dendritic and axonal targeting of type 5 metabotropic glutamate receptor is regulated by Homer1 proteins and neuronal excitation. *The Journal of Neuroscience, 20*, 8710–8716.

Ango, F., Prezeau, L., Muller, T., Tu, J. C., Xiao, B., Worley, P. F., Pin, J. P., Bockaert, J., & Fagni, L. (2001). Agonist-independent activation of metabotropic glutamate receptors by the intracellular protein Homer. *Nature, 411*, 962–965.

Ango, F., Robbe, D., Tu, J. C., Xiao, B., Worley, P. F., Pin, J. P., Bockaert, J., & Fagni, L. (2002). Homer-dependent cell surface expression of metabotropic glutamate receptor type 5 in neurons. *Molecular and Cellular Neuroscience, 20*, 323–329.

Azdad, K., Gall, D., Woods, A. S., Ledent, C., Ferre, S., & Schiffmann, S. N. (2009). Dopamine D2 and adenosine A2A receptors regulate NMDA-mediated excitation in accumbens neurons through A2A-D2 receptor heteromerization. *Neuropsychopharmacology, 34*, 972–986.

Bao, W. L., Williams, A. J., Faden, A. I., & Tortella, F. C. (2001). Selective mGlu5 receptor antagonist or agonist provides neuroprotection in a rat model of focal cerebral ischemia. *Brain Research, 922*, 173–179.

Bear, M. F., Huber, K. M., & Warren, S. T. (2004). The mGluR theory of fragile X mental retardation. *Trends in Neurosciences, 27*, 370–377.

Benneyworth, M. A., Smith, R. L., & Sanders-Bush, E. (2008). Chronic phenethylamine hallucinogen treatment alters behavioral sensitivity to a metabotropic glutamate 2/3 receptor agonist. *Neuropsychopharmacology, 33*, 2206–2216.

Benneyworth, M. A., Xiang, Z., Smith, R. L., Garcia, E. E., Conn, P. J., & Sanders-Bush, E. (2007). A selective positive allosteric modulator of metabotropic glutamate receptor subtype 2 blocks a hallucinogenic drug model of psychosis. *Molecular Pharmacology, 72*, 477–484.

Bertaso, F., Lill, Y., Airas, J. M., Espeut, J., Blahos, J., Bockaert, J., Fagni, L., Betz, H., & El-Far, O. (2006). MacMARCKS interacts with the metabotropic glutamate receptor type 7 and modulates G protein-mediated constitutive inhibition of calcium channels. *Journal of Neurochemistry, 99*, 288–298.

Bertaso, F., Zhang, C., Scheschonka, A., De Bock, F., Fontanaud, P., Marin, P., Huganir, R. L., Betz, H., Bockaert, J., Fagni, L., & Lerner-Natoli, M. (2008). PICK1 uncoupling from mGluR7a causes absence-like seizures. *Nature Neuroscience, 7*, 211–224.

Blumcke, I., Becker, A. J., Klein, C., Scheiwe, C., Lie, A. A., Beck, H., Waha, A., Friedl, M. G., Kuhn, R., Emson, P., Elger, C., & Wiestler, O. D. (2000). Temporal lobe epilepsy associated up-regulation of metabotropic glutamate receptors: Correlated changes in mGluR1 mRNA and protein expression in experimental animals and human patients. *Journal of Neuropathology and Experimental Neurology, 59*, 1–10.

Brakeman, P. R., Lanahan, A. A., O'Brien, R., Roche, K., Barnes, C. A., Huganir, R. L., & Worley, P. F. (1997). Homer: A protein that selectively binds metabotropic glutamate receptors. *Nature, 386*, 284–288.

Burgueno, J., Canela, E. I., Mallol, J., Lluis, C., Franco, R., & Ciruela, F. (2004). Mutual regulation between metabotropic glutamate type 1alpha receptor and caveolin proteins: From traffick to constitutive activity. *Experimental Cell Research, 300*, 23–34.

Cabello, N., Gandia, J., Bertarelli, D. C., Watanabe, M., Lluis, C., Franco, R., Ferre, S., Lujan, R., & Ciruela, F. (2009). Metabotropic glutamate type 5, dopamine D2 and adenosine A2a receptors form higher-order oligomers in living cells. *Journal of Neurochemistry, 109*, 1497–1507.

Cabello, N., Remelli, R., Canela, L., Soriguera, A., Mallol, J., Canela, E. I., Robbins, M. J., Lluis, C., Franco, R., McIlhinney, R. A., & Ciruela, F. (2007). Actin-binding protein alpha-actinin-1 interacts with the metabotropic glutamate receptor type 5b and modulates the cell surface expression and function of the receptor. *The Journal of Biological Chemistry, 282*, 12143–12153.

Cai, Z., Saugstad, J. A., Sorensen, S. D., Ciombor, K. J., Zhang, C., Schaffhauser, H., Hubalek, F., Pohl, J., Duvoisin, R. M., & Conn, P. J. (2001). Cyclic AMP-dependent protein kinase phosphorylates group III metabotropic glutamate receptors and inhibits their function as presynaptic receptors. *Journal of Neurochemistry, 78*, 756–766.

Canela, L., Fernandez-Duenas, V., Albergaria, C., Watanabe, M., Lluis, C., Mallol, J., Canela, E. I., Franco, R., Lujan, R., & Ciruela, F. (2009). The association of metabotropic glutamate receptor type 5 with the neuronal Ca^{2+}-binding protein 2 modulates receptor function. *Journal of Neurochemistry, 111*, 555–567.

Cheng, S., Zhang, J., Zhu, P., Ma, Y., Xiong, Y., Sun, L., Xu, J., Zhang, H., & He, J. (2010). The PDZ domain protein CAL interacts with mGluR5a and modulates receptor expression. *Journal of Neurochemistry, 112*, 588–598.

Ciruela, F., Burgueno, J., Casado, V., Canals, M., Marcellino, D., Goldberg, S. R., Bader, M., Fuxe, K., Agnati, L. F., Lluis, C., Franco, R., Ferre, S., & Woods, A. S. (2004). Combining mass spectrometry and pull-down techniques for the study of receptor heteromerization. Direct epitope-epitope electrostatic interactions between adenosine A2A and dopamine D2 receptors. *Analytical Chemistry, 76*, 5354–5363.

Ciruela, F., Escriche, M., Burgueno, J., Angulo, E., Casado, V., Soloviev, M. M., Canela, E. I., Mallol, J., Chan, W. Y., Lluis, C., McIlhinney, R. A., & Franco, R. (2001). Metabotropic glutamate 1alpha and adenosine A1 receptors assemble into functionally interacting complexes. *The Journal of Biological Chemistry, 276*, 18345–18351.

Dev, K. K., Nakajima, Y., Kitano, J., Braithwaite, S. P., Henley, J. M., & Nakanishi, S. (2000). PICK1 interacts with and regulates PKC phosphorylation of mGLUR7. *The Journal of Neuroscience, 20*, 7252–7257.

Dhami, G. K., Babwah, A. V., Sterne-Marr, R., & Ferguson, S. S. (2005). Phosphorylation-independent regulation of metabotropic glutamate receptor 1 signaling requires g protein-coupled receptor kinase 2 binding to the second intracellular loop. *The Journal of Biological Chemistry, 280*, 24420–24427.

Doumazane, E., Scholler, P., Zwier, J. M., Trinquet, E., Rondard, P., & Pin, J. P. (2010). A new approach to analyze cell surface protein complexes reveals specific heterodimeric metabotropic glutamate receptors. *FASEB Journal.*

El Far, O., Airas, J., Wischmeyer, E., Nehring, R. B., Karschin, A., & Betz, H. (2000). Interaction of the C-terminal tail region of the metabotropic glutamate receptor 7 with the protein kinase C substrate PICK1. *The European Journal of Neuroscience, 12*, 4215–4221.

El Far, O., Bofill-Cardona, E., Airas, J. M., O'Connor, V., Boehm, S., Freissmuth, M., Nanoff, C., & Betz, H. (2001). Mapping of calmodulin and Gbetagamma binding domains within the C-terminal region of the metabotropic glutamate receptor 7A. *The Journal of Biological Chemistry, 276*, 30662–30669.

Enz, R. (2002a). The actin-binding protein Filamin-A interacts with the metabotropic glutamate receptor type 7. *FEBS Letters, 514*, 184–188.

Enz, R. (2002b). The metabotropic glutamate receptor mGluR7b binds to the catalytic gamma-subunit of protein phosphatase 1. *Journal of Neurochemistry, 81*, 1130–1140.

Fagni, L., Ango, F., Perroy, J., & Bockaert, J. (2004). Identification and functional roles of metabotropic glutamate receptor-interacting proteins. *Seminars in Cell and Developmental Biology, 15*, 289–298.

Ferre, S., Ciruela, F., Woods, A. S., Lluis, C., & Franco, R. (2007). Functional relevance of neurotransmitter receptor heteromers in the central nervous system. *Trends in Neurosciences, 30*, 440–446.

Ferreira, L. T., Dale, L. B., Ribeiro, F. M., Babwah, A. V., Pampillo, M., & Ferguson, S. S. (2009). Calcineurin inhibitor protein (CAIN) attenuates Group I metabotropic glutamate receptor endocytosis and signaling. *The Journal of Biological Chemistry, 284*, 28986–28994.

Flajolet, M., Rakhilin, S., Wang, H., Starkova, N., Nuangchamnong, N., Nairn, A. C., & Greengard, P. (2003). Protein phosphatase 2C binds selectively to and dephosphorylates metabotropic glutamate receptor 3. *Proceedings of the National Academy of Sciences of the United States of America, 100*, 16006–16011.

Francesconi, A., Kumari, R., & Zukin, R. S. (2009a). Proteomic analysis reveals novel binding partners of metabotropic glutamate receptor 1. *Journal of Neurochemistry, 108*, 1515–1525.

Francesconi, A., Kumari, R., & Zukin, R. S. (2009b). Regulation of group I metabotropic glutamate receptor trafficking and signaling by the caveolar/lipid raft pathway. *Journal of Neuroscience, 29*, 3590–3602.

Friedman, R. A., Van Laer, L., Huentelman, M. J., Sheth, S. S., Van Eyken, E., Corneveaux, J. J., Tembe, W. D., Halperin, R. F., Thorburn, A. Q., Thys, S., Bonneux, S., Fransen, E., Huyghe, J., Pyykko, I., Cremers, C. W., Kremer, H., Dhooge, I., Stephens, D., Orzan, E., Pfister, M., Bille, M., Parving, A., Sorri, M., Van de Heyning, P. H., Makmura, L., Ohmen, J. D., Linthicum, F. H., Jr., Fayad, J. N., Pearson, J. V., Craig, D. W., Stephan, D. A., & Van Camp, G. (2009). GRM7 variants confer susceptibility to age-related hearing impairment. *Human Molecular Genetics, 18*, 785–796.

Gama, L., Wilt, S. G., & Breitwieser, G. E. (2001). Heterodimerization of calcium sensing receptors with metabotropic glutamate receptors in neurons. *The Journal of Biological Chemistry, 276*, 39053–39059.

Geurts, J. J., Wolswijk, G., Bo, L., Redeker, S., Ramkema, M., Troost, D., & Aronica, E. (2005). Expression patterns of Group III metabotropic glutamate receptors mGluR4 and mGluR8 in multiple sclerosis lesions. *Journal of Neuroimmunology, 158*, 182–190.

Giuffrida, R., Musumeci, S., D'Antoni, S., Bonaccorso, C. M., Giuffrida-Stella, A. M., Oostra, B. A., & Catania, M. V. (2005). A reduced number of metabotropic glutamate subtype 5 receptors are associated with constitutive Homer proteins in a mouse model of fragile X syndrome. *The Journal of Neuroscience, 25*, 8908–8916.

Gonzalez-Maeso, J., Ang, R. L., Yuen, T., Chan, P., Weisstaub, N. V., Lopez-Gimenez, J. F., Zhou, M., Okawa, Y., Callado, L. F., Milligan, G., Gingrich, J. A., Filizola, M., Meana, J. J., & Sealfon, S. C. (2008). Identification of a serotonin/glutamate receptor complex implicated in psychosis. *Nature, 452*, 93–97.

Hirbec, H., Perestenko, O., Nishimune, A., Meyer, G., Nakanishi, S., Henley, J. M., & Dev, K. K. (2002). The PDZ proteins PICK1, GRIP, and syntenin bind multiple glutamate receptor subtypes. Analysis of PDZ binding motifs. *The Journal of Biological Chemistry, 277*, 15221–15224.

Huang, G., Kim, J. Y., Dehoff, M., Mizuno, Y., Kamm, K. E., Worley, P. F., Muallem, S., & Zeng, W. (2007). Ca^{2+} signaling in microdomains: Homer1 mediates the interaction between RyR2 and Cav1.2 to regulate excitation-contraction coupling. *The Journal of Biological Chemistry, 282*, 14283–14290.

Ishikawa, K., Nash, S. R., Nishimune, A., Neki, A., Kaneko, S., & Nakanishi, S. (1999). Competitive interaction of seven in absentia homolog-1A and Ca^{2+}/calmodulin with the cytoplasmic tail of group 1 metabotropic glutamate receptors. *Genes to Cells, 4*, 381–390.

Jess, U., El Far, O., Kirsch, J., & Betz, H. (2002). Interaction of the C-terminal region of the rat serotonin transporter with MacMARCKS modulates 5-HT uptake regulation by protein kinase C. *Biochemical and Biophysical Research Communications, 294*, 272–279.

Karim, F., Wang, C. C., & Gereau, R. Wt. (2001). Metabotropic glutamate receptor subtypes 1 and 5 are activators of extracellular signal-regulated kinase signaling required for inflammatory pain in mice. *The Journal of Neuroscience, 21*, 3771–3779.

Kato, A., Ozawa, F., Saitoh, Y., Hirai, K., & Inokuchi, K. (1997). vesl, a gene encoding VASP/Ena family related protein, is upregulated during seizure, long-term potentiation and synaptogenesis. *FEBS Letters, 412*, 183–189.

Kitano, J., Kimura, K., Yamazaki, Y., Soda, T., Shigemoto, R., Nakajima, Y., & Nakanishi, S. (2002). Tamalin, a PDZ domain-containing protein, links a protein complex formation of group 1 metabotropic glutamate receptors and the guanine nucleotide exchange factor cytohesins. *The Journal of Neuroscience, 22*, 1280–1289.

Kitano, J., Nishida, M., Itsukaichi, Y., Minami, I., Ogawa, M., Hirano, T., Mori, Y., & Nakanishi, S. (2003). Direct interaction and functional coupling between metabotropic glutamate receptor subtype 1 and voltage-sensitive Cav2.1 Ca^{2+} channel. *The Journal of Biological Chemistry, 278*, 25101–25108.

Kunishima, N., Shimada, Y., Tsuji, Y., Sato, T., Yamamoto, M., Kumasaka, T., Nakanishi, S., Jingami, H., & Morikawa, K. (2000). Structural basis of glutamate recognition by a dimeric metabotropic glutamate receptor. *Nature, 407*, 971–977.

Lee, H. G., Ogawa, O., Zhu, X., O'Neill, M. J., Petersen, R. B., Castellani, R. J., Ghanbari, H., Perry, G., & Smith, M. A. (2004). Aberrant expression of metabotropic glutamate receptor 2 in the vulnerable neurons of Alzheimer's disease. *Acta Neuropathologica, 107*, 365–371.

Lu, D., Yan, H., Othman, T., & Rivkees, S. A. (2004). Cytoskeletal protein 4.1G is a binding partner of the metabotropic glutamate receptor subtype 1 alpha. *Journal of Neuroscience Research, 78*, 49–55.

Mameli, M., Balland, B., Lujan, R., & Luscher, C. (2007). Rapid synthesis and synaptic insertion of GluR2 for mGluR-LTD in the ventral tegmental area. *Science, 317*, 530–533.

Mao, L. M., Liu, X. Y., Zhang, G. C., Chu, X. P., Fibuch, E. E., Wang, L. S., Liu, Z., & Wang, J. Q. (2008). Phosphorylation of group I metabotropic glutamate receptors (mGluR1/5) in vitro and in vivo. *Neuropharmacology, 55*, 403–408.

Masugi, M., Yokoi, M., Shigemoto, R., Muguruma, K., Watanabe, Y., Sansig, G., van der Putten, H., & Nakanishi, S. (1999). Metabotropic glutamate receptor subtype 7 ablation causes deficit in fear response and conditioned taste aversion. *The Journal of Neuroscience, 19*, 955–963.

Minakami, R., Jinnai, N., & Sugiyama, H. (1997). Phosphorylation and calmodulin binding of the metabotropic glutamate receptor subtype 5 (mGluR5) are antagonistic in vitro. *The Journal of Biological Chemistry, 272*, 20291–20298.

Moritz, A., Scheschonka, A., Beckhaus, T., Karas, M., & Betz, H. (2009). Metabotropic glutamate receptor 4 interacts with microtubule-associated protein 1B. *Biochemical and Biophysical Research Communications, 390*, 82–86.

Nakajima, Y., Mochida, S., Okawa, K., & Nakanishi, S. (2009). Ca^{2+}-dependent release of Munc18-1 from presynaptic mGluRs in short-term facilitation. *Proceedings of the National Academy of Sciences of the United States of America, 106*, 18385–18389.

O'Connor, V., El Far, O., Bofill-Cardona, E., Nanoff, C., Freissmuth, M., Karschin, A., Airas, J. M., Betz, H., & Boehm, S. (1999). Calmodulin dependence of presynaptic metabotropic glutamate receptor signaling. *Science, 286*, 1180–1184.

Paquet, M., Asay, M. J., Fam, S. R., Inuzuka, H., Castleberry, A. M., Oller, H., Smith, Y., Yun, C. C., Traynelis, S. F., & Hall, R. A. (2006). The PDZ scaffold NHERF-2 interacts with mGluR5 and regulates receptor activity. *The Journal of Biological Chemistry, 281*, 29949–29961.

Perroy, J., El Far, O., Bertaso, F., Pin, J. P., Betz, H., Bockaert, J., & Fagni, L. (2002). PICK1 is required for the control of synaptic transmission by the metabotropic glutamate receptor 7. *EMBO Journal, 21*, 2990–2999.

Perroy, J., Raynaud, F., Homburger, V., Rousset, M., Telley, L., Bockaert, J., & Fagni, L. (2008). Direct interaction enables crosstalk between ionotropic and group-I metabotropic glutamate receptors. *The Journal of Biological Chemistry, 283*, 6799–6805.

Pisani, A., Bonsi, P., Centonze, D., Gubellini, P., Bernardi, G., & Calabresi, P. (2003). Targeting striatal cholinergic interneurons in Parkinson's disease: Focus on metabotropic glutamate receptors. *Neuropharmacology, 45*, 45–56.

Porter, R. H., Jaeschke, G., Spooren, W., Ballard, T. M., Buttelmann, B., Kolczewski, S., Peters, J. U., Prinssen, E., Wichmann, J., Vieira, E., Muhlemann, A., Gatti, S., Mutel, V., & Malherbe, P. (2005). Fenobam: A clinically validated nonbenzodiazepine anxiolytic is a potent, selective, and noncompetitive mGlu5 receptor antagonist with inverse agonist activity. *The Journal of Pharmacology and Experimental Therapeutics, 315*, 711–721.

Poulopoulou, C., Davaki, P., Koliaraki, V., Kolovou, D., Markakis, I., & Vassilopoulos, D. (2005). Reduced expression of metabotropic glutamate receptor 2mRNA in T cells of ALS patients. *Annals of Neurology, 58*, 946–949.

Prezeau, L., Rives, M. L., Comps-Agrar, L., Maurel, D., Kniazeff, J., & Pin, J. P. (2010). Functional crosstalk between GPCRs: With or without oligomerization. *Current Opinion in Pharmacology, 10*, 6–13.

Rao, A. M., Hatcher, J. F., & Dempsey, R. J. (2000). Neuroprotection by group I metabotropic glutamate receptor antagonists in forebrain ischemia of gerbil. *Neuroscience Letters, 293*, 1–4.

Renner, M., Lacor, P. N., Velasco, P. T., Xu, J., Contractor, A., Klein, W. L., & Triller, A. (2010). Deleterious effects of amyloid beta oligomers acting as an extracellular scaffold for mGluR5. *Neuron, 66*, 739–754.

Rose, M., Dutting, E., & Enz, R. (2008). Band 4.1 proteins are expressed in the retina and interact with both isoforms of the metabotropic glutamate receptor type 8. *Journal of Neurochemistry, 105*, 2375–2387.

Roth, B. L., Berry, S. A., Kroeze, W. K., Willins, D. L., & Kristiansen, K. (1998). Serotonin 5-HT2A receptors: Molecular biology and mechanisms of regulation. *Critical Reviews in Neurobiology, 12*, 319–338.

Scherer, S. W., Duvoisin, R. M., Kuhn, R., Heng, H. H., Belloni, E., & Tsui, L. C. (1996). Localization of two metabotropic glutamate receptor genes, GRM3 and GRM8, to human chromosome 7q. *Genomics, 31*, 230–233.

Scherer, S. W., Soder, S., Duvoisin, R. M., Huizenga, J. J., & Tsui, L. C. (1997). The human metabotropic glutamate receptor 8 (GRM8) gene: A disproportionately large gene located at 7q31.3-q32.1. *Genomics, 44*, 232–236.

Schroder, H., Wu, D. F., Seifert, A., Rankovic, M., Schulz, S., Hollt, V., & Koch, T. (2009). Allosteric modulation of metabotropic glutamate receptor 5 affects phosphorylation, internalization, and desensitization of the micro-opioid receptor. *Neuropharmacology, 56*, 768–778.

Schwendt, M., & McGinty, J. F. (2007). Regulator of G-protein signaling 4 interacts with metabotropic glutamate receptor subtype 5 in rat striatum: Relevance to amphetamine behavioral sensitization. *The Journal of Pharmacology and Experimental Therapeutics, 323*, 650–657.

Seebahn, A., Rose, M., & Enz, R. (2008). RanBPM is expressed in synaptic layers of the mammalian retina and binds to metabotropic glutamate receptors. *FEBS Letters, 582*, 2453–2457.

Swanson, C. J., Bures, M., Johnson, M. P., Linden, A. M., Monn, J. A., & Schoepp, D. D. (2005). Metabotropic glutamate receptors as novel targets for anxiety and stress disorders. *Nature Reviews: Drug Discovery, 4*, 131–144.

Szumlinski, K. K., Lominac, K. D., Kleschen, M. J., Oleson, E. B., Dehoff, M. H., Schwarz, M. K., Seeburg, P. H., Worley, P. F., & Kalivas, P. W. (2005). Behavioral and neurochemical phenotyping of Homer1 mutant mice: Possible relevance to schizophrenia. *Genes, Brain and Behavior, 4*, 273–288.

Tabata, T., Araishi, K., Hashimoto, K., Hashimotodani, Y., van der Putten, H., Bettler, B., & Kano, M. (2004). Ca^{2+} activity at GABAB receptors constitutively promotes metabotropic glutamate signaling in the absence of GABA. *Proceedings of the National Academy of Sciences of the United States of America, 101*, 16952–16957.

Tang, F. R., Chia, S. C., Chen, P. M., Gao, H., Lee, W. L., Yeo, T. S., Burgunder, J. M., Probst, A., Sim, M. K., & Ling, E. A. (2004). Metabotropic glutamate receptor 2/3 in the hippocampus of patients with mesial temporal lobe epilepsy, and of rats and mice after pilocarpine-induced status epilepticus. *Epilepsy Research, 59*, 167–180.

Tang, Z., El Far, O., Betz, H., & Scheschonka, A. (2005). Pias1 interaction and sumoylation of metabotropic glutamate receptor 8. *The Journal of Biological Chemistry, 280*, 38153–38159.

Tappe, A., Klugmann, M., Luo, C., Hirlinger, D., Agarwal, N., Benrath, J., Ehrengruber, M. U., During, M. J., & Kuner, R. (2006). Synaptic scaffolding protein Homer1a protects against chronic inflammatory pain. *Nature Medicine, 12*, 677–681.

Tu, J. C., Xiao, B., Yuan, J. P., Lanahan, A. A., Leoffert, K., Li, M., Linden, D. J., & Worley, P. F. (1998). Homer binds a novel proline-rich motif and links group 1 metabotropic glutamate receptors with IP3 receptors. *Neuron, 21*, 717–726.

Valerio, A., Ferrario, M., Paterlini, M., Liberini, P., Moretto, G., Cairns, N. J., Pizzi, M., & Spano, P. (2002). Spinal cord mGlu1a receptors: Possible target for amyotrophic lateral sclerosis therapy. *Pharmacology, Biochemistry, and Behavior, 73*, 447–454.

Wang, H., Westin, L., Nong, Y., Birnbaum, S., Bendor, J., Brismar, H., Nestler, E., Aperia, A., Flajolet, M., & Greengard, P. (2009). Norbin is an endogenous regulator of metabotropic glutamate receptor 5 signaling. *Science, 326*, 1554–1557.

Xiao, B., Tu, J. C., Petralia, R. S., Yuan, J. P., Doan, A., Breder, C. D., Ruggiero, A., Lanahan, A. A., Wenthold, R. J., & Worley, P. F. (1998). Homer regulates the association of group 1 metabotropic glutamate receptors with multivalent complexes of Homer-related, synaptic proteins. *Neuron, 21*, 707–716.

Yuan, J. P., Kiselyov, K., Shin, D. M., Chen, J., Shcheynikov, N., Kang, S. H., Dehoff, M. H., Schwarz, M. K., Seeburg, P. H., Muallem, S., & Worley, P. F. (2003). Homer binds to TRPC family channels and is required for gating of TRPC1 by IP3 receptors. *Cell, 114*, 777–789.

Zhang, J., Cheng, S., Xiong, Y., Ma, Y., Luo, D., Jeromin, A., Zhang, H., & He, J. (2008). A novel association of mGluR1a with the PDZ scaffold protein CAL modulates receptor activity. *FEBS Letters, 582*, 4117–4124.

Chapter 4
Regulation of the Actin Cytoskeleton in Dendritic Spines

Peter Penzes and Igor Rafalovich

Abstract Spine morphogenesis is largely dependent on the remodeling of the actin cytoskeleton. Actin dynamics within spines is regulated by a complex network of signaling molecules, which relay signals from synaptic receptors, through small GTPases and their regulators, to actin-binding proteins. In this chapter, we will discuss molecules involved in dendritic spine plasticity beginning with actin and moving upstream toward neuromodulators and trophic factors that initiate signaling involved in these plasticity events. We will place special emphasis on small GTPase pathways, as they have an established importance in dendritic spine plasticity and pathology. Finally, we will discuss some epigenetic mechanisms that control spine morphogenesis.

Keywords Actin binding proteins • GTPase activating proteins • Guanine exchange factors • Small GTPases • Spine synapses

4.1 Introduction

More than a century ago, after observing dendritic spines on Purkinje cell dendrites, Santiago Ramon y Cajal proposed that "such spines could be the points where electrical charge or current is received." This hypothesis has proved to be correct,

P. Penzes (✉)
Department of Physiology, Northwestern University Feinberg School of Medicine, 303 E. Chicago Avenue, 60611 Chicago, IL, USA

Department of Psychiatry and Behavioral Sciences, Northwestern University Feinberg School of Medicine, 303 E. Chicago Avenue, 60611 Chicago, IL, USA
e-mail: p-penzes@northwestern.edu

I. Rafalovich
Department of Physiology, Northwestern University Feinberg School of Medicine, 303 E. Chicago Avenue, 60611 Chicago, IL, USA

M.R. Kreutz and C. Sala (eds.), *Synaptic Plasticity*,
Advances in Experimental Medicine and Biology 970,
DOI 10.1007/978-3-7091-0932-8_4, © Springer-Verlag/Wien 2012

and it has been demonstrated that most excitatory synapses are formed between axon terminals and these "dendritic spines."

Dendritic spines are heterogeneous in their shape and size. However, their morphology can be classified into long-thin, stubby, and mushroom-shaped. Their shape often reflects their stability and the strength of the synapse, the latter presumably due to AMPA receptor levels. Often, the destabilization of spines – their shrinkage from mushroom-shaped to long and thin leads to their disappearance (Alvarez and Sabatini 2007). Conversely, spines that appear as long, thin filopodia might increase in area in an activity-dependent manner (Bonhoeffer and Yuste 2002). The change in spine morphology and spine morphogenesis is mainly dependent on the remodeling of β- and γ-actin, the main isoforms of actin present in neurons (Schubert and Dotti 2007). In this chapter, we will discuss molecules involved in dendritic spine plasticity beginning with actin and moving upstream toward neuromodulators and trophic factors that initiate signaling involved in these plasticity events. We will place special emphasis on small GTPase pathways, as they have an established importance in dendritic spine plasticity and pathology.

Actin is found as soluble monomeric G-actin and polymerized F-actin filaments, the latter likely conferring the characteristic spine morphology. The polymerization of free G-actin is subject to regulation by numerous pathways activated by various surface receptors (Cingolani and Goda 2008). Most notably, activations of N-methyl-D-aspartic acid (NMDA) receptors lead to the aforementioned changes. The pathways that act as transducers of these changes are subject to modulation by converging pathways giving rise to a complex molecular network.

4.2 Actin-Binding Proteins

The conversion of soluble G-actin into F-actin is a highly dynamic and reversible process that is regulated through interactions with actin-binding proteins (ABPs). The differential effect of ABPs on actin (some favor polymerization while others depolymerization) confers intricate regulation of the cytoskeletal remodeling at the synapse. The actin-related proteins 2 and 3 (Arp2/3) complex is a major component of actin remodeling that is localized to dendritic spines of hippocampal neurons (Racz and Weinberg 2008). Upon activation, Arp2/3 binds existing acting filaments, nucleating them into a branched network of actin filaments (Goley and Welch 2006). Recent knockdown studies of Arp2/3 in hippocampal neurons have revealed its importance for dendritic spine formation (Wegner et al. 2008). An interesting consideration is that the Arp2/3 complex is the target of many converging pathways involved in dendritic spine morphogenesis. For example, the F-actin-binding protein cortactin binds Arp2/3, activating and localizing it to dendritic spines (Weaver et al. 2001; Hering and Sheng 2003). Another crucial Arp2/3 activator is WAVE-1 (Wiskott-Aldrich syndrome protein family member 1). WAVE-1 serves as a signal transducer between the Rho GTPase Rac1 and Arp2/3. Knockdown studies of WAVE-1 have revealed its importance in spine morphology (Soderling et al. 2007). Depletion of other Arp2/3 activators including Abi2,

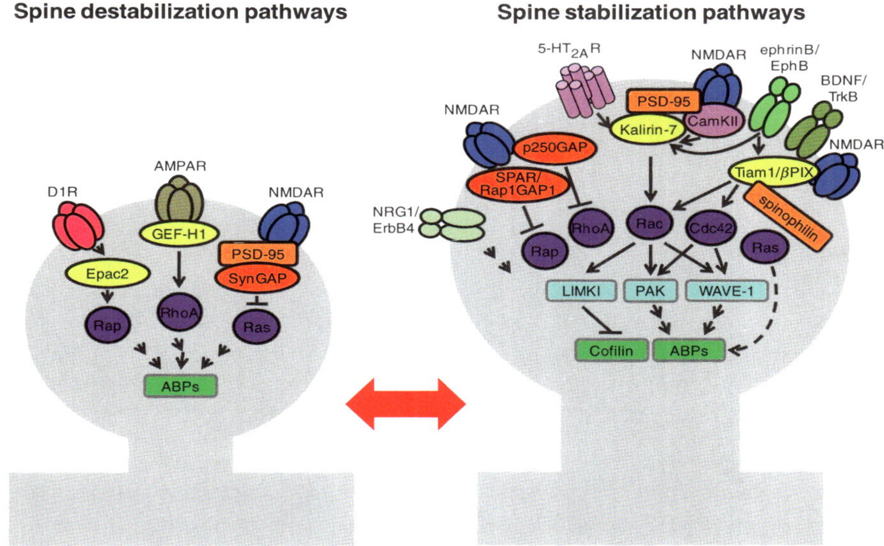

Fig. 4.1 Schematic representation of the molecules that regulate bidirectional dendritic spine plasticity. Pathways that promote spine enlargement and stabilization are on the *right*; pathways promoting spine shrinkage and destabilization are on the *left*

N-WASP, and Abp1 alters morphology and number of spines (Grove et al. 2004; Haeckel et al. 2008; Wegner et al. 2008) (Fig. 4.1).

Profilin is another key player in actin polymerization that targets to dendritic spines upon chemical or electrical stimulation of hippocampal neurons (Witke et al. 2001; Neuhoff et al. 2005). Experiments utilizing a peptide competitor of profilin prevented profilin targeting and destabilized dendritic spines (Ackermann and Matus 2003). Concordantly, it has been observed that profilin translocates from the dendritic shaft into the dendritic spine in the amygdala after fear conditioning (Lamprecht et al. 2006).

Another important promoter of actin polymerization is drebrin. Drebrin is an F-actin-binding protein that is highly concentrated in dendritic spines, where it associates with actin filaments (Hayashi et al. 1996; Hayashi and Shirao 1999). Studies have shown that drebrin accumulates in dendritic spines prior to PSD-95 during spine formation. Knockdown of drebrin with siRNA disrupts accumulation of PSD-95 in spines. These studies suggest that drebrin's role is to promote actin assembly and the clustering of PSD-95 in synaptic spines (Takahashi et al. 2003).

Gelsolin is another actin-binding protein whose actin-binding activity is Ca^{2+} dependent. In the presence of high Ca^{2+} concentration, gelsolin binds to the ends of actin filaments and prevents further elongation. This action also serves to stabilize the actin filaments during synaptic plasticity (Star et al. 2002).

Spinophilin, named after its prominent localization to dendritic spines, targets protein phosphatase 1 (PP1) to dendritic spines. Spinophilin promotes the phosphatase activity of PP1 (Allen et al. 1997; Hsieh-Wilson et al. 1999). Spinophilin's actin binding is modulated by protein kinase A (PKA) and Ca^{2+}/calmodulin-dependent kinase II

(CaMKII), allowing for its activity-dependent regulation (Grossman et al. 2004). Additionally, spinophilin has been shown to serve as a Rac1 regulator through its interaction with the Rac1 guanine exchange factor (GEF) Tiam1 (Buchsbaum et al. 2003).

The balance between G- and F-actin is also controlled by the actin depolymerizing factor (ADF)-related protein cofilin. Depending on phosphorylation state, cofilin can either disassemble filaments or sever them providing a barbed end of actin assembly (Carlier et al. 1997). Knocking down cofilin in neuronal culture results in a reduction of F-actin turnover and a loss of dendritic spine density (Hotulainen et al. 2009). Phosphorylation of cofilin on serine 3 by LIM kinases inhibits its function. LIM kinase I (LIMK-1) is an ADF/cofilin-specific kinase enriched in dendritic spines. Hippocampal neurons cultured from LIMK-1 mice show reduced cofilin phosphorylation and aberrant F-actin accumulation in spines (Meng et al. 2003). Although the roles of the above molecules have predominantly been identified as ABPs, many of these have other roles within the synaptic plasticity network.

4.3 Small GTPases

Small GTPases are crucial for the remodeling of the actin cytoskeleton. Since GTPases can exist in two states, an active GTP-bound and an inactive GDP-bound, they serve as "on" or "off" binary molecular switches. These small molecules are regulated by guanine exchange factors (GEFs) which catalyze the exchange of GDP for GTP, resulting in activation of the GTPase. Conversely, GTPase-activating proteins (GAPs) enhance the hydrolysis of GTP into GDP, inactivating the GTPase. There are a few different families of GTPases involved in spine morphogenesis. The most widely studied is the Rho family, which includes RhoA, Rac1, Cdc42, and others which are not as well characterized. Tight regulation of these molecules is necessary for proper spine formation and function. It is generally accepted that Rac1 activation stimulates F-actin polymerization and stabilizes dendritic spines through the activation of downstream effectors p21-activated kinase (PAK), LIM kinase I (LIMKI), and the actin-binding protein cofilin (Meng et al. 2002; Govek et al. 2004; Zhang et al. 2005) (Fig. 4.1). Luo and colleagues showed that transgenic mice expressing a mutant form of Rac1, lacking the ability to hydrolyze GTP thus remaining constitutively active, resulted in increased spine density at the expense of spine size in cerebellar Purkinje cells (Luo et al. 1996). Experiments where constitutively active Rac1 was overexpressed in hippocampal cultured neurons and slices documented the formation of irregularly shaped protrusions and impairment of synapse formation which contrasts with in vivo data (Luo et al. 1996; Tashiro et al. 2000; Zhang et al. 2003). Overexpression of a dominant-negative form of Rac1, incapable of interacting with GEFs, drastically decreases the number of spines and synapses in cultured hippocampal slices and dissociated neurons (Nakayama et al. 2000; Penzes et al. 2003; Zhang et al. 2003). These data suggest that an optimal level of Rac1 activation is required for proper maintenance of dendritic spines, and only small fluctuations in Rac1 activity are responsible for spine morphogenesis.

Another small GTPase, Cdc42, plays a role in spine morphogenesis as demonstrated in experiments where a dominant-negative version prevented morphology changes in cultured hippocampal neurons (Irie and Yamaguchi 2002). In contrast to Rac1 and Cdc42, the RhoA GTPase seems to promote spine destabilization, shrinkage, and reduction in density. For example, overexpression of constitutively active RhoA in hippocampal slice cultures fosters spine retraction and elimination (Tashiro et al. 2000).

Ras and Rap are a pair of closely related GTPases in the Ras subfamily that share many common regulators and effectors but exert contrasting actions on dendritic spines. Whereas Ras has been shown to stabilize synapses and traffic AMPA receptors into spines in a phosphoinositide 3-kinase (PI3K)/Akt-dependent manor (Zhu et al. 2002), Rap has been shown to destabilize spines through B-Raf signaling.

In general, the consensus seems to be that Rac1, Cdc42, and Ras promote spine formation/stability, while RhoA and Rap promote spine destabilization, shrinkage, and elimination.

4.4 Guanine Exchange Factors and GTPase-Activating Proteins

Through catalyzing the exchange of the GTPase-bound GDP to GTP, guanine exchange factors (GEFs) serve to activate GTPases. Kalirin-7 is one such GEF, regulating the activity of Rac1. Kalirin-7 is especially unique due to the fact that it is the only known Rac1 GEF expressed in the cortex of adult mice (Penzes et al. 2008). Overexpression of this GEF in cortical cultures leads to an increase in spine head area and density. Concomitantly, knockdown of kalirin-7 through an RNAi approach reduces the spine area and density (Xie et al. 2007). In the hippocampus, the role of kalirin-7 is obscured due to the presence of two other Rac1 GTPases, Tiam1 and β-PIX (Zhang et al. 2005; Tolias et al. 2007). Interestingly, mice in which the kalirin gene has been deleted exhibit many phenotypes reminiscent of schizophrenia including deficits in working memory as well as reduced dendritic spine density in the cortex (Cahill et al. 2009).

In the hippocampus, Tiam1 is regulated by NMDA receptor activation and has also been implicated in EphB receptor–dependent dendritic spine development (Tolias et al. 2005, 2007). Likewise, the Rac1 GEF β-PIX, a downstream target of NMDA receptors, has been shown to be regulated by CaM kinase kinase and CaM kinase I (Zhang et al. 2005).

RhoA is associated with spine shrinkage and destabilization. Thus, GEFs that activate this GTPase have similar effects on dendritic spine morphology. For example, the recently identified GEF-H1 has been shown to colocalize with the AMPA receptor complex and negatively regulate spine density and length through a RhoA signaling cascade (Kang et al. 2009). Another GEF involved in the destabilization and shrinkage of spines is Epac2. This Rap1 GEF is activated by cAMP and leads to reduced spine AMPA receptor content, depressed excitatory transmission, as well as spine destabilization as demonstrated by live imaging studies. Conversely, inhibition of Epac2 leads to spine enlargement and stabilization (Woolfrey et al. 2009). Recent studies have associated Epac2 with autism. Thus, further studies centered around this GEF may shed light on the pathology of this complex disorder (Bacchelli et al. 2003).

Despite their name, GTPase-activating proteins (GAPs), serve to inhibit GTPase activity by increasing the rate at which the GTPase-bound GTP is hydrolyzed to GDP. For example, SPAR1 has been found to be enriched in dendritic spines of cultured hippocampal neurons. Here it interacts with the PSD-95 and the NMDA receptor complex to dampen Rap activity and enlargement of dendritic spines (Pak et al. 2001). In medium spiny neurons, Rap1GAP1 serves an analogous role to SPAR1 in pyramidal neurons. McAvoy et al. showed that overexpression of this GAP leads to increased spine area (McAvoy et al. 2009). p250GAP is a RhoA GAP associated with the NMDA receptors. Studies where p250GAP was knocked down in primary hippocampal neurons show an increase in dendritic spine width as well as elevated RhoA activity (Nakazawa et al. 2008) (Fig. 4.1).

GAPs are not limited to the regulation of GTPases associated with spine destabilization and shrinkage. SynGAP is a Ras/Rap GAP associated with trafficking of glutamate receptors to the synapse. Heterozygous deletion of SynGAP was sufficient to result in an elevated number of mushroom-shaped spines. In addition, both Ras and Rac activation was decreased in the forebrain of these heterozygous animals. Activation of NMDA receptors in neurons cultured from SynGAP-knockout animals resulted in aberrant cofilin function. Finally, normal EPSPs were also disrupted in hippocampal slices cultured from heterozygous animals (Carlisle et al. 2008). GEFs and GAPS are very large molecules often incorporating various domains and motifs including lipid-binding domains, cAMP-binding domains, PDZ-binding motifs, and others (Rossman et al. 2005). Together, GEFs add another layer of GTPase control.

4.5 Trophic Factors

A few trophic factor signaling pathways have been identified to feed into and modulate the abovementioned pathways involved in dendritic spine morphogenesis. The trophic factor neuregulin-1 (NRG1), polymorphisms in which are associated with schizophrenia, binds to ErbB4, the postsynaptic receptor tyrosine kinase. Long-term treatment of hippocampal pyramidal neurons with NRG1 has been shown to increase spine density as well as increasing the proportion of spines with a mature phenotype (Barros et al. 2009). Furthermore, mouse models in which ErbB4 was knocked out from the CNS show a deficit in dendritic spine density in both the hippocampus and cortex (Barros et al. 2009). Another study has demonstrated that the overexpression of ErbB4 leads to an increase in dendritic spine density (Li et al. 2007). Interestingly, this signaling interaction has recently been associated with schizophrenia (Jaaro-Peled et al. 2009).

The trophic factor, brain-derived neurotrophic factor (BDNF), and its high-affinity receptor, tropomyocin-related kinase B (TrkB), have long been associated with synaptic formation and plasticity (Luikart and Parada 2006; Lu et al. 2008). Numerous studies have reported BDNF-induced changes in dendritic spine density and morphology in a variety of neuron populations (Luikart and Parada 2006; Lu et al. 2008). Consistently, TrkB-deficient mice have significantly fewer dendritic spines on CA1 hippocampal neurons (Luikart et al. 2005). The TrkB receptor has also been shown to interact with the Rac1 GEF Tiam1 (Miyamoto et al. 2006). This finding further implicates BDNF/TrkB signaling in dendritic spine plasticity. Of all trophic factors, the effects of NRG1 and BDNF on dendritic spine morphogenesis have been best described. However, future studies will identify undoubtedly other factors involved in regulating dendritic spine plasticity.

4.6 Neurotransmitter Signaling Regulating Dendritic Spine Plasticity

The predominant receptor in regulating dendritic spine plasticity is the NMDA receptor. Numerous studies have shown that many of the signaling molecules mentioned earlier either interact with or are downstream of NMDA receptors. Upon activation of this receptor, the dendritic spine undergoes a transient increase in calcium concentration (Sobczyk and Svoboda 2007). This rise in calcium activates the calcium-sensing calmodulin (CaM). Calcium-bound CaM activates the CaMK family of serine/threonine kinases including CaMKI, CaMKII, and CaMKIV (Hook and Means 2001). These kinases go on to phosphorylate a variety of targets involved in spine structural plasticity including kalirin-7, as well as other signaling and scaffolding proteins involved in plasticity (Soderling 2000; Xie et al. 2007).

Aside from glutamate, other neurotransmitters have been shown to modulate dendritic spine plasticity. Activation of 5-HT2A receptors in pyramidal neurons increased spine size through a kalirin-7-Rac1-PAK-dependent mechanism (Jones et al. 2009). This study is of particular importance as it provides a direct link between serotonergic signaling and dendritic spine morphogenesis, both implicated in schizophrenia.

Another important neurotransmitter implicated in dendritic spines is dopamine. A study examining dendritic spine density in the prelimbic cortex of rats treated with 6-hydroxydopamine, a neurotoxin that selectively ablates dopaminergic and noradrenergic neurons, found a decrease in spine density in this region 3 weeks after toxin administration (Solis et al. 2007). Similar findings were also reported in another study (Wang and Deutch 2008). At the molecular level, activation of the D1/D5 receptors with the selective agonist SKF-38393 leads to spine shrinkage through activation of the Rap GEF Epac2 (Woolfrey et al. 2009).

Ablation of the cholinergic system using 192 IgG-saporin has been shown to decrease dendritic spine density in layer V cortical pyramidal neurons (Sherren and Pappas 2005). Furthermore, muscarinic acetylcholine receptors have recently been localized to the extrasynaptic membrane of pyramidal neurons (Yamasaki et al. 2010); however, their exact role in spine morphogenesis has not been determined. In vivo evidence has demonstrated that deletion of the $\beta2$ subunit of the acetylcholine nicotine receptor leads to reduction of spines in the higher-order association areas (Ballesteros-Yanez et al. 2010).

Classically defined as a hormone, estrogen has recently come into the spotlight as an important modulator of dendritic spine plasticity. A study in 2008 by Srivastava et al. demonstrated the nonlinearity of signaling pathways. In this study, treatment of cortical cultures with estradiol increased spine density while decreasing the AMPA receptor content of spines. These "silent synapses" were potentiated by activation of NMDA receptors, reminiscent of activity-dependent maturation of silent synapses during development (Srivastava et al. 2008). These effects were mediated by the Rap/AF-6/ERK1/2 signaling pathways (Srivastava et al. 2008). Additionally, recent studies have demonstrated that treatment of rat cortical cultures leads to phosphorylation of WAVE1 and its targeting to spines, leading to the polymerization of actin (Sanchez et al. 2009). Similar findings have been reported in hippocampal cultured neurons. Here, treatment of hippocampal cultures resulted in increased number of synapses in addition to increased kalirin-7 localization in dendritic spines (Ma et al. 2010). These actions of estradiol seem to be mediated through the ER-β receptor as ER-β, but not ER-α, agonists are able to recapitulate these effects (Ma et al. 2010; Srivastava et al. 2010).

Taken together, these findings demonstrate the importance of modulatory neurotransmitter signaling in dendritic spine plasticity.

4.7 Epigenetic Mechanisms in Dendritic Spine Plasticity

Epigenetics is the study of inherited changes in phenotype caused by mechanisms other than changes in the underlying DNA sequence. A field still in its infancy, epigenetic research has only begun to amalgamate with neuroscience in general and synaptic and spine plasticity specifically. Nevertheless, a handful of studies have begun to elucidate the role of epigenetics in dendritic spine plasticity. The acetylation and deacetylation of histone proteins has been associated with regulation of gene transcription through the loosening of heterochromatin. A study by Guan et al. has shown the role of histone deacetylase 2 (HDAC2) in synaptic plasticity. In this study, the authors showed that neuron-specific overexpression of HDAC2 decreased spine density, synapse number, and enhanced learning. On the other hand, HDAC2 deficits resulted in an increased synapse and spine number. The same was observed in mice treated with HDAC2 inhibitors (Guan et al. 2009).

DNA methylation is another type of epigenetic modification. Emerging evidence is beginning to implicate this process in the formation of dendritic spines. Repeated exposure of rodents to cocaine increases spine density in the nucleus accumbens. However, this effect is mitigated by lack of lysine dimethyltransferase G9a, the enzyme involved in transferring methyl groups to DNA cysteine residues (Maze et al. 2010).

The modulation of compact chromatin and nucleosomes, by deacetylases, kinases, phosphatases, and other chromatin-modifying enzymes is crucial for the binding of transcription factors (TFs) and initiation of transcription. The effort toward an understanding of epigenetic mechanisms has spilled over into synaptic and dendritic spine plasticity leading to an identification of transcription factors involved in these processes. As such, a recent study has identified two transcription factors, Cux1 and Cux2, in the control of dendritic spine morphology (Cubelos et al. 2010). Cubelos et al. showed that mice deficient in either one of these factors are deficient in dendritic spine density in layer 2/3 but not layer 5 cortical pyramidal neurons (Cubelos et al. 2010).

Whereas the abovementioned transcription factors are crucial for dendritic spine maintenance, the TF MEF2 is needed for proper synapse elimination. Flavell et al. showed that MEF2 is involved in synapse elimination in hippocampal neuron cultures. Activity-dependent dephosphorylation of MEF2 leads to the expression of activity-regulated cytoskeletal-associated protein (Arc) and SynGAP. Since Arc has been associated with AMPA receptor endocytosis (Chowdhury et al. 2006), dephosphorylation of MEF2 induces destabilization of the synapse leading to spine elimination. These in vitro observations have recently been confirmed in vivo where Mef2c deficient mice showed an elevation in dendritic spine number compared to control (Flavell et al. 2006; Pfeiffer et al. 2010). Additionally, MEF2 has also been shown to be involved in the spine elimination of striatal medium spiny neurons (Tian et al. 2010).

After TFs initiate transcription, microRNAs (miRNAs) regulate protein expression by binding to mRNA and suppressing its translation. miRNAs that are involved in dendritic spine formation and stabilization are just begging to emerge. Morgan Sheng's group has recently discovered two miRNAs, miR-125b and miR-132, which associate with the fragile X mental retardation protein (FMRP) as knockdown of this protein ameliorates the effect of these miRNAs on spine morphology (Edbauer et al. 2010). In this study, they demonstrated that the NR2A subunit of the NMDA receptor is the target for both of these miRNA; however, their effect on spine morphology is conflicting. Whereas overexpression of miR-125b induced long, narrow spines correlated with reduced synaptic transmission, overexpression of miR-132 lead to enlarged spine heads (Edbauer et al. 2010). MiR-132 has also been associated with another mechanism. Through suppression of p250GAP translation, miR-132 induces spine formation. A similar effect is seen with knockdown of p250GAP. Inhibition of miR-132 results in smaller dendritic spines and reduced EPSCs (Impey et al. 2010).

Another molecule that is regulated at the miRNA level is LIMK1. Schratt et al. found that miR-134 suppresses LIMK1; however, exposure of cultured cells to BDNF relieves this suppression and contributes to spine stabilization (Schratt et al. 2006).

A novel and interesting way by which miRNAs regulate dendritic spine plasticity is by inhibiting the translation of a protein involved in palmitoylation. miR-138 has been shown to negatively control dendritic spine size in rat hippocampal neurons through the regulation of acyl protein thioesterase 1 (APT1) expression. Because APT1 is involved in the palmitoylation of proteins at the synapse, its modulation may effect dendritic spine morphogenesis through regulating the targeting of synaptic proteins to the synapse (Siegel et al. 2009).

Another miRNA target recently identified to be involved in regulating dendritic spines is the ubiquitin ligase mind bomb-1 (MB1). Smart et al. have shown that suppression of MB1 by miR-137 results in a decrease in dendritic spine density in hippocampal cultured neurons (Smart et al. 2010).

Although epigenetic control of dendritic spine plasticity has only begun to emerge, the importance of control at the RNA level has already been demonstrated. Undoubtedly, more miRNA targets involved in the dendritic spine regulation will soon be elucidated.

4.8 Conclusions

Dendritic spine plasticity is a relatively nascent field. Recent technological advancement, particularly imaging technologies, has allowed for the study of the above molecules and their roles in these plasticity events. As fluorescent probes and delivery techniques continue to be developed, studies demonstrating the dynamic nature of signaling events will begin to phase out the currently used static approaches. For example, two seminal studies recently utilized time-lapse imaging to elucidate the molecular mechanisms in spine plasticity (Srivastava et al. 2008; Woolfrey et al. 2009). Furthermore, development of 2-photon imaging has already revolutionized our understanding of the structural and functional dynamics of spines, through in vivo studies of the intact cortex (Grutzendler et al. 2002; Holtmaat et al. 2005; Yang et al. 2009). As we move through the twenty-first century, computational analysis and modeling of molecular pathway networks will provide a more comprehensive understanding of the nonlinear molecular interactions regulating spine plasticity. This is particularly important as many proteins involved in spine plasticity have been implicated in various psychiatric disorders. Of particular importance are studies that show disease-related phenotypes in concord with dendritic spine aberrations such as the study by Cahill et al. (2009). Additionally, there exists a dearth of studies in which an observed anatomical phenotype in correlation with a molecular abnormality is modeled in a laboratory setting. In conclusion, understanding the functional relationships between different signaling molecules associated with a particular disorder will undoubtedly shed light on the underpinnings of pathology as well as identify possible targets for treatment.

References

Ackermann, M., & Matus, A. (2003). Activity-induced targeting of profilin and stabilization of dendritic spine morphology. *Nature Neuroscience, 6*, 1194–1200.

Allen, P. B., Ouimet, C. C., & Greengard, P. (1997). Spinophilin, a novel protein phosphatase 1 binding protein localized to dendritic spines. *Proceedings of the National Academy of Sciences of the United States of America, 94*, 9956–9961.

Alvarez, V. A., & Sabatini, B. L. (2007). Anatomical and physiological plasticity of dendritic spines. *Annual Review of Neuroscience, 30*, 79–97.

Bacchelli, E., Blasi, F., Biondolillo, M., Lamb, J. A., Bonora, E., Barnby, G., Parr, J., Beyer, K. S., Klauck, S. M., Poustka, A., Bailey, A. J., Monaco, A. P., & Maestrini, E. (2003). Screening of nine candidate genes for autism on chromosome 2q reveals rare nonsynonymous variants in the cAMP-GEFII gene. *Molecular Psychiatry, 8*, 916–924.

Ballesteros-Yanez, I., Benavides-Piccione, R., Bourgeois, J. P., Changeux, J. P., & DeFelipe, J. (2010). Alterations of cortical pyramidal neurons in mice lacking high-affinity nicotinic receptors. *Proceedings of the National Academy of Sciences of the United States of America, 107*, 11567–11572.

Barros, C. S., Calabrese, B., Chamero, P., Roberts, A. J., Korzus, E., Lloyd, K., Stowers, L., Mayford, M., Halpain, S., & Muller, U. (2009). Impaired maturation of dendritic spines without disorganization of cortical cell layers in mice lacking NRG1/ErbB signaling in the central nervous system. *Proceedings of the National Academy of Sciences of the United States of America, 106*, 4507–4512.

Bonhoeffer, T., & Yuste, R. (2002). Spine motility. Phenomenology, mechanisms, and function. *Neuron, 35*, 1019–1027.

Buchsbaum, R. J., Connolly, B. A., & Feig, L. A. (2003). Regulation of p70 S6 kinase by complex formation between the Rac guanine nucleotide exchange factor (Rac-GEF) Tiam1 and the scaffold spinophilin. *Journal of Biological Chemistry, 278*, 18833–18841.

Cahill, M. E., Xie, Z., Day, M., Photowala, H., Barbolina, M. V., Miller, C. A., Weiss, C., Radulovic, J., Sweatt, J. D., Disterhoft, J. F., Surmeier, D. J., & Penzes, P. (2009). Kalirin regulates cortical spine morphogenesis and disease-related behavioral phenotypes. *Proceedings of the National Academy of Sciences of the United States of America, 106*, 13058–13063.

Carlier, M. F., Laurent, V., Santolini, J., Melki, R., Didry, D., Xia, G. X., Hong, Y., Chua, N. H., & Pantaloni, D. (1997). Actin depolymerizing factor (ADF/cofilin) enhances the rate of filament turnover: Implication in actin-based motility. *The Journal of Cell Biology, 136*, 1307–1322.

Carlisle, H. J., Manzerra, P., Marcora, E., & Kennedy, M. B. (2008). SynGAP regulates steady-state and activity-dependent phosphorylation of cofilin. *Journal of Neuroscience, 28*, 13673–13683.

Chowdhury, S., Shepherd, J. D., Okuno, H., Lyford, G., Petralia, R. S., Plath, N., Kuhl, D., Huganir, R. L., & Worley, P. F. (2006). Arc/Arg3.1 interacts with the endocytic machinery to regulate AMPA receptor trafficking. *Neuron, 52*, 445–459.

Cingolani, L. A., & Goda, Y. (2008). Actin in action: The interplay between the actin cytoskeleton and synaptic efficacy. *Nature Reviews Neuroscience, 9*, 344–356.

Cubelos, B., Sebastian-Serrano, A., Beccari, L., Calcagnotto, M. E., Cisneros, E., Kim, S., Dopazo, A., Alvarez-Dolado, M., Redondo, J. M., Bovolenta, P., Walsh, C. A., & Nieto, M. (2010). Cux1 and Cux2 regulate dendritic branching, spine morphology, and synapses of the upper layer neurons of the cortex. *Neuron, 66*, 523–535.

Edbauer, D., Neilson, J. R., Foster, K. A., Wang, C. F., Seeburg, D. P., Batterton, M. N., Tada, T., Dolan, B. M., Sharp, P. A., & Sheng, M. (2010). Regulation of synaptic structure and function by FMRP-associated microRNAs miR-125b and miR-132. *Neuron, 65*, 373–384.

Flavell, S. W., Cowan, C. W., Kim, T. K., Greer, P. L., Lin, Y., Paradis, S., Griffith, E. C., Hu, L. S., Chen, C., & Greenberg, M. E. (2006). Activity-dependent regulation of MEF2 transcription factors suppresses excitatory synapse number. *Science, 311*, 1008–1012.

Goley, E. D., & Welch, M. D. (2006). The ARP2/3 complex: An actin nucleator comes of age. *Nature Reviews: Molecular Cell Biology, 7*, 713–726.

Govek, E. E., Newey, S. E., Akerman, C. J., Cross, J. R., Van der Veken, L., & Van Aelst, L. (2004). The X-linked mental retardation protein oligophrenin-1 is required for dendritic spine morphogenesis. *Nature Neuroscience, 7*, 364–372.

Grossman, S. D., Futter, M., Snyder, G. L., Allen, P. B., Nairn, A. C., Greengard, P., & Hsieh-Wilson, L. C. (2004). Spinophilin is phosphorylated by Ca^{2+}/calmodulin-dependent protein kinase II resulting in regulation of its binding to F-actin. *Journal of Neurochemistry, 90*, 317–324.

Grove, M., Demyanenko, G., Echarri, A., Zipfel, P. A., Quiroz, M. E., Rodriguiz, R. M., Playford, M., Martensen, S. A., Robinson, M. R., Wetsel, W. C., Maness, P. F., & Pendergast, A. M. (2004). ABI2-deficient mice exhibit defective cell migration, aberrant dendritic spine morphogenesis, and deficits in learning and memory. *Molecular and Cellular Biology, 24*, 10905–10922.

Grutzendler, J., Kasthuri, N., & Gan, W. B. (2002). Long-term dendritic spine stability in the adult cortex. *Nature, 420*, 812–816.

Guan, J. S., Haggarty, S. J., Giacometti, E., Dannenberg, J. H., Joseph, N., Gao, J., Nieland, T. J., Zhou, Y., Wang, X., Mazitschek, R., Bradner, J. E., DePinho, R. A., Jaenisch, R., & Tsai, L. H. (2009). HDAC2 negatively regulates memory formation and synaptic plasticity. *Nature, 459*, 55–60.

Haeckel, A., Ahuja, R., Gundelfinger, E. D., Qualmann, B., & Kessels, M. M. (2008). The actin-binding protein Abp1 controls dendritic spine morphology and is important for spine head and synapse formation. *Journal of Neuroscience, 28*, 10031–10044.

Hayashi, K., & Shirao, T. (1999). Change in the shape of dendritic spines caused by overexpression of drebrin in cultured cortical neurons. *Journal of Neuroscience, 19*, 3918–3925.

Hayashi, K., Ishikawa, R., Ye, L. H., He, X. L., Takata, K., Kohama, K., & Shirao, T. (1996). Modulatory role of drebrin on the cytoskeleton within dendritic spines in the rat cerebral cortex. *Journal of Neuroscience, 16*, 7161–7170.

Hering, H., & Sheng, M. (2003). Activity-dependent redistribution and essential role of cortactin in dendritic spine morphogenesis. *Journal of Neuroscience, 23*, 11759–11769.

Holtmaat, A. J., Trachtenberg, J. T., Wilbrecht, L., Shepherd, G. M., Zhang, X., Knott, G. W., & Svoboda, K. (2005). Transient and persistent dendritic spines in the neocortex in vivo. *Neuron, 45*, 279–291.

Hook, S. S., & Means, A. R. (2001). Ca(2+)/CaM-dependent kinases: From activation to function. *Annual Review of Pharmacology and Toxicology, 41*, 471–505.

Hotulainen, P., Llano, O., Smirnov, S., Tanhuanpaa, K., Faix, J., Rivera, C., & Lappalainen, P. (2009). Defining mechanisms of actin polymerization and depolymerization during dendritic spine morphogenesis. *The Journal of Cell Biology, 185*, 323–339.

Hsieh-Wilson, L. C., Allen, P. B., Watanabe, T., Nairn, A. C., & Greengard, P. (1999). Characterization of the neuronal targeting protein spinophilin and its interactions with protein phosphatase-1. *Biochemistry, 38*, 4365–4373.

Impey, S., Davare, M., Lasiek, A., Fortin, D., Ando, H., Varlamova, O., Obrietan, K., Soderling, T. R., Goodman, R. H., & Wayman, G. A. (2010). An activity-induced microRNA controls dendritic spine formation by regulating Rac1-PAK signaling. *Molecular and Cellular Neuroscience, 43*, 146–156.

Irie, F., & Yamaguchi, Y. (2002). EphB receptors regulate dendritic spine development via intersectin, Cdc42 and N-WASP. *Nature Neuroscience, 5*, 1117–1118.

Jaaro-Peled, H., Hayashi-Takagi, A., Seshadri, S., Kamiya, A., Brandon, N. J., & Sawa, A. (2009). Neurodevelopmental mechanisms of schizophrenia: Understanding disturbed postnatal brain maturation through neuregulin-1-ErbB4 and DISC1. *Trends in Neurosciences, 32*, 485–495.

Jones, K. A., Srivastava, D. P., Allen, J. A., Strachan, R. T., Roth, B. L., & Penzes, P. (2009). Rapid modulation of spine morphology by the 5-HT2A serotonin receptor through kalirin-7 signaling. *Proceedings of the National Academy of Sciences of the United States of America, 106*, 19575–19580.

Kang, M. G., Guo, Y., & Huganir, R. L. (2009). AMPA receptor and GEF-H1/Lfc complex regulates dendritic spine development through RhoA signaling cascade. *Proceedings of the National Academy of Sciences of the United States of America, 106*, 3549–3554.

Lamprecht, R., Farb, C. R., Rodrigues, S. M., & LeDoux, J. E. (2006). Fear conditioning drives profilin into amygdala dendritic spines. *Nature Neuroscience, 9*, 481–483.

Li, B., Woo, R. S., Mei, L., & Malinow, R. (2007). The neuregulin-1 receptor erbB4 controls glutamatergic synapse maturation and plasticity. *Neuron, 54*, 583–597.

Lu, Y., Christian, K., & Lu, B. (2008). BDNF: A key regulator for protein synthesis-dependent LTP and long-term memory? *Neurobiology of Learning and Memory, 89*, 312–323.

Luikart, B. W., & Parada, L. F. (2006). Receptor tyrosine kinase B-mediated excitatory synaptogenesis. *Progress in Brain Research, 157*, 15–24.

Luikart, B. W., Nef, S., Virmani, T., Lush, M. E., Liu, Y., Kavalali, E. T., & Parada, L. F. (2005). TrkB has a cell-autonomous role in the establishment of hippocampal Schaffer collateral synapses. *Journal of Neuroscience, 25*, 3774–3786.

Luo, L., Hensch, T. K., Ackerman, L., Barbel, S., Jan, L. Y., & Jan, Y. N. (1996). Differential effects of the Rac GTPase on Purkinje cell axons and dendritic trunks and spines. *Nature, 379*, 837–840.

Ma, X. M., Huang, J. P., Kim, E. J., Zhu, Q., Kuchel, G. A., Mains, R. E., & Eipper, B. A. (2010). Kalirin-7, an important component of excitatory synapses, is regulated by estradiol in hippocampal neurons. *Hippocampus, 21*, 661–677.

Maze, I., Covington, H. E., 3rd, Dietz, D. M., LaPlant, Q., Renthal, W., Russo, S. J., Mechanic, M., Mouzon, E., Neve, R. L., Haggarty, S. J., Ren, Y., Sampath, S. C., Hurd, Y. L., Greengard, P., Tarakhovsky, A., Schaefer, A., & Nestler, E. J. (2010). Essential role of the histone methyltransferase G9a in cocaine-induced plasticity. *Science, 327*, 213–216.

McAvoy, T., Zhou, M. M., Greengard, P., & Nairn, A. C. (2009). Phosphorylation of Rap1GAP, a striatally enriched protein, by protein kinase A controls Rap1 activity and dendritic spine morphology. *Proceedings of the National Academy of Sciences of the United States of America, 106*, 3531–3536.

Meng, Y., Zhang, Y., Tregoubov, V., Janus, C., Cruz, L., Jackson, M., Lu, W. Y., MacDonald, J. F., Wang, J. Y., Falls, D. L., & Jia, Z. (2002). Abnormal spine morphology and enhanced LTP in LIMK-1 knockout mice. *Neuron, 35*, 121–133.

Meng, Y., Zhang, Y., Tregoubov, V., Falls, D. L., & Jia, Z. (2003). Regulation of spine morphology and synaptic function by LIMK and the actin cytoskeleton. *Reviews in the Neurosciences, 14*, 233–240.

Miyamoto, Y., Yamauchi, J., Tanoue, A., Wu, C., & Mobley, W. C. (2006). TrkB binds and tyrosine-phosphorylates Tiam1, leading to activation of Rac1 and induction of changes in cellular morphology. *Proceedings of the National Academy of Sciences of the United States of America, 103*, 10444–10449.

Nakayama, A. Y., Harms, M. B., & Luo, L. (2000). Small GTPases Rac and Rho in the maintenance of dendritic spines and branches in hippocampal pyramidal neurons. *Journal of Neuroscience, 20*, 5329–5338.

Nakazawa, T., Kuriu, T., Tezuka, T., Umemori, H., Okabe, S., & Yamamoto, T. (2008). Regulation of dendritic spine morphology by an NMDA receptor-associated Rho GTPase-activating protein, p250GAP. *Journal of Neurochemistry, 105*, 1384–1393.

Neuhoff, H., Sassoe-Pognetto, M., Panzanelli, P., Maas, C., Witke, W., & Kneussel, M. (2005). The actin-binding protein profilin I is localized at synaptic sites in an activity-regulated manner. *European Journal of Neuroscience, 21*, 15–25.

Pak, D. T., Yang, S., Rudolph-Correia, S., Kim, E., & Sheng, M. (2001). Regulation of dendritic spine morphology by SPAR, a PSD-95-associated RapGAP. *Neuron, 31*, 289–303.

Penzes, P., Beeser, A., Chernoff, J., Schiller, M. R., Eipper, B. A., Mains, R. E., & Huganir, R. L. (2003). Rapid induction of dendritic spine morphogenesis by trans-synaptic ephrinB-EphB receptor activation of the Rho-GEF kalirin. *Neuron, 37,* 263–274.

Penzes, P., Cahill, M. E., Jones, K. A., & Srivastava, D. P. (2008). Convergent CaMK and RacGEF signals control dendritic structure and function. *Trends in Cell Biology, 18,* 405–413.

Pfeiffer, B. E., Zang, T., Wilkerson, J. R., Taniguchi, M., Maksimova, M. A., Smith, L. N., Cowan, C. W., & Huber, K. M. (2010). Fragile X mental retardation protein is required for synapse elimination by the activity-dependent transcription factor MEF2. *Neuron, 66,* 191–197.

Racz, B., & Weinberg, R. J. (2008). Organization of the Arp2/3 complex in hippocampal spines. *Journal of Neuroscience, 28,* 5654–5659.

Rossman, K. L., Der, C. J., & Sondek, J. (2005). GEF means go: Turning on RHO GTPases with guanine nucleotide-exchange factors. *Nature Reviews: Molecular Cell Biology, 6,* 167–180.

Sanchez, A. M., Flamini, M. I., Fu, X. D., Mannella, P., Giretti, M. S., Goglia, L., Genazzani, A. R., & Simoncini, T. (2009). Rapid signaling of estrogen to WAVE1 and moesin controls neuronal spine formation via the actin cytoskeleton. *Molecular Endocrinology, 23,* 1193–1202.

Schratt, G. M., Tuebing, F., Nigh, E. A., Kane, C. G., Sabatini, M. E., Kiebler, M., & Greenberg, M. E. (2006). A brain-specific microRNA regulates dendritic spine development. *Nature, 439,* 283–289.

Schubert, V., & Dotti, C. G. (2007). Transmitting on actin: Synaptic control of dendritic architecture. *Journal of Cell Science, 120,* 205–212.

Sherren, N., & Pappas, B. A. (2005). Selective acetylcholine and dopamine lesions in neonatal rats produce distinct patterns of cortical dendritic atrophy in adulthood. *Neuroscience, 136,* 445–456.

Siegel, G., Obernosterer, G., Fiore, R., Oehmen, M., Bicker, S., Christensen, M., Khudayberdiev, S., Leuschner, P. F., Busch, C. J., Kane, C., Hubel, K., Dekker, F., Hedberg, C., Rengarajan, B., Drepper, C., Waldmann, H., Kauppinen, S., Greenberg, M. E., Draguhn, A., Rehmsmeier, M., Martinez, J., & Schratt, G. M. (2009). A functional screen implicates microRNA-138-dependent regulation of the depalmitoylation enzyme APT1 in dendritic spine morphogenesis. *Nature Cell Biology, 11,* 705–716.

Smrt, R. D., Szulwach, K. E., Pfeiffer, R. L., Li, X., Guo, W., Pathania, M., Teng, Z. Q., Luo, Y., Peng, J., Bordey, A., Jin, P., & Zhao, X. (2010). MicroRNA miR-137 regulates neuronal maturation by targeting ubiquitin ligase mind bomb-1. *Stem Cells, 28,* 1060–1070.

Sobczyk, A., & Svoboda, K. (2007). Activity-dependent plasticity of the NMDA-receptor fractional Ca^{2+} current. *Neuron, 53,* 17–24.

Soderling, T. R. (2000). CaM-kinases: Modulators of synaptic plasticity. *Current Opinion in Neurobiology, 10,* 375–380.

Soderling, S. H., Guire, E. S., Kaech, S., White, J., Zhang, F., Schutz, K., Langeberg, L. K., Banker, G., Raber, J., & Scott, J. D. (2007). A WAVE-1 and WRP signaling complex regulates spine density, synaptic plasticity, and memory. *Journal of Neuroscience, 27,* 355–365.

Solis, O., Limon, D. I., Flores-Hernandez, J., & Flores, G. (2007). Alterations in dendritic morphology of the prefrontal cortical and striatum neurons in the unilateral 6-OHDA-rat model of Parkinson's disease. *Synapse, 61,* 450–458.

Srivastava, D. P., Woolfrey, K. M., Jones, K. A., Shum, C. Y., Lash, L. L., Swanson, G. T., & Penzes, P. (2008). Rapid enhancement of two-step wiring plasticity by estrogen and NMDA receptor activity. *Proceedings of the National Academy of Sciences of the United States of America, 105,* 14650–14655.

Srivastava, D. P., Woolfrey, K. M., Liu, F., Brandon, N. J., & Penzes, P. (2010). Estrogen receptor ss activity modulates synaptic signaling and structure. *Journal of Neuroscience, 30,* 13454–13460.

Star, E. N., Kwiatkowski, D. J., & Murthy, V. N. (2002). Rapid turnover of actin in dendritic spines and its regulation by activity. *Nature Neuroscience, 5,* 239–246.

Takahashi, H., Sekino, Y., Tanaka, S., Mizui, T., Kishi, S., & Shirao, T. (2003). Drebrin-dependent actin clustering in dendritic filopodia governs synaptic targeting of postsynaptic density-95 and dendritic spine morphogenesis. *Journal of Neuroscience, 23*, 6586–6595.

Tashiro, A., Minden, A., & Yuste, R. (2000). Regulation of dendritic spine morphology by the rho family of small GTPases: Antagonistic roles of Rac and Rho. *Cerebral Cortex, 10*, 927–938.

Tian, X., Kai, L., Hockberger, P. E., Wokosin, D. L., & Surmeier, D. J. (2010). MEF-2 regulates activity-dependent spine loss in striatopallidal medium spiny neurons. *Molecular and Cellular Neuroscience, 44*, 94–108.

Tolias, K. F., Bikoff, J. B., Burette, A., Paradis, S., Harrar, D., Tavazoie, S., Weinberg, R. J., & Greenberg, M. E. (2005). The Rac1-GEF Tiam1 couples the NMDA receptor to the activity-dependent development of dendritic arbors and spines. *Neuron, 45*, 525–538.

Tolias, K. F., Bikoff, J. B., Kane, C. G., Tolias, C. S., Hu, L., & Greenberg, M. E. (2007). The Rac1 guanine nucleotide exchange factor Tiam1 mediates EphB receptor-dependent dendritic spine development. *Proceedings of the National Academy of Sciences of the United States of America, 104*, 7265–7270.

Wang, H. D., & Deutch, A. Y. (2008). Dopamine depletion of the prefrontal cortex induces dendritic spine loss: Reversal by atypical antipsychotic drug treatment. *Neuropsychopharmacology, 33*, 1276–1286.

Weaver, A. M., Karginov, A. V., Kinley, A. W., Weed, S. A., Li, Y., Parsons, J. T., & Cooper, J. A. (2001). Cortactin promotes and stabilizes Arp2/3-induced actin filament network formation. *Current Biology, 11*, 370–374.

Wegner, A. M., Nebhan, C. A., Hu, L., Majumdar, D., Meier, K. M., Weaver, A. M., & Webb, D. J. (2008). N-wasp and the arp2/3 complex are critical regulators of actin in the development of dendritic spines and synapses. *Journal of Biological Chemistry, 283*, 15912–15920.

Witke, W., Sutherland, J. D., Sharpe, A., Arai, M., & Kwiatkowski, D. J. (2001). Profilin I is essential for cell survival and cell division in early mouse development. *Proceedings of the National Academy of Sciences of the United States of America, 98*, 3832–3836.

Woolfrey, K. M., Srivastava, D. P., Photowala, H., Yamashita, M., Barbolina, M. V., Cahill, M. E., Xie, Z., Jones, K. A., Quilliam, L. A., Prakriya, M., & Penzes, P. (2009). Epac2 induces synapse remodeling and depression and its disease-associated forms alter spines. *Nature Neuroscience, 12*, 1275–1284.

Xie, Z., Srivastava, D. P., Photowala, H., Kai, L., Cahill, M. E., Woolfrey, K. M., Shum, C. Y., Surmeier, D. J., & Penzes, P. (2007). Kalirin-7 controls activity-dependent structural and functional plasticity of dendritic spines. *Neuron, 56*, 640–656.

Yamasaki, M., Matsui, M., & Watanabe, M. (2010). Preferential localization of muscarinic M1 receptor on dendritic shaft and spine of cortical pyramidal cells and its anatomical evidence for volume transmission. *Journal of Neuroscience, 30*, 4408–4418.

Yang, G., Pan, F., & Gan, W. B. (2009). Stably maintained dendritic spines are associated with lifelong memories. *Nature, 462*, 920–924.

Zhang, B., Zhang, Y., & Shacter, E. (2003). Caspase 3-mediated inactivation of rac GTPases promotes drug-induced apoptosis in human lymphoma cells. *Molecular and Cellular Biology, 23*, 5716–5725.

Zhang, H., Webb, D. J., Asmussen, H., Niu, S., & Horwitz, A. F. (2005). A GIT1/PIX/Rac/PAK signaling module regulates spine morphogenesis and synapse formation through MLC. *Journal of Neuroscience, 25*, 3379–3388.

Zhu, J. J., Qin, Y., Zhao, M., Van Aelst, L., & Malinow, R. (2002). Ras and Rap control AMPA receptor trafficking during synaptic plasticity. *Cell, 110*, 443–455.

Chapter 5
Synaptic Cell Adhesion Molecules

Olena Bukalo and Alexander Dityatev

Abstract During development of the nervous system following axon pathfinding, synaptic connections are established between neurons. Specific cell adhesion molecules (CAMs) accumulate at pre- and postsynaptic sites and trigger synaptic differentiation through interactions with intra- and extracellular scaffolds. These interactions are important to align pre- and postsynaptic transduction machineries and to couple the sites of cell-to-cell adhesion to the cytoskeleton and signaling complexes necessary to accumulate and recycle presynaptic vesicles, components of exo- and endocytic zones, and postsynaptic receptors. In mature brains, CAMs contribute to regulation of synaptic efficacy and plasticity, partially via direct interactions with postsynaptic neurotransmitter receptors and presynaptic voltage-gated ion channels. This chapter is to highlight the major classes of synaptic CAMs, their multiple functions, and the multistage concerted interactions between different CAMs and other components of synapses.

Keywords Cell adhesion • NCAM • Scaffold • Synaptic plasticity • Synaptogenesis

O. Bukalo
National Institute of Child Health and Human Development, National Institutes of Health, 20892 Bethesda, MD, USA
e-mail: bukalool@mail.nih.gov

A. Dityatev (✉)
Department of Neuroscience and Brain Technologies, Istituto Italiano di Tecnologia, Via Morego 30, 16163 Genova, Italy

Laboratory for Brain Extracellular Matrix Research, University of Nizhny Novgorod, 603950 Nizhny Novgorod, Russia
e-mail: alexander.dityatev@iit.it

M.R. Kreutz and C. Sala (eds.), *Synaptic Plasticity*,
Advances in Experimental Medicine and Biology 970,
DOI 10.1007/978-3-7091-0932-8_5, © Springer-Verlag/Wien 2012

5.1 Introduction

CAMs in the brain enable cell–cell recognition and are responsible for mechanical stabilization of synaptic contacts, as well as for synapse organization through assembling signaling molecules, neurotransmitter receptors, and actin cytoskeleton. Numerous studies indicate that synaptically localized CAMs are not just involved in physical adhesion but can control synapse formation, regulate dendritic spine morphology, and modify synaptic receptor function in an activity-dependent manner.

Considerable progress has been made in the characterization of several CAM families at both developing and mature synapses (Fig. 5.1). These include cadherins, immunoglobulin-containing cell adhesion molecules (Ig-CAMs), neurexins and neuroligins, ephrins, and Eph receptors. Recent data have highlighted the synaptic functions of SynCAMs and IgLONs that belong to Ig-CAMs, and leucine-rich repeat (LRR)–containing CAMs, such as NGLs (netrin-G ligands), LRRTMs (leucine-rich repeat transmembrane neuronal), and SALMs (synaptic adhesion–like molecules). Each of these synaptic CAMs differs in terms of homo-/heterophilic adhesion, calcium sensitivity, and synaptic/extrasynaptic localization and is thought to act in different processes, such as recognition of specific target domains within a neuron, synaptic differentiation, synaptic stability, and synaptic plasticity. In this chapter, we provide an overview of the current data on how each of the CAM families influences these diverse pre- and postsynaptic functions (see also Tables 5.1 and 5.2). Furthermore, we focus on recent studies that start to shed light on the molecular interactions by which mammalian CAMs shape the developing synapse and determine the molecular organization of mature synaptic contacts.

5.2 N-cadherin

Cadherins form a large superfamily of Ca^{2+}-dependent CAMs mediating mostly homophilic interactions and are grouped into subfamilies of classic cadherins and protocadherins. Strong cadherin adhesion is dependent on their ability to dimerize in *cis* orientation (between molecules presented on the same cell), which then binds in *trans* (molecules from different cells). N-cadherin, the most extensively studied classic cadherin, is expressed at excitatory synapses. It localizes at active zones of developing synapses, and synapse maturation is accompanied by the clustering of N-cadherin at *puncta adherentia* junctions (PAJs), the region flanking the active zone (Benson et al. 1998; Tallafuss et al. 2010). Several lines of evidence indicate that N-cadherin homophilic adhesion contributes to synapse stabilization both during early development (synapse formation) and in adult (synaptic plasticity), providing structural support to the synaptic complex and/or by activating intracellular signaling that regulates neuronal physiology. Multiple features of immature spines, such as reduced spine number, more filopodia-like spines, thinner spines, or

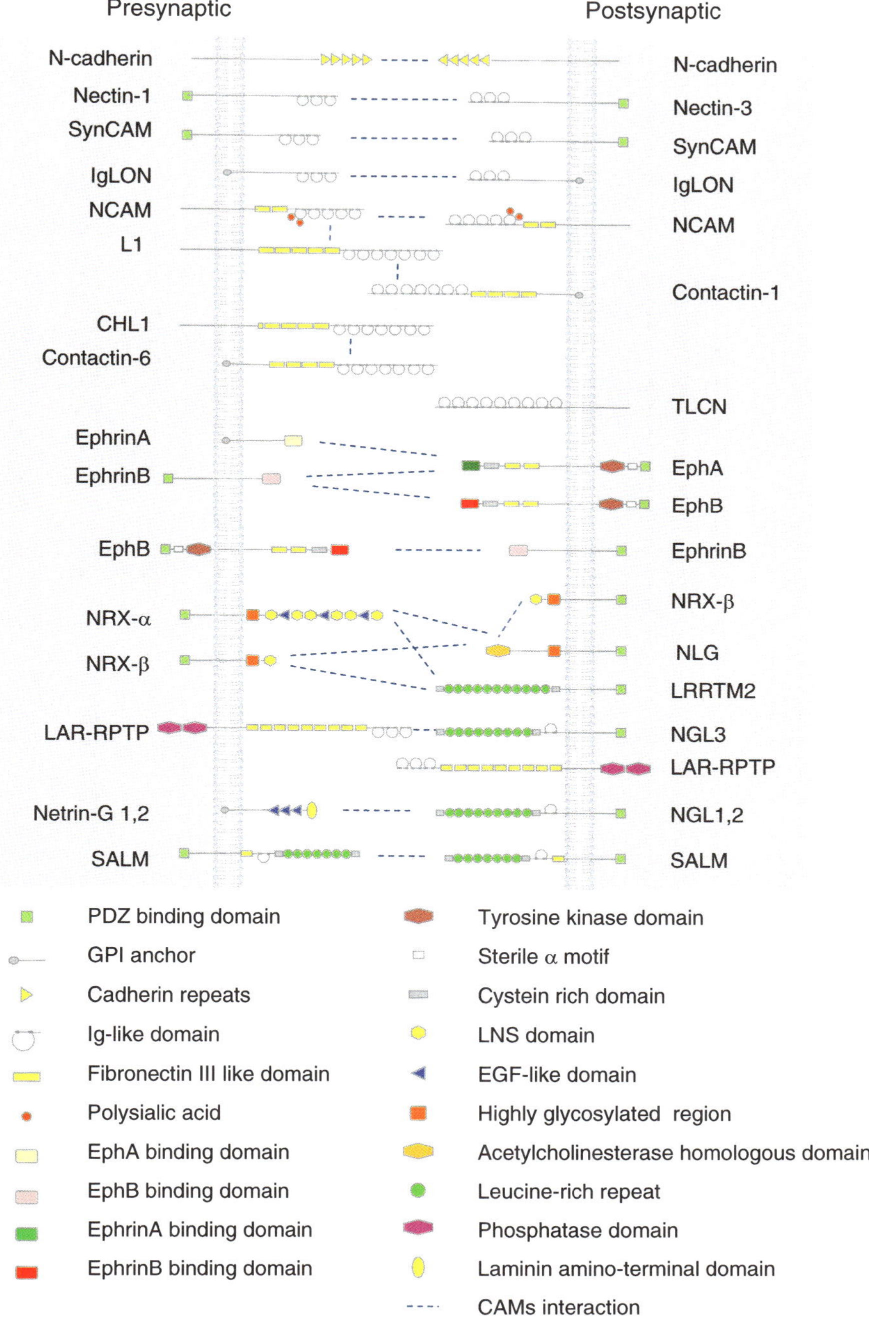

Fig. 5.1 Domain structure of synaptic CAMs and schematic diagram of the protein–protein interactions between synaptic CAMs expressed on pre- and postsynaptic sites, as discussed in this chapter. Abbreviations: *CHL1* close homolog of L1, *GPI* glycosylphosphatidylinositol, *Ig* immunoglobulin, *NCAM* neural cell adhesion molecule, *NLG* neuroligin, *NRX* neurexin, *NGL* netrin-G ligand, *LAR-RPTP* leukocyte antigen-related family protein tyrosine phosphatase, *LRRTM* leucine-rich repeat transmembrane gene family. Protein–protein interactions between CAMs are shown by *dotted lines*

Table 5.1 Presynaptic scaffold and functions mediated by CAMs

Molecules	Function	Interactions and possible mechanisms	References
N-cadherin	Docked vesicle localization, ↑vesicle accumulation and recycling, ↑STP, ↑release probability, ↑mEPSC frequency	*Trans*-synaptic retrograde regulation via N-cadherin, NLG1, and S-SCAM	Jungling et al. (2006), Stan et al. (2010)
N-cadherin	Modulation of presynaptic calcium influx	p120-catenin, presynaptic VDCCs, RhoA GTPase, and myosin–actin	Marrs et al. (2009)
NCAM	↑vesicle recycling, ↑PPF at NMJ	Via C-terminal to MLCK, MLC, and myosin II	Polo-Parada et al. (2005), Rafuse et al. (2000)
CHL1	Vesicle localization at inhibitory synapses, ↑vesicle recycling	Association in complex with Hsc70/SGT, CSP, and SNARE	Nikonenko et al. (2006), Andreyeva et al. (2010)
Contactin-1	↑PPF	ND	Murai et al. (2002)
SynCAM1	↑mEPSC frequency, ↑vesicle recycling	Association with CASK via PDZ-binding domain	Biederer et al. (2002)
EphrinB1	↑mEPSC frequency, ↑PPF, ↑vesicle recycling, ↑LTP in the optic tectum	Reverse signaling via postsynaptic EphB2	Lim et al. (2008)
NLG1	Docked vesicle localization	ND	Dahlhaus et al. (2010)
NRXβ	↑release probability	*Trans*-synaptic retrograde modulation via NLG1, PSD-95 Binding via PDZ domain to CASK/Mint1and CASK/caskin1, protein 4.1, and actin	Futai et al. (2007), Tabuchi et al. (2002), Biederer and Sudhof (2001)
NRXα	↑Ca^{2+} triggered transmitter release, ↑mIPSC frequency, ↑vesicle recycling, ↑presynaptic Ca^{2+} current	Regulation of presynaptic N- and P/Q-type VDCCs activation via extracellular domain of NRXα	Missler et al. (2003), Zhang et al. (2005)

STP short-term plasticity, *mEPSCs* miniature excitatory postsynaptic currents, *mIPSCs* miniature inhibitory postsynaptic currents, *PPF* paired-pulse facilitation, *VDCCs* voltage-dependent calcium channels, *CHL1* close homolog of L1, *NCAM* neural cell adhesion molecule, *NLG* neuroligin, *NRX* neurexin, ↑ increase, *ND* not determined

spines with smaller heads, are detected in cultures deficient in N-cadherin or associated β-catenin that mediates cadherin interaction with the cytoskeleton (Mendez et al. 2010; Okuda et al. 2007; Saglietti et al. 2007). Adhesion mediated by N-cadherin is not static, but may be regulated in an activity-dependent manner either by dimerization of N-cadherin molecules, leading to more association of β-catenin with N-cadherin; or by redistribution of N-cadherin and β-catenin from dendritic shafts to spines; or by the protocadherin arcadlin–mediated N-cadherin

Table 5.2 Postsynaptic scaffolding and functions mediated by CAMs

Molecules	Function	Interactions and possible mechanisms	References
N-cadherin	Postsynaptic scaffolding	δ-Catenin PDZ domain binds to GluA2 via ABP and GRIP	Silverman et al. (2007)
N-cadherin	↑AMPAR-mediated current	Extracellular interaction with GluA1 and GluA2	Nuriya and Huganir (2006), Saglietti et al. (2007)
N-cadherin	↑LTP, not LTD in CA1	Dendritic spine enlargement	Bozdagi et al. (2010)
NCAM	Postsynaptic scaffolding	Associated with β-spectrin and accumulation of PSD95, GluN1, GluN2B, CaMKII	Sytnyk et al. (2006)
NCAM	↑AMPAR-mediated current	By PSA in immature neurons	Vaithianathan et al. (2004)
NCAM	↓GluN2B-mediated current, ↑LTP in CA1	PSA inhibition of GluN2B, activation of p38 MAPK	Hammond et al. (2006), Kochlamazashvili et al. (2010)
TLCN	↓LTP in CA1	Dendritic spine retraction, interaction with ERM proteins and F-actin	Nakamura et al. (2001), Tian et al. (2007)
SynCAM1	Postsynaptic scaffolding, NMDAR trafficking, ↑NMDAR-mediated current, ↓LTD, not LTP in CA1	FERM-binding domain to 4.1B protein, CASK, syntenin1, GluN1/GluN2B accumulation	Hoy et al. (2009), Robbins et al. (2010)
EphA4	↓LTP in amygdala	EphrinB3-activated forward signaling, Rab5-GEF Rin1–mediated EphA4 endocytosis	Deininger et al. (2008)
EphB2	AMPAR trafficking	Via PDZ domain to GluA2/3	Kayser et al. (2006)
EphB2	↑ NMDAR-mediated current, ↑LTD and late LTP in CA1	Extracellular binding to GluN1; activation of Erk1/2, Fyn, and Src tyrosine kinases; GluN2A/GluN2B phosphorylation	Dalva et al. (2000), Grunwald et al. (2001), Takasu et al. (2002), Henderson et al. (2001)
EphrinB2	↑LTP in CA3	Retrograde signaling via postsynaptic EphB, GluA2 clustering by GRIP and PICK	Contractor et al. (2002)
EphrinB3	Postsynaptic scaffolding, ↑excitatory synaptic transmission, ↑LTP and LTD in CA1	EphA4, reverse signaling. Interactions with GRIP1 via PDZ-binding domain	Aoto et al. (2007), Grunwald et al. (2004), Rodenas-Ruano et al. (2006)

(continued)

Table 5.2 (continued)

Molecules	Function	Interactions and possible mechanisms	References
NLG1	AMPAR trafficking	GluA2 recruitment to PSD-95	Heine et al. (2008)
NLG1	Postsynaptic scaffolding, NMDAR trafficking, ↑NMDAR-mediated current, ↑LTP and STDP in amygdala	GluN1 recruitment via PDZ-binding to PSD-95 and other scaffolding protein	Barrow et al. (2009), Kim et al. (2008), Jung et al. (2010)
NLG2	Postsynaptic scaffolding, ↑GABAergic and glycinergic transmission	Gephyrin, collybistin, $GABA_A$Rs	Poulopoulos et al. (2009)
NLG3	AMPAR (GluA2/3) trafficking during spine remodeling	Epac2 recruitment, Rap-GEF activation	Woolfrey et al. (2009)
NLG1/3	Activity-dependent targeting to postsynaptic membrane	↑ surface NLG1/3 after chem-LTP ↓ surface NLG1/3 after chem-LTD PSD-95 binding, microtubule and dynein motor complex	Schapitz et al. (2010)
NRXβ	Alignment of pre- and postsynaptic sites	Retrograde transsynaptic interaction with GluD2 via cerebellin1 precursor protein	Uemura et al. (2010)
NRXβ	↓$GABA_A$R-mediated transmission	Independent of NLG1 and NLG2; extracellular binding to $GABA_A\alpha$R1	Zhang et al. (2010)
NGLs	Postsynaptic scaffolding	Associated with PSD-95, GKAP, Shank, GluN2A, GluN1, GluA2	Kim et al. (2006), Woo et al. (2009)
LAR-RPTPs	AMPAR trafficking, ↑AMPAR-mediated current	LAR-mediated dephosphorylation of β-catenin, complex with GRIP/α-liprin/GluA2/3 and cadherin	Dunah et al. (2005)
LRRTM2	AMPAR trafficking, ↑excitatory synaptic transmission	Extracellular binding to GluA1, GluA2 via LAR domain, interaction with PSD-95 via PDZ-domain, NRXα, NRXβ	De Wit et al. (2009)
SALM1	Postsynaptic scaffolding, NMDAR trafficking	Associated with PSD-95, PDZ-dependent interaction with GluN1	Wang et al. (2006)

(continued)

Table 5.2 (continued)

Molecules	Function	Interactions and possible mechanisms	References
SALM2	Postsynaptic scaffolding	Induces clustering of PSD-95, GKAP, GluA1 LAR-mediated interaction with reticulon3 within ER	Ko et al. (2006), Chang et al. (2010)
SALM3	Postsynaptic scaffolding	Associated/clustered with PSD-95	Mah et al. (2010)

chem chemical, *NCAM* neural cell adhesion molecule, *PSA* polysialic acid, *TLCN* telencephalin, *NLG* neuroligin, *NRX* neurexin, *NGLs* netrin-G ligands, *LAR-RPTPs* leukocyte antigen-related family protein tyrosine phosphatases, *LRRTM* leucine-rich repeat transmembrane gene family, *ER* endoplasmic reticulum, ↓ reduction, ↑ increase, *ND* not determined

endocytosis (Murase et al. 2002; Tanaka et al. 2000; Yasuda et al. 2007). Therefore, in response to activity, changes in N-cadherin-mediating adhesion may affect dendritic spine morphology and synaptic transmission. Indeed, activity-dependent stabilization of spines by N-cadherin has been demonstrated recently. Firstly, in conditional N-cadherin deficient mice despite normal spine morphogenesis and basal synaptic transmission, the magnitude of LTP and LTP-associated spine enlargement are prominently reduced in the CA1 region (Bozdagi et al. 2010). Secondly, interfering with N-cadherin functions either by dominant-negative protein or by siRNA prevents plasticity-induced stabilization of spines in organotypic hippocampal cultures (Mendez et al. 2010).

Disrupting N-cadherin homophilic interactions influences both the pre- and postsynapses, by affecting accumulation of synaptic proteins, synaptic vesicle recycling, and potentiation of neurotransmission (Bozdagi et al. 2000; Okuda et al. 2007; Tanaka et al. 2000). Accumulating evidence supports the role of the retrograde transsynaptic signaling by postsynaptic N-cadherin that in turn may regulate presynaptic vesicle exocytosis (Saglietti et al. 2007) or vesicle recycling under enhanced synaptic activity (Jungling et al. 2006). This signaling is possibly initiated by formation of a postsynaptic complex containing N-cadherin, neuroligin-1, and a scaffolding molecule S-SCAM (Stan et al. 2010). Another mechanism, independent of cadherin binding, implicates presynaptic β-catenin in regulation of the reserve pool of synaptic vesicles via binding of the β-catenin PDZ domain to the Veli protein, which is involved in synaptic vesicle exocytosis (Bamji et al. 2003). In mice with conditionally ablated β-catenin, a decrease in the reserve but not docked pool of synaptic vesicles and an impaired response to prolonged repetitive stimulation was shown (Bamji et al. 2003). Presynaptic function, measured as the prepulse facilitation ratio (PPF), is also impaired in hippocampal slices from mice deficient in another member of the catenin family δ-catenin (Israely et al. 2004). In contrast to β-catenin that regulates presynaptic release in a cadherin-independent manner, the δ-catenin effects on PPF may be mediated via N-cadherin since a selective loss of N-cadherin has been reported in the brain of δ-catenin deficient mice (Israely et al. 2004). Interestingly, homophilic N-cadherin binding controls presynaptic voltage-activated Ca^{2+} influx by interacting with p120-catenin and regulating the RhoA GTPase activity and myosin–actin interactions downstream of RhoA (Marrs et al. 2009). Differences in the above described mechanisms may rely on the distinct engagement of catenins to the actin cytoskeleton either by direct binding (as for p120 catenin and δ-catenin) or via intermediate α-catenin (as for β-catenin), thereby linking different pools of synaptic vesicles and affecting presynaptic function. At the postsynapse, N-cadherin is involved in AMPA receptor (AMPAR) trafficking associated with the spine growth and regulation of excitatory synaptic transmission. Published data suggest that N-cadherin may bind via its extracellular domain to the extracellular part of AMPARs (GluA1 and GluA2), thereby regulating function and availability of AMPARs at the postsynaptic membrane (Nuriya and Huganir 2006; Saglietti et al. 2007). Also, N-cadherin-induced intracellular signaling may regulate AMPAR trafficking and modulate excitatory synaptic transmission. In this scenario, the availability of GluA2 on the synaptic surface is regulated by N-cadherin's interaction with δ-catenin and additional recruitment of AMPAR binding proteins ABP and GRIP (Silverman et al. 2007). An increase in GluA1 content at the postsynaptic membrane is associated with N-cadherin-mediated spine growth via interaction of its PDZ domain to Rac-guanine exchange factor (GEF) kalirin-7, resulting in activation of Rac-1 and p21-activated kinase (PAK) (Xie et al. 2008).

5.3   NCAM

The neural cell adhesion molecule (NCAM), a member of the Ig-CAM family, is expressed at both the pre- and postsynaptic membrane, where it can interact with NCAM and other membrane molecules in *cis* or *trans*. NCAM regulates synapse formation, maturation, and function through homo- and heterophilic interactions. During the early stages of synapse formation, NCAM is clustered at nascent synaptic contacts, where it interacts with β1-spectrin, and recruits *trans*-Golgi network-derived organelles, transforming these contacts into functional synapses (Sytnyk et al. 2002). In heterotypic hippocampal cultures, obtained from wild-type and NCAM-deficient mice, a reduced number of synapses, decreased amplitude of excitatory postsynaptic current, and abolished potentiation after glutamate application are observed selectively in cells with ablated postsynaptic NCAM. This suggests that excitatory synapse formation and function is regulated by postsynaptic NCAM (Dityatev et al. 2000). Further, in vitro studies revealed a decrease in PSD size, as well as reduced expression of PSD-associated molecules such as PSD-95, GluN1, GluN2B, CaMKII, and α-actinin in NCAM-deficient hippocampal cultures. Activity-dependent translocation of CaMKII to the PSD in response to glutamate was abolished in this system (Sytnyk et al. 2006). These results are in line with data obtained in vivo demonstrating increased NCAM180 expression and its colocalization with GluN2A in the PSD after induction of LTP in the dentate gyrus (Fux et al. 2003; Schuster et al. 1998). In addition, NMDAR-dependent forms of synaptic plasticity are impaired in NCAM-deficient mice, including LTP and LTD in CA1 (Bukalo et al. 2004; Muller et al. 1996) and LTP in the dentate gyrus (Stoenica et al. 2006). It appears that NCAM-mediated targeting of NMDARs to the PSD and redistribution of PSD components in response to activity are required for both synapse formation and synaptic plasticity.

In the mammalian brain, NCAM is a predominant carrier of the unusual long-chain polyanionic carbohydrate, polysialic acid (PSA) that is able to modify NCAM-mediated adhesion and is involved in synaptogenesis and synaptic plasticity (Rutishauser 2008). It has been demonstrated that removals of PSA by endoneuraminidase N (endoN) or interfering by NCAM-PSA-Fc impairs preferential formation of synapses on NCAM-expressing neurons in hippocampal cultures (Dityatev et al. 2004). The mechanism by which PSA-NCAM regulates synaptogenesis requires an interaction with heparan sulfate proteoglycans and activation of NMDA and FGF receptors (Dityatev et al. 2004). In cultures, soluble PSA increases AMPAR-mediated current in immature, but not mature cells (Vaithianathan et al. 2004) and modulates NMDAR-mediated currents in a concentration-dependent manner (Hammond et al. 2006). LTP, LTD, and activity-dependent formation of perforated spines in the CA1 region are impaired after enzymatic cleavage of PSA with endoN (Becker et al. 1996; Dityatev et al. 2004; Muller et al. 1996), as well as by exogenous application of PSA (Muller et al. 2000; Senkov et al. 2006). Experiments in mice deficient in polysialyl-transferase ST8SiaIV/PST, the enzyme that is responsible for PSA synthesis at late developmental stages, demonstrated impaired LTP and LTD in CA1 (Eckhardt et al. 2000). Recently, it has been shown that a deficiency in NCAM or PSA upregulates GluN2B-mediated transmission and phosphorylation of p38 MAPK (Kochlamazashvili et al. 2010). In this study, CA1 LTP recordings in mice deficient in NCAM or PSA could be restored by suppression of GluN2B or p38 MAPK activity and by ablation of Ras-GRF1 that transduces signaling between GluN2B and p38 MAPK. Furthermore, CA1 LTP could be restored by a glutamate scavenger. The latter observation and the fact that PSA potentiates only GluN2B-mediated currents that are activated by low micromolar concentrations of glutamate suggest that PSA restrains signaling through extrasynaptic GluN2B receptors and by this mechanism controls synaptic plasticity (Kochlamazashvili et al. 2010). Available data demonstrate that postsynaptically expressed NCAM and PSA are important components of synapse organization and function at different developmental stages by being involved in the reorganization of the PSD and function of NMDARs. Recently, a fraction of SynCAM1 has been demonstrated to be polysialylated on a glial cell subpopulation in the early postnatal brain (Galuska et al. 2010), and it will be interesting to determine to which extent PSA expressed on these cells may contribute to synapse formation and function.

NCAM also has presynaptic functions. NCAM-deficient neuromuscular junctions (NMJs) are lacking paired-pulse facilitation and fail to maintain transmitter output with repetitive stimuli (Rafuse et al. 2000). Furthermore, clusters of dye-loaded vesicles are observed not only at the end plate but also at the pretermi-nal part of the axon, as it has been found previously in immature axons

(Polo-Parada et al. 2001). A highly conserved C-terminal domain of NCAM is required to maintain effective transmission via a pathway involving myosin light-chain kinase (MLCK) and probably MLC and myosin II (Polo-Parada et al. 2005). This pathway is necessary to replenish synaptic vesicles during high levels of exocytosis by facilitating myosin-driven delivery of synaptic vesicles to active zones for subsequent exocytosis. In addition, NCAM$-$/$-$ mice exhibit deficits in catecholamine granule trafficking between the readily releasable pool and the immediately releasable pool (Chan et al. 2005). Thus, NCAM appears to play a fundamental role in the transmitter release mechanism at least in neuroendocrine cells and at the neuromuscular junction.

5.4 L1-CAMs

The L1 family of neural cell adhesion molecules (L1-CAMs) contains transmembrane receptors with critical functions in neurodevelopment. It has been demonstrated that the L1 intracellular domain binds to ankyrin, FERM, and 14-3-3 proteins, suggesting that L1 may be involved in synaptic organization (Hortsch et al. 2009; Ramser et al. 2010). Ultrastructural studies performed in the hippocampus of L1-deficient mice revealed a decreased density of perisomatic synapses and structural abnormalities in the presynaptic terminals (Saghatelyan et al. 2004). A more diffuse distribution of synaptic vesicles is accompanied by decreased mIPSC frequency and by increased activity-dependent disinhibition in hippocampal slices from juvenile L1-deficient mice (Saghatelyan et al. 2004). Although in conditional L1 knockout mice enhanced basal synaptic transmission is possibly due to abnormalities in inhibitory currents, no structural abnormalities in morphology of inhibitory or excitatory synapses are detected in these mice (Law et al. 2003). Recently, it has been demonstrated that a loss of the L1–ankyrin interaction impaired branching of GABAergic interneurons, including basket cells, and reduced the number of perisomatic synapses in the cingulate cortex in mice (Guan and Maness 2010). Interestingly, another L1-CAM, neurofascin, has also been implicated in development of GABAergic innervation of cerebellar Purkinje cells, and its function also depends on the interaction with ankyrin (Ango et al. 2004). Thus, at least two L1-CAMs are involved in the formation of inhibitory perisomatic synapses during development via interactions with ankyrin.

With regard to the importance of L1 in synaptic plasticity, no changes have been reported for the CA3–CA1 synapses in L1-deficient mice (Bliss et al. 2000; Law et al. 2003; Saghatelyan et al. 2004), but systematic analysis of seven major subtypes of excitatory synapses in the hippocampus revealed a specific impairment of LTP at synapses made by perforant path axons on distal dendrites of pyramidal cells (Lepsveridze, Dityatev, Schachner, unpublished data).

Another member of the L1 subfamily, CHL1 (close homolog of L1), is also involved in synaptogenesis of inhibitory interneurons. In the hippocampus of juvenile CHL1-deficient mice, the density and total number of perisomatic interneurons are increased (Nikonenko et al. 2006). Ultrastructural studies demonstrated an increased length and linear density of active zones in inhibitory synapses that was accompanied by increased inhibitory currents recorded from CHL1-deficient neurons (Nikonenko et al. 2006). Thus, CHL1 has an effect opposite to L1 on structure and function of inhibitory synapses in the juvenile hippocampus. The increase in inhibition results in a reduced LTP in CA1 in juvenile CHL1-deficient mice, which can be abrogated by suppression of GABAergic inhibition (Nikonenko et al. 2006). However, in young adult CHL1-deficient mice, LTP in seven major excitatory connections in the hippocampus is normal, indicating a compensation of developmental synaptic abnormalities in the CA1 region. At this age, however, basal synaptic transmission in lateral and medial perforant path projections to the dentate gyrus is elevated in CHL1-deficient mice, correlating with a reduced reactivity to environmental stimuli and reduced expression of social behaviors (Morellini et al. 2007).

The underlying mechanisms may involve interaction between the intracellular domain of CHL1 and ATPase 70 kDa heat shock cognate protein (Hsc70) and SNARE complex components, which are involved in clathrin-mediated vesicles endocytosis (Andreyeva et al. 2010; Leshchyns'ka et al. 2006). In response to synaptic activity, CHL1 recruits Hsc70 and SNARE complex component to the presynaptic vesicles (Leshchyns'ka et al. 2006). In CHL1-deficient mice, Hsc70-mediated chaperone activity in synapses is reduced and, therefore, the SNARE complex is unable to sustain vesicle recycling during prolonged synaptic activity (Andreyeva et al. 2010) (Fig. 5.1).

5.5 Telencephalin

Telencephalin (TLCN) is a member of the ICAM subfamily of Ig-CAMs with expression restricted to the soma and dendrites of neurons (Oka et al. 1990). TLCN facilitates the formation, elongation, and maintenance of dendritic filopodia and thereby slows spine maturation in hippocampal neurons, as demonstrated in experiments using TLCN-deficient mice (Matsuno et al. 2006). TLCN is linked to the actin cytoskeleton via its intracellular domain that interacts with α-actin in the dendritic shafts. In dendritic filopodia, TLCN interacts with the phosphorylated, active form of ERM (ezrin/radixin/moesin) proteins (Furutani et al. 2007). It has been shown that in response to synaptic activity, membrane-associated TLCN is cleaved by matrix metalloproteases and is dissociated from F-actin, which results in increased number and length of filopodia in wild-type neurons (Tian et al. 2007). Accordingly, retraction of spine heads is shown in TLCN-deficient hippocampal neurons (Tian et al. 2007). Ablation of TLCN leads to the enhancement of CA1 LTP and increased saturation level for LTP (Nakamura et al. 2001). These data suggest that synapse "rejuvenation" by TLCN deficiency may increase LTP and the dynamic range of synaptic plasticity; in this respect, abnormalities in LTD can be predicted in TLCN mutants. Altogether, these data indicate that TLCN-mediated adhesion controls structural synaptic plasticity by counteracting adhesion systems that are facilitating spine maturation.

5.6 Contactins

Contactins are GPI-linked Ig-CAMs that exert heterophilic binding activities, interacting with members of the L1-CAMs and extracellular matrix components (Shimoda and Watanabe 2009). Electron microscopy studies detected contactin-1 as a component of the PSD in CA1 pyramidal cells (Murai et al. 2002). The development of synaptic ultrastructure, basal excitatory synaptic transmission, NMDA receptor function, LTP, and depotentiation in the CA1 region are intact in the mutant. Contactin-1-deficient mice are specifically impaired in LTD in CA1 pyramidal cells, and it has been proposed that the mutation appears to affect the stabilization of LTD rather than its induction (Murai et al. 2002). Although contactin-1 is localized at the postsynaptic membrane, impaired PPF was detected in contactin-1-deficient mice, suggesting that contactin-1 either decreases postsynaptic expression of GluA2-containing AMPARs or regulates presynaptic function via transsynaptic interactions. Recently, a presynaptic localization of another member of contactin family, contactin-6, was shown in hippocampus and cerebellum (Sakurai et al. 2009, 2010). A reduced number of excitatory presynaptic terminals were detected in these brain regions in contactin-6-deficient mice (Sakurai et al. 2009, 2010). If contactin-6/contactin-1 may function as the transsynaptic system and play a role in synapse, development and function remain to be investigated.

5.7 Nectins

Nectins are transmembrane Ig-CAMs that interact in *trans*, in a Ca^{2+}-independent manner through their extracellular domains with each other or with other Ig-CAMs. Nectins through their PDZ domain bind to the actin-binding protein afadin that interacts with α-catenin, thereby anchoring nectins to the actin cytoskeleton and to the cadherin/catenin adhesion complex (Giagtzoglou et al. 2009). Interestingly, at an early developmental stage, nectin-1/afadin complex is found at both excitatory and inhibitory synapses in association with F-actin (Benson et al. 1998; Lim et al. 2008). However, following synaptic maturation, nectin-1 localization is restricted to excitatory synapses similar to N-cadherin (Benson et al. 1998; Lim et al. 2008). In the hippocampal CA3 region of adult brain, nectin-1 and -3 localize at pre- and postsynaptic sites, respectively, whereas afadin is symmetrically present at both sites (Mizoguchi et al. 2002). Inhibiting the function of nectin-1 in hippocampal cell culture results in a reduction in the size of synapses, which is accompanied by an increased number of synapses (Mizoguchi et al. 2002), suggesting a role of nectins in synaptogenesis. On the other hand, analysis of spine morphology in the hippocampus of nectin-1- and nectin-3-deficient mice revealed no abnormalities in these dendritic structures, despite a reduced number of PAJs and aberrantly localized N-cadherin and afadin (Honda et al. 2006). Conditional ablation of afadin results in reduced expression of nectin-1, nectin-3, N-cadherin and β-catenin, a reduction of PAJs, and an increased number of perforated synapses in the CA3 region (Majima et al. 2009). Collectively, these results indicate that afadin is important for the accumulation of not only nectins but also N-cadherin and β-catenin at synapses and plays a key role in the formation of PAJs. Nectin-1 and afadin also colocalize with the scaffolding protein S-SCAM (Yamada et al. 2003) that associates with N-cadherin via the cell adhesion molecule neuroligin 1 (Stan et al. 2010), indicating that several cell adhesion systems may operate cooperatively in the formation of junctional complexes. In spite of aberrant mossy fiber projections, basic synaptic transmission and LTP at the mossy fiber synapse of nectin-1 knockout mice are not distinguished from wild-type mice (Honda et al. 2006).

5.8 SynCAMs

Synaptic cell adhesion molecules (SynCAM) represent a family of four Ca^{2+}-independent transmembrane Ig-CAMs that are prominently expressed throughout the brain at both excitatory and inhibitory neurons (Thomas et al. 2008). At early developmental stages, SynCAM1 is involved in the contact-mediated differentiation of synapses (Stagi et al. 2010). During synapse maturation, SynCAM proteins are enriched in pre- and postsynaptic plasma membranes and are engaged in specific homo- and heterophilic adhesive interactions, representing *trans*-interacting adhesion system (Biederer et al. 2002; Fogel et al. 2007). Interestingly, heterophilic adhesion between SynCAM1 and SynCAM2 is affected by their N-glycosylation pattern (Fogel et al. 2007). Functionally, the heterophilic partners SynCAM1 and SynCAM2 increase the number of presynaptic terminals and enhance excitatory neurotransmission in cultured neurons, suggesting a role of SynCAMs in presynapse formation and function (Biederer et al. 2002; Fogel et al. 2007; Sara et al. 2005). Through its PDZ domain, SynCAM1 binds to the scaffolding proteins syntenin and CASK and recruits them from the cytosol to the membrane (Biederer et al. 2002; Meyer et al. 2004). In addition, SynCAM1 is able to interact with the FERM domain of proteins 4.1B and 4.1N, recruiting NMDAR and AMPAR, respectively, to the postsynaptic membrane (Hoy et al. 2009). Manipulating 4.1B expression in hippocampal cultures and coexpressing SynCAM with GluN1/GluN2B lead to enhanced synaptic localization of NMDARs, NMDAR-mediated current, and synaptogenesis (Hoy et al. 2009), suggesting that 4.1B is a SynCAM1 effector molecule that influences postsynaptic development. In vivo experiments in SynCAM1-deficient and overexpressing mice have convincingly demonstrated that the organization of excitatory synapses is the key developmental role of SynCAM1 (Robbins et al. 2010). In adult brains, LTP in the CA1 remains intact in both transgenic lines, whereas LTD is abolished in overexpressing mice and enhanced in mice with ablated SynCAM1, suggesting that SynCAM1 is involved in regulation of synaptic plasticity via restricting LTD (Robbins et al. 2010).

5.9 IgLONs

The IgLON subfamily of Ig-CAMs consists of four highly glycosylated proteins that are attached to membrane lipid rafts via a GPI anchor. LAMP (limbic system–associated membrane protein), OBCAM (opioid-binding cell adhesion molecule), and Ntm (neurotrimin), but not Kilon, are capable of homophilic interactions both in *trans* and *cis* (Lodge et al. 2000). LAMP, OBCAM, and Ntm bind also in a heterophilic manner to each other and to Kilon (Lodge et al. 2000; Miyata et al. 2003). IgLON proteins are abundant at synapses of the limbic system and cerebral cortex (Miyata et al. 2003); however, their subcellular localization is developmentally regulated. In fetal neurons, LAMP is widely expressed on somata, dendrites, and axons, whereas in mature neurons, it is restricted to postsynaptic sites (Pimenta et al. 1996). Kilon expression is confined to axons and presynaptic terminals at early developmental stages, but is also localized at dendritic spines and somatic synapses later on (Hashimoto et al. 2008; Miyata et al. 2003). OBCAM was shown to have a much more restricted distribution pattern with a selective localization at dendritic spines in both immature and mature brains (Miyata et al. 2003). Studies performed in cell culture suggest that LAMP, Kilon, and OBCAM are important for controlling synapse number. Overexpression or suppression of LAMP and OBCAM in cultured neurons results in corresponding changes in synapse numbers (Hashimoto et al. 2009; Yamada et al. 2007). Early in development, Kilon is not anchored at lipid rafts and decreases the number of synapses after overexpression. However, with synapse maturation, Kilon associates with lipid rafts and then promotes synapse formation (Hashimoto et al. 2008). Furthermore, in response to increased neuronal activity, OBCAM at the cell surface is internalized via a raft-dependent pathway (Yamada et al. 2007). These data suggest that control of the IgLONs association with lipid rafts may be implicated in synapse formation and function. It has been demonstrated that synaptic lipid rafts are important for maintenance of postsynaptic structures (Hering et al. 2003) and retention of potassium channel Kir3 by NCAM (Kleene et al. 2010). The functional significance of raft associated IgLONs and their relevance to modulation of synaptic functions in vivo remain to be investigated.

5.10 Ephrins and Eph Receptors

Eph receptors represent a family of receptor tyrosine kinases which have been divided into EphA and EphB subclasses. EphA receptors typically bind to GPI-anchored ephrinA, and EphBs bind to transmembrane ephrinB ligands; the exception is EphA4, which binds to both classes of ephrins. Because of the signaling capability of ephrins, Eph–ephrin binding leads to bidirectional signal transduction into both the receptor expressing cell (forward signaling) and the ligand expressing cell (reverse signaling). Eph and ephrins can be expressed either at pre- and postsynaptic membrane, as well as extrasynaptically; EphA4 and Ephrin A3 expression has also been detected on astrocytes (Klein 2009).

Several lines of evidence indicate that activation of EphAs by ephrinAs is involved in spine retraction in hippocampal neurons (Fu et al. 2007; Murai et al. 2003). Regulation of small Rho family GTPases that control the actin cytoskeleton has been proposed as the underlying mechanism. EphA4 stimulation leads to inactivation of Rap1 GAP, with following inhibition of integrin signaling and reduced adhesion to the extracellular matrix (Bourgin et al. 2007; Richter et al. 2007) or activation of serine/threonine kinase Cdk5, RhoA-specific GEF ephexin1, and RhoA (Fu et al. 2007). The regulation of spine length and retraction of spines may also involve the activation of PLCγ1 by ephrinA3 that leads to reduced membrane association of the actin depolymerization factor cofilin (Zhou et al. 2007). The significance of EphA4 forward signaling in maturation of dendritic spines is confirmed in EphA4-deficient mice, which have longer, irregular shaped dendritic spines with disorganized appearance (Murai et al. 2003). Apparently, synaptic adhesion mediated by EphA–ephrinA is also important in regulation of spines dynamic and function in the adult brain because a deficit in hippocampal CA1 LTP and LTD is observed in EphA4-deficient mice, independent of EphA4 kinase activity (Grunwald et al. 2004). In mature neurons of the amygdala, EphA4 forward signaling activated by ephrinB3 is required for regulation of Rab5-GEF Rin1, which is involved in EphA4 internalization and restricting LTP (Deininger et al. 2008).

In contrast to EphA–ephrinA, activation of EphB–ephrinB signaling promotes excitatory synaptogenesis, as it has been demonstrated both in vitro and in vivo for EphB1–B3 and ephrinB1–B3 (Aoto et al. 2007; Bouzioukh et al. 2007; Henkemeyer et al. 2003; Kayser et al. 2008; Kayser et al. 2006; Penzes et al. 2003). EphBs role in synaptogenesis is temporally restricted, being the most critical during contact initiation, when filopodia are most abundant and motile (Kayser et al. 2008). Interfering with EphBs expression in mature culture resulted in fewer spines and synapses but also in more filopodia (Henkemeyer et al. 2003; Penzes et al. 2003). EphB forward signaling leads to activation of PAK and increased filopodia motility (Kayser et al. 2008), suggesting that signaling initiated by EphBs may lead to the reorganization of the actin cytoskeleton by Rho GTPases. EphB forward signaling leads to phosphorylation of the Rho GEF Tiam1 and kalirin-7, which activate the Rho family GTPases Rac1 and Cdc42 (Irie and Yamaguchi 2002; Penzes et al. 2003; Tolias et al. 2007). EphB2 also phosphorylates the transmembrane heparan sulfate proteoglycan syndecan-2 (Ethell et al. 2001), which activates Cdc42 (Irie and Yamaguchi 2002).

EphB signaling is also involved in the regulation of glutamatergic receptor trafficking and function. In the hippocampus of triple EphB1/EphB2/EphB3 knockout mice, less dendritic NMDAR and AMPAR clusters were detected (Kayser et al. 2006). Distinct modes of EphB2 association with neurotransmitter receptors may regulate synapse maturation and function. EphB2 controls, in a kinase-dependent manner, the localization of GluA2/3 through PDZ-binding domain interactions and triggers presynaptic differentiation via its extracellular ephrin-binding domain (Kayser et al. 2006). Modulation of mature spine number by presynaptic mechanisms is demonstrated, as an increase in release sites by long-term incubation with ephrinB1–Fc or EphB2–Fc (Dalva et al. 2000; Penzes et al. 2003) and a reduction in mEPSC frequency by EphB2 siRNA (Kayser et al. 2006). Clustering of GluA2-containing AMPARs by glutamate receptor–binding protein (GRIP), protein interacting with C kinase (PICK), and EphB2 at postsynaptic sites initiates the reverse signaling to enhance the increase in glutamate release underlying mossy fiber LTP in the CA3 region (Contractor et al. 2002). Also PDZ-mediated interactions of ephrinB3 with GRIP1 are critical for shaft synapse formation (Aoto et al. 2007). Deficiency in ephrinB3 expression affects the number of presynaptic terminals, frequency of mEPSC, and mossy fiber LTP (Aoto et al. 2007; Armstrong et al. 2006), suggesting that reverse signaling by ephrinB3 is involved in regulation of synapse formation and function.

EphB2 directly binds to the extracellular domain of GluN1, and this interaction is implicated in EphB2-mediated synapse formation (Dalva et al. 2000) and synaptic plasticity (Grunwald et al. 2001). Activation of EphB2 kinase activity by ephrinB1/ephrinB2 binding results in activation of Erk1/2, Fyn, and Src tyrosine kinases, phosphorylation of GluN2A/GluN2B, and Ca^{2+} influx via NMDARs (Grunwald et al. 2001; Takasu et al. 2002). Impaired LTP in the dentate gyrus of EphB2-deficient mice is accompanied by reduced synaptic NMDAR-mediated current, whereas total NMDAR- and AMPAR-mediated currents are unchanged (Henderson et al. 2001). In the CA1 region of EphB2 knockout mice, there are less GluN1/EphB2 clusters, impaired LTD, decreased late LTP, and normal early LTP (Grunwald et al. 2001). EphB2, as EphA4, is required for synaptic plasticity in a kinase domain-independent fashion, suggesting that EphB2 may be either required in the dendrite, interacting in *cis* with other postsynaptic proteins (including NMDARs), or in the axon terminal, where it transsynaptically interacts with postsynaptic ephrinBs. Reduced LTP and LTD in CA1 are recorded in ephrinB2- and ephrinB3-deficient mice (Bouzioukh et al. 2007; Grunwald et al. 2004; Rodenas-Ruano et al. 2006).

In summary, transsynaptic ephrin–Eph adhesion is involved in the dynamic formation and retraction of spines during development and in the adult brain by regulating transmitter release, clustering of neurotransmitter receptors in the PSD, regulation of NMDAR's function, and signaling through activation of tyrosine kinases and small GTPases activity.

5.11 Neurexins and Neuroligins

Neuroligins (NLGs) constitute a family of cell adhesion proteins that transsynaptically interact with neurexins (NRXs) in a Ca^{2+}-dependent manner. Four neuroligin genes (NLG1–4) and three neurexin genes (NRX1–3) are widely expressed in mouse brains.

Alternative promoter choice generates two transcripts per NRX gene (NRXα and NRXβ), and alternative splicing at five splicing sites generates >1,000 neurexins isoforms. Also, several isoforms of NLGs are generated by alternative splicing at two sites in the extracellular domain. Ultrastructural studies indicate a postsynaptic localization of NLGs, whereas NRXs are found both pre- and postsynaptically (Tallafuss et al. 2010).

Numerous studies performed in cell culture suggest that the NLG–NRX interaction may control the formation of excitatory and inhibitory synapses (Chih et al. 2005; Dean et al. 2003; Graf et al. 2004). In contrast, in vivo studies in triple NLG1–NLG2–NLG3 knockout mice have shown that elimination of NLGs does not affect synapse numbers in the brain, although it alters the recruitment of postsynaptic receptors to glutamatergic, GABAergic, and glycinergic synapses (Varoqueaux et al. 2006). Overexpressing NLG1 in vivo increases proportion of dendritic spines with mature morphology in comparison to wild-type mice (Dahlhaus et al. 2010). In triple NRXα-deficient mice, a normal excitatory synapse number and morphology, but a decreased number of inhibitory synapses, have been demonstrated (Missler et al. 2003). These findings suggest that NLGs and NRXs are essential for proper synapse maturation by stabilization of transient synaptic contacts. Another important aspect of the NLG–NRX adhesion system is its ability to determine excitatory or inhibitory synapse specification in an activity-dependent manner (Chubykin et al. 2007; Graf et al. 2004). Overexpression of NLG1 and NLG2 results in increased clustering of presynaptic vesicles in glutamatergic and GABAergic synapses in culture (Graf et al. 2004; Prange et al. 2004). Using NLG1- and NLG2-deficient mice, it was shown that NLG1 selectively increases excitatory synaptic responses, whereas NLG2 enhances inhibitory transmission (Chubykin et al. 2007), suggesting that the relative expression levels of NLG1 and NLG2 may be involved in the regulation of the excitation/inhibition ratio. In fact, the small extracellular splice insertions A and B restrict the function of NLG1 and NLG2 to glutamatergic versus GABAergic synapses and alter interaction with presynaptic neurexins. The NLG isoforms associated with GABAergic synapses (containing the splice insert A) bind to NRX1α and a subset of NRX1βs, with a potency to induce GABAergic but not glutamatergic postsynaptic differentiation (Chih et al. 2006).

NLG–NRX signaling is involved in presynaptic organization of excitatory and inhibitory synapses. The PDZ domain of NRXβ interacts with several proteins of the synaptic release machinery, including synapsins, CASK–Mint1, and CASK–caskin1, linking NRXβ to the actin cytoskeleton (Biederer and Sudhof 2001; Dean et al. 2003; Tabuchi et al. 2002). The functional significance of these interactions has been further verified in cultures overexpressing PSD-95, in which an increase in AMPAR-mediated current is explained by an increase in release probability due to the transsynaptic NLG1–NRXβ interaction (Futai et al. 2007). An impaired presynaptic function, measured as a reduction of quantal content and an increase in the failure rate, is responsible for reduction in inhibitory transmission in fast-spiking interneurons of NLG1–NLG2 double knockout or NLG2 single knockout mice (Gibson et al. 2009; Poulopoulos et al. 2009). Abnormalities in presynaptic release are also observed in triple NRXα mice that show severely depressed Ca^{2+}-triggered excitatory and inhibitory neurotransmitter release measured as a reduced frequency of mEPSC and mIPSC. Additionally, an increased failure rate and decreased whole-cell Ca^{2+} currents are seen (Missler et al. 2003). Later studies using both NRXα-deficient and transgenically rescued mice show that the described changes in synaptic properties are due to selective alterations in N- and P-/Q-type VDCC-mediated currents (Zhang et al. 2005), suggesting an involvement of NRXα in the regulation of presynaptic Ca^{2+} influx through these channels.

At the postsynaptic membrane, NLG–NRX signaling is involved in neurotransmitter receptor targeting to the cell surface. In mice with altered NLG1, NLG2, NLG3, and NRXα expression (either single or in combination), NMDAR-mediated synaptic transmission and NMDAR-dependent synaptic plasticity are impaired (Chubykin et al. 2007; Dahlhaus et al. 2010; Jung et al. 2010; Kim et al. 2008; Varoqueaux et al. 2006). Interaction of NLG1 with NRXβ results in increased clustering of PSD-95 and accumulation of NMDA and AMPA receptors within the PSD (Chih et al. 2005; Graf et al. 2004; Heine et al. 2008; Nam and Chen 2005). Interestingly, distinct mechanisms are shown to recruit NMDARs and PSD-95 to dendritic NLG1 clusters. During early synaptogenesis NLG1 clusters are able to associate with GluN1 via PDZ-containing adapter proteins, other than PSD-95, and aggregate on the membrane within a few minutes after formation of axon–dendritic contact (Barrow et al. 2009). Additional NMDARs can be recruited to the established synaptic sites during spine maturation via a slower NLG-dependent recruitment of GluN1 that is associated with PSD-95 and requires a palmitoylation step (Barrow et al. 2009). NRXβ-induced NLG1 clustering results in rapid constitutive accumulation of GluA2 at PSD-95 clusters during early synaptogenesis (Heine et al. 2008). The membrane surface level of NLGs can be modulated by activity also in the adult brain. Chemical stimulation that induces either LTP or LTD in hippocampal neurons leads to a corresponding membrane accumulation or endocytosis of NGL1/3, which occurs in a microtubule- and dynein-dependent manner (Schapitz et al. 2010). These results suggest that activity-dependent NLGs turnover through active cytoskeletal transport is coupled to neurotransmitter receptors delivery or removal. In cultured rat cortical neurons, NLG3 is associated with and is able to activate Epac2, a PKA-independent cAMP target and Rap-GEF. Epac2 activation, likely by D1/D5 G protein–coupled receptors, induces spine shrinkage, increases spine motility, removal of synaptic GluA2/3, and depression of excitatory transmission, whereas its inhibition promotes spine enlargement and stabilization (Woolfrey et al. 2009).

At inhibitory synapses formed by perisomatic interneurons, GABA$_A$Rs clustering is regulated by PDZ-mediated NGL2 binding to inhibitory postsynaptic scaffolding proteins gephyrin and collybistin, affecting GABAergic and glycinergic transmission (Poulopoulos et al. 2009). Overexpression of NRXβs or addition of recombinant NRXβ to cell cultures has been shown to reduce surface GABA$_A$$\alpha$1R expression and to prevent the normal developmental increase in GABAergic transmission without decreasing the synapse number (Zhang et al. 2010). Notably, these effects are not mediated via interactions with NLGs, but rather due to direct extracellular interaction of NRXβs with GABA$_A$α1R (Zhang et al. 2010). Whether this interaction occurs in *cis* or *trans* is still unclear. In mature brains, NRXβs are detected at the presynaptic specialization, although a small fraction are also localized postsynaptically (Berninghausen et al. 2007), resulting in inactivation of NLG1 and the destabilization of synapses (Taniguchi et al. 2007). Interestingly, NRXs redistribute from a dendritic localization to the axon shaft in cultured neurons as a result of signaling from astrocytes (Barker and Ullian 2008), suggesting that interactions of NRXs with their binding partners may depend on the developmental stage and the microenvironment. Additionally, binding of presynaptic NRXβ to postsynaptic neurotransmitter receptors via secreted extracellular proteins can be essential for synapse formation, as it was demonstrated for GluD2 that transsynaptically interacts with NRXβ1 via the cerebellin 1 precursor protein (Uemura et al. 2010).

5.12 NGLs

NGL (netrin-G ligand) proteins form a family of leucine-rich repeat (LRR)–containing CAMs, which have three members NGL-1, NGL-2, and NGL-3. NGL-1 and NGL-2 bind the GPI-anchored netrin-G1 or netrin-G2 in an isoform-specific manner (Kim et al. 2006), while NGL-3 interacts with receptor tyrosine phosphatases family proteins (LAR-RPTPs), including LAR, PTPσ, and PTPδ (Kwon et al. 2010; Woo et al. 2009). NGLs are mainly detected at postsynaptic sites of excitatory synapses, where they interact with PSD-95 (Kim et al. 2006). Because overexpression of NGLs promotes pre- and postsynaptic differentiation and knockdown of NGLs reduces the number and functions of excitatory synapses (Kim et al. 2006; Woo et al. 2009), the role of NGL-mediated adhesion in synaptogenesis has been proposed. Also, ablating LAR expression in hippocampal cultures leads to a decreased number and function of excitatory synapses (Dunah et al. 2005); however, a normal density of PSD-95 clusters was observed in the hippocampus of netrin-G1 and netrin-G2 knockout mice (Nishimura-Akiyoshi et al. 2007).

The precise mechanism how NGLs may be involved in synaptogenesis is still unknown. The aggregation of NGLs on the surface of dendrites recruits postsynaptic proteins to the dendritic surface, including scaffolding proteins PSD-95, GKAP, and Shank, but not gephyrin (Kim et al. 2006; Woo et al. 2009). In addition, both NLG-2 and NLG-3 are able to aggregate GluN2A and GluN1 subunits of NMDARs, respectively, whereas GluA2 is clustered selectively by NGL-3 (Kim et al. 2006; Woo et al. 2009). One should also note that LAR-RPTPs, via their intracellular domain, interact with liprin-α (Pulido et al. 1995), a cytoplasmic adaptor protein that is important for both presynaptic and postsynaptic development and function. At the presynapse, liprin-α is coupled to presynaptic active zone proteins, including RIM, ELKS/ERC, and CASK/LIN (Ko et al. 2003; Olsen et al.

2005; Schoch et al. 2002), similar to β-catenins and NRXβ (Bamji et al. 2003; Tabuchi et al. 2002), suggesting that LAR-RPTPs may converge onto liprin-α to promote synaptic differentiation. At postsynaptic sites, LAR binding to liprin-α is involved in clustering of GRIP1 and GluA2/3 (Dunah et al. 2005). LAR-RPTPs are associated with β-catenin in neurons, and LAR phosphatase activity leads to β-catenin dephosphorylation and reduces its targeting to the dendritic spine (Dunah et al. 2005). LTD induced by activation of muscarinic acetylcholine receptors involves interactions between GluA2, GRIP, and liprin-α that are likely to be important for the subsequent GluA2 dephosphorylation and endocytosis (Dickinson et al. 2009). Although presynaptically localized LAR is involved in synapse formation (Woo et al. 2009), *cis* interactions between LAR and NGL-3 at postsynaptic membrane may be important in mature synapses to support synaptic function and plasticity. However, the functional significance of interactions between NGLs and their binding partners, netrin-Gs and LAR-RPTPs, remains to be investigated in vivo.

5.13 LRRTMs

The LRRTM (leucine-rich repeat transmembrane neuronal) proteins form a family of four LRR containing cell surface receptors that are enriched in the PSD fraction and that are dynamically expressed in the developing and adult nervous systems (Lauren et al. 2003). Although all LRRTMs expressed in cell lines are able to initiate the formation of presynaptic terminals in cocultured hippocampal neurons, LRRTM1 and LRRTM2 exhibit the most potent synaptogenic activity (Linhoff et al. 2009). The LRR (extracellular) domain of LRRTM2 is necessary and sufficient to induce excitatory presynaptic differentiation without any contribution from other factors (de Wit et al. 2009; Linhoff et al. 2009). Knocking down the expression of LRRTM2 results in a reduction (de Wit et al. 2009) and, conversely, overexpression results in an increase in the number of excitatory synapses (Ko et al. 2009). These effects are shown to be solely dependent on the extracellular regions of LRRTM2 and to occur in *trans*, suggesting the presence of a presynaptic extracellular ligand for LRRTM2 that was subsequently able to instruct the formation of a presynaptic terminal. Two independent studies (de Wit et al. 2009; Ko et al. 2009) demonstrate that both NRXα1 and NRXβ1 are receptors for LRRTM2 and that all four LRRTMs can bind NRXβ1. LRRTM binding to NRXs is Ca^{2+} dependent and competitive with those of NLGs (Siddiqui et al. 2010). Overexpressing both NLGs and LRRTMs enhances the recruitment of the presynaptic proteins bassoon and synaptophysin to presynaptic terminals, suggesting that these receptors function in a cooperative manner to promote glutamatergic synapse development (Siddiqui et al. 2010). Similar to NLG-deficient mice, LRRTM1-deficient mice exhibit subtle morphological abnormalities, demonstrating a selective increase in the size of presynaptic terminals in hippocampal CA1 region

(Linhoff et al. 2009). In support of the in vivo function of LRRTMs, knockdown of LRRTM2 results in a reduction of AMPAR- and NMDAR-mediated currents in hippocampal slices (de Wit et al. 2009).

5.14 SALMs

Synaptic adhesion–like molecules (SALM1–SALM5) are transmembrane CAMs containing both LRR and Ig-like domains in their extracellular region and a PDZ-binding motif that interacts with PSD-95 at the C-terminal. In an overexpression system, SALMs 4 and 5 form homomeric *trans* interactions, through their extracellular domains, whereas other SALMs do not (Seabold et al. 2008). Members of the SALM family regulate excitatory and inhibitory synapse formation through distinct mechanisms. SALM1 via its PDZ domain is able to recruit NMDARs (GluN1, GluN2B, and GluN2A) and PSD-95 to the postsynaptic membrane, but not AMPARs. In addition, the SALM1 extracellular domain was shown to bind directly to the extracellular domain of GluN1 within the endoplasmic reticulum (ER) (Wang et al. 2006) and to ER protein reticulon-3 that is involved in protein trafficking (Chang et al. 2010). Although the functional meaning of this interaction remains to be studied, targeting of NMDARs to the postsynaptic membrane by SALM1 is a plausible mechanism. SALM2 is also involved in excitatory synaptogenesis by regulating postsynaptic differentiation and function, possibly via PSD-95, GKAP, GluA1, and to a lesser extent GluN1 clustering at the postsynaptic membrane surface (Ko et al. 2006). Unlike SALM1, SALM2 effects on neurons are more prominent at later developmental stages, suggesting that SALM2 may be involved in the maturation of excitatory synapses through mechanisms including synaptic enrichment of AMPARs. On the other hand, SALM3 and SALM5, but not other members of the SALM family, are capable of inducing both excitatory and inhibitory presynaptic differentiations in contacting axons. SALM3 and SALM5 induce clustering of presynaptic proteins, including the excitatory and inhibitory presynaptic markers VGluT and VGAT, the presynaptic vesicle protein synaptophysin, and the presynaptic active zone protein Piccolo (Mah et al. 2010). Aggregation of SALM3 on the dendritic surface leads to the subsequent clustering of PSD-95, similar to SALM1 and SALM2 (Mah et al. 2010). Knockdown of SALM5 expression with siRNA results in reduced excitatory and inhibitory synapse number and function (Mah et al. 2010).

5.15 Summary and Outlook

In vitro data provided strong evidence for the role of numerous CAMs in synaptogenesis, which, however, not always were confirmed by in vivo studies, in which synaptic changes due to manipulation of single or even multiple CAMs are typically rather subtle and often even undetectable. The most likely reason is redundancy between CAMs within the same class/group and complex interaction between multiple cell adhesion systems. No single pair of CAMs seems to be necessary or sufficient to organize all aspects of synapse development, indicating that CAMs might have overlapping functions and act together at synaptic sites. This concerted action is necessary to support the dynamic nature of synapse formation (different stages involve distinct CAMs, adhesion modulations, redistribution of CAMs, a switch from *trans* to *cis* interaction, etc.).

The diversity of the synaptogenic CAMs and their isoforms also corresponds to the vast heterogeneity of synapses (in terms of presynaptic axons and postsynaptic targets, and hence morphology, molecular composition, and even the type of neurotransmitter released). Further studies are warranted to delineate specific contributions of distinct CAMs and interconnected complexes of CAMs in formation, maintenance, and use-dependent plasticity of specific subtypes of excitatory and inhibitory synapses.

As described in this chapter, numerous studies demonstrate the importance of interactions between CAMs and intracellular scaffold molecules (Fig. 5.1). In addition, accumulating evidences emphasizes the role of ECM scaffolds in transsynaptic signaling at central synapses (Dityatev and Schachner 2003; Dityatev and Rusakov 2011). Induction and maintenance of functional compartments from the extracellular space can be advantageous in comparison to that via intracellular scaffolding proteins because cross-linking of pre- and postsynaptic CAMs by ECM molecules can result in a well-defined stoichiometric relationship between pre- and postsynaptic components. To some degree, this is also achieved by direct transsynaptic interactions between CAMs. However, as ECM scaffolds may have binding sites for multiple CAMs and other membrane-associated molecules, a higher order of coordination between pre- and postsynaptic composition might be gained. Thus, the impact of interplay between (peri)synaptic ECM and CAMs should be further elucidated.

Among other important aspects, which have not received enough attention in the past but could deepen our understanding of mechanisms mediated by synaptic CAMs, are as follows:

- Direct effects of CAMs and associated glycans on the activity of neurotransmitter receptors
- Transsynaptic bidirectional signaling via CAMs as a basis for coordinated changes at the pre- and postsynaptic compartments during synaptogenesis and synaptic plasticity
- Mechanisms of activity-dependent regulation of CAM synaptic expression, such as control of endo- and exocytosis, proteolytic degradation, and lateral diffusion

At the systemic level, more research is necessary to understand the importance of synaptic CAMs for learning and memory and their roles in synaptopathies. In fact, there are exciting examples implicating diverse CAMs in mental retardation, autism, and schizophrenia. Still, studies on the impact of mutations/genetic variability of CAMs in humans are in their infancy, not saying about the development of small molecules that would interfere with CAM-mediated interactions. Such molecules, however, could direct formation or plasticity of synapses and be potentially used for improvement of cognitive functions in human patients suffering from synaptopathies.

Acknowledgments The work in the authors' laboratories is supported by the Italian Institute of Technology, San Paolo Foundation, and the Government of the Russian Federation (AD) and by NICHD funding for intramural research (OB). We thank Dr. Philip Lee for his careful and critical reading of this manuscript. We sincerely apologize to all those colleagues whose work is not cited here because of space considerations.

References

Andreyeva, A., Leshchyns'ka, I., Knepper, M., Betzel, C., Redecke, L., Sytnyk, V., & Schachner, M. (2010). CHL1 is a selective organizer of the presynaptic machinery chaperoning the SNARE complex. *PloS One, 5*, e12018.

Ango, F., di Cristo, G., Higashiyama, H., Bennett, V., Wu, P., & Huang, Z. J. (2004). Ankyrin-based subcellular gradient of neurofascin, an immunoglobulin family protein, directs GABAergic innervation at purkinje axon initial segment. *Cell, 119*, 257–272.

Aoto, J., Ting, P., Maghsoodi, B., Xu, N., Henkemeyer, M., & Chen, L. (2007). Postsynaptic ephrinB3 promotes shaft glutamatergic synapse formation. *The Journal of Neuroscience, 27*, 7508–7519.

Armstrong, J. N., Saganich, M. J., Xu, N. J., Henkemeyer, M., Heinemann, S. F., & Contractor, A. (2006). B-ephrin reverse signaling is required for NMDA-independent long-term potentiation of mossy fibers in the hippocampus. *The Journal of Neuroscience, 26*, 3474–3481.

Bamji, S. X., Shimazu, K., Kimes, N., Huelsken, J., Birchmeier, W., Lu, B., & Reichardt, L. F. (2003). Role of beta-catenin in synaptic vesicle localization and presynaptic assembly. *Neuron, 40*, 719–731.

Barker, A. J., & Ullian, E. M. (2008). New roles for astrocytes in developing synaptic circuits. *Communicative & Integrative Biology, 1*, 207–211.

Barrow, S. L., Constable, J. R., Clark, E., El-Sabeawy, F., McAllister, A. K., & Washbourne, P. (2009). Neuroligin1: A cell adhesion molecule that recruits PSD-95 and NMDA receptors by distinct mechanisms during synaptogenesis. *Neural Development, 4*, 17.

Becker, C. G., Artola, A., Gerardy-Schahn, R., Becker, T., Welzl, H., & Schachner, M. (1996). The polysialic acid modification of the neural cell adhesion molecule is involved in spatial learning and hippocampal long-term potentiation. *Journal of Neuroscience Research, 45*, 143–152.

Benson, D. L., Yoshihara, Y., & Mori, K. (1998). Polarized distribution and cell type-specific localization of telencephalin, an intercellular adhesion molecule. *Journal of Neuroscience Research, 52*, 43–53.

Berninghausen, O., Rahman, M. A., Silva, J. P., Davletov, B., Hopkins, C., & Ushkaryov, Y. A. (2007). Neurexin Ibeta and neuroligin are localized on opposite membranes in mature central synapses. *Journal of Neurochemistry, 103*, 1855–1863.

Biederer, T., & Sudhof, T. C. (2001). CASK and protein 4.1 support F-actin nucleation on neurexins. *Journal of Biological Chemistry, 276*, 47869–47876.

Biederer, T., Sara, Y., Mozhayeva, M., Atasoy, D., Liu, X., Kavalali, E. T., & Sudhof, T. C. (2002). SynCAM, a synaptic adhesion molecule that drives synapse assembly. *Science, 297*, 1525–1531.

Bliss, T., Errington, M., Fransen, E., Godfraind, J. M., Kauer, J. A., Kooy, R. F., Maness, P. F., & Furley, A. J. (2000). Long-term potentiation in mice lacking the neural cell adhesion molecule L1. *Current Biology, 10*, 1607–1610.

Bourgin, C., Murai, K. K., Richter, M., & Pasquale, E. B. (2007). The EphA4 receptor regulates dendritic spine remodeling by affecting beta1-integrin signaling pathways. *The Journal of Cell Biology, 178*, 1295–1307.

Bouzioukh, F., Wilkinson, G. A., Adelmann, G., Frotscher, M., Stein, V., & Klein, R. (2007). Tyrosine phosphorylation sites in ephrinB2 are required for hippocampal long-term potentiation but not long-term depression. *The Journal of Neuroscience, 27*, 11279–11288.

Bozdagi, O., Shan, W., Tanaka, H., Benson, D. L., & Huntley, G. W. (2000). Increasing numbers of synaptic puncta during late-phase LTP: N-cadherin is synthesized, recruited to synaptic sites, and required for potentiation. *Neuron, 28*, 245–259.

Bozdagi, O., Wang, X. B., Nikitczuk, J. S., Anderson, T. R., Bloss, E. B., Radice, G. L., Zhou, Q., Benson, D. L., & Huntley, G. W. (2010). Persistence of coordinated long-term potentiation and dendritic spine enlargement at mature hippocampal CA1 synapses requires N-cadherin. *The Journal of Neuroscience, 30*, 9984–9989.

Bukalo, O., Fentrop, N., Lee, A. Y., Salmen, B., Law, J. W., Wotjak, C. T., Schweizer, M., Dityatev, A., & Schachner, M. (2004). Conditional ablation of the neural cell adhesion molecule reduces precision of spatial learning, long-term potentiation, and depression in the CA1 subfield of mouse.hippocampus. *The Journal of Neuroscience, 24*, 1565–1577.

Chan, S. A., Polo-Parada, L., Landmesser, L. T., & Smith, C. (2005). Adrenal chromaffin cells exhibit impaired granule trafficking in NCAM knockout mice. *Journal of Neurophysiology, 94*, 1037–1047.

Chang, K., Seabold, G. K., Wang, C. Y., & Wenthold, R. J. (2010). Reticulon 3 is an interacting partner of the SALM family of adhesion molecules. *Journal of Neuroscience Research, 88*, 266–274.

Chih, B., Engelman, H., & Scheiffele, P. (2005). Control of excitatory and inhibitory synapse formation by neuroligins. *Science, 307*, 1324–1328.

Chih, B., Gollan, L., & Scheiffele, P. (2006). Alternative splicing controls selective trans-synaptic interactions of the neuroligin-neurexin complex. *Neuron, 51*, 171–178.

Chubykin, A. A., Atasoy, D., Etherton, M. R., Brose, N., Kavalali, E. T., Gibson, J. R., & Sudhof, T. C. (2007). Activity-dependent validation of excitatory versus inhibitory synapses by neuroligin-1 versus neuroligin-2. *Neuron, 54*, 919–931.

Contractor, A., Rogers, C., Maron, C., Henkemeyer, M., Swanson, G. T., & Heinemann, S. F. (2002). Trans-synaptic Eph receptor-ephrin signaling in hippocampal mossy fiber LTP. *Science, 296*, 1864–1869.

Dahlhaus, R., Hines, R. M., Eadie, B. D., Kannangara, T. S., Hines, D. J., Brown, C. E., Christie, B. R., & El-Husseini, A. (2010). Overexpression of the cell adhesion protein neuroligin-1 induces learning deficits and impairs synaptic plasticity by altering the ratio of excitation to inhibition in the hippocampus. *Hippocampus, 20*, 305–322.

Dalva, M. B., Takasu, M. A., Lin, M. Z., Shamah, S. M., Hu, L., Gale, N. W., & Greenberg, M. E. (2000). EphB receptors interact with NMDA receptors and regulate excitatory synapse formation. *Cell, 103*, 945–956.

de Wit, J., Sylwestrak, E., O'Sullivan, M. L., Otto, S., Tiglio, K., Savas, J. N., Yates, J. R., 3rd, Comoletti, D., Taylor, P., & Ghosh, A. (2009). LRRTM2 interacts with neurexin1 and regulates excitatory synapse formation. *Neuron, 64*, 799–806.

Dean, C., Scholl, F. G., Choih, J., DeMaria, S., Berger, J., Isacoff, E., & Scheiffele, P. (2003). Neurexin mediates the assembly of presynaptic terminals. *Nature Neuroscience, 6*, 708–716.

Deininger, K., Eder, M., Kramer, E. R., Zieglgansberger, W., Dodt, H. U., Dornmair, K., Colicelli, J., & Klein, R. (2008). The Rab5 guanylate exchange factor Rin1 regulates endocytosis of the EphA4 receptor in mature excitatory neurons. *Proceedings of the National Academy of Sciences of the United States of America, 105*, 12539–12544.

Dickinson, B. A., Jo, J., Seok, H., Son, G. H., Whitcomb, D. J., Davies, C. H., Sheng, M., Collingridge, G. L., & Cho, K. (2009). A novel mechanism of hippocampal LTD involving muscarinic receptor-triggered interactions between AMPARs. GRIP and liprin-alpha. *Molecular Brain, 2*, 18.

Dityatev, A., & Rusakov, D. A. (2011). Molecular signals of plasticity at the tetrapartite synapse. *Current Opinion in Neurobiology, 21*, 353–359.

Dityatev, A., & Schachner, M. (2003). Extracellular matrix molecules and synaptic plasticity. *Nature Reviews Neuroscience, 4*, 456–468.

Dityatev, A., Dityateva, G., & Schachner, M. (2000). Synaptic strength as a function of post-versus presynaptic expression of the neural cell adhesion molecule NCAM. *Neuron, 26*, 207–217.

Dityatev, A., Dityateva, G., Sytnyk, V., Delling, M., Toni, N., Nikonenko, I., Muller, D., & Schachner, M. (2004). Polysialylated neural cell adhesion molecule promotes remodeling and formation of hippocampal synapses. *The Journal of Neuroscience, 24*, 9372–9382.

Dunah, A. W., Hueske, E., Wyszynski, M., Hoogenraad, C. C., Jaworski, J., Pak, D. T., Simonetta, A., Liu, G., & Sheng, M. (2005). LAR receptor protein tyrosine phosphatases in the development and maintenance of excitatory synapses. *Nature Neuroscience, 8*, 458–467.

Eckhardt, M., Bukalo, O., Chazal, G., Wang, L., Goridis, C., Schachner, M., Gerardy-Schahn, R., Cremer, H., & Dityatev, A. (2000). Mice deficient in the polysialyltransferase ST8SiaIV/PST-1 allow discrimination of the roles of neural cell adhesion molecule protein and polysialic acid in neural development and synaptic plasticity. *The Journal of Neuroscience, 20*, 5234–5244.

Ethell, I. M., Irie, F., Kalo, M. S., Couchman, J. R., Pasquale, E. B., & Yamaguchi, Y. (2001). EphB/syndecan-2 signaling in dendritic spine morphogenesis. *Neuron, 31*, 1001–1013.

Fogel, A. I., Akins, M. R., Krupp, A. J., Stagi, M., Stein, V., & Biederer, T. (2007). SynCAMs organize synapses through heterophilic adhesion. *The Journal of Neuroscience, 27*, 12516–12530.

Fu, W. Y., et al. (2007). Cdk5 regulates EphA4-mediated dendritic spine retraction through an ephexin1-dependent mechanism. *Nature Neuroscience, 10*, 67–76.

Furutani, Y., Matsuno, H., Kawasaki, M., Sasaki, T., Mori, K., & Yoshihara, Y. (2007). Interaction between telencephalin and ERM family proteins mediates dendritic filopodia formation. *The Journal of Neuroscience, 27*, 8866–8876.

Futai, K., Kim, M. J., Hashikawa, T., Scheiffele, P., Sheng, M., & Hayashi, Y. (2007). Retrograde modulation of presynaptic release probability through signaling mediated by PSD-95-neuroligin. *Nature Neuroscience, 10*, 186–195.

Fux, C. M., Krug, M., Dityatev, A., Schuster, T., & Schachner, M. (2003). NCAM180 and glutamate receptor subtypes in potentiated spine synapses: An immunogold electron microscopic study. *Molecular and Cellular Neuroscience, 24*, 939–950.

Galuska, S. P., et al. (2010). Synaptic cell adhesion molecule SynCAM 1 is a target for polysialylation in postnatal mouse brain. *Proceedings of the National Academy of Sciences of the United States of America, 107*, 10250–10255.

Giagtzoglou, N., Ly, C. V., & Bellen, H. J. (2009). Cell adhesion, the backbone of the synapse: "vertebrate" and "invertebrate" perspectives. *Cold Spring Harbor Perspectives in Biology, 1*, a003079.

Gibson, J. R., Huber, K. M., & Sudhof, T. C. (2009). Neuroligin-2 deletion selectively decreases inhibitory synaptic transmission originating from fast-spiking but not from somatostatin-positive interneurons. *The Journal of Neuroscience, 29*, 13883–13897.

Graf, E. R., Zhang, X., Jin, S. X., Linhoff, M. W., & Craig, A. M. (2004). Neurexins induce differentiation of GABA and glutamate postsynaptic specializations via neuroligins. *Cell, 119*, 1013–1026.

Grunwald, I. C., Korte, M., Wolfer, D., Wilkinson, G. A., Unsicker, K., Lipp, H. P., Bonhoeffer, T., & Klein, R. (2001). Kinase-independent requirement of EphB2 receptors in hippocampal synaptic plasticity. *Neuron, 32*, 1027–1040.

Grunwald, I. C., Korte, M., Adelmann, G., Plueck, A., Kullander, K., Adams, R. H., Frotscher, M., Bonhoeffer, T., & Klein, R. (2004). Hippocampal plasticity requires postsynaptic ephrinBs. *Nature Neuroscience, 7*, 33–40.

Guan, H., & Maness, P. F. (2010). Perisomatic GABAergic innervation in prefrontal cortex is regulated by ankyrin interaction with the L1 cell adhesion molecule. *Cerebral Cortex, 20*, 2684–2693.

Hammond, M. S., Sims, C., Parameshwaran, K., Suppiramaniam, V., Schachner, M., & Dityatev, A. (2006). Neural cell adhesion molecule-associated polysialic acid inhibits NR2B-containing N-methyl-D-aspartate receptors and prevents glutamate-induced cell death. *Journal of Biological Chemistry, 281*, 34859–34869.

Hashimoto, T., Yamada, M., Maekawa, S., Nakashima, T., & Miyata, S. (2008). IgLON cell adhesion molecule Kilon is a crucial modulator for synapse number in hippocampal neurons. *Brain Research, 1224*, 1–11.

Hashimoto, T., Maekawa, S., & Miyata, S. (2009). IgLON cell adhesion molecules regulate synaptogenesis in hippocampal neurons. *Cell Biochemistry and Function, 27*, 496–498.

Heine, M., Thoumine, O., Mondin, M., Tessier, B., Giannone, G., & Choquet, D. (2008). Activity-independent and subunit-specific recruitment of functional AMPA receptors at neurexin/neuroligin contacts. *Proceedings of the National Academy of Sciences of the United States of America, 105*, 20947–20952.

Henderson, J. T., Georgiou, J., Jia, Z., Robertson, J., Elowe, S., Roder, J. C., & Pawson, T. (2001). The receptor tyrosine kinase EphB2 regulates NMDA-dependent synaptic function. *Neuron, 32*, 1041–1056.

Henkemeyer, M., Itkis, O. S., Ngo, M., Hickmott, P. W., & Ethell, I. M. (2003). Multiple EphB receptor tyrosine kinases shape dendritic spines in the hippocampus. *The Journal of Cell Biology, 163*, 1313–1326.

Hering, H., Lin, C. C., & Sheng, M. (2003). Lipid rafts in the maintenance of synapses, dendritic spines, and surface AMPA receptor stability. *The Journal of Neuroscience, 23*, 3262–3271.

Honda, T., et al. (2006). Involvement of nectins in the formation of puncta adherentia junctions and the mossy fiber trajectory in the mouse hippocampus. *Molecular and Cellular Neuroscience, 31*, 315–325.

Hortsch, M., Nagaraj, K., & Godenschwege, T. A. (2009). The interaction between L1-type proteins and ankyrins–a master switch for L1-type CAM function. *Cellular and Molecular Biology Letters, 14*, 57–69.

Hoy, J. L., Constable, J. R., Vicini, S., Fu, Z., & Washbourne, P. (2009). SynCAM1 recruits NMDA receptors via protein 4.1B. *Molecular and Cellular Neuroscience, 42*, 466–483.

Irie, F., & Yamaguchi, Y. (2002). EphB receptors regulate dendritic spine development via intersectin, Cdc42 and N-WASP. *Nature Neuroscience, 5*, 1117–1118.

Israely, I., Costa, R. M., Xie, C. W., Silva, A. J., Kosik, K. S., & Liu, X. (2004). Deletion of the neuron-specific protein delta-catenin leads to severe cognitive and synaptic dysfunction. *Current Biology, 14*, 1657–1663.

Jung, S. Y., et al. (2010). Input-specific synaptic plasticity in the amygdala is regulated by neuroligin-1 via postsynaptic NMDA receptors. *Proceedings of the National Academy of Sciences of the United States of America, 107*, 4710–4715.

Jungling, K., Eulenburg, V., Moore, R., Kemler, R., Lessmann, V., & Gottmann, K. (2006). N-cadherin transsynaptically regulates short-term plasticity at glutamatergic synapses in embryonic stem cell-derived neurons. *The Journal of Neuroscience, 26*, 6968–6978.

Kayser, M. S., McClelland, A. C., Hughes, E. G., & Dalva, M. B. (2006). Intracellular and trans-synaptic regulation of glutamatergic synaptogenesis by EphB receptors. *The Journal of Neuroscience, 26*, 12152–12164.

Kayser, M. S., Nolt, M. J., & Dalva, M. B. (2008). EphB receptors couple dendritic filopodia motility to synapse formation. *Neuron, 59*, 56–69.

Kim, S., Burette, A., Chung, H. S., Kwon, S. K., Woo, J., Lee, H. W., Kim, K., Kim, H., Weinberg, R. J., & Kim, E. (2006). NGL family PSD-95-interacting adhesion molecules regulate excit-atory synapse formation. *Nature Neuroscience, 9*, 1294–1301.

Kim, J., et al. (2008). Neuroligin-1 is required for normal expression of LTP and associative fear memory in the amygdala of adult animals. *Proceedings of the National Academy of Sciences of the United States of America, 105*, 9087–9092.

Kleene, R., Cassens, C., Bahring, R., Theis, T., Xiao, M. F., Dityatev, A., Schafer-Nielsen, C., Doring, F., Wischmeyer, E., & Schachner, M. (2010). Functional consequences of the interactions among the neural cell adhesion molecule NCAM, the receptor tyrosine kinase TrkB, and the inwardly rectifying K+ channel KIR3.3. *Journal of Biological Chemistry, 285*, 28968–28979.

Klein, R. (2009). Bidirectional modulation of synaptic functions by Eph/ephrin signaling. *Nature Neuroscience, 12*, 15–20.

Ko, J., Na, M., Kim, S., Lee, J. R., & Kim, E. (2003). Interaction of the ERC family of RIM-binding proteins with the liprin-alpha family of multidomain proteins. *Journal of Biological Chemistry, 278*, 42377–42385.

Ko, J., Kim, S., Chung, H. S., Kim, K., Han, K., Kim, H., Jun, H., Kaang, B. K., & Kim, E. (2006). SALM synaptic cell adhesion-like molecules regulate the differentiation of excitatory synapses. *Neuron, 50*, 233–245.

Ko, J., Fuccillo, M. V., Malenka, R. C., & Sudhof, T. C. (2009). LRRTM2 functions as a neurexin ligand in promoting excitatory synapse formation. *Neuron, 64*, 791–798.

Kochlamazashvili, G., et al. (2010). Neural cell adhesion molecule-associated polysialic acid regulates synaptic plasticity and learning by restraining the signaling through GluN2B-containing NMDA receptors. *The Journal of Neuroscience, 30*, 4171–4183.

Kwon, S. K., Woo, J., Kim, S. Y., Kim, H., & Kim, E. (2010). Trans-synaptic adhesions between netrin-G ligand-3 (NGL-3) and receptor tyrosine phosphatases LAR, protein-tyrosine phosphatase delta (PTPdelta), and PTPsigma via specific domains regulate excitatory synapse formation. *Journal of Biological Chemistry, 285*, 13966–13978.

Lauren, J., Airaksinen, M. S., Saarma, M., & Timmusk, T. (2003). A novel gene family encoding leucine-rich repeat transmembrane proteins differentially expressed in the nervous system. *Genomics, 81*, 411–421.

Law, J. W., Lee, A. Y., Sun, M., Nikonenko, A. G., Chung, S. K., Dityatev, A., Schachner, M., & Morellini, F. (2003). Decreased anxiety, altered place learning, and increased CA1 basal excitatory synaptic transmission in mice with conditional ablation of the neural cell adhesion molecule L1. *The Journal of Neuroscience, 23*, 10419–10432.

Leshchyns'ka, I., Sytnyk, V., Richter, M., Andreyeva, A., Puchkov, D., & Schachner, M. (2006). The adhesion molecule CHL1 regulates uncoating of clathrin-coated synaptic vesicles. *Neuron, 52*, 1011–1025.

Lim, S. T., Lim, K. C., Giuliano, R. E., & Federoff, H. J. (2008). Temporal and spatial localization of nectin-1 and l-afadin during synaptogenesis in hippocampal neurons. *The Journal of Comparative Neurology, 507*, 1228–1244.

Linhoff, M. W., Lauren, J., Cassidy, R. M., Dobie, F. A., Takahashi, H., Nygaard, H. B., Airaksinen, M. S., Strittmatter, S. M., & Craig, A. M. (2009). An unbiased expression screen for synaptogenic proteins identifies the LRRTM protein family as synaptic organizers. *Neuron, 61*, 734–749.

Lodge, A. P., Howard, M. R., McNamee, C. J., & Moss, D. J. (2000). Co-localisation, heterophilic interactions and regulated expression of IgLON family proteins in the chick nervous system. *Molecular Brain Research, 82*, 84–94.

Mah, W., Ko, J., Nam, J., Han, K., Chung, W. S., & Kim, E. (2010). Selected SALM (synaptic adhesion-like molecule) family proteins regulate synapse formation. *The Journal of Neuroscience, 30*, 5559–5568.

Majima, T., Ogita, H., Yamada, T., Amano, H., Togashi, H., Sakisaka, T., Tanaka-Okamoto, M., Ishizaki, H., Miyoshi, J., & Takai, Y. (2009). Involvement of afadin in the formation and remodeling of synapses in the hippocampus. *Biochemical and Biophysical Research Communications, 385*, 539–544.

Marrs, G. S., Theisen, C. S., & Bruses, J. L. (2009). N-cadherin modulates voltage activated calcium influx via RhoA, p120-catenin, and myosin-actin interaction. *Molecular and Cellular Neuroscience, 40*, 390–400.

Matsuno, H., Okabe, S., Mishina, M., Yanagida, T., Mori, K., & Yoshihara, Y. (2006). Telencephalin slows spine maturation. *The Journal of Neuroscience, 26,* 1776–1786.

Mendez, P., De Roo, M., Poglia, L., Klauser, P., & Muller, D. (2010). N-cadherin mediates plasticity-induced long-term spine stabilization. *The Journal of Cell Biology, 189,* 589–600.

Meyer, G., Varoqueaux, F., Neeb, A., Oschlies, M., & Brose, N. (2004). The complexity of PDZ domain-mediated interactions at glutamatergic synapses: A case study on neuroligin. *Neuropharmacology, 47,* 724–733.

Missler, M., Zhang, W., Rohlmann, A., Kattenstroth, G., Hammer, R. E., Gottmann, K., & Sudhof, T. C. (2003). Alpha-neurexins couple Ca^{2+} channels to synaptic vesicle exocytosis. *Nature, 423,* 939–948.

Miyata, S., Matsumoto, N., Taguchi, K., Akagi, A., Iino, T., Funatsu, N., & Maekawa, S. (2003). Biochemical and ultrastructural analyses of IgLON cell adhesion molecules, Kilon and OBCAM in the rat brain. *Neuroscience, 117,* 645–658.

Mizoguchi, A., et al. (2002). Nectin: An adhesion molecule involved in formation of synapses. *The Journal of Cell Biology, 156,* 555–565.

Morellini, F., Lepsveridze, E., Kahler, B., Dityatev, A., & Schachner, M. (2007). Reduced reactivity to novelty, impaired social behavior, and enhanced basal synaptic excitatory activity in perforant path projections to the dentate gyrus in young adult mice deficient in the neural cell adhesion molecule CHL1. *Molecular and Cellular Neuroscience, 34,* 121–136.

Muller, D., Wang, C., Skibo, G., Toni, N., Cremer, H., Calaora, V., Rougon, G., & Kiss, J. Z. (1996). PSA-NCAM is required for activity-induced synaptic plasticity. *Neuron, 17,* 413–422.

Muller, D., Djebbara-Hannas, Z., Jourdain, P., Vutskits, L., Durbec, P., Rougon, G., & Kiss, J. Z. (2000). Brain-derived neurotrophic factor restores long-term potentiation in polysialic acid-neural cell adhesion molecule-deficient hippocampus. *Proceedings of the National Academy of Sciences of the United States of America, 97,* 4315–4320.

Murai, K. K., Misner, D., & Ranscht, B. (2002). Contactin supports synaptic plasticity associated with hippocampal long-term depression but not potentiation. *Current Biology, 12,* 181–190.

Murai, K. K., Nguyen, L. N., Koolpe, M., McLennan, R., Krull, C. E., & Pasquale, E. B. (2003). Targeting the EphA4 receptor in the nervous system with biologically active peptides. *Molecular and Cellular Neuroscience, 24,* 1000–1011.

Murase, S., Mosser, E., & Schuman, E. M. (2002). Depolarization drives beta-Catenin into neuronal spines promoting changes in synaptic structure and function. *Neuron, 35,* 91–105.

Nakamura, K., et al. (2001). Enhancement of hippocampal LTP, reference memory and sensorimotor gating in mutant mice lacking a telencephalon-specific cell adhesion molecule. *European Journal of Neuroscience, 13,* 179–189.

Nam, C. I., & Chen, L. (2005). Postsynaptic assembly induced by neurexin-neuroligin interaction and neurotransmitter. *Proceedings of the National Academy of Sciences of the United States of America, 102,* 6137–6142.

Nikonenko, A. G., Sun, M., Lepsveridze, E., Apostolova, I., Petrova, I., Irintchev, A., Dityatev, A., & Schachner, M. (2006). Enhanced perisomatic inhibition and impaired long-term potentiation in the CA1 region of juvenile CHL1-deficient mice. *European Journal of Neuroscience, 23,* 1839–1852.

Nishimura-Akiyoshi, S., Niimi, K., Nakashiba, T., & Itohara, S. (2007). Axonal netrin-Gs transneuronally determine lamina-specific subdendritic segments. *Proceedings of the National Academy of Sciences of the United States of America, 104,* 14801–14806.

Nuriya, M., & Huganir, R. L. (2006). Regulation of AMPA receptor trafficking by N-cadherin. *Journal of Neurochemistry, 97,* 652–661.

Oka, S., Mori, K., & Watanabe, Y. (1990). Mammalian telencephalic neurons express a segment-specific membrane glycoprotein, telencephalin. *Neuroscience, 35,* 93–103.

Okuda, T., Yu, L. M., Cingolani, L. A., Kemler, R., & Goda, Y. (2007). Beta-Catenin regulates excitatory postsynaptic strength at hippocampal synapses. *Proceedings of the National Academy of Sciences of the United States of America, 104,* 13479–13484.

Olsen, O., et al. (2005). Neurotransmitter release regulated by a MALS-liprin-alpha presynaptic complex. *The Journal of Cell Biology, 170*, 1127–1134.

Penzes, P., Beeser, A., Chernoff, J., Schiller, M. R., Eipper, B. A., Mains, R. E., & Huganir, R. L. (2003). Rapid induction of dendritic spine morphogenesis by trans-synaptic ephrinB-EphB receptor activation of the Rho-GEF kalirin. *Neuron, 37*, 263–274.

Pimenta, A. F., Reinoso, B. S., & Levitt, P. (1996). Expression of the mRNAs encoding the limbic system-associated membrane protein (LAMP): II. Fetal rat brain. *The Journal of Comparative Neurology, 375*, 289–302.

Polo-Parada, L., Bose, C. M., & Landmesser, L. T. (2001). Alterations in transmission, vesicle dynamics, and transmitter release machinery at NCAM-deficient neuromuscular junctions. *Neuron, 32*, 815–828.

Polo-Parada, L., Plattner, F., Bose, C., & Landmesser, L. T. (2005). NCAM 180 acting via a conserved C-terminal domain and MLCK is essential for effective transmission with repetitive stimulation. *Neuron, 46*, 917–931.

Poulopoulos, A., et al. (2009). Neuroligin 2 drives postsynaptic assembly at perisomatic inhibitory synapses through gephyrin and collybistin. *Neuron, 63*, 628–642.

Prange, O., Wong, T. P., Gerrow, K., Wang, Y. T., & El-Husseini, A. (2004). A balance between excitatory and inhibitory synapses is controlled by PSD-95 and neuroligin. *Proceedings of the National Academy of Sciences of the United States of America, 101*, 13915–13920.

Pulido, R., Serra-Pages, C., Tang, M., & Streuli, M. (1995). The LAR/PTP delta/PTP sigma subfamily of transmembrane protein-tyrosine-phosphatases: Multiple human LAR, PTP delta, and PTP sigma isoforms are expressed in a tissue-specific manner and associate with the LAR-interacting protein LIP.1. *Proceedings of the National Academy of Sciences of the United States of America, 92*, 11686–11690.

Rafuse, V. F., Polo-Parada, L., & Landmesser, L. T. (2000). Structural and functional alterations of neuromuscular junctions in NCAM-deficient mice. *The Journal of Neuroscience, 20*, 6529–6539.

Ramser, E. M., Wolters, G., Dityateva, G., Dityatev, A., Schachner, M., & Tilling, T. (2010). The 14-3-3zeta protein binds to the cell adhesion molecule L1, promotes L1 phosphorylation by CKII and influences L1-dependent neurite outgrowth. *PloS One, 5*, e13462.

Richter, M., Murai, K. K., Bourgin, C., Pak, D. T., & Pasquale, E. B. (2007). The EphA4 receptor regulates neuronal morphology through SPAR-mediated inactivation of Rap GTPases. *The Journal of Neuroscience, 27*, 14205–14215.

Robbins, E. M., Krupp, A. J., Perez de Arce, K., Ghosh, A. K., Fogel, A. I., Boucard, A., Sudhof, T. C., Stein, V., & Biederer, T. (2010). SynCAM 1 adhesion dynamically regulates synapse number and impacts plasticity and learning. *Neuron, 68*, 894–906.

Rodenas-Ruano, A., Perez-Pinzon, M. A., Green, E. J., Henkemeyer, M., & Liebl, D. J. (2006). Distinct roles for ephrinB3 in the formation and function of hippocampal synapses. *Developmental Biology, 292*, 34–45.

Rutishauser, U. (2008). Polysialic acid in the plasticity of the developing and adult vertebrate nervous system. *Nature Reviews Neuroscience, 9*, 26–35.

Saghatelyan, A. K., Nikonenko, A. G., Sun, M., Rolf, B., Putthoff, P., Kutsche, M., Bartsch, U., Dityatev, A., & Schachner, M. (2004). Reduced GABAergic transmission and number of hippocampal perisomatic inhibitory synapses in juvenile mice deficient in the neural cell adhesion molecule L1. *Molecular and Cellular Neuroscience, 26*, 191–203.

Saglietti, L., et al. (2007). Extracellular interactions between GluR2 and N-cadherin in spine regulation. *Neuron, 54*, 461–477.

Sakurai, K., Toyoshima, M., Ueda, H., Matsubara, K., Takeda, Y., Karagogeos, D., Shimoda, Y., & Watanabe, K. (2009). Contribution of the neural cell recognition molecule NB-3 to synapse formation between parallel fibers and Purkinje cells in mouse. *Developmental Neurobiology, 69*, 811–824.

Sakurai, K., Toyoshima, M., Takeda, Y., Shimoda, Y., & Watanabe, K. (2010). Synaptic formation in subsets of glutamatergic terminals in the mouse hippocampal formation is affected by a deficiency in the neural cell recognition molecule NB-3. *Neuroscience Letters, 473*, 102–106.

Sara, Y., Biederer, T., Atasoy, D., Chubykin, A., Mozhayeva, M. G., Sudhof, T. C., & Kavalali, E. T. (2005). Selective capability of SynCAM and neuroligin for functional synapse assembly. *The Journal of Neuroscience, 25*, 260–270.

Schapitz, I. U., et al. (2010). Neuroligin 1 is dynamically exchanged at postsynaptic sites. *The Journal of Neuroscience, 30*, 12733–12744.

Schoch, S., Castillo, P. E., Jo, T., Mukherjee, K., Geppert, M., Wang, Y., Schmitz, F., Malenka, R. C., & Sudhof, T. C. (2002). RIM1alpha forms a protein scaffold for regulating neurotransmitter release at the active zone. *Nature, 415*, 321–326.

Schuster, T., Krug, M., Hassan, H., & Schachner, M. (1998). Increase in proportion of hippocampal spine synapses expressing neural cell adhesion molecule NCAM180 following long-term potentiation. *Journal of Neurobiology, 37*, 359–372.

Seabold, G. K., Wang, P. Y., Chang, K., Wang, C. Y., Wang, Y. X., Petralia, R. S., & Wenthold, R. J. (2008). The SALM family of adhesion-like molecules forms heteromeric and homomeric complexes. *Journal of Biological Chemistry, 283*, 8395–8405.

Senkov, O., Sun, M., Weinhold, B., Gerardy-Schahn, R., Schachner, M., & Dityatev, A. (2006). Polysialylated neural cell adhesion molecule is involved in induction of long-term potentiation and memory acquisition and consolidation in a fear-conditioning paradigm. *The Journal of Neuroscience, 26*, 10888–109898.

Shimoda, Y., & Watanabe, K. (2009). Contactins: Emerging key roles in the development and function of the nervous system. *Cell Adhesion & Migration, 3*, 64–70.

Siddiqui, T. J., Pancaroglu, R., Kang, Y., Rooyakkers, A., & Craig, A. M. (2010). LRRTMs and neuroligins bind neurexins with a differential code to cooperate in glutamate synapse development. *The Journal of Neuroscience, 30*, 7495–7506.

Silverman, J. B., Restituito, S., Lu, W., Lee-Edwards, L., Khatri, L., & Ziff, E. B. (2007). Synaptic anchorage of AMPA receptors by cadherins through neural plakophilin-related arm protein AMPA receptor-binding protein complexes. *The Journal of Neuroscience, 27*, 8505–8516.

Stagi, M., Fogel, A. I., & Biederer, T. (2010). SynCAM 1 participates in axo-dendritic contact assembly and shapes neuronal growth cones. *Proceedings of the National Academy of Sciences of the United States of America, 107*, 7568–7573.

Stan, A., Pielarski, K. N., Brigadski, T., Wittenmayer, N., Fedorchenko, O., Gohla, A., Lessmann, V., Dresbach, T., & Gottmann, K. (2010). Essential cooperation of N-cadherin and neuroligin-1 in the transsynaptic control of vesicle accumulation. *Proceedings of the National Academy of Sciences of the United States of America, 107*, 11116–11121.

Stoenica, L., Senkov, O., Gerardy-Schahn, R., Weinhold, B., Schachner, M., & Dityatev, A. (2006). In vivo synaptic plasticity in the dentate gyrus of mice deficient in the neural cell adhesion molecule NCAM or its polysialic acid. *European Journal of Neuroscience, 23*, 2255–2264.

Sytnyk, V., Leshchyns'ka, I., Delling, M., Dityateva, G., Dityatev, A., & Schachner, M. (2002). Neural cell adhesion molecule promotes accumulation of TGN organelles at sites of neuron-to-neuron contacts. *The Journal of Cell Biology, 159*, 649–661.

Sytnyk, V., Leshchyns'ka, I., Nikonenko, A. G., & Schachner, M. (2006). NCAM promotes assembly and activity-dependent remodeling of the postsynaptic signaling complex. *The Journal of Cell Biology, 174*, 1071–1085.

Tabuchi, K., Biederer, T., Butz, S., & Sudhof, T. C. (2002). CASK participates in alternative tripartite complexes in which Mint 1 competes for binding with caskin 1, a novel CASK-binding protein. *The Journal of Neuroscience, 22*, 4264–4273.

Takasu, M. A., Dalva, M. B., Zigmond, R. E., & Greenberg, M. E. (2002). Modulation of NMDA receptor-dependent calcium influx and gene expression through EphB receptors. *Science, 295*, 491–495.

Tallafuss, A., Constable, J. R., & Washbourne, P. (2010). Organization of central synapses by adhesion molecules. *European Journal of Neuroscience, 32*, 198–206.

Tanaka, H., Shan, W., Phillips, G. R., Arndt, K., Bozdagi, O., Shapiro, L., Huntley, G. W., Benson, D. L., & Colman, D. R. (2000). Molecular modification of N-cadherin in response to synaptic activity. *Neuron, 25*, 93–107.

Taniguchi, H., Gollan, L., Scholl, F. G., Mahadomrongkul, V., Dobler, E., Limthong, N., Peck, M., Aoki, C., & Scheiffele, P. (2007). Silencing of neuroligin function by postsynaptic neurexins. *The Journal of Neuroscience, 27*, 2815–2824.

Thomas, L. A., Akins, M. R., & Biederer, T. (2008). Expression and adhesion profiles of SynCAM molecules indicate distinct neuronal functions. *The Journal of Comparative Neurology, 510*, 47–67.

Tian, L., Stefanidakis, M., Ning, L., Van Lint, P., Nyman-Huttunen, H., Libert, C., Itohara, S., Mishina, M., Rauvala, H., & Gahmberg, C. G. (2007). Activation of NMDA receptors promotes dendritic spine development through MMP-mediated ICAM-5 cleavage. *The Journal of Cell Biology, 178*, 687–700.

Tolias, K. F., Bikoff, J. B., Kane, C. G., Tolias, C. S., Hu, L., & Greenberg, M. E. (2007). The Rac1 guanine nucleotide exchange factor Tiam1 mediates EphB receptor-dependent dendritic spine development. *Proceedings of the National Academy of Sciences of the United States of America, 104*, 7265–7270.

Uemura, T., Lee, S. J., Yasumura, M., Takeuchi, T., Yoshida, T., Ra, M., Taguchi, R., Sakimura, K., & Mishina, M. (2010). Trans-synaptic interaction of GluRdelta2 and Neurexin through Cbln1 mediates synapse formation in the cerebellum. *Cell, 141*, 1068–1079.

Vaithianathan, T., Matthias, K., Bahr, B., Schachner, M., Suppiramaniam, V., Dityatev, A., & Steinhauser, C. (2004). Neural cell adhesion molecule-associated polysialic acid potentiates alpha-amino-3-hydroxy-5-methylisoxazole-4-propionic acid receptor currents. *Journal of Biological Chemistry, 279*, 47975–47984.

Varoqueaux, F., Aramuni, G., Rawson, R. L., Mohrmann, R., Missler, M., Gottmann, K., Zhang, W., Sudhof, T. C., & Brose, N. (2006). Neuroligins determine synapse maturation and function. *Neuron, 51*, 741–754.

Wang, C. Y., Chang, K., Petralia, R. S., Wang, Y. X., Seabold, G. K., & Wenthold, R. J. (2006). A novel family of adhesion-like molecules that interacts with the NMDA receptor. *The Journal of Neuroscience, 26*, 2174–2183.

Woo, J., Kwon, S. K., Choi, S., Kim, S., Lee, J. R., Dunah, A. W., Sheng, M., & Kim, E. (2009). Trans-synaptic adhesion between NGL-3 and LAR regulates the formation of excitatory synapses. *Nature Neuroscience, 12*, 428–437.

Woolfrey, K. M., et al. (2009). Epac2 induces synapse remodeling and depression and its disease-associated forms alter spines. *Nature Neuroscience, 12*, 1275–1284.

Xie, Z., Photowala, H., Cahill, M. E., Srivastava, D. P., Woolfrey, K. M., Shum, C. Y., Huganir, R. L., & Penzes, P. (2008). Coordination of synaptic adhesion with dendritic spine remodeling by AF-6 and kalirin-7. *The Journal of Neuroscience, 28*, 6079–6091.

Yamada, A., Irie, K., Deguchi-Tawarada, M., Ohtsuka, T., & Takai, Y. (2003). Nectin-dependent localization of synaptic scaffolding molecule (S-SCAM) at the puncta adherentia junctions formed between the mossy fibre terminals and the dendrites of pyramidal cells in the CA3 area of the mouse hippocampus. *Genes to Cells, 8*, 985–994.

Yamada, M., Hashimoto, T., Hayashi, N., Higuchi, M., Murakami, A., Nakashima, T., Maekawa, S., & Miyata, S. (2007). Synaptic adhesion molecule OBCAM; synaptogenesis and dynamic internalization. *Brain Research, 1165*, 5–14.

Yasuda, S., et al. (2007). Activity-induced protocadherin arcadlin regulates dendritic spine number by triggering N-cadherin endocytosis via TAO2beta and p38 MAP kinases. *Neuron, 56*, 456–471.

Zhang, W., Rohlmann, A., Sargsyan, V., Aramuni, G., Hammer, R. E., Sudhof, T. C., & Missler, M. (2005). Extracellular domains of alpha-neurexins participate in regulating synaptic transmission by selectively affecting N- and P/Q-type Ca^{2+} channels. *The Journal of Neuroscience, 25*, 4330–4342.

Zhang, C., Atasoy, D., Arac, D., Yang, X., Fucillo, M. V., Robison, A. J., Ko, J., Brunger, A. T., & Sudhof, T. C. (2010). Neurexins physically and functionally interact with GABA(A) receptors. *Neuron, 66*, 403–416.

Zhou, L., Martinez, S. J., Haber, M., Jones, E. V., Bouvier, D., Doucet, G., Corera, A. T., Fon, E. A., Zisch, A. H., & Murai, K. K. (2007). EphA4 signaling regulates phospholipase Cgamma1 activation, cofilin membrane association, and dendritic spine morphology. *The Journal of Neuroscience, 27*, 5127–5138.

Part II
Molecular Dynamics of the Synapse

Chapter 6
Molecular Dynamics of the Excitatory Synapse

Shigeo Okabe

Abstract Molecular dynamics of synapses are one of the most important factors that control the remodeling of synaptic connection and efficacy of transmission. This chapter focuses on the dynamics of postsynaptic molecular machinery and describes the imaging technologies important for quantitative analyses of synapses, their application to the postsynaptic molecules, and the insights obtained from these analyses. New visualization techniques, such as super-resolution microscopy, will become an indispensable approach to reveal submicron changes of synaptic molecules. New methods of monitoring protein interactions will also be integrated with experimental paradigms of synaptic plasticity. Cell biological analyses, together with cutting-edge imaging technologies, have been applied to the studies of nascent synapse formation, synapse maintenance, and activity-dependent synapse remodeling. From these studies, a variety of new concepts emerged, such as local assembly of postsynaptic scaffolds, presence of "transport packets" of postsynaptic receptors, heterogeneity of actin movement within spines, and activity-free fluctuation of PSD/spine sizes. These new concepts are useful in understanding specific properties of postsynaptic functions and should be integrated in future to build a realistic model of the postsynaptic organization that can explain its remarkable stability and tunability.

Keywords AMPA-receptors • NMDA-receptors • Postsynaptic density • Scaffolding molecules • Spine synapses

S. Okabe (✉)
Department of Cellular Neurobiology, Graduate School of Medicine, The University of Tokyo, 7-3-1 Hongo, Bunkyo-ku, 113-0033 Tokyo, Japan
e-mail: okabe@m.u-tokyo.ac.jp

M.R. Kreutz and C. Sala (eds.), *Synaptic Plasticity*,
Advances in Experimental Medicine and Biology 970,
DOI 10.1007/978-3-7091-0932-8_6, © Springer-Verlag/Wien 2012

6.1 Introduction

Excitatory synapses, the major sites of communication between neurons in the mammalian CNS, are composed of two distinct components: the presynaptic exocytotic machinery that releases neurotransmitters into the synaptic cleft and the postsynaptic structure specialized for the signal transduction initiated by the binding of neurotransmitters to their membrane receptors. To increase the efficacy of detecting neurotransmitters released to the synaptic cleft and to regulate the signal transduction, specialized postsynaptic structures are differentiated, such as dendritic spines, postsynaptic densities (PSDs), and spine apparatuses. Dynamics of functional molecules in the postsynaptic compartment are one of the most important factors controlling the remodeling of synaptic connection and efficacy of synaptic transmission. In this chapter, I will focus on the dynamics of the postsynaptic molecular machinery and describe the imaging technologies for the detection of molecular dynamics, their application to the postsynaptic molecules, and the insights obtained from these analyses.

6.2 Microstructure of the Excitatory Postsynaptic Cytoplasm

Most excitatory synapses are formed onto dendritic spines in the mammalian forebrain. Dendritic spines contain several unique cytoplasmic structures, including the PSD (Palade and Palay 1954), meshwork of the actin cytoskeleton (Matus 2000), spine apparatus (Spacek 1985), endosomal membranes (Park et al. 2006), and mitochondria (Li et al. 2004) (Fig. 6.1). The PSD is located at the plasma membrane of the dendritic spine and apposed to the presynaptic active zone. The typical PSD has a disk-like structure with a diameter of 200–500 nm and a thickness

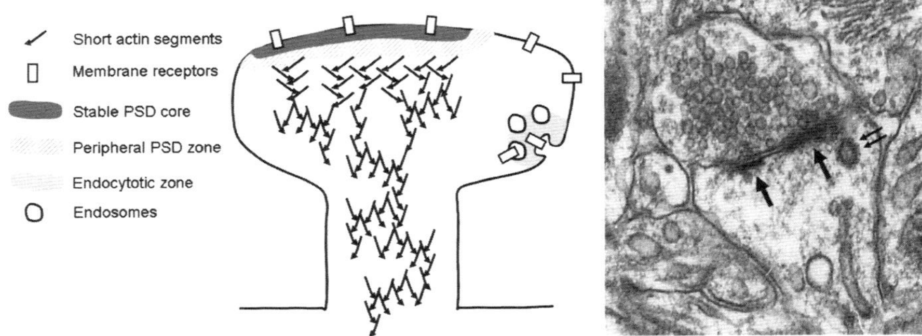

Fig. 6.1 *Microstructure of the excitatory postsynaptic cytoplasm.* The excitatory synapses are mainly formed onto the head of dendritic spines. The spine cytoplasm contains the PSD (*thick arrows* in *right* EM image), the endocytotic machinery (*double thin arrows* in *right* EM image), and F-actin meshwork. Fast-growing ends of F-actin are oriented toward the plasma membrane of the spine. However, recent reports indicate that this polarity orientation is not strict

of 30–60 nm (Harris et al. 1992; Spacek and Harris 1998). The PSD is a highly organized network of membrane proteins and cytoplasmic interacting proteins. Neurotransmitter receptors, such as AMPA-type and NMDA-type glutamate receptors, are accumulated at the PSD. Cell adhesion molecules are also present in the PSD. The PSD also plays a role as a platform for the recruitment of protein kinases and phosphatases. Their interaction with PSD scaffold is often regulated by synaptic activity. Recycling endosomes and endocytic zones are found within spines or at their base (Blanpied et al. 2002; Racz et al. 2004). The spine cytoplasm is enriched with the actin cytoskeleton (Hirokawa 1989; Landis and Reese 1983), and other cytoskeletal filaments are relatively scarce.

Spine morphology is regulated by the combination of local synaptic transmission and global activity of the postsynaptic neuron. Experimental procedures that induce long-lasting change in synaptic efficacy have been shown to affect spine structure. When synaptic efficacy is increased (long-term potentiation, or LTP), the spine is enlarged (Matsuzaki et al. 2004). In turn, when synaptic transmission is depressed (long-term depression, or LTD), the spine reduces its size (Nagerl et al. 2004; Zhou et al. 2004). These experiments illustrate the importance of dynamic behavior of PSD molecules and spine cytoskeletal elements in fine tuning of the synaptic transmission.

6.3 Molecular Composition of the PSD

High-sensitivity mass spectrometry identified more than 400 proteins in highly purified PSD preparations (Husi et al. 2000; Jordan et al. 2004; Yoshimura et al. 2004). The list includes proteins belonging to classes with their major subcellular localization distinct from the PSD, such as those involved in signaling to the nucleus, RNA trafficking, and protein translation. Imaging studies of PSD molecules have been mainly focused on proteins abundant in the PSD, such as glutamate receptors and scaffolding proteins. Both AMPA-type and NMDA-type glutamate receptors are abundant in the purified PSDs (Cheng et al. 2006). Immunoelectron microscopy also confirmed relative enrichment of AMPA and NMDA receptors in the postsynaptic membrane (Nusser 1999; Nusser et al. 1998; Petralia et al. 1994a, b; Tanaka et al. 2005). AMPA and NMDA receptors are essential functional elements of the fast synaptic transmission. Recruitment and turnover of these receptors both in nascent and mature synapses have been studied extensively, and previous imaging studies illustrated activity-dependent regulation of receptor mobility.

PSD scaffolding molecules, such as PSD-95 and Homer, are highly enriched at postsynaptic sites. PSD-95 belongs to the membrane-associated guanylate kinase (MAGUK) protein family (Cho et al. 1992). PSD-95 directly binds to NR2 subunits of NMDA receptors (Kornau et al. 1995) and indirectly with AMPA receptors via interaction with transmembrane AMPA receptor regulatory proteins (TARPs), an auxiliary component of the native AMPA receptor complex (Tomita et al. 2005).

PSD-95 also interacts with a variety of cell adhesion molecules, including neuroligins (Irie et al. 1997; Meyer et al. 2004). Homer proteins interact with group I metabotropic glutamate receptors (mGluRs) (Brakeman et al. 1997) and other synaptic molecules including a scaffolding protein Shank (Tu et al. 1999). Absolute numbers of PSD scaffolding proteins per synapse were rigorously estimated to be in the range of 100–500 (Chen et al. 2005; Sugiyama et al. 2005). They are thought to provide the framework of the PSD through their interactions with both membrane proteins and the cytoplasmic cytoskeletal components (Okabe 2007).

Synapses are specialized sites of cell-to-cell contact and cell adhesion molecules should be important in both formation of synaptic contacts and maintenance of assembled synaptic structures. A variety of cell adhesion molecules have been identified to be localized in the postsynaptic membrane, and many of them have been shown to play important roles in synapse formation and maintenance. Among these, neuroligins are the molecules that function as "synapse organizers" (Scheiffele et al. 2000), and their roles in both synapse formation and regulation of assembled synaptic junctions have been extensively studied (Brose 2009). Because neuroligin binds to the presynaptic receptor neurexins and induce differentiation of the presynaptic structure by itself, the molecular dynamics of neuroligins in the postsynaptic compartment should be clarified. This point will be discussed in the following chapter.

6.4 Cytoskeletal Organization of the Spine Cytoplasm

Abundance of actin in the spine cytoplasm is one of the remarkable features of the excitatory postsynaptic sites. Local actin content is already high in immature filopodia lacking contact with axons, indicating that reorganization of actin cytoskeleton precedes differentiation of the postsynaptic specialization. Possibly, this cell autonomous event alters the dendritic cytoplasm to be able to respond to the presynaptic contact. Organization of actin filaments in dendritic filopodia (Korobova and Svitkina 2010) shows similarity with that in non-neuronal filopodia (Okabe and Hirokawa 1989; Svitkina et al. 2003). In both cases, actin filaments show uniform polarity orientation with their barbed ends (fast-growing ends) enriched at the distal ends of protrusions. After contact with the presynaptic component, dendritic filopodia change their shape into mature spines with enlarged heads (Dailey and Smith 1996; Yasumatsu et al. 2008). This expansion of the heads is a unique feature of neuronal filopodia/spines, and specific regulatory mechanism should be present. Branched actin meshwork has been shown to be present in the mature spine heads (Korobova and Svitkina 2010), together with regulatory proteins involved in branch formation of actin filaments, such as Arp2/3 complex (Hotulainen et al. 2009) and cortactin (Hering and Sheng 2003; Iki et al. 2005).

6.5 Technologies of Monitoring Protein Dynamics in Synapses

Studies on protein dynamics in synapses heavily rely on imaging technologies that can visualize and quantitate mobility of proteins within the small compartment of synapses. The sizes of the spine and PSD are close to the resolution of a high-numerical-aperture objective lens. Therefore, conventional imaging technologies are not useful to extract information on dynamics and redistribution of proteins within the spine and PSDs. Quantitative techniques of fluorescence microscopy (such as fluorescence recovery after photobleaching (FRAP), photoactivation (PA)/photoconversion) and detection of single molecules (single receptor tracking by quantum dots (QDs) and photoactivated localization microscopy (PALM)/stochastic optical reconstruction microscopy (STORM)) are the methods applicable to the study of submicron-scale analysis of postsynaptic structures.

6.5.1 Fluorescence Recovery After Photobleaching (FRAP)

FRAP is a technique of rapidly eliminating fluorescence signals from a defined region of cellular structures by exposure to the intense excitation light and measuring the recovery of fluorescence (Axelrod et al. 1976). This technique has been applied to monitor dynamics of a variety of proteins in different cellular components, including the cytoplasm, plasma membrane, nucleoplasm, cytoskeleton, and specific organelles (Reits and Neefjes 2001). In general, FRAP measurements can provide two parameters of protein dynamics: the mobile fraction of molecules and the rate of mobility (Fig. 6.2a). The mobile fraction can be calculated from three parameters: the fluorescence in the region of interest before bleaching (Fi), immediately after bleaching (Fb), and also after full recovery (Fr). The mobile fraction (R) is calculated as follows:

$$R = (Fr - Fb)/(Fi - Fb)$$

The rate of mobility can be determined by fitting the FRAP recovery curve with theoretical curves. The shapes of theoretical curves are dependent on the models of molecular diffusion/mobility. Usually, unrestricted two-dimensional diffusion or three-dimensional diffusion models are applied to membrane proteins or cytoplasmic protein, respectively.

FRAP measurements of membrane molecules and cytoplasmic proteins within spines are possible, but analyses of data require considerations of the specific geometry of spines (Fig. 6.2b). Fluorescence recovery within the spine head is determined by the rate of molecular diffusion through the spine neck. By using a simple compartment model of the spine, which has volume V and concentration of the fluorophore Cs and is connected to a large dendritic shaft (infinite volume and

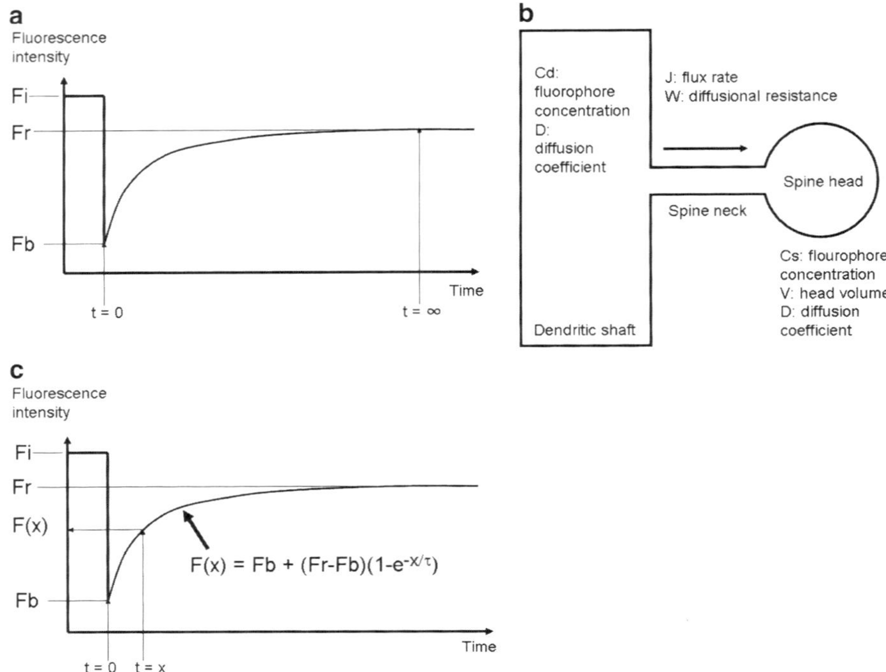

Fig. 6.2 *FRAP analysis of the spine.* (**a**) FRAP recovery curve can determine the amount of mobile fraction (Fr − Fb)/(Fi − Fb) without information of the initial concentration of the fluorophore. (**b**) A simple two compartment model of the molecular mobility between the spine and the dendrite. (**c**) Estimation of time constant τ from the FRAP recovery curve by fitting with a single exponential

fluorophore concentration of Cd) with a thin neck (diffusional resistance of W), it is possible to relate the concentration of the fluorophore in the spine (Cs) with the rate of fluorophore flux through the neck (J) as follows:

$$J = D(Cs - Cd)/W$$

where D is the diffusion coefficient of the fluorophore in the cytoplasm (Svoboda et al. 1996). This equation is similar to the one that describes the simple one-dimensional diffusion model. Therefore, fluorescence recovery of freely diffusible molecules within spines is expected to follow a single exponential with a time constant τ (Fig. 6.2c), which can be related to diffusion coefficient as follows:

$$\tau = WV/D$$

Because D can be assumed to be similar throughout the cytoplasm, estimation of D from measurements within the dendritic shaft or cell body is possible (Kuriu et al. 2006; Majewska et al. 2000). Estimated τ from spine FRAP, combined with measurement of D in the dendritic shaft and estimation of spine volume from the fluorescence image, can be utilized to calculate W, which directly reflects the structural resistance of the spine neck.

Selection of fluorophores for FRAP measurement is important. First, fluorophores should be resistant to the excitation light during the measurement of fluorescence recovery, but should be sensitive enough to allow rapid photobleaching. Second, fluorophores should not generate reactive oxygen species (ROS) by photobleaching. This is occasionally problematic with proteins labeled by conventional chemical fluorophores, such as fluorescein (Okabe and Hirokawa 1993). Indeed, chromophore-assisted light inactivation (CALI)/fluorophore-assisted light inactivation (FALI) are the techniques based on light-induced generation of ROS to selectively inactivate functional proteins labeled by fluorophores (Hoffman-Kim et al. 2007). Fortunately, GFP and related fluorescent proteins are not efficient in ROS production (Rajfur et al. 2002), and the side effects of FRAP is usually negligible if the irradiation dose is set to be minimal for photobleaching. Third, fluorophores should not show reversible photobleaching, which is the recovery of fluorescence by photochemical processes independent of fluorophore mobility. Reversible photobleaching occurs if fluorophores transiently stay in a triplet state. Wild-type GFP shows a complex photochromism, but this feature is largely suppressed in engineered GFP variants such as eGFP. However, eGFP together with eCFP and eYFP still shows reversible recovery to some extent with time constants of 30–60 s (Sinnecker et al. 2005). This property should be taken into account in the interpretation of FRAP data, especially when small differences in the recovery curve should be evaluated.

6.5.2 Photoactivation (PA)/Photoconversion

Local photoactivation (PA) of fluorescence and subsequent monitoring of fluorescence intensity and distribution is a technique complementary to FRAP. Originally, this technique was applied to monitor microtubule dynamics in mitotic cells and in growing axons by using caged fluorescein-labeled tubulin (Mitchison 1989; Okabe and Hirokawa 1992). Development of photoactivatable GFP variants (Patterson and Lippincott-Schwartz 2002) accelerated application of this technique to a variety of biological samples, including fully differentiated neurons with synaptic connections. An obvious advantage of PA strategy in the study of protein dynamics is its low background. Detection of the stable bleached molecules in the background of fluorescence from newly incorporated molecules is very difficult in FRAP measurements. In contrast, photoactivated molecules stably incorporated into synaptic structures can be easily detected and quantitated after local PA. Another advantage of the PA technique is the relatively low dose of irradiation required for PA of either caged fluorescein or PA-GFP variants compared with the irradiation dose required for photobleaching. Further reduction of phototoxicity can be achieved by using two-photon excitation with longer wavelength laser pulses

(720–840 nm) in comparison with single-photon excitation (usually with a 405 nm diode laser) (Schneider et al. 2005).

Imaging of photoactivatable GFP/RFP variants needs identification of the nonfluorescent target structure before activation. This is usually achieved by co-expression of other fluorescent markers. Discovery of photoconvertible fluorescent proteins, such as Kaede (Ando et al. 2002), enabled researchers to identify the target structure by fluorescence from the photoconvertible molecules of their original color and subsequently mark tagged molecules by changing their color with exposure to short-wavelength light. As in the case of photoactivatable fluorescent proteins, the irradiation dose required for photoconversion is much less than that for photobleaching.

6.5.3 Tracking of Single Membrane Molecules by Labeling with Semiconductor Quantum Dots (QDs) or Fluorescent Dyes

FRAP, PA, and photoconversion technologies are useful in measuring the ensemble behavior of fluorescent molecules in synapses. However, the heterogeneous behavior of individual molecules and discrimination of mobility within substructures of synapses cannot be obtained by these techniques. Tracking of single membrane molecules by labeling with QDs or fluorescent dyes is a powerful method to extract the dynamic properties of single molecules on the spine surface (Fujiwara et al. 2002; Groc et al. 2004). This technique has been applied to the analysis of both excitatory and inhibitory synapses by scientists in Paris and Bordeaux (Triller and Choquet 2008), and detailed discussions of this topic are present in other chapters of this book. Essential points are as follows: (1) Cell surface receptors, such as glutamate receptors, GABA receptors, and glycine receptors, can be labeled by antibodies coupled to QDs or fluorescent dyes. (2) Signals from single QDs or fluorescent dyes can be identified on the cell surface. (3) Movement of single QDs and fluorescent dyes can be tracked for a time period sufficient to calculate their kinetic parameters, such as instantaneous diffusion coefficient, the degree of confinement, and residence time within specific compartment, such as the PSD area (Triller and Choquet 2008). These parameters can be obtained only from the detection of single membrane receptors on the spine surface, illustrating the uniqueness of this approach. The only pitfall of the technique is the possible restriction of labeled membrane receptors by other membrane proteins and extracellular components present within the synaptic cleft. This possibility was evaluated by Groc et al. (2007), and the results indicated that only very fast diffusing molecules may be slowed down by attaching bulky probes such as QDs coated with antibodies, but the average of diffusion coefficients for all particles within synapses were not different. This result indicates that average mobility of receptors within synapses can be reliably monitored by the QD-based visualization technique.

6.5.4 Photoactivated Localization Microscopy (PALM)/Stochastic Optical Reconstruction Microscopy (STORM)

Single molecule tracking is a powerful technique to measure dynamics of proteins in small domains of cells. Antibody labeling methods applied to cell surface receptors is difficult to extend to cytoplasmic components, such as PSD scaffolds and actin filaments. The recent development of the PALM/STORM technique and its application to living cells using photoactivatable fluorescent proteins opened the door into live imaging of single cytoplasmic proteins in the subcompartments of spines (Betzig et al. 2006; Rust et al. 2006). In PALM/STORM imaging, only a small number of photoactivatable fluorescent molecules are activated by weak irradiation with a UV-violet laser, and their images were recorded by illumination with a second laser with appropriate excitation wavelengths. The positions of single activated fluorescent molecules are estimated by using nonlinear least squares fitting or maximum-likelihood fitting with assumed Gaussian point-spread functions. After repeating cycles of activation recording of single molecules, the data of molecular positions are utilized to reconstruct high-resolution images or precise trajectories of individual molecules. There has been a rapid technical development associated with this reconstruction method. PALM/STORM imaging of multiple fluorophores (Bates et al. 2007; Shroff et al. 2007) and applications of PALM/STORM to three-dimensional imaging (Huang et al. 2008) have been reported.

6.6 Molecular Dynamics in Nascent Synapses

Precise roles of filopodia protruding from immature dendritic shafts are not yet clarified (Portera-Cailliau et al. 2003). Dendritic filopodia are highly motile and transient. Only a small proportion of filopodia may be stabilized and start to form synaptic contacts (Okabe et al. 2001a). This means that the main role of filopodia in immature dendrites may not be to generate spines, but to search the surrounding tissue for appropriate targets. In this sense, dendritic filopodia may be classified into two categories: One is transient protrusions for probing the environment, and the other is spine precursors. A recent report suggested that actin filaments within dendritic filopodia are not bundled but form meshwork of branched and linear actin filaments (Korobova and Svitkina 2010). This organization is different from bundled straight F-actin in conventional filopodia of non-neuronal cells (Okabe and Hirokawa 1989; Svitkina et al. 2003), suggesting importance of distinct actin organization in different types of filopodia. FRAP analysis of dendritic filopodia revealed unique incorporation pattern of GFP-actin (Hotulainen et al. 2009). In non-neuronal filopodia, incorporation of G-actin is restricted to their tips. However, dendritic filopodia show addition of new actin also at their roots. This additional site of actin incorporation may be responsible for more flexible structural change of dendritic filopodia, which may be associated with their activity of searching nearby axons.

A subset of motile filopodia interacts with axons and starts to form stable synaptic contacts. Target recognition is one of the most important events in synapse formation, but information about recruitment of membrane proteins and submembranous scaffolds to the contact sites is not sufficient to build a model of molecular assembly during target recognition. Initial recognition of synaptic partners should be mediated by trans-interaction of cell adhesion molecules. N-cadherin is a prominent component of developing synapses and may play a critical role in the initial contact (Togashi et al. 2002). In vivo trafficking of N-cadherin in growing axons of zebra fish embryos revealed the presence of highly mobile transport packets and rapid formation (<1 h) of stable N-cadherin clusters in the wake of growth cones (Jontes et al. 2004). Regulated delivery of cadherin-containing vesicles should also be important in postsynaptic functions of cadherins.

Neuroligins are heterophilic postsynaptic cell adhesion molecules interacting with a presynaptic partner neurexins (Brose 2009). Neurexin expressed in non-neuronal cells can induce postsynaptic differentiation in dendrites within a few days (Graf et al. 2004). These artificial postsynaptic sites contain PSD scaffolding protein PSD-95, GKAP, and synGAP, together with NMDA receptor subunits, but not AMPA receptor subunits (Graf et al. 2004; Nam and Chen 2005). The ability of postsynaptic neuroligins to accumulate multiple PSD proteins after interaction with presynaptic neurexins indicates possible roles of neuroligins to recognize presynaptic partners and initiate postsynaptic molecular cascades of synaptogenesis. Time-lapse imaging of GFP-/CFP-tagged neuroligin 1 in cultured cortical neurons at 4–5 days in vitro revealed appearance of neuroligin 1 clusters at sites of axodendritic contact within 10 min (Barrow et al. 2009). At the same developmental stage, mobile neuroligin 1 clusters were also present both in dendritic shafts and filopodia. Although time course of neuroligin clustering in cultured cortical neurons is faster than that of cadherin in zebra fish embryos, neuroligin clustering may need cadherin functions, as N-cadherin knockout neurons show marked reduction of neuroligin 1 clusters in dendrites (Stan et al. 2010).

Induction of artificial postsynaptic specialization by contact with neurexin-expressing cells revealed accumulation of both PSD-95 and NMDA receptors at the sites of neuroligin clusters, but the route of this recruitment is not clear, and imaging approaches are necessary to resolve this issue. Imaging of immature hippocampal neurons in culture revealed the presence of two distinct populations of postsynaptic structures (Gerrow et al. 2006). One is a mobile nonsynaptic complex of multiple scaffolding proteins, including PSD-95, GKAP, and Shank, but without neuroligin 1. The other is also positive with PSD-95, GKAP, and Shank, but stationary and contains neuroligin 1. The presence of these two populations of clusters may indicate marking of synaptic contact sites by neuroligin clustering and subsequent recruitment of mobile packets to predetermined sites. However, several other reports postulate an alternative mechanism for the recruitment of postsynaptic scaffolds to the synaptic contact sites. First, artificially induced patches of neuroligin 1 on the surface of immature neurons accumulate PSD-95 gradually, suggesting palmitoylation-dependent recruitment of individual molecules (Barrow et al. 2009). Second, FRAP analysis of PSD-95, GKAP, Shank, and Homer molecules indicates exchange between assembled and diffusible pools (Gray et al. 2006; Kuriu et al. 2006; Okabe et al. 2001b). Third, time-lapse imaging of individual PSDs by PSD-95 or Shank tagged with GFP showed gradual increase of total fluorescence signals in single PSDs, further supporting addition of new molecules from a diffusible cytoplasmic pool (Bresler et al. 2001, 2004). These data collectively indicate local assembly of PSD scaffolds is a predominant recruitment pathway. On the other hand, recruitment of NMDA receptors to nascent synapses should depend on transport vesicles, and this process was successfully visualized in immature neurons in culture (Washbourne et al. 2002). The relative importance and the amount of scaffolding proteins provided from two distinct sources (preassembled packets and diffusible pool) may be dependent on several factors, such as maturity of neurons, location along dendrites, and stage of synapse development. Comparison of the same culture system at different stages of maturation should be performed more extensively to clarify the routes of transport.

Both filopodia protrusion and their contact with presynaptic membranes via cell adhesion molecules are important events in synapse formation, but molecular mechanisms linking these two events have not yet been clarified. Filopodia may recruit clusters of cell adhesion molecules to increase their ability to form stable contacts with axons, or clustering of cell adhesion molecules may initiate filopodia formation. Cell adhesion and filopodia formation may be coordinated by cell adhesion molecules that can regulate the actin cytoskeleton. Telencephalin, a cell adhesion molecule expressed in the forebrain neurons (Matsuno et al. 2006; Yoshihara et al. 1994), is unique in its activity to interact with ERM (exrin/radixin/moesin) family proteins with its cytoplasmic domain (Furutani et al. 2007). ERM proteins are adaptor molecules between membrane proteins and F-actin, and enriched in membrane-protruding structures such as microvilli. It is not yet clear if telencephalin recruitment to the contact sites initiates interaction between the plasma membrane and F-actin locally. Alternatively, synapse organizers, such as neuroligin, may be recruited to telencephalin-enriched filopodia by cis-interaction. These possibilities should be tested by real-time imaging of cell adhesion molecules at nascent synapses in future.

6.7 Molecular Dynamics in Spine Synapse Maintenance

Stability of synaptic junctions is fundamentally important in the maintenance of neuron network connectivity and preservation of memory. Recent in vivo imaging studies of dendritic spines in the neocortex revealed remarkable stability of individual spines, which can be preserved throughout the entire life of an animal (Zuo et al. 2005) (Fig. 6.3a). In contrast to the stability of spines, turnover of molecules building postsynaptic structure has been shown to be very rapid. In culture preparations, FRAP measurements revealed that glutamate receptors, PSD scaffolding proteins, and cytoskeletal proteins exchange with retention times of minutes to several hours (Kuriu et al. 2006; Sharma et al. 2006; Star et al. 2002). Rapid

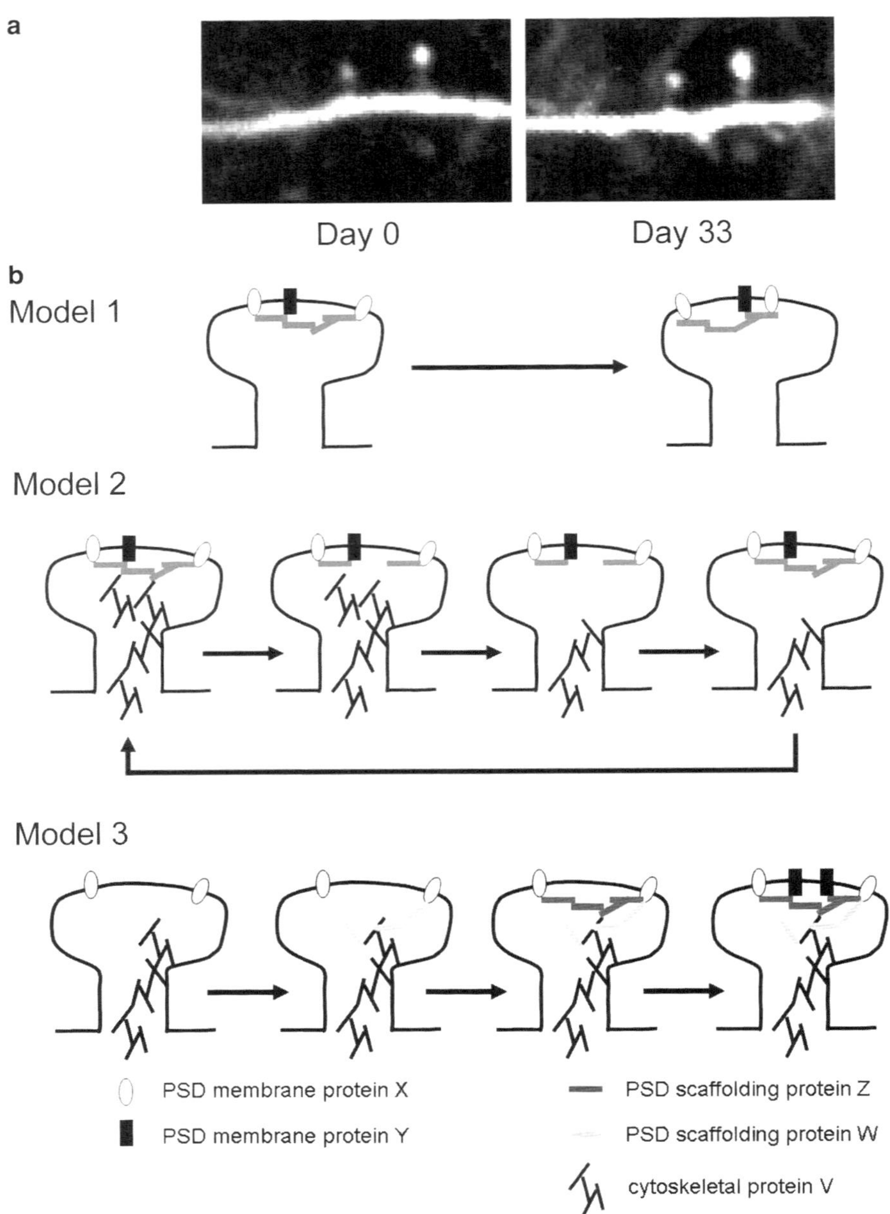

Fig. 6.3 *Models of PSD/spine stability.* (**a**) In vivo two-photon microscopy of dendritic spines from a transgenic mouse expressing GFP. These two spines of the pyramidal neuron in the somatosensory cortex are stable at least for 33 days. (**b**) Three models of PSD/spine stability. Model 1: Hypothetical PSD membrane protein Y, which shows prolonged residence in the PSD, supports the PSD stability. Model 2: Feedback regulation between PSD scaffolding protein Z and the cytoskeletal protein V supports the PSD stability. If the content of protein Z spontaneously

turnover of PSD-95 was confirmed by in vivo two-photon microscopy of PA-GFP-labeled PSD-95 (Gray et al. 2006). Even with these rapid exchange rates, structure of PSDs can be maintained for a prolonged period. The molecular mechanisms underlying this stability are currently unknown. A simple model is to hypothesize the existence of yet unidentified "core" proteins that are persistently present in the PSD structure (Model 1 in Fig. 6.3b). Indeed, proteins forming nuclear pore complexes can disassemble during mitosis and reassemble into the newly forming nuclei, but do not turn over in differentiated cells (D'Angelo et al. 2009). Existence of similar "core" proteins in the stable PSDs can explain the integrity of the structure for years. Second possibility is the existence of a feedback mechanism that maintains the molecular content of PSDs (Model 2 in Fig. 6.3b). If rapid actin turnover in spines facilitate elimination of PSD scaffolding molecules, but decrease of PSD scaffolds secondarily induces reduction of molecules involved in actin nucleation, this will create a feedback loop to maintain the constant amount of scaffolding molecules and F-actin. This model can explain short-term regulation of synaptic molecular content, but the setting point of the feedback loop should be sensitive to other factors, such as the spine volume. The third possible explanation of PSD stability is to hypothesize that there exist multiple steps in the growth process of PSDs (Model 3 in Fig. 6.3b). Even if transition between adjacent steps is relatively rapid in the order of hours, transition through tens of the steps will require a prolonged period, possibly in the scale of months. Previous FRAP analyses of multiple PSD scaffolding proteins revealed their differential dynamics (Kuriu et al. 2006), suggesting differential contribution of individual molecules for the PSD stability. Consistent with this observation, immunoelectron microscopic analysis indicated that individual PSD molecules have distinct patterns of tangential distribution along the PSD (Valtschanoff and Weinberg 2001). If the amounts of multiple PSD molecules encode the current step of PSD stabilization and increase of stable scaffolds triggers the transition to the next step, this mechanism may function to confer prolonged stability to the PSDs. Indeed, gradual increase of scaffolding protein content per individual synapses along neuronal differentiation was reported (Sugiyama et al. 2005), suggesting correlation between PSD stability and the content of scaffolding proteins. This model is theoretically possible, but may not explain rapid acquisition of new information by changing the state of PSDs. Quantitative imaging of PSD molecular contents will be essential to test possible models of the PSD stability.

Live imaging of spines revealed continual change of morphology while they are preserving contact with presynaptic components. This structural change is on the

---

Fig. 6.3 (continued) fluctuates downward, it will induce suppression of the cytoskeletal system regulated by protein V. This suppression of the cytoskeletal system will subsequently result in enhanced incorporation of protein Z back to the initial level. Model 3: Multiple steps from the nascent to mature spines prevent sudden elimination of the PSD. If there exist tens of intermediate steps from nascent synapses to the fully stabilized synapses and transition between the adjacent steps is regulated by addition/elimination of specific PSD molecules, elimination of PSD in fully stabilized spines should be difficult and require prolonged transition time

order of minutes and based on local actin polymerization (Fischer et al. 1998). Actin polymerization is also coupled to structural changes of the PSDs (Blanpied et al. 2008), and rapid elimination of spine actin filaments triggers partial disassembly of the PSD scaffolds (Kuriu et al. 2006), suggesting active roles of actin reorganization in the control of PSD function and molecular content. In turn, PSDs contain molecules that can interact and regulate actin cytoskeleton. For example, cortactin binds to both actin filaments and Arp2/3 complex and is enriched in the PSD via its interaction with Shank molecules (Hering and Sheng 2003; Iki et al. 2005). Multiple interactions between PSDs and actin network should exist, and physiological roles of individual pathways in synaptic function should be clarified. The presence of retrograde actin flow in the spine head was successfully visualized by using two-photon activation of PA-GFP-labeled actin (Honkura et al. 2008). Recent application of high-resolution microscopic techniques (PALM/STORM) to the behavior of single actin molecules in spines confirmed the presence of actin flow (Frost et al. 2010; Tatavarty et al. 2009). In addition, these PALM/STORM studies revealed more complex behavior of actin filaments in the spine cytoplasm. They reported that velocity of individual actin molecules incorporated into the network is heterogeneous even within a single spine. In the microdomain close to the PSD, actin velocity is specifically enhanced. In contrast, the endocytic zone showed no enhancement of actin velocity, indicating clear functional distinction of actin network between spine microdomains (Fig. 6.1). Bidirectional translocation of single actin molecules also exists within spines, indicating the presence of actin filaments with their pointed ends toward the spine head. These reports illustrate heterogeneity in actin movement and relatively short distance of their net flow, suggesting that overall organization of actin filaments in spines is based on short filaments with less aligned orientation (Fig. 6.1). This model perfectly matches with the snapshots of spine actin network obtained by using platinum replica electron microscopy (Korobova and Svitkina 2010).

Maintenance of glutamate receptors on the postsynaptic membrane is achieved by surface delivery and retrieval of receptors and also by local two-dimensional mobility of receptors and subsequent confinement within the postsynaptic membrane. To monitor exocytosis of AMPA receptor-containing vesicles, tagging of AMPA receptors with a pH-sensitive form of GFP (SEP) has been widely utilized in combination with two-photon microscopy of neurons in slice culture (Kopec et al. 2006). In addition to the detection of total surface AMPA receptors, elimination of SEP signals from preexisting surface AMPA receptor by photobleaching is possible, allowing direct monitoring of AMPA receptors newly incorporated into the plasma membrane. By tracking single AMPA receptors labeled with QDs, the trajectories of single AMPA receptors exposed to the dendritic surface can be traced and analyzed quantitatively (Triller and Choquet 2008). General agreement is that, in a resting state, AMPA receptors are continuously cycling between cell surface and the intracellular pool, and this process is required for the supply of sufficient surface AMPA receptors that can be trapped at the synaptic sites (Petrini et al. 2009).

6.8 Molecular Dynamics in Response to Activity

Synaptic transmission, postsynaptic depolarization, and elevation of intracellular calcium can induce multiple changes in molecular composition, structure, and function of dendritic spines. Among different forms of synapse plasticity, LTP has provided insight into the mechanisms of memory formation triggered by sensory stimuli in animals. The cellular and molecular mechanisms of LTP have been extensively studied, but still not yet fully understood. Because excellent reviews on the topic of activity-dependent synaptic plasticity and molecular dynamics already exist (Newpher and Ehlers 2008; Shepherd and Huganir 2007), I will introduce several new concepts related to this topic and discuss future directions.

6.8.1 Identification of New Rules of Synapse Remodeling

Models of synaptic plasticity are relatively limited. LTP and homeostatic plasticity are two major paradigms widely used to trigger changes in synaptic molecules. However, it is unclear how many different forms of synaptic plasticity exist. In order to identify new rules of synapse remodeling, long-term imaging of single spines/PSDs should be a very powerful approach. Yasumatsu et al. reported intrinsic fluctuations of spine volume in the absence of synaptic activity and proposed this instability is essential in ensemble behavior of the spine population and also establishment of a small number of large stable spines (Yasumatsu et al. 2008). This report is unique in identification of activity-independent mechanisms that support variability of the spine shape and spontaneous appearance of large spines (activity-free fluctuation). Another experiment performed by Minerbi et al. illustrates the long-term effects of activity blockade on the distribution of PSD sizes by using real-time imaging of PSD-95-GFP in cultured neurons (Minerbi et al. 2009). Their first conclusion is similar to the activity-free fluctuation proposed by Yasumatsu et al.: Neurons have an activity-independent intrinsic mechanism to increase the heterogeneity of PSD size. An interesting point is that, in addition to the first rule, they proposed a second rule: Activity induces PSD sizes to be more uniform, by suppressing further growth of large PSDs and enhancing growth of small PSDs. This second rule ("democratic" rule) is distinct from both LTP/LTD and homeostatic plasticity. LTP/LTD should regulate individual synapses, but the "democratic" rule applies to the population of synapses. Homeostatic plasticity predicts increase of the average of synaptic strength after activity elimination, but the "democratic" rule does not affect the population average. An important point is whether these new rules may operate in a physiological context or not. To this end, long-term imaging of spines or PSDs should be performed in combination with a strategy to suppress neuronal activity in vivo.

6.8.2   Plasticity-Dependent Remodeling of Cell Adhesion Molecules

Recent researches in activity-dependent redistribution of AMPA receptors revealed importance of recycling endosomes within spines (Petrini et al. 2009; Wang et al. 2008). These analyses suggest that recycling of membrane proteins in general may be enhanced in spines after synaptic activity. Cell adhesion molecules are important in the establishment of initial target recognition at the beginning of synapse formation, but may also be involved in activity-dependent changes of mature spines. Indeed, Schapitz et al. reported that a treatment triggering LTD in cultured neurons also accelerates internalization of neuroligin 1 at synaptic sites (Schapitz et al. 2010). Reduction of neuroligin 1 in the postsynaptic sites may affect synaptic contacts and impair AMPA receptor-mediated synaptic transmission via inhibition of PSD-95. A new technology of visualizing interaction between two protein molecules by site-specific biotin ligation was developed and applied to detect interactions between neurexin and neuroligin in hippocampal neuron culture (Thyagarajan and Ting 2010). Transsynaptic interaction between neurexin and neuroligin was enhanced after acute increase of synaptic transmission, suggesting importance of cell adhesion itself in the expression of synaptic plasticity. Importance of N-cadherin in spines was also confirmed by using single synapse imaging in slice culture preparations (Mendez et al. 2010). In control slices, there was a large difference in spine lifetime when they were classified by the presence of N-cadherin clusters. Application of LTP-inducing stimuli promoted formation of N-cadherin-GFP positive spines, which were also more stable than spines without N-cadherins. A variety of "synapse organizers" have been identified, and all of these molecules can potentially be involved in molecular mechanisms of LTP. It is important to know how targeting of cell adhesion molecules to the synaptic junctions is regulated in both a resting state and after induction of synaptic plasticity. Unexpected interactions between cell adhesion molecules and ionotropic glutamate receptors have been identified, and their roles in synaptic plasticity should also be evaluated (Saglietti et al. 2007).

6.8.3   Remodeling and Actin-Microtubule Interaction

It is well established that actin reorganization is associated with synaptic plasticity, and stabilization of actin network underlies structural enlargement of spines. Although actin filaments are persistently present within spines and interact with molecules present within PSDs, the spine cytoplasm is usually devoid of microtubules and they are only infrequently enter into spines. A recent report illustrated the importance of microtubule entry in spine enlargement, via the interaction of microtubule tip-tracking protein EB3, postsynaptic protein p140Cap, and actin-binding protein cortactin (Jaworski et al. 2009). This molecular link indicates importance of coordination between microtubules and F-actin in regulated spine enlargement. Several other cross-linking systems between microtubules and F-actin have been reported, and their roles in regulation of spine morphology should be tested (Bielas et al. 2007).

6.9 Conclusion

In this chapter, I described the technologies important for the analyses of molecular dynamics, their application to the postsynaptic molecules, and the insights obtained from these analyses. New techniques for visualization of synaptic molecules are expanding rapidly. Among these techniques, high-resolution microscopy should become an indispensable approach to reveal molecular distribution in submicron domains of spines. New methods of monitoring protein interactions will also be integrated with experimental paradigms of synaptic plasticity. These technical advancements will reveal the complex interactions among receptors, cell adhesion molecules, PSD scaffolds, and actin network in spines. The remarkable consistency and tunability of the postsynaptic functions can only be understood through these bottom-up approaches.

References

Ando, R., Hama, H., Yamamoto-Hino, M., Mizuno, H., & Miyawaki, A. (2002). An optical marker based on the UV-induced green-to-red photoconversion of a fluorescent protein. *Proceedings of the National Academy of Sciences of the United States of America, 99*, 12651–12656.

Axelrod, D., Koppel, D. E., Schlessinger, J., Elson, E., & Webb, W. W. (1976). Mobility measurement by analysis of fluorescence photobleaching recovery kinetics. *Biophysical Journal, 16*, 1055–1069.

Barrow, S. L., Constable, J. R., Clark, E., El-Sabeawy, F., McAllister, A. K., & Washbourne, P. (2009). Neuroligin1: A cell adhesion molecule that recruits PSD-95 and NMDA receptors by distinct mechanisms during synaptogenesis. *Neural Development, 4*, 17.

Bates, M., Huang, B., Dempsey, G. T., & Zhuang, X. (2007). Multicolor super-resolution imaging with photo-switchable fluorescent probes. *Science, 317*, 1749–1753.

Betzig, E., Patterson, G. H., Sougrat, R., Lindwasser, O. W., Olenych, S., Bonifacino, J. S., Davidson, M. W., Lippincott-Schwartz, J., & Hess, H. F. (2006). Imaging intracellular fluorescent proteins at nanometer resolution. *Science, 313*, 1642–1645.

Bielas, S. L., Serneo, F. F., Chechlacz, M., Deerinck, T. J., Perkins, G. A., Allen, P. B., Ellisman, M. H., & Gleeson, J. G. (2007). Spinophilin facilitates dephosphorylation of doublecortin by PP1 to mediate microtubule bundling at the axonal wrist. *Cell, 129*, 579–591.

Blanpied, T. A., Kerr, J. M., & Ehlers, M. D. (2008). Structural plasticity with preserved topology in the postsynaptic protein network. *Proceedings of the National Academy of Sciences of the United States of America, 105*, 12587–12592.

Blanpied, T. A., Scott, D. B., & Ehlers, M. D. (2002). Dynamics and regulation of clathrin coats at specialized endocytic zones of dendrites and spines. *Neuron, 36*, 435–449.

Brakeman, P. R., Lanahan, A. A., O'Brien, R., Roche, K., Barnes, C. A., Huganir, R. L., & Worley, P. F. (1997). Homer: A protein that selectively binds metabotropic glutamate receptors. *Nature, 386,* 284–288.

Bresler, T., Ramati, Y., Zamorano, P. L., Zhai, R., Garner, C. C., & Ziv, N. E. (2001). The dynamics of SAP90/PSD-95 recruitment to new synaptic junctions. *Molecular and Cellular Neuroscience, 18,* 149–167.

Bresler, T., Shapira, M., Boeckers, T., Dresbach, T., Futter, M., Garner, C. C., Rosenblum, K., Gundelfinger, E. D., & Ziv, N. E. (2004). Postsynaptic density assembly is fundamentally different from presynaptic active zone assembly. *Journal of Neuroscience, 24,* 1507–1520.

Brose, N. (2009). Synaptogenic proteins and synaptic organizers: "Many hands make light work". *Neuron, 61,* 650–652.

Chen, X., Vinade, L., Leapman, R. D., Petersen, J. D., Nakagawa, T., Phillips, T. M., Sheng, M., & Reese, T. S. (2005). Mass of the postsynaptic density and enumeration of three key molecules. *Proceedings of the National Academy of Sciences of the United States of America, 102,* 11551–11556.

Cheng, D., Hoogenraad, C. C., Rush, J., Ramm, E., Schlager, M. A., Duong, D. M., Xu, P., Wijayawardana, S. R., Hanfelt, J., Nakagawa, T., et al. (2006). Relative and absolute quantification of postsynaptic density proteome isolated from rat forebrain and cerebellum. *Molecular & Cellular Proteomics, 5,* 1158–1170.

Cho, K. O., Hunt, C. A., & Kennedy, M. B. (1992). The rat brain postsynaptic density fraction contains a homolog of the Drosophila discs-large tumor suppressor protein. *Neuron, 9,* 929–942.

D'Angelo, M. A., Raices, M., Panowski, S. H., & Hetzer, M. W. (2009). Age-dependent deterioration of nuclear pore complexes causes a loss of nuclear integrity in postmitotic cells. *Cell, 136,* 284–295.

Dailey, M. E., & Smith, S. J. (1996). The dynamics of dendritic structure in developing hippocampal slices. *Journal of Neuroscience, 16,* 2983–2994.

Fischer, M., Kaech, S., Knutti, D., & Matus, A. (1998). Rapid actin-based plasticity in dendritic spines. *Neuron, 20,* 847–854.

Frost, N. A., Shroff, H., Kong, H., Betzig, E., & Blanpied, T. A. (2010). Single-molecule discrimination of discrete perisynaptic and distributed sites of actin filament assembly within dendritic spines. *Neuron, 67,* 86–99.

Fujiwara, T., Ritchie, K., Murakoshi, H., Jacobson, K., & Kusumi, A. (2002). Phospholipids undergo hop diffusion in compartmentalized cell membrane. *The Journal of Cell Biology, 157,* 1071–1081.

Furutani, Y., Matsuno, H., Kawasaki, M., Sasaki, T., Mori, K., & Yoshihara, Y. (2007). Interaction between telencephalin and ERM family proteins mediates dendritic filopodia formation. *Journal of Neuroscience, 27,* 8866–8876.

Gerrow, K., Romorini, S., Nabi, S. M., Colicos, M. A., Sala, C., & El-Husseini, A. (2006). A preformed complex of postsynaptic proteins is involved in excitatory synapse development. *Neuron, 49,* 547–562.

Graf, E. R., Zhang, X., Jin, S. X., Linhoff, M. W., & Craig, A. M. (2004). Neurexins induce differentiation of GABA and glutamate postsynaptic specializations via neuroligins. *Cell, 119,* 1013–1026.

Gray, N. W., Weimer, R. M., Bureau, I., & Svoboda, K. (2006). Rapid redistribution of synaptic PSD-95 in the neocortex in vivo. *PLoS Biology, 4,* e370.

Groc, L., Heine, M., Cognet, L., Brickley, K., Stephenson, F. A., Lounis, B., & Choquet, D. (2004). Differential activity-dependent regulation of the lateral mobilities of AMPA and NMDA receptors. *Nature Neuroscience, 7,* 695–696.

Groc, L., Lafourcade, M., Heine, M., Renner, M., Racine, V., Sibarita, J. B., Lounis, B., Choquet, D., & Cognet, L. (2007). Surface trafficking of neurotransmitter receptor: Comparison between single-molecule/quantum dot strategies. *Journal of Neuroscience, 27,* 12433–12437.

Harris, K. M., Jensen, F. E., & Tsao, B. (1992). Three-dimensional structure of dendritic spines and synapses in rat hippocampus (CA1) at postnatal day 15 and adult ages: Implications for the maturation of synaptic physiology and long-term potentiation. *Journal of Neuroscience, 12*, 2685–2705.

Hering, H., & Sheng, M. (2003). Activity-dependent redistribution and essential role of cortactin in dendritic spine morphogenesis. *Journal of Neuroscience, 23*, 11759–11769.

Hirokawa, N. (1989). The arrangement of actin filaments in the postsynaptic cytoplasm of the cerebellar cortex revealed by quick-freeze deep-etch electron microscopy. *Neuroscience Research, 6*, 269–275.

Hoffman-Kim, D., Diefenbach, T. J., Eustace, B. K., & Jay, D. G. (2007). Chromophore-assisted laser inactivation. *Methods in Cell Biology, 82*, 335–354.

Honkura, N., Matsuzaki, M., Noguchi, J., Ellis-Davies, G. C., & Kasai, H. (2008). The subspine organization of actin fibers regulates the structure and plasticity of dendritic spines. *Neuron, 57*, 719–729.

Hotulainen, P., Llano, O., Smirnov, S., Tanhuanpaa, K., Faix, J., Rivera, C., & Lappalainen, P. (2009). Defining mechanisms of actin polymerization and depolymerization during dendritic spine morphogenesis. *The Journal of Cell Biology, 185*, 323–339.

Huang, B., Wang, W., Bates, M., & Zhuang, X. (2008). Three-dimensional super-resolution imaging by stochastic optical reconstruction microscopy. *Science, 319*, 810–813.

Husi, H., Ward, M. A., Choudhary, J. S., Blackstock, W. P., & Grant, S. G. (2000). Proteomic analysis of NMDA receptor-adhesion protein signaling complexes. *Nature Neuroscience, 3*, 661–669.

Iki, J., Inoue, A., Bito, H., & Okabe, S. (2005). Bi-directional regulation of postsynaptic cortactin distribution by BDNF and NMDA receptor activity. *European Journal of Neuroscience, 22*, 2985–2994.

Irie, M., Hata, Y., Takeuchi, M., Ichtchenko, K., Toyoda, A., Hirao, K., Takai, Y., Rosahl, T. W., & Sudhof, T. C. (1997). Binding of neuroligins to PSD-95. *Science, 277*, 1511–1515.

Jaworski, J., Kapitein, L. C., Gouveia, S. M., Dortland, B. R., Wulf, P. S., Grigoriev, I., Camera, P., Spangler, S. A., Di Stefano, P., Demmers, J., et al. (2009). Dynamic microtubules regulate dendritic spine morphology and synaptic plasticity. *Neuron, 61*, 85–100.

Jontes, J. D., Emond, M. R., & Smith, S. J. (2004). In vivo trafficking and targeting of N-cadherin to nascent presynaptic terminals. *Journal of Neuroscience, 24*, 9027–9034.

Jordan, B. A., Fernholz, B. D., Boussac, M., Xu, C., Grigorean, G., Ziff, E. B., & Neubert, T. A. (2004). Identification and verification of novel rodent postsynaptic density proteins. *Molecular & Cellular Proteomics, 3*, 857–871.

Kopec, C. D., Li, B., Wei, W., Boehm, J., & Malinow, R. (2006). Glutamate receptor exocytosis and spine enlargement during chemically induced long-term potentiation. *Journal of Neuroscience, 26*, 2000–2009.

Kornau, H. C., Schenker, L. T., Kennedy, M. B., & Seeburg, P. H. (1995). Domain interaction between NMDA receptor subunits and the postsynaptic density protein PSD-95. *Science, 269*, 1737–1740.

Korobova, F., & Svitkina, T. (2010). Molecular architecture of synaptic actin cytoskeleton in hippocampal neurons reveals a mechanism of dendritic spine morphogenesis. *Molecular Biology of the Cell, 21*, 165–176.

Kuriu, T., Inoue, A., Bito, H., Sobue, K., & Okabe, S. (2006). Differential control of postsynaptic density scaffolds via actin-dependent and -independent mechanisms. *Journal of Neuroscience, 26*, 7693–7706.

Landis, D. M., & Reese, T. S. (1983). Cytoplasmic organization in cerebellar dendritic spines. *The Journal of Cell Biology, 97*, 1169–1178.

Li, Z., Okamoto, K., Hayashi, Y., & Sheng, M. (2004). The importance of dendritic mitochondria in the morphogenesis and plasticity of spines and synapses. *Cell, 119*, 873–887.

Majewska, A., Tashiro, A., & Yuste, R. (2000). Regulation of spine calcium dynamics by rapid spine motility. *Journal of Neuroscience, 20*, 8262–8268.

Matsuno, H., Okabe, S., Mishina, M., Yanagida, T., Mori, K., & Yoshihara, Y. (2006). Telencephalin slows spine maturation. *Journal of Neuroscience, 26*, 1776–1786.

Matsuzaki, M., Honkura, N., Ellis-Davies, G. C., & Kasai, H. (2004). Structural basis of long-term potentiation in single dendritic spines. *Nature, 429*, 761–766.

Matus, A. (2000). Actin-based plasticity in dendritic spines. *Science, 290*, 754–758.

Mendez, P., De Roo, M., Poglia, L., Klauser, P., & Muller, D. (2010). N-cadherin mediates plasticity-induced long-term spine stabilization. *The Journal of Cell Biology, 189*, 589–600.

Meyer, G., Varoqueaux, F., Neeb, A., Oschlies, M., & Brose, N. (2004). The complexity of PDZ domain-mediated interactions at glutamatergic synapses: A case study on neuroligin. *Neuropharmacology, 47*, 724–733.

Minerbi, A., Kahana, R., Goldfeld, L., Kaufman, M., Marom, S., & Ziv, N. E. (2009). Long-term relationships between synaptic tenacity, synaptic remodeling, and network activity. *PLoS Biology, 7*, e1000136.

Mitchison, T. J. (1989). Polewards microtubule flux in the mitotic spindle: Evidence from photoactivation of fluorescence. *The Journal of Cell Biology, 109*, 637–652.

Nagerl, U. V., Eberhorn, N., Cambridge, S. B., & Bonhoeffer, T. (2004). Bidirectional activity-dependent morphological plasticity in hippocampal neurons. *Neuron, 44*, 759–767.

Nam, C. I., & Chen, L. (2005). Postsynaptic assembly induced by neurexin-neuroligin interaction and neurotransmitter. *Proceedings of the National Academy of Sciences of the United States of America, 102*, 6137–6142.

Newpher, T. M., & Ehlers, M. D. (2008). Glutamate receptor dynamics in dendritic microdomains. *Neuron, 58*, 472–497.

Nusser, Z. (1999). A new approach to estimate the number, density and variability of receptors at central synapses. *European Journal of Neuroscience, 11*, 745–752.

Nusser, Z., Lujan, R., Laube, G., Roberts, J. D., Molnar, E., & Somogyi, P. (1998). Cell type and pathway dependence of synaptic AMPA receptor number and variability in the hippocampus. *Neuron, 21*, 545–559.

Okabe, S. (2007). Molecular anatomy of the postsynaptic density. *Molecular and Cellular Neuroscience, 34*, 503–518.

Okabe, S., & Hirokawa, N. (1989). Incorporation and turnover of biotin-labeled actin microinjected into fibroblastic cells: An immunoelectron microscopic study. *The Journal of Cell Biology, 109*, 1581–1595.

Okabe, S., & Hirokawa, N. (1992). Differential behavior of photoactivated microtubules in growing axons of mouse and frog neurons. *The Journal of Cell Biology, 117*, 105–120.

Okabe, S., & Hirokawa, N. (1993). Do photobleached fluorescent microtubules move?: Re-evaluation of fluorescence laser photobleaching both in vitro and in growing Xenopus axon. *The Journal of Cell Biology, 120*, 1177–1186.

Okabe, S., Miwa, A., & Okado, H. (2001a). Spine formation and correlated assembly of presynaptic and postsynaptic molecules. *Journal of Neuroscience, 21*, 6105–6114.

Okabe, S., Urushido, T., Konno, D., Okado, H., & Sobue, K. (2001b). Rapid redistribution of the postsynaptic density protein PSD-Zip45 (Homer 1c) and its differential regulation by NMDA receptors and calcium channels. *Journal of Neuroscience, 21*, 9561–9571.

Palade, G. E., & Palay, S. L. (1954). Electron microscopic observations of interneuronal and neuromuscular synapses. *Anatomical Record, 118*, 335–336.

Park, M., Salgado, J. M., Ostroff, L., Helton, T. D., Robinson, C. G., Harris, K. M., & Ehlers, M. D. (2006). Plasticity-induced growth of dendritic spines by exocytic trafficking from recycling endosomes. *Neuron, 52*, 817–830.

Patterson, G. H., & Lippincott-Schwartz, J. (2002). A photoactivatable GFP for selective photolabeling of proteins and cells. *Science, 297*, 1873–1877.

Petralia, R. S., Wang, Y. X., & Wenthold, R. J. (1994a). The NMDA receptor subunits NR2A and NR2B show histological and ultrastructural localization patterns similar to those of NR1. *Journal of Neuroscience, 14*, 6102–6120.

Petralia, R. S., Yokotani, N., & Wenthold, R. J. (1994b). Light and electron microscope distributions of the NMDA receptor subunit NMDAR1 in the rat nervous system using a selective anti-peptide antibody. *Journal of Neuroscience, 14*, 667–696.

Petrini, E. M., Lu, J., Cognet, L., Lounis, B., Ehlers, M. D., & Choquet, D. (2009). Endocytic trafficking and recycling maintain a pool of mobile surface AMPA receptors required for synaptic potentiation. *Neuron, 63*, 92–105.

Portera-Cailliau, C., Pan, D. T., & Yuste, R. (2003). Activity-regulated dynamic behavior of early dendritic protrusions: Evidence for different types of dendritic filopodia. *Journal of Neuroscience, 23*, 7129–7142.

Racz, B., Blanpied, T. A., Ehlers, M. D., & Weinberg, R. J. (2004). Lateral organization of endocytic machinery in dendritic spines. *Nature Neuroscience, 7*, 917–918.

Rajfur, Z., Roy, P., Otey, C., Romer, L., & Jacobson, K. (2002). Dissecting the link between stress fibres and focal adhesions by CALI with EGFP fusion proteins. *Nature Cell Biology, 4*, 286–293.

Reits, E. A., & Neefjes, J. J. (2001). From fixed to FRAP: Measuring protein mobility and activity in living cells. *Nature Cell Biology, 3*, E145–E147.

Rust, M. J., Bates, M., & Zhuang, X. (2006). Sub-diffraction-limit imaging by stochastic optical reconstruction microscopy (STORM). *Nature Methods, 3*, 793–795.

Saglietti, L., Dequidt, C., Kamieniarz, K., Rousset, M. C., Valnegri, P., Thoumine, O., Beretta, F., Fagni, L., Choquet, D., Sala, C., et al. (2007). Extracellular interactions between GluR2 and N-cadherin in spine regulation. *Neuron, 54*, 461–477.

Schapitz, I. U., Behrend, B., Pechmann, Y., Lappe-Siefke, C., Kneussel, S. J., Wallace, K. E., Stempel, A. V., Buck, F., Grant, S. G., Schweizer, M., et al. (2010). Neuroligin 1 is dynamically exchanged at postsynaptic sites. *Journal of Neuroscience, 30*, 12733–12744.

Scheiffele, P., Fan, J., Choih, J., Fetter, R., & Serafini, T. (2000). Neuroligin expressed in nonneuronal cells triggers presynaptic development in contacting axons. *Cell, 101*, 657–669.

Schneider, M., Barozzi, S., Testa, I., Faretta, M., & Diaspro, A. (2005). Two-photon activation and excitation properties of PA-GFP in the 720–920-nm region. *Biophysical Journal, 89*, 1346–1352.

Sharma, K., Fong, D. K., & Craig, A. M. (2006). Postsynaptic protein mobility in dendritic spines: Long-term regulation by synaptic NMDA receptor activation. *Molecular and Cellular Neuroscience, 31*, 702–712.

Shepherd, J. D., & Huganir, R. L. (2007). The cell biology of synaptic plasticity: AMPA receptor trafficking. *Annual Review of Cell and Developmental Biology, 23*, 613–643.

Shroff, H., Galbraith, C. G., Galbraith, J. A., White, H., Gillette, J., Olenych, S., Davidson, M. W., & Betzig, E. (2007). Dual-color superresolution imaging of genetically expressed probes within individual adhesion complexes. *Proceedings of the National Academy of Sciences of the United States of America, 104*, 20308–20313.

Sinnecker, D., Voigt, P., Hellwig, N., & Schaefer, M. (2005). Reversible photobleaching of enhanced green fluorescent proteins. *Biochemistry, 44*, 7085–7094.

Spacek, J. (1985). Three-dimensional analysis of dendritic spines. II. Spine apparatus and other cytoplasmic components. *Anatomy and Embryology, 171*, 235–243.

Spacek, J., & Harris, K. M. (1998). Three-dimensional organization of cell adhesion junctions at synapses and dendritic spines in area CA1 of the rat hippocampus. *The Journal of Comparative Neurology, 393*, 58–68.

Stan, A., Pielarski, K. N., Brigadski, T., Wittenmayer, N., Fedorchenko, O., Gohla, A., Lessmann, V., Dresbach, T., & Gottmann, K. (2010). Essential cooperation of N-cadherin and neuroligin-1 in the transsynaptic control of vesicle accumulation. *Proceedings of the National Academy of Sciences of the United States of America, 107*, 11116–11121.

Star, E. N., Kwiatkowski, D. J., & Murthy, V. N. (2002). Rapid turnover of actin in dendritic spines and its regulation by activity. *Nature Neuroscience, 5*, 239–246.

Sugiyama, Y., Kawabata, I., Sobue, K., & Okabe, S. (2005). Determination of absolute protein numbers in single synapses by a GFP-based calibration technique. *Nature Methods, 2*, 677–684.

Svitkina, T. M., Bulanova, E. A., Chaga, O. Y., Vignjevic, D. M., Kojima, S., Vasiliev, J. M., & Borisy, G. G. (2003). Mechanism of filopodia initiation by reorganization of a dendritic network. *The Journal of Cell Biology, 160*, 409–421.

Svoboda, K., Tank, D. W., & Denk, W. (1996). Direct measurement of coupling between dendritic spines and shafts. *Science, 272*, 716–719.

Tanaka, J., Matsuzaki, M., Tarusawa, E., Momiyama, A., Molnar, E., Kasai, H., & Shigemoto, R. (2005). Number and density of AMPA receptors in single synapses in immature cerebellum. *Journal of Neuroscience, 25*, 799–807.

Tatavarty, V., Kim, E. J., Rodionov, V., & Yu, J. (2009). Investigating sub-spine actin dynamics in rat hippocampal neurons with super-resolution optical imaging. *PloS One, 4*, e7724.

Thyagarajan, A., & Ting, A. Y. (2010). Imaging activity-dependent regulation of neurexin-neuroligin interactions using trans-synaptic enzymatic biotinylation. *Cell, 143*, 456–469.

Togashi, H., Abe, K., Mizoguchi, A., Takaoka, K., Chisaka, O., & Takeichi, M. (2002). Cadherin regulates dendritic spine morphogenesis. *Neuron, 35*, 77–89.

Tomita, S., Adesnik, H., Sekiguchi, M., Zhang, W., Wada, K., Howe, J. R., Nicoll, R. A., & Bredt, D. S. (2005). Stargazin modulates AMPA receptor gating and trafficking by distinct domains. *Nature, 435*, 1052–1058.

Triller, A., & Choquet, D. (2008). New concepts in synaptic biology derived from single-molecule imaging. *Neuron, 59*, 359–374.

Tu, J. C., Xiao, B., Naisbitt, S., Yuan, J. P., Petralia, R. S., Brakeman, P., Doan, A., Aakalu, V. K., Lanahan, A. A., Sheng, M., & Worley, P. F. (1999). Coupling of mGluR/Homer and PSD-95 complexes by the Shank family of postsynaptic density proteins. *Neuron, 23*, 583–592.

Valtschanoff, J. G., & Weinberg, R. J. (2001). Laminar organization of the NMDA receptor complex within the postsynaptic density. *Journal of Neuroscience, 21*, 1211–1217.

Wang, Z., Edwards, J. G., Riley, N., Provance, D. W., Jr., Karcher, R., Li, X. D., Davison, I. G., Ikebe, M., Mercer, J. A., Kauer, J. A., & Ehlers, M. D. (2008). Myosin Vb mobilizes recycling endosomes and AMPA receptors for postsynaptic plasticity. *Cell, 135*, 535–548.

Washbourne, P., Bennett, J. E., & McAllister, A. K. (2002). Rapid recruitment of NMDA receptor transport packets to nascent synapses. *Nature Neuroscience, 5*, 751–759.

Yasumatsu, N., Matsuzaki, M., Miyazaki, T., Noguchi, J., & Kasai, H. (2008). Principles of long-term dynamics of dendritic spines. *Journal of Neuroscience, 28*, 13592–13608.

Yoshihara, Y., Oka, S., Nemoto, Y., Watanabe, Y., Nagata, S., Kagamiyama, H., & Mori, K. (1994). An ICAM-related neuronal glycoprotein, telencephalin, with brain segment-specific expression. *Neuron, 12*, 541–553.

Yoshimura, Y., Yamauchi, Y., Shinkawa, T., Taoka, M., Donai, H., Takahashi, N., Isobe, T., & Yamauchi, T. (2004). Molecular constituents of the postsynaptic density fraction revealed by proteomic analysis using multidimensional liquid chromatography-tandem mass spectrometry. *Journal of Neurochemistry, 88*, 759–768.

Zhou, Q., Homma, K. J., & Poo, M. M. (2004). Shrinkage of dendritic spines associated with long-term depression of hippocampal synapses. *Neuron, 44*, 749–757.

Zuo, Y., Lin, A., Chang, P., & Gan, W. B. (2005). Development of long-term dendritic spine stability in diverse regions of cerebral cortex. *Neuron, 46*, 181–189.

Chapter 7
The Brain's Extracellular Matrix and Its Role in Synaptic Plasticity

Renato Frischknecht and Eckart D. Gundelfinger

Abstract The extracellular matrix (ECM) of the brain has important roles in regulating synaptic function and plasticity. A juvenile ECM supports the wiring of neuronal networks, synaptogenesis, and synaptic maturation. The closure of critical periods for experience-dependent shaping of neuronal circuits coincides with the implementation of a mature form of ECM that is characterized by highly elaborate hyaluronan-based structures, the perineuronal nets (PNN), and PNN-like perisynaptic ECM specializations. In this chapter, we will focus on some recently reported aspects of ECM functions in brain plasticity. These include (a) the discovery that the ECM can act as a passive diffusion barrier for cell surface molecules including neurotransmitter receptors and in this way compartmentalize cell surfaces, (b) the specific functions of ECM components in actively regulating synaptic plasticity and homeostasis, and (c) the shaping processes of the ECM by extracellular proteases and in turn the activation particular signaling pathways.

Keywords Cell surface molecules • Extracellular matrix • Perineuronal nets • Proteases • Synaptic plasticity

7.1 Introduction

The extracellular matrix (ECM) wrapping neural cells in the brain is produced by both neurons and glial cells. During postnatal maturation of neuronal circuits, this originally rather diffuse ECM condenses into a netlike structure around a subclass of neurons. These structures are termed perineuronal nets (PNN) and have been discovered by the pioneers of brain cell biology including Camillo Golgi and Santiago Ramon y Cajal (for review, see Celio et al. 1998). Nonetheless, the

R. Frischknecht (✉) • E.D. Gundelfinger
Leibniz Institute for Neurobiology (LIN), Brenneckestr. 6, 39118 Magdeburg, Germany
e-mail: rfrischk@ifn-magdeburg.de; gundelfi@ifn-magdeburg.de

M.R. Kreutz and C. Sala (eds.), *Synaptic Plasticity*,
Advances in Experimental Medicine and Biology 970,
DOI 10.1007/978-3-7091-0932-8_7, © Springer-Verlag/Wien 2012

occurrence of an ECM in the brain has only been generally accepted in the 1970s (for review, see Zimmermann and Dours-Zimmermann 2008). The unbranched polysaccharide hyaluronic acid is the core component of the ECM in the brain. It acts as a kind of backbone to recruit proteoglycans and glycoproteins into ECM structures (Bandtlow and Zimmermann 2000; Rauch 2004; Frischknecht and Seidenbecher 2008). Major components of the hyaluronic acid–based ECM are chondroitin sulfate proteoglycans (CSPGs) of the lectican family (also named hyalecticans), tenascins, and so-called link proteins (Bandtlow and Zimmermann 2000; Yamaguchi 2000; Rauch 2004). In addition, a variety of other glycoproteins and proteoglycans contribute to the brain's ECM. These include laminins, pentraxins, pleiotrophin/HB-GAM, phosphocan, reelin, thrombospondins, and the heparan-sulfate proteoglycan (HSPG) agrin or cell surface-bound HSPGs of the syndecan and glypican families. Moreover, matrix-shaping enzymes, like proteases and hyaluronidases, are found in the brain ECM (Bandtlow and Zimmermann 2000; Dityatev and Schachner 2003; Christopherson et al. 2005; Dityatev and Fellin 2008; Frischknecht and Seidenbecher 2008).

During CNS development, the ECM undergoes significant changes. Initially, a juvenile ECM is synthesized during late embryonic and early postnatal development of the mammalian brain. The lectican neurocan and tenascin-C are prominent constituents of this juvenile matrix (Carulli et al. 2007). The adult ECM is characterized by downregulation of these components and the upregulation of other components including tenascin-R, brevican, aggrecan (Fig. 7.1), or particular versican isoforms (Milev et al. 1998; Carulli et al. 2007; Zimmermann and Dours-Zimmermann 2008; Carulli et al. 2010). A systematic biochemical and immuno-histochemical investigation revealed differentially extractable ECM fractions from the adult brain (Deepa et al. 2006). Most of the material is loosely associated with

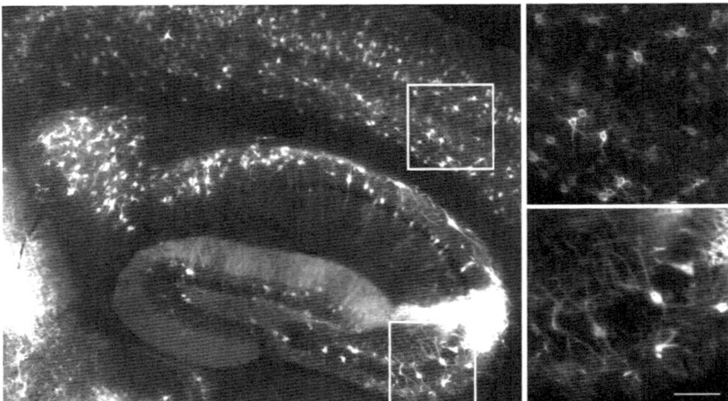

Fig. 7.1 Aggrecan is a major component of perineuronal nets (*PNN*). Staining of brain sections from 6-month-old mice with anti-aggrecan antibodies identifies numerous interneurons in the hippocampus and cortex surrounded by a PNN. Higher magnifications of cortical and hippocampal CA3 areas are shown in the upper and the lower box, respectively (scale bar: 100 μm). Of note, the massive aggrecan immunoreactivity of the CA2 region in the hippocampus

brain membranes. Nonionic detergents and salt can solubilize another fraction of ECM material that is thought to be more tightly associated with neural cell membranes. A final fraction comprising roughly a quarter of the CSPG material can only be extracted with urea. This fraction includes aggrecan, versican V2, neurocan, brevican, as well as phosphacan, and is not present in the young brain before closure of the critical period in the visual cortex (Deepa et al. 2006). The fraction is thought to represent cartilage-like ECM material forming the PNN (Fawcett 2009a). The PNN-like material can be removed from brain tissue entirely with the hyaluronic acid hydrolyzing enzyme hyaluronidase and partly with chondroitinase ABC, an enzyme that removes glycosaminoglycan chains from CSPGs (Deepa et al. 2006).

PNN are most prominent on GABAergic interneurons expressing the calcium buffer protein parvalbumin (Celio et al. 1998; Hartig et al. 1999). However, recent studies revealed that PNN are highly heterogeneous and that they occur on various types of neurons including excitatory principal neurons and inhibitory neurons throughout the CNS (Bruckner et al. 2000; Matthews et al. 2002; Wegner et al. 2003; Alpar et al. 2006). Mouse mutants for brevican, aggrecan, cartilage link protein Crtl1 (Hapln1) and tenascin-R display abnormal PNN indicating the importance of lecticans and tenascins for these ECM specializations (Bruckner et al. 2000; Brakebusch et al. 2002; Giamanco et al. 2010). After prolonged time in culture, PNN-like structures form also in primary neuronal cultures of various CNS areas (Miyata et al. 2005; John et al. 2006; Dityatev et al. 2007). Also here, GABAergic neurons first accumulate ECM material on their surfaces (Dityatev et al. 2007); however, virtually all neurons including their neurites are more or less densely covered within netlike structures after about 3 weeks in culture (John et al. 2006). This hyaluronan-based ECM tightly wraps synapses and is interspersed between neurons and astrocytes. Formation of cartilage-like cell surface structures resembling PNN can be triggered by heterologous overexpression of hyaluronan synthase (HAS), Crtl1, and aggrecan in human embryonic kidney cells (Kwok et al. 2010).

7.2 The Tetrapartite Synapse

The concept of a tripartite synapse implying that not only the canonical pre- and postsynaptic elements of two adjacent neurons but also the endfeet of glial cells contribute to the structure and function of brain synapses has been developed more than a decade ago and is nowadays widely accepted (for a review, see Araque et al. 1999; Haydon 2001; Slezak and Pfrieger 2003; Faissner et al. 2010). The term tetrapartite synapse or "synaptic quadriga" has been coined to indicate that, in addition to these three cellular parts, ECM structures produced by both astrocytes and neurons contribute to the functional synaptic complex (Dityatev et al. 2006; John et al. 2006; Dityatev et al. 2010b). A well-studied example of such a quadriga is the vertebrate neuromuscular junction where nonmyelinating Schwann cells tightly wrap the presynaptic terminal, and a prominent agrin–laminin-based basal lamina fills the synaptic cleft between the motorneuron ending and the postsynaptic membrane of the muscle (for review, see Sanes and Lichtman 2001; Faissner et al. 2010). At CNS synapses, the synaptic cleft is much less wide as compared to the neuromuscular synapse (20 vs. 50 microns, respectively), and it does not contain a basal lamina. Nonetheless, within the synaptic cleft of excitatory CNS synapses, regularly assembled ECM structures are found, the biochemical identity of which is currently unknown (Zuber et al. 2005). Maybe, like at the neuromuscular junction, an agrin-based ECM is found also in the cleft of central synapses. Consistent with this hypothesis, processing of agrin by the extracellular protease neurotrypsin is involved in the development and plasticity of CNS synapses (Matsumoto-Miyai et al. 2009; see below).

The ECM of the synaptic cleft seems to be clearly distinct from the perisynaptic ECM wrapping CNS synapses. For example, biochemical fractionation has demonstrated that ECM components of the mature brain, like brevican, tightly associate with synaptic protein preparations (Seidenbecher et al. 1995, 2002; Li et al. 2004). However, at the ultrastructural level, brevican immunoreactivity is strictly perisynaptic and not found within the synaptic cleft (Seidenbecher et al. 1997). Brevican is primarily synthesized by astrocytes (Yamada et al. 1997; John et al. 2006) confirming that glia-derived components contribute to the PNN-like ECM tightly associated with the synaptic complex. As will be discussed in detail below, both passive and active functions might be assigned to the perisynaptic ECM. Passive functions include the trapping of trophic factors like neurotrophins, fibroblast growth factors, midkine, or pleiotrophin in the vicinity of their cognate high affinity receptors (Celio and Blumcke 1994; Galtrey and Fawcett 2007; Fawcett 2009a), as well as the formation of diffusion barriers for cell membrane proteins (Frischknecht et al. 2009, see below). Active functions include specific interactions of ECM components with synaptic proteins and pathways as well as the generation of signaling components via proteolysis of ECM components (see below).

Astrocyte-derived ECM components serve important functions already during synaptogenesis and synaptic maturation. For example, thrombospondins (TSPs), oligomeric ECM proteins secreted by astrocytes in the brain, have been identified as a major factor for synapse development inducing synapse formation and maturation of the presynaptic bouton. However, these synapses were postsynaptically silent (Christopherson et al. 2005). Mice lacking both TSP1 and 2 exhibited a significantly decreased number of excitatory synapses in the cerebral cortex (Christopherson et al. 2005). Another member of the thrombospondin family (TSP4) is strongly expressed in adult astrocytes and is a candidate for regulating synaptic plasticity in the CNS (Eroglu 2009). The postsynaptic partner of TSPs is the non-channel-forming $\alpha2\delta$-1 subunit of voltage-gated calcium channels, which also acts as receptor for the antiepileptic analgesic drug gabapentin. Binding of TSPs to the gabapentin receptor is also required for postsynaptic activation of excitatory synapses (Eroglu et al. 2009). Moreover, the interaction of TSP1 with the postsynaptic cell adhesion molecule neuroligin 1 seems to accelerate the process of synaptogenesis in hippocampal primary cultures (Xu et al. 2010).

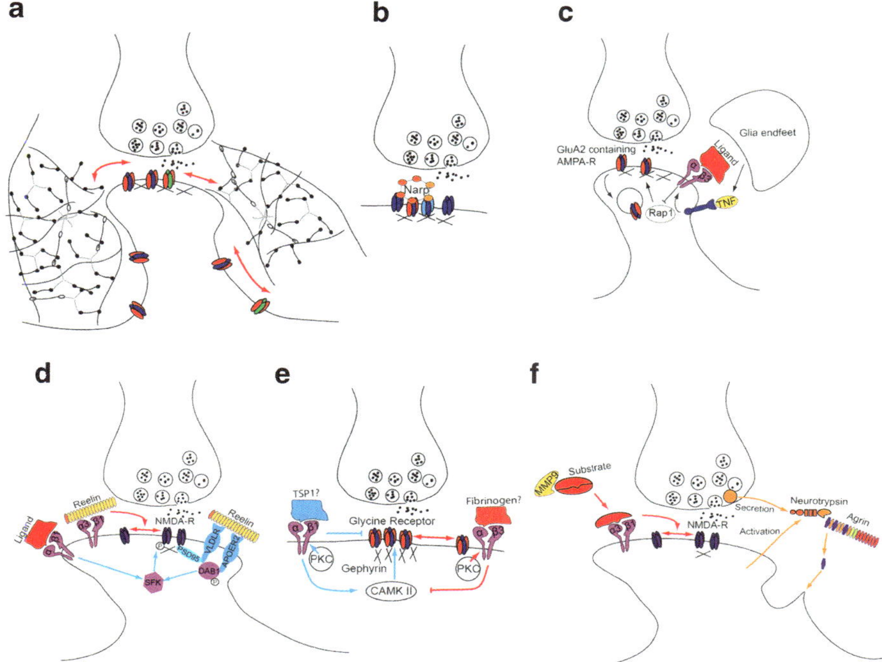

Fig. 7.2 Examples of ECM function in the modulation of synaptic transmission. (**a**) The perisynaptic ECM represents a physical barrier for lateral diffusion of neurotransmitter receptors (*red arrows*). Surface receptors such as AMPA-receptors are limited in lateral diffusion by the ECM, which reduces the exchange of synaptic and extrasynaptic receptors in the adult brain (Frischknecht et al. 2009). (**b**) On parvalbumin-positive inhibitory neurons multimers of the neuronal pentraxins NP1 and narp bind to and cluster AMPA receptors in response to elevated synaptic activity and thus regulate homeostatic scaling (Chang et al. 2010). (**c**) ECM-integrin signaling reduces AMPA receptor internalization and thus contributes to synaptic scaling. Reduced synaptic activity is thought to induce the release of tumor necrosis factor alpha (*TNF*) from glial cells, which in turn enhances surface expression of β3-containing integrins. β3-Integrins inhibit the small GTPase Rap1, which in its active state is responsible for removal of GluA2-containg AMPA receptors from the cell surface (for review, see Pozo and Goda 2010; Dityatev et al. 2010a). (**d**) Reelin signaling alters activity (*blue arrows*) and lateral diffusion of NMDA receptors. Upon binding, reelin clusters its receptors, the very-low-density lipoprotein receptor (*VLDLR*) and the apolipoprotein E receptor type 2 (*ApoER2*), which is in contact with NMDA receptors through binding to the scaffold protein PSD-95. Co-clustering of the intracellular adaptor protein disabled 1 (*DAB1*), which binds to both receptors induces activation Src family tyrosine kinases (*SFK*) and to tyrosine phosphorylation of NMDA receptors leading to increased receptor activity. Similarly, SFK mediate NMDA receptor tyrosine phosphorylation after activation of β1-containing integrins (*blue arrows*). Reelin signaling further enhances lateral diffusion of GluN2B but not GluN2A containing NMDA-receptors and thus may contributes to the decrease of synaptic GluN2B receptors during late development (*red arrows*). Hypothetically, in this case reelin signaling may act through direct binding to α3β1-containing integrins (for review, see Herz and Chen 2006; Dityatev et al. 2010a; Frotscher 2010). (**e**) Integrins control surface trafficking of glycine receptors on spinal cord neurons. β1 – (*blue arrows*) and β3 – (*red arrows*) containing integrins act on glycine receptor diffusion in an antagonistic manner. Both integrins require protein kinase C (*PKC*) for their activity. However, activation of β1-integrins by trombospondin 1 also

At inhibitory synapses of the spinal cord, TSP1 via its integrin $\beta1$ receptor slows the mobility of extrasynaptic glycine receptors (GlyR) and stabilizes these receptors in synapses (Charrier et al. 2010, see Fig. 7.2e).

7.3 Formation of the Adult ECM: Switch from Developmental to Mature Forms of Synaptic Plasticity

One of the most striking aspects of adult ECM function is its appearance at the end of the critical period of circuit wiring suggesting that it is involved in the implementation of adult plasticity modes (Mataga et al. 2002; Pizzorusso et al. 2002; Mataga et al. 2004; Oray et al. 2004). This seems to occur at the expense of the regenerative potential of the central nervous system (Fawcett 2009b). Initial findings derive from studies on ocular dominance (OD) plasticity in the visual cortex of cats, which, since its discovery (Wiesel and Hubel 1963), has represented a valuable model for studying experience-dependent plasticity in vivo. In the visual cortex, certain groups of neurons respond preferentially to one but not the other eye. Visual deprivation of one eye during a developmental period, the so-called critical period (postnatal days P21–P25 in rodents), leads to a drastic change in the neuronal circuits in the visual cortex. There the number of neurons responding to the nondeprived eye increases at the expense of those responding to the deprived eye. In contrast to what is observed during the critical period, visual deprivation performed in adult animals results in little or no plasticity (Pizzorusso et al. 2002; Berardi et al. 2003). Experiments by Pizzorusso and colleagues (2002) have demonstrated that removal of the hyaluronan-based ECM from the visual cortex can restore OD plasticity in the adult stage. Injection of chondroitinase ABC into the visual cortex and subsequent monocular deprivation resulted in a shift of ocular dominance. Utilizing this treatment, it was even possible to rescue visual acuity in adult animals reared with one long-term deprived eye and thus suffering from strongly asymmetric ocular dominance (Pizzorusso et al. 2006). Investigations on a mouse mutant for the cartilage link protein 1 (Crtl1) revealed that it does not develop

---

Fig. 7.2 (continued) activates Ca^{2+}/calmodulin-dependent protein kinase II (*CaMKII*) and reduces glycine receptor diffusion and accumulation of the receptor at the synapse. In contrast, activation of $\beta1$-integrins by fibronectin blocks of CaMKII and increases lateral mobility and in turn the loss of receptors from the synaptic membrane (Charrier et al. 2010). (**f**) Effects of proteolytic cleavage of ECM molecules on synaptic function. Cleavage of an unknown substrate in the ECM by MMP9 unmasks an RGD signal for $\beta1$-containing integrins. Subsequently GluN2A-containing NMDA receptors exhibit increased lateral diffusion within the neuronal membrane (Michaluk et al. 2009). Neurotrypsin is released from presynaptic vesicles. Activation of neurotrypsin requires concomitant activation of the postsynapse (*orange arrows*). Active neurotrypsin processes agrin and releases a 22-kD stable fragment harboring a single laminin G3 domain. This fragment is able to induce dendritic filopodia that may give rise to new synapses (Frischknecht et al. 2008; Matsumoto-Miyai et al. 2009)

normal PNN, retains juvenile levels of OD plasticity, and the visual acuity remains sensitive to visual deprivation (Carulli et al. 2010). In these mice, also plasticity of Purkinje cell axon terminals in the deep cerebellar nuclei is enhanced (Foscarin et al. 2011). In rodents, formation of PNN in the visual cortex is delayed upon dark rearing (Pizzorusso et al. 2002; Carulli et al. 2010) and is reduced in deep cerebellar nuclei by rearing in enriched environment (Foscarin et al. 2011) suggesting an interplay of external stimuli and the synthesis and/or maintenance of PNN.

Another study by Gogolla and colleagues (2009) suggests that similar mechanisms involving the ECM may make particular memories, in this case fear memories, resistant to erasure. Conditioned fear memories can be erased permanently in young rats while animals older than 3–4 weeks are largely resistant to this fear extinction. Fear extinction in both adult and young rats is amygdala dependent. In this brain structure, PNN develop between postnatal days P16 and P21. After this critical period, fear memory can be reduced by repeated exposure to the conditioned stimulus in the absence of the aversive fear-provoking stimulus. However, in contrast to young animals, fear response is reinstated in adult rats when the aversive stimulus is presented again. Similar to the experiments in the visual cortex, removal of the hyaluronan/CSPG-based ECM achieved a rapid and permanent erasure of newly acquired fear memories. Extinction did not take place when fear experience was made before enzyme application indicating that CSPGs are essential for protecting fear memories from erasure during the acquisition phase (Gogolla et al. 2009; Pizzorusso 2009).

7.4 The Hyaluronic Acid–Based ECM Is Indispensible for Mature Forms of Synaptic Plasticity

In the adult mouse brain, the lack of components of the ECM leads to impaired synaptic plasticity (for a comprehensive review, see Dityatev and Fellin 2008). For instance, TNR-deficient mice exhibit impaired long-term potentiation (LTP) but normal long-term depression (LTD). Further, a lack of TNR leads to disinhibition of the CA1 region of the hippocampus and to shift in the threshold for induction of LTP (Bukalo et al. 2007). Mice lacking the CSPGs brevican or neurocan also show impairments in LTP 30 min and 2 h after induction, respectively (Zhou et al. 2001; Brakebusch et al. 2002). Removal of chondroitin sulfates results in the degradation of perineuronal nets, the increase in excitability of perisomatic interneurons (Dityatev et al. 2007), and in the $GABA_A$ receptor-dependent inhibition of LTP induction (Bukalo et al. 2001). Interestingly, it has recently been suggested that hyaluronic acid is not only a fundamental structural element of the PNN but also has a direct impact on synaptic plasticity (Kochlamazashvili et al. 2010). According to this report, hyaluronic acid binds directly to and modulates L-type voltage-dependent calcium channels (L-VDCC; Cav1.2). Acute enzymatic removal of hyaluronic acid with hyaluronidase reduced nifedipine-sensitive Ca^{2+} currents in dendrites and spines of hippocampal neurons in slices from 2- to 3-month-old mice. Furthermore, it abolished an L-VDCC-dependent component of LTP at the CA3–CA1 Schaffer collateral synapses in the hippocampus. This deficit was completely rescued by the application of exogenous hyaluronic acid. In a heterologous expression system, exogenous HA rapidly increased currents mediated by Cav1.2, but not Cav1.3 subunit-containing L-VDCCs indicating a direct binding of hyaluronic acid to the channel. At the systemic level, intrahippocampal injection of hyaluronidase impaired contextual fear conditioning. Thus, the perisynaptic extracellular matrix influences use-dependent synaptic plasticity through regulation of dendritic Ca^{2+} channels (Kochlamazashvili et al. 2010).

7.5 Role of ECM in Control of Neurotransmitter Receptors

7.5.1 The ECM as a Passive Diffusion Barrier for Cell Surface Molecules

A large pool of surface molecules is highly mobile within the plasma membrane due to lateral Brownian diffusion (Kusumi et al. 1993; Triller and Choquet 2008). However, most surface molecules are restricted in lateral mobility by obstacles (pickets and corrals) compartmentalizing the cell surface. These obstacles are most likely formed by the underlying cytoskeleton or by rigid membrane structures (Kusumi et al. 1993; Choquet and Triller 2003; Kusumi et al. 2005). Synapses represent one of the most important surface compartments and are the sites of neurotransmitter release and detection for interneuronal communication. Interestingly neurotransmitter receptors, such as AMPA-type and NMDA-type glutamate receptors or $GABA_A$ receptors, not only are present in synaptic areas but are also found extrasynaptically. Lateral diffusion of receptors in extrasynaptic and synaptic domains has been investigated intensely in the past years (Triller and Choquet 2008). In general, diffusion of these receptors is more confined in the synaptic compartment as compared to extrasynaptic areas. However, receptors are steadily exchanging between synaptic and extrasynaptic pools. This steady replacement probably constitutes a fundamental mechanism for the maintenance of synaptic receptor pools as the exchange between cell surface and intracellular receptors through exo- and endocytosis occurs outside the synaptic membrane (Newpher and Ehlers 2008; Petrini et al. 2009). In addition, studies on hippocampal slices and primary hippocampal neurons have revealed that lateral diffusion may account for the exchange of desensitized synaptic AMPA receptors, which emerge during high-frequency firing, for naive extrasynaptic ones (Heine et al. 2008). Blockade of lateral diffusion, e.g., by cross-linking with antibodies, resulted in strong paired-pulse depression (PPD) presumably caused by the accumulation of desensitized receptors under the release site. These results demonstrated that the lateral diffusion of AMPA receptors was a novel postsynaptic mechanism influencing short-term plasticity of individual synapses.

Interestingly, the diffusion rates of AMPA receptors on dissociated hippocampal neurons decreased during synapse maturation, between the second and third week in vitro (Borgdorff and Choquet 2002). During this time period, a hyaluronan–CSPG-based ECM resembling the perisynaptic netlike ECM of the adult CNS is formed in these cultures (John et al. 2006). Similar to the in vivo situation, the netlike structure divides the neuronal surface into multiple compartments of variable size. These ECM-derived cell surface structures restrict the lateral diffusion of extrasynaptic AMPA receptors (Frischknecht et al. 2009; Fig. 7.2a). Removal of the ECM with the enzyme hyaluronidase increased diffusion rates of extrasynaptic but not of synaptic receptors. The exchange between receptors at synapses and extrasynaptic compartments was also increased. Extrasynaptic diffusion rates after hyaluronidase treatment resembled the "juvenile" situation before the ECM is established in the cultures at day 10 in vitro.

An electrophysiological assessment revealed that removal of ECM from dissociated hippocampal neurons and, under certain conditions, also from hippocampal slices affected short-term synaptic plasticity (Frischknecht et al. 2009; Kochlamazashvili et al. 2010). In the presence of the ECM, PPD seems to be much stronger than after hyaluronidase treatment when basically no PPD was observed. Thus, ECM-derived surface compartments can influence short-term plasticity of neurons by controlling lateral diffusion and thus control the synaptic availability of naive AMPA receptors. It should be noted here that ECM nets are not impermeable barriers for diffusing surface proteins. They rather have to be considered as viscous structures that reduce the surface mobility of proteins through weak, transient interactions, or simply as passive obstacles. Accordingly, the size and shape of the extracellular domains of surface-exposed membrane proteins influence the mobility shift by the ECM (Frischknecht et al. 2009). Along this line, the recent characterization of the full crystal structure of AMPA receptors points to their very large extracellular domain, protruding over 10 nm into the extracellular space (Sobolevsky et al. 2009) and, thus, likely to bump into these ECM components.

7.5.2 Specific Effects of ECM Elements on Neurotransmitter Receptor Regulation and Synaptic Plasticity

In addition, there are ECM components that interact specifically with surface molecules and thereby modify synaptic function. Especially the neuronal activity–regulated pentraxin (Narp), an immediate early gene that is upregulated by neuronal activity, has been implicated in activity-dependent synapse formation and synaptic scaling (O'Brien et al. 1999; Chang et al. 2010). Narp is a secreted calcium-dependent lectin that forms a covalent heteromeric complex with the neuronal pentraxin 1 (NP1) (Tsui et al. 1996; Xu et al. 2003). Narp complexes are enriched at excitatory synapses especially on parvalbumin (PV)-expressing interneurons; they have been suggested to associate with the GluA4 subunit of AMPA receptors and to regulate their synaptic clustering (O'Brien et al. 1999; Sia et al. 2007; Chang et al. 2010). Digestion of the PNN using chondroitinase ABC abolishes Narp clustering on PV-positive neurons indicating that the formation of Narp clusters depends on hyaluronan/CSPG-based ECM structure (Chang et al. 2010). It has been further reported that presynaptically released Narp is required for homeostatic synaptic scaling in PV cells (Chang et al. 2010; Fig. 7.2b). Increased network activity drives Narp expression, which in turn is required for GluA4 upregulation and enhanced mEPSC amplitudes on PV cells after network silencing using TTX. PV-expressing cells from Narp$^{-/-}$ mice show neither increased GluA4 expression nor any form of synaptic scaling (Chang et al. 2010). It is known that PV-positive cells are key players in the control of network activity (Sohal et al. 2009), and the inhibitory network plays an important role in suppression of seizures. Consistently, Narp$^{-/-}$ mice are more sensitive to kindling-induced seizures (Chang et al. 2010).

It is known for long that integrins as classical ECM receptors are involved in synaptic function and plasticity (e.g. Staubli et al. 1998; Chavis and Westbrook 2001; Dityatev and Schachner 2003). Different integrin heterodimers seem to have differential functions in the induction and maintenance of synaptic plasticity including LTP (Chan et al. 2006; for review, see Dityatev et al. 2010a). One particular example is the antagonistic regulation of synaptic accumulations of GlyRs in rat spinal cord neurons (Charrier et al. 2010). Here, β1-containing integrins seem to increase GlyR and gephyrin clusters in synapses, whereas β3-containing integrins have the opposite effect. Potential ligands are TSP1 for β1-containing integrins and fibrinogen for β3-containing integrins. Signaling occurs both via protein kinase C, what affects GlyR mobility inside and outside of synapses, and CaM kinase II that is activated and inhibited by β1 and β3 integrins, respectively, and has opposing effects on synaptic GlyRs (Fig. 7.2e). Thus, active CaMKII is responsible for keeping GlyRs within the synapse. Integrin signaling cooperates with other signaling pathways to regulate synaptic functions. For instance, β1-integrin signaling seems to interact with reelin-induced signaling pathways to regulate NMDA receptor composition and mobility (see below). Interestingly, matrix metalloproteases, like MMP9, can unmask binding motifs on integrin ligands upon proteolytic cleavage and in this way induce integrin-mediated regulation of NMDA receptors (Wang et al. 2008; Michaluk et al. 2009, see below).

Finally, integrins are also involved in homeostatic processes like synaptic scaling. For example, β3-containing integrin complexes regulate the synaptic availability of GluA2-containing AMPA receptors via the small GTPase Rap1 (Cingolani et al. 2008). It is thought that the upregulation of β3-integrins by tumor necrosis factor-α (TNFα) released from astrocytes is responsible for a reduction of GluA2 endocytosis and accordingly a synaptic upscaling in response to tetrodotoxin-induced suppression of network activity (Pozo and Goda 2010; Fig. 7.2c). Accordingly, Steinmetz and Turrigiano (2010) showed that addition of TNFα to hippocampal cultures leads to increased AMPA amplitude at control synapses. However, addition of the factor to prescaled cultures had exactly the opposite effect suggesting that TNFα is a critical factor for maintaining synapses in a plastic range within which scaling can be accomplished (Steinmetz and Turrigiano 2010).

Reelin is a 400-kDa ECM protein that is well appreciated for its function during development, where it is involved in controlling the migration and laminar arrangement of neurons in various structures including the neocortex, the hippocampus, the cerebellum, and the spinal cord. This molecule has also been implicated in the maintenance of neuronal networks (Frotscher 2010) and in mechanisms of synaptic plasticity, e.g., by controlling NMDA receptor function (Herz and Chen 2006). Reelin mediates its function in the adult CNS as well as during development by binding to its cell surface receptors the very-low-density lipoprotein receptor (VLDLR) and ApoE2 receptor (ApoE2R) and in turn the downstream adaptor protein Dab1 (see Fig. 7.2d for details on reelin signaling pathways). Reelin signaling eventually stabilizes F-actin via inducing n-cofilin phosphorylation (Chai et al. 2009). Reelin-mediated stabilization of F-actin is not only crucial for directional migration processes during cortex development, but in addition may be essential for the maintenance of the adult brain and thus was hypothesized to act as a mediator between stability and plasticity in the adult brain (Frotscher 2010).

Moreover, the reelin receptors VLDLR and ApoE2 seem to be directly in contact with synaptic NMDA receptors via the membrane-associated guanylate kinase homologue PSD-95, and reelin signaling is closely connected to NMDA receptor signaling and thereby regulates synaptic plasticity (Herz and Chen 2006; Rogers and Weeber 2008). Reelin has also been implicated in the control of the subunit composition of somatic NMDA receptors during hippocampal maturation (Sinagra et al. 2005), and reelin secreted by GABAergic interneurons is responsible for maintaining the adult NMDA receptor composition. Blockade of reelin secretion reversibly increases the fraction of juvenile GluN2B-containing NMDA receptors, and addition of exogenous reelin can rescue this effect (Campo et al. 2009). In addition, reelin controls the surface trafficking of GluN2B-containing NMDA receptors (Fig. 7.2d). As shown by single-particle tracking, inhibition of reelin function reduced the surface mobility of these receptors and increased their synaptic dwell time in an integrin-dependent manner (Groc et al. 2007). β1-containing integrin receptors are supposed to cooperate with ApoE2Rs and/ or VLDLRs in this context.

Reelin has also been discussed as a serine protease that is able to digest fibronectin, laminin, and to a lesser extent also collagen IV. Recently, also Caspr (contactin-associated protein), a molecule known to be required for the formation of axoglial paranodal junctions surrounding the nodes of Ranvier in myelinated axons, has been added as a substrate of reelin (Devanathan et al. 2010). Caspr inhibits neurite outgrowth of cerebellar neurons, and it has been proposed that shedding of Caspr by proteolytic action of reelin counteracts its repulsive function (Devanathan et al. 2010). Further, it was proposed that binding of the prion protein (PrP) to Caspr protects Caspr from proteolysis and thus supports its repulsive function during neurite outgrowth (Devanathan et al. 2010). However, it should be noted that the protease activity of reelin is controversially discussed in the literature (Kohno and Hattori 2010).

7.6 Proteolysis of the ECM and the Generation of Synaptic Signals

ECM-modulating enzymes with major impact on brain development and synapse function are the large family of the matrix metalloproteases (Ethell and Ethell 2007). Probably, the best-studied member of the MMP family in the nervous system is MMP9. Increased neuronal activity enhances expression of MMP9 and leads to increased proteolysis of β-dystroglycan (Szklarczyk et al. 2002; Michaluk et al. 2007). Depletion of MMP9 results in an impairment of LTP at hippocampal synapses (Nagy et al. 2006). Application of MMP9 to neuronal primary cultures affects lateral diffusion of NMDA receptors without changing the mobility of AMPA receptors or the structure of the hyaluronic acid–based ECM (Michaluk et al. 2009). Rather, extracellular MMP9 proteolysis induced β1-integrin-dependent signaling, which then led to the mobilization of NMDA receptors (Fig. 7.2f). β1-Integrin signaling was also identified as being responsible for MMP9-induced spine enlargement and synaptic potentiation (Wang et al. 2008). To date, the target of MMP9 proteolysis within the ECM that induces integrin signaling remains unknown. Recently, the effect of an enriched environment on synapse morphology, ECM structure, and MMP9 and MMP2 activity in the deep cerebellar nuclei (DCN) of adult mice has been examined (Foscarin et al. 2011). Enriched environment leads to an increased size of Purkinje cell axon termini on DCN neurons. At the same time, a reduction of *Wisteria floribunda* agglutinin (WFA) staining, which labels chondroitin sulfates, and hyaluronic acid labeling was observed indicating a downregulation of PNN. This may happen through a downregulation of components of the PNN after environmental enrichment as it was observed on the level of transcripts for aggrecan and Crtl1. Alternatively, the ECM may be degraded by MMP9 and its close relative MMP2, which exhibited an increased activity after environmental enrichment (Foscarin et al. 2011).

Another matrix metalloprotease, MMP3, processes agrin at the basal lamina of the neuromuscular junction in an activity-dependent manner (Werle and VanSaun 2003). Agrin plays a pivotal role in the development and maintenance of the neuromuscular junction. It induces acetylcholine receptor clustering through the muscle-specific receptor tyrosine kinase (MuSK) and its coreceptor low-density lipoprotein receptor-related protein 4 (LRP4) (Sanes and Lichtman 2001; Dityatev et al. 2010a). It has been suggested that agrin processing by MMP3 is indispensable for normal neuromuscular junction development. In line with this view, MMP3 knockout mice exhibit abnormal neuromuscular junction morphology and acetylcholine receptor distribution (VanSaun et al. 2003).

Also the brain-specific serine protease neurotrypsin has been reported to specifically process agrin (Reif et al. 2008). Actually, neurotrypsin seems to be the sole agrin-cleaving protease in the CNS while, in contrast to MMP3, it is not relevant for agrin cleavage at the neuromuscular junction (Bolliger et al. 2010). Like MMP3, neurotrypsin cleaves agrin at two highly conserved sites releasing a 90-kDa and a 22-kDa fragment. Both agrin fragments are absent in the brain of neurotrypsin-deficient mice indicating that in vivo cleavage of agrin in the brain depends on neurotrypsin (Reif et al. 2007; Reif et al. 2008). The 22-kDa fragment resembles a laminin G3 domain. Indeed, it has been shown that the naturally occurring cleavage product forms a stable, well-folded domain while shorter constructs were aberrantly folded (Tidow et al. 2011). Neurotrypsin has been identified as essential for cognitive functions in the human brain. Deletion mutation in the coding region resulting in a truncated protein without protease domain leads to severe mental retardation (Molinari et al. 2002). Furthermore, neurotrypsin is recruited and released at synapses in an activity-dependent manner (Frischknecht et al. 2008). Activation of neurotrypsin proteolytic activity requires concomitant activation of the postsynaptic neuron (Matsumoto-Miyai et al. 2009). It has been demonstrated that the proteolytic fragment of 22 kDa acquired filopodia-inducing signaling properties in hippocampal slice cultures after induction of synaptic LTP (Matsumoto-Miyai et al. 2009). Similar to MMP9, proteolytic cleavage of ECM components by neurotrypsin unmasks a signaling molecule (Fig. 7.2f), which in turn induces alterations in spine morphology and even the generation of new synapses. These examples suggest that the ECM contains a variety of hidden instructive signals that can be unmasked by specific proteolytic enzymes.

Similar to chondroitinase ABC treatment, the topical application of the serine protease tissue-type plasminogen activator (tPA) can prolong the critical period in the visual cortex (Mataga et al. 2004; Oray et al. 2004). Moreover, tPA expression is significantly increased after monocular deprivation during the critical period of development (Mataga et al. 2002). In line with this observation, tPA knockout mice display reduced OD plasticity during the critical period and repetitive application of tPA to the cortex rescues normal plasticity in mutant mice (Mataga et al. 2002). In adult animals, where no OD plasticity occurs, monocular deprivation leads to no significant increase in tPA activity (Mataga et al. 2004). Application of tPA to cortical slices increases spine motility, and a 2-day period of monocular deprivation, which leads to tPA release, has a similar effect (Mataga et al. 2004). The effect of exogenous tPA is prevented in slice from monocularly deprived animals, suggesting that tPA is the mediator of the increased spine motility after monocular deprivation (Mataga et al. 2004). Moreover, tPA activity is also responsible for a change in spine density that follows monocular deprivation (Oray et al. 2004) with a transient loss of spines in the binocular region of the visual cortex. This effect is absent in the tPA knockout mouse and can be rescued by exogenous application of tPA (Oray et al. 2004). Thus, tPA release after monocular deprivation is developmentally regulated, and by degrading ECM proteins, tPA might be a crucial determinant of experience-dependent plasticity *in vivo*.

7.7 Outlook

After their discovery, PNN have been enigmatic for more than a century. Research activities of the last decade strongly imply that one important function of PNN is the restriction of juvenile plasticity in the adult brain at the end of critical periods. This occurs at the expense of the regenerative potential of the mature vertebrate CNS. Moreover, while PNN are most prominent on a particular class of GABAergic interneurons, PNN-like structures are found throughout the brain where they influence and regulate synaptic plasticity processes. There they have passive and active roles, the molecular mechanisms of which are just about to be discovered. The knowledge of these mechanisms will open new avenues for a better basic understanding of brain function. In longer terms this may also provide opportunities for curative intervention, e.g., by keeping the critical periods for brain wiring open for longer time or even reopening them therapeutically to correct miswiring.

Acknowledgments Work in the authors' laboratory was supported by the Deutsche Forschungsgemeinschaft (GU 230/5-3) and the German Federal Minster for Education and Science BMBF via EraNET NEURON (Moddifsyn).

References

Alpar, A., Gartner, U., Hartig, W., & Bruckner, G. (2006). Distribution of pyramidal cells associated with perineuronal nets in the neocortex of rat. *Brain Research, 1120*, 13–22.

Araque, A., Parpura, V., Sanzgiri, R. P., & Haydon, P. G. (1999). Tripartite synapses: Glia, the unacknowledged partner. *Trends in Neurosciences, 22*, 208–215.

Bandtlow, C. E., & Zimmermann, D. R. (2000). Proteoglycans in the developing brain: New conceptual insights for old proteins. *Physiological Reviews, 80*, 1267–1290.

Berardi, N., Pizzorusso, T., Ratto, G. M., & Maffei, L. (2003). Molecular basis of plasticity in the visual cortex. *Trends in Neurosciences, 26*, 369–378.

Bolliger, M. F., Zurlinden, A., Luscher, D., Butikofer, L., Shakhova, O., Francolini, M., Kozlov, S. V., Cinelli, P., Stephan, A., Kistler, A. D., Rulicke, T., Pelczar, P., Ledermann, B., Fumagalli, G., Gloor, S. M., Kunz, B., & Sonderegger, P. (2010). Specific proteolytic cleavage of agrin regulates maturation of the neuromuscular junction. *Journal of Cell Science, 123*, 3944–3955.

Borgdorff, A. J., & Choquet, D. (2002). Regulation of AMPA receptor lateral movements. *Nature, 417*, 649–653.

Brakebusch, C., Seidenbecher, C. I., Asztely, F., Rauch, U., Matthies, H., Meyer, H., Krug, M., Bockers, T. M., Zhou, X., Kreutz, M. R., Montag, D., Gundelfinger, E. D., & Fassler, R. (2002). Brevican-deficient mice display impaired hippocampal CA1 long-term potentiation but show no obvious deficits in learning and memory. *Molecular and Cellular Biology, 22*, 7417–7427.

Bruckner, G., Grosche, J., Schmidt, S., Hartig, W., Margolis, R. U., Delpech, B., Seidenbecher, C. I., Czaniera, R., & Schachner, M. (2000). Postnatal development of perineuronal nets in wild-type mice and in a mutant deficient in tenascin-R. *The Journal of Comparative Neurology, 428*, 616–629.

Bukalo, O., Schachner, M., & Dityatev, A. (2001). Modification of extracellular matrix by enzymatic removal of chondroitin sulfate and by lack of tenascin-R differentially affects several forms of synaptic plasticity in the hippocampus. *Neuroscience, 104*, 359–369.

Bukalo, O., Schachner, M., & Dityatev, A. (2007). Hippocampal metaplasticity induced by deficiency in the extracellular matrix glycoprotein tenascin-R. *Journal of Neuroscience, 27*, 6019–6028.

Campo, C. G., Sinagra, M., Verrier, D., Manzoni, O. J., & Chavis, P. (2009). Reelin secreted by GABAergic neurons regulates glutamate receptor homeostasis. *PloS One, 4*, e5505.

Carulli, D., Pizzorusso, T., Kwok, J. C., Putignano, E., Poli, A., Forostyak, S., Andrews, M. R., Deepa, S. S., Glant, T. T., & Fawcett, J. W. (2010). Animals lacking link protein have attenuated perineuronal nets and persistent plasticity. *Brain, 133*, 2331–2347.

Carulli, D., Rhodes, K. E., & Fawcett, J. W. (2007). Upregulation of aggrecan, link protein 1, and hyaluronan synthases during formation of perineuronal nets in the rat cerebellum. *The Journal of Comparative Neurology, 501*, 83–94.

Celio, M. R., & Blumcke, I. (1994). Perineuronal nets – A specialized form of extracellular matrix in the adult nervous system. *Brain Research: Brain Research Reviews, 19*, 128–145.

Celio, M. R., Spreafico, R., De Biasi, S., & Vitellaro-Zuccarello, L. (1998). Perineuronal nets: Past and present. *Trends in Neurosciences, 21*, 510–515.

Chai, X., Forster, E., Zhao, S., Bock, H. H., & Frotscher, M. (2009). Reelin stabilizes the actin cytoskeleton of neuronal processes by inducing n-cofilin phosphorylation at serine3. *Journal of Neuroscience, 29*, 288–299.

Chan, C. S., Weeber, E. J., Zong, L., Fuchs, E., Sweatt, J. D., & Davis, R. L. (2006). Beta1-integrins are required for hippocampal AMPA receptor-dependent synaptic transmission, synaptic plasticity, and working memory. *Journal of Neuroscience, 26*, 223–232.

Chang, M. C., Park, J. M., Pelkey, K. A., Grabenstatter, H. L., Xu, D., Linden, D. J., Sutula, T. P., McBain, C. J., & Worley, P. F. (2010). Narp regulates homeostatic scaling of excitatory synapses on parvalbumin-expressing interneurons. *Nature Neuroscience, 13*, 1090–1097.

Charrier, C., Machado, P., Tweedie-Cullen, R. Y., Rutishauser, D., Mansuy, I. M., & Triller, A. (2010). A crosstalk between beta1 and beta3 integrins controls glycine receptor and gephyrin trafficking at synapses. *Nature Neuroscience, 13*, 1388–1395.

Chavis, P., & Westbrook, G. (2001). Integrins mediate functional pre- and postsynaptic maturation at a hippocampal synapse. *Nature, 411*, 317–321.

Choquet, D., & Triller, A. (2003). The role of receptor diffusion in the organization of the postsynaptic membrane. *Nature Reviews Neuroscience, 4*, 251–265.

Christopherson, K. S., Ullian, E. M., Stokes, C. C., Mullowney, C. E., Hell, J. W., Agah, A., Lawler, J., Mosher, D. F., Bornstein, P., & Barres, B. A. (2005). Thrombospondins are astrocyte-secreted proteins that promote CNS synaptogenesis. *Cell, 120*, 421–433.

Cingolani, L. A., Thalhammer, A., Yu, L. M., Catalano, M., Ramos, T., Colicos, M. A., & Goda, Y. (2008). Activity-dependent regulation of synaptic AMPA receptor composition and abundance by beta3 integrins. *Neuron, 58*, 749–762.

Deepa, S. S., Carulli, D., Galtrey, C., Rhodes, K., Fukuda, J., Mikami, T., Sugahara, K., & Fawcett, J. W. (2006). Composition of perineuronal net extracellular matrix in rat brain: A different disaccharide composition for the net-associated proteoglycans. *Journal of Biological Chemistry, 281*, 17789–17800.

Devanathan, V., Jakovcevski, I., Santuccione, A., Li, S., Lee, H. J., Peles, E., Leshchyns'ka, I., Sytnyk, V., & Schachner, M. (2010). Cellular form of prion protein inhibits reelin-mediated shedding of Caspr from the neuronal cell surface to potentiate Caspr-mediated inhibition of neurite outgrowth. *Journal of Neuroscience, 30*, 9292–9305.

Dityatev, A., Bruckner, G., Dityateva, G., Grosche, J., Kleene, R., & Schachner, M. (2007). Activity-dependent formation and functions of chondroitin sulfate-rich extracellular matrix of perineuronal nets. *Developmental Neurobiology, 67*, 570–588.

Dityatev, A., & Fellin, T. (2008). Extracellular matrix in plasticity and epileptogenesis. *Neuron Glia Biology, 4*, 235–247.

Dityatev, A., Frischknecht, R., & Seidenbecher, C. I. (2006). Extracellular matrix and synaptic functions. *Results and Problems in Cell Differentiation, 43*, 69–97.

Dityatev, A., & Schachner, M. (2003). Extracellular matrix molecules and synaptic plasticity. *Nature Reviews Neuroscience, 4*, 456–468.

Dityatev, A., Schachner, M., & Sonderegger, P. (2010a). The dual role of the extracellular matrix in synaptic plasticity and homeostasis. *Nature Reviews Neuroscience, 11*, 735–746.

Dityatev, A., Seidenbecher, C. I., & Schachner, M. (2010b). Compartmentalization from the outside: The extracellular matrix and functional microdomains in the brain. *Trends in Neurosciences, 33*, 503–512.

Eroglu, C. (2009). The role of astrocyte-secreted matricellular proteins in central nervous system development and function. *Journal of Cell Communication Signaling, 3*, 167–176.

Eroglu, C., Allen, N. J., Susman, M. W., O'Rourke, N. A., Park, C. Y., Ozkan, E., Chakraborty, C., Mulinyawe, S. B., Annis, D. S., Huberman, A. D., Green, E. M., Lawler, J., Dolmetsch, R., Garcia, K. C., Smith, S. J., Luo, Z. D., Rosenthal, A., Mosher, D. F., & Barres, B. A. (2009). Gabapentin receptor alpha2delta-1 is a neuronal thrombospondin receptor responsible for excitatory CNS synaptogenesis. *Cell, 139*, 380–392.

Ethell, I. M., & Ethell, D. W. (2007). Matrix metalloproteinases in brain development and remodeling: Synaptic functions and targets. *Journal of Neuroscience Research, 85*, 2813–2823.

Faissner, A., Pyka, M., Geissler, M., Sobik, T., Frischknecht, R., Gundelfinger, E. D., & Seidenbecher, C. (2010). Contributions of astrocytes to synapse formation and maturation – Potential functions of the perisynaptic extracellular matrix. *Brain Research Reviews, 63*, 26–38.

Fawcett, J. (2009a). Molecular control of brain plasticity and repair. *Progress in Brain Research, 175*, 501–509.

Fawcett, J. W. (2009b). Recovery from spinal cord injury: Regeneration, plasticity and rehabilitation. *Brain, 132*, 1417–1418.

Foscarin, S., Ponchione, D., Pajaj, E., Leto, K., Gawlak, M., Wilczynski, G. M., Rossi, F., & Carulli, D. (2011). Experience-dependent plasticity and modulation of growth regulatory molecules at central synapses. *PloS One, 6*, e16666.

Frischknecht, R., Fejtova, A., Viesti, M., Stephan, A., & Sonderegger, P. (2008). Activity-induced synaptic capture and exocytosis of the neuronal serine protease neurotrypsin. *Journal of Neuroscience, 28*, 1568–1579.

Frischknecht, R., Heine, M., Perrais, D., Seidenbecher, C. I., Choquet, D., & Gundelfinger, E. D. (2009). Brain extracellular matrix affects AMPA receptor lateral mobility and short-term synaptic plasticity. *Nature Neuroscience, 12*, 897–904.

Frischknecht, R., & Seidenbecher, C. I. (2008). The crosstalk of hyaluronan-based extracellular matrix and synapses. *Neuron Glia Biology, 4*, 249–257.

Frotscher, M. (2010). Role for reelin in stabilizing cortical architecture. *Trends in Neurosciences, 33*, 407–414.

Galtrey, C. M., & Fawcett, J. W. (2007). The role of chondroitin sulfate proteoglycans in regeneration and plasticity in the central nervous system. *Brain Research Reviews, 54*, 1–18.

Giamanco, K. A., Morawski, M., & Matthews, R. T. (2010). Perineuronal net formation and structure in aggrecan knockout mice. *Neuroscience, 170*, 1314–1327.

Gogolla, N., Caroni, P., Luthi, A., & Herry, C. (2009). Perineuronal nets protect fear memories from erasure. *Science, 325*, 1258–1261.

Groc, L., Choquet, D., Stephenson, F. A., Verrier, D., Manzoni, O. J., & Chavis, P. (2007). NMDA receptor surface trafficking and synaptic subunit composition are developmentally regulated by the extracellular matrix protein reelin. *Journal of Neuroscience, 27*, 10165–10175.

Hartig, W., Derouiche, A., Welt, K., Brauer, K., Grosche, J., Mader, M., Reichenbach, A., & Bruckner, G. (1999). Cortical neurons immunoreactive for the potassium channel Kv3.1b subunit are predominantly surrounded by perineuronal nets presumed as a buffering system for cations. *Brain Research, 842*, 15–29.

Haydon, P. G. (2001). GLIA: Listening and talking to the synapse. *Nature Reviews Neuroscience, 2*, 185–193.

Heine, M., Groc, L., Frischknecht, R., Beique, J. C., Lounis, B., Rumbaugh, G., Huganir, R. L., Cognet, L., & Choquet, D. (2008). Surface mobility of postsynaptic AMPARs tunes synaptic transmission. *Science, 320*, 201–205.

Herz, J., & Chen, Y. (2006). Reelin, lipoprotein receptors and synaptic plasticity. *Nature Reviews Neuroscience, 7*, 850–859.

John, N., Krugel, H., Frischknecht, R., Smalla, K. H., Schultz, C., Kreutz, M. R., Gundelfinger, E. D., & Seidenbecher, C. I. (2006). Brevican-containing perineuronal nets of extracellular matrix in dissociated hippocampal primary cultures. *Molecular and Cellular Neuroscience, 31*, 774–784.

Kochlamazashvili, G., Henneberger, C., Bukalo, O., Dvoretskova, E., Senkov, O., Lievens, P. M., Westenbroek, R., Engel, A. K., Catterall, W. A., Rusakov, D. A., Schachner, M., & Dityatev, A. (2010). The extracellular matrix molecule hyaluronic acid regulates hippocampal synaptic plasticity by modulating postsynaptic L-type Ca(2+) channels. *Neuron, 67*, 116–128.

Kohno, T., & Hattori, M. (2010). Re-evaluation of protease activity of reelin. *Biological and Pharmaceutical Bulletin, 33*, 1047–1049.

Kusumi, A., Ike, H., Nakada, C., Murase, K., & Fujiwara, T. (2005). Single-molecule tracking of membrane molecules: Plasma membrane compartmentalization and dynamic assembly of raft-philic signaling molecules. *Seminars in Immunology, 17*, 3–21.

Kusumi, A., Sako, Y., & Yamamoto, M. (1993). Confined lateral diffusion of membrane receptors as studied by single particle tracking (nanovid microscopy). Effects of calcium-induced differentiation in cultured epithelial cells. *Biophysical Journal, 65*, 2021–2040.

Kwok, J. C., Carulli, D., & Fawcett, J. W. (2010). In vitro modeling of perineuronal nets: Hyaluronan synthase and link protein are necessary for their formation and integrity. *Journal of Neurochemistry, 114*, 1447–1459.

Li, K. W., Hornshaw, M. P., Van Der Schors, R. C., Watson, R., Tate, S., Casetta, B., Jimenez, C. R., Gouwenberg, Y., Gundelfinger, E. D., Smalla, K. H., & Smit, A. B. (2004). Proteomics analysis of rat brain postsynaptic density. Implications of the diverse protein functional groups for the integration of synaptic physiology. *Journal of Biological Chemistry, 279*, 987–1002.

Mataga, N., Mizuguchi, Y., & Hensch, T. K. (2004). Experience-dependent pruning of dendritic spines in visual cortex by tissue plasminogen activator. *Neuron, 44*, 1031–1041.

Mataga, N., Nagai, N., & Hensch, T. K. (2002). Permissive proteolytic activity for visual cortical plasticity. *Proceedings of the National Academy of Sciences of the United States of America, 99*, 7717–7721.

Matsumoto-Miyai, K., Sokolowska, E., Zurlinden, A., Gee, C. E., Luscher, D., Hettwer, S., Wolfel, J., Ladner, A. P., Ster, J., Gerber, U., Rulicke, T., Kunz, B., & Sonderegger, P. (2009). Coincident pre- and postsynaptic activation induces dendritic filopodia via neurotrypsin-dependent agrin cleavage. *Cell, 136*, 1161–1171.

Matthews, R. T., Kelly, G. M., Zerillo, C. A., Gray, G., Tiemeyer, M., & Hockfield, S. (2002). Aggrecan glycoforms contribute to the molecular heterogeneity of perineuronal nets. *Journal of Neuroscience, 22*, 7536–7547.

Michaluk, P., Kolodziej, L., Mioduszewska, B., Wilczynski, G. M., Dzwonek, J., Jaworski, J., Gorecki, D. C., Ottersen, O. P., & Kaczmarek, L. (2007). Beta-dystroglycan as a target for MMP-9, in response to enhanced neuronal activity. *Journal of Biological Chemistry, 282*, 16036–16041.

Michaluk, P., Mikasova, L., Groc, L., Frischknecht, R., Choquet, D., & Kaczmarek, L. (2009). Matrix metalloproteinase-9 controls NMDA receptor surface diffusion through integrin beta1 signaling. *Journal of Neuroscience, 29*, 6007–6012.

Milev, P., Maurel, P., Chiba, A., Mevissen, M., Popp, S., Yamaguchi, Y., Margolis, R. K., & Margolis, R. U. (1998). Differential regulation of expression of hyaluronan-binding proteoglycans in developing brain: Aggrecan, versican, neurocan, and brevican. *Biochemical and Biophysical Research Communications, 247*, 207–212.

Miyata, S., Nishimura, Y., Hayashi, N., & Oohira, A. (2005). Construction of perineuronal net-like structure by cortical neurons in culture. *Neuroscience, 136*, 95–104.

Molinari, F., Rio, M., Meskenaite, V., Encha-Razavi, F., Auge, J., Bacq, D., Briault, S., Vekemans, M., Munnich, A., Attie-Bitach, T., Sonderegger, P., & Colleaux, L. (2002). Truncating neurotrypsin mutation in autosomal recessive nonsyndromic mental retardation. *Science, 298*, 1779–1781.

Nagy, V., Bozdagi, O., Matynia, A., Balcerzyk, M., Okulski, P., Dzwonek, J., Costa, R. M., Silva, A. J., Kaczmarek, L., & Huntley, G. W. (2006). Matrix metalloproteinase-9 is required for hippocampal late-phase long-term potentiation and memory. *Journal of Neuroscience, 26*, 1923–1934.

Newpher, T. M., & Ehlers, M. D. (2008). Glutamate receptor dynamics in dendritic microdomains. *Neuron, 58*, 472–497.

O'Brien, R. J., Xu, D., Petralia, R. S., Steward, O., Huganir, R. L., & Worley, P. (1999). Synaptic clustering of AMPA receptors by the extracellular immediate-early gene product Narp. *Neuron, 23*, 309–323.

Oray, S., Majewska, A., & Sur, M. (2004). Dendritic spine dynamics are regulated by monocular deprivation and extracellular matrix degradation. *Neuron, 44*, 1021–1030.

Petrini, E. M., Lu, J., Cognet, L., Lounis, B., Ehlers, M. D., & Choquet, D. (2009). Endocytic trafficking and recycling maintain a pool of mobile surface AMPA receptors required for synaptic potentiation. *Neuron, 63*, 92–105.

Pizzorusso, T. (2009). Neuroscience. Erasing fear memories. *Science, 325*, 1214–1215.

Pizzorusso, T., Medini, P., Berardi, N., Chierzi, S., Fawcett, J. W., & Maffei, L. (2002). Reactivation of ocular dominance plasticity in the adult visual cortex. *Science, 298*, 1248–1251.

Pizzorusso, T., Medini, P., Landi, S., Baldini, S., Berardi, N., & Maffei, L. (2006). Structural and functional recovery from early monocular deprivation in adult rats. *Proceedings of the National Academy of Sciences of the United States of America, 103*, 8517–8522.

Pozo, K., & Goda, Y. (2010). Unraveling mechanisms of homeostatic synaptic plasticity. *Neuron, 66*, 337–351.

Rauch, U. (2004). Extracellular matrix components associated with remodeling processes in brain. *Cellular and Molecular Life Sciences, 61*, 2031–2045.

Reif, R., Sales, S., Dreier, B., Luscher, D., Wolfel, J., Gisler, C., Baici, A., Kunz, B., & Sonderegger, P. (2008). Purification and enzymological characterization of murine neurotrypsin. *Protein Expression and Purification, 61*, 13–21.

Reif, R., Sales, S., Hettwer, S., Dreier, B., Gisler, C., Wolfel, J., Luscher, D., Zurlinden, A., Stephan, A., Ahmed, S., Baici, A., Ledermann, B., Kunz, B., & Sonderegger, P. (2007). Specific cleavage of agrin by neurotrypsin, a synaptic protease linked to mental retardation. *The FASEB Journal, 21*, 3468–3478.

Rogers, J. T., & Weeber, E. J. (2008). Reelin and apoE actions on signal transduction, synaptic function and memory formation. *Neuron Glia Biology, 4*, 259–270.

Sanes, J. R., & Lichtman, J. W. (2001). Induction, assembly, maturation and maintenance of a postsynaptic apparatus. *Nature Reviews Neuroscience, 2*, 791–805.

Seidenbecher, C., Richter, K., & Gundelfinger, E. D. (1997). Brevican, a conditional proteoglycan from rat brain: Characterization of secreted and GPI-anchored isoforms. In A. W. Teelken & J. Korf (Eds.), *Neurochemistry* (pp. 901–904). New York: Plenum Press.

Seidenbecher, C. I., Richter, K., Rauch, U., Fassler, R., Garner, C. C., & Gundelfinger, E. D. (1995). Brevican, a chondroitin sulfate proteoglycan of rat brain, occurs as secreted and cell surface glycosylphosphatidylinositol-anchored isoforms. *Journal of Biological Chemistry, 270*, 27206–27212.

Seidenbecher, C. I., Smalla, K. H., Fischer, N., Gundelfinger, E. D., & Kreutz, M. R. (2002). Brevican isoforms associate with neural membranes. *Journal of Neurochemistry, 83*, 738–746.

Sia, G. M., Beique, J. C., Rumbaugh, G., Cho, R., Worley, P. F., & Huganir, R. L. (2007). Interaction of the N-terminal domain of the AMPA receptor GluR4 subunit with the neuronal pentraxin NP1 mediates GluR4 synaptic recruitment. *Neuron, 55*, 87–102.

Sinagra, M., Verrier, D., Frankova, D., Korwek, K. M., Blahos, J., Weeber, E. J., Manzoni, O. J., & Chavis, P. (2005). Reelin, very-low-density lipoprotein receptor, and apolipoprotein E receptor 2 control somatic NMDA receptor composition during hippocampal maturation in vitro. *Journal of Neuroscience, 25*, 6127–6136.

Slezak, M., & Pfrieger, F. W. (2003). New roles for astrocytes: Regulation of CNS synaptogenesis. *Trends in Neurosciences, 26*, 531–535.

Sobolevsky, A. I., Rosconi, M. P., & Gouaux, E. (2009). X-ray structure, symmetry and mechanism of an AMPA-subtype glutamate receptor. *Nature, 462*, 745–756.

Sohal, V. S., Zhang, F., Yizhar, O., & Deisseroth, K. (2009). Parvalbumin neurons and gamma rhythms enhance cortical circuit performance. *Nature, 459*, 698–702.

Staubli, U., Chun, D., & Lynch, G. (1998). Time-dependent reversal of long-term potentiation by an integrin antagonist. *Journal of Neuroscience, 18*, 3460–3469.

Steinmetz, C. C., & Turrigiano, G. G. (2010). Tumor necrosis factor-alpha signaling maintains the ability of cortical synapses to express synaptic scaling. *Journal of Neuroscience, 30*, 14685–14690.

Szklarczyk, A., Lapinska, J., Rylski, M., McKay, R. D., & Kaczmarek, L. (2002). Matrix metalloproteinase-9 undergoes expression and activation during dendritic remodeling in adult hippocampus. *Journal of Neuroscience, 22*, 920–930.

Tidow, H., Mattle, D., & Nissen, P. (2011). Structural and biophysical characterisation of agrin laminin G3 domain constructs. *Protein Engineering, Design & Selection, 24*, 219–224.

Triller, A., & Choquet, D. (2008). New concepts in synaptic biology derived from single-molecule imaging. *Neuron, 59*, 359–374.

Tsui, C. C., Copeland, N. G., Gilbert, D. J., Jenkins, N. A., Barnes, C., & Worley, P. F. (1996). Narp, a novel member of the pentraxin family, promotes neurite outgrowth and is dynamically regulated by neuronal activity. *Journal of Neuroscience, 16*, 2463–2478.

VanSaun, M., Herrera, A. A., & Werle, M. J. (2003). Structural alterations at the neuromuscular junctions of matrix metalloproteinase 3 null mutant mice. *Journal of Neurocytology, 32*, 1129–1142.

Wang, X. B., Bozdagi, O., Nikitczuk, J. S., Zhai, Z. W., Zhou, Q., & Huntley, G. W. (2008). Extracellular proteolysis by matrix metalloproteinase-9 drives dendritic spine enlargement and long-term potentiation coordinately. *Proceedings of the National Academy of Sciences of the United States of America, 105*, 19520–19525.

Wegner, F., Hartig, W., Bringmann, A., Grosche, J., Wohlfarth, K., Zuschratter, W., & Bruckner, G. (2003). Diffuse perineuronal nets and modified pyramidal cells immunoreactive for glutamate and the GABA(A) receptor alpha1 subunit form a unique entity in rat cerebral cortex. *Experimental Neurology, 184*, 705–714.

Werle, M. J., & VanSaun, M. (2003). Activity dependent removal of agrin from synaptic basal lamina by matrix metalloproteinase 3. *Journal of Neurocytology, 32*, 905–913.

Wiesel, T. N., & Hubel, D. H. (1963). Effects of visual deprivation on morphology and physiology of cells in the cats lateral geniculate body. *Journal of Neurophysiology, 26*, 978–993.

Xu, D., Hopf, C., Reddy, R., Cho, R. W., Guo, L., Lanahan, A., Petralia, R. S., Wenthold, R. J., O'Brien, R. J., & Worley, P. (2003). Narp and NP1 form heterocomplexes that function in developmental and activity-dependent synaptic plasticity. *Neuron, 39*, 513–528.

Xu, J., Xiao, N., & Xia, J. (2010). Thrombospondin 1 accelerates synaptogenesis in hippocampal neurons through neuroligin 1. *Nature Neuroscience, 13*, 22–24.

Yamada, H., Fredette, B., Shitara, K., Hagihara, K., Miura, R., Ranscht, B., Stallcup, W. B., & Yamaguchi, Y. (1997). The brain chondroitin sulfate proteoglycan brevican associates with astrocytes ensheathing cerebellar glomeruli and inhibits neurite outgrowth from granule neurons. *Journal of Neuroscience, 17*, 7784–7795.

Yamaguchi, Y. (2000). Lecticans: Organizers of the brain extracellular matrix. *Cellular and Molecular Life Sciences, 57*, 276–289.

Zhou, X. H., Brakebusch, C., Matthies, H. et al. (2001). Neurocan is dispensable for brain development. *Mol Cell Biol, 21*, 5970–5978.

Zimmermann, D. R., & Dours-Zimmermann, M. T. (2008). Extracellular matrix of the central nervous system: From neglect to challenge. *Histochemistry and Cell Biology, 130*, 635–653.

Zuber, B., Nikonenko, I., Klauser, P., Muller, D., & Dubochet, J. (2005). The mammalian central nervous synaptic cleft contains a high density of periodically organized complexes. *Proceedings of the National Academy of Sciences of the United States of America, 102*, 19192–19197.

Chapter 8
Molecular Motors in Cargo Trafficking and Synapse Assembly

Robert van den Berg and Casper C. Hoogenraad

Abstract Every production process, be it cellular or industrial, depends on a constant supply of energy and resources. Synapses, specialized junctions in the central nervous system through which neurons signal to each other, are no exception to this rule. In order to form new synapses and alter the strength of synaptic transmission, neurons need a regulatory mechanism to deliver and remove synaptic proteins at synaptic sites. Neurons make use of active transport driven by molecular motor proteins to move synaptic cargo over either microtubules (kinesin, dynein) or actin filaments (myosin) to their specific site of action. These mechanisms are crucial for the initial establishment of synaptic specializations during synaptogenesis and for activity-dependent changes in synaptic strength during plasticity. In this chapter, we address the organization of the neuronal cytoskeleton, focus on synaptic cargo transport activities that operate in axons and dendrites, and discuss the spatial and temporal regulation of motor protein-based transport.

Keywords Actin • Axon • Cytoskeleton • Dendrite • Disease • Dynein • Kinesin • Microtubule • Myosin • Neuron • Synapse • Synaptic plasticity

8.1 Introduction

One apparent feature of neurons is that once the axon and dendrites have grown out, they establish synaptic contacts forming neuronal networks that propagate information in a unidirectional fashion. Excitatory synaptic signaling in the brain occurs by releasing glutamate from "sending" neurons and activating glutamate receptors

R. van den Berg • C.C. Hoogenraad (⊠)
Cell Biology, Faculty of Science, Utrecht University, Padualaan 8, 3584CH, Utrecht, The Netherlands
e-mail: c.hoogenraad@uu.nl

M.R. Kreutz and C. Sala (eds.), *Synaptic Plasticity*,
Advances in Experimental Medicine and Biology 970,
DOI 10.1007/978-3-7091-0932-8_8, © Springer-Verlag/Wien 2012

at "receiving" neurons (Sheng and Hoogenraad 2007; Südhof 2004). Structurally, synapses are divided into two specialized domains: the presynaptic bouton on the axon side of the "sending neuron" and the postsynaptic compartment on the dendrite of the "receiving" neuron. The directional nature of signal relay requires that synaptic contacts are morphologically asymmetric with distinct protein components. Recent studies have identified the molecular components of synapses, particularly by using genetic and proteomic strategies, and have revealed that the specification of synaptic function, for example, excitatory or inhibitory, at both pre- and postsynapses is achieved via the recruitment and assembly of very distinct synaptic complexes (Jin and Garner 2008; Kim and Sheng 2004; Margeta et al. 2008; Sheng and Hoogenraad 2007). Proper arrangement of pre- and postsynaptic membranes and organization of pre- and postsynaptic compartments is essential for accurate synaptic signaling, neural network activity, and cognitive processes such as learning and memory formation (Kasai et al. 2010; Lisman et al. 2007; Yuste and Bonhoeffer 2001).

Most of the synaptic cargos, such as neurotransmitter receptors, ion channels, integral membrane proteins, signaling complexes, mRNAs, synaptic vesicle precursors, or mitochondria, are made and preassembled in the cell soma and need to be transported to the proper synaptic destinations. Studies on intracellular trafficking have demonstrated various mechanisms for compartment-specific localization (Goldstein and Yang 2000; Hirokawa and Takemura 2005; Hoogenraad and Bradke 2009; Winckler and Mellman 2010). For example, several synaptic cargos are nonspecifically transported to both axons and dendrites and are then selectively retained at the required compartments (Bel et al. 2009; Garrido et al. 2001; Leterrier et al. 2006; Sampo et al. 2003; Wisco et al. 2003). Alternatively, many presynaptic cargos are correctly sorted into axons (Kaether et al. 2000; Pennuto et al. 2003), whereas postsynaptic components move specifically into dendritic branches and spines, which are specialized dendritic protrusions that mediate most of the excitatory synaptic transmission (Craig et al. 1993; Ruberti and Dotti 2000; Stowell and Craig 1999; Wang et al. 2008). In consequence, synaptic precursor vesicles need to be steered into axons in order to reach the presynaptic terminals, and glutamate receptors need to be transported into dendrites to be correctly inserted in the postsynaptic membrane. Importantly, several neurological diseases are linked to abnormalities in the machinery that controls synaptic cargo trafficking (Chevalier-Larsen and Holzbaur 2006; Gunawardena and Goldstein 2004; Lau and Zukin 2007; Shepherd and Huganir 2007; van Spronsen and Hoogenraad 2010).

Most intracellular cargo transport is driven by molecular motor proteins that move along two types of cytoskeletal elements: actin filaments and microtubules (Schliwa and Woehlke 2003; Vale 2003). Actin facilitates motility of motor proteins of the myosin superfamily, whereas microtubules serve as tracks for two families of motor proteins, kinesin and dynein. While many different motor proteins have been found to participate in neuronal cargo trafficking (Goldstein and Yang 2000; Hirokawa and Takemura 2005), for many of these their precise contribution to synaptic cargo transport has remained unclear. Most current models for neuronal trafficking rely heavily on microtubule plus-end-directed kinesin family members (Hirokawa and Takemura 2005); however, recent work reported important roles for dynein and myosin in synaptic cargo transport (Kapitein et al. 2010; Lewis et al. 2009; Zheng et al. 2008). Further evidence suggests that the docking of molecular motors to synaptic cargo vesicles via adaptor molecules is an important mechanism to achieve transport specificity (Akhmanova and Hammer 2010; Schlager and Hoogenraad 2009).

It has intrigued scientists for a long time how synaptic cargo can be sorted in neurons along the cytoskeleton network to ensure precise cargo delivery. How are motor-cargo complexes able to choose between transport routes to the axon or dendrites? Here, it is important to consider that the actin and microtubule filaments themselves are intrinsically polarized structures with two functionally distinct ends, a "plus" and "minus" end. The polarity of the cytoskeleton filaments exists not only at the two ends but also all along the length of its lattice, which is critical for the directional movement of molecular motor proteins. For example, dynein transports cargo toward the minus end of microtubules, while kinesins motor proteins move toward the plus end of microtubules. In this way, local polarity patterns of microtubules and actin filaments in axon and dendrites can direct motor-driven cargo trafficking within neurons. Recent evidence suggests that a well-organized cytoskeleton network exists in neurons that can facilitate directional motor-driven cargo trafficking and establish asymmetric distributions of specific synaptic proteins (Kapitein and Hoogenraad 2010). Moreover, variations in cytoskeleton density, binding proteins, and posttranslational modifications could also drive synaptic cargo transport in specific directions. Thus, knowing the polarity and modification pattern of microtubules and actin in axon and dendrites is an instrumental piece of information for understanding how molecular motors direct synaptic cargo traffic.

In this chapter, we aim to give an overview of the molecular trafficking mechanisms important for the delivery of synaptic proteins. We will review current knowledge about the organization of the neuronal cytoskeleton, focus on synaptic cargo sorting and trafficking into axons and dendrites, and discuss the spatial and temporal regulation of motor protein–based transport. Studying the basic cellular machinery for synaptic cargo trafficking will help us to understand fundamental principles of synapse formation, function, and plasticity.

8.2 Microtubule and Actin Cytoskeleton in Neurons

The cytoskeletal organization in neurons is specialized in several ways, involving intracellular variations in density, orientations, binding proteins, and posttranslational modifications. Recently, it has become increasingly clear that these specific cytoskeletal properties directly modulate the activity of specific molecular motor proteins. In this section, we will review current knowledge about the structure, organization, and modifications of the microtubule and actin cytoskeleton (Fig. 8.1).

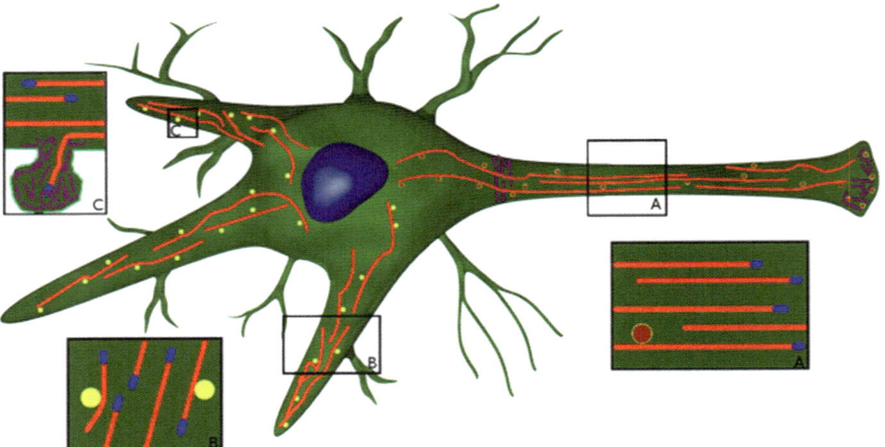

Fig. 8.1 Cytoskeleton organization in neurons. Microtubules (*red*) are present in both axons and dendritic shaft, while actin (*purple*) is enriched in the axon initial segment and dendritic spines. From the cell soma, synaptic cargo vesicles are sorted into the dendrites (*yellow vesicles*) or the axon (*red vesicles*). (**a**) In the axon, all microtubules are oriented with their plus ends (*blue comets*) pointing outward. (**b**) In dendrites, microtubules have a mixed orientation. (**c**) Dynamic microtubules occasionally enter spines

8.2.1   Actin and Microtubule Structure and Dynamics

The main transport infrastructure in eukaryotic cells is formed by the microtubules and actin cytoskeleton. To serve this function, the cytoskeleton must be organized into a wide variety of configurations, ranging from the higher-order actin-based networks in dendritic spines to the dense antiparallel microtubule array in dendritic shaft. Without the cytoskeleton, neurons would not be able to maintain their complex axonal and dendritic architectures and synaptic organization. Micro-tubules are noncovalent polymers of α- and β-tubulin dimers that form a hollow cylinder with a diameter of 25 nm and actin filaments consist of monomers of the protein actin that polymerizes to form 8-nm fibers (Howard and Hyman 2009; Pollard and Cooper 2009). There are multiple actin and α- and β-tubulin genes that are highly conserved among and between species and might be utilized for distinct neuronal functions. With the recent discovery of several congenital neurological disorders that result from mutations in genes that encode different α- and β-tubulin isotypes, novel paradigms have emerged to assess how selective perturbations in microtubule subunits affect neuronal functioning (Baran et al. 2010; Keays et al. 2007). For example, a series of heterozygous missense mutations in TUBB3, encoding the neuron-specific ß-tubulin isotype III, have been described that result in a spectrum of human nervous system disorders (Tischfield et al. 2010). A knock-in disease mouse model reveals axon guidance defects without evidence of cortical cell migration abnormalities. Importantly, it was demonstrated that the disease-associated mutations impair tubulin heterodimer formation and disrupt the interaction with kinesin motors.

Microtubules and actin filaments are intrinsically polar structures and contain two distinct ends, a "plus end" favored for assembly/disassembly and a "minus end" which is less favored for these dynamics. Minus ends of microtubules are often, but not always, attached to the centrosome from which the microtubule is nucleated. It was recently found that the centrosome loses its function as a microtubule-organizing center during development of hippocampal neurons (Stiess et al. 2010). It is well possible that acentrosomal microtubule nucleation arranges the neuronal microtubule cytoskeleton in mature neurons and is responsible for the extended and complex morphology of neurons. In living cells, microtubules and actin filaments are highly dynamic, and their dominant kinetic behavior is known as dynamic instability, where the individual ends alternate between bouts of growth ("polymerization") and shrinkage ("depolymerization") (Mitchison and Kirschner 1984). Microtubules and actin filaments may also undergo treadmilling, a phenomenon in which filament length remains approximately constant, while monomers add at the plus end and dissociate from the minus end (Kueh and Mitchison 2009).

Regulation of microtubule and actin dynamics and turnover plays an important role in neuronal development and synaptic plasticity (Conde and Caceres 2009; Dent and Gertler 2003; Frost et al. 2010a; Hotulainen and Hoogenraad 2010). It is not surprising that recently, a lot of attention has been given to the plus ends of the microtubule, the site where most dynamics takes place. The microtubule plus end can grow, then undergo a shrinking event ("catastrophe"), pause, and grow again ("rescue"), all in a matter of seconds. The fate of the microtubule is determined by a large number of plus-end tracking proteins, most of them only found on the microtubule tip, that can control microtubule dynamics (Akhmanova and Steinmetz 2008; Gouveia and Akhmanova 2010). Plus-end tracking proteins regulate different aspects of neuronal architecture by mediating the cross talk between microtubule ends, the actin cytoskeleton, and the cell cortex, and participate in transport and positioning of signaling factors and membrane organelles (Hoogenraad and Bradke 2009; Jaworski et al. 2008). For example, it was recently found that the microtubule plus-end binding protein EB3 regulates actin dynamics in dendritic spines and is involved in spine morphology and synaptic plasticity (Jaworski et al. 2009).

8.2.2 Actin Organization in Neurons

Microtubules and actin filaments are present throughout the cell body and axonal and dendritic compartments (Cingolani and Goda 2008; Conde and Caceres 2009; Hotulainen and Hoogenraad 2010). In mature neurons, actin is the most prominent cytoskeletal protein at synapses, being present at both the pre- and the postsynaptic terminals but highly enriched at dendritic spines (Landis et al. 1988; Landis and

Reese 1983). Neurons treated with latrunculin, an inhibitor of actin polymerization, showed that the actin cytoskeleton is necessary for synapse formation, stability, and normal synaptic activity (Krucker et al. 2000; Okamoto et al. 2004). At the presynaptic terminal, actin filaments are important in organizing and recycling of synaptic vesicle (Cingolani and Goda 2008), while at the postsynaptic site, actin is involved in receptor trafficking and spine plasticity (Frost et al. 2010a; Hotulainen and Hoogenraad 2010). Spines exhibit a continuous network of both branched and long, linear actin filaments (Korobova and Svitkina 2010). In the spine heads, actin filaments are oriented with their plus ends to the postsynaptic density and synaptic membrane and their minus ends toward the dendritic shaft. The most likely role of actin in mature spines is to stabilize postsynaptic proteins and modulate spine head structure in response to postsynaptic signaling (Frost et al. 2010a; Hotulainen and Hoogenraad 2010). Several recent studies using advanced microscopy techniques present new insights in the organization and molecular composition of actin cytoskeleton in dendritic spines (Frost et al. 2010b; Honkura et al. 2008; Korobova and Svitkina 2010). For example, electron microscopy revealed that actin filaments in the neck of dendritic spines were shown to exhibit mixed polarity, although the plus ends are predominantly oriented away from the dendritic shaft (Hotulainen et al. 2009; Korobova and Svitkina 2010). Mutations in genes that code for actin regulatory proteins, like Rho GTPases, are commonly associated with mental retardation (Govek et al. 2004). Subtle variations in spine size and shape, mediated by the actin cytoskeleton, are associated with a variety of neurological and psychiatric disorders like schizophrenia and drug addiction (Blanpied and Ehlers 2004; van Spronsen and Hoogenraad 2010).

Apart from its organization in synapses and spines, little is known about the arrangement of the actin cytoskeleton in the axonal and dendritic shafts in mature neurons (Kapitein and Hoogenraad 2010). Evidence suggests that actin is present as short, branched filaments in the axon oriented perpendicular to the plasma membrane and is sometimes aligned with microtubules (Bearer and Reese 1999). Several papers showed that actin is enriched in the axon initial segment (Nakada et al. 2003; Winckler et al. 1999). Here, actin is an important regulator of Na^+-channel stability at the initial segment membrane and is actively involved in the maintenance of neuronal polarity (Rasband 2010). The actin network in the initial segment could also function as a selective barrier for cargo transport to enter the axon. In fact, a recent paper presented evidence for an actin-based molecular sieve that prevents the diffusional entry of large macromolecules at the initial segment and isolates the axon from the cell body (Song et al. 2009). In contrast, other work proposed that the actin organization in axons promotes retrograde cargo transport by myosin Va motors (Lewis et al. 2009), which would require actin filaments of uniform polarity oriented parallel to the axon long axis. To better understand actin-based transport activity in neurons, detailed studies are required to reveal the actin cytoskeleton organization in axons, dendrites, and at the initial segment.

8.2.3 Microtubule Organization in Neurons

In mature neurons, most neuronal microtubules are not attached to the centrosome (Stiess et al. 2010) and form dense bundles running along the length of axons and dendrites (Fig. 8.1). Individual microtubules do not extend along the entire length of neuronal processes; instead, microtubule fragments stabilized at their minus ends form regularly spaced longitudinal arrays cross-linked by microtubule-associated proteins (MAPs) (Chen et al. 1992). Two abundant neuronal MAPs whose distributions are polarized between axons and dendrites are tau and MAP2 (Dehmelt and Halpain 2005). Since tau and MAP2 induce the formation of longitudinal microtubule bundles with a distinct spacing, it is plausible that MAPs are directly involved in the organizing of polarized cargo trafficking in axons and dendrites. So far, there is no evidence that this different spacing directly influences the transport machinery. In contrast, the presence of MAPs on the microtubule lattice seems to affect motor protein motility in vitro (Dixit et al. 2008). MAPs can cause decreased attachment and/or increased detachment of kinesin-1 motors and/or vesicles (Verhey and Hammond 2009). However, a direct role of MAPs in guiding polarized transport seems unlikely, as tau and MAP2 knockout mice show relatively mild phenotypes (Dehmelt and Halpain 2005).

Microtubule arrays within axon and dendrites are highly organized with respect to their intrinsic polarity (Hoogenraad and Bradke 2009; Kapitein and Hoogenraad 2010). In the late 1980s, an elegant hook-decoration technique was used to determine the orientation of microtubules in axons and dendrites by electron microscopy (Baas et al. 1988; Burton 1988). Surprisingly, microtubules in the axon are generally long and uniformly oriented, with their plus ends facing outward, while the microtubules in the dendrites are oriented in both directions (Fig. 8.1). Thinner, more distal dendrites, however, contain unipolar microtubules oriented the same way as axonal ones (Baas et al. 1988, 1989). This specialized microtubule organization has also been captured in action by visualizing EB3-GFP in living neuronal cells (Jaworski et al. 2009; Morrison et al. 2002; Stepanova et al. 2003). Stepanova et al. showed that in axons and distal dendrites, all dynamic microtubule plus ends point toward growth cones, while in proximal dendrites, significant EB3-GFP displacement was directed toward the cell body (Stepanova et al. 2003). Remarkably, the use of EB1-GFP tracking in *Drosophila* neurons provided strong evidence that axons and dendrites both have microtubule arrays of uniform polarity orientation. However, while axonal microtubules were uniformly plus-end distal, dendritic microtubules were of the opposite orientation, nearly all minus-end out (Rolls et al. 2007; Stone et al. 2008). These results show a striking difference in microtubule orientations in axons and dendrites and further prompt the idea that these distinct patterns might underlie polarized sorting of synaptic cargos (Fig. 8.2). Distinct patterns of microtubule orientation can directly generate asymmetries in the composition of each neuronal compartment by directing specific motor protein transport into axons or dendrites. Indeed, recent work in vertebrates, worms, and flies reported a specific role for plus- and minus-end-directed motor proteins in steering

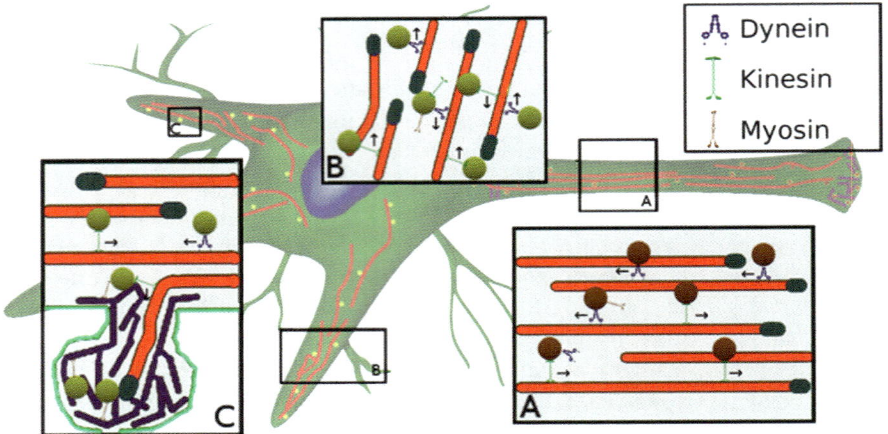

Fig. 8.2 Essential components of the synaptic cargo transport machinery. Kinesin, dynein, and myosin are the three classes of motor proteins that form the workforce of the transport system. (**a**) In the axon, dynein typically moves its cargo retrograde, toward the cell body, while kinesins move outward toward the synapse. Multiple active and inactive motors can be simultaneously attached to one synaptic cargo vesicle. (**b**) The situation in the dendrite is more complex, as both kinesins and dynein can move in antero- and retrograde direction, depending on the orientation of the underlying microtubule. (**c**) Upon arrival at the base of the spine, cargo vesicles are transported along the actin network by myosin motors

synaptic cargo transport into either axons or dendrites (Kapitein et al. 2010; Zheng et al. 2008) and found that the cyclin-dependent kinase pathway regulates polarized trafficking of presynaptic components (Ou et al. 2010).

8.2.4 Microtubule Modifications in Neurons

One other way to directly influence synaptic cargo transport is to generate functional diversity by modifying the cytoskeleton; motor proteins recognize the various spatial cues and establish specific synaptic cargo trafficking routes. It has also recently been demonstrated that posttranslational modification of microtubules can alter their stability and motor protein–binding characteristics (Verhey and Hammond 2009). Stable microtubules in neurons typically accumulate a variety of posttranslational modifications, like acetylation, detyrosination, and (poly) glutamylation. Although both dendrites and axons have high levels of acetylated microtubules, acetylated microtubules are abundantly present in axons (Witte et al. 2008). The selective enrichment of acetylated microtubules in axons can be abolished by inhibition of a known α-tubulin deacetylase histone deacetylase 6 (HDAC6), suggesting that in normal situations, the activity of tubulin-modifying enzymes differs between axons and dendrites rather than that the acetylation reaction is restricted to the axon (Janke and Kneussel 2010). Biochemical evidence revealed that kinesin-1 has an increased affinity for acetylated microtubules, consistent with the observation that in fibroblast cells, kinesin-1 motility occurs predominantly over modified microtubules (Cai et al. 2009). It was proposed that acetylation of microtubules was the major determinant for the selective motility of kinesin-1 motors into specific neurites. Kinesin-1 also preferentially binds detyrosinated microtubules (Liao and Gundersen 1998), which are also enriched in the axon. Recent work identified a specific region in kinesin-1, termed β5-L8, to be responsible for this preference (Konishi and Setou 2009). Interestingly, upon knockdown of tubulin tyrosine ligase (TTL), the fraction of dendrites that contained kinesin-1 increased. Recently, synaptic activity has been shown to modify microtubules posttranslational modifications (Maas et al. 2009). Treatment of neurons with strychnine, an inhibitor of the glycine receptor, increases neuronal activity, leads to increased polyglutamylation, and influences synaptic cargo transport. In this way, microtubule modifications are intimately related to synaptic changes and synaptic cargo transport.

8.3 Microtubule- and Actin-Based Motor Proteins

In general, cargos are transported over long distances along microtubules before transferring to the actin cytoskeleton for the final part of their journey. A common feature of both actin- and microtubule-based transport is that the force required for cargo transport is generated by molecular motor proteins through ATP hydrolysis (Kardon and Vale 2009; Schliwa and Woehlke 2003; Vale 2003; Woolner and Bement 2009). The motor proteins that use microtubules as tracks are the minus-end-directed dynein and plus-end-directed kinesins, whereas myosins move along actin filaments. Neuronal cargo trafficking is achieved by the concerted efforts of both microtubule-based and actin-based motors (Hirokawa and Takemura 2005; Schlager and Hoogenraad 2009). Several classes of myosin motors participate in synaptic cargo transport in axon and dendrites—most commonly used are myosin V and myosin VI. Although the basic machinery for microtubule- and actin-dependent transport in neurons is well established, how synaptic cargos achieve specificity and directionality to their site of action is an emerging field of investigation. In this section, we will introduce the actin and microtubule transport system and its main components. We will also explain the role the actin network plays in selecting the destination of cargo and its role in synaptic function.

8.3.1 Myosin Motor Proteins in Neurons

Myosins were the first molecular motors to be discovered and comprise a large (~25 classes) evolutionarily conserved family of actin-based motor proteins (Conti and Adelstein 2008). Early studies focused on the role of myosins as force generators in muscle tissue. The head region in the myosin heavy chain contains the motor domain; this is the site of ATP hydrolysis, which leads to a force generating conformational change (Conti and Adelstein 2008). It was long thought that the sole function of myosin was to generate force in muscles; however, with the discovery of nonmuscle (unconventional) myosin, new roles for myosin motors were uncovered. Several myosin motors can move directionally along actin filaments, such as myosin V toward the plus end and myosin VI toward the minus end (Sweeney and Houdusse 2010; Woolner and Bement 2009). Myosin motors have been implicated in short-range transport of synaptic cargos, especially in the areas of the neuron where there is hardly any microtubule network, like dendritic spines and presynaptic terminals. For example, myosin V and VI regulate the mobility of synaptic vesicles at the presynapse (Cingolani and Goda 2008) and AMPA receptor–containing recycling endosomes in dendritic spines (Nash et al. 2010; Osterweil et al. 2005; Rudolf et al. 2010; Wang et al. 2008). Moreover, myosin motors can also be involved in regulating microtubule-based cargo transport, either by making direct physical contact with kinesin motors and enhancing each other's processivity (Ali et al. 2008; Huang et al. 1999) or by cooperative actions of actin- and microtubule-based motors on a single cargo (Gross et al. 2007). These motor-motor interactions may represent a mechanism by which the transition of vesicles from microtubules to actin filaments or vice versa is regulated. In contrast, recent data suggest that myosin V and VI can facilitate organelle docking by opposing, rather than complementing, microtubule-based movements (Pathak et al. 2010). Emerging data show that myosin motors are not only important for transport of cargo, they also regulate the secretion of exocytotic vesicles by docking them in actin-rich areas (Desnos et al. 2007). Moreover, myosin II can also regulate actin dynamics in spines in response to synaptic stimulation (Rex et al. 2010; Ryu et al. 2006). All these different actions of myosin motors, such as regulating synapse shape and stability, transporting and docking synaptic vesicles, and influencing actin dynamics, are important for synaptic function and plasticity (Hotulainen and Hoogenraad 2010).

8.3.2 Kinesin Motor Proteins in Neurons

Kinesin-1 was the neuronal transport motor to be identified based on assays of vesicular motility along microtubules in extruded axoplasm (Vale et al. 1985). Similar to the myosin superfamily nowadays, many different kinesin motors have been found. It is thought that approximately 45 different mammalian kinesin genes exist which, by virtue of alternative splicing, code for as many as 90 different kinesin proteins (Hirokawa et al. 2009). The vast majority of kinesin proteins share a number of structural characteristics: a highly conserved kinesin motor domain, responsible for microtubule binding and force generation by ATP hydrolysis; one or more coiled-coil domains for protein dimerization; and a cargo-binding domain (Hirokawa and Noda 2008; Lawrence et al. 2004; Schliwa and Woehlke 2003;

Vale 2003). The motor domain is ATP bound and upon attachment to the microtubule, ATP is hydrolyzed to ADP, resulting in a conformational change of the kinesin protein. This conformational change effectively results in a step of 8 nm along the microtubule (Schnitzer and Block 1997). Since most kinesin form a homodimer, the two heads of the kinesin molecule move in a hand-over-hand mechanism along the microtubule (Yildiz et al. 2004). Kinesin movement is highly processive, meaning that once bound to the microtubule, the motor will move prior to detaching, allowing it to transport cargo over long distances in axons and dendrites. The position of the kinesin motor domain determines the direction in which it moves: kinesin proteins with an N-terminal motor domain (the most common layout) move to the microtubule plus end, whereas kinesins with a C-terminal motor domain move toward the minus end. Kinesins with a motor domain in the middle are involved in regulating microtubule dynamics (Hirokawa and Noda 2008; Verhey and Hammond 2009). The nonmotor regions of kinesin motor proteins are poorly conserved and have been shown to regulate both motor function (by intramolecular folding and inhibition of ATPase activity) (Verhey and Hammond 2009) and cargo binding (by interacting with adaptor proteins) (Schlager and Hoogenraad 2009).

Most cargos bound to kinesin-1 motor proteins, such as amyloid precursor protein (APP) and Reelin receptor AporER2, interact indirectly via kinesin-1 light chains (Hirokawa et al. 2010). However, some cargos bind directly to the kinesin heavy chain, such as the AMPA receptor, which binds via adaptor protein GRIP1 (Setou et al. 2002), and mitochondria, which bind via adaptor protein Milton (Glater et al. 2006). A recent study showed that kinesin-1 binds synaptic precursor vesicles via syntabulin and syntaxin-1 (Cai et al. 2007). Knockdown of syntabulin impairs the anterograde transport of synaptic vesicle precursors. Members of the kinesin-3 family, including KIF1A and KIF1Bβ, also transport synaptic vesicle precursors in the axon. Both KIF1A and KIF1Bβ knockout mice exhibit defects in sensory and motor neurons, including a loss of synaptic vesicles (Hirokawa et al. 2010). The delivery of GABAA receptors to synapses is mediated by the kinesin-HAP-1 complex and is disrupted by mutant huntingtin (Twelvetrees et al. 2010). Several kinesin motors are now critically involved in neuronal disease pathogenesis (Goldstein 2001; Hirokawa and Noda 2008; Mandelkow and Mandelkow 2002; Morfini et al. 2009). For example, mutations in kinesin-1/KIF5A are associated with spastic paraplegia (Ebbing et al. 2008), truncating mutations in KIF17 are associated with schizophrenia (Tarabeux et al. 2010), and heterozygous missense mutations in KIF21A have been found to cause congenital fibrosis of the extraocular muscles (CFEOM), a rare congenital eye movement disorder that results from the dysfunction of the oculomotor nerve (Yamada et al. 2003).

8.3.3 *Dynein Motor Proteins in Neurons*

Similar to kinesin motors, the motor proteins of the dynein family move along microtubule structures. However, while most kinesin motors move toward the microtubule plus end, dyneins move toward the minus end of the microtubule. In most species, more than 15 dynein encoding genes have been identified, where the majority of dynein motor proteins are involved in nontransport actions but drive flagellar beating (Kardon and Vale 2009). In stark contrast to kinesin proteins where many different motor types perform many different tasks, only one dynein motor, cytoplasmic dynein 1, is responsible for the bulk of the minus-end-directed cargo transport. In other respects, dynein is very different from kinesin and myosin motor proteins. One of the most obvious differences is the sheer size of the dynein motor complex: with a mass of 1–2 MDa, it is several times larger than a typical myosin or kinesin motor. At the core of the motor complex is the dynein heavy chain (DHC1), a polypeptide of over 500 kDa, which is essential for the motor activity consisting of three different domains: the stalk, motor, and tail domains. While in kinesin and myosin motors, the polymer-binding site and catalytic site are integrated within a single globular motor domain; in dynein, the microtubule-binding domain is separated from the motor domain by a ~15-nm stalk (Carter et al. 2008). The stalk is a coiled-coil structure that extends directly from the motor domain and is thought to function as a lever (Houdusse and Carter 2009; Kardon and Vale 2009). The circular motor domain consists of six ATPase domains, the tail domains mediate dimerization and form the interaction site for five additional dynein subunits. The intermediate chain (IC) and light intermediate chain (LIC) bind directly to the heavy chain tail, whereas light chain 8 (LC8), light chain 7 (LC7 or roadblock), and T-complex testis-specific protein 1 (Tctex1) bind to the intermediate chain (Kardon and Vale 2009). The dynein subunits are essential in determining the binding of dynein to specific cargos, the cellular localization, and even intrinsic properties of dynein, like its processivity. Interestingly, missense point mutations in the tail domain of cytoplasmic dynein heavy chain have been shown to result in progressive motor neuron degeneration in mice (Hafezparast et al. 2003). Recent data show that the mutations in the nonmotor part of the protein inhibit dynein motor run length and significantly alter motor domain coordination (Ori-McKenney et al. 2010). These results suggest a potential role for the dynein tail in motor function and provide direct evidence for a link between single-motor processivity and disease.

The efficient function of cytoplasmic dynein critically depends on the dynactin (dynein activator) accessory complex. Most dynein-dependent processes from yeast and filamentous fungi to invertebrates and mammals require dynactin (Schroer 2004). Dynactin has been shown to regulate dynein transport in several ways; it is involved in targeting of dynein, functions as adaptor protein, and regulates processive dynein movement (Kardon and Vale 2009; Schroer 2004). Dynactin is a large complex that contains 11 different subunits, and since some subunits are present in multiple copies, the complete assembly can be comprised of as many as 20 proteins. Detailed EM studies have revealed that dynactin basically consists of two parts, a rod domain and arm domain projecting from the rod (Schroer 2004). Mutations in the p150glued gene, coding for one of the large dynactin subunits, are found in a family with motor neuron disease (Puls et al. 2003). Affected patients develop adult-onset vocal fold paralysis, facial weakness, and distal limb muscle weakness, mainly caused by the selective loss of motor neurons. Mutant mice with impaired dynein/dynactin function showed disrupted retrograde axonal transport and develop motor neuron disease similar to amyotrophic lateral sclerosis (ALS) (LaMonte et al. 2002; Teuling et al. 2008). This illustrates both the importance of subunits in the dynactin complex as well as the crucial role of dynein-based transport in the nervous system.

Cytoplasmic dynein transports neurotrophic tyrosine kinase receptor family (Trk) (Ha et al. 2008), Rab6-positive neuropeptide-containing secretory vesicles (Colin et al. 2008; Schlager et al. 2010), the piccolo/bassoon complex (Fejtova et al. 2009), and mitochondria (Hollenbeck and Saxton 2005) retrogradely in the axon, while in the dendrites, the cargos carried by cytoplasmic dynein include glycine receptor vesicles (Maas et al. 2006), messenger ribonucleoproteins (mRNPs) (Villace et al. 2004), Rab5 endosomes (Satoh et al. 2008), and AMPA receptor vesicles (Kapitein et al. 2010). In these cases, the cargos bind to the dynein complex either directly or through adaptor proteins such as gephyrin (glycine receptor) or bicaudal-D and related proteins (Rab6, BDNF, NPY) (Fig. 8.3). Regulation of the binding is controlled via phosphorylation, GTP hydrolysis of the small Rab-GTPases, or Ca^{2+} signaling (Schlager and Hoogenraad 2009).

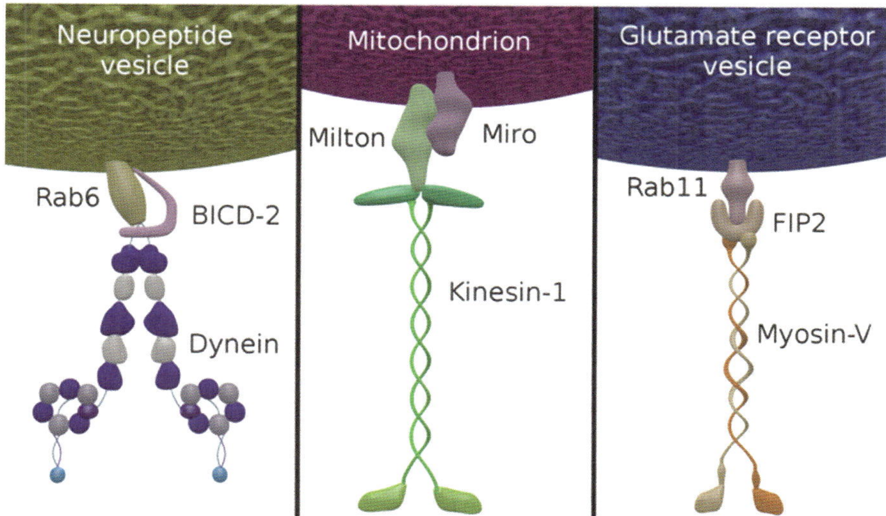

Fig. 8.3 Motor-adaptor-cargo transport complexes. Three typical examples of motor-adaptor-cargo transport machineries. Each motor protein connects to its cargo via adaptor molecules. Although the motor proteins, adaptors, and cargos vary, the essential building blocks are the same in each case. (**a**) Dynein motor complex connected to neuropeptide containing Rab6-positive vesicles via adaptor BICD-2. (**b**) Kinesin-1 is connected to mitochondria via Miro and Milton. This complex is regulated by calcium. (**c**) Myosin V transports glutamate receptor–containing vesicles by connecting to Rab11-FIP2

8.4 Regulating Motor Protein–Based Transport

Precise regulation of motor-based transport is essential to ensure precise cargo delivery to synapses. The list of molecules that are known to link specific transport motors to synaptic cargos is rapidly expanding (Hirokawa et al. 2010; Schlager and Hoogenraad 2009). Biochemical and proteomics approaches and high-throughput yeast two-hybrid screens have identified more than 100 proteins that bind to kinesin-1 in mammals, flies, and worms (Gindhart 2006). Most of these proteins act as cargo molecules themselves or function as motor-adaptor proteins (scaffolding proteins, Rab GTPases, signaling proteins); some are regulators of motor activity. Emerging data from several organisms and different experimental systems suggest that transport motors can be regulated at several points, including motor-cargo binding, motor activation, motor switching, microtubule track selection, cargo release at the destination, and the recycling of motors (Schlager and Hoogenraad 2009;Verhey and Hammond 2009). It is becoming increasingly clear that motor-adaptor-cargo interactions play a key role in identifying synaptic cargos and regulating synaptic cargo trafficking. In this section, we discuss recent work that has shed light on the regulation of the synaptic cargo-motor interactions. To illustrate the common layout and components of motor-adaptor-cargo complexes, we will focus here on two calcium-regulated motor adaptors, the endosomal myosin-V-FIP2 and mitochondrial kinesin-1-Milton-Miro complexes (Fig. 8.3).

8.4.1 Motor-Adaptor-Cargo Interactions

In the large majority of cases, motor proteins do not bind directly to vesicles or synaptic proteins, but they interact with cargo via so-called adaptor proteins. The main role of adaptor proteins is to provide an additional layer of regulation for transport specificity and selectivity. Adaptors can be single proteins, such as GRIP linking KIF5 to AMPA receptors (Hoogenraad et al. 2005; Setou et al. 2002), or protein complexes such as the Mint/CASK/MALS linking KIF17 to NR2B subunits of NMDA receptors (Setou et al. 2000). Interestingly, recent findings show that CaMKII activity is correlated with regulated cargo release near the postsynaptic membrane. Here, CaMKII has been shown to phosphorylate KIF17, which induces the dissociation of the Mint scaffolding protein complex and release of NMDA receptor–containing cargo near the postsynaptic membrane (Guillaud et al. 2008). In this way, regulated CaMKII activity provides an attractive mechanism for targeting NMDA receptor complexes to active synapses where an activity regulated kinase is switched "on." The significance of KIF17 function to brain function is further illustrated by the observation that transgenic mice overexpressing KIF17 show enhanced spatial and working memory (Wong et al. 2002). Additional examples of regulated adaptor proteins are the DENN/MADD adaptor protein that binds KIF1A and KIF1Bß and interacts with Rab3 vesicles (Niwa et al. 2008), liprin family proteins as adaptors that link KIF1A to synaptic vesicle precursors (Miller et al. 2005; Wagner et al. 2009), and bicaudal-D family proteins connecting dynein motors to Rab6-positive neuropeptide secretory vesicles (Grigoriev et al. 2007; Matanis et al. 2002; Schlager et al. 2010). Interestingly, these Rab6-positive secretory vesicles also contain semaphorin 3A and BDNF and are anterogradely transported in axons by kinesin-3 motors (Barkus et al. 2008; de Wit et al. 2006; Schlager et al. 2010). Thus, regulation of cargo binding can be controlled via phosphorylation or GTP hydrolysis of the small Rab-GTPases (Schlager and Hoogenraad 2009).

A third way to control intracellular trafficking is to regulate motor-cargo interactions by responding to changes in local ion concentrations. It is well known that activation of NMDA receptors causes a rapid influx of Ca^{2+} in dendritic spines. A recent study shows that myosin Vb is a "Ca^{2+} sensor" for actin-based postsynaptic AMPA receptor trafficking (Wang et al. 2008). Increased Ca^{2+} levels lead to unfolding of myosin Vb motors and allows for binding to Rab11-FIP2 adaptors on recycling endosomes (Schlager and Hoogenraad 2009) (Fig. 8.3). The association of myosin Vb with Rab11-FIP2 transports AMPA receptor–containing recycling endosomes into the actin-rich spines. Thus, elevated Ca^{2+} levels in spines promote local postsynaptic trafficking. On the other hand, Ca^{2+} influx reduces mitochondrial motility (Boldogh and Pon 2007). The Milton-Miro complex was identified as an adaptor between kinesin-1 and mitochondria and a candidate for Ca^{2+}-dependent regulation of mitochondrial transport. Indeed, recent studies suggest that the mitochondrial Miro-Milton adaptor complex is important for the Ca^{2+}-dependent regulation of mitochondria trafficking (Wang and Schwarz 2009) (Fig. 8.3). Elevated Ca^{2+} levels permit Miro to interact directly with the motor domain of kinesin-1. The interesting aspect of this model is that kinesin-1 remains associated with mitochondria regardless of whether they are moving or stationary. In the "moving" state, kinesin-1 is bound to mitochondria by binding to Milton, which in turn interacts with Miro on the mitochondrial surface. In the "stationary" state, in the presence of high Ca^{2+} levels, the kinesin-1 motor domain interacts directly with Miro and prevents microtubule interactions. In contrast, another recent paper showed that the presence of Ca^{2+} inhibits the Miro1/kinesin-1 protein-protein interaction and that the motor is dissociated from mitochondria yielding arrested movement (Macaskill et al. 2009). Both findings imply the existence of "Ca^{2+} sensors" that detect neuronal activity stimuli and convert Ca^{2+} influx into mechanisms regulating cargo trafficking.

8.5 Conclusion and Future Directions

A typical fully differentiated neuron within an active neuronal circuitry faces an enormous logistical challenge. Synaptic cargo needs to be sorted into dendrites and axons both during basal neuronal activity and changes in activity, such during firing of action potentials. Not all synapses are equally active, and their requirements can vary greatly; some may require a constant flow of receptors and neurotransmitters, while others undergo depression and mainly need transport out of the synapse into a reserve pool or back to the cell body. Moreover, the distance between the cell soma and the most distant synapse can be huge, for example, up to 1 m for a motor neuron. And all these transported proteins, mitochondria, neurotransmitters, and synaptic vesicle precursors flow through an axon of only 5 micrometer in diameter. Therefore, neurons are equipped with a well-balanced and meticulously regulated transport system in order to facilitate synapse formation, function, and plasticity. First of all, the actin and microtubule cytoskeleton play a pivotal role in synaptic plasticity—together they determine synaptic architecture, organize subcellular compartments, and transport intracellular synaptic constituents. Second, the characteristic dynamics, polarity, and modifications patterns of cytoskeleton elements are instrumental for establishing and maintaining the structural and compositional polarity of synapses. Third, this highly specialized microtubule and actin cytoskeletal organization facilitates local, polarized transport by guiding specific motor proteins to specific directions. Fourth, synaptic activity may regulate the cytoskeleton organization and motor protein transport in neurons. All these mechanisms occur simultaneous and can influence each other, creating a highly dynamic infrastructure that is able to rebuild itself in order to adapt to changes in the cellular environment. Without this highly dynamic cytoskeleton system, synaptic plasticity and cognitive brain functions would be impossible.

The relationship between cytoskeleton, motor protein transport, and synaptic signaling is never more apparent than when the brain becomes dysfunctional. Molecular motor proteins, especially kinesin proteins, are prime candidates to be involved in several psychiatric and neurological disorders (Goldstein 2001; Mandelkow and Mandelkow 2002), ranging from schizophrenia (Tarabeux et al. 2010) to spastic paraplegia (Ebbing et al. 2008). Both KIF5 and dynein may be involved in the pathology of Huntington and other polyQ diseases (Colin et al. 2008; Twelvetrees et al. 2010). The role of dynein and dynactin in neurological diseases is best described in the motor neuron disease ALS (Chevalier-Larsen and Holzbaur 2006), and mutations in tubulin isotypes have been observed in patients with severe neurodevelopmental disorders (Keays et al. 2007; Tischfield et al. 2010). There is also a strong link between the activity of tubulin modifying enzymes and neuronal abnormalities. Mutant mice that lack the gene for tubulin tyrosine ligase (TTL), the enzyme that catalyzes the addition of a C-terminal tyrosine residue to α-tubulin in the tubulin tyrosination cycle, die shortly after birth because of neuronal disorganization, including premature axon specialization (Erck et al. 2005). On the other hand, mice that lack functional cytosolic carboxypeptidase (CCP1), the enzymes catalyzing deglutamylation, have increased microtubule hyperglutamylation and Purkinje cell degeneration (Rogowski et al. 2010). However, a number of key mechanistic questions remain to be answered. What are the downstream effects of transport deficits that lead to neurodegeneration? In some cases, it may be a failure to supply new material to the distal axons and dendrites, so that synapses degrade over time. Consistently, defects in both axon and dendritic transport in various organisms can lead to neuropathies (Hirokawa et al. 2010).

In other cases, neuronal degeneration may result from the accumulation of toxic substances in the processes. The molecular motor machinery itself could be critically involved in removing toxic waste in neurons, but these protein accumulations may also lead to traffic jams and disrupt normal synaptic trafficking routes.

In this chapter, we have shown how the cellular infrastructure is essential for neuronal development and plasticity. This highly adaptive network of filaments, motor, adaptor, and cargo proteins is able to answer the ever-changing demands of neuronal networks in action. As a large number of neurological diseases illustrate, there is little room for error. If this fascinating intracellular transport system is not working at peak efficiency, neurons are not able to function properly and will eventually degenerate and die. Future studies on neuronal cytoskeletal dynamics and the synaptic transport machinery may lead to new insights and, hopefully, new treatments for neurological and psychiatric disorders.

Acknowledgments R.v.d.B. is supported by a grant from the "Stichting MS Research." C.C.H. is supported by the Netherlands Organization for Scientific Research (NWO-ALW and NWO-CW), the Netherlands Organization for Health Research and Development (ZonMW-VIDI and ZonMW-TOP), the European Science Foundation (EURYI), EMBO Young Investigators Program (YIP), and the Human Frontier Science Program (HFSP-CDA).

References

Akhmanova, A., & Hammer, J. A., 3rd. (2010). Linking molecular motors to membrane cargo. *Current Opinion in Cell Biology, 22*, 479–487.

Akhmanova, A., & Steinmetz, M. O. (2008). Tracking the ends: A dynamic protein network controls the fate of microtubule tips. *Nature Reviews Molecular Cell Biology, 9*, 309–322.

Ali, M. Y., Lu, H., Bookwalter, C. S., Warshaw, D. M., & Trybus, K. M. (2008). Myosin V and kinesin act as tethers to enhance each others' processivity. *Proceedings of the National Academy of Sciences of the United States of America, 105*, 4691–4696.

Baas, P. W., Black, M. M., & Banker, G. A. (1989). Changes in microtubule polarity orientation during the development of hippocampal neurons in culture. *The Journal of Cell Biology, 109*, 3085–3094.

Baas, P. W., Deitch, J. S., Black, M. M., & Banker, G. A. (1988). Polarity orientation of microtubules in hippocampal neurons: Uniformity in the axon and nonuniformity in the dendrite. *Proceedings of the National Academy of Sciences of the United States of America, 85*, 8335–8339.

Baran, R., Castelblanco, L., Tang, G., Shapiro, I., Goncharov, A., & Jin, Y. (2010). Motor neuron synapse and axon defects in a *C. elegans* alpha-tubulin mutant. *PLoS One, 5*, e9655.

Barkus, R. V., Klyachko, O., Horiuchi, D., Dickson, B. J., & Saxton, W. M. (2008). Identification of an axonal kinesin-3 motor for fast anterograde vesicle transport that facilitates retrograde transport of neuropeptides. *Molecular Biology of the Cell, 19*, 274–283.

Bearer, E. L., & Reese, T. S. (1999). Association of actin filaments with axonal microtubule tracts. *Journal of Neurocytology, 28*, 85–98.

Bel, C., Oguievetskaia, K., Pitaval, C., Goutebroze, L., & Faivre-Sarrailh, C. (2009). Axonal targeting of Caspr2 in hippocampal neurons via selective somatodendritic endocytosis. *Journal of Cell Science, 122*, 3403–3413.

Blanpied, T. A., & Ehlers, M. D. (2004). Microanatomy of dendritic spines: Emerging principles of synaptic pathology in psychiatric and neurological disease. *Biological Psychiatry, 55*, 1121–1127.

Boldogh, I. R., & Pon, L. A. (2007). Mitochondria on the move. *Trends in Cell Biology, 17*, 502–510.

Burton, P. R. (1988). Dendrites of mitral cell neurons contain microtubules of opposite polarity. *Brain Research, 473*, 107–115.

Cai, D., McEwen, D. P., Martens, J. R., Meyhofer, E., & Verhey, K. J. (2009). Single molecule imaging reveals differences in microtubule track selection between Kinesin motors. *PLoS Biology, 7*, e1000216.

Cai, Q., Pan, P. Y., & Sheng, Z. H. (2007). Syntabulin-kinesin-1 family member 5B-mediated axonal transport contributes to activity-dependent presynaptic assembly. *Journal of Neuroscience, 27*, 7284–7296.

Carter, A. P., Garbarino, J. E., Wilson-Kubalek, E. M., Shipley, W. E., Cho, C., Milligan, R. A., Vale, R. D., & Gibbons, I. R. (2008). Structure and functional role of dynein's microtubule-binding domain. *Science, 322*, 1691–1695.

Chen, J., Kanai, Y., Cowan, N. J., & Hirokawa, N. (1992). Projection domains of MAP2 and tau determine spacings between microtubules in dendrites and axons. *Nature, 360*, 674–677.

Chevalier-Larsen, E., & Holzbaur, E. L. (2006). Axonal transport and neurodegenerative disease. *Biochimica et Biophysica Acta, 1762*, 1094–1108.

Cingolani, L. A., & Goda, Y. (2008). Actin in action: The interplay between the actin cytoskeleton and synaptic efficacy. *Nature Reviews Neuroscience, 9*, 344–356.

Colin, E., Zala, D., Liot, G., Rangone, H., Borrell-Pages, M., Li, X. J., Saudou, F., & Humbert, S. (2008). Huntingtin phosphorylation acts as a molecular switch for anterograde/retrograde transport in neurons. *EMBO Journal, 27*, 2124–2134.

Conde, C., & Caceres, A. (2009). Microtubule assembly, organization and dynamics in axons and dendrites. *Nature Reviews Neuroscience, 10*, 319–332.

Conti, M. A., & Adelstein, R. S. (2008). Nonmuscle myosin II moves in new directions. *Journal of Cell Science, 121*, 11–18.

Craig, A. M., Blackstone, C. D., Huganir, R. L., & Banker, G. (1993). The distribution of glutamate receptors in cultured rat hippocampal neurons: Postsynaptic clustering of AMPA-selective subunits. *Neuron, 10*, 1055–1068.

de Wit, J., Toonen, R. F., Verhaagen, J., & Verhage, M. (2006). Vesicular trafficking of semaphorin 3A is activity-dependent and differs between axons and dendrites. *Traffic, 7*, 1060–1077.

Dehmelt, L., & Halpain, S. (2005). The MAP2/Tau family of microtubule-associated proteins. *Genome Biology, 6*, 204.

Dent, E. W., & Gertler, F. B. (2003). Cytoskeletal dynamics and transport in growth cone motility and axon guidance. *Neuron, 40*, 209–227.

Desnos, C., Huet, S., Fanget, I., Chapuis, C., Bottiger, C., Racine, V., Sibarita, J. B., Henry, J. P., & Darchen, F. (2007). Myosin va mediates docking of secretory granules at the plasma membrane. *Journal of Neuroscience, 27*, 10636–10645.

Dixit, R., Ross, J. L., Goldman, Y. E., & Holzbaur, E. L. (2008). Differential regulation of dynein and kinesin motor proteins by tau. *Science, 319*, 1086–1089.

Ebbing, B., Mann, K., Starosta, A., Jaud, J., Schols, L., Schule, R., & Woehlke, G. (2008). Effect of spastic paraplegia mutations in KIF5A kinesin on transport activity. *Human Molecular Genetics, 17*, 1245–1252.

Erck, C., Peris, L., Andrieux, A., Meissirel, C., Gruber, A. D., Vernet, M., Schweitzer, A., Saoudi, Y., Pointu, H., Bosc, C., et al. (2005). A vital role of tubulin-tyrosine-ligase for neuronal organization. *Proceedings of the National Academy of Sciences of the United States of America, 102*, 7853–7858.

Fejtova, A., Davydova, D., Bischof, F., Lazarevic, V., Altrock, W. D., Romorini, S., Schone, C., Zuschratter, W., Kreutz, M. R., Garner, C. C., et al. (2009). Dynein light chain regulates axonal trafficking and synaptic levels of Bassoon. *The Journal of Cell Biology, 185*, 341–355.

Frost, N. A., Kerr, J. M., Lu, H. E., & Blanpied, T. A. (2010a). A network of networks: Cytoskeletal control of compartmentalized function within dendritic spines. *Current Opinion in Neurobiology, 20*, 578–587.

Frost, N. A., Shroff, H., Kong, H., Betzig, E., & Blanpied, T. A. (2010b). Single-molecule discrimination of discrete perisynaptic and distributed sites of actin filament assembly within dendritic spines. *Neuron, 67*, 86–99.

Garrido, J. J., Fernandes, F., Giraud, P., Mouret, I., Pasqualini, E., Fache, M. P., Jullien, F., & Dargent, B. (2001). Identification of an axonal determinant in the C-terminus of the sodium channel Na(v)1.2. *EMBO Journal, 20*, 5950–5961.

Gindhart, J. G. (2006). Towards an understanding of kinesin-1 dependent transport pathways through the study of protein–protein interactions. *Briefings in Functional Genomics & Proteomics, 5*, 74–86.

Glater, E. E., Megeath, L. J., Stowers, R. S., & Schwarz, T. L. (2006). Axonal transport of mitochondria requires Milton to recruit kinesin heavy chain and is light chain independent. *The Journal of Cell Biology, 173*, 545–557.

Goldstein, L. S. (2001). Kinesin molecular motors: Transport pathways, receptors, and human disease. *Proceedings of the National Academy of Sciences of the United States of America, 98*, 6999–7003.

Goldstein, L. S., & Yang, Z. (2000). Microtubule-based transport systems in neurons: The roles of kinesins and dyneins. *Annual Review of Neuroscience, 23*, 39–71.

Gouveia, S. M., & Akhmanova, A. (2010). Cell and molecular biology of microtubule plus end tracking proteins: End binding proteins and their partners. *International Review of Cell and Molecular Biology, 285*, 1–74.

Govek, E. E., Newey, S. E., Akerman, C. J., Cross, J. R., Van der Veken, L., & Van Aelst, L. (2004). The X-linked mental retardation protein oligophrenin-1 is required for dendritic spine morphogenesis. *Nature Neuroscience, 7*, 364–372.

Grigoriev, I., Splinter, D., Keijzer, N., Wulf, P. S., Demmers, J., Ohtsuka, T., Modesti, M., Maly, I. V., Grosveld, F., Hoogenraad, C. C., et al. (2007). Rab6 regulates transport and targeting of exocytotic carriers. *Developmental Cell, 13*, 305–314.

Gross, S. P., Vershinin, M., & Shubeita, G. T. (2007). Cargo transport: Two motors are sometimes better than one. *Current Biology, 17*, R478–R486.

Guillaud, L., Wong, R., & Hirokawa, N. (2008). Disruption of KIF17-Mint1 interaction by CaMKII-dependent phosphorylation: A molecular model of kinesin-cargo release. *Nature Cell Biology, 10*, 19–29.

Gunawardena, S., & Goldstein, L. S. (2004). Cargo-carrying motor vehicles on the neuronal highway: Transport pathways and neurodegenerative disease. *Journal of Neurobiology, 58*, 258–271.

Ha, J., Lo, K. W., Myers, K. R., Carr, T. M., Humsi, M. K., Rasoul, B. A., Segal, R. A., & Pfister, K. K. (2008). A neuron-specific cytoplasmic dynein isoform preferentially transports TrkB signaling endosomes. *The Journal of Cell Biology, 181*, 1027–1039.

Hafezparast, M., Klocke, R., Ruhrberg, C., Marquardt, A., Ahmad-Annuar, A., Bowen, S., Lalli, G., Witherden, A. S., Hummerich, H., Nicholson, S., et al. (2003). Mutations in dynein link motor neuron degeneration to defects in retrograde transport. *Science, 300*, 808–812.

Hirokawa, N., Niwa, S., & Tanaka, Y. (2010). Molecular motors in neurons: Transport mechanisms and roles in brain function, development, and disease. *Neuron, 68*, 610–638.

Hirokawa, N., & Noda, Y. (2008). Intracellular transport and kinesin superfamily proteins, KIFs: Structure, function, and dynamics. *Physiological Reviews, 88*, 1089–1118.

Hirokawa, N., Noda, Y., Tanaka, Y., & Niwa, S. (2009). Kinesin superfamily motor proteins and intracellular transport. *Nature Reviews Molecular Cell Biology, 10*, 682–696.

Hirokawa, N., & Takemura, R. (2005). Molecular motors and mechanisms of directional transport in neurons. *Nature Reviews Neuroscience, 6*, 201–214.

Hollenbeck, P. J., & Saxton, W. M. (2005). The axonal transport of mitochondria. *Journal of Cell Science, 118*, 5411–5419.

Honkura, N., Matsuzaki, M., Noguchi, J., Ellis-Davies, G. C., & Kasai, H. (2008). The subspine organization of actin fibers regulates the structure and plasticity of dendritic spines. *Neuron, 57*, 719–729.

Hoogenraad, C. C., & Bradke, F. (2009). Control of neuronal polarity and plasticity—A renaissance for microtubules? *Trends in Cell Biology, 19*, 669–676.

Hoogenraad, C. C., Milstein, A. D., Ethell, I. M., Henkemeyer, M., & Sheng, M. (2005). GRIP1 controls dendrite morphogenesis by regulating EphB receptor trafficking. *Nature Neuroscience, 8*, 906–915.

Hotulainen, P., & Hoogenraad, C. C. (2010). Actin in dendritic spines: Connecting dynamics to function. *The Journal of Cell Biology, 189*, 619–629.

Hotulainen, P., Llano, O., Smirnov, S., Tanhuanpaa, K., Faix, J., Rivera, C., & Lappalainen, P. (2009). Defining mechanisms of actin polymerization and depolymerization during dendritic spine morphogenesis. *The Journal of Cell Biology, 185*, 323–339.

Houdusse, A., & Carter, A. P. (2009). Dynein swings into action. *Cell, 136*, 395–396.

Howard, J., & Hyman, A. A. (2009). Growth, fluctuation and switching at microtubule plus ends. *Nature Reviews Molecular Cell Biology, 10*, 569–574.

Huang, J. D., Brady, S. T., Richards, B. W., Stenolen, D., Resau, J. H., Copeland, N. G., & Jenkins, N. A. (1999). Direct interaction of microtubule- and actin-based transport motors. *Nature, 397*, 267–270.

Janke, C., & Kneussel, M. (2010). Tubulin post-translational modifications: Encoding functions on the neuronal microtubule cytoskeleton. *Trends in Neurosciences, 33*, 362–372.

Jaworski, J., Hoogenraad, C. C., & Akhmanova, A. (2008). Microtubule plus-end tracking proteins in differentiated mammalian cells. *The International Journal of Biochemistry & Cell Biology, 40*, 619–637.

Jaworski, J., Kapitein, L. C., Gouveia, S. M., Dortland, B. R., Wulf, P. S., Grigoriev, I., Camera, P., Spangler, S. A., Di Stefano, P., Demmers, J., et al. (2009). Dynamic microtubules regulate dendritic spine morphology and synaptic plasticity. *Neuron, 61*, 85–100.

Jin, Y., & Garner, C. C. (2008). Molecular mechanisms of presynaptic differentiation. *Annual Review of Cell and Developmental Biology, 24*, 237–262.

Kaether, C., Skehel, P., & Dotti, C. G. (2000). Axonal membrane proteins are transported in distinct carriers: A two-color video microscopy study in cultured hippocampal neurons. *Molecular Biology of the Cell, 11*, 1213–1224.

Kapitein, L. C., & Hoogenraad, C. C. (2010). Which way to go? Cytoskeletal organization and polarized transport in neurons. *Molecular Cell Neuroscience, 46(1)*, 9–20.

Kapitein, L. C., Schlager, M. A., Kuijpers, M., Wulf, P. S., van Spronsen, M., MacKintosh, F. C., & Hoogenraad, C. C. (2010). Mixed microtubules steer dynein-driven cargo transport into dendrites. *Current Biology, 20*, 290–299.

Kapitein L. C., & Hoogenraad C. C. (2011). Which way to go? Cytoskeletal organization and polarized transport in neurons. *Mol Cell Neurosci 46*, 9–20

Kardon, J. R., & Vale, R. D. (2009). Regulators of the cytoplasmic dynein motor. *Nature Reviews Molecular Cell Biology, 10*, 854–865.

Kasai, H., Fukuda, M., Watanabe, S., Hayashi-Takagi, A., & Noguchi, J. (2010). Structural dynamics of dendritic spines in memory and cognition. *Trends in Neurosciences, 33*, 121–129.

Keays, D. A., Tian, G., Poirier, K., Huang, G. J., Siebold, C., Cleak, J., Oliver, P. L., Fray, M., Harvey, R. J., Molnar, Z., et al. (2007). Mutations in alpha-tubulin cause abnormal neuronal migration in mice and lissencephaly in humans. *Cell, 128*, 45–57.

Kim, E., & Sheng, M. (2004). PDZ domain proteins of synapses. *Nature Reviews Neuroscience, 5*, 771–781.

Konishi, Y., & Setou, M. (2009). Tubulin tyrosination navigates the kinesin-1 motor domain to axons. *Nature Neuroscience, 12*, 559–567.

Korobova, F., & Svitkina, T. (2010). Molecular architecture of synaptic actin cytoskeleton in hippocampal neurons reveals a mechanism of dendritic spine morphogenesis. *Molecular Biology of the Cell, 21*, 165–176.

Krucker, T., Siggins, G. R., & Halpain, S. (2000). Dynamic actin filaments are required for stable long-term potentiation (LTP) in area CA1 of the hippocampus. *Proceedings of the National Academy of Sciences of the United States of America, 97,* 6856–6861.

Kueh, H. Y., & Mitchison, T. J. (2009). Structural plasticity in actin and tubulin polymer dynamics. *Science, 325,* 960–963.

LaMonte, B. H., Wallace, K. E., Holloway, B. A., Shelly, S. S., Ascano, J., Tokito, M., Van Winkle, T., Howland, D. S., & Holzbaur, E. L. (2002). Disruption of dynein/dynactin inhibits axonal transport in motor neurons causing late-onset progressive degeneration. *Neuron, 34,* 715–727.

Landis, D. M., Hall, A. K., Weinstein, L. A., & Reese, T. S. (1988). The organization of cytoplasm at the presynaptic active zone of a central nervous system synapse. *Neuron, 1,* 201–209.

Landis, D. M., & Reese, T. S. (1983). Cytoplasmic organization in cerebellar dendritic spines. *The Journal of Cell Biology, 97,* 1169–1178.

Lau, C. G., & Zukin, R. S. (2007). NMDA receptor trafficking in synaptic plasticity and neuropsychiatric disorders. *Nature Reviews Neuroscience, 8,* 413–426.

Lawrence, C. J., Dawe, R. K., Christie, K. R., Cleveland, D. W., Dawson, S. C., Endow, S. A., Goldstein, L. S., Goodson, H. V., Hirokawa, N., Howard, J., et al. (2004). A standardized kinesin nomenclature. *The Journal of Cell Biology, 167,* 19–22.

Leterrier, C., Laine, J., Darmon, M., Boudin, H., Rossier, J., & Lenkei, Z. (2006). Constitutive activation drives compartment-selective endocytosis and axonal targeting of type 1 cannabinoid receptors. *Journal of Neuroscience, 26,* 3141–3153.

Lewis, T. L., Jr., Mao, T., Svoboda, K., & Arnold, D. B. (2009). Myosin-dependent targeting of transmembrane proteins to neuronal dendrites. *Nature Neuroscience, 12,* 568–576.

Liao, G., & Gundersen, G. G. (1998). Kinesin is a candidate for cross-bridging microtubules and intermediate filaments. Selective binding of kinesin to detyrosinated tubulin and vimentin. *Journal of Biological Chemistry, 273,* 9797–9803.

Lisman, J. E., Raghavachari, S., & Tsien, R. W. (2007). The sequence of events that underlie quantal transmission at central glutamatergic synapses. *Nature Reviews Neuroscience, 8,* 597–609.

Maas, C., Belgardt, D., Lee, H. K., Heisler, F. F., Lappe-Siefke, C., Magiera, M. M., van Dijk, J., Hausrat, T. J., Janke, C., & Kneussel, M. (2009). Synaptic activation modifies microtubules underlying transport of postsynaptic cargo. *Proceedings of the National Academy of Sciences of the United States of America, 106,* 8731–8736.

Maas, C., Tagnaouti, N., Loebrich, S., Behrend, B., Lappe-Siefke, C., & Kneussel, M. (2006). Neuronal cotransport of glycine receptor and the scaffold protein gephyrin. *The Journal of Cell Biology, 172,* 441–451.

Macaskill, A. F., Rinholm, J. E., Twelvetrees, A. E., Arancibia-Carcamo, I. L., Muir, J., Fransson, A., Aspenstrom, P., Attwell, D., & Kittler, J. T. (2009). Miro1 is a calcium sensor for glutamate receptor-dependent localization of mitochondria at synapses. *Neuron, 61,* 541–555.

Mandelkow, E., & Mandelkow, E. M. (2002). Kinesin motors and disease. *Trends in Cell Biology, 12,* 585–591.

Margeta, M. A., Shen, K., & Grill, B. (2008). Building a synapse: Lessons on synaptic specificity and presynaptic assembly from the nematode *C. elegans. Current Opinion in Neurobiology, 18,* 69–76.

Matanis, T., Akhmanova, A., Wulf, P., Del Nery, E., Weide, T., Stepanova, T., Galjart, N., Grosveld, F., Goud, B., De Zeeuw, C. I., et al. (2002). Bicaudal-D regulates COPI-independent Golgi-ER transport by recruiting the dynein-dynactin motor complex. *Nature Cell Biology, 4,* 986–992.

Miller, K. E., DeProto, J., Kaufmann, N., Patel, B. N., Duckworth, A., & Van Vactor, D. (2005). Direct observation demonstrates that Liprin-alpha is required for trafficking of synaptic vesicles. *Current Biology, 15,* 684–689.

Mitchison, T., & Kirschner, M. (1984). Dynamic instability of microtubule growth. *Nature, 312,* 237–242.

Morfini, G. A., Burns, M., Binder, L. I., Kanaan, N. M., LaPointe, N., Bosco, D. A., Brown, R. H., Jr., Brown, H., Tiwari, A., Hayward, L., et al. (2009). Axonal transport defects in neurodegenerative diseases. *Journal of Neuroscience, 29*, 12776–12786.

Morrison, E. E., Moncur, P. M., & Askham, J. M. (2002). EB1 identifies sites of microtubule polymerisation during neurite development. *Brain Research: Molecular Brain Research, 98*, 145–152.

Nakada, C., Ritchie, K., Oba, Y., Nakamura, M., Hotta, Y., Iino, R., Kasai, R. S., Yamaguchi, K., Fujiwara, T., & Kusumi, A. (2003). Accumulation of anchored proteins forms membrane diffusion barriers during neuronal polarization. *Nature Cell Biology, 5*, 626–632.

Nash, J. E., Appleby, V. J., Correa, S. A., Wu, H., Fitzjohn, S. M., Garner, C. C., Collingridge, G. L., & Molnar, E. (2010). Disruption of the interaction between myosin VI and SAP97 is associated with a reduction in the number of AMPARs at hippocampal synapses. *Journal of Neurochemistry, 112*, 677–690.

Niwa, S., Tanaka, Y., & Hirokawa, N. (2008). KIF1Bbeta- and KIF1A-mediated axonal transport of presynaptic regulator Rab3 occurs in a GTP-dependent manner through DENN/MADD. *Nature Cell Biology, 10*, 1269–1279.

Okamoto, K., Nagai, T., Miyawaki, A., & Hayashi, Y. (2004). Rapid and persistent modulation of actin dynamics regulates postsynaptic reorganization underlying bidirectional plasticity. *Nature Neuroscience, 7*, 1104–1112.

Ori-McKenney, K. M., Xu, J., Gross, S. P., & Vallee, R. B. (2010). A cytoplasmic dynein tail mutation impairs motor processivity. *Nature Cell Biology, 12*, 1228–1234.

Osterweil, E., Wells, D. G., & Mooseker, M. S. (2005). A role for myosin VI in postsynaptic structure and glutamate receptor endocytosis. *The Journal of Cell Biology, 168*, 329–338.

Ou, C. Y., Poon, V. Y., Maeder, C. I., Watanabe, S., Lehrman, E. K., Fu, A. K., Park, M., Fu, W. Y., Jorgensen, E. M., Ip, N. Y., et al. (2010). Two cyclin-dependent kinase pathways are essential for polarized trafficking of presynaptic components. *Cell, 141*, 846–858.

Pathak, D., Sepp, K. J., & Hollenbeck, P. J. (2010). Evidence that myosin activity opposes microtubule-based axonal transport of mitochondria. *Journal of Neuroscience, 30*, 8984–8992.

Pennuto, M., Bonanomi, D., Benfenati, F., & Valtorta, F. (2003). Synaptophysin I controls the targeting of VAMP2/synaptobrevin II to synaptic vesicles. *Molecular Biology of the Cell, 14*, 4909–4919.

Pollard, T. D., & Cooper, J. A. (2009). Actin, a central player in cell shape and movement. *Science, 326*, 1208–1212.

Puls, I., Jonnakuty, C., LaMonte, B. H., Holzbaur, E. L., Tokito, M., Mann, E., Floeter, M. K., Bidus, K., Drayna, D., Oh, S. J., et al. (2003). Mutant dynactin in motor neuron disease. *Nature Genetics, 33*, 455–456.

Rasband, M. N. (2010). The axon initial segment and the maintenance of neuronal polarity. *Nature Reviews Neuroscience, 11*, 552–562.

Rex, C. S., Gavin, C. F., Rubio, M. D., Kramar, E. A., Chen, L. Y., Jia, Y., Huganir, R. L., Muzyczka, N., Gall, C. M., Miller, C. A., et al. (2010). Myosin IIb regulates actin dynamics during synaptic plasticity and memory formation. *Neuron, 67*, 603–617.

Rogowski, K., van Dijk, J., Magiera, M. M., Bosc, C., Deloulme, J. C., Bosson, A., Peris, L., Gold, N. D., Lacroix, B., Grau, M. B., et al. (2010). A family of protein-deglutamylating enzymes associated with neurodegeneration. *Cell, 143*, 564–578.

Rolls, M. M., Satoh, D., Clyne, P. J., Henner, A. L., Uemura, T., & Doe, C. Q. (2007). Polarity and intracellular compartmentalization of Drosophila neurons. *Neural Development, 2*, 7.

Ruberti, F., & Dotti, C. G. (2000). Involvement of the proximal C terminus of the AMPA receptor subunit GluR1 in dendritic sorting. *Journal of Neuroscience, 20*, RC78.

Rudolf, R., Bittins, C. M., & Gerdes, H. H. (2010). The role of myosin V in exocytosis and synaptic plasticity. *Journal of Neurochemistry, 116(2)*, 177–191.

Rudolf, R., Bittins, C. M., Gerdes, H. H. (2011) The role of myosin V in exocytosis and synaptic plasticity. *J Neurochem 116*, 177–191

Ryu, J., Liu, L., Wong, T. P., Wu, D. C., Burette, A., Weinberg, R., Wang, Y. T., & Sheng, M. (2006). A critical role for myosin IIb in dendritic spine morphology and synaptic function. *Neuron, 49*, 175–182.

Sampo, B., Kaech, S., Kunz, S., & Banker, G. (2003). Two distinct mechanisms target membrane proteins to the axonal surface. *Neuron, 37*, 611–624.

Satoh, D., Sato, D., Tsuyama, T., Saito, M., Ohkura, H., Rolls, M. M., Ishikawa, F., & Uemura, T. (2008). Spatial control of branching within dendritic arbors by dynein-dependent transport of Rab5-endosomes. *Nature Cell Biology, 10*, 1164–1171.

Schlager, M. A., & Hoogenraad, C. C. (2009). Basic mechanisms for recognition and transport of synaptic cargos. *Molecular Brain, 2*, 25.

Schlager, M. A., Kapitein, L. C., Grigoriev, I., Burzynski, G. M., Wulf, P. S., Keijzer, N., de Graaff, E., Fukuda, M., Shepherd, I. T., Akhmanova, A., et al. (2010). Pericentrosomal targeting of Rab6 secretory vesicles by Bicaudal-D-related protein 1 (BICDR-1) regulates neuritogenesis. *EMBO Journal, 29*, 1637–1651.

Schliwa, M., & Woehlke, G. (2003). Molecular motors. *Nature, 422*, 759–765.

Schnitzer, M. J., & Block, S. M. (1997). Kinesin hydrolyses one ATP per 8-nm step. *Nature, 388*, 386–390.

Schroer, T. A. (2004). Dynactin. *Annual Review of Cell and Developmental Biology, 20*, 759–779.

Setou, M., Nakagawa, T., Seog, D. H., & Hirokawa, N. (2000). Kinesin superfamily motor protein KIF17 and mLin-10 in NMDA receptor-containing vesicle transport. *Science, 288*, 1796–1802.

Setou, M., Seog, D. H., Tanaka, Y., Kanai, Y., Takei, Y., Kawagishi, M., & Hirokawa, N. (2002). Glutamate-receptor-interacting protein GRIP1 directly steers kinesin to dendrites. *Nature, 417*, 83–87.

Sheng, M., & Hoogenraad, C. C. (2007). The postsynaptic architecture of excitatory synapses: A more quantitative view. *Annual Review of Biochemistry, 76*, 823–847.

Shepherd, J. D., & Huganir, R. L. (2007). The cell biology of synaptic plasticity: AMPA receptor trafficking. *Annual Review of Cell and Developmental Biology, 23*, 613–643.

Song, A. H., Wang, D., Chen, G., Li, Y., Luo, J., Duan, S., & Poo, M. M. (2009). A selective filter for cytoplasmic transport at the axon initial segment. *Cell, 136*, 1148–1160.

Stepanova, T., Slemmer, J., Hoogenraad, C. C., Lansbergen, G., Dortland, B., De Zeeuw, C. I., Grosveld, F., van Cappellen, G., Akhmanova, A., & Galjart, N. (2003). Visualization of microtubule growth in cultured neurons via the use of EB3-GFP (end-binding protein 3-green fluorescent protein). *Journal of Neuroscience, 23*, 2655–2664.

Stiess, M., Maghelli, N., Kapitein, L. C., Gomis-Ruth, S., Wilsch-Brauninger, M., Hoogenraad, C. C., Tolic-Norrelykke, I. M., & Bradke, F. (2010). Axon extension occurs independently of centrosomal microtubule nucleation. *Science, 327*, 704–707.

Stone, M. C., Roegiers, F., & Rolls, M. M. (2008). Microtubules have opposite orientation in axons and dendrites of Drosophila neurons. *Molecular Biology of the Cell, 19*, 4122–4129.

Stowell, J. N., & Craig, A. M. (1999). Axon/dendrite targeting of metabotropic glutamate receptors by their cytoplasmic carboxy-terminal domains. *Neuron, 22*, 525–536.

Südhof, T. C. (2004). The synaptic vesicle cycle. *Annual Review of Neuroscience, 27*, 509–547.

Sweeney, H. L., & Houdusse, A. (2010). Myosin VI rewrites the rules for myosin motors. *Cell, 141*, 573–582.

Tarabeux, J., Champagne, N., Brustein, E., Hamdan, F. F., Gauthier, J., Lapointe, M., Maios, C., Piton, A., Spiegelman, D., Henrion, E., et al. (2010). De novo truncating mutation in Kinesin 17 associated with schizophrenia. *Biological Psychiatry, 68*, 649–656.

Teuling, E., van Dis, V., Wulf, P. S., Haasdijk, E. D., Akhmanova, A., Hoogenraad, C. C., & Jaarsma, D. (2008). A novel mouse model with impaired dynein/dynactin function develops amyotrophic lateral sclerosis (ALS)-like features in motor neurons and improves lifespan in SOD1-ALS mice. *Human Molecular Genetics, 17*, 2849–2862.

Tischfield, M. A., Baris, H. N., Wu, C., Rudolph, G., Van Maldergem, L., He, W., Chan, W. M., Andrews, C., Demer, J. L., Robertson, R. L., et al. (2010). Human TUBB3 mutations perturb microtubule dynamics, kinesin interactions, and axon guidance. *Cell, 140*, 74–87.

Twelvetrees, A. E., Yuen, E. Y., Arancibia-Carcamo, I. L., MacAskill, A. F., Rostaing, P., Lumb, M. J., Humbert, S., Triller, A., Saudou, F., Yan, Z., et al. (2010). Delivery of GABAARs to synapses is mediated by HAP1-KIF5 and disrupted by mutant huntingtin. *Neuron, 65*, 53–65.

Vale, R. D. (2003). The molecular motor toolbox for intracellular transport. *Cell, 112*, 467–480.

Vale, R. D., Schnapp, B. J., Reese, T. S., & Sheetz, M. P. (1985). Movement of organelles along filaments dissociated from the axoplasm of the squid giant axon. *Cell, 40*, 449–454.

van Spronsen, M., & Hoogenraad, C. C. (2010). Synapse pathology in psychiatric and neurologic disease. *Current Neurology and Neuroscience Reports, 10*, 207–214.

Verhey, K. J., & Hammond, J. W. (2009). Traffic control: Regulation of kinesin motors. *Nature Reviews Molecular Cell Biology, 10*, 765–777.

Villace, P., Marion, R. M., & Ortin, J. (2004). The composition of Staufen-containing RNA granules from human cells indicates their role in the regulated transport and translation of messenger RNAs. *Nucleic Acids Research, 32*, 2411–2420.

Wagner, O. I., Esposito, A., Kohler, B., Chen, C. W., Shen, C. P., Wu, G. H., Butkevich, E., Mandalapu, S., Wenzel, D., Wouters, F. S., et al. (2009). Synaptic scaffolding protein SYD-2 clusters and activates kinesin-3 UNC-104 in *C. elegans*. *Proceedings of the National Academy of Sciences of the United States of America, 106*, 19605–19610.

Wang, Z., Edwards, J. G., Riley, N., Provance, D. W., Jr., Karcher, R., Li, X. D., Davison, I. G., Ikebe, M., Mercer, J. A., Kauer, J. A., et al. (2008). Myosin Vb mobilizes recycling endosomes and AMPA receptors for postsynaptic plasticity. *Cell, 135*, 535–548.

Wang, X., & Schwarz, T. L. (2009). The mechanism of Ca^{2+}-dependent regulation of kinesin-mediated mitochondrial motility. *Cell, 136*, 163–174.

Winckler, B., Forscher, P., & Mellman, I. (1999). A diffusion barrier maintains distribution of membrane proteins in polarized neurons. *Nature, 397*, 698–701.

Winckler, B., & Mellman, I. (2010). Trafficking guidance receptors. *Cold Spring Harbor Perspectives in Biology, 2*, a001826.

Wisco, D., Anderson, E. D., Chang, M. C., Norden, C., Boiko, T., Folsch, H., & Winckler, B. (2003). Uncovering multiple axonal targeting pathways in hippocampal neurons. *The Journal of Cell Biology, 162*, 1317–1328.

Witte, H., Neukirchen, D., & Bradke, F. (2008). Microtubule stabilization specifies initial neuronal polarization. *The Journal of Cell Biology, 180*, 619–632.

Wong, R. W., Setou, M., Teng, J., Takei, Y., & Hirokawa, N. (2002). Overexpression of motor protein KIF17 enhances spatial and working memory in transgenic mice. *Proceedings of the National Academy of Sciences of the United States of America, 99*, 14500–14505.

Woolner, S., & Bement, W. M. (2009). Unconventional myosins acting unconventionally. *Trends in Cell Biology, 19*, 245–252.

Yamada, K., Andrews, C., Chan, W. M., McKeown, C. A., Magli, A., de Berardinis, T., Loewenstein, A., Lazar, M., O'Keefe, M., Letson, R., et al. (2003). Heterozygous mutations of the kinesin KIF21A in congenital fibrosis of the extraocular muscles type 1 (CFEOM1). *Nature Genetics, 35*, 318–321.

Yildiz, A., Tomishige, M., Vale, R. D., & Selvin, P. R. (2004). Kinesin walks hand-over-hand. *Science, 303*, 676–678.

Yuste, R., & Bonhoeffer, T. (2001). Morphological changes in dendritic spines associated with long-term synaptic plasticity. *Annual Review of Neuroscience, 24*, 1071–1089.

Zheng, Y., Wildonger, J., Ye, B., Zhang, Y., Kita, A., Younger, S. H., Zimmerman, S., Jan, L. Y., & Jan, Y. N. (2008). Dynein is required for polarized dendritic transport and uniform microtubule orientation in axons. *Nature Cell Biology, 10*, 1172–1180.

Chapter 9
Surface Traffic in Synaptic Membranes

Martin Heine

Abstract The precision of signal transmission in chemical synapses is highly dependent on the structural alignment between pre- and postsynaptic components. The thermal agitation of transmembrane signaling molecules by surrounding lipid molecules and activity-driven changes in the local protein interaction affinities indicate a dynamic molecular traffic of molecules within synapses. The observation of local protein surface dynamics starts to be a useful tool to determine the contribution of intracellular and extracellular structures in organizing a plastic synapse. Local rearrangements by lateral diffusion in the synaptic and perisynaptic membrane induce fast density changes of signaling molecules and enable the synapse to change efficacy in short time scales. The degree of lateral mobility is restricted by many passive and active interactions inside and outside the membrane. AMPAR at the glutamatergic synapse are the best explored receptors in this respect and reviewed here as an example molecule. In addition, transsynaptic adhesion molecule complexes also appear highly dynamically in the synapse and do further support the importance of local surface traffic in subcellular compartments like synapses.

Keywords AMPA-receptors • Endocytosis • Exocytosis • Lateral diffusion • Quantum dots

M. Heine (✉)
Research Group Molecular Physiology, Leibniz Institute for Neurobiology, 39118 Magdeburg, Germany
e-mail: Martin.Heine@ifn-magdeburg.de

M.R. Kreutz and C. Sala (eds.), *Synaptic Plasticity*,
Advances in Experimental Medicine and Biology 970,
DOI 10.1007/978-3-7091-0932-8_9, © Springer-Verlag/Wien 2012

9.1 Introduction

The local number, composition, and density of signaling molecules at synapses are important determinants of synaptic plasticity. Changes in synaptic protein content were identified to be basic features of synaptic plasticity and memory formation (Kessels and Malinow 2009). Mechanisms like the endo-exocytotic cycle of synaptic membrane proteins, intracellular transport, and local synthesis of new molecules were shown to change local number, density, and composition of proteins within a time window of minutes to hours (Newpher and Ehlers 2008). Initialization of short- and long-term changes in synaptic plasticity takes place in a few milliseconds and is known to depend on interplay between pre- and postsynaptic mechanisms. Kinetic properties of pre- and postsynaptic molecules are identified to play a major role within this very first moment of activity changes, including calcium-induced facilitation or depression of presynaptic release properties (Catterall and Few 2008; Neher and Sakaba 2008) or postsynaptic receptor saturation and desensitization (Trussell et al. 1993; Xu-Friedman and Regehr 2003). In order to weaken or strengthen synaptic transmission, the alignment between pre- and postsynaptic elements has been shown to be crucial (Franks et al. 2003; Raghavachari and Lisman 2004; Shouval 2005; Xie et al. 1997). In this respect, the view of the synapse as a highly ordered compartment of signaling molecules has to be combined with a high molecular dynamic. The self-agitation of lipids and transmembrane molecules in biological membranes predicts that the position and local density of signaling molecules might be highly variable (Fig. 9.1). Within the membrane, all components can undergo rotational, translational, and lateral motions (Edidin 2003; Saxton and Jacobson 1997). These provoked questions as how is the synaptic molecular structure and density established and maintained? Is lateral mobility controlled or even used as a mechanism for synaptic organization and plasticity? Which quantities of molecules are involved? What is the turnover of crucial molecules?

Here, I will focus on functional implications and consequences of the surface traffic of proteins within and around pre- and postsynaptic membranes. Several mechanisms of dynamic protein stabilization were proposed (Poo 1985, Fig. 1) and recently revisited:

- Diffusion-Mediated trapping by intracellular anchors (Triller, Choquet 2008)
- Interactions with the extracellular matrix (Dityatev et al. 2010; Gundelfinger et al. 2010)
- Changes of membrane viscosity by different lipid composition (Renner et al. 2009a)
- Iontophoretic forces (Fromherz 1988; Savtchenko et al. 2000; Sylantyev et al. 2008).

Plasticity concepts based on a dynamic surface organization within both membrane compartments have been proposed (Franks et al. 2003; Raghavachari and Lisman 2004; Shouval 2005; Xie et al. 1997) but have not been experimentally

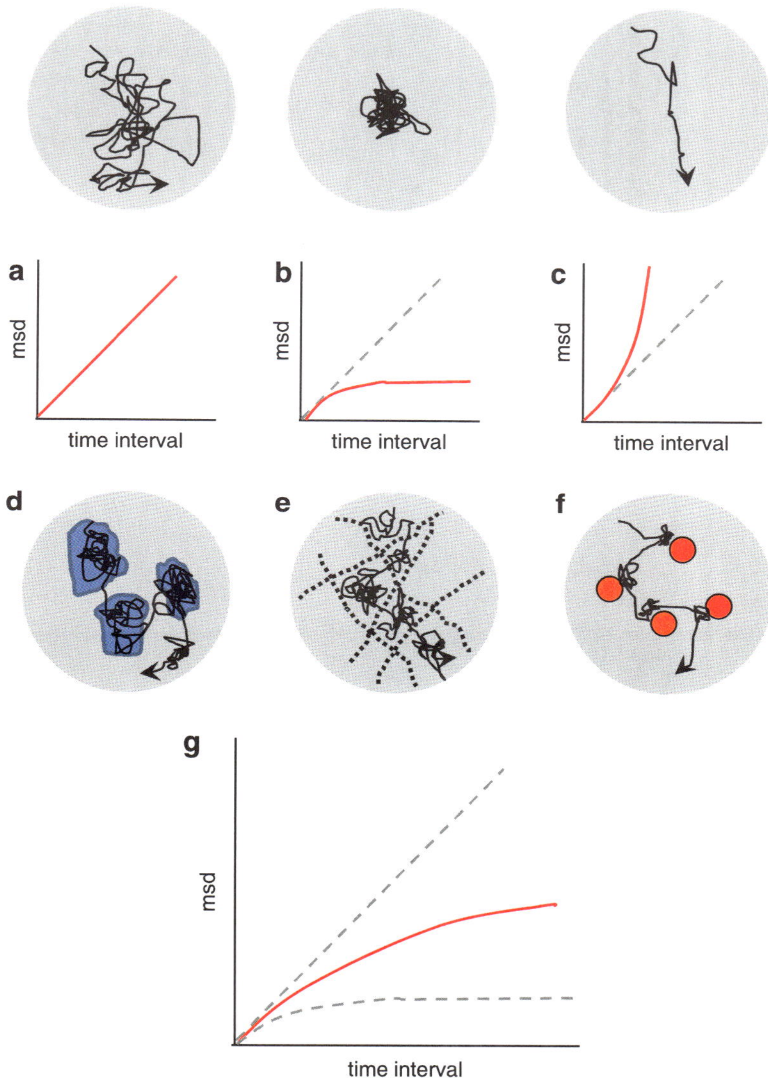

Fig. 9.1 The pictograms should demonstrate the different modes of diffusive behavior in the membrane and possible changes by structural elements within or attached to the membrane. The mean square displacement (msd) over time interval is the basal function for distinct diffusive properties as demonstrated for the extreme cases in **a–c**. (**a**) Free Brownian diffusion within the membrane is characterized by a linear msd versus time interval plot. The slope of this relationship determines the diffusion coefficient. (**b**) Confined motions within a defined area are reflected in the curved msd plot. The slope of the first points reflects the diffusion coefficient within the area. (**c**) The combination of diffusion and directed motion will result in a parabolic msd plot and can be separated from purely diffusive behavior. **d–f** are examples for anomalous diffusion, which represents the most abandon case of diffusion behavior in neuronal synapses and is an intermediate situation between free and confined diffusion. Several possible structural elements

addressed in much detail. The development and application of imaging techniques as FRAP (Axelrod et al. 1976), single particle tracking (SPT) (Saxton and Jacobson 1997), and STED microscopy (Dyba et al. 2003; Toomre and Bewersdorf 2010) to synaptic structures initiated the visualization of the molecular organization inside and around synapses. Recent developments in the field of microscopy to probe membrane dynamics have been reviewed extensively and will be not introduced here in detail (Marguet et al. 2006; Thoumine et al. 2008).

Application of these techniques has evaluated theoretical concepts of synaptic surface organization and there dynamics in such crowded and tiny compartments (diameter of 500 nm for a cortical synapse). In the past years, many experimental data support the view that the molecular noise induced by the thermal self-agitation of molecules, membrane potential changes, and intermolecular repulsive charges in the neuronal membrane are important variables to describe synaptic function as a dynamic network of signaling molecules. In the following sections, I will review the dynamics of molecules that are examples for the dynamic organization of the synaptic signaling apparatus.

9.2 Diffusion-Mediated Trapping of Postsynaptic Receptors

The first report about surface mobility of synaptic proteins came from FRAP experiments of acetylcholine receptors in the muscular membrane (Edidin and Fambrough 1973). Such experiments gave indications about the proportion of mobile and immobile acetylcholine receptors. With the application of SPT to neuronal cells (Borgdorff and Choquet 2002; Dahan et al. 2003; Meier et al. 2001), many more synaptic molecules were described to be mobile in the plasma membrane (see Table 9.1). The use of SPT provides us with detailed information about the location (accuracy of 10–50 nm), the temporal diffusion properties, relative dwell times, and relative population size of exchanging molecules within small membrane compartments. Diffusive properties can be quantified by the mean square displacement over time as an indication of the explored surface and give a quantitative measure for the apparent viscosity around the molecule of interest (for review, see Kusumi et al. 2005; Saxton and Jacobson 1997).

Fig. 9.1 (continued) are discussed in the text and illustrated here. (**d**) Changes in the membrane viscosity (*blue patches*) can introduce a confinement to free diffusive molecules in the membrane. (**e**) Intracellular as well as extracellular meshwork of structural elements (*dotted lines*) of the cytoskeleton or the extracellular matrix, respectively, can confine free diffusive molecules. (**f**) Transmembrane obstacles (*red dots*) can induce a transient confinement of diffusive molecules. (**g**) The curved msd plot is an often observed phenomenon of diffusive molecules in the membrane. Correlation of the trajectory with structural elements, for example, postsynaptic density markers, can be used to segregate the diffusive behavior in different membrane domains of the cell that are defined in time and space

Table 9.1 Median of diffusion coefficients from FRAP and SPT experiments

Molecule	$D_{synaptic}$ (μm/s)	$D_{extrasynaptic}$ (μm/s)	Mobile fraction synaptic (%)	Cell type	References
Ionotropic neurotransmitter receptor subunits					
GluA1	0.05	0.1	50–60 (D > 0.0075 μm/s)	Hippocampal neurons 14–21 DIV	Ehlers et al. (2007), Heine et al. (2008a)
GluA2	0.01	0.087–0.11	40–50 (D > 0.0075 μm/s)	Hippocampal neurons 10–14 DIV	Groc et al. (2004), Tardin et al. (2003)
TARP (γ2)	~0.01	0.075	25 (D > 0.0075 μm/s)	Hippocampal neurons culture (7–10 DIV)	Bats et al. (2007)
GluN1	0.02	0.037	70 (D > 0.0075 μm/s)	Hippocampal neurons 10–12 DIV	Groc et al. (2004)
GluN2A	0.0002	0.00075	25 (D > 0.0075 μm/s)	Hippocampal neurons 10–14 DIV	Groc et al. (2006)
GluN2B	0.05	0.025	71 (D > 0.0075 μm/s)	Hippocampal neurons 10–14 DIV	Groc et al. (2006)
GABAAR (γ2)	0.024	0.012	–	Hippocampal neurons	Bannai et al. (2009)
GlycinR	~0.005	~0.07	–	DRG neurons	Charrier et al. (2006)
nAchR (α3)	0.07	0.18	34 (total population)	Chick sympathetic ganglion neurons	Fernandes et al. (2010)
nAchR (α7)	0.067	0.18	61 (total population)	Chick sympathetic ganglion neurons	Fernandes et al. (2010)
Metabotropic neurotransmitter receptors					
mGluR5	0.014	0.14	>95 (D > 0.0001 μm/s)	Hippocampal neurons culture 21–27 DIV	Renner et al. (2010)
GABABR	–	–	50 (FRAP)	Hippocampal neurons culture 7 DIV	Pooler and McIlhinney (2007)
Dopamine receptor D1	–	0.7 (FRAP)	65 (FRAP)	Striatal organotypic culture	Scott et al. (2006)
Adhesion molecules					
α-Neurexin	–	–	100 (FRAP)	Parvalbumin-positive cells(organotypic slice)	Fu and Huang (2010)
β-Neurexin	–	~0.07 axon	100 (FRAP)	Parvalbumin-positive cells(organotypic slice) hippocampal neurons culture (6–8 DIV)	Fu and Huang (2010), Saint-Michel et al. (2009)
NCam	–	~0.13	–	Hippocampal neurons culture 10–14 DIV	Bard et al. (2008), Opazo et al. (2010)
SynCam	–	–	–	Hippocampal neurons	Breillat et al. (2007)

(continued)

Table 9.1 (continued)

Molecule	$D_{synaptic}$ (μm/s)	$D_{extrasynaptic}$ (μm/s)	Mobile fraction synaptic (%)	Cell type	References
N-cadherin	–	0.08 growth cone	n.d.	Hippocampal neurons culture (1–2 DIV)	Bard et al. (2008)
L1	–	0.25 axon	70 (D > 0.05 μm/s)	Hippocampal neurons culture (4–5 DIV)	Dequidt et al. (2007)
Membrane lipids					
DOPI	~0.06	–	42	Hippocampal neurons culture (14–20 DIV)	Renner et al. (2009b)
GPI	0.023[a] 0.046[b]	0.17	>99 (D > 0.0001 μm/s)	Hippocampal neurons culture (14–20 DIV)	Renner et al. (2009a)
Cholera toxin (G_{M1})	0.015[a] 0.029[b]	0.18	>99 (D > 0.0001 μm/s)	Hippocampal neurons culture (14–20 DIV)	Renner et al. (2009a)

[a]Median of diff.-coeff in inhibitory neurons

[b]Median of diff.-coeff in excitatory neurons; in some cases, data were only given as graphs indicated by (~)

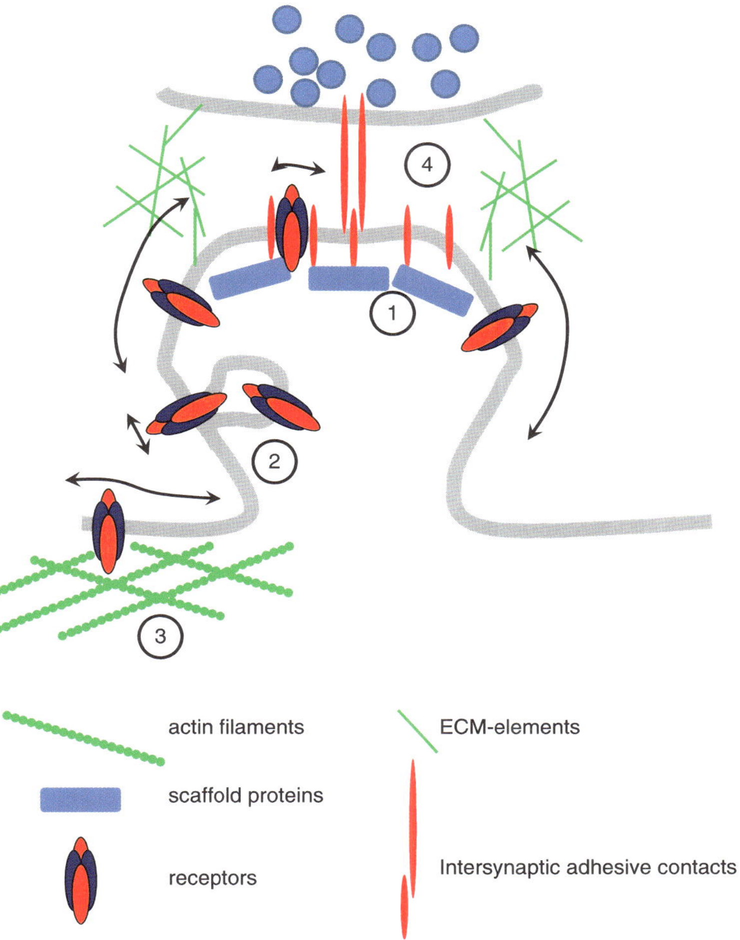

Fig. 9.2 Sketch of the dynamic organization of a spiny postsynaptic density: (*1*) intracellular scaffold proteins can interact with receptors and postsynaptic elements of adhesive contacts, (*2*) specialized zones like endocytotic pits do have a stabilization effect on diffusive receptors and will hinder them to escape from the synapse, (*3*) intracellular cytoskeleton is an important structure to regulate the surface traffic outside the postsynaptic density, and (*4*) intercellular structures like the extracellular matrix of adhesive contacts will have a passive, and, perhaps, also an active, impact on the mobility of receptors in and out of the PSD

In general, the macroscopic mobility of synaptic transmembrane molecules fulfills the characteristics of free and anomalous Brownian diffusion as evidenced by the linear or sublinear relationship of the mean square displacement (MSD) plots over time interval, respectively. For ionotropic receptors of inhibitory and excitatory synapses, it has been demonstrated by SPT (see Table 9.1) as well as electrophysiological measurements (Adesnik et al. 2005; Heine et al. 2008a; Thomas et al. 2005; Tovar and Westbrook 2002; Young and Poo 1983) that they are in a continuous exchange between the synaptic and extrasynaptic membrane. Therefore, the number and density of those crucial postsynaptic molecules depend on the regulation of diffusion properties and the capacity of a synapse to trap these receptors inside the postsynaptic density (PSD, Fig. 9.2). Molecular interactions between receptors, scaffold proteins below the inner surface of the membrane, and elements of the cytoskeleton are mechanisms to enrich molecules in the PSD. In order to clarify the contribution of a particular interaction to the stabilization of the molecule outside and inside the synapse, the surface mobility is a sensitive readout. The cytoskeleton has been identified to control the diffusion of proteins as probed by the disruption of microtubules and actin filaments (Charrier et al. 2010; Gu et al. 2010; Lee et al. 2009; Renner et al. 2009a; Rust et al. 2010). Inside synapses, major scaffold proteins are identified for AMPA-, NMDA-, GABA-, and Gly-receptors, which contribute to the local organization of these receptors in the synaptic membrane (Choquet 2010; Newpher and Ehlers 2008; Triller and Choquet 2008). As an example, interacting partners for AMPAR in the glutamatergic synapse and their surface dynamic as a functional variable of synaptic transmission are discussed below (see also Choquet 2010; Gerrow and Triller 2010).

AMPARs are composed of four subunits (GluA1–4) and expressed as heteromeric tetramers in the neuronal membrane (Collingridge et al. 2004; Hollmann and Heinemann 1994). The different length of the intracellular C-terminus, the identification of PDZ-binding domains at the end of those termini, and the different types of PDZ-binding domains, type 1 GluA1 and type 2 for GluA2 (Shi et al. 2001), suggested a direct link between AMPARs and intracellular scaffold proteins with multiple PDZ domains like PSD95, GRIP, and PICK (Feng and Zhang 2009; Kessels and Malinow 2009). The different splice variants and subunits of AMPAR, in particular, GluA1 and GluA2, are often assembled as heterodimers and crucial for plastic changes as studied extensively in hippocampal synapses by the use of plasticity protocols to induce long-term potentiation or depression (LTP, LTD) (Kessels and Malinow 2009). Supported by a number of experimental findings, this has led to the hypothesis that the capacity of the postsynaptic side is defined by the number of PDZ domains and the composition and phosphorylation state of the AMPARs that predicts their affinity to the intracellular scaffold (Malenka 2003). Knockout of the GluA1-subunit impairs the induction of long-term potentiation (LTP) in classical high-frequency stimulation paradigm (Zamanillo et al. 1999). GluA1-subunit containing AMPARs are preferentially in cooperated into the synapse during LTP induction as compared to GluA2/GluA3 heterodimers with a shorter C-terminus and a different PDZ-binding domain (Shi et al. 2001). In line, a point mutation within the PDZ-binding domain of GluA1 to disrupt the affinity to the PDZ domain impairs LTP (Hayashi et al. 2000). After LTP induction, synapses are reconsolidated by the gradual loss of GluA1/GluA2-containing receptors which are replaced by GluA2/GluA3 receptors. The disruption of the endo-exocytosis of AMPA-receptors leads to a reduction of postsynaptic receptor population and an impairment of LTP (Ehlers 2000; Park et al. 2004). In parallel to these findings, a transient increase of calcium or induction of chemical LTP was demonstrated to immobilize receptors in the PSD (Borgdorff and Choquet 2002; Tardin et al. 2003).

Contradictory to these experiments, studies in knockout mice lacking GluA2- and GluA3-subunits show no impairment in LTP or LTD (Meng et al. 2003). Furthermore, the specific deletion of the complete PDZ-binding domain of the GluA1- or GluA2-subunit did not disturb basal transmission or LTP induction (Kim et al. 2005; Panicker et al. 2008). Both findings speak against a model that subunit-specific PDZ-binding domains of AMPARs alone are critical for the traffic of the receptor to the synapse and plastic changes of their strength.

The discovery of additional subunits of AMPAR, named transmembrane AMPA receptor regulatory proteins (TARPs, Chen et al. 2000) helped to resolve this discrepant results. Initial investigations of the stargazer mutant mouse lead to the discovery of stargazin, which was believed to be a subunit of calcium channels but has much more dominant function in the traffic of AMPAR in the cerebellum (Chen et al. 2000). TARPs enhance the surface expression of AMPAR (Priel et al. 2005; Tomita et al. 2005), bind directly to PSD95 (Schnell et al. 2002), change pharmacological properties of AMPAR (Milstein and Nicoll 2008), alter kinetic properties of AMPAR (Kato et al. 2010; Milstein et al. 2007; Morimoto-Tomita et al. 2009; Priel et al. 2005; Tomita et al. 2005), and modulate the diffusion properties in the synaptic membrane (Bats et al. 2007; Opazo et al. 2010). The knockout of the dominant TARP isoform $\gamma 8$ in the hippocampus results in impaired LTP similar to those seen for the GluA1 knockout (Rouach et al. 2005). Following the surface mobility of TARPs together with AMPARs by SPT revealed that AMPARs without this additional subunit stay much shorter inside the PSD and spontaneous postsynaptic currents almost disappear. In addition, the interruption of the PDZ-binding domain of TARP ($\gamma 2$, stargazin) but not of the GluA2-subunit disrupts the confinement of receptors in the synapse (Bats et al. 2007). By tagging stargazin, the diffusion coefficients were not different in comparison to AMPAR, which implicates that a large population of AMPAR is associated with stargazin (Bats et al. 2007). In order to prove the idea that TARPs stabilize AMPAR in the PSD, Sainlos et al. (2011) developed cell-permeable biomimetic divalent ligands to disrupt AMPAR stabilization in the PSD. Those ligands do specifically bind to PSD95 type 1 PDZ domains and similar PDZ domains of PSD95 like MAGUKS (SAP102, SAP97). Those peptides do indeed disrupt the stabilization of an AMPAR subpopulation but only transient. Application of the biomimetic ligands has a time-dependent effect on AMPAR diffusion but reduces EPSC amplitude to about 40% of the control. Monomeric ligands were not effective (Sainlos et al. 2011). The temporal effect on diffusion can be seen as a loss of stabilization sides in the synapse and a readjustment of the tightly controlled equilibrium between surface and intracellular pool of receptors controlled by endocytotic zones close to the PSD (Blanpied et al. 2002; Petrini et al. 2009; Racz et al. 2004). Endocytic zones have been identified to be a stabilization side for mobile receptors in the periphery of the PSD controlling the local surface population of AMPAR (Petrini et al. 2009). The milder effects of these biomimetic tools point to the multivalent interactions within the PSD to control receptor mobility. The transient change of receptor diffusion therefore supports the trap diffusion model and indicates the dynamic organization of AMPAR within the PSD.

To be plastic, such local trapping of receptors inside the synapse must be modular. Global manipulation of synaptic activity has profound effects on the receptor mobility in the neuronal membrane (Groc et al. 2004; Tardin et al. 2003). A disruption of presynaptic transmitter release revealed that receptor stabilization inside the PSD is activity dependent (Ehlers et al. 2007). Artificial modulation of the mobile population of synaptic AMPAR in parallel to electrophysiological measurements of synaptic responses demonstrated a significant contribution of mobile receptors to basal synaptic transmission (Choquet 2010; Heine et al. 2008a). In case of fast repetitive activation (>10 Hz) of the postsynaptic receptors, the responsive population declines faster if receptors are immobilized in the postsynaptic membrane. Liberating receptors by digestion of the extracellular matrix has opposite effects (Frischknecht et al. 2009; see also Frischknecht and Gundelfinger within this book). Recordings of synaptic responses confirmed that receptor mobility is a variable of short-term plasticity (Choquet 2010; Heine et al. 2008a). The kinetic properties of AMPAR, low affinity to glutamate (Featherstone and Shippy 2008), fast desensitization, and slow recovery from desensitization (Jonas et al. 1993), do support the finding that a fast exchange of glutamate-bound and glutamate-unbound receptors will influence the postsynaptic responsiveness to high-frequency transmitter release. In addition, it has been shown that the synaptic population of AMPARs is not saturated by a single transmitter vesicle (Liu et al. 1999; McAllister and Stevens 2000), supporting the possibility of a fast (within ms) exchange of glutamate-bound receptors by free receptors within the synapse.

Another kinetic property of AMPAR is the steady state desensitization in the presence of micromolar concentrations of glutamate (Featherstone and Shippy 2008; Raman and Trussell 1995). Investigating the interaction between AMPAR and stargazin, Morimoto-Tomita et al. (2009) propose that the association of GluA1 and stargazin is critical for a steady state current evoked by micromolar ambient glutamate concentrations. Furthermore, the dissociation of the receptor-TARP complex occurs at glutamate concentrations above 100 μM and leads to a faster and more complete desensitization as well as slower recovery from this conformational state. Hence, the described dissociation of the complex within a few ms proposes a more complex picture as seen by only looking at AMPAR diffusion. Liberated AMPAR may diffuse away from the scaffold-bound stargazin and stick to their next neighbor TARPs outside the focal plane of glutamate release. With the invention of new imaging techniques, we will be enabled to gain dynamic information for a large population of receptors simultaneously (Giannone et al. 2010; Manley et al. 2008). One might postulate that different activation states of AMPAR will show different diffusion kinetics and will allow further insights in dynamic association and dissociation of receptor complexes.

Focusing on a key molecule for plasticity in glutamatergic synapses, the calcium/calmodulin-dependent protein kinase II (CaMKII; Lisman et al. 2002), Opazo et al. (2010) have demonstrated that CaMKII induces a phosphorylation-dependent stabilization of stargazin that dominates the stability of AMPAR in the PSD and that has similar physiological consequences like artificial cross-linking of AMPAR in the postsynaptic membrane (Heine et al. 2008a). Taken together, this underscores the importance of local molecular dynamics in the plasma membrane for synaptic transmission and plasticity.

The discovery of other subunits associated with AMPAR in different brain regions, as cornichons (Schwenk et al. 2009), CKAMP44 (von Engelhardt et al. 2010), and SynDIG1 (Kalashnikova et al. 2010), will probably further extend the dynamic view of the AMPAR as an association point for many interacting molecules that might tune the surface mobility and kinetic properties of AMPAR.

Despite the strong interactions of surface molecules with the PSD as discussed above for AMPARs and seen for GABAA-, Glycin-, and NMDA-receptors as well (Bannai et al. 2009; Bard et al. 2010; Charrier et al. 2010; Dumoulin et al. 2009; Jacob et al. 2005; Muir et al. 2010; Tretter et al. 2008; Tretter and Moss 2008) other factors inside and outside the membrane contribute to a dynamic surface organization of signaling molecules (Fig. 9.1). The transient stabilization of receptors in the synapse leads to the idea that other mechanisms in addition to the intracellular anchors will contribute to maintain the concentration of signaling molecules within the synapse. Such mechanisms could include a different lipid composition (Allen et al. 2007), extracellular structures, like components of the extracellular matrix (Dityatev et al. 2010; Gundelfinger et al. 2010, Frischknecht and Gundelfinger in this book), repulsive or attractive intermolecular forces induced by charge differences (electrodiffusion (Savtchenko et al. 2000) or stable transmembrane molecules acting as diffusion obstacles (Kusumi et al. 2005)).

9.3 Molecular Crowding Confines Molecules Inside the Synapse

The existence of cholesterol/sphingolipid microdomains (lipid rafts) within the dendritic membrane was proposed to influence the membrane stabilization of AMPAR at synapses (Hering et al. 2003). Characterization of the diffusion properties of two lipid raft markers, glycophosphatidylinositol-anchored green fluorescent protein (GFP-GPI) and cholera toxin (beta-subunit binds to G_{M1}) in the postsynaptic membrane revealed confined diffusion inside the synapse (Renner et al. 2009a, b). Both raft markers are not enriched in the postsynaptic membrane of excitatory or inhibitory synapses, despite the confined diffusion within the synapse. Interestingly, lipid diffusion is twofold slower in inhibitory than in excitatory synapses, indicating a stiffer organization of inhibitory postsynaptic membranes (Renner et al. 2009a). Manipulation of the cytoskeletal integrity by depolymerization of F-actin leads to an even higher diffusion of fast-diffusing lipids in the postsynaptic membrane. The acceleration of lipid diffusion in the outer leaflet of the membrane can be explained as a decrease in the apparent viscosity of the membrane resulting from a weaker stabilization of transmembrane proteins in the absence of F-actin filaments. The molecular manipulation of actin polymerization by the actin depolymerization factor (ADF) n-cofilin had no effect on cholera toxin diffusion, confirming the rather indirect effect of the actin skeleton within the PSD (Rust et al. 2010).

A partial extraction of cholesterol from the membrane does not dramatically change receptor or lipid diffusion inside synapses. The confined diffusion without enrichment of lipids inside synapses speaks in favor for a situation where the density of obstacles is the main source of confinement for small molecules like lipids as seen in the axonal initial segment and synapse (Nakada et al. 2003; Renner et al. 2009a, b), rather than a specific population of saturated lipids like those described in lipid rafts. The noncovalent interactions of synaptic receptors with the PSD, as described above, have a significant contribution to the crowded environment keeping a homeostatic concentration of receptors (Santamaria et al. 2010; Shouval 2005). Other forces beside lateral diffusion and intracellular binding affinities manipulating local molecular density are electrostatical interactions between molecules and transient electric fields generated by the opening of receptors after neurotransmitter binding (Fromherz 1988; Poo et al. 1979). For small cortical synapses, the contribution of electric forces has only been investigated theoretically, suggesting that AMPAR will be clustered during trains of high-frequency stimulation (>20 Hz) in opposite to the presynaptic release side (Savtchenko et al. 2000). In order to test this idea experimentally, SPT with high spatial and temporal resolution could give an answer (Kusumi et al. 2005). However, the mainly used video-rate acquisition (30 Hz) is not sufficient to follow fast motions. Increasing the temporal resolution is often paid by the loss of spatial resolution and needs further methodological development.

9.4 Membrane Structures Contribute to Local Confinement

Apart from the scaffold protein interaction and packed organization within the postsynaptic membrane, the question occurs how the receptor concentration is maintained within the synapse if each dissociation from an interaction partner results in increased mobility. It appears therefore reasonable to assume that specific structures around the synapse will prevent the escape of proteins over time. Models of the synaptic membrane organization have suggested barriers in the periphery of the PSD (Holcman and Triller 2006; Schuss et al. 2007) with restricted numbers of open gates that allow both the flux of molecules and the retaining of a critical local population inside the synapse. Investigations of the endo-exocytotic cycle and diffusional membrane organization made clear that first the source and sink of receptors are located outside the PSD (Blanpied et al. 2002; Jacob et al. 2009; Jaskolski et al. 2009; Kennedy et al. 2010; Lu et al. 2007; Yudowski et al. 2006), the distance between the PSD and clathrin-coated endocytotic pits does regulate the number of available receptors (Lu et al. 2007), exocytosis occurs within a specific membrane compartment close to the PSD (Kennedy et al. 2010) or within the dendritic shaft and soma (Adesnik et al. 2005; Jaskolski et al. 2009; Makino and Malinow 2009; Yudowski et al. 2006), and transient receptor stabilization in clathrin-coated endocytotic pits modifies the synaptic responsiveness to activity changes (Jacob et al. 2009; Petrini et al. 2009).

The microscale organization of the perisynaptic membrane has profound consequences for long-term synaptic plasticity. Dendritic exocytosis of AMPAR has been identified as essential for LTP maintenance (Makino and Malinow 2009). The molecular players were recently identified and will help to further explore the molecular network underlying changes in synaptic plasticity. Receptor exocytosis occurs within local clusters of the t-SNARE protein syntaxin4 (Kennedy et al. 2010). Syntaxin4-mediated exocytosis is regulated by the interaction with F-actin filaments (Band et al. 2002), which do control the association with the vesicular SNARE protein synaptobrevin2 (VAMP2) (Jewell et al. 2008). The mobile organization of syntaxin4 aggregates shortly before exocytosis, and the interaction with the vesicular SNARE protein is controlled by the local depolimerization of F-actin (Gu et al. 2010; Kennedy et al. 2010). As mentioned above, the actin depolymerization factor n-cofilin has little effect for the synaptic membrane organization but influences spine morphology, exocytosis, and extrasynaptic surface mobility of AMPAR which seems also to be regulated by the phosphorylation status of n-cofilin (Gu et al. 2010; Rust et al. 2010). Similar mechanisms were reported for acetylcholine receptors in the *Xenopus* neuromuscular junction (Lee et al. 2009). Once exocytosed, the majority of receptors are confined close to the synaptic density (Kennedy et al. 2010). Membrane curvature and the intracellular condensation state of the actin filaments are proposed to hinder diffusion escape from the PSD (Holcman and Triller 2006) and hence allow receptors to become incorporated in the PSD. ADF/cofilins are responsible for the dynamic surface organization of syntaxin4-containing exocytotic zones around the synapse. The structures that cause such restricted mobility still need to be identified. Septins are very likely candidates as demonstrated for the developmental switch from microdomains to nanodomains in the presynaptic terminal of the calyx of Held (Yang et al. 2010).

Knockout mice for n-cofilin show no difference in synaptic transmission and short-term plasticity but have impaired late LTP and LTD as well as deficits in associative learning (Rust et al. 2010). The concept of the membrane microdomain organization around the synapse is interconnected by lateral surface diffusion of proteins between these domains that regulate synaptic plasticity and contributes to the spatial isolation of individual synapses.

At the presynaptic side, the very efficient and fast endo-exocytotic coupling is discussed to be a directed surface diffusion/flow of vesicular proteins (Haucke et al. 2011) within the presynaptic membrane. Its role would be to bridge the space between vesicle fusion and vesicle retrieval in the periphery of the active zone and to prevent an enlargement of the presynaptic terminal. At the postsynaptic side, volume changes in dendritic spines are reported to be associated with the insertion of new AMPAR and the induction of LTP (Kopec et al. 2007; Makino and Malinow 2009), suggesting a less tight coupling between membrane insertion and membrane retrieval. Such structural plasticity may interfere with the adhesion between pre- and postsynaptic membranes. Adhesion molecule surface organization may provide an insight in such dynamic processes.

9.5   Surface Dynamic of Adhesion Molecules as Modulator of Synaptic Molecular Organization

As shown in other cellular systems, the formation and maintenance of focal intercellular contacts, like synaptic junctions, depend on the density and mobility of adhesive partners on both membranes (Chan et al. 1991). It is reasonable to assume that primarily the density of cell adhesive molecules (CAM) will determine the strength of the formed contact. However, the mobility of CAMs in both membranes strongly accelerate the formation of the contact formation (Chan et al. 1991).

Many synaptic CAMs have been found, and the list of their intra- and extracellular binding partners is still growing (for further review, see Dalva et al. 2007; Tallafuss et al. 2010). Beside their known function to tether pre- and postsynaptic membranes, the capacity of transsynaptic signaling of CAMs is an important variable of synaptic plasticity (Futai et al. 2007; Stan et al. 2010). During synaptogenesis, the lateral recruitment of adhesion molecules to focal intercellular contacts is well accepted and used in many approaches to demonstrate the adhesive function of the molecule by the capacity to corecruit other synaptic molecules. For example, a prominent pair of synaptic adhesion molecules, neuroligins (postsynaptic) and neurexins (presynaptic), has been studied by the use HEK cells expressing either neuroligin (Scheiffele et al. 2000) or neurexin (Dean et al. 2003; Graf et al. 2004) cocultured with neurons, demonstrating the binding ability and recruitment of pre- and postsynaptic elements like transmitter vesicles (Scheiffele et al. 2000) or postsynaptic scaffold proteins (PSD95; Irie et al. 1997), Gephrin; (Poulopoulos et al. 2009), NMDA- and AMPA-receptors, respectively. Another assay was to use neurexin-coated beads to recruit neuroligin (Heine et al. 2008b; Nam and Chen 2005) or the simple application of Fc-tagged β-neurexins and subsequential cross-linking by Fc-specific antibodies (Barrow et al. 2009). It can be assumed that at least during synaptogenesis, adhesive molecules are expressed in the outer membrane and able to form preliminary intercellular contacts that will be consolidated following their establishment. However, the direct investigation of the surface distribution and dynamic of adhesive molecules as well as their intracellular trafficking is little understood, due to the methodological problem of interference with the intercellular binding partners by fluorescent labeling without interference with the function of these proteins.

Using transsynaptic enzymatic biotinylation to ensure functional surface labeling of heterophilic contacts between β-neurexin and neuroligin 1, Thyagarajan and Ting (2010) proposed a very local synapse-specific recycling of both molecules, which is modulated by neuronal activity. A similar local trafficking has been proposed for N-cadherin endocytosis (Tai et al. 2007).

Recent work by Fu and Huang (2010) directly addressed the question of subcellular distribution and surface accumulation of neurexins in the axon of parvalbumin-positive interneurons in cultured organotypic brain slices. Using pHluorin (pH-sensitive variant of GFP, Miesenbock et al. 1998) tagged α- and β-neurexins within organotypic slices, they conducted FRAP experiments to probe the dynamic of neurexins in the axonal membrane. β-Neurexins have been seen mostly clustered within the axonal terminal, whereas α-neurexins are nearly equally distributed in the synaptic and axonal membrane and exchange between neighboring synapses. Despite the differences in surface distribution, both probes recover to 100% after local photobleaching within several minutes. Modulating presynaptic activity by the blocking of sodium channels with TTX or disrupting transmitter release with tetanus toxin did decrease the mobility of β-neurexins but did not influence the mobility of α-neurexins (Fu and Huang 2010). The described functions of these two forms of Neurexins are different and seem to be reflected in the surface dynamic. β-Neurexins possess a short extrasynaptic domain and bind to postsynaptic neuroligin 1 and 2 involved in the establishment and maintenance of specific synaptic contacts depending on the splice isoforms (Dalva et al. 2007; Tallafuss et al. 2010). Whereas, α-neurexins are recognized to organize the number or localization of presynaptic calcium channels (Missler et al. 2003; Zhang et al. 2005) and have only moderate binding affinities to neuroligin 2 in GABAergic synapses. The side of interaction between α-neurexin and N- and P/Q-type channels is still not known. It is conceivable that the differences in surface mobility could be due to a difference in function of neurexin isoforms and splice variants in synapse maturation and plasticity. Another explanation comes from the discovery of new postsynaptic binding partners of β-neurexins, suggesting a stronger anchoring of β-neurexins due to the multivalent interactions with other adhesion molecules (de Wit et al. 2009; Ko et al. 2009) or subunits (α1) of GABAA-receptors (Zhang et al. 2010). The application of SPT to investigate the mobility of adhesion molecules in the synapse will contribute to the validation of biochemically identified binding partners and might help to measure association and dissociation constants in a more physiological cellular environment (Saint-Michel et al. 2009).

9.6 Conclusions

The molecular noise within biological membranes has been documented as an important variable for immunological synapses, which have a lifetime of a few hours at most. The function of neuronal synapses as structural elements for learning and memory suggests a much more stable organization. However, molecular lifetime, nonequal distribution of binding partners, activity-driven changes of binding affinities, and concentrations argue against a rigid structural organization. The advent of imaging techniques that allow resolving single-molecule level shed new light in the nm-scale organization of synapses. The combination of such imaging techniques with physiological assays will clarify the functional impact of molecular motion in processes like learning and memory. As reviewed here for AMPAR synaptic transmission and short-term plasticity, initial changes of synaptic integration in the millisecond time window depend on the mobility of signaling molecules. Therefore, molecular flexibility, which is partially reflected in the differential diffusion of transmitter receptors, has a fundamental impact on neuronal network function.

References

Adesnik, H., Nicoll, R. A., & England, P. M. (2005). Photoinactivation of native AMPA receptors reveals their real-time trafficking. *Neuron, 48*, 977–985.

Allen, J. A., Halverson-Tamboli, R. A., & Rasenick, M. M. (2007). Lipid raft microdomains and neurotransmitter signalling. *Nature Reviews Neuroscience, 8*, 128–140.

Axelrod, D., Koppel, D. E., Schlessinger, J., Elson, E., & Webb, W. W. (1976). Mobility measurement by analysis of fluorescence photobleaching recovery kinetics. *Biophysical Journal, 16*, 1055–1069.

Band, A. M., Ali, H., Vartiainen, M. K., Welti, S., Lappalainen, P., Olkkonen, V. M., & Kuismanen, E. (2002). Endogenous plasma membrane t-SNARE syntaxin 4 is present in rab11 positive endosomal membranes and associates with cortical actin cytoskeleton. *FEBS Letters, 531*, 513–519.

Bannai, H., Levi, S., Schweizer, C., Inoue, T., Launey, T., Racine, V., Sibarita, J. B., Mikoshiba, K., & Triller, A. (2009). Activity-dependent tuning of inhibitory neurotransmission based on GABAAR diffusion dynamics. *Neuron, 62*, 670–682.

Bard, L., Boscher, C., Lambert, M., Mege, R. M., Choquet, D., & Thoumine, O. (2008). A molecular clutch between the actin flow and N-cadherin adhesions drives growth cone migration. *Journal of Neuroscience, 28*, 5879–5890.

Bard, L., Sainlos, M., Bouchet, D., Cousins, S., Mikasova, L., Breillat, C., Stephenson, F. A., Imperiali, B., Choquet, D., & Groc, L. (2010). Dynamic and specific interaction between synaptic NR2-NMDA receptor and PDZ proteins. *Proceedings of the National Academy of Sciences of the United States of America, 107*, 19561–19566.

Barrow, S. L., Constable, J. R., Clark, E., El-Sabeawy, F., McAllister, A. K., & Washbourne, P. (2009). Neuroligin1: A cell adhesion molecule that recruits PSD-95 and NMDA receptors by distinct mechanisms during synaptogenesis. *Neural Development, 4*, 17.

Bats, C., Groc, L., & Choquet, D. (2007). The interaction between Stargazin and PSD-95 regulates AMPA receptor surface trafficking. *Neuron, 53*, 719–734.

Blanpied, T. A., Scott, D. B., & Ehlers, M. D. (2002). Dynamics and regulation of clathrin coats at specialized endocytic zones of dendrites and spines. *Neuron, 36*, 435–449.

Borgdorff, A. J., & Choquet, D. (2002). Regulation of AMPA receptor lateral movements. *Nature, 417*, 649–653.

Breillat, C., Thoumine, O., & Choquet, D. (2007). Characterization of SynCAM surface trafficking using a SynCAM derived ligand with high homophilic binding affinity. *Biochemical and Biophysical Research Communications, 359*, 655–659.

Catterall, W. A., & Few, A. P. (2008). Calcium channel regulation and presynaptic plasticity. *Neuron, 59*, 882–901.

Chan, P. Y., Lawrence, M. B., Dustin, M. L., Ferguson, L. M., Golan, D. E., & Springer, T. A. (1991). Influence of receptor lateral mobility on adhesion strengthening between membranes containing LFA-3 and CD2. *The Journal of Cell Biology, 115*, 245–255.

Charrier, C., Ehrensperger, M. V., Dahan, M., Levi, S., & Triller, A. (2006). Cytoskeleton regulation of glycine receptor number at synapses and diffusion in the plasma membrane. *Journal of Neuroscience, 26*, 8502–8511.

Charrier, C., Machado, P., Tweedie-Cullen, R. Y., Rutishauser, D., Mansuy, I. M., & Triller, A. (2010). A crosstalk between beta1 and beta3 integrins controls glycine receptor and gephyrin trafficking at synapses. *Nature Neuroscience, 13*, 1388–1395.

Chen, L., Chetkovich, D. M., Petralia, R. S., Sweeney, N. T., Kawasaki, Y., Wenthold, R. J., Bredt, D. S., & Nicoll, R. A. (2000). Stargazin regulates synaptic targeting of AMPA receptors by two distinct mechanisms. *Nature, 408*, 936–943.

Choquet, D. (2010). Fast AMPAR trafficking for a high-frequency synaptic transmission. *European Journal of Neuroscience, 32*, 250–260.

Collingridge, G. L., Isaac, J. T., & Wang, Y. T. (2004). Receptor trafficking and synaptic plasticity. *Nature Reviews Neuroscience, 5*, 952–962.

Dahan, M., Levi, S., Luccardini, C., Rostaing, P., Riveau, B., & Triller, A. (2003). Diffusion dynamics of glycine receptors revealed by single-quantum dot tracking. *Science, 302*, 442–445.

Dalva, M. B., McClelland, A. C., & Kayser, M. S. (2007). Cell adhesion molecules: Signalling functions at the synapse. *Nature Reviews Neuroscience, 8*, 206–220.

de Wit, J., Sylwestrak, E., O'Sullivan, M. L., Otto, S., Tiglio, K., Savas, J. N., Yates, J. R., III, Comoletti, D., Taylor, P., & Ghosh, A. (2009). LRRTM2 interacts with Neurexin1 and regulates excitatory synapse formation. *Neuron, 64*, 799–806.

Dean, C., Scholl, F. G., Choih, J., DeMaria, S., Berger, J., Isacoff, E., & Scheiffele, P. (2003). Neurexin mediates the assembly of presynaptic terminals. *Nature Neuroscience, 6*, 708–716.

Dequidt, C., Danglot, L., Alberts, P., Galli, T., Choquet, D., & Thoumine, O. (2007). Fast turnover of L1 adhesions in neuronal growth cones involving both surface diffusion and exo/endocytosis of L1 molecules. *Molecular Biology of the Cell, 18*, 3131–3143.

Dityatev, A., Schachner, M., & Sonderegger, P. (2010). The dual role of the extracellular matrix in synaptic plasticity and homeostasis. *Nature Reviews Neuroscience, 11*, 735–746.

Dumoulin, A., Triller, A., & Kneussel, M. (2009). Cellular transport and membrane dynamics of the glycine receptor. *Frontiers in Molecular Neuroscience, 2*, 28.

Dyba, M., Jakobs, S., & Hell, S. W. (2003). Immunofluorescence stimulated emission depletion microscopy. *Nature Biotechnology, 21*, 1303–1304.

Edidin, M. (2003). The state of lipid rafts: From model membranes to cells. *Annual Review of Biophysics and Biomolecular Structure, 32*, 257–283.

Edidin, M., & Fambrough, D. (1973). Fluidity of the surface of cultured muscle fibers. Rapid lateral diffusion of marked surface antigens. *The Journal of Cell Biology, 57*, 27–37.

Ehlers, M. D. (2000). Reinsertion or degradation of AMPA receptors determined by activity-dependent endocytic sorting. *Neuron, 28*, 511–525.

Ehlers, M. D., Heine, M., Groc, L., Lee, M. C., & Choquet, D. (2007). Diffusional trapping of GluR1 AMPA receptors by input-specific synaptic activity. *Neuron, 54*, 447–460.

Featherstone, D. E., & Shippy, S. A. (2008). Regulation of synaptic transmission by ambient extracellular glutamate. *The Neuroscientist, 14*, 171–181.

Feng, W., & Zhang, M. (2009). Organization and dynamics of PDZ-domain-related supramodules in the postsynaptic density. *Nature Reviews Neuroscience, 10*, 87–99.

Fernandes, C. C., Berg, D. K., & Gomez-Varela, D. (2010). Lateral mobility of nicotinic acetylcholine receptors on neurons is determined by receptor composition, local domain, and cell type. *Journal of Neuroscience, 30*, 8841–8851.

Franks, K. M., Stevens, C. F., & Sejnowski, T. J. (2003). Independent sources of quantal variability at single glutamatergic synapses. *Journal of Neuroscience, 23*, 3186–3195.

Frischknecht, R., Heine, M., Perrais, D., Seidenbecher, C. I., Choquet, D., & Gundelfinger, E. D. (2009). Brain extracellular matrix affects AMPA receptor lateral mobility and short-term synaptic plasticity. *Nature Neuroscience, 12*, 897–904.

Fromherz, P. (1988). Self-organization of the fluid mosaic of charged channel proteins in membranes. *Proceedings of the National Academy of Sciences of the United States of America, 85*, 6353–6357.

Fu, Y., & Huang, Z. J. (2010). Differential dynamics and activity-dependent regulation of alpha- and beta-neurexins at developing GABAergic synapses. *Proceedings of the National Academy of Sciences of the United States of America, 107*, 22699–22704.

Futai, K., Kim, M. J., Hashikawa, T., Scheiffele, P., Sheng, M., & Hayashi, Y. (2007). Retrograde modulation of presynaptic release probability through signaling mediated by PSD-95-neuroligin. *Nature Neuroscience, 10*, 186–195.

Gerrow, K., & Triller, A. (2010). Synaptic stability and plasticity in a floating world. *Current Opinion in Neurobiology, 20*, 631–639.

Giannone, G., Hosy, E., Levet, F., Constals, A., Schulze, K., Sobolevsky, A. I., Rosconi, M. P., Gouaux, E., Tampe, R., Choquet, D., & Cognet, L. (2010). Dynamic superresolution imaging of endogenous proteins on living cells at ultra-high density. *Biophysical Journal, 99*, 1303–1310.

Graf, E. R., Zhang, X., Jin, S. X., Linhoff, M. W., & Craig, A. M. (2004). Neurexins induce differentiation of GABA and glutamate postsynaptic specializations via neuroligins. *Cell, 119*, 1013–1026.

Groc, L., Heine, M., Cognet, L., Brickley, K., Stephenson, F. A., Lounis, B., & Choquet, D. (2004). Differential activity-dependent regulation of the lateral mobilities of AMPA and NMDA receptors. *Nature Neuroscience, 7*, 695–696.

Groc, L., Heine, M., Cousins, S. L., Stephenson, F. A., Lounis, B., Cognet, L., & Choquet, D. (2006). NMDA receptor surface mobility depends on NR2A-2B subunits. *Proceedings of the National Academy of Sciences of the United States of America, 103*, 18769–18774.

Gu, J., Lee, C. W., Fan, Y., Komlos, D., Tang, X., Sun, C., Yu, K., Hartzell, H. C., Chen, G., Bamburg, J. R., & Zheng, J. Q. (2010). ADF/cofilin-mediated actin dynamics regulate AMPA receptor trafficking during synaptic plasticity. *Nature Neuroscience, 13*, 1208–1215.

Gundelfinger, E. D., Frischknecht, R., Choquet, D., & Heine, M. (2010). Converting juvenile into adult plasticity: a role for the brain's extracellular matrix. *European Journal of Neuroscience, 31*, 2156–2165.

Haucke, V., Neher, E., & Sigrist, S. J. (2011). Protein scaffolds in the coupling of synaptic exocytosis and endocytosis. *Nature Reviews Neuroscience, 12*, 127–138.

Hayashi, Y., Shi, S. H., Esteban, J. A., Piccini, A., Poncer, J. C., & Malinow, R. (2000). Driving AMPA receptors into synapses by LTP and CaMKII: Requirement for GluR1 and PDZ domain interaction. *Science, 287*, 2262–2267.

Heine, M., Groc, L., Frischknecht, R., Beique, J. C., Lounis, B., Rumbaugh, G., Huganir, R. L., Cognet, L., & Choquet, D. (2008a). Surface mobility of postsynaptic AMPARs tunes synaptic transmission. *Science, 320*, 201–205.

Heine, M., Thoumine, O., Mondin, M., Tessier, B., Giannone, G., & Choquet, D. (2008b). Activity-independent and subunit-specific recruitment of functional AMPA receptors at neurexin/neuroligin contacts. *Proceedings of the National Academy of Sciences of the United States of America, 105*, 20947–20952.

Hering, H., Lin, C. C., & Sheng, M. (2003). Lipid rafts in the maintenance of synapses, dendritic spines, and surface AMPA receptor stability. *Journal of Neuroscience, 23*, 3262–3271.

Holcman, D., & Triller, A. (2006). Modeling synaptic dynamics driven by receptor lateral diffusion. *Biophysical Journal, 91*, 2405–2415.

Hollmann, M., & Heinemann, S. (1994). Cloned glutamate receptors. *Annual Review of Neuroscience, 17*, 31–108.

Irie, M., Hata, Y., Takeuchi, M., Ichtchenko, K., Toyoda, A., Hirao, K., Takai, Y., Rosahl, T. W., & Sudhof, T. C. (1997). Binding of neuroligins to PSD-95. *Science, 277*, 1511–1515.

Jacob, T. C., Bogdanov, Y. D., Magnus, C., Saliba, R. S., Kittler, J. T., Haydon, P. G., & Moss, S. J. (2005). Gephyrin regulates the cell surface dynamics of synaptic GABAA receptors. *Journal of Neuroscience, 25*, 10469–10478.

Jacob, T. C., Wan, Q., Vithlani, M., Saliba, R. S., Succol, F., Pangalos, M. N., & Moss, S. J. (2009). GABA(A) receptor membrane trafficking regulates spine maturity. *Proceedings of the National Academy of Sciences of the United States of America, 106*, 12500–12505.

Jaskolski, F., Mayo-Martin, B., Jane, D., & Henley, J. M. (2009). Dynamin-dependent membrane drift recruits AMPA receptors to dendritic spines. *Journal of Biological Chemistry, 284*, 12491–12503.

Jewell, J. L., Luo, W., Oh, E., Wang, Z., & Thurmond, D. C. (2008). Filamentous actin regulates insulin exocytosis through direct interaction with syntaxin 4. *Journal of Biological Chemistry, 283*, 10716–10726.

Jonas, P., Major, G., & Sakmann, B. (1993). Quantal components of unitary EPSCs at the mossy fibre synapse on CA3 pyramidal cells of rat hippocampus. *The Journal of Physiology, 472*, 615–663.

Kalashnikova, E., Lorca, R. A., Kaur, I., Barisone, G. A., Li, B., Ishimaru, T., Trimmer, J. S., Mohapatra, D. P., & Diaz, E. (2010). SynDIG1: An activity-regulated, AMPA-receptor-interacting transmembrane protein that regulates excitatory synapse development. *Neuron, 65*, 80–93.

Kato, A. S., Gill, M. B., Ho, M. T., Yu, H., Tu, Y., Siuda, E. R., Wang, H., Qian, Y. W., Nisenbaum, E. S., Tomita, S., & Bredt, D. S. (2010). Hippocampal AMPA receptor gating controlled by both TARP and cornichon proteins. *Neuron, 68*, 1082–1096.

Kennedy, M. J., Davison, I. G., Robinson, C. G., & Ehlers, M. D. (2010). Syntaxin-4 defines a domain for activity-dependent exocytosis in dendritic spines. *Cell, 141*, 524–535.

Kessels, H. W., & Malinow, R. (2009). Synaptic AMPA receptor plasticity and behavior. *Neuron, 61*, 340–350.

Kim, C. H., Takamiya, K., Petralia, R. S., Sattler, R., Yu, S., Zhou, W., Kalb, R., Wenthold, R., & Huganir, R. (2005). Persistent hippocampal CA1 LTP in mice lacking the C-terminal PDZ ligand of GluR1. *Nature Neuroscience, 8*, 985–987.

Ko, J., Fuccillo, M. V., Malenka, R. C., & Sudhof, T. C. (2009). LRRTM2 functions as a neurexin ligand in promoting excitatory synapse formation. *Neuron, 64*, 791–798.

Kopec, C. D., Real, E., Kessels, H. W., & Malinow, R. (2007). GluR1 links structural and functional plasticity at excitatory synapses. *Journal of Neuroscience, 27*, 13706–13718.

Kusumi, A., Nakada, C., Ritchie, K., Murase, K., Suzuki, K., Murakoshi, H., Kasai, R. S., Kondo, J., & Fujiwara, T. (2005). Paradigm shift of the plasma membrane concept from the two-dimensional continuum fluid to the partitioned fluid: High-speed single-molecule tracking of membrane molecules. *Annual Review of Biophysics and Biomolecular Structure, 34*, 351–378.

Lee, C. W., Han, J., Bamburg, J. R., Han, L., Lynn, R., & Zheng, J. Q. (2009). Regulation of acetylcholine receptor clustering by ADF/cofilin-directed vesicular trafficking. *Nature Neuroscience, 12*, 848–856.

Lisman, J., Schulman, H., & Cline, H. (2002). The molecular basis of CaMKII function in synaptic and behavioural memory. *Nature Reviews Neuroscience, 3*, 175–190.

Liu, G., Choi, S., & Tsien, R. W. (1999). Variability of neurotransmitter concentration and nonsaturation of postsynaptic AMPA receptors at synapses in hippocampal cultures and slices. *Neuron, 22*, 395–409.

Lu, J., Helton, T. D., Blanpied, T. A., Racz, B., Newpher, T. M., Weinberg, R. J., & Ehlers, M. D. (2007). Postsynaptic positioning of endocytic zones and AMPA receptor cycling by physical coupling of dynamin-3 to Homer. *Neuron, 55*, 874–889.

Makino, H., & Malinow, R. (2009). AMPA receptor incorporation into synapses during LTP: The role of lateral movement and exocytosis. *Neuron, 64*, 381–390.

Malenka, R. C. (2003). Synaptic plasticity and AMPA receptor trafficking. *Annals of the New York Academy of Sciences, 1003*, 1–11.

Manley, S., Gillette, J. M., Patterson, G. H., Shroff, H., Hess, H. F., Betzig, E., & Lippincott-Schwartz, J. (2008). High-density mapping of single-molecule trajectories with photoactivated localization microscopy. *Nature Methods, 5*, 155–157.

Marguet, D., Lenne, P. F., Rigneault, H., & He, H. T. (2006). Dynamics in the plasma membrane: How to combine fluidity and order. *EMBO Journal, 25*, 3446–3457.

McAllister, A. K., & Stevens, C. F. (2000). Nonsaturation of AMPA and NMDA receptors at hippocampal synapses. *Proceedings of the National Academy of Sciences of the United States of America, 97*, 6173–6178.

Meier, J., Vannier, C., Serge, A., Triller, A., Choquet, D., & M-247* (2001). Fast and reversible trapping of surface glycine receptors by gephyrin. *Nature Neuroscience, 4*, 253–260.

Meng, Y., Zhang, Y., Jia, Z., & M-508* (2003). Synaptic transmission and plasticity in the absence of AMPA glutamate receptor GluR2 and GluR3. *Neuron, 39*, 163–176.

Miesenbock, G., De Angelis, D. A., & Rothman, J. E. (1998). Visualizing secretion and synaptic transmission with pH-sensitive green fluorescent proteins. *Nature, 394*, 192–195.

Milstein, A. D., & Nicoll, R. A. (2008). Regulation of AMPA receptor gating and pharmacology by TARP auxiliary subunits. *Trends in Pharmacological Sciences, 29*, 333–339.

Milstein, A. D., Zhou, W., Karimzadegan, S., Bredt, D. S., & Nicoll, R. A. (2007). TARP subtypes differentially and dose-dependently control synaptic AMPA receptor gating. *Neuron, 55*, 905–918.

Missler, M., Zhang, W., Rohlmann, A., Kattenstroth, G., Hammer, R. E., Gottmann, K., & Sudhof, T. C. (2003). Alpha-neurexins couple Ca^{2+} channels to synaptic vesicle exocytosis. *Nature, 423,* 939–948.

Morimoto-Tomita, M., Zhang, W., Straub, C., Cho, C. H., Kim, K. S., Howe, J. R., & Tomita, S. (2009). Autoinactivation of neuronal AMPA receptors via glutamate-regulated TARP interaction. *Neuron, 61,* 101–112.

Muir, J., Arancibia-Carcamo, I. L., MacAskill, A. F., Smith, K. R., Griffin, L. D., & Kittler, J. T. (2010). NMDA receptors regulate GABAA receptor lateral mobility and clustering at inhibitory synapses through serine 327 on the gamma2 subunit. *Proceedings of the National Academy of Sciences of the United States of America, 107,* 16679–16684.

Nakada, C., Ritchie, K., Oba, Y., Nakamura, M., Hotta, Y., Iino, R., Kasai, R. S., Yamaguchi, K., Fujiwara, T., & Kusumi, A. (2003). Accumulation of anchored proteins forms membrane diffusion barriers during neuronal polarization. *Nature Cell Biology, 5,* 626–632.

Nam, C. I., & Chen, L. (2005). Postsynaptic assembly induced by neurexin-neuroligin interaction and neurotransmitter. *Proceedings of the National Academy of Sciences of the United States of America, 102,* 6137–6142.

Neher, E., & Sakaba, T. (2008). Multiple roles of calcium ions in the regulation of neurotransmitter release. *Neuron, 59,* 861–872.

Newpher, T. M., & Ehlers, M. D. (2008). Glutamate receptor dynamics in dendritic microdomains. *Neuron, 58,* 472–497.

Opazo, P., Labrecque, S., Tigaret, C. M., Frouin, A., Wiseman, P. W., De Koninck, P., & Choquet, D. (2010). CaMKII triggers the diffusional trapping of surface AMPARs through phosphorylation of stargazin. *Neuron, 67,* 239–252.

Panicker, S., Brown, K., & Nicoll, R. A. (2008). Synaptic AMPA receptor subunit trafficking is independent of the C terminus in the GluR2-lacking mouse. *Proceedings of the National Academy of Sciences of the United States of America, 105,* 1032–1037.

Park, M., Penick, E. C., Edwards, J. G., Kauer, J. A., & Ehlers, M. D. (2004). Recycling endosomes supply AMPA receptors for LTP. *Science, 305,* 1972–1975.

Petrini, E. M., Lu, J., Cognet, L., Lounis, B., Ehlers, M. D., & Choquet, D. (2009). Endocytic trafficking and recycling maintain a pool of mobile surface AMPA receptors required for synaptic potentiation. *Neuron, 63,* 92–105.

Poo, M. M. (1985). Mobility and localization of proteins in excitable membranes. *Annual Review of Neuroscience, 8,* 369–406.

Poo, M., Lam, J. W., Orida, N., & Chao, A. W. (1979). Electrophoresis and diffusion in the plane of the cell membrane. *Biophysical Journal, 26,* 1–21.

Pooler, A. M., & McIlhinney, R. A. (2007). Lateral diffusion of the GABAB receptor is regulated by the GABAB2 C terminus. *Journal of Biological Chemistry, 282,* 25349–25356.

Poulopoulos, A., Aramuni, G., Meyer, G., Soykan, T., Hoon, M., Papadopoulos, T., Zhang, M., Paarmann, I., Fuchs, C., Harvey, K., Jedlicka, P., Schwarzacher, S. W., Betz, H., Harvey, R. J., Brose, N., Zhang, W., & Varoqueaux, F. (2009). Neuroligin 2 drives postsynaptic assembly at perisomatic inhibitory synapses through gephyrin and collybistin. *Neuron, 63,* 628–642.

Priel, A., Kolleker, A., Ayalon, G., Gillor, M., Osten, P., & Stern-Bach, Y. (2005). Stargazin reduces desensitization and slows deactivation of the AMPA-type glutamate receptors. *Journal of Neuroscience, 25,* 2682–2686.

Racz, B., Blanpied, T. A., Ehlers, M. D., & Weinberg, R. J. (2004). Lateral organization of endocytic machinery in dendritic spines. *Nature Neuroscience, 7,* 917–918.

Raghavachari, S., & Lisman, J. E. (2004). Properties of quantal transmission at CA1 synapses. *Journal of Neurophysiology, 92,* 2456–2467.

Raman, I. M., & Trussell, L. O. (1995). The mechanism of alpha-amino-3-hydroxy-5-methyl-4-isoxazolepropionate receptor desensitization after removal of glutamate. *Biophysical Journal, 68,* 137–146.

Renner, M., Choquet, D., & Triller, A. (2009a). Control of the postsynaptic membrane viscosity. *Journal of Neuroscience, 29,* 2926–2937.

Renner, M. L., Cognet, L., Lounis, B., Triller, A., & Choquet, D. (2009b). The excitatory postsynaptic density is a size exclusion diffusion environment. *Neuropharmacology, 56*, 30–36.

Renner, M., Lacor, P. N., Velasco, P. T., Xu, J., Contractor, A., Klein, W. L., & Triller, A. (2010). Deleterious effects of amyloid beta oligomers acting as an extracellular scaffold for mGluR5. *Neuron, 66*, 739–754.

Rouach, N., Byrd, K., Petralia, R. S., Elias, G. M., Adesnik, H., Tomita, S., Karimzadegan, S., Kealey, C., Bredt, D. S., Nicoll, R. A. (2005). TARP gamma-8 controls hippocampal AMPA receptor number, distribution and synaptic plasticity. *Nature Neuroscience, 8*, 1525–1533.

Rust, M. B., Gurniak, C. B., Renner, M., Vara, H., Morando, L., Gorlich, A., Sassoe-Pognetto, M., Banchaabouchi, M. A., Giustetto, M., Triller, A., Choquet, D., & Witke, W. (2010). Learning, AMPA receptor mobility and synaptic plasticity depend on n-cofilin-mediated actin dynamics. *EMBO Journal, 29*, 1889–1902.

Sainlos, M., Tigaret, C., Poujol, C., Olivier, N. B., Bard, L., Breillat, C., Thiolon, K., Choquet, D., & Imperiali, B. (2011). Biomimetic divalent ligands for the acute disruption of synaptic AMPAR stabilization. *Nature Chemical Biology, 7*, 81–91.

Saint-Michel, E., Giannone, G., Choquet, D., & Thoumine, O. (2009). Neurexin/neuroligin interaction kinetics characterized by counting single cell-surface attached quantum dots. *Biophysical Journal, 97*, 480–489.

Santamaria, F., Gonzalez, J., Augustine, G. J., & Raghavachari, S. (2010). Quantifying the effects of elastic collisions and non-covalent binding on glutamate receptor trafficking in the post-synaptic density. *PLoS Computational Biology, 6*, e1000780.

Savtchenko, L. P., Korogod, S. M., & Rusakov, D. A. (2000). Electrodiffusion of synaptic receptors: A mechanism to modify synaptic efficacy? *Synapse, 35*, 26–38.

Saxton, M. J., & Jacobson, K. (1997). Single-particle tracking: Applications to membrane dynamics. *Annual Review of Biophysics and Biomolecular Structure, 26*, 373–399.

Scheiffele, P., Fan, J., Choih, J., Fetter, R., & Serafini, T. (2000). Neuroligin expressed in nonneuronal cells triggers presynaptic development in contacting axons. *Cell, 101*, 657–669.

Schnell, E., Sizemore, M., Karimzadegan, S., Chen, L., Bredt, D. S., Nicoll, R. A., & M-421* (2002). Direct interactions between PSD-95 and stargazin control synaptic AMPA receptor number. *Proceedings of the National Academy of Sciences of the United States of America, 99*, 13902–13907. Epub 12002 Oct 13901.

Schuss, Z., Singer, A., & Holcman, D. (2007). The narrow escape problem for diffusion in cellular microdomains. *Proceedings of the National Academy of Sciences of the United States of America, 104*, 16098–16103.

Schwenk, J., Harmel, N., Zolles, G., Bildl, W., Kulik, A., Heimrich, B., Chisaka, O., Jonas, P., Schulte, U., Fakler, B., & Klocker, N. (2009). Functional proteomics identify cornichon proteins as auxiliary subunits of AMPA receptors. *Science, 323*, 1313–1319.

Scott, L., Zelenin, S., Malmersjo, S., Kowalewski, J. M., Markus, E. Z., Nairn, A. C., Greengard, P., Brismar, H., & Aperia, A. (2006). Allosteric changes of the NMDA receptor trap diffusible dopamine 1 receptors in spines. *Proceedings of the National Academy of Sciences of the United States of America, 103*, 762–767.

Shi, S., Hayashi, Y., Esteban, J. A., & Malinow, R. (2001). Subunit-specific rules governing AMPA receptor trafficking to synapses in hippocampal pyramidal neurons. *Cell, 105*, 331–343.

Shouval, H. Z. (2005). Clusters of interacting receptors can stabilize synaptic efficacies. *Proceedings of the National Academy of Sciences of the United States of America, 102*, 14440–14445.

Stan, A., Pielarski, K. N., Brigadski, T., Wittenmayer, N., Fedorchenko, O., Gohla, A., Lessmann, V., Dresbach, T., & Gottmann, K. (2010). Essential cooperation of N-cadherin and neuroligin-1 in the transsynaptic control of vesicle accumulation. *Proceedings of the National Academy of Sciences of the United States of America, 107*, 11116–11121.

Sylantyev, S., Savtchenko, L. P., Niu, Y. P., Ivanov, A. I., Jensen, T. P., Kullmann, D. M., Xiao, M. Y., & Rusakov, D. A. (2008). Electric fields due to synaptic currents sharpen excitatory transmission. *Science, 319*, 1845–1849.

Tai, C. Y., Mysore, S. P., Chiu, C., & Schuman, E. M. (2007). Activity-regulated N-cadherin endocytosis. *Neuron, 54*, 771–785.

Tallafuss, A., Constable, J. R., & Washbourne, P. (2010). Organization of central synapses by adhesion molecules. *European Journal of Neuroscience, 32*, 198–206.

Tardin, C., Cognet, L., Bats, C., Lounis, B., & Choquet, D. (2003). Direct imaging of lateral movements of AMPA receptors inside synapses. *EMBO Journal, 22*, 4656–4665.

Thomas, P., Mortensen, M., Hosie, A. M., Smart, T. G. (2005). Dynamic mobility of functional GABAA receptors at inhibitory synapses. *Nature Neuroscience, 8*, 889–897.

Thoumine, O., Ewers, H., Heine, M., Groc, L., Frischknecht, R., Giannone, G., Poujol, C., Legros, P., Lounis, B., Cognet, L., & Choquet, D. (2008). Probing the dynamics of protein-protein interactions at neuronal contacts by optical imaging. *Chemical Reviews, 108*, 1565–1587.

Thyagarajan, A., & Ting, A. Y. (2010). Imaging activity-dependent regulation of neurexin-neuroligin interactions using trans-synaptic enzymatic biotinylation. *Cell, 143*, 456–469.

Tomita, S., Adesnik, H., Sekiguchi, M., Zhang, W., Wada, K., Howe, J. R., Nicoll, R. A., & Bredt, D. S. (2005). Stargazin modulates AMPA receptor gating and trafficking by distinct domains. *Nature, 435*, 1052–1058.

Toomre, D., & Bewersdorf, J. (2010). A new wave of cellular imaging. *Annual Review of Cell and Developmental Biology, 26*, 285–314.

Tovar, K. R., & Westbrook, G. L. (2002). Mobile NMDA receptors at hippocampal synapses. *Neuron, 34*, 255–264.

Tretter, V., Jacob, T. C., Mukherjee, J., Fritschy, J. M., Pangalos, M. N., & Moss, S. J. (2008). The clustering of GABA(A) receptor subtypes at inhibitory synapses is facilitated via the direct binding of receptor alpha 2 subunits to gephyrin. *Journal of Neuroscience, 28*, 1356–1365.

Tretter, V., & Moss, S. J. (2008). GABA(A) receptor dynamics and constructing GABAergic synapses. *Frontiers in Molecular Neuroscience, 1*, 7.

Triller, A., & Choquet, D. (2008). New concepts in synaptic biology derived from single-molecule imaging. *Neuron, 59*, 359–374.

Trussell, L. O., Zhang, S., & Raman, I. M. (1993). Desensitization of AMPA receptors upon multiquantal neurotransmitter release. *Neuron, 10*, 1185–1196.

von Engelhardt, J., Mack, V., Sprengel, R., Kavenstock, N., Li, K. W., Stern-Bach, Y., Smit, A. B., Seeburg, P. H., & Monyer, H. (2010). CKAMP44: A brain-specific protein attenuating short-term synaptic plasticity in the dentate gyrus. *Science, 327*, 1518–1522.

Xie, X., Liaw, J. S., Baudry, M., & Berger, T. W. (1997). Novel expression mechanism for synaptic potentiation: Alignment of presynaptic release site and postsynaptic receptor. *Proceedings of the National Academy of Sciences of the United States of America, 94*, 6983–6988.

Xu-Friedman, M. A., & Regehr, W. G. (2003). Ultrastructural contributions to desensitization at cerebellar mossy fiber to granule cell synapses. *Journal of Neuroscience, 23*, 2182–2192.

Yang, Y. M., Fedchyshyn, M. J., Grande, G., Aitoubah, J., Tsang, C. W., Xie, H., Ackerley, C. A., Trimble, W. S., & Wang, L. Y. (2010). Septins regulate developmental switching from microdomain to nanodomain coupling of Ca(2+) influx to neurotransmitter release at a central synapse. *Neuron, 67*, 100–115.

Young, S. H., & Poo, M. M. (1983). Topographical rearrangement of acetylcholine receptors alters channel kinetics. *Nature, 304*, 161–163.

Yudowski, G. A., Puthenveedu, M. A., & von Zastrow, M. (2006). Distinct modes of regulated receptor insertion to the somatodendritic plasma membrane. *Nature Neuroscience, 9*, 622–627.

Zamanillo, D., Sprengel, R., Hvalby, O., Jensen, V., Burnashev, N., Rozov, A., Kaiser, K. M., Koster, H. J., Borchardt, T., Worley, P., Lubke, J., Frotscher, M., Kelly, P. H., Sommer, B., Andersen, P., Seeburg, P. H., & Sakmann, B. (1999). Importance of AMPA receptors for hippocampal synaptic plasticity but not for spatial learning. *Science, 284*, 1805–1811.

Zhang, C., Atasoy, D., Arac, D., Yang, X., Fucillo, M. V., Robison, A. J., Ko, J., Brunger, A. T., & Sudhof, T. C. (2010). Neurexins physically and functionally interact with GABA(A) receptors. *Neuron, 66*, 403–416.

Zhang, W., Rohlmann, A., Sargsyan, V., Aramuni, G., Hammer, R. E., Sudhof, T. C., & Missler, M. (2005). Extracellular domains of alpha-neurexins participate in regulating synaptic transmission by selectively affecting N- and P/Q-type Ca^{2+} channels. *Journal of Neuroscience, 25*, 4330–4342.

Chapter 10
Synaptic Protein Degradation in Memory Reorganization

Bong-Kiun Kaang and Jun-Hyeok Choi

Abstract The ubiquitin-proteasome system (UPS) is a ubiquitous, major pathway of protein degradation that is involved in most cellular processes by regulating the abundance of certain proteins. Accumulating evidence indicates a role for the UPS in specific functions of neurons. In this chapter, we first introduce the role of the UPS in neuronal function and the mechanism of UPS regulation following synaptic activity. Then, we focus on the recently revealed, distinct role of the UPS in the destabilization of a reactivated memory. Finally, we discuss the physiological role of this destabilization process. The reactivated memory may undergo modification from the initial memory depending on the context in which the memory is reactivated, which we will term memory reorganization. We will introduce the role of the *protein degradation–dependent destabilization process* for memory reorganization and suggest a hypothetical model combining the recent findings.

Keywords E3 ubiquitin ligase • Long-term memory • Spine • Ubiquitin proteasome system

B.-K. Kaang (⊠)
National Creative Research Initiative Center for Memory, Department of Biological Sciences, College of Natural Sciences, Seoul National University, 151-742 Seoul, South Korea

Department of Brain and Cognitive Sciences, College of Natural Sciences, Seoul National University, 151-742 Seoul, South Korea
e-mail: kaang@snu.ac.kr

J.-H. Choi
National Creative Research Initiative Center for Memory, Department of Biological Sciences, College of Natural Sciences, Seoul National University, 151-742 Seoul, South Korea
e-mail: jhchoi15@gmail.com

M.R. Kreutz and C. Sala (eds.), *Synaptic Plasticity*,
Advances in Experimental Medicine and Biology 970,
DOI 10.1007/978-3-7091-0932-8_10, © Springer-Verlag/Wien 2012

10.1 Introduction

The ubiquitin-proteasome system (UPS) is a ubiquitous, major pathway of protein degradation that governs the turnover of proteins, thereby inevitably affecting every process in which proteins are involved. In the UPS, the small protein ubiquitin is covalently conjugated to a substrate protein by the serial action of the E1 ubiquitin-activating enzyme, the E2 ubiquitin-conjugating enzyme, and the E3 ubiquitin ligase. After a serial reaction to produce a polyubiquitin chain on the substrate, the polyubiquitinated substrate is directed to a large proteasome complex that manages the degradation. E3 ubiquitin ligase seems to be the major component that determines substrate specificity (Fig. 10.1). Emerging evidence indicates the critical involvement of protein degradation in specialized functions of the neurons. Ubiquitin-proteasome-dependent degradation is known to play important roles in the regulation of synaptogenesis and the elimination of synapses in the development (DiAntonio et al. 2001; Ding et al. 2007; Liao et al. 2004; Schaefer et al. 2000; van Roessel et al. 2004; Wan et al. 2000), maintenance, and modulation of neurotransmission functions (Arancibia-Carcamo et al. 2009; Bedford et al. 2001; Burbea et al. 2002; Colledge et al. 2003; Dreier et al. 2005; Haas et al. 2007; Juo and Kaplan 2004; Kato et al. 2005; Patrick et al. 2003; Speese et al. 2003; Tada et al. 2010; van Roessel et al. 2004; Willeumier et al. 2006; Yao et al. 2007) and the structural remodeling of the synapse (Cartier et al. 2009; Colledge et al. 2003; Hoogenraad et al. 2007; Hung et al. 2010; Pak and Sheng 2003). Also, recent findings indicate that the UPS can be regulated by neuronal activity, suggesting a specific role for the UPS in plastic changes of synaptic strength (Ehlers 2003; Bingol et al. 2010;

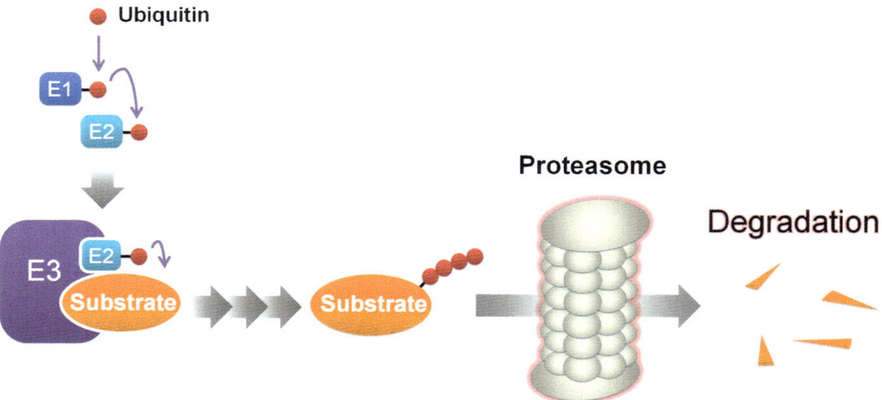

Fig. 10.1 *Mechanism of the ubiquitin–proteasome system.* Ubiquitin is first conjugated to E1 ubiquitin-activating enzyme (E1) in an ATP-dependent manner. The conjugated ubiquitin is then transferred to E2 ubiquitin-conjugating enzyme (E2). E3 ubiquitin ligase (E3) recognizes specific target proteins (substrates) and transfers and conjugates the ubiquitin from E2 to the substrate. E2 and E3 may also transfer the ubiquitin to a previously conjugated ubiquitin. After a serial reaction to produce a polyubiquitin chain on the substrate, the polyubiquitinated substrate is directed to a large proteasome complex that manages the degradation

Colledge et al. 2003; Deng and Lei 2007; Hou et al. 2006; Karpova et al. 2006; Kato et al. 2005; Pak and Sheng 2003; Patrick et al. 2003; Bingol and Schuman 2006; Djakovic et al. 2009; Fonseca et al. 2006; Shen et al. 2007).

In accordance with the findings on the role of the UPS in synaptic plasticity *in vitro*, recent *in vivo* studies show an involvement of the UPS in memory (Merlo and Romano 2007; Artinian et al. 2008; Wood et al. 2005; Lee et al. 2008; Lee 2008; Choi et al. 2010). Some of these findings suggest a distinct role of protein degradation in a specific step of reconsolidation (Lee et al. 2008; Lee 2008). Nader and colleagues (Nader et al. 2000) demonstrated that after a memory is retrieved, the previously consolidated memory becomes "labile" or sensitive to the amnesic effect of *protein synthesis inhibitors*, for a certain period of time. This indicates that the reactivated memory may have undergone an active destabilization process followed by a restabilization process, and this is termed reconsolidation. The early studies on reconsolidation focused on the consolidation-like restabilization process, which is mainly protein synthesis dependent (reviewed in Tronson and Taylor (2007), Nader and Hardt (2009), Dudai (2006)). However, the destabilization process is now demonstrated to rely on ubiquitin-proteasome-dependent degradation (Lee et al. 2008) (for a brief review, see Kaang et al. (2009)).

In this chapter, we will first discuss the specific role of the UPS in neuronal function and the mechanism for regulating the UPS following neuronal activity. Then, we will focus on recent studies exploring the distinct role of protein degradation as a mechanism of destabilization induced by the reactivation of a previously consolidated memory and also the significance of this process in memory reorganization.

10.2 The Ubiquitin-Proteasome System in Neurons

10.2.1 Regulation of Synapse Formation, Elimination, and Function by the UPS

Specific genes involved in the UPS are required for axon growth, synapse formation, and elimination. In *C. elegans*, Rpm-1, which is a subunit of the SCF ubiquitin ligase complex, is involved in axon growth and synaptogenesis (Schaefer et al. 2000). A mutant for this gene showed disorganized axon morphologies and presynaptic structures, while these phenotypes were rescued by expressing Rpm-1. FSN-1, another subunit of the SCF complex in *C. elegans,* was also shown to be involved in synapse formation (Liao et al. 2004). In the *Drosophila* neuromuscular junction (NMJ), a mutant for Highwire (a *Drosophila* homologue of Rpm-1) resulted in synapse outgrowth and expanded the extent of branches and the number of boutons (Wan et al. 2000). The overexpression of deubiquitinating protease fat facets resulted in a similar phenotype as the Highwire mutant in the *Drosophila* NMJ, suggesting that synapse formation may be regulated by the balance between positive and negative regulators of ubiquitination (DiAntonio et al. 2001). APC, another E3 ligase complex, has also been shown to be involved in synapse formation in the NMJ of *Drosophila* by regulating the degradation of the scaffold protein liprin-α (van Roessel et al. 2004). Disrupting the functions of SCF complex subunits SKR-1, Cullin and SEL-10 in *C. elegans* also caused defects in synapse elimination. SKR-1-binding protein SYG-1 is shown to protect synapses from elimination by inhibiting the association between SKR-1 and SEL-10 (Ding et al. 2007).

There are also studies demonstrating that the UPS modulates presynaptic neurotransmission function. In the *Drosophila* NMJ, the UPS components are shown to regulate the level of the presynaptic and essential synaptic vesicle-priming protein DUNC-13. An inhibition of proteasome activity resulted in an accumulation of DUNC-13 and an increased presynaptic efficacy (Speese et al. 2003). Pharmacological inhibition of proteasome activity has demonstrated that the UPS also plays an important role in regulating synaptic transmission in mammalian presynaptic terminals. Using a fluorescent dye in a hippocampal neuron culture, it was shown that a 2-hour inhibition of proteasome activity increased the recycling pool of vesicles by 76%, with no change in the rate or total amount of dye release (Willeumier et al. 2006). SCRAPPER, a synapse-localized E3 ubiquitin ligase, was shown to bind and ubiquitinate RIM1, a modulator of presynaptic plasticity. Neurons from SCRAPPER-knockout mice showed an increased frequency of miniature excitatory postsynaptic currents that was rescued by the expression of exogenous SCRAPPER or the knockdown of RIM1 (Yao et al. 2007). A novel ubiquitin ligase, Fbxo45, selectively expressed in the nervous system, was demonstrated to regulate neurotransmission, likely by modulating the synaptic vesicle-priming factor Munc13-1 at the synapse (Tada et al. 2010).

Several studies have demonstrated that the level of GLR-1 glutamate receptor is regulated by the UPS in *C. elegans* (Burbea et al. 2002; Dreier et al. 2005; Juo and Kaplan 2004; van Roessel et al. 2004). By expressing the dominant-negative form of proteasome subunits postsynaptically in the *Drosophila* NMJ, it was shown that the proteasome regulates the abundance of GluRIIB-containing glutamate receptors, limiting the synaptic strength (Haas et al. 2007). Agonist-induced AMPA receptor internalization was also regulated by the ubiquitin-proteasome-dependent degradation of PSD-95 in mammalian neurons (Patrick et al. 2003; Colledge et al. 2003). By expressing a dominant-negative form of Fbx2 that directs the ubiquitination of NR1 in hippocampal neuron, increased NR1 levels and NMDA receptor currents were seen in an activity-dependent manner, suggesting that the UPS is involved in the homeostatic control of synaptic NR1 (Kato et al. 2005). There is also evidence showing that the level of $GABA_A$ receptor, the key receptor for inhibitory transmission, is regulated by the UPS (Arancibia-Carcamo et al. 2009; Bedford et al. 2001).

Besides the regulation of receptors that directly mediates synaptic transmission, UPS also regulates the architectural components of the synapse. Serum-inducible kinase (SNK) was induced in hippocampal neurons by synaptic activity and was targeted to dendritic spines. Then SNK phosphorylated spine-associated Rap guanosine triphosphatase–activating protein (SPAR, a postsynaptic actin regulatory protein), which was then subjected to ubiquitin-proteasome-dependent degradation, thereby affecting the morphological change in the spines. The activation of SNK was dependent on the activities of the NMDA receptor, the AMPA receptor, and the

L-type voltage-gated calcium channel (LVGCC) (Pak and Sheng 2003). The activity of ubiquitin C-terminal hydrolase L1 (UCH-L1), a deubiquitinating enzyme, was rapidly regulated by NMDA receptor activation, affecting the synaptic protein distribution and spine morphology, size, and density, indirectly showing that the UPS is involved in activity-dependent structural remodeling (Cartier et al. 2009). Also, scaffolding proteins such as *Shank, GKAP, AKAP79/150, PSD-95, and liprin-α* have been demonstrated to be regulated by the UPS in an activity-dependent manner (Ehlers 2003; Colledge et al. 2003; Hoogenraad et al. 2007). Among these proteins, the specific E3 ligases for GKAP and PSD-95 were identified as TRIM3 and Mdm2, respectively (Colledge et al. 2003; Hung et al. 2010). Given the role of these scaffolding proteins in mediating multiple protein-protein interactions in synapse architecture and function, the UPS may be one of the pathways regulating activity-driven synapse remodeling.

10.2.2 Synaptic Activity–Dependent Regulation of the UPS

Long-lasting synaptic plasticity requires the incorporation of newly synthesized proteins. Protein degradation provides another mechanism for regulating the protein profile in activated neurons. Chronic inhibition or upregulation of synaptic activity in cultured neurons results in a changed protein profile: the levels of some proteins increase with the upregulation of activity and decrease upon the inhibition of activity, some are inversely regulated, and some are maintained at stable levels (Ehlers 2003). Some of the changes in synapse structure and function mediated by the UPS, as mentioned in the previous section, were induced in an activity-dependent manner (Colledge et al. 2003; Patrick et al. 2003; Kato et al. 2005; Pak and Sheng 2003; Cartier et al. 2009). In addition, the UPS also modulates rapid, activity-induced plasticity, long-term potentiation (LTP), and long-term depression (LTD) (Hou et al. 2006; Colledge et al. 2003; Fonseca et al. 2006; Karpova et al. 2006; Deng and Lei 2007).

Polyribosomes are transported to dendritic spines during LTP, and there is a body of evidence showing that proteins are locally synthesized in the activated sites (Aakalu et al. 2001; Ostroff et al. 2002; Pfeiffer and Huber 2006). Similar to this local protein synthesis, several studies demonstrated that proteasomes are transported from the dendritic shaft to the synaptic spines after synaptic activity, suggesting the possibility of local protein degradation (Bingol and Schuman 2006; Shen et al. 2007; Bingol et al. 2010). Synaptic activity enhanced the proteasome entry rate by ~1.5-fold while dramatically reducing the exit rate by at least sixfold, likely induced by an association with the actin cytoskeleton (Bingol and Schuman 2006). Another report has shown that NAC1, a cocaine-regulated transcriptional protein that associates with subunits of the proteasome complex, is cotranslocated with the proteasome from the nucleus into the dendritic spines by enhanced synaptic activity (Shen et al. 2007). Translocation of the proteasome can be blocked either by the depletion of NAC1 or by the expression of a dominant-negative mutant lacking the proteasome binding domain. A recent report demonstrated that calcium-/calmodulin-dependent protein kinase IIα (CaMKIIα) acts as a scaffold responsible for the activity-dependent translocation of the proteasome to dendritic spines (Bingol et al. 2010). CaMKIIα showed a biochemical association with the proteasome in the brain and also showed colocalization with the proteasome in a hippocampal culture. Activity-dependent translocation of CaMKIIα in hippocampal culture was necessary and sufficient for the translocation of the proteasome. This process required autophosphorylation of CaMKIIα, while kinase activity itself was not necessary. This evidence supports the possibility of activity-dependent local protein degradation, which may serve as one of the mechanisms controlling the local protein composition at synapses after stimulation (Fig. 10.2).

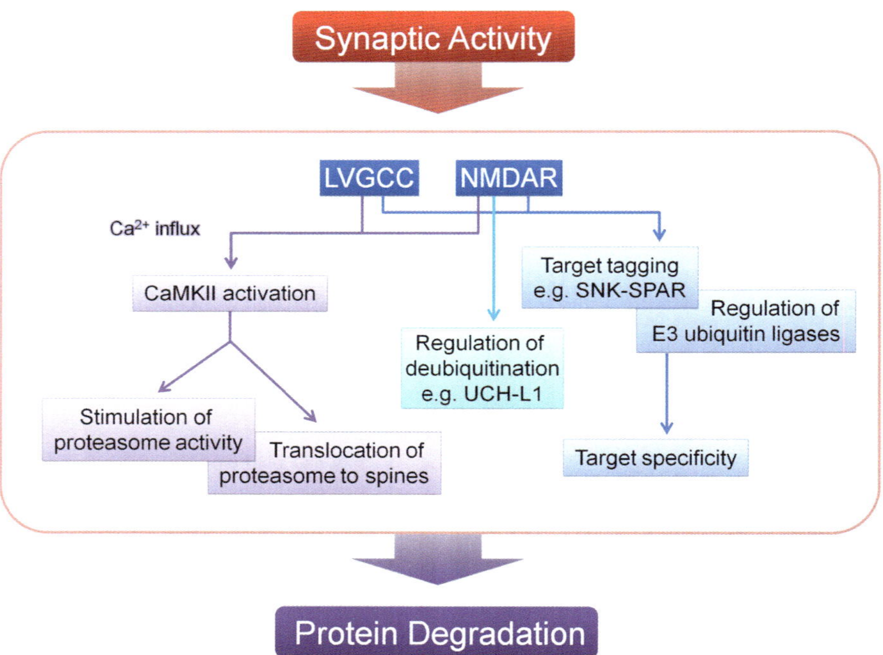

Fig. 10.2 *Regulation of the UPS by synaptic activity.* Roughly three pathways that affect protein degradation are regulated by synaptic activity. When there is a synaptic activity, NMDA receptors and LVGCCs are activated, resulting in an influx of external calcium ions. These calcium ions in turn activate CaMKII, which may then phosphorylate a proteasome complex subunit, thereby upregulating general proteasome activity. Autophosphorylated CaMKII also works as a scaffold for the translocation of the proteasome from the dendritic shaft to the synaptic spines. As postsynaptic proteins seem to be differentially regulated by neuronal activity in vitro or by retrieval in vivo, another pathway should regulate this target-specific differential turnover ratio. Although the upstream members of this pathway are not well characterized, the pathway is likely to involve regulation of either target tagging or specific E3 ligases' activity, which governs the target-specific protein degradation by regulating polyubiquitination. For example, SNK phosphorylates a specific protein SPAR, leading to the degradation of this protein. This process is dependent on NMDA receptor and LVGCC. Deubiquitination, an opposite process of ubiquitination, is also regulated by synaptic activity. UCH-L1 is a deubiquitinating enzyme regulated by NMDA receptor activation

A recent study revealed one of the upstream pathways that may regulate neuronal activity-induced proteasome stimulation (Djakovic et al. 2009). Blockade or upregulation of neuronal activity induced rapid inhibition or enhancement of proteasome activity, respectively. This regulation of proteasome activity is dependent on NMDA receptors and LVGCCs and also requires CaMKII activity, which phosphorylates a subunit of the proteasome complex, Rpt6. As external calcium entry and CaMKII activation are crucial molecular requirements for synaptic plasticity, the regulation of proteasome activity by this pathway may provide a mechanism for remodeling the synaptic composition and strength via protein degradation. However, as many synaptic molecules are differentially regulated, i.e., some are increased by an upregulation of activity, while others are decreased (Ehlers 2003), there should be mechanisms to differentially regulate the degradation of each protein, probably by differentially regulating various E3 ligases. This synaptic activity–induced regulation of specific E3 ligases is largely unknown so far (Fig. 10.2).

10.3 Role of the UPS in the Destabilization of Retrieved Memory

10.3.1 Protein Degradation as a Mechanism of Postretrieval Destabilization

Recently, Lee et al. suggested protein degradation as a mechanism of destabilizing memory after it is activated (Lee et al. 2008). Overall, polyubiquitination of synaptic proteins in the hippocampus was specifically increased after the retrieval of consolidated contextual fear conditioning, which induces *protein synthesis–dependent reconsolidation*. As polyubiquitination is a key step of the ubiquitin-proteasome-dependent protein degradation pathway, this result suggests that total ubiquitin-proteasome-dependent protein degradation of synaptic proteins is increased under this condition. This result is in accordance with reports showing that activity regulates postsynaptic protein composition through the ubiquitin-proteasome system mentioned in the previous section (Ehlers 2003), likely providing a mechanism for the activity-driven functional reorganization of synapses in culture systems.

The retrieval-induced degradation of synaptic proteins seems to be target specific. For example, the polyubiquitination of specific synaptic proteins, including Shank and GKAP, was increased, whereas that of PSD-95 was stable. This pattern resembles the results acquired in culture systems. Notably, the endogenous level of Shank in the synaptosomal fraction of the hippocampus decreased after retrieval, reaching the lowest level 2 h after retrieval and recovering to basal levels at 6 h after retrieval. This retrieval-induced decrease in the endogenous Shank level was blocked by clasto-lactacystin-β-lactone (β-lactone), a specific proteasome inhibitor, strongly suggesting that specific synaptic proteins are destabilized after retrieval through the ubiquitin-proteasome-dependent degradation pathway.

The inhibition of proteasome activity in the hippocampus after retrieval seems to prevent the destabilization of memory. Postretrieval anisomycin treatment leads to impairment of the previously formed memory. However, local treatment of proteasome inhibitor β-lactone along with anisomycin in the hippocampus after the retrieval of contextual fear memory prevented the amnesic effect of anisomycin. β-lactone treatment alone did not affect memory. These results suggest that ubiquitin-proteasome-dependent protein degradation underlies the destabilization of a previously formed memory after it is retrieved. On the other hand, β-lactone treatment immediately after conditioning did not prevent the amnesic effect of anisomycin on consolidation. This result demonstrates that β-lactone does not have a critical role in the consolidation process of this fear memory and that the effect of β-lactone cannot be attributed to a direct compensation of the effects of anisomycin. This supports the hypothesis that protein degradation plays a critical role in the destabilization of previously formed memories after retrieval, rather than in the consolidation-like restabilization process. However, another study demonstrated that both consolidation and reconsolidation of spatial memory in a water maze task were impaired by the inhibition of proteasome activity (Artinian et al. 2008), and the consolidation of learning in the crab *Chasmagnathus* was also interfered with by UPS inhibition (Merlo and Romano 2007). These indicate that the involvement of proteasome-dependent degradation may differ between species and memory types.

There are also reports suggesting a critical role for proteasome activity in LTP (Karpova et al. 2006; Fonseca et al. 2006), though the treatment of proteasome inhibitor started more than 30 min before LTP induction and might possibly have affected the protein profile before the induction of LTP (which is different from the research of Lee et al., where the drug was injected after the memory task) (Lee et al. 2008). It is also possible that the effect of proteasome inhibition on consolidation was simply not detected in the relatively strong conditioning protocol in the research of Lee et al. Meanwhile, the involvement of the UPS in LTD might have some relationship with the role of the UPS in the destabilization of reactivated memory (Colledge et al. 2003; Deng and Lei 2007; Hou et al. 2006). This destabilization process shows a similar outcome as depotentiation, the reversal of potentiation that shares some mechanisms with LTD.

10.3.2 Molecules Involved in Postretrieval Destabilization

Several molecules, including the NMDA receptor, are also involved in the destabilization of reactivated memory (Ben Mamou et al. 2006). NMDA receptor antagonist AP5, as well as NR2B selective inhibitor ifenprodil, locally applied in the amygdala before the retrieval of cued fear conditioning prevented the amnesic effect of postretrieval anisomycin injection. On the other hand, AMPA receptor antagonist CNQX did not interfere with the blocking effect of anisomycin. However, several studies have shown that the NMDA receptor antagonist itself has an amnesic effect when the previously formed memory is retrieved (Brown et al. 2008; Itzhak 2008; Lee and Everitt 2008; Milton et al. 2008; Suzuki et al. 2004; Lee et al. 2006). Systemic treatment with the NMDA antagonist MK-801 produced an amnesic effect on the reconsolidation of contextual and cued fear conditioning, odor-reward association, and drug-associated memories. Intra-amygdala NMDA receptor antagonism by AP5 also prevented the reconsolidation of drug-associated memory. These results demonstrate that the effect of NMDA receptor inhibition differs among various memory paradigms and treatment methods and also that NMDA receptors may be required for the restabilization of destabilized memory under certain conditions.

LVGCC and central cannabinoid receptor 1 (CB1 receptor) are also involved in the destabilization of reactivated contextual fear memory (Suzuki et al. 2008). Systemic and hippocampal treatments of LVGCC or CB1 receptor inhibitors prevented the amnesic effect of anisomycin after the retrieval of contextual fear memory. Systemic blockade of LVGCCs also protected reactivated memories against the amnesic effects of CREB activity inhibition. As LVGCCs and CB1 receptors are also required for memory extinction (Suzuki et al. 2004, 2008), there may be overlap between the initial destabilization mechanisms during reconsolidation and extinction.

These molecules may work as upstream factors in the protein degradation pathway after memory is reactivated. As mentioned in the previous section, NMDA receptor and LVGCC-dependent external calcium entry, and the resulting activation of CaMKII, constitute a pathway that regulates proteasome activity *in vitro* (Djakovic et al. 2009). Autophosphorylation of CaMKIIα and its translocation are also responsible for the regulation of proteasome translocation. Studies of the relationships among these molecules and the protein degradation induced by memory reactivation are required to fully understand the mechanism of destabilization induced by memory reactivation.

10.4 Memory Reorganization

10.4.1 Weakening the Reactivated Memory

Although memory can be stably stored for a long time, it sometimes has to be updated as circumstances change. The idea that reconsolidation may be an updating mechanism was hypothesized years ago (Dudai and Eisenberg 2004), and accumulating evidence suggests that this is indeed the case (Garcia-DeLaTorre et al. 2009; Lee 2008; Lee et al. 2008; Morris et al. 2006; Rodriguez-Ortiz et al. 2005; Rodriguez-Ortiz et al. 2008; Rossato et al. 2007; Winters et al. 2009).

The idea that reconsolidation is required for reorganization assumes that new information is incorporated during the labile state, leading to the stabilization of new information together with restabilization of the reactivated initial memory. Although there are some differences between the restabilization process of the reactivated memory and the *consolidation process* of the initially encoded memory, these two share many molecular mechanisms. The pharmacological treatments that can block restabilization of the reactivated memory usually also block the consolidation of a new memory. Even if these treatments block the incorporation of new information after the previously formed memory is reactivated, the results can be interpreted as the inhibition of either the independent consolidation of new information or the reconsolidation-based updating mechanism. Furthermore, even if there were treatments that exclusively impaired reconsolidation, such treatments would also lead to impaired initial memory. As the new information is related to the initial memory, it is hard to determine whether the incorporation of new information is actually impaired or whether it is simply not expressed due to an impairment of the initial memory that may be required for the expression of the updated component. Pioneering studies of the destabilization mechanism after the reactivation of a previously formed memory have provided a breakthrough regarding the role of reconsolidation as an updating mechanism. If the destabilization process is critical for the incorporation of the new information into the previously formed memory, pharmacological treatments that block the destabilization of the reactivated memory should impair the updating procedure while preserving the previous memory.

The strength of a previously formed memory may be weakened as one realizes that the memory of the initial situation is no longer valid. Extinction is an example of this kind of learning paradigm. In extinction of classical conditioning, for example, the subjects are extensively exposed to the conditioned stimulus (CS) without unconditioned stimulus (US), leading to a weaker conditioned response (CR) to the CS (Fig. 10.3a). This type of learning paradigm may be considered the modification and reorganization of the original memory in conjunction with the new information, i.e., that the CS is no longer associated with the US. In the paper reporting protein degradation as a mechanism of destabilization in reactivated memory, Lee et al. also confirmed that blocking protein degradation results in impaired contextual fear memory extinction (Lee et al. 2008). Local treatment with protein degradation inhibitors in the hippocampus after the extinction trial blocked the decrement of freezing the next day, whereas the vehicle group showed normal memory extinction. This result indicates that *protein degradation– dependent destabilization* of the reactivated memory is required for further reorganization or specifically for weakening of the initial memory. Similar results were found when considering the putative upstream molecules of destabilization, the LVGCCs and CB1 receptors (Suzuki et al. 2008). Other than their role in destabilization within the reconsolidation process, these molecules are also required for extinction (Suzuki et al. 2004; Suzuki et al. 2008). These results are in accordance with the findings on protein degradation inhibition, although the possibility remains that these molecules have unique roles in extinction learning.

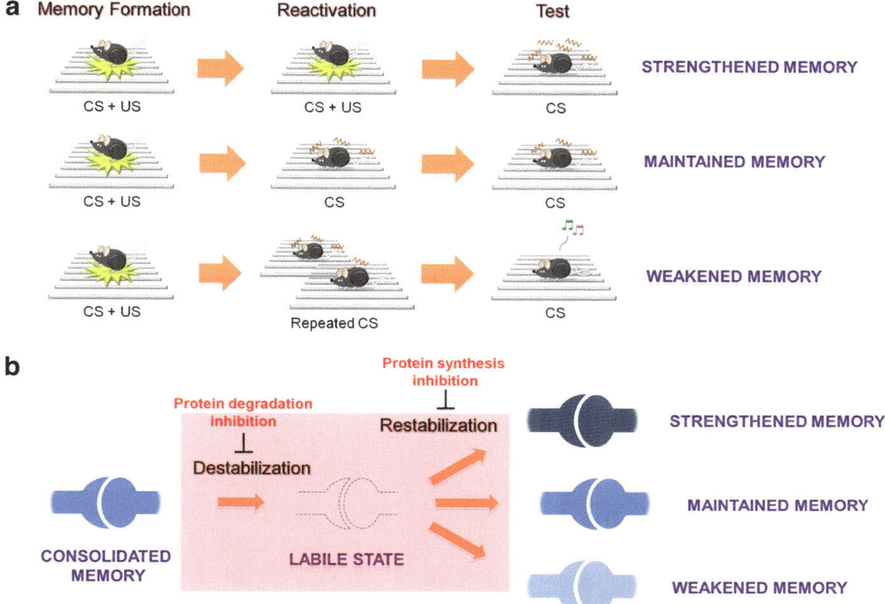

Fig. 10.3 *A model for memory reorganization – strengthening, maintaining, and weakening.* (**a**) Cartoons of the behavioral scheme used to reveal the mechanism underlying memory strengthening, maintaining, and weakening. After the original contextual fear conditioning, the memory is reactivated in various situations. In the scheme for memory strengthening, the animal receives an additional US shock. In the scheme for memory maintaining, it is exposed to the training context (CS) for a few minutes. In the scheme for memory weakening, it is repeatedly exposed to the training context (CS). Drugs are applied after memory reactivation, and the memory level is tested on the next day. (**b**) The diagram represents the state of the memory during the strengthening, maintaining, and weakening of the consolidated memory. Although the diagrams are shown with a single synapse, note that this is a simple symbolic representation

Although reconsolidation and extinction have been considered distinct processes thus far, the results described above demonstrate that reconsolidation and extinction share a common molecular mechanism, at least in the initial stages after the reactivation of the memory. Based on this interpretation, it may be possible to consider reconsolidation and extinction under a unified model in the reorganization of preexisting memory. After the consolidated memory is reactivated, it undergoes a destabilization process, which involves active degradation of scaffolding proteins such as Shank and GKAP in the spines, followed by restabilization either to recover the initial memory (reconsolidation) via protein synthesis or to maintain the destabilized state (extinction) with either minimal protein synthesis or active suppressive memory formation. Although some reports support the "unlearning" paradigm of extinction (Kim et al. 2007, 2009), active relearning of the CS-"no US" association (which is dependent on protein synthesis) is also a well-known mechanism of extinction (reviewed in (Lattal et al. 2006; Quirk and Mueller 2008)). It is not yet clear whether the *protein degradation–dependent destabilization* process is the initial part of either the unlearning or the relearning mechanism of extinction. It is also possible that different independent mechanisms cooperatively work toward the result of extinction.

10.4.2   Strengthening the Reactivated Memory

In some learning paradigms, one learning trial leads to robust memory that can be saturated, but in most cases, repeated learning leads to a gradual strengthening of memory. Several experiments utilized this gradual strengthening of memory to demonstrate that reconsolidation occurs when there is new information. Additionally, several studies have indicated that the application of certain drugs during reconsolidation can enhance the strength of memory, suggesting that reconsolidation can be potentially associated with an increase in memory strength (Lee et al. 2006; Tronson et al. 2006). However, these studies do not provide direct evidence for the hypothesis that the reconsolidation process is required for updating and increasing memory strength.

Following the report that protein degradation underlies the weakening of reactivated memory, another study demonstrated that destabilization of reactivated memory is also required for the strengthening of contextual fear memory (Lee 2008). The author first demonstrated that contextual fear memory can be further strengthened by repeated conditioning with a relatively weak, aversive US (Fig. 10.3a). Given that the consolidation and reconsolidation of contextual fear have different molecular requirements (Lee et al. 2004), Lee showed that the strengthening of a consolidated memory that occurs upon second training does not match the molecular mechanism of consolidation, as the treatment that has an amnesic effect exclusively upon consolidation had no effect. The author also demonstrated that a treatment that has an amnesic effect exclusively on reconsolidation could impair the strengthening as well as the reactivated memory. However, the amnesic treatment of reconsolidation impairs and ablates the original reactivated memory and thus also impairs strengthening, no matter whether the strengthening mechanism actually relies on reconsolidation. To more directly demonstrate the requirement for the reconsolidation mechanism in memory strengthening, the author locally applied a *protein degradation inhibitor* to the hippocampus after the second training. If *protein degradation–dependent destabilization* was required to strengthen the reactivated memory, the *protein degradation inhibitor* would block further enhancement of the memory, leaving it at the level of initially consolidated memory. This was what the author observed (Lee 2008).

10.4.3 Hypothetical Model for Memory Reorganization

The fact that strengthening reactivated memories requires *protein degradation–dependent destabilization*, together with the evidence that reconsolidation and extinction partly share a common mechanism, indicates that the maintenance, weakening, and strengthening of a reactivated original memory may be interpreted under a unified model of reorganization (Fig. 10.3b). After a memory is consolidated, it can be retrieved by certain situations that include one or more components related to the original memory. These situations may be quite diverse and can determine the fate of the retrieved memory. In some cases, the memory seems to be maintained without being reactivated. When the memory retrieval is very brief, or when the memory is saturated by overtraining, it is not susceptible to the amnesic effect of *protein synthesis inhibitors*, even though the memory is well retrieved (Suzuki et al. 2004; Rodriguez-Ortiz et al. 2005, 2008; Garcia-DeLaTorre et al. 2009; Wang et al. 2009). On the other hand, when a memory, usually unsaturated, is retrieved for more than a very brief period, it can be reactivated and reorganized. The reactivated memory first becomes destabilized by a mechanism that is likely initiated by the NMDA receptor, LVGCC, or CB1 receptor and involves protein degradation. The fate of the destabilized memory depends on the incoming information specific to the situation. In cases where the original memory is no longer valid, the destabilized memory will either passively remain in a destabilized state or the extinction information will be actively encoded, weakening the memory. In cases where the original memory should be strengthened by additional training, the destabilized memory is restabilized into a stronger memory. Finally, in cases where there is no additional training, but there is not sufficient information to conclude that the original memory is no longer valid, the destabilized memory is restabilized to a similar level as the original memory.

The underlying molecular pathway of this reorganization mechanism is still under investigation. The destabilization process seems to be initiated by activation of NMDA receptors, LVGCC, and CB1 receptors (Suzuki et al. 2008; Ben Mamou et al. 2006). The UPS seems to have a critical role in this process, though the direct links of the upstream molecules have not been demonstrated *in vivo*. In vitro studies show the possibility that NMDA receptors and LVGCC can activate CaMKII, which in turn activates and translocates the proteasome to the synaptic spines along with the autophosphorylated CaMKII (Bingol et al. 2010; Bingol and Schuman 2006). The increase of degradation in the synaptosomal fraction can be well explained by this pathway (Lee et al. 2008). However, the pathway that links synaptic activity to the specificity of the substrate for degradation is unknown. Two substrates demonstrated to be actively degraded during the destabilization step are Shank and GKAP, both of which have been proven to be regulated by synaptic activity *in vitro* (Ehlers 2003), where GKAP is especially ubiquitinated by TRIM3 ubiquitin ligase (Hung et al. 2010). Given the role of these proteins as scaffolding proteins of the synaptic spine, in which Shank specifically acts as a "master" scaffolding protein that holds together intermediate scaffolding proteins such as

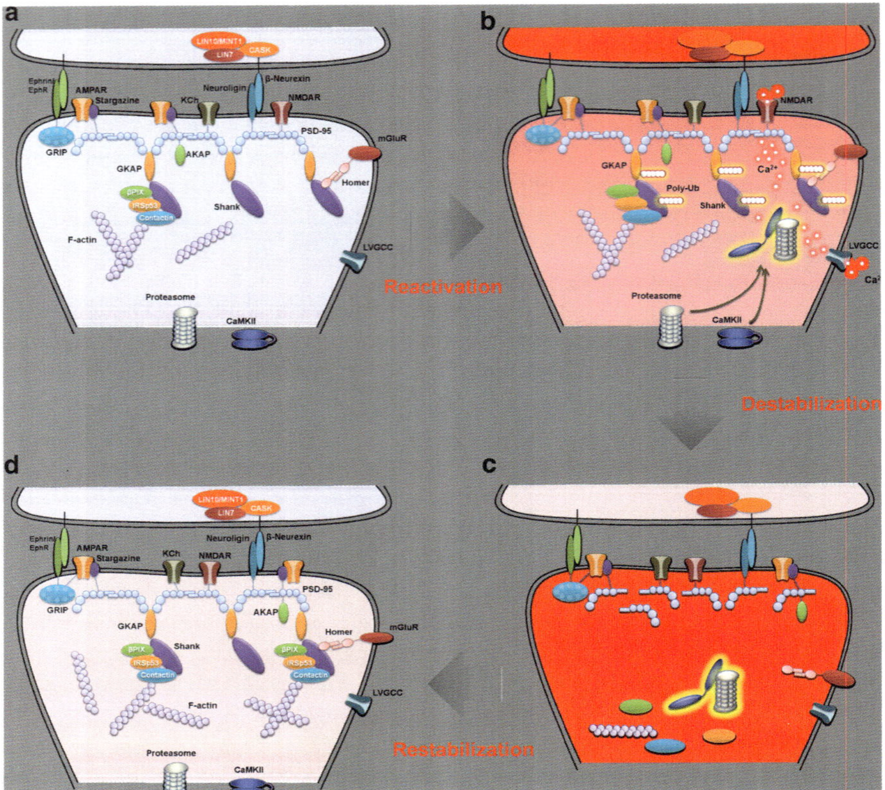

Fig. 10.4 *A model for memory reorganization – synaptic remodeling.* (**a**) Structure of a synapse encoding memory. (**b**) When the memory is reactivated, NMDA receptor and LVGCC are opened, allowing calcium influx to the spine. These calcium ions activate CaMKII, which then phosphorylates the proteasome to increase the activity. The activated CaMKII may undergo autophosphorylation and can associate with and translocate the proteasome from the dendritic shaft to the spine. Meanwhile, target proteins are polyubiquitinated by the specific action of E3 ligases and other proteins. The known proteins that undergo polyubiquitination after memory reactivation are Shank and GKAP, as indicated. (**c**) The recruited active proteasomes degrade these specifically polyubiquitinated targets. Since the targets here are scaffolding proteins, it is a likely consideration that this spine undergoes structural remodeling. (**d**) A protein synthesis–dependent process restabilizes the synapse either to a state similar to the initial state or to a modified state

GKAP and PSD-95, and also considering the fact that the UPS is involved in activity-dependent synaptic remodeling (Pak and Sheng 2003; Cartier et al. 2009), it seems that during reconsolidation, reactivated synapses undergo synaptic remodeling, first being disassembled during the destabilization step and then being recovered to a state similar to the initial one or becoming stabilized as a modified state. This process might accompany morphological changes as well. Restabilization is basically protein synthesis dependent and shares many molecular mechanisms with the original consolidation, although some differences exist. The process of restabilization may be the key step that governs the fate of the memory (Fig. 10.4). The reorganization process investigated so far is focused on the postsynaptic site. The role of protein degradation on the presynaptic site is largely unknown.

The model here is based on the reorganization of memory strength. However, it is noteworthy that there are other types of reorganization where the memory content is changed rather than the quantitative extent of the memory. A recent study demonstrated that partial modification of an object-place associative memory requires both protein synthesis and degradation (Choi et al. 2010). One day after the animal had initially formed object-place associative memory for four objects placed in a context, it was exposed to a context where two of the objects positions were changed. Without any treatment, the animal would reorganize the initial memory in order to learn the changed position of the objects. However, when either a *protein synthesis inhibitor* or a *proteasome inhibitor* was treated right after the second exposure, the animal could not appropriately reorganize the memory. Although more research is required to clearly reveal the memory reorganization process in this situation, the requirement of both protein synthesis and degradation matches the suggested model.

10.5 Conclusions and Future Directions

As for many basic cellular processes, neuronal functions are also under the influence of the UPS. Recent studies indicate that the UPS can be regulated in response to synaptic activity, suggesting a role for the UPS in synaptic plasticity and memory. The role of protein degradation in the destabilization step of reconsolidation shows that the UPS may serve a very specific role, more than simply maintaining proteins at an appropriate level.

Since reconsolidation was first demonstrated, many studies have focused on the mechanism of the restabilization step of the full process. However, the destabilization that occurs prior to restabilization is also a unique and important process. Recent studies focusing on the destabilization process of reactivated memory have not only revealed the underlying mechanism of this process but also given insight into important aspects of the fate of reactivated memory. In contrast to the *protein synthesis–dependent restabilization process*, destabilization of reactivated memory seems to be dependent on protein degradation. There are several molecules that may work in a putative upstream pathway to regulate protein degradation. As synaptic protein turnover rates are differentially regulated by neuronal activity, more studies are required to elucidate the target-specific regulation of each protein. Using a proteasome inhibitor as a tool to block destabilization of reactivated memory, it was shown that this reactivation-induced destabilization is required for the reorganization of the reactivated memory, a process that includes maintaining, weakening, and strengthening the memory. These results suggest a unified model of reorganization, beginning with the destabilization of reactivated memory and followed by stabilization of appropriate information, depending on the situation.

The studies based on culture systems and molecular analyses after behavioral processes suggest that the protein degradation–based mechanism may work on a synaptic level. However, there is no direct evidence as to whether each synapse that is involved in the memory behaves according to the memory state, i.e., destabilization followed by restabilization, which is an important issue. Another important issue is whether this protein degradation–dependent reorganization mechanism can be applied to systems-level changes such as systems consolidation and systems reconsolidation (Debiec et al. 2002; Frankland and Bontempi 2005). More studies are required to reveal the details of the mechanism and also to apply studies on *in vitro* systems to the *in vivo* destabilization process. Studies utilizing more selective targeting of a specifically regulated proteasome function would also be valuable compared to those using a general, pharmacological inhibition of proteasome activity. It is also important to determine the range of memory tasks and animal models to which this model can be applied.

Acknowledgments This work was supported by the National Creative Research Initiative Program and WCU program of the Korean Ministry of Science and Technology.

References

Aakalu, G., Smith, W. B., Nguyen, N., Jiang, C., & Schuman, E. M. (2001). Dynamic visualization of local protein synthesis in hippocampal neurons. *Neuron, 30*(2), 489–502. doi:S0896-6273 (01)00295-1 [pii].

Arancibia-Carcamo, I. L., Yuen, E. Y., Muir, J., Lumb, M. J., Michels, G., Saliba, R. S., Smart, T. G., Yan, Z., Kittler, J. T., & Moss, S. J. (2009). Ubiquitin-dependent lysosomal targeting of GABA(A) receptors regulates neuronal inhibition. *Proceedings of the National Academy of Sciences of the United States of America, 106*(41), 17552–17557. doi:0905502106 [pii] 10.1073/pnas.0905502106.

Artinian, J., McGauran, A. M., De Jaeger, X., Mouledous, L., Frances, B., & Roullet, P. (2008). Protein degradation, as with protein synthesis, is required during not only long-term spatial memory consolidation but also reconsolidation. *The European Journal of Neuroscience, 27* (11), 3009–3019. doi:EJN6262 [pii] 10.1111/j.1460-9568.2008.06262.x.

Bedford, F. K., Kittler, J. T., Muller, E., Thomas, P., Uren, J. M., Merlo, D., Wisden, W., Triller, A., Smart, T. G., & Moss, S. J. (2001). GABA(A) receptor cell surface number and subunit stability are regulated by the ubiquitin-like protein Plic-1. *Nature Neuroscience, 4*(9), 908–916. doi:10.1038/nn0901-908 nn0901-908 [pii].

Ben Mamou, C., Gamache, K., & Nader, K. (2006). NMDA receptors are critical for unleashing consolidated auditory fear memories. *Nature Neuroscience, 9*(10), 1237–1239. doi:nn1778 [pii] 10.1038/nn1778.

Bingol, B., & Schuman, E. M. (2006). Activity-dependent dynamics and sequestration of proteasomes in dendritic spines. *Nature, 441*(7097), 1144–1148. doi:nature04769 [pii] 10.1038/nature04769.

Bingol, B., Wang, C. F., Arnott, D., Cheng, D., Peng, J., & Sheng, M. (2010). Autophosphorylated CaMKIIalpha acts as a scaffold to recruit proteasomes to dendritic spines. *Cell, 140*(4), 567–578. doi:S0092-8674(10)00059-0 [pii] 10.1016/j.cell.2010.01.024.

Brown, T. E., Lee, B. R., & Sorg, B. A. (2008). The NMDA antagonist MK-801 disrupts reconsolidation of a cocaine-associated memory for conditioned place preference but not for self-administration in rats. *Learning and Memory, 15*(12), 857–865. doi:15/12/857 [pii] 10.1101/lm.1152808.

Burbea, M., Dreier, L., Dittman, J. S., Grunwald, M. E., & Kaplan, J. M. (2002). Ubiquitin and AP180 regulate the abundance of GLR-1 glutamate receptors at postsynaptic elements in *C. elegans. Neuron, 35*(1), 107–120. doi:doi:S0896627302007493 [pii].

Cartier, A. E., Djakovic, S. N., Salehi, A., Wilson, S. M., Masliah, E., & Patrick, G. N. (2009). Regulation of synaptic structure by ubiquitin C-terminal hydrolase L1. *The Journal of Neuroscience, 29*(24), 7857–7868. doi:29/24/7857 [pii] 10.1523/JNEUROSCI.1817-09.2009.

Choi, J. H., Kim, J. E., & Kaang, B. K. (2010). Protein synthesis and degradation are required for the incorporation of modified information into the pre-existing object-location memory. *Molecular Brain, 3*, 1. doi:1756-6606-3-1 [pii] 10.1186/1756-6606-3-1.

Colledge, M., Snyder, E. M., Crozier, R. A., Soderling, J. A., Jin, Y., Langeberg, L. K., Lu, H., Bear, M. F., & Scott, J. D. (2003). Ubiquitination regulates PSD-95 degradation and AMPA receptor surface expression. *Neuron, 40*(3), 595–607. doi:S0896627303006871 [pii].

Debiec, J., LeDoux, J. E., & Nader, K. (2002). Cellular and systems reconsolidation in the hippocampus. *Neuron, 36*(3), 527–538. doi:S0896627302010012 [pii].

Deng, P. Y., & Lei, S. (2007). Long-term depression in identified stellate neurons of juvenile rat entorhinal cortex. *Journal of Neurophysiology, 97*(1), 727–737. doi:01089.2006 [pii] 10.1152/jn.01089.2006.

DiAntonio, A., Haghighi, A. P., Portman, S. L., Lee, J. D., Amaranto, A. M., & Goodman, C. S. (2001). Ubiquitination-dependent mechanisms regulate synaptic growth and function. *Nature, 412*(6845), 449–452. doi:10.1038/35086595 35086595 [pii].

Ding, M., Chao, D., Wang, G., & Shen, K. (2007). Spatial regulation of an E3 ubiquitin ligase directs selective synapse elimination. *Science, 317*(5840), 947–951. doi:1145727 [pii] 10.1126/science.1145727.

Djakovic, S. N., Schwarz, L. A., Barylko, B., DeMartino, G. N., & Patrick, G. N. (2009). Regulation of the proteasome by neuronal activity and calcium/calmodulin-dependent protein kinase II. *The Journal of Biological Chemistry, 284*(39), 26655–26665. doi:M109.021956 [pii] 10.1074/jbc.M109.021956.

Dreier, L., Burbea, M., & Kaplan, J. M. (2005). LIN-23-mediated degradation of beta-catenin regulates the abundance of GLR-1 glutamate receptors in the ventral nerve cord of *C. elegans. Neuron, 46*(1), 51–64. doi:doi:S0896-6273(05)00161-3 [pii] 10.1016/j.neuron.2004.12.058.

Dudai, Y. (2006). Reconsolidation: The advantage of being refocused. *Current Opinion in Neurobiology, 16*(2), 174–178. doi:S0959-4388(06)00035-3 [pii] 10.1016/j.conb.2006.03.010.

Dudai, Y., & Eisenberg, M. (2004). Rites of passage of the engram: Reconsolidation and the lingering consolidation hypothesis. *Neuron, 44*(1), 93–100. doi:10.1016/j.neuron.2004.09.003 S0896627304005720 [pii].

Ehlers, M. D. (2003). Activity level controls postsynaptic composition and signaling via the ubiquitin-proteasome system. *Nature Neuroscience, 6*(3), 231–242. doi:10.1038/nn1013 nn1013 [pii].

Fonseca, R., Vabulas, R. M., Hartl, F. U., Bonhoeffer, T., & Nagerl, U. V. (2006). A balance of protein synthesis and proteasome-dependent degradation determines the maintenance of LTP. *Neuron, 52*(2), 239–245. doi:S0896-6273(06)00637-4 [pii] 10.1016/j.neuron.2006.08.015.

Frankland, P. W., & Bontempi, B. (2005). The organization of recent and remote memories. *Nature Reviews Neuroscience, 6*(2), 119–130. doi:nrn1607 [pii] 10.1038/nrn1607.

Garcia-DeLaTorre, P., Rodriguez-Ortiz, C. J., Arreguin-Martinez, J. L., Cruz-Castaneda, P., & Bermudez-Rattoni, F. (2009). Simultaneous but not independent anisomycin infusions in insular cortex and amygdala hinder stabilization of taste memory when updated. *Learning and Memory, 16*(9), 514–519. doi:16/9/514 [pii] 10.1101/lm.1356509.

Haas, K. F., Miller, S. L., Friedman, D. B., & Broadie, K. (2007). The ubiquitin-proteasome system postsynaptically regulates glutamatergic synaptic function. *Molecular and Cellular Neuroscience, 35*(1), 64–75. doi:S1044-7431(07)00027-9 [pii] 10.1016/j.mcn.2007.02.002.

Hoogenraad, C. C., Feliu-Mojer, M. I., Spangler, S. A., Milstein, A. D., Dunah, A. W., Hung, A. Y., & Sheng, M. (2007). Liprinalpha1 degradation by calcium/calmodulin-dependent protein kinase II regulates LAR receptor tyrosine phosphatase distribution and dendrite development. *Developmental Cell, 12*(4), 587–602. doi:S1534-5807(07)00056-1 [pii] 10.1016/j.devcel.2007.02.006.

Hou, L., Antion, M. D., Hu, D., Spencer, C. M., Paylor, R., & Klann, E. (2006). Dynamic translational and proteasomal regulation of fragile X mental retardation protein controls mGluR-dependent long-term depression. *Neuron, 51*(4), 441–454. doi:S0896-6273(06)00545-9 [pii] 10.1016/j.neuron.2006.07.005.

Hung, A. Y., Sung, C. C., Brito, I. L., & Sheng, M. (2010). Degradation of postsynaptic scaffold GKAP and regulation of dendritic spine morphology by the TRIM3 ubiquitin ligase in rat hippocampal neurons. *PLoS One, 5*(3), e9842. doi:10.1371/journal.pone.0009842.

Itzhak, Y. (2008). Role of the NMDA receptor and nitric oxide in memory reconsolidation of cocaine-induced conditioned place preference in mice. *Annals of the New York Academy of Sciences, 1139*, 350–357. doi:NYAS1139051 [pii] 10.1196/annals.1432.051.

Juo, P., & Kaplan, J. M. (2004). The anaphase-promoting complex regulates the abundance of GLR-1 glutamate receptors in the ventral nerve cord of *C. elegans*. *Current Biology, 14*(22), 2057–2062. doi:doi:S0960982204008802 [pii] 10.1016/j.cub.2004.11.010.

Kaang, B. K., Lee, S. H., & Kim, H. (2009). Synaptic protein degradation as a mechanism in memory reorganization. *The Neuroscientist, 15*(5), 430–435. doi:1073858408331374 [pii] 10.1177/1073858408331374.

Karpova, A., Mikhaylova, M., Thomas, U., Knopfel, T., & Behnisch, T. (2006). Involvement of protein synthesis and degradation in long-term potentiation of Schaffer collateral CA1 synapses. *The Journal of Neuroscience, 26*(18), 4949–4955. doi:26/18/4949 [pii] 10.1523/JNEUROSCI.4573-05.2006.

Kato, A., Rouach, N., Nicoll, R. A., & Bredt, D. S. (2005). Activity-dependent NMDA receptor degradation mediated by retrotranslocation and ubiquitination. *Proceedings of the National Academy of Sciences of the United States of America, 102*(15), 5600–5605. doi:0501769102 [pii] 10.1073/pnas.0501769102.

Kim, J., Lee, S., Park, K., Hong, I., Song, B., Son, G., Park, H., Kim, W. R., Park, E., Choe, H. K., Kim, H., Lee, C., Sun, W., Kim, K., Shin, K. S., & Choi, S. (2007). Amygdala depotentiation and fear extinction. *Proceedings of the National Academy of Sciences of the United States of America, 104*(52), 20955–20960. doi:0710548105 [pii] 10.1073/pnas.0710548105.

Kim, J., Park, S., Lee, S., & Choi, S. (2009). Amygdala depotentiation ex vivo requires mitogen-activated protein kinases and protein synthesis. *Neuroreport, 20*(5), 517–520. doi:10.1097/WNR.0b013e328329412d.

Lattal, K. M., Radulovic, J., & Lukowiak, K. (2006). Extinction: [corrected] does it or doesn't it? The requirement of altered gene activity and new protein synthesis. *Biological Psychiatry, 60*(4), 344–351. doi:S0006-3223(06)00766-9 [pii] 10.1016/j.biopsych.2006.05.038.

Lee, J. L. (2008). Memory reconsolidation mediates the strengthening of memories by additional learning. *Nature Neuroscience, 11*(11), 1264–1266. doi:nn.2205 [pii] 10.1038/nn.2205.

Lee, S. H., Choi, J. H., Lee, N., Lee, H. R., Kim, J. I., Yu, N. K., Choi, S. L., Kim, H., & Kaang, B. K. (2008). Synaptic protein degradation underlies destabilization of retrieved fear memory. *Science, 319*(5867), 1253–1256. doi:1150541 [pii] 10.1126/science.1150541.

Lee, J. L., & Everitt, B. J. (2008). Appetitive memory reconsolidation depends upon NMDA receptor-mediated neurotransmission. *Neurobiology of Learning and Memory, 90*(1), 147–154. doi:S1074-7427(08)00031-2 [pii] 10.1016/j.nlm.2008.02.004.

Lee, J. L., Everitt, B. J., & Thomas, K. L. (2004). Independent cellular processes for hippocampal memory consolidation and reconsolidation. *Science, 304*(5672), 839–843. doi:10.1126/science.1095760 1095760 [pii].

Lee, J. L., Milton, A. L., & Everitt, B. J. (2006). Reconsolidation and extinction of conditioned fear: Inhibition and potentiation. *The Journal of Neuroscience, 26*(39), 10051–10056. doi:26/39/10051 [pii] 10.1523/JNEUROSCI.2466-06.2006.

Liao, E. H., Hung, W., Abrams, B., & Zhen, M. (2004). An SCF-like ubiquitin ligase complex that controls presynaptic differentiation. *Nature, 430*(6997), 345–350. doi:10.1038/nature02647 nature02647 [pii].

Merlo, E., & Romano, A. (2007). Long-term memory consolidation depends on proteasome activity in the crab Chasmagnathus. *Neuroscience, 147*(1), 46–52. doi:S0306-4522(07) 00520-9 [pii] 10.1016/j.neuroscience.2007.04.022.

Milton, A. L., Lee, J. L., Butler, V. J., Gardner, R., & Everitt, B. J. (2008). Intra-amygdala and systemic antagonism of NMDA receptors prevents the reconsolidation of drug-associated memory and impairs subsequently both novel and previously acquired drug-seeking behaviors. *The Journal of Neuroscience, 28*(33), 8230–8237. doi:28/33/8230 [pii] 10.1523/ JNEUROSCI.1723-08.2008.

Morris, R. G., Inglis, J., Ainge, J. A., Olverman, H. J., Tulloch, J., Dudai, Y., & Kelly, P. A. (2006). Memory reconsolidation: Sensitivity of spatial memory to inhibition of protein synthesis in dorsal hippocampus during encoding and retrieval. *Neuron, 50*(3), 479–489. doi:S0896-6273 (06)00280-7 [pii] 10.1016/j.neuron.2006.04.012.

Nader, K., & Hardt, O. (2009). A single standard for memory: The case for reconsolidation. *Nature Reviews Neuroscience, 10*(3), 224–234. doi:nrn2590 [pii] 10.1038/nrn2590.

Nader, K., Schafe, G. E., & Le Doux, J. E. (2000). Fear memories require protein synthesis in the amygdala for reconsolidation after retrieval. *Nature, 406*(6797), 722–726. doi:10.1038/ 35021052.

Ostroff, L. E., Fiala, J. C., Allwardt, B., & Harris, K. M. (2002). Polyribosomes redistribute from dendritic shafts into spines with enlarged synapses during LTP in developing rat hippocampal slices. *Neuron, 35*(3), 535–545. doi:S0896627302007857 [pii].

Pak, D. T., & Sheng, M. (2003). Targeted protein degradation and synapse remodeling by an inducible protein kinase. *Science, 302*(5649), 1368–1373. doi:10.1126/science.1082475 1082475 [pii].

Patrick, G. N., Bingol, B., Weld, H. A., & Schuman, E. M. (2003). Ubiquitin-mediated proteasome activity is required for agonist-induced endocytosis of GluRs. *Current Biology, 13*(23), 2073–2081. doi:S0960982203007826 [pii].

Pfeiffer, B. E., & Huber, K. M. (2006). Current advances in local protein synthesis and synaptic plasticity. *The Journal of Neuroscience, 26*(27), 7147–7150. doi:26/27/7147 [pii] 10.1523/ JNEUROSCI.1797-06.2006.

Quirk, G. J., & Mueller, D. (2008). Neural mechanisms of extinction learning and retrieval. *Neuropsychopharmacology, 33*(1), 56–72. doi:1301555 [pii] 10.1038/sj.npp. 1301555.

Rodriguez-Ortiz, C. J., De la Cruz, V., Gutierrez, R., & Bermudez-Rattoni, F. (2005). Protein synthesis underlies post-retrieval memory consolidation to a restricted degree only when updated information is obtained. *Learning and Memory, 12*(5), 533–537. doi:lm.94505 [pii] 10.1101/lm.94505.

Rodriguez-Ortiz, C. J., Garcia-DeLaTorre, P., Benavidez, E., Ballesteros, M. A., & Bermudez-Rattoni, F. (2008). Intrahippocampal anisomycin infusions disrupt previously consolidated spatial memory only when memory is updated. *Neurobiology of Learning and Memory, 89*(3), 352–359. doi:S1074-7427(07)00168-2 [pii] 10.1016/j.nlm.2007.10.004.

Rossato, J. I., Bevilaqua, L. R., Myskiw, J. C., Medina, J. H., Izquierdo, I., & Cammarota, M. (2007). On the role of hippocampal protein synthesis in the consolidation and reconsolidation of object recognition memory. *Learning and Memory, 14*(1), 36–46. doi:14/1/36 [pii] 10.1101/ lm.422607.

Schaefer, A. M., Hadwiger, G. D., & Nonet, M. L. (2000). rpm-1, a conserved neuronal gene that regulates targeting and synaptogenesis in *C. elegans*. *Neuron, 26*(2), 345–356. doi:doi:S0896-6273(00)81168-X [pii].

Shen, H., Korutla, L., Champtiaux, N., Toda, S., LaLumiere, R., Vallone, J., Klugmann, M., Blendy, J. A., Mackler, S. A., & Kalivas, P. W. (2007). NAC1 regulates the recruitment of the proteasome complex into dendritic spines. *The Journal of Neuroscience, 27*(33), 8903–8913. doi:27/33/8903 [pii] 10.1523/JNEUROSCI.1571-07.2007.

Speese, S. D., Trotta, N., Rodesch, C. K., Aravamudan, B., & Broadie, K. (2003). The ubiquitin proteasome system acutely regulates presynaptic protein turnover and synaptic efficacy. *Current Biology, 13*(11), 899–910. doi:S0960982203003385 [pii].

Suzuki, A., Josselyn, S. A., Frankland, P. W., Masushige, S., Silva, A. J., & Kida, S. (2004). Memory reconsolidation and extinction have distinct temporal and biochemical signatures. *The Journal of Neuroscience, 24*(20), 4787–4795. doi:10.1523/JNEUROSCI.5491-03.2004 24/20/4787 [pii].

Suzuki, A., Mukawa, T., Tsukagoshi, A., Frankland, P. W., & Kida, S. (2008). Activation of LVGCCs and CB1 receptors required for destabilization of reactivated contextual fear memories. *Learning and Memory, 15*(6), 426–433. doi:15/6/426 [pii] 10.1101/lm.888808.

Tada, H., Okano, H. J., Takagi, H., Shibata, S., Yao, I., Matsumoto, M., Saiga, T., Nakayama, K. I., Kashima, H., Takahashi, T., Setou, M., & Okano, H. (2010). Fbxo45, a novel ubiquitin ligase, regulates synaptic activity. *The Journal of Biological Chemistry, 285*(6), 3840–3849. doi: M109.046284 [pii] 10.1074/jbc.M109.046284.

Tronson, N. C., & Taylor, J. R. (2007). Molecular mechanisms of memory reconsolidation. *Nature Reviews Neuroscience, 8*(4), 262–275. doi:nrn2090 [pii] 10.1038/nrn2090.

Tronson, N. C., Wiseman, S. L., Olausson, P., & Taylor, J. R. (2006). Bidirectional behavioral plasticity of memory reconsolidation depends on amygdalar protein kinase A. *Nature Neuroscience, 9*(2), 167–169. doi:nn1628 [pii] 10.1038/nn1628.

van Roessel, P., Elliott, D. A., Robinson, I. M., Prokop, A., & Brand, A. H. (2004). Independent regulation of synaptic size and activity by the anaphase-promoting complex. *Cell, 119*(5), 707–718. doi:S0092867404010967 [pii] 10.1016/j.cell.2004.11.028.

Wan, H. I., DiAntonio, A., Fetter, R. D., Bergstrom, K., Strauss, R., & Goodman, C. S. (2000). Highwire regulates synaptic growth in Drosophila. *Neuron, 26*(2), 313–329. doi:S0896-6273(00)81166-6 [pii].

Wang, S. H., de Oliveira, A. L., & Nader, K. (2009). Cellular and systems mechanisms of memory strength as a constraint on auditory fear reconsolidation. *Nature Neuroscience, 12*(7), 905–912. doi:nn.2350 [pii] 10.1038/nn.2350.

Willeumier, K., Pulst, S. M., & Schweizer, F. E. (2006). Proteasome inhibition triggers activity-dependent increase in the size of the recycling vesicle pool in cultured hippocampal neurons. *The Journal of Neuroscience, 26*(44), 11333–11341. doi:26/44/11333 [pii] 10.1523/JNEUROSCI.1684-06.2006.

Winters, B. D., Tucci, M. C., & DaCosta-Furtado, M. (2009). Older and stronger object memories are selectively destabilized by reactivation in the presence of new information. *Learning and Memory, 16*(9), 545–553. doi:16/9/545 [pii] 10.1101/lm.1509909.

Wood, M. A., Kaplan, M. P., Brensinger, C. M., Guo, W., & Abel, T. (2005). Ubiquitin C-terminal hydrolase L3 (Uchl3) is involved in working memory. *Hippocampus, 15*(5), 610–621. doi:10.1002/hipo.20082.

Yao, I., Takagi, H., Ageta, H., Kahyo, T., Sato, S., Hatanaka, K., Fukuda, Y., Chiba, T., Morone, N., Yuasa, S., Inokuchi, K., Ohtsuka, T., Macgregor, G. R., Tanaka, K., & Setou, M. (2007). SCRAPPER-dependent ubiquitination of active zone protein RIM1 regulates synaptic vesicle release. *Cell, 130*(5), 943–957. doi:S0092-8674(07)00902-6 [pii] 10.1016/j.cell.2007.06.052.

Chapter 11
AMPA Receptor Assembly: Atomic Determinants and Built-In Modulators

Madhav Sukumaran, Andrew C. Penn, and Ingo H. Greger

Abstract Glutamate-gated ion channels (iGluRs) predominantly operate as heterotetramers to mediate excitatory neurotransmission at glutamatergic synapses. The subunit composition of the receptors determines their targeting to synaptic sites and signalling properties and is therefore a fundamental parameter for neuronal computations. iGluRs assemble as obligatory or preferential heteromers; the mechanisms underlying this selective assembly are only starting to emerge. Here we review recent work in the field and provide an in-depth update on atomic determinants in the assembly domains, which have been facilitated by recent advances in iGluR structural biology. We also discuss the role of alternative RNA processing in the ligand-binding domain, which modulates a central subunit interface and has the capacity to modulate receptor formation in response to external cues. Finally, we review the emerging physiological significance of signalling via distinct iGluR heterotetramers and provide examples of how recruitment of functionally diverse receptors modulates excitatory neurotransmission under physiological and pathological conditions.

M. Sukumaran
Laboratory of Cellular and Synaptic MRC LMB and Neurophysiology, National Institutes of Health, Eunice Kennedy Shriver National Institute of Child Health and Human Development, Bethesda, MD 20892, USA

A.C. Penn
Univ. de Bordeaux, Interdisciplinary Institute for Neuroscience, UMR 5297, F-33000, Bordeaux, France

CNRS, Interdisciplinary Institute for Neuroscience, UMR 5297, F-33000 Bordeaux, France

I.H. Greger (✉)
Neurobiology Division, MRC Laboratory of Molecular Biology, Hills Road, CB2 0QH Cambridge, UK
e-mail: ig@mrc-lmb.cam.ac.uk

M.R. Kreutz and C. Sala (eds.), *Synaptic Plasticity*,
Advances in Experimental Medicine and Biology 970,
DOI 10.1007/978-3-7091-0932-8_11, © Springer-Verlag/Wien 2012

Keywords Alternative splicing • AMPA receptor biosynthesis • Endoplasmic reticulum • Ligand binding domain • N-(amino) terminal domain • RNA editing • Synaptic plasticity

11.1 From Polysome to Receptor Oligomer

Cell surface receptors, such as ion channels and G-protein-coupled receptors, prominently operate as hetero-oligomers. Assembly from a pool of different subunits increases the versatility and plasticity of signal transmission and is under complex cellular control. Ionotropic glutamate receptors (iGluRs) provide a dramatic example of how functionally diverse receptor stoichiometries shape an essential cellular process. iGluRs mediate excitatory neurotransmission in vertebrate nervous systems. This process involves three distinct iGluR subfamilies (AMPA-, NMDA- and kainate types), differentially expressed and regulated subunits within each subfamily, and a multitude of accessory subunits (Hollmann and Heinemann 1994; Traynelis et al. 2010). The result of this rich variety of assembly substituents is a combinatorial diversity of receptor expression, which impact such receptor properties as gating kinetics (which can operate on time scales spanning four orders of magnitude), ion conductance, pharmacology and synaptic trafficking; all of these properties are dependent upon the receptor's subunit stoichiometry, in both vertebrates (Cull-Candy et al. 2006; Greger et al. 2007; Traynelis et al. 2010) and invertebrates (Abuin et al. 2011; Rasse et al. 2005; Qin et al. 2005). The resulting diversity of possible receptor properties will ultimately shape synaptic transmission and in turn the operation of neuronal networks.

Like a multitude of other post-synaptic signalling components, iGluRs are embedded in the post-synaptic density (PSD), a sub-synaptic anchoring platform which concentrates and positions receptors directly opposite presynaptic release sites (discussed in Chap. 3) (Sheng and Hoogenraad 2007). In addition to synaptic trafficking, positioning and anchorage in the PSD can be determined by the subunit composition of the receptor. In the case of NMDA-type iGluRs, receptors containing the NR2B subunit locate to the edge of the PSD (i.e. extrasynaptically) whereas receptors harbouring NR2A are concentrated more centrally (Tovar and Westbrook 1999; Rumbaugh and Vicini 1999). This location dependence arises from sequence determinants within cytosolic carboxy-termini and will ultimately impact signal transmission (Steigerwald et al. 2000).

In the three main iGluR subfamilies, assembly into heteromers is either obligatory (NMDA-type and GluK4 and GluK5-containing kainate iGluRs) or preferential (AMPA-type and GluK1–3 kainate receptors). Due to the less stringent assembly rules, AMPA- and low-affinity kainate receptors (GluK1–3) can also exist as homotetramers. Signalling through AMPA receptor (AMPAR) homomers, which in the absence of the GluA2 subunit are Ca^{2+} permeable (CP) (Jia et al. 1996; Isaac et al. 2007), modulates synaptic physiology; recruitment of CP-AMPARs appears to be dynamically regulated in a number of neurons, rendering these synapses more plastic (see below) (Cull-Candy et al. 2006; Kauer and Malenka 2007). Determinants underlying these different assembly routes are starting to emerge.

Assembly into heteromers unlikely occurs by default. As a result of translation from polyribosomes, identical subunits, synthesized from an individual mRNA molecule, will be spatially and temporally concentrated on a patch of endoplasmic reticulum (ER) membrane (Fig. 11.1a). For example, *GRIA2* mRNA, encoding the GluA2 AMPA receptor subunit, with a length of ~3,000 base pairs could be translated by up to 30 ribosomes (Staehlin et al. 1964), resulting in ~30 nascent GluA2 polypeptides in close proximity; this local concentration is expected to promote assembly into homomers, which are not commonly observed. Parameters such as (1) diffusion in the plane of the ER membrane, (2) relative affinities of inter-subunit contacts as well as (3) the concentration of assembly partners in the ER are expected to determine the rate and extent of heteromeric assembly (summarized in Fig. 11.1a). We shall discuss the latter two parameters and how they are expected to affect different stages of iGluR biogenesis. We will focus on AMPA-type receptors; however, emerging principles will be generally applicable to the other iGluR subfamilies.

11.2 Dimer Formation

AMPARs form in two steps – subunits first dimerize, followed by assembly of dimers into tetramers. The dimer-of-dimers assembly has been observed at various levels. The crystal structure of the isolated L-glutamate ligand-binding domain (LBD) of GluA2 revealed a twofold symmetrical homodimeric complex (Armstrong and Gouaux 2000). Similarly, the second extracellular portion, the N-terminal domain (NTD), crystallized as a dimer, in both AMPA and kainate receptors (Fig. 11.1b) (Clayton et al. 2009; Jin et al. 2009; Kumar et al. 2009; Karakas et al. 2009; Kumar and Mayer 2010; Sukumaran et al. 2011). This overall twofold symmetry of the extracellular portion is also observed at the level of the intact receptor, whereas the ion channel adopts fourfold symmetry (Sobolevsky et al. 2009). Secondly, dimers (together with monomers and tetramers but not trimers) were also apparent on native gels (Penn et al. 2008; Greger et al. 2003). Dimers form first and are readily isolated from GluA2-expressing HEK293 cells for subsequent structural analysis (Shanks et al. 2010). Monomers are barely detected, thus dimers are the first stable assembly intermediate (Greger et al. 2003; Shanks et al. 2010). Dimer formation will be driven by the NTD (see below). This domain at the extreme N-terminus encompasses ~50% of primary sequence and is expected to fold first once threaded through the translocon into the ER lumen (Netzer and Hartl 1997). It is conceivable that subunit contacts via the NTD take place co-translationally, i.e. prior to folding of the remaining nascent chain (Fig. 11.1a), which would explain the paucity of monomers in biochemical experiments (see above). Accordingly, the NTD will initiate receptor formation. Recent crystal structures have provided atomic resolution of this critical assembly interface.

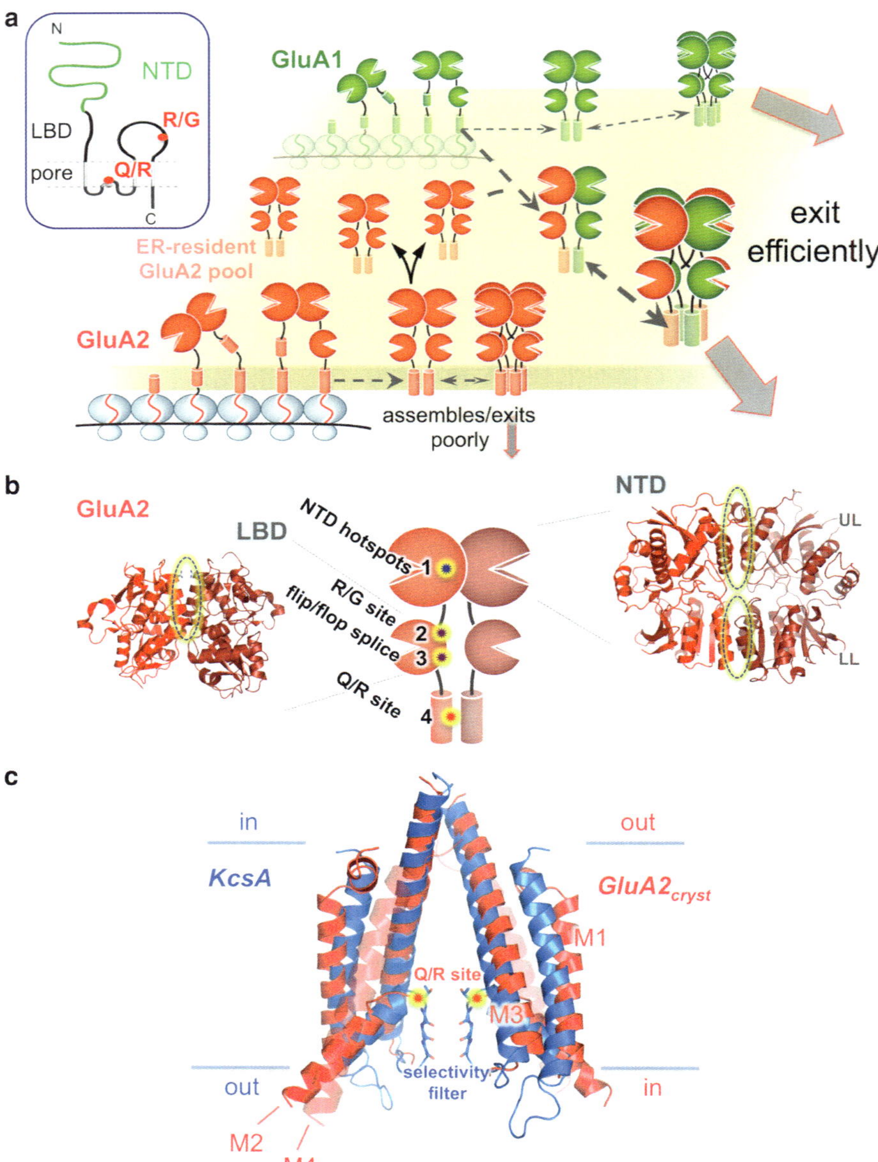

Fig. 11.1 AMPA receptors assemble into tetramers in the endoplasmic reticulum, with selective assembly modulated by domain-specific and subunit-specific determinants. (**a**) *Inset*: topology of an individual GluA2 AMPA receptor subunit. The extracellular N-terminal domain (NTD) and ligand-binding domain (LBD) are shown as *green* and *black* lines, respectively. Transmembrane helices that constitute the ion-channel pore are denoted as *grey* cylinders. Amino acid changes corresponding to RNA-editing sites are also denoted; the R/G site is conserved between GluA2, GluA3 and GluA4, whereas the Q/R site on a re-entrant pore loop is unique to GluA2. Note that the NTD is continuous in primary sequence, while the LBD is interrupted by multiple transmembrane segments.

11.2.1 The N-Terminal Domain Assembly Surface

The NTD is a hallmark of metazoan iGluRs but is absent in the prokaryotic GluR0-type channels (Chen et al. 1999) and in vertebrate kainate-binding proteins (Henley 1994). The function of this domain in non-NMDARs (AMPA and kainate receptors) has not been fully resolved. In NMDARs, powerful allosteric modulation of the channel via the NTD is well established, where channel open probability is reduced in response to binding of Zn^{2+} and other ligands (Mony et al. 2009). An allosteric role in the non-NMDAR NTD has not been described to date but cannot be ruled out (Sukumaran et al. 2011; Jensen et al. 2011). In all iGluRs, the NTD is implicated in subunit assembly (Hansen et al. 2010). Multiple iGluR genes are found in higher eukaryotes, which is not generally the case in prokaryotic genomes. Therefore, the need for a more sophisticated assembly determinant, orchestrating a fine balance of associations between homo- and heterotetramers *within* subfamilies in addition to preventing co-assembly *between* subfamilies, may explain the appearance of the NTD later in evolution concomitant with the radiation of iGluR paralogs by gene duplication and subsequent mutation. This distal segment appears to also play a role in synapse formation (Passafaro et al. 2003) and provides a binding site, both for presynaptic elements and for soluble factors released upon

←―――

Fig. 11.1 (continued) Main figure: A schematic of assembly in the endoplasmic reticulum (ER), highlighting different steps of AMPA receptor assembly. Nascent GluA2 (*red*) and GluA1 (*green*) polypeptides are shown emerging from polyribosomes (*grey*), translating into the ER lumen. NTD dimerization likely occurs co-translationally, due to its location at the extreme N-terminus. Because of the high local concentration of identical subunits, due to nearby ribosomes translating polypeptide in close proximity, homodimerization likely dominates at this stage. After folding is complete, dimeric subunits then subsequently assemble into tetramers. Due to its subunit-specific set of assembly determinants, native GluA2 assembles poorly into tetramers and inefficiently exits the ER; therefore, GluA2 likely forms a stable, ER-resident pool of dimers (*solid arrow*). This relatively higher concentration of dimers, concomitant with GluA2's favourable heteromerization capability, allows heterodimers to be formed efficiently upon translation of GluA1 subunits. Heterodimers assemble into heterotetramers and exit the ER efficiently; GluA1 is also capable of efficient homotetramerization and ER exit. (**b**) Sites of assembly determinants within a GluA2 subunit. A GluA2 dimer is depicted, with individual protomers coloured dark and light red, respectively. Homodimeric crystal structures of the isolated GluA2 domains are also shown (pdb codes 3hsy and 2uxa), with subunit dimer-interfaces depicted as ovals. Sites of assembly - specifying determinants are shown on the schematic: (*1*) hotspots in the NTD mediate homo- and heterodimerization, (*2*) the R/G site in GluA2–4 and (*3*) alternative splicing in GluA1-4 modulate overall heteromerization and ER-exit competence, and (*4*) the Q/R site in the GluA2 transmembrane domain renders GluA2 homotetramerization unfavourable versus heterotetramerization, leading to the stable ER-resident pool of dimers mentioned above. (**c**) The transmembrane domain of GluA2 is similar to potassium channels. A superposition of the transmembrane domains of the crystallized full-length GluA2 construct, GluA$_{cryst}$ (*red*; pdb code 3kg2), and the KcsA potassium channel (*blue*; pdb code 1r3j) shows that both are highly similar, albeit in opposite topological orientations. The assembly-critical Q/R site of GluA2 is shown, located on top of the selectivity filter. Transmembrane domains that make contacts in the tetrameric structure (*M1, M3* and *M4*) and the re-entrant pore loop (*M2*) are denoted

intense synaptic activity (O'Brien et al. 1999; Hansen et al. 2010). The latter functions will only be relevant in species with nervous systems, whereas in prokaryotes ion homeostasis is likely to be the main function of the NTD-lacking, homomeric GluR0 channel types.

Recent structural data on the intact GluA2 homomer together with high-resolution structures of isolated domains provide an overview of inter-subunit interactions along the major axis of the receptor (Sobolevsky et al. 2009). Dimeric contacts are prominent at the level of the NTD, whereas tetrameric packing is mediated by the transmembrane helices of the ion channel. A similar principle has been inferred for kainate receptors (Das et al. 2010). NTDs form extensive twofold symmetrical dimeric assemblies (Fig. 11.2) which, when isolated from the receptor, can also be measured in solution (Clayton et al. 2009; Jin et al. 2009; Kumar et al. 2009; Rossmann et al. 2011), in contrast to the LBDs, which are largely monomeric (in the case of RNA-edited GluA2-flop; Sun et al. 2002). Interestingly, in receptors assembling as obligatory heteromers, homodimeric NTD contacts are either absent (NR2B; Karakas et al. 2009) or severely reduced (GluK5; Kumar and Mayer 2010). A similar observation holds for the GluA3 subunit, which exhibits 'obligatory' heteromeric assembly behaviour within the AMPAR family (see below) (Rossmann et al. 2011). These observations underline the key role of the NTD in driving distinct assembly routes.

The NTD protomer adopts a fold analogous to prokaryotic type I periplasmic-binding proteins (PBPs; Quiocho and Ledvina 1996), where two globular lobes (the upper and lower lobes; Fig. 11.1b) are connected by three short hinges. Each lobe contributes to the bipartite NTD dimer interface (Fig. 11.2a). Contacts between the upper lobes are tight and evolutionarily conserved, thus bearing the hallmarks of a functionally relevant interface, whereas packing across the lower lobes is looser. This functional division is apparent in GluA2 and GluA3 (Sukumaran et al. 2011), with GluA3 providing the most striking example: in the most commonly observed dimeric form (Sukumaran et al. 2011) an unfavourable electrostatic potential between the lower lobes results in lobe separation (PDB 3O21; chains CD), to a degree seen in the analogous metabotropic GluR (mGluR) ligand-binding cores (Kunishima et al. 2000). In GluK2, the upper and lower lobes form similar interfaces that are both less compact than the upper lobe interface of the AMPAR subfamily.

Analytical ultracentrifugation with fluorescence detection (AU-FDS) provided a sensitive tool facilitating measurements of NTD associations at high resolution, in the sub-nanomolar range (MacGregor et al. 2004) and, more importantly, permits measurement of heteromeric assemblies (Rossmann et al. 2011). A surprising range of affinities among AMPAR NTDs, covering almost three orders of magnitude, could be discerned: GluA2 and GluA3 lie at the functional extremes with dimer dissociation constants (K_ds) of 1.8 and 1,200 nM, respectively; values for GluA1 and GluA4 NTD dimers were intermediate (Fig. 11.3a) (Rossmann et al. 2011). The relatively unstable homodimeric GluA3 contacts are most certainly a result of the 'unzipped' lower lobe interface (Fig. 11.2a), which underlies the 'obligatory' assembly behaviour of GluA3. First measurements of heteromeric assemblies revealed that AMPAR NTDs preferentially heterodimerize (Rossmann et al. 2011). Contrasting with the relatively poor homodimeric affinity, GluA3 produced tight heterodimers ($K_d \sim 1.3$ nM).

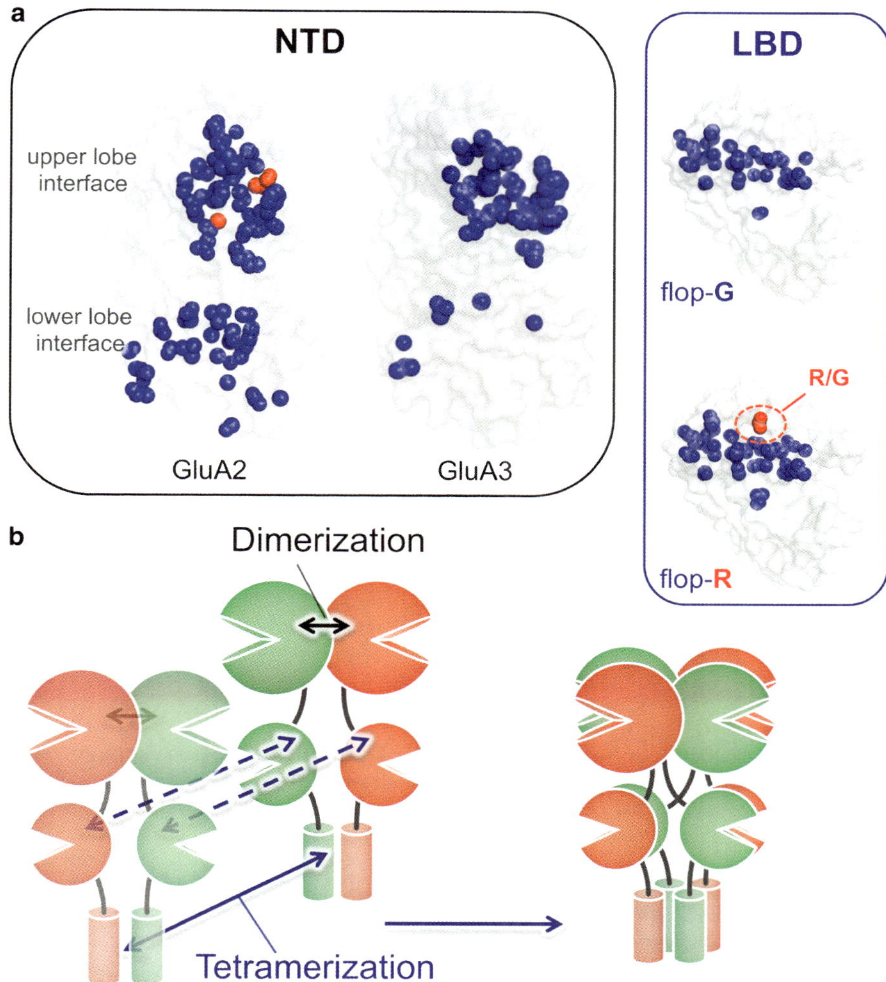

Fig. 11.2 Critical assembly determinants in the extracellular domains are located at subunit interfaces. (**a**) Assembly surfaces of the AMPAR extracellular domains. Atoms that make contacts across the dimer interfaces are shown as *blue* spheres, with specific assembly 'hotspots' and determinants highlighted in red. The GluA2 NTD (pdb code 3hsy) shows extensive dimerization contacts across both upper and lower lobe interfaces; however, GluA3 (pdb code 3o21) shows markedly less interface contacts in the lower lobe, suggesting differential, subunit-specific assembly behaviour for the NTD (Reproduced from (Sukumaran et al. 2011) with permission from Nature Publishing Group). In the case of the LBD (*flop-G*: pdb code 1ftj; *flop-R*: unpublished), interface contacts are largely uniform across subunits, but the interfaces are modulated at the level of RNA editing and alternative splicing; different interfaces for the edited G and unedited R forms are shown, with the Arg shown in *red*. Note a minor increase in subunit contacts with the unedited Arg, perhaps due to the favourable symmetrical arginine-arginine contacts across the homomer interface; therefore, editing to glycine reduces the homomeric LBD affinity and favours heteromerization. (**b**) The assembly steps from Fig. 11.1a, shown in detail. Dimerization (*black arrows*) is mediated co-translationally by the NTDs, whereas tetramerization (*blue arrows*) is mediated by determinants in the transmembrane domains. LBD 'dimerization' observed crystallographically for isolated LBDs only occurs upon tetramerization in the context of the full receptor

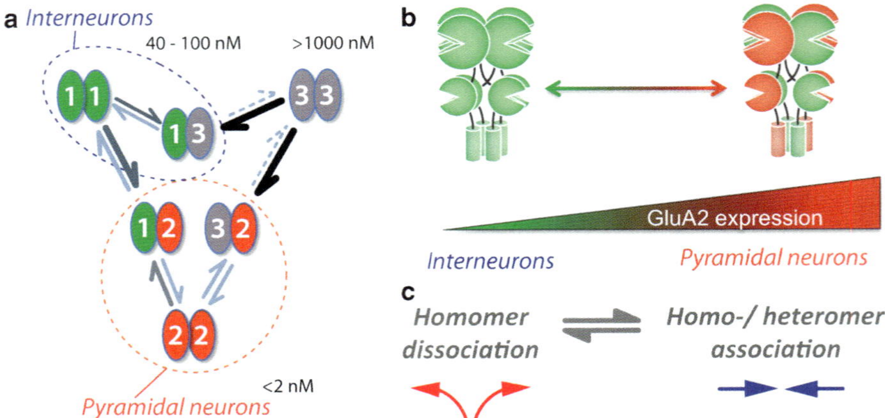

Fig. 11.3 Differential AMPA receptor assembly will be a balance between subunit affinities and subunit expression levels. (**a**) Measured association affinities between GluA1–3 NTDs define specific assembly regimes. Measured K_ds of homodimer and heterodimer dissociation by AU-FDS are shown for GluA1, GluA2 and GluA3 homomers and heteromers. These K_ds span three orders of magnitude, from very tight (<2 nM, *bottom*), for assembly driven by GluA2, to relatively loose, as in the case of GluA3 homomers (>1,000 nM, *top right*). The relationship between these homomeric and heteromeric affinities defines different 'regimes' dependent on cellular expression profile; example regimes are given for GluA2-expressing hippocampal neurons (*red outline*), which will efficiently incorporate GluA2 into receptors, versus GluA2-lacking hippocampal interneurons (*blue outline*), which will express GluA1/3 heteromers and GluA1 homomers but no GluA3 homomers. Due to its poor homomerization capability, GluA3 will 'obligatorily' form heteromers in the presence of the other subunits (Reproduced from (Rossmann et al. 2011 with permission from Nature Publishing Group). (**b**) Titration of subunit expression allows neurons to modulate channel properties. Despite GluA2's dominant assembly and functional phenotypes, neurons may be able to express functionally different receptors by modulating the ratio of expressed GluA2 versus GluA1. The example expression regimes outlined in panel A (hippocampal pyramidal neurons vs. interneurons) are denoted. (**c**) Heteromerization requires at least two steps: homomer dissociation and subsequent association of heteromers. For the second step, heteromerization will have to compete with re-association of homomers

Accordingly, GluA3 homomers are only expected to form under conditions of substoichiometric expression of other assembly partners (Fig. 11.3a, b).

A number of concepts emerged from these results (Rossmann et al. 2011). First, AMPARs preferentially heteromerize at the level of the dimer. Preferential heterodimerization will bear upon subunit stoichiometry, spatial arrangement of subunits within tetramers, and will allow for formation of tri-heteromeric AMPARs. Secondly, GluA2, which restricts Ca^{2+} flux through AMPARs, is dominantly incorporated into heterodimers. This property of the GluA2 NTD together with assembly determinants in the LBD interface and the channel pore (see below) (Greger et al. 2007) likely explain the dominant expression of GluA2-containing heteromers throughout the brain (Isaac et al. 2007). Third, GluA1 and GluA4 exhibit a more 'balanced' assembly between homo- and heteromeric modes. This property likely underlies the existence of Ca^{2+}-permeable GluA1 homomers, which are detected in selected neurons under certain conditions (Cull-Candy et al. 2006; Carlezon and Nestler 2002). In sum, the data imply the existence of two assembly routes for AMPAR heteromers, termed 'preferential' and 'obligatory' (Fig. 11.3a) (Rossmann et al. 2011). The affinity network shown in Fig. 11.3a together with relative expression levels of assembly partners in the ER (Fig. 11.3b) will ultimately determine the nature of receptor oligomer subpopulations in a given neuronal type.

These data support the previously recognized role of the NTD in assembly (Ayalon and Stern-Bach 2001; Leuschner and Hoch 1999) but reveal a dominant organizing function, which turns out to be more sophisticated. In fact, individual assembly determinants or 'hotspots' encoded in the highly conserved upper lobe dimer interface have been detected in GluA2 (Rossmann et al. 2011). These evolutionarily variable residues help explain how the tight GluA2 homodimeric contacts, which presumably will form co-translationally (Fig. 11.1a), allow formation of heteromers. A biophysical dissection of these 'hotspots' alluded to a principle whereby assembly is driven by two parameters: dissociation of homodimers prior to associations of heterodimers. If re-association is energetically favoured, the equilibrium will be shifted towards the newly formed heterodimer (Fig. 11.3c).

11.2.2 The Modulatory LBD Dimer Interface

The ligand-binding domain (LBD) overall resembles the bilobate fold of the NTD; however, its role in signal transmission and assembly is vastly different. Relatively loose dimer associations, mediated by the LBD upper lobes (Fig. 11.2), facilitate inter-subunit flexibility, a likely requirement for gating transitions (Mayer and Armstrong 2004). Like the NTD, the LBD has been crystallized as a twofold symmetrical dimer, which is evident for the isolated domain (for all iGluRs) and in the complete GluA2 AMPAR (Sobolevsky et al. 2009). LBD dimers are not detected in solution, except when stabilized by mutation or by allosteric modulators (Sun et al. 2002; Jin et al. 2005). Surprisingly, in the full-length receptor structure, LBDs swap to form '*trans*-dimers', i.e. the twofold symmetrical dimer observed for the isolated LBD is formed only upon tetramerization. Moreover, characterization of GluA2 dimers by single-particle electron microscopy revealed that the two LBDs are separated, and are 'held together' by the NTD contacts at the top and the transmembrane sector at the bottom (Shanks et al. 2010; Nakagawa 2010). These findings suggest that the LBD forms the crystallographically described twofold symmetrical dimer interface only in the tetrameric context, with LBDs associating between (rather than within) subunit dimers (Fig. 11.2b). Whether this represents the only accessible conformation remains to be seen. For example, heteromeric AMPARs or the AMPAR-TARP complex may give rise to '*cis*-dimers', featuring a closed LBD dimer interface within a subunit dimer analogous to the L483Y mutant dimer (Shanks et al. 2010). These may be energetically less favourable in the absence of a heteromeric or TARP partner and were, thus, not seen under the conditions used for crystallization and single-particle electron microscopy. Furthermore, a recent study has suggested that the LBD dimer interface is only formed upon ligand binding and channel opening, with individual LBDs decoupled from each other in the resting (unliganded, closed channel) and desensitized (liganded, closed channel) states (Gonzalez et al. 2010). However, as this study was conducted in GluA4 constructs lacking the entire NTD, more work is required to fully understand the exact oligomeric conformation(s) of LBDs in the physiological context.

In stark contrast to the NTD, the LBD sequence is well conserved between AMPAR paralogs. Versatility is introduced post-transcriptionally by adenosine-to-inosine RNA editing and by alternative splicing (Seeburg 1996). All AMPAR subunits harbour the alternative flip/flop exons (Sommer et al. 1990), whereas RNA editing only targets GluA2–4 (Figs. 11.1a and 11.2a), resulting in a switch from a genomically encoded arginine (R) to a glycine (G) at the R/G site (Lomeli et al. 1994). These alternative RNA processing sites line the dimer interface (Fig. 11.2a) and alter assembly and secretory traffic (Greger et al. 2006; Greger et al. 2007; Coleman et al. 2006, 2010). This adds another facet to the assembly process: in addition to preferential heteromerization between subunit paralogs, different alternatively processed homologs (e.g. GluA1-flip + GluA1-flop) preferably co-assemble (Brorson et al. 2004). Whether these switches are affinity determinants or purely operate by altering the dwell time of assembly intermediates is not fully resolved. Clearly, the fact that editing at both the R/G and the Q/R site reduces secretory traffic of GluA2 will increase availability (i.e. the concentration) of this critical subunit in the ER (Greger et al. 2002, 2006), which will facilitate its uptake into heterodimers (Fig. 11.1a). The interplay between tight NTD- and looser LBD interactions in co-ordinating assembly are currently unclear.

Alternative splicing can be regulated by external cues resulting, for example, in changes of intracellular Ca^{2+} (Xie 2008; Stamm 2002). Similarly, editing by the editases ADAR1 and 2 can be reprogrammed (Schmauss 2005). Due to their strategic location at subunit interfaces (Fig. 11.2a), these 'built-in' modulators are primed to remodel assembly and secretory traffic of AMPARs, providing a homeostatic control hub for adjusting receptor type and number in response to altered neuronal activity (Penn et al. in review; Penn and Greger 2009).

11.3 Tetramer Formation

As discussed above, recent studies have indicated that AMPARs preferentially assemble as heterodimers (Rossmann et al. 2011), but whether there are similar mechanisms of preferential assembly at the tetramerization step is currently unclear. Earlier data indicate that tetramerization also follows specific pathways when bringing together different populations of homo- and heterodimers (Mansour et al. 2001). In the ion-channel sector, two major determinants will drive assembly: (1) packing between three transmembrane helices M1, M2 and M4 (Fig. 11.4a) and (2) the pore loop (M2), which forms a fourfold symmetrical contact point (Figs. 11.1c and 11.4a).

Helical packing in the transmembrane domain (TMD) of the AMPA receptor largely mirrors an inverted K^+ channel (Fig. 11.1c), which contains extensive contacts between transmembrane helices and provides the largest packing interface in the tetrameric context (Long et al. 2005), suggesting that the TMDs contribute

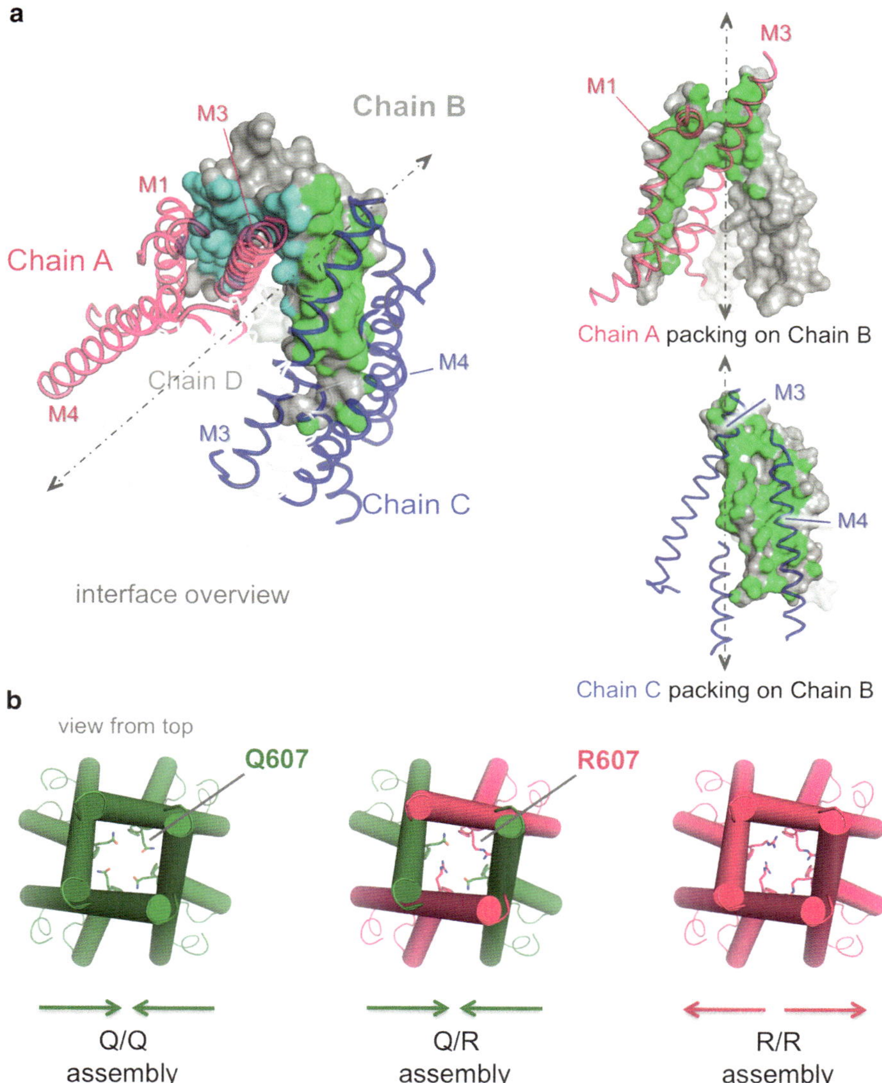

Fig. 11.4 Assembly determinants in the transmembrane domain. (**a**) Overview of the packing of the transmembrane domains (TMDs) of individual subunits against each other in the GluA2$_{cryst}$ homotetramer (pdb code 3kg2). In the interface overview, the tetrameric ion-channel domains of all four chains (A–D) are shown, with chain B (*grey*) shown as a molecular surface and chains A (*magenta*), C (*blue*) and D (*white*) shown in ribbon. Both chain A and chain C pack against chain B, forming two distinct assembly interfaces; the interface formed with chain A is shown in cyan on chain B, with the interface formed with chain C shown in green. Roughly, the M4 helix from each chain packs against a 'groove' formed by helices M1 and M3 from a partner chain; chain B forms the 'groove' for M4 from chain C, whereas the M4 from chain B packs against the M1/M3 groove from chain A. Views onto the individual A–B and B–C interfaces are also shown. (**b**) A key assembly determinant in the transmembrane domain is the Q/R site in GluA2. Q/R editing in

the major tetramerization drive. In further analogy, K^+ channels also exhibit a dimer-of-dimers assembly pattern (Tu and Deutsch 1999; Deutsch 2002).

In GluA2, the TMD contains an extensive packing interface (Sobolevsky et al. 2009). As outlined in Fig. 11.4, each protomer contributes three distinct contact points: the first interface is largely external to the aqueous ion-channel pore and consists of M4 packing against a 'groove' formed by M1 and M3 of the neighbouring subunit. The second, larger interface is the 'groove' provided by M1 and M3 of the other subunit in the dimer. As M4 is a novel insertion in eukaryotic iGluRs versus prokaryotic GluR0-type receptors and kainate-binding proteins, these two interfaces are expected to contribute additional assembly determinants in metazoa.

The third 'interface' contact is internal to the ion channel and is provided by residues in the re-entrant pore loop (M2) packing against copies of itself in the other protomers, as well as near a narrow constriction, where the four M3 helices from each subunit align with one another, putatively forming the channel gate (Fig. 11.1c). As the re-entrant pore loop is built incompletely in the currently available crystal structure (PDB: 3KG2), the packing of the pore loops is unknown at this point. As contacts in this region shape the aqueous vestibules of the pore upon channel assembly, regions around these interfaces will transition from lipid-exposed to solvent-exposed, and these transfer reactions could also contribute thermodynamically to channel assembly.

Sequence conservation among AMPAR paralogs in the transmembrane region is high; however, variability is introduced by RNA editing in the apex of the pore loop of GluA2 and GluK1 and GluK2 (Seeburg 1996). Recoding at the GluA2 Q/R site plays a critical role in the dimer-to-tetramer transition by destabilizing edited tetramers, thereby providing a strategically positioned assembly determinant (Greger et al. 2003) (Figs. 11.1c and 11.4b). An analogous case has been described recently for GluK2 (Ball et al. 2010) and most likely results from unfavourable electrostatics from the approximation of four arginines during pore formation (Fig. 11.4b). It is worth pointing out that Q/R-edited GluA2-R channels can form (Swanson et al. 1997), in particular when expressed with auxiliary factors (Yamazaki et al. 2004), albeit far less efficiently than unedited GluA2-Q (Greger et al. 2003). Therefore, in the presence of other subunit partners in the ER, the energetically preferred Q/R heteromers are likely to prevail over R-pore homomers. To draw another parallel between iGluRs and K^+ channels, an analogous position in the pore loop of the KCNQ3 potassium channel also slows channel transit from the ER relative to KCNQ2 (Gomez-Posada et al. 2010), suggesting that this site is critical to the general tetrameric channel fold.

Fig. 11.4 (continued) GluA2 results in a switch from Q607 (*green*), which allows efficient homomeric and heteromeric assembly, to R607 (*magenta*), which, due to charge repulsion of the arginines, allows heteromic assembly with Q but disfavours homomeric assembly and in turn ER export, giving rise to a stable pool of ER-resident GluA2 (Fig. 11.1a). Note that GluA1, GluA3 and GluA4 express Q at the analogous position, resulting in a favoured heteromeric Q/R pore stoichiometry in native AMPA receptors. Q607 and R607 TMD structures are homology models of the full-length pore modelled against the KcsA pore (Greger et al. 2003)

In addition to the pore sector, the extracellular domains also come into contact in the context of the crystallized GluA2$_{cryst}$ tetramer. The lower lobes of the NTD pack in the region of V209, and the LBDs form a small packing interface between two symmetrical helices (Sobolevsky et al. 2009). However, both of these inter-dimer contacts are only seen in one subunit pair. Also, different three-dimensional structures of the overexpressed GluA2 homotetramer and purified native AMPAR complexes reconstructed by electron microscopy show different inter-dimer arrangements and also show that the extracellular domains adopt different quaternary arrangements dependent on subunit inclusion and gating status (Midgett and Madden 2008; Nakagawa et al. 2005). Therefore, the inter-dimer interfaces in the extracellular domains might be transient and are unlikely the primary drivers of tetrameric assembly. Moreover, as discussed above, the LBD 'dimer' interface is putatively formed only upon channel tetramerization (Fig. 11.2b) (Nakagawa 2010). This suggests that variability in the LBD interface (flip/flop splicing and R/G editing) may also contribute to differential channel assembly at the level of tetramerization; as yet this question has not been explored experimentally and LBD dimer variability affecting tetrameric assembly is still speculative at this point.

AMPARs co-assemble with a variety of different auxiliary and modulatory subunits. At what stage these auxiliary factors, including TARPs (transmembrane AMPA receptor regulatory proteins), cornichons and CKAMP-44 (reviewed in Guzman and Jonas 2010; Jackson et al. 2011), complex with the core receptor is not fully resolved. TARPs, the best-studied auxiliary factors, do not co-purify stably with dimeric assembly intermediates, suggesting that they bind to already-assembled, tetrameric AMPARs (Shanks et al. 2010). Moreover, TARPs appear to associate in varying stoichiometries, depending on expression levels (Kim et al. 2010; Shi et al. 2009; Kato et al. 2010). Since TARP expression is not homogenous across neuronal populations, cell-type-specific receptor modulation by these cofactors can be expected (see below).

11.4 Why Heteromers?

The ability for receptors to assemble as heteromers imparts significant diversity and flexibility in the regulation of nervous system function. Much of what is inferred about the oligomeric state of native receptors comes from single-cell profiling of subunit mRNA expression, *in situ* localization of subunit mRNA or protein, and subunit-specific pharmacology or genetic manipulations. Overall, the findings reveal a predominance of heteromeric receptors, which vary in abundance during development and in a tissue- and neuron-specific manner. As discussed above, in the principal neurons of the vertebrate CNS, GluA2-containing AMPAR heteromers prevail, whereas interneurons often express a large population of GluA2-lacking receptors.

11.4.1   The Functional Dominance of Q/R-Edited GluA2

Heteromerization can have important consequences on AMPAR functional properties, trafficking and subcellular localization. Of particular significance, the incorporation of the Q/R-edited GluA2 subunit renders the channel pore imperme-able to divalent cations, thus disarming them as an activator of calcium-dependent signalling cascades (Cull-Candy et al. 2006; Isaac et al. 2007). In addition, GluA2 inhibits the voltage-dependent block of the pore by intracellular polyamines, which can have consequences for short-term synaptic plasticity (Rozov and Burnashev 1999; Rozov et al. 1998) and redefine the rules for long-term changes in synaptic strength (Kullmann and Lamsa 2008). Furthermore, the attenuation of single-channel conductance and desensitization by GluA2 provides a means to regulate synaptic strength (Liu and Cull-Candy 2000) and shape synaptic transmission (Zhu 2009; Gardner et al. 2001a). Recently, it has also been suggested that incorporation of GluA2 can alter the capacity for regulation of receptor function by type II TARPs (Kato et al. 2008; Soto et al. 2009). These various means by which Q/R-edited GluA2 can have potent effects on receptor properties have prompted it to be referred to as the 'functionally dominant' subunit.

11.4.2   Subunit-Specific Accessory Factors

The formation of heteromeric receptors provides a means to expand the repertoire of interactions with scaffolding proteins and thus regulate AMPAR trafficking (Shepherd and Huganir 2007). GluA1 and GluA3 have long and short cytoplasmic carboxy-terminal domains (CTDs) respectively, whereas GluA2 and GluA4 can undergo alternative splicing to include either long or short CTDs (Bredt and Nicoll 2003). These AMPAR CTDs can participate in a variety of different protein interactions to regulate receptor trafficking. Long CTD subunit variants of GluA1 and GluA4 interact with band 4.1 protein, to could stabilize AMPAR surface expression via the spectrin-actin cytoskeleton (Coleman et al. 2003; Shen et al. 2000; Lin et al. 2009). In addition, GluA1 has a unique motif conforming to the general consensus for type I PDZ interactions to enable SAP97-mediated secretory trafficking (Sans et al. 2001). Short CTD subunits, on the other hand, have a clathrin-adaptor AP2 interaction site critical for clathrin-mediated endocytosis and central to the recycling of synaptic AMPARs (Sheng and Hoogenraad 2007). This site overlaps with the NSF-binding site in GluA2, which plays a critical role in the membrane fusion and synaptic expression of AMPA receptors (Steinberg et al. 2004; Luthi et al. 1999; Huang et al. 2005). In addition, the short CTD terminus has a type II PDZ ligand with phosphory-lation-modulated PDZ interactions that are central to sorting of synaptic AMPA receptors and for synaptic plasticity in the cerebellum (Cull-Candy et al. 2006; Chung et al. 2000; Xia et al. 2000). Also of interest, it has recently been demonstrated that incorporation of subunits with short CTDs into heteromeric receptors blocks the effects of a type II TARP auxiliary subunit (Soto et al. 2009). In general, factors controlling the subunit specificity or stoichiometry of auxiliary subunits to AMPARs could tune channel gating and potentially modify interactions with scaffolding proteins (Kim et al. 2010; Shi et al. 2009; Milstein and Nicoll 2008). Finally, multiple subunit-specific sites for post-translational modifications such as phosphorylation and palmitoylation further expand the repertoire of mechanisms regulating AMPA receptor function and trafficking (Bredt and Nicoll 2003; Shepherd and Huganir 2007). Therefore, combinatorial protein interactions resulting from the formation of heteromeric receptors is likely fundamental in tuning the regulation of AMPA receptors and the plasticity of synaptic transmission.

The same concepts for combinatorial CTD protein interactions likely also apply to the extracellular domain of the receptor. For example, the NTD of GluA2 (but not GluA1) directly interacts with the synaptic adhesion molecule N-cadherin to stimulate presynaptic development (Saglietti et al. 2007). In addition, AMPA receptor clustering and excitatory synaptogenesis is facilitated by trans-synaptic interactions of the secreted pentraxin NARP with the NTDs of GluA1–3 (but not GluA4) (O'Brien et al. 2002). In contrast, synaptic recruitment of GluA4 receptors is mediated by another pentraxin, NP1 (Sia et al. 2007). Together, combinatorial subunit-protein interactions might shed light on input-specific recruitment of AMPA receptors differing in oligomeric state (Zhu 2009; Good and Lupica 2010; Kielland et al. 2009; Soler-Llavina and Sabatini 2006; Harms et al. 2005; Gardner et al. 2001b; Toth and McBain 1998).

11.4.3 Differential Activity Patterns Result in Synapse-Specific Heteromer Recruitment

The recruitment of different AMPA receptor oligomers to synapses by activity or external cues has been demonstrated at various synapses in the nervous system. In the cerebellar cortex, stimulation of parallel fibre-derived stellate synapses induced a switch to GluA2-containing receptors and a concomitant decrease in the size of the synaptic current (Liu and Cull-Candy 2000), an effect more recently shown to be synapse specific (Soler-Llavina and Sabatini 2006). In a related *in vivo* study, noradrenaline release caused by fear-inducing stimulus was also shown to trigger a switch in AMPA receptor composition at these synapses by boosting expression of the GluA2 subunit (Liu et al. 2010). Plasticity at stellate cell synapses may provide a means to regulate this inhibitory network and optimize cerebellar learning (Jorntell et al. 2010). Strong emotional cues have also been shown to trigger changes in the AMPA receptor composition at synapses in the lateral amygdala. Here, fear conditioning has been shown to trigger both potentiation at thalamic inputs and a slow but transient incorporation of GluA2-lacking receptors, which impart the capacity for fear memory erasure by long-term depression (Clem and Huganir 2010).

Plasticity of the oligomeric state of AMPA receptors onto cortical pyramidal neurons has also been demonstrated. Whisker experience induced pathway-specific strengthening of spared excitatory inputs onto layer 2/3 pyramidal cells of barrel cortex by recruitment of GluA2-lacking receptors (Clem and Barth 2006). In addition, dark-rearing rodents resulted in reversible incorporation of GluA2-lacking receptors and an increase in the size of synaptic currents of analogous connections in the visual cortex (Goel et al. 2006), a finding that is consistent with *in vitro* models of activity deprivation (Thiagarajan et al. 2005). Transient incorporation of GluA2-lacking receptors has also been detected by some groups at CA1 Schaffer collateral inputs in hippocampus, and that their calcium-permeability may be required for consolidating LTP of synaptic strength (Plant et al. 2006; but see Gray et al. 2007; Adesnik and Nicoll 2007).

In addition to these examples, a developmental switch from GluA2-lacking receptors to heteromers containing this subunit has been demonstrated throughout the nervous system including the retina (Osswald et al. 2007), neocortex (Kumar et al. 2002) and hippocampus (Ho et al. 2007; Stubblefield and Benke 2010), where the calcium-signalling capacity may play a role in juvenile forms of synaptic plasticity (Jensen et al. 2003). In summary, these are a few examples where heteromerization provides an important physiological means to regulate calcium signalling through AMPA receptors and synaptic transmission.

11.4.4 Failure to Heteromerize Is Associated with Neuropathology

There are also a number of conditions associated with insufficient heteromerization resulting in pathological synaptic expression of calcium-permeable AMPA receptors. Prolonged withdrawal from cocaine leads to the incorporation of GluA2-lacking receptors at excitatory inputs onto neurons of the nucleus accumbens (Conrad et al. 2008). In contrast, single cocaine administration caused a redistribution of GluA2-lacking receptors to synapses on dopaminergic cells of the ventral tegmental area (VTA); this effect was reversed by stimulating LTD with mGluR1 enhancers (Bellone and Luscher 2006). Interestingly, disruption of mGluR1 function reduced the level of cocaine exposure required to induce the switch in synaptic AMPA receptors in the nucleus accumbens (Mameli et al. 2009).

In the hippocampus, repeated morphine administration has been shown to increase synaptic expression of GluA2-lacking receptors and reduce the magnitude of LTD (Billa et al. 2010). Interestingly, recent findings show that the introduction of GluA2-lacking receptors at CA1 synaptic inputs adds the capacity for anti-Hebbian LTP and attenuation of NMDAR-dependent learning (Wiltgen et al. 2010). Finally, toxic expression of calcium-permeable AMPA receptors has been reported in a number of disease states including at CA1 neurons following ischemia and in motor neurons of patients with sporadic amyotrophic lateral sclerosis, where calcium-permeable AMPA receptor blockers or expression of the edited GluA2 subunit is neuroprotective (Liu and Zukin 2007; Peng et al. 2006; Hideyama et al.

2010; Noh et al. 2005). In summary, the plasticity of AMPA receptor expression and oligomerization is a key mediator in the pathology of various diseases.

Given the significance of forming AMPA receptor heteromers containing the GluA2 subunit in protecting certain neuronal types from pathological demise, modulation of AMPA receptor assembly may prove an appropriate therapeutic approach. Recently, it has become apparent that dimerization properties of the NTD impact the competence of subunits to heteromerize (Rossmann et al. 2011). Furthermore, clamshell movements of some NTDs may alter dimer affinities, thus making the putative capacity of these periplasmic binding protein homologs to bind ligand an attractive drug target (Sukumaran et al. 2011). The emerging structures of AMPA receptor NTDs and methods described to purify the soluble domain and quantify their dimer affinities provide a potential platform for drug discovery.

References

Abuin, L., Bargeton, B., Ulbrich, M. H., Isacoff, E. Y., Kellenberger, S., & Benton, R. (2011). Functional architecture of olfactory ionotropic glutamate receptors. *Neuron, 69*(1), 44–60. doi: S0896-6273(10)00984-0 [pii] 10.1016/j.neuron.2010.11.042.

Adesnik, H., & Nicoll, R. A. (2007). Conservation of glutamate receptor 2-containing AMPA receptors during long-term potentiation. *Journal of Neuroscience, 27*(17), 4598–4602. doi:27/17/4598 [pii] 10.1523/JNEUROSCI.0325-07.2007.

Armstrong, N., & Gouaux, E. (2000). Mechanisms for activation and antagonism of an AMPA-sensitive glutamate receptor: Crystal structures of the GluR2 ligand binding core. *Neuron, 28*, 165–181.

Ayalon, G., & Stern-Bach, Y. (2001). Functional assembly of AMPA and kainate receptors is mediated by several discrete protein–protein interactions. *Neuron, 31*, 103–113.

Ball, S. M., Atlason, P. T., Shittu-Balogun, O. O., & Molnar, E. (2010). Assembly and intracellular distribution of kainate receptors is determined by RNA editing and subunit composition. *Journal of Neurochemistry, 114*(6), 1805–1818. doi:JNC6895 [pii] 10.1111/j.1471-4159.2010.06895.x.

Bellone, C., & Luscher, C. (2006). Cocaine triggered AMPA receptor redistribution is reversed in vivo by mGluR-dependent long-term depression. *Nature Neuroscience, 9*(5), 636–641. doi: nn1682 [pii] 10.1038/nn1682.

Billa, S. K., Liu, J., Bjorklund, N. L., Sinha, N., Fu, Y., Shinnick-Gallagher, P., & Moron, J. A. (2010). Increased insertion of glutamate receptor 2-lacking alpha-amino-3-hydroxy-5-methyl-4-isoxazole propionic acid (AMPA) receptors at hippocampal synapses upon repeated morphine administration. *Molecular Pharmacology, 77*(5), 874–883. doi:mol.109.060301 [pii] 10.1124/mol.109.060301.

Bredt, D. S., & Nicoll, R. A. (2003). AMPA receptor trafficking at excitatory synapses. *Neuron, 40*, 361–379.

Brorson, J. R., Li, D., & Suzuki, T. (2004). Selective expression of heteromeric AMPA receptors driven by flip-flop differences. *Journal of Neuroscience, 24*(14), 3461–3470.

Carlezon, W. A., Jr., & Nestler, E. J. (2002). Elevated levels of GluR1 in the midbrain: A trigger for sensitization to drugs of abuse? *Trends in Neurosciences, 25*(12), 610–615. doi: S0166223602022890 [pii].

Chen, G. Q., Cui, C., Mayer, M. L., & Gouaux, E. (1999). Functional characterization of a potassium-selective prokaryotic glutamate receptor. *Nature, 402*(6763), 817–821. doi:10.1038/45568.

Chung, H. J., Xia, J., Scannevin, R. H., Zhang, X., & Huganir, R. L. (2000). Phosphorylation of the AMPA receptor subunit GluR2 differentially regulates its interaction with PDZ domain-containing proteins. *Journal of Neuroscience, 20*(19), 7258–7267.

Clayton, A., Siebold, C., Gilbert, R. J., Sutton, G. C., Harlos, K., McIlhinney, R. A., Jones, E. Y., & Aricescu, A. R. (2009). Crystal structure of the GluR2 amino-terminal domain provides insights into the architecture and assembly of ionotropic glutamate receptors. *Journal of Molecular Biology, 392*(5), 1125–1132. doi:S0022-2836(09)00973-5 [pii] 10.1016/j.jmb.2009.07.082.

Clem, R. L., & Barth, A. (2006). Pathway-specific trafficking of native AMPARs by in vivo experience. *Neuron, 49*(5), 663–670. doi:S0896-6273(06)00072-9 [pii] 10.1016/j.neuron.2006.01.019.

Clem, R. L., & Huganir, R. L. (2010). Calcium-permeable AMPA receptor dynamics mediate fear memory erasure. *Science, 330*(6007), 1108–1112. doi:science.1195298 [pii] 10.1126/science.1195298.

Coleman, S. K., Cai, C., Mottershead, D. G., Haapalahti, J. P., & Keinanen, K. (2003). Surface expression of GluR-D AMPA receptor is dependent on an interaction between its C-terminal domain and a 4.1 Protein. *Journal of Neuroscience, 23*(3), 798–806. doi:23/3/798 [pii].

Coleman, S. K., Möykkynen, T., Cai, C., von Ossowski, L., Kuismanen, E., Korpi, E. R., & Keinänen, K. (2006). Isoform-specific early trafficking of AMPA receptor flip and flop variants. *Journal of Neuroscience, 26*, 11220–11229.

Coleman, S. K., Moykkynen, T., Hinkkuri, S., Vaahtera, L., Korpi, E. R., Pentikainen, O. T., & Keinanen, K. (2010). Ligand-binding domain determines endoplasmic reticulum exit of AMPA receptors. *Journal of Biological Chemistry, 285*(46), 36032–36039. doi:M110.156943 [pii] 10.1074/jbc.M110.156943.

Conrad, K. L., Tseng, K. Y., Uejima, J. L., Reimers, J. M., Heng, L. J., Shaham, Y., Marinelli, M., & Wolf, M. E. (2008). Formation of accumbens GluR2-lacking AMPA receptors mediates incubation of cocaine craving. *Nature, 454*(7200), 118–121. doi:nature06995 [pii] 10.1038/nature06995.

Cull-Candy, S., Kelly, L., & Farrant, M. (2006). Regulation of Ca^{2+}-permeable AMPA receptors: Synaptic plasticity and beyond. *Current Opinion in Neurobiology, 16*, 288–297.

Das, U., Kumar, J., Mayer, M. L., & Plested, A. J. (2010). Domain organization and function in GluK2 subtype kainate receptors. *Proceedings of the National Academy of Sciences of the United States of America, 107*(18), 8463–8468. doi:1000838107 [pii] 10.1073/pnas.1000838107.

Gardner, S. M., Trussell, L. O., & Oertel, D. (2001). Correlation of AMPA receptor subunit composition with synaptic input in the mammalian cochlear nuclei. *Journal of Neuroscience, 21*(18), 7428–7437. doi:21/18/7428 [pii].

Goel, A., Jiang, B., Xu, L. W., Song, L., Kirkwood, A., & Lee, H. K. (2006). Cross-modal regulation of synaptic AMPA receptors in primary sensory cortices by visual experience. *Nature Neuroscience, 9*(8), 1001–1003. doi:nn1725 [pii] 10.1038/nn1725.

Gomez-Posada, J. C., Etxeberria, A., Roura-Ferrer, M., Areso, P., Masin, M., Murrell-Lagnado, R. D., & Villarroel, A. (2010). A pore residue of the KCNQ3 potassium M-channel subunit controls surface expression. *Journal of Neuroscience, 30*(27), 9316–9323. doi:30/27/9316 [pii] 10.1523/JNEUROSCI.0851-10.2010.

Gonzalez, J., Du, M., Parameshwaran, K., Suppiramaniam, V., & Jayaraman, V. (2010). Role of dimer interface in activation and desensitization in AMPA receptors. *Proceedings of the National Academy of Sciences of the United States of America, 107*(21), 9891–9896. doi:0911854107 [pii] 10.1073/pnas.0911854107.

Good, C. H., & Lupica, C. R. (2010). Afferent-specific AMPA receptor subunit composition and regulation of synaptic plasticity in midbrain dopamine neurons by abused drugs. *Journal of Neuroscience, 30*(23), 7900–7909. doi:30/23/7900 [pii] 10.1523/JNEUROSCI.1507-10.2010.

Gray, E. E., Fink, A. E., Sarinana, J., Vissel, B., & O'Dell, T. J. (2007). Long-term potentiation in the hippocampal CA1 region does not require insertion and activation of GluR2-lacking AMPA receptors. *Journal of Neurophysiology, 98*(4), 2488–2492. doi:00473.2007 [pii] 10.1152/jn.00473.2007.

Greger, I. H., Akamine, P., Khatri, L., & Ziff, E. B. (2006). Developmentally regulated, combinatorial RNA processing modulates AMPA receptor biogenesis. *Neuron, 51*, 85–97.

Greger, I. H., Khatri, L., Kong, X., & Ziff, E. B. (2003). AMPA receptor tetramerization is mediated by Q/R editing. *Neuron, 40*, 763–774.

Greger, I. H., Khatri, L., & Ziff, E. B. (2002). RNA editing at Arg607 controls AMPA receptor exit from the endoplasmic reticulum. *Neuron, 34*, 759–772.

Greger, I. H., Ziff, E. B., & Penn, A. C. (2007). Molecular determinants of AMPA receptor subunit assembly. *Trends in Neurosciences, 30*, 407–416.

Guzman, S. J., & Jonas, P. (2010). Beyond TARPs: The growing list of auxiliary AMPAR subunits. *Neuron, 66*(1), 8–10. doi:S0896-6273(10)00243-6 [pii] 10.1016/j.neuron.2010.04.003.

Hansen, K. B., Furukawa, H., & Traynelis, S. F. (2010). Control of assembly and function of glutamate receptors by the amino-terminal domain. *Molecular Pharmacology, 78*(4), 535–549. doi:mol.110.067157 [pii] 10.1124/mol.110.067157.

Harms, K. J., Tovar, K. R., & Craig, A. M. (2005). Synapse-specific regulation of AMPA receptor subunit composition by activity. *Journal of Neuroscience, 25*(27), 6379–6388. doi:25/27/6379 [pii] 10.1523/JNEUROSCI.0302-05.2005.

Henley, J. M. (1994). Kainate-binding proteins: Phylogeny, structures and possible functions. *Trends in Pharmacological Sciences, 15*(6), 182–190.

Hideyama, T., Yamashita, T., Suzuki, T., Tsuji, S., Higuchi, M., Seeburg, P. H., Takahashi, R., Misawa, H., & Kwak, S. (2010). Induced loss of ADAR2 engenders slow death of motor neurons from Q/R site-unedited GluR2. *Journal of Neuroscience, 30*(36), 11917–11925. doi:30/36/11917 [pii] 10.1523/JNEUROSCI.2021-10.2010.

Ho, M. T., Pelkey, K. A., Topolnik, L., Petralia, R. S., Takamiya, K., Xia, J., Huganir, R. L., Lacaille, J. C., & McBain, C. J. (2007). Developmental expression of Ca^{2+}−permeable AMPA receptors underlies depolarization-induced long-term depression at mossy fiber CA3 pyramid synapses. *Journal of Neuroscience, 27*(43), 11651–11662. doi:27/43/11651 [pii] 10.1523/JNEUROSCI.2671-07.2007.

Hollmann, M., & Heinemann, S. (1994) Cloned glutamate receptors. *Annu. Rev. Neurosci. 17*, 31–108.

Huang, Y., Man, H. Y., Sekine-Aizawa, Y., Han, Y., Juluri, K., Luo, H., Cheah, J., Lowenstein, C., Huganir, R. L., & Snyder, S. H. (2005). S-nitrosylation of N-ethylmaleimide sensitive factor mediates surface expression of AMPA receptors. *Neuron, 46*(4), 533–540. doi:S0896-6273(05) 00313-2 [pii] 10.1016/j.neuron.2005.03.028.

Isaac, J. T., Ashby, M., & McBain, C. J. (2007). The role of the GluR2 subunit in AMPA receptor function and synaptic plasticity. *Neuron, 54*, 859–871.

Jensen, V., Kaiser, K. M., Borchardt, T., Adelmann, G., Rozov, A., Burnashev, N., Brix, C., Frotscher, M., Andersen, P., Hvalby, O., Sakmann, B., Seeburg, P. H., & Sprengel, R. (2003). A juvenile form of postsynaptic hippocampal long-term potentiation in mice deficient for the AMPA receptor subunit GluR-a. *The Journal of Physiology, 553*(Pt 3), 843–856. doi:10.1113/ jphysiol.2003.053637 jphysiol.2003.053637 [pii].

Jensen, M. H., Sukumaran, M., Johnson, C. M., Greger, I. H., & Neuweiler, H. (2011) Intrinsic motions in the N-terminal domain of an ionotropic glutamate receptor detected by fluorescence correlation spectroscopy. *J. Mol. Biol. 414*, 96–105.

Jia, Z., Agopyan, N., Miu, P., Xiong, Z., Henderson, J., Gerlai, R., Taverna, F. A., Velumian, A., MacDonald, J., Carlen, P., Abramow-Newerly, W., & Roder, J. (1996). Enhanced LTP in mice deficient in the AMPA receptor GluR2. *Neuron, 17*(5), 945–956. doi:S0896-6273(00)80225-1 [pii].

Jin, R., Clark, S., Weeks, A. M., Dudman, J. T., Gouaux, E., & Partin, K. M. (2005). Mechanism of positive allosteric modulators acting on AMPA receptors. *Journal of Neuroscience, 25*(39), 9027–9036. doi:25/39/9027 [pii] 10.1523/JNEUROSCI.2567-05.2005.

Jin, R., Singh, S. K., Gu, S., Furukawa, H., Sobolevsky, A. I., Zhou, J., Jin, Y., & Gouaux, E. (2009). Crystal structure and association behaviour of the GluR2 amino-terminal domain. *EMBO Journal, 28*(12), 1812–1823. doi:emboj2009140 [pii] 10.1038/emboj.2009.140.

Jorntell, H., Bengtsson, F., Schonewille, M., & De Zeeuw, C. I. (2010). Cerebellar molecular layer interneurons – Computational properties and roles in learning. *Trends in Neurosciences, 33* (11), 524–532. doi:S0166-2236(10)00119-0 [pii] 10.1016/j.tins.2010.08.004.

Karakas, E., Simorowski, N., & Furukawa, H. (2009). Structure of the zinc-bound amino-terminal domain of the NMDA receptor NR2B subunit. *EMBO Journal*. doi:emboj2009338 [pii] 10.1038/emboj.2009.338.

Kato, A. S., Siuda, E. R., Nisenbaum, E. S., & Bredt, D. S. (2008). AMPA receptor subunit-specific regulation by a distinct family of type II TARPs. *Neuron, 59*(6), 986–996. doi:S0896-6273(08)00633-8 [pii] 10.1016/j.neuron.2008.07.034.

Kauer, J. A., & Malenka, R. C. (2007). Synaptic plasticity and addiction. *Nature Reviews Neuroscience, 8*(11), 844–858. doi:nrn2234 [pii] 10.1038/nrn2234.

Kielland, A., Bochorishvili, G., Corson, J., Zhang, L., Rosin, D. L., Heggelund, P., & Zhu, J. J. (2009). Activity patterns govern synapse-specific AMPA receptor trafficking between deliverable and synaptic pools. *Neuron, 62*(1), 84–101. doi:S0896-6273(09)00198-6 [pii] 10.1016/j. neuron.2009.03.001.

Kim, K. S., Yan, D., & Tomita, S. (2010). Assembly and stoichiometry of the AMPA receptor and transmembrane AMPA receptor regulatory protein complex. *Journal of Neuroscience, 30*(3), 1064–1072. doi:30/3/1064 [pii] 10.1523/JNEUROSCI.3909-09.2010.

Kullmann, D. M., & Lamsa, K. (2008). Roles of distinct glutamate receptors in induction of anti-hebbian long-term potentiation. *The Journal of Physiology, 586*(6), 1481–1486. doi: jphysiol.2007.148064 [pii] 10.1113/jphysiol.2007.148064.

Kumar, S. S., Bacci, A., Kharazia, V., & Huguenard, J. R. (2002). A developmental switch of AMPA receptor subunits in neocortical pyramidal neurons. *Journal of Neuroscience, 22*(8), 3005–3015. doi:20026285 22/8/3005 [pii].

Kumar, J., & Mayer, M. L. (2010). Crystal structures of the glutamate receptor ion channel GluK3 and GluK5 amino-terminal domains. *Journal of Molecular Biology, 404*(4), 680–696. doi: S0022-2836(10)01103-4 [pii] 10.1016/j.jmb.2010.10.006.

Kumar, J., Schuck, P., Jin, R., & Mayer, M. L. (2009). The N-terminal domain of GluR6-subtype glutamate receptor ion channels. *Nature Structural & Molecular Biology, 16*(6), 631–638. doi: nsmb.1613 [pii] 10.1038/nsmb.1613.

Kunishima, N., Shimada, Y., Tsuji, Y., Sato, T., Yamamoto, M., Kumasaka, T., Nakanishi, S., Jingami, H., & Morikawa, K. (2000). Structural basis of glutamate recognition by a dimeric metabotropic glutamate receptor. *Nature, 407*(6807), 971–977. doi:10.1038/35039564.

Leuschner, W. D., & Hoch, W. (1999). Subtype-specific assembly of alpha-amino-3-hydroxy-5-methyl-4-isoxazole propionic acid receptor subunits is mediated by their n-terminal domains. *Journal of Biological Chemistry, 274*(24), 16907–16916.

Lin, D. T., Makino, Y., Sharma, K., Hayashi, T., Neve, R., Takamiya, K., & Huganir, R. L. (2009). Regulation of AMPA receptor extrasynaptic insertion by 4.1N, phosphorylation and palmitoylation. *Nature Neuroscience, 12*(7), 879–887. doi:nn.2351 [pii] 10.1038/nn.2351.

Liu, S. Q., & Cull-Candy, S. G. (2000). Synaptic activity at calcium-permeable AMPA receptors induces a switch in receptor subtype. *Nature, 405*(6785), 454–458. doi:10.1038/35013064.

Liu, Y., Formisano, L., Savtchouk, I., Takayasu, Y., Szabo, G., Zukin, R. S., & Liu, S. J. (2010). A single fear-inducing stimulus induces a transcription-dependent switch in synaptic AMPAR phenotype. *Nature Neuroscience, 13*(2), 223–231. doi:nn.2474 [pii] 10.1038/nn.2474.

Liu, S. J., & Zukin, R. S. (2007). Ca^{2+}−permeable AMPA receptors in synaptic plasticity and neuronal death. *Trends in Neurosciences, 30*, 126–134.

Lomeli, H., Mosbacher, J., Melcher, T., Hoger, T., Geiger, J. R., Kuner, T., Monyer, H., Higuchi, M., Bach, A., & Seeburg, P. H. (1994). Control of kinetic properties of AMPA receptor channels by nuclear RNA editing. *Science, 266*, 1709–1713.

Long, S. B., Campbell, E. B., & Mackinnon, R. (2005). Crystal structure of a mammalian voltage-dependent shaker family K+ channel. *Science, 309*(5736), 897–903. doi:1116269 [pii] 10.1126/science.1116269.

Luthi, A., Chittajallu, R., Duprat, F., Palmer, M. J., Benke, T. A., Kidd, F. L., Henley, J. M., Isaac, J. T., & Collingridge, G. L. (1999). Hippocampal LTD expression involves a pool of AMPARs regulated by the NSF-GluR2 interaction. *Neuron, 24*(2), 389–399. doi:S0896-6273(00)80852-1 [pii].

MacGregor, I. K., Anderson, A. L., & Laue, T. M. (2004). Fluorescence detection for the XLI analytical ultracentrifuge. *Biophysical Chemistry, 108*(1–3), 165–185. doi:10.1016/j. bpc.2003.10.018 S0301462203003041 [pii].

Mameli, M., Halbout, B., Creton, C., Engblom, D., Parkitna, J. R., Spanagel, R., & Luscher, C. (2009). Cocaine-evoked synaptic plasticity: Persistence in the VTA triggers adaptations in the NAc. *Nature Neuroscience, 12*(8), 1036–1041. doi:nn.2367 [pii] 10.1038/nn.2367.

Mansour, M., Nagarajan, N., Nehring, R. B., Clements, J. D., & Rosenmund, C. (2001). Heteromeric AMPA receptors assemble with a preferred subunit stoichiometry and spatial arrangement. *Neuron, 32*(5), 841–853. doi:S0896-6273(01)00520-7 [pii].

Mayer, M. L., & Armstrong, N. (2004). Structure and function of glutamate receptor ion channels. *Annual Review of Physiology, 66*, 161–181.

Midgett, C. R., & Madden, D. R. (2008). The quaternary structure of a calcium-permeable AMPA receptor: Conservation of shape and symmetry across functionally distinct subunit assemblies. *Journal of Molecular Biology, 382*(3), 578–584. doi:S0022-2836(08)00863-2 [pii] 10.1016/j.jmb.2008.07.021.

Milstein, A. D., & Nicoll, R. A. (2008). Regulation of AMPA receptor gating and pharmacology by TARP auxiliary subunits. *Trends in Pharmacological Sciences, 29*(7), 333–339. doi:S0165-6147(08)00109-0 [pii] 10.1016/j.tips.2008.04.004.

Mony, L., Kew, J. N., Gunthorpe, M. J., & Paoletti, P. (2009). Allosteric modulators of NR2B-containing NMDA receptors: Molecular mechanisms and therapeutic potential. *British Journal of Pharmacology, 157*(8), 1301–1317. doi:BPH304 [pii] 10.1111/j.1476-5381.2009.00304.x.

Nakagawa, T. (2010). The biochemistry, ultrastructure, and subunit assembly mechanism of AMPA receptors. *Molecular Neurobiology, 42*(3), 161–184. doi:10.1007/s12035-010-8149-x.

Nakagawa, T., Cheng, Y., Ramm, E., Sheng, M., & Walz, T. (2005). Structure and different conformational states of native AMPA receptor complexes. *Nature, 433*(7025), 545–549. doi:nature03328 [pii] 10.1038/nature03328.

Netzer, W. J., & Hartl, F. U. (1997). Recombination of protein domains facilitated by co-translational folding in eukaryotes. *Nature, 388*(6640), 343–349. doi:10.1038/41024.

Noh, K. M., Yokota, H., Mashiko, T., Castillo, P. E., Zukin, R. S., & Bennett, M. V. (2005). Blockade of calcium-permeable AMPA receptors protects hippocampal neurons against global ischemia-induced death. *Proceedings of the National Academy of Sciences of the United States of America, 102*(34), 12230–12235. doi:0505408102 [pii] 10.1073/pnas.0505408102.

O'Brien, R. J., Xu, D., Petralia, R. S., Steward, O., Huganir, R. L., & Worley, P. (1999) Synaptic clustering of AMPA receptors by the extracellular immediate-early gene product Narp. *Neuron 23*, 309–323.

O'Brien, R., Xu, D., Mi, R., Tang, X., Hopf, C., & Worley, P. (2002). Synaptically targeted narp plays an essential role in the aggregation of AMPA receptors at excitatory synapses in cultured spinal neurons. *Journal of Neuroscience, 22*(11), 4487–4498. doi:20026354 22/11/4487 [pii].

Osswald, I. K., Galan, A., & Bowie, D. (2007). Light triggers expression of philanthotoxin-insensitive Ca^{2+}–permeable AMPA receptors in the developing rat retina. *The Journal of Physiology, 582* (Pt 1), 95–111. doi:jphysiol.2007.127894 [pii] 10.1113/jphysiol.2007.127894.

Passafaro, M., Nakagawa, T., Sala, C., & Sheng, M. (2003). Induction of dendritic spines by an extracellular domain of AMPA receptor subunit GluR2. *Nature, 424*(6949), 677–681. doi:10.1038/nature01781 nature01781 [pii].

Peng, P. L., Zhong, X., Tu, W., Soundarapandian, M. M., Molner, P., Zhu, D., Lau, L., Liu, S., Liu, F., & Lu, Y. (2006). ADAR2-dependent RNA editing of AMPA receptor subunit GluR2 determines vulnerability of neurons in forebrain ischemia. *Neuron, 49*(5), 719–733. doi:S0896-6273(06)00079-1 [pii] 10.1016/j.neuron.2006.01.025.

Penn, A. C., & Greger, I. H. (2009). Sculpting AMPA receptor formation and function by alternative RNA processing. *RNA Biology, 6*(5), 517–522. doi:9552 [pii].

Penn, A. C., Williams, S. R., & Greger, I. H. (2008). Gating motions underlie AMPA receptor secretion from the endoplasmic reticulum. *EMBO Journal, 27*(22), 3056–3068. doi:emboj2008222 [pii] 10.1038/emboj.2008.222.

Plant, K., Pelkey, K. A., Bortolotto, Z. A., Morita, D., Terashima, A., McBain, C. J., Collingridge, G. L., & Isaac, J. T. (2006). Transient incorporation of native GluR2-lacking AMPA receptors during hippocampal long-term potentiation. *Nature Neuroscience, 9*(5), 602–604. doi:nn1678 [pii] 10.1038/nn1678.

Qin, G., Schwarz, T., Kittel, R. J., Schmid, A., Rasse, T. M., Kappei, D., Ponimaskin, E., Heckmann, M., & Sigrist, S. J. (2005). Four different subunits are essential for expressing the synaptic glutamate receptor at neuromuscular junctions of Drosophila. *Journal of Neuroscience, 25*(12), 3209–3218. doi:25/12/3209 [pii] 10.1523/JNEUROSCI.4194-04.2005.

Quiocho, F. A., & Ledvina, P. S. (1996). Atomic structure and specificity of bacterial periplasmic receptors for active transport and chemotaxis: Variation of common themes. *Molecular Microbiology, 20*, 17–25.

Rasse, T. M., Fouquet, W., Schmid, A., Kittel, R. J., Mertel, S., Sigrist, C. B., Schmidt, M., Guzman, A., Merino, C., Qin, G., Quentin, C., Madeo, F. F., Heckmann, M., & Sigrist, S. J. (2005). Glutamate receptor dynamics organizing synapse formation in vivo. *Nature Neuroscience, 8*(7), 898–905.

Rossmann, M., Sukumaran, M., Penn, A. C., Veprintsev, D. B., Babu, M. M., & Greger, I. H. (2011). Subunit-selective N-terminal domain associations organize the formation of AMPA receptor heteromers. *EMBO Journal.* doi:emboj2011.16.

Rozov, A., & Burnashev, N. (1999). Polyamine-dependent facilitation of postsynaptic AMPA receptors counteracts paired-pulse depression. *Nature, 401*(6753), 594–598. doi:10.1038/44151.

Rozov, A., Zilberter, Y., Wollmuth, L. P., & Burnashev, N. (1998). Facilitation of currents through rat Ca^{2+}–permeable AMPA receptor channels by activity-dependent relief from polyamine block. *The Journal of Physiology, 511*(Pt 2), 361–377.

Rumbaugh, G., & Vicini, S. (1999). Distinct synaptic and extrasynaptic NMDA receptors in developing cerebellar granule neurons. *Journal of Neuroscience, 19*(24), 10603–10610.

Saglietti, L., Dequidt, C., Kamieniarz, K., Rousset, M. C., Valnegri, P., Thoumine, O., Beretta, F., Fagni, L., Choquet, D., Sala, C., Sheng, M., & Passafaro, M. (2007). Extracellular interactions between GluR2 and N-cadherin in spine regulation. *Neuron, 54*(3), 461–477. doi:S0896-6273 (07)00291-7 [pii] 10.1016/j.neuron.2007.04.012.

Sans, N., Racca, C., Petralia, R. S., Wang, Y. X., McCallum, J., & Wenthold, R. J. (2001). Synapse-associated protein 97 selectively associates with a subset of AMPA receptors early in their biosynthetic pathway. *Journal of Neuroscience, 21*(19), 7506–7516. doi:21/19/7506 [pii].

Schmauss, C. (2005). Regulation of serotonin 2C receptor pre-mRNA editing by serotonin. *International Review of Neurobiology, 63*, 83–100. doi:S00747742205630048 [pii] 10.1016/S0074-7742(05)63004-8.

Seeburg, P. H. (1996). The role of RNA editing in controlling glutamate receptor channel properties. *Journal of Neurochemistry, 66*, 1–5.

Shanks, N. F., Maruo, T., Farina, A. N., Ellisman, M. H., & Nakagawa, T. (2010). Contribution of the global subunit structure and stargazin on the maturation of AMPA receptors. *Journal of Neuroscience, 30*(7), 2728–2740. doi:30/7/2728 [pii] 10.1523/JNEUROSCI.5146-09.2010.

Shen, L., Liang, F., Walensky, L. D., & Huganir, R. L. (2000). Regulation of AMPA receptor GluR1 subunit surface expression by a 4. 1N-linked actin cytoskeletal association. *Journal of Neuroscience, 20*(21), 7932–7940. doi:20/21/7932 [pii].

Sheng, M., & Hoogenraad, C. C. (2007). The postsynaptic architecture of excitatory synapses: A more quantitative view. *Annual Review of Biochemistry, 76*, 823–847. doi:10.1146/annurev. biochem.76.060805.160029.

Shepherd, J.D., Huganir, R.L. (2007). The cell biology of synaptic plasticity: AMPA receptor trafficking. *Annu. Rev. Cell. Dev. Biol. 23*, 613–643.

Shi, Y., Lu, W., Milstein, A. D., & Nicoll, R. A. (2009). The stoichiometry of AMPA receptors and TARPs varies by neuronal cell type. *Neuron, 62*(5), 633–640. doi:S0896-6273(09)00391-2 [pii] 10.1016/j.neuron.2009.05.016.

Sia, G. M., Beique, J. C., Rumbaugh, G., Cho, R., Worley, P. F., & Huganir, R. L. (2007). Interaction of the N-terminal domain of the AMPA receptor GluR4 subunit with the neuronal pentraxin NP1 mediates GluR4 synaptic recruitment. *Neuron, 55*(1), 87–102. doi:S0896-6273 (07)00453-9 [pii] 10.1016/j.neuron.2007.06.020.

Sobolevsky, A. I., Rosconi, M. P., & Gouaux, E. (2009). X-ray structure, symmetry and mechanism of an AMPA-subtype glutamate receptor. *Nature, 462*, 745–756. doi:nature08624 [pii] 10.1038/nature08624.

Soler-Llavina, G. J., & Sabatini, B. L. (2006). Synapse-specific plasticity and compartmentalized signaling in cerebellar stellate cells. *Nature Neuroscience, 9*(6), 798–806. doi:nn1698 [pii] 10.1038/nn1698.

Sommer, B., Keinanen, K., Verdoorn, T. A., Wisden, W., Burnashev, N., Herb, A., Kohler, M., Takagi, T., Sakmann, B., & Seeburg, P. H. (1990). Flip and flop: A cell-specific functional switch in glutamate-operated channels of the CNS. *Science, 249*, 1580–1585.

Soto, D., Coombs, I. D., Renzi, M., Zonouzi, M., Farrant, M., & Cull-Candy, S. G. (2009). Selective regulation of long-form calcium-permeable AMPA receptors by an atypical TARP, gamma-5. *Nature Neuroscience, 12*(3), 277–285. doi:nn.2266 [pii] 10.1038/nn.2266.

Staehlin, T., Wettstein, F. O., Oura, H., & Noll, H. (1964). Determination of the coding ratio based on molecular weight of messenger ribonucleic acid associated with ergosomes of different aggregate size. *Nature, 201*, 264–270.

Stamm, S. (2002). Signals and their transduction pathways regulating alternative splicing: A new dimension of the human genome. *Human Molecular Genetics, 11*(20), 2409–2416.

Steigerwald, F., Schulz, T. W., Schenker, L. T., Kennedy, M. B., Seeburg, P. H., & Kohr, G. (2000). C-terminal truncation of NR2A subunits impairs synaptic but not extrasynaptic localization of NMDA receptors. *Journal of Neuroscience, 20*(12), 4573–4581. doi:20/12/4573 [pii].

Steinberg, J. P., Huganir, R. L., & Linden, D. J. (2004). N-ethylmaleimide-sensitive factor is required for the synaptic incorporation and removal of AMPA receptors during cerebellar long-term depression. *Proceedings of the National Academy of Sciences of the United States of America, 101*(52), 18212–18216. doi:0408278102 [pii] 10.1073/pnas.0408278102.

Stubblefield, E. A., & Benke, T. A. (2010). Distinct AMPA-type glutamatergic synapses in developing rat CA1 hippocampus. *Journal of Neurophysiology, 104*(4), 1899–1912. doi:jn.00099.2010 [pii] 10.1152/jn.00099.2010.

Sukumaran, M., Rossmann, M., Shrivastava, I., Dutta, A., Bahar, I., & Greger, I. H. (2011). Dynamics and allosteric potential of the AMPA receptor N-terminal domain. *EMBO Journal*. doi:emboj2011.17.

Sun, Y., Olson, R., Horning, M., Armstrong, N., Mayer, M., & Gouaux, E. (2002). Mechanism of glutamate receptor desensitization. *Nature, 417*, 245–253.

Swanson, G. T., Kamboj, S. K., & Cull-Candy, S. G. (1997). Single-channel properties of recombinant AMPA receptors depend on RNA editing, splice variation, and subunit composition. *Journal of Neuroscience, 17*(1), 58–69.

Thiagarajan, T. C., Lindskog, M., & Tsien, R. W. (2005). Adaptation to synaptic inactivity in hippocampal neurons. *Neuron, 47*(5), 725–737. doi:S0896-6273(05)00647-1 [pii] 10.1016/j.neuron.2005.06.037.

Toth, K., & McBain, C. J. (1998). Afferent-specific innervation of two distinct AMPA receptor subtypes on single hippocampal interneurons. *Nature Neuroscience, 1*(7), 572–578. doi:10.1038/2807.

Tovar, K. R., & Westbrook, G. L. (1999). The incorporation of NMDA receptors with a distinct subunit composition at nascent hippocampal synapses in vitro. *Journal of Neuroscience, 19*(10), 4180–4188.

Traynelis, S. F., Wollmuth, L. P., McBain, C. J., Menniti, F. S., Vance, K. M., Ogden, K. K., Hansen, K. B., Yuan, H., Myers, S. J., Dingledine, R., & Sibley, D. (2010). Glutamate receptor ion channels: Structure, regulation, and function. *Pharmacological Reviews, 62*(3), 405–496. doi:62/3/405 [pii] 10.1124/pr.109.002451.

Tu, L., & Deutsch, C. (1999). Evidence for dimerization of dimers in K+ channel assembly. *Biophysical Journal, 76*(4), 2004–2017. doi:S0006-3495(99)77358-3 [pii] 10.1016/S0006-3495(99)77358-3.

Wiltgen, B. J., Royle, G. A., Gray, E. E., Abdipranoto, A., Thangthaeng, N., Jacobs, N., Saab, F., Tonegawa, S., Heinemann, S. F., O'Dell, T. J., Fanselow, M. S., & Vissel, B. (2010). A role for calcium-permeable AMPA receptors in synaptic plasticity and learning. *PloS One, 5*(9). doi:10.1371/journal.pone.0012818.

Xia, J., Chung, H. J., Wihler, C., Huganir, R. L., & Linden, D. J. (2000). Cerebellar long-term depression requires PKC-regulated interactions between GluR2/3 and PDZ domain-containing proteins. *Neuron, 28*(2), 499–510. doi:S0896-6273(00)00128-8 [pii].

Xie, J. (2008). Control of alternative pre-mRNA splicing by Ca(++) signals. *Biochimica et Biophysica Acta, 1779*(8), 438–452. doi:S1874-9399(08)00032-1 [pii] 10.1016/j.bbagrm.2008.01.003.

Yamazaki, M., Ohno-Shosaku, T., Fukaya, M., Kano, M., Watanabe, M., & Sakimura, K. (2004). A novel action of stargazin as an enhancer of AMPA receptor activity. *Neuroscience Research, 50*(4), 369–374. doi:S0168-0102(04)00258-5 [pii] 10.1016/j.neures.2004.10.002.

Zhu, J. J. (2009). Activity level-dependent synapse-specific AMPA receptor trafficking regulates transmission kinetics. *Journal of Neuroscience, 29*(19), 6320–6335. doi:29/19/6320 [pii] 10.1523/JNEUROSCI.4630-08.2009.

Chapter 12
Developmental Plasticity of the Dendritic Compartment: Focus on the Cytoskeleton

Malgorzata Urbanska, Lukasz Swiech, and Jacek Jaworski

Abstract Plasticity, the ability to undergo lasting changes in response to a stimulus, is an important attribute of neurons. It allows proper development and underlies learning, memory, and the recovery of the nervous system after severe injuries. Often, an outcome of neuronal plasticity is a structural plasticity manifested as a change of neuronal morphology. In this chapter, we focus on the structural plasticity of dendritic arbors and spines during development. Dendrites receive and compute synaptic inputs from other neurons. The number of dendrites and their branching pattern are strictly correlated with the function of a particular neuron and the geometry of the connections it receives. The development of proper dendritic tree morphology depends on the interplay between genetic programming and extracellular signals. Spines are tiny actin-rich dendritic protrusions that harbor excitatory synapses. No consensus has been reached regarding how dendritic spines form, and several models of spine morphogenesis exist. Nevertheless, most researchers agree that spinogenesis is an important target for structural plasticity. In this chapter, we discuss examples of such plasticity and describe the principles and molecular mechanisms underlying this process, focusing mostly on the regulation of the cytoskeleton during dendrito- and spinogenesis.

Keywords Actin dynamics • Cytoskeleton • Dendritogenesis • Dendritic spines • Microtubules

M. Urbanska • L. Swiech • J. Jaworski (✉)
International Institute of Molecular and Cell Biology, 4 Ks. Trojdena St., 02-109 Warsaw, Poland
e-mail: jaworski@iimcb.gov.pl

M.R. Kreutz and C. Sala (eds.), *Synaptic Plasticity*,
Advances in Experimental Medicine and Biology 970,
DOI 10.1007/978-3-7091-0932-8_12, © Springer-Verlag/Wien 2012

12.1 Introduction

Plasticity, the ability to undergo lasting changes in response to a stimulus, is an important attribute of neurons. It allows proper development and underlies learning, memory, and the recovery of the nervous system after severe injuries. Improper neuronal plasticity accompanies several central nervous system disorders, including mental retardation, autism spectrum disorders, schizophrenia, epilepsy, and neurodegenerative disorders. Long-lasting adjustments of neuronal functions have been postulated and subsequently proven to require changes in cell morphology (Kasai et al. 2010a; Stuart et al. 2007; Yuste 2010). Such changes are referred to as structural plasticity. During development, neurons rapidly change their morphology, but structural plasticity subsequently becomes greatly reduced, and only specialized dendritic protrusions, referred to as spines, remain plastic throughout life. In this chapter, we focus on the structural plasticity of dendritic arbors and spines during development. We discuss examples of such plasticity and describe the principles and molecular mechanisms underlying this process, focusing mostly on the regulation of the cytoskeleton.

12.2 Structural Plasticity of Dendrites and Dendritic Spines During Development

Dendrites receive and compute synaptic inputs from other neurons (Stuart et al. 2007). The number of dendrites and their branching pattern are strictly correlated with the function of a particular neuron and the geometry of the connections it receives (Stuart et al. 2007). Imaging neurons cultured in vitro or in vivo in the brain (Dotti et al. 1988; Mizrahi 2007; Stuart et al. 2007; Wu et al. 1999) revealed that dendritic arbor development is a multistage process (Fig. 12.1) that begins shortly after axon specification. Initially, dendrites undergo extensive elongation without new branch formation. Dendrites then start to branch to effectively cover appropriate dendritic fields, a process which is correlated with the formation of synaptic contacts. How dendrites branch mechanistically is still unclear. Dendrite elongation is driven by dendritic growth cones, and some observations suggest that dendritic branching occurs by dendritic growth cone splitting (Acebes and Ferrus 2000; Portera-Cailliau et al. 2003). Dendritic growth cones of Purkinje and pyramidal neurons form filopodia that search the extracellular environment for attractive axonal partners (Acebes and Ferrus 2000; Bradley and Berry 1976; Laxson and King 1983; Portera-Cailliau et al. 2003). Growth cone filopodia that succeed and become stabilized form dendritic branches. The final shape of a dendritic arbor, however, is not achieved exclusively by the robust addition of new branches. It is rather reached through repetitive rounds of branch additions and retractions that occur in response to a variety of extracellular signals (Stuart et al. 2007), which makes this stage of development the most plastic. Both in vitro and in vivo imaging

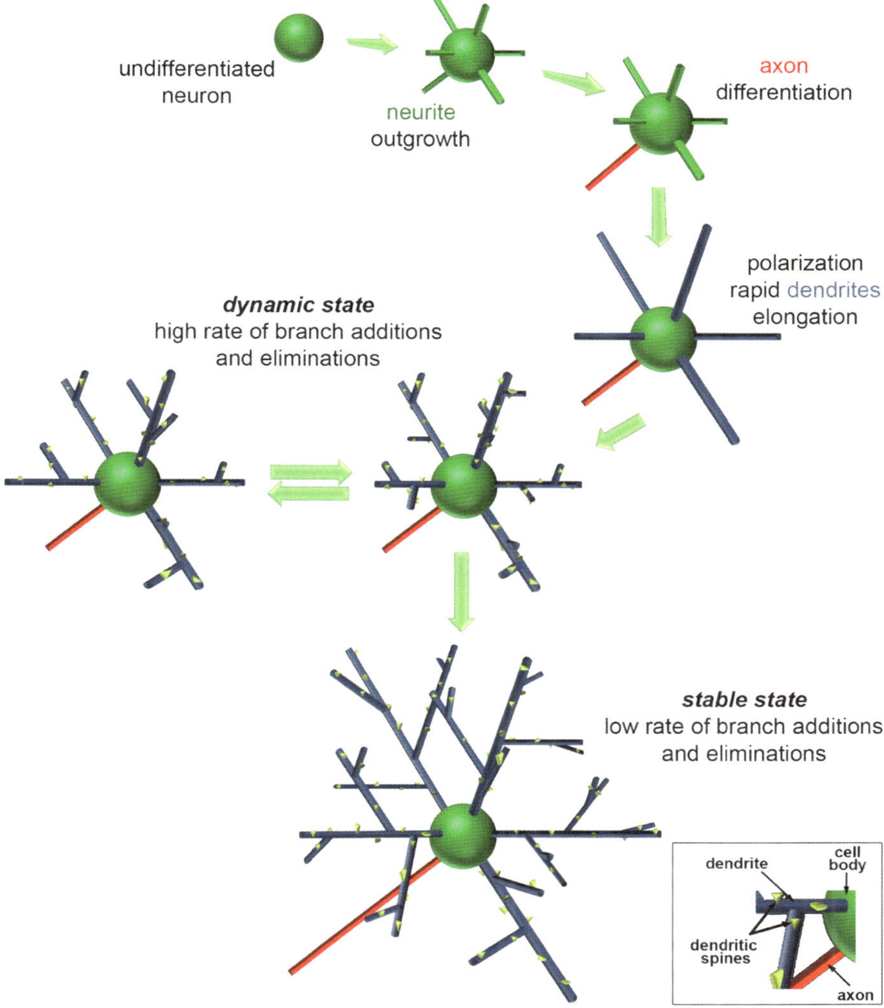

Fig. 12.1 *Dendritic arbor development is a multistage process.* Dendritogenesis starts shortly after axon specification. Dendrites first elongate. Next, they start to branch dynamically. The final shape of a dendritic arbor is achieved through repetitive rounds of branch additions and retractions that occur in response to a variety of extracellular signals. Once the final shape is reached, the dendritic tree becomes stable, and structural plasticity occurs quite rarely under basal conditions

of neurons confirm that after this period, the dendritic tree becomes stable, and structural plasticity occurs quite rarely under basal conditions (Mizrahi 2007; Mizrahi and Katz 2003; Wu et al. 1999).

Spines are tiny actin-rich dendritic protrusions that harbor excitatory synapses. Not all neurons have dendritic spines, and not all functions of spines have been fully established (Harris 1999; Yuste 2010). Nevertheless, the spine has been generally

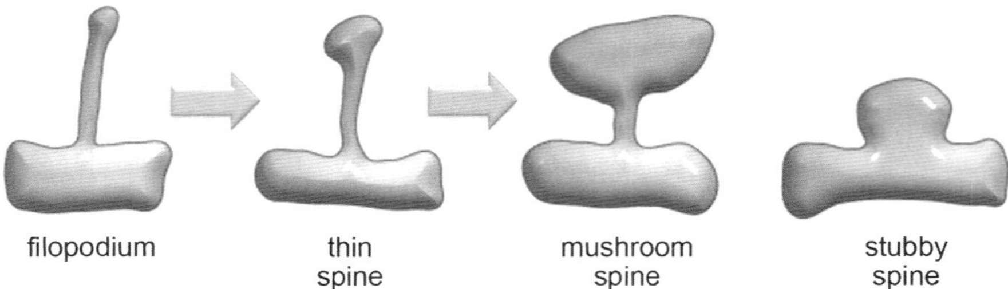

Fig. 12.2 *Dendritic spines can have different morphology.* Spines are tiny actin-rich dendritic protrusions that harbor excitatory synapses. Number of spines and their shape are hypothesized to be important for information processing. Shapes represent a continuum rather than separate classes, but spines are usually categorized as thin, stubby, or mushroom. As depicted, dendritic filopodia are considered precursors of dendritic spines. Arrows indicate direction of spine shape changes upon spine maturation

accepted to constitute a small, separate, biochemical compartment that allows clear synaptic input separation and computation (Harris 1999; Hayashi and Majewska 2005; Yuste 2010). Therefore, both the number of spines and their shape are hypothesized to be important for information processing (Harris 1999; Kasai et al. 2010a, b). Indeed, both parameters are changed in several brain diseases that are characterized by cognitive deficits (Fiala et al. 2002; van Spronsen and Hoogenraad 2010). Spines can have multiple shapes (Arellano et al. 2007), and although these shapes represent a continuum rather than separate classes, spines are usually categorized as thin, stubby, mushroom, or multishaped (Harris 1999; Hering and Sheng 2001; Yuste 2010) (Fig. 12.2). A prototypical spine has a relatively thin neck and bulbous head. Spines have been repeatedly demonstrated to change neck and head dimensions (i.e., switch categories) (Bourne and Harris 2007; Kasai et al. 2010a; Yasumatsu et al. 2008). A current view purports that the shape of a particular spine mirrors its history or plastic potential. Thin spines with long, thin necks and small-volume heads are considered relatively immature and plastic (Kasai et al. 2010a). In contrast, mushroom spines have shorter, wider necks and bigger heads, pass more current, and are considered mature and less prone to change, however, have been shown to shrink under certain conditions (Kasai et al. 2010a; Okamoto et al. 2004). In addition to dendritic spines, filopodia growing directly from a dendritic shaft have been observed (Portera-Cailliau et al. 2003; Yuste 2010) (Fig. 12.2). These protrusions, although called the same, are different from ones grown by growth cones and described above. They also differ, as we will discuss later, from identically named cytoplasmic projections of nonneuronal cells. Thus, in this chapter, we will refer to filopodia growing directly from dendritic shaft as to dendritic filopodia while those formed by growth cones or by nonneuronal cells will be called growth cone or conventional filopodia, respectively. The dendritic filopodia differ from spines because of the lack of the head, longer length, and higher motility. Dendritic filopodia often lack a presynaptic contact and are therefore not functional sites of neurotransmission.

Observations that the number and shape of spines change during development lead to several models of spinogenesis (Harris 1999; Portera-Cailliau et al. 2003; Yuste 2010; Yuste and Bonhoeffer 2004). The most popular one assumes that dendritic filopodia are precursors of spines, which, because of their length and motility, sample the neuron's environment for a contact with an axon. Once a connection is formed and stabilized by synaptic contact, the dendritic filopodium turns into a spine. This conversion involves growing the spine head and shortening the neck. The further progression of such changes results in a mushroom spine. Indeed, conversion of some dendritic filopodia into spines has been observed with two-photon microscopy (Marrs et al. 2001).

Spinogenesis is an important target for structural plasticity, and below, we discuss several extracellular signals that affect it. However, one must remember that developmental pruning of excessive spines is equally important for the refinement of neuronal networks (Yuste and Bonhoeffer 2004). Finally, long-term imaging of spines in vivo revealed that spine motility decreases with neuronal maturation, most likely because of the stabilization of successful synaptic contacts (Knott and Holtmaat 2008; Majewska and Sur 2003; Yuste 2010).

12.3 Molecular Mechanisms of Structural Plasticity of Dendrites and Dendritic Spines

The development of proper dendritic tree morphology depends on the interplay between genetic programming and extracellular signals (Jan and Jan 2010; Parrish et al. 2007; Urbanska et al. 2008). Inside the cell, these two instructions must be combined and properly executed by effector mechanisms (Jan and Jan 2010; Parrish et al. 2007; Urbanska et al. 2008). Genetic regulation of dendritic growth has been studied mostly in *Drosophila* and was reviewed extensively (Gao and Bogert 2003; Jan and Jan 2010; Parrish et al. 2007). Therefore, we will not discuss this topic in depth and focus instead on external signals that induce dendritic growth plasticity. Prior to the arrival of axons and formation of synaptic connections, dendritic arbor development is regulated by diffusible cues. Several of them have been identified, including brain-derived neurotrophic factor (BDNF), bone morphogenetic protein (BMP) family members, semaphorins, and Slits (Jan and Jan 2010; Urbanska et al. 2008). Diffusible cues have also been reported to be important for the number, distribution, and shape of dendritic spines. For example, a lack of proper semaphorin 3F signaling leads to the formation of supernumerary enlarged spines, suggesting the importance of this trophic factor for spine pruning (Tran et al. 2009). BDNF application, in contrast, leads to very diverse effects, depending on the mode of application. Acute, short-term administration of a high dose increases spine head volume, whereas a gradually increasing dose induces spine elongation (Ji et al. 2010).

Once dendrites and axons begin to contact each other, synaptic activity and physical interactions of surface proteins begin to play an important role in dendrito- and spinogenesis. Neuronal transmission can either increase or decrease dendritic arborization (McAllister 2000). A very clear demonstration of the positive effect of neuronal activity on dendrite growth comes from in vivo time-lapse imaging of developing neurons in the optic tectum of *Xenopus laevis*. In this model, visual stimulation resulted in substantial increases in dendrite growth dynamics and total dendritic length (Sin et al. 2002), which required α-amino-3-hydroxy-5-methyl-4-isoxazolepropionic acid (AMPA) and *N*-methyl-D-aspartate (NMDA) receptor-mediated glutamatergic transmission. NMDA receptors (NMDA-R) are also important for proper dendritic field coverage. NR2B knockout stellate neurons of the barrel cortex could not restrict dendrite expansion to a single barrel (Espinosa et al. 2009).

In addition to excitatory transmission, depolarization of hippocampal, cortical, and cerebellar neurons cultured in vitro in the presence of high KCl leads to increased dendritic growth (Chen et al. 2005; Gaudilliere et al. 2004; Redmond et al. 2002; Wayman et al. 2006; Wu et al. 2007; Yu and Malenka 2003). Depolarization-induced growth was inhibited by nimodipine, an L-type voltage-gated calcium channel blocker, and the effects of neuronal activity on dendritic arborization are widely accepted to be attributable to elevated cellular calcium concentrations (Konur and Ghosh 2005; Redmond and Ghosh 2005). Indeed, Lohman et al. (2002) directly confirmed the importance of calcium ions for dendritic development, showing that Ca^{2+}-induced Ca^{2+} release (CICR) locally in dendrites prevented the retraction of dendrites of chick retinal ganglion neurons.

Neuronal activity is postulated to be a crucial factor for spinogenesis and morphological spine changes in adult neurons (Holtmaat et al. 2006; Knott and Holtmaat 2008; Matsuzaki et al. 2004; Okamoto et al. 2004). However, several researchers disagree with the absolute requirement of neuronal activity for developmental spinogenesis (Segal 2001; Yuste 2010). The leading example is one in which Purkinje neuron spines make synaptic contacts with parallel fibers, and these particular spines can develop without synaptic input (Yuste and Bonhoeffer 2004). If neuronal transmission is important for spinogenesis, then knockout of neurotransmitter receptors should have dramatic impacts on synaptogenesis. In contrast to this assumption mice with conditional NR1-receptor knockout in cortical and hippocampal pyramidal neurons, although lacking NMDAR transmission, still developed spines, and clear differences in spine number were only evident after postnatal day 10 (Ultanir et al. 2007). Removal of the NR2B subunit from forebrain neurons also decreased dendritic spine density but did not prevent dendritic spine formation (Brigman et al. 2010). However, in both cases, the shape of the spines differed between mutant and control animals, suggesting that neuronal activity is needed more for maturation than for the formation of dendritic spines. The role of neuronal activity was also demonstrated for developmental spine pruning and inhibition of spine motility (Knott and Holtmaat 2008; Segal 2001; Yuste 2010).

The interaction between proteins at the cell surface is another force that shapes the dendritic arbor and drives spinogenesis. Several examples are derived from different types of neurons and species. Molecules such as Delta, Notch, ephrinB, EphB, and cadherins can either accelerate or inhibit dendrite growth, branching, and stability (Jan and Jan 2010; Urbanska et al. 2008). The contacts of dendrites and axons via cell surface proteins also play a crucial role in dendritic spine development and plasticity (Calabrese et al. 2006; Yoshihara et al. 2009). Interactions via surface proteins, such as cadherins, neurexins, ephrin receptors, syndecan-2, nectins, and SALMs, have rather permissive effects on the dendritic filopodia-spine transition and spine maturation (Calabrese et al. 2006; Yoshihara et al. 2009). Additionally, most surface proteins that induce dendritic filopodia support their transition to spines. But some of them, for example, telencephalin (TLCN), a cell adhesion molecule abundantly present in dendritic filopodia, induce their formation but prevent their conversion to spines (Furutani et al. 2007).

Extracellular factors require intracellular messengers (signal transduction proteins, e.g. kinases, phosphatases, small GTPases and transcription factors) and executors to affect the dendritic arbor and spine development (Jan and Jan 2010; Tada and Sheng 2006; Urbanska et al. 2008). The effectors identified to date are involved in cytoskeleton dynamics, membrane trafficking, macromolecule synthesis and degradation (Jan and Jan 2010; Tada and Sheng 2006; Urbanska et al. 2008). In the case of spine morphogenesis, scaffold proteins of the postsynaptic density also need to be considered (Sala et al. 2008). In the following sections, we will focus on the role of the cytoskeleton in dendrito- and spinogenesis.

12.4 The Cytoskeleton of the Synaptodendritic Compartment

Three components of the cytoskeleton are present in neurons: microfilaments, intermediate filaments, and microtubules. Although each of these play particular roles in the cell, they cooperate to ensure proper cell shape, efficient intracellular transport, and resistance to mechanical stress. Moreover, the ability of actin and microtubules to undergo very fast dynamic changes allows the movements and morphological adaptation of the dendritic compartment, which are crucial for structural neuronal plasticity. Microtubules, but not microfilaments, constitute the major cytoskeletal component of a dendrite. Their leading role in shaping the dendritic arbor was long assumed. Quite the opposite, however, dendritic spines are rich in actin, and microtubules were reported absent in this compartment. Consequently, most work focused on actin dynamics, thus overlooking the potential role of microtubules in the structural plasticity of spines. However, over the past decade, actin was proven to be crucial for dendritic growth, and dynamic microtubules were shown to invade spines (Hoogenraad and Bradke 2009; Urbanska et al. 2008) (Fig. 12.3). Thus, both microfilaments and microtubules are clearly important for the structural plasticity of the synaptodendritic compartment, regardless of which cytoskeletal element predominates as a major component.

Microtubules are composed of α- and β-tubulin heterodimers connected in a head-to-tail fashion. Microtubules are polarized structures, meaning that their ends are not equal and have different characteristics. In vivo, microtubules grow and shrink on their plus end, whereas the minus end remains relatively stable because of the protection

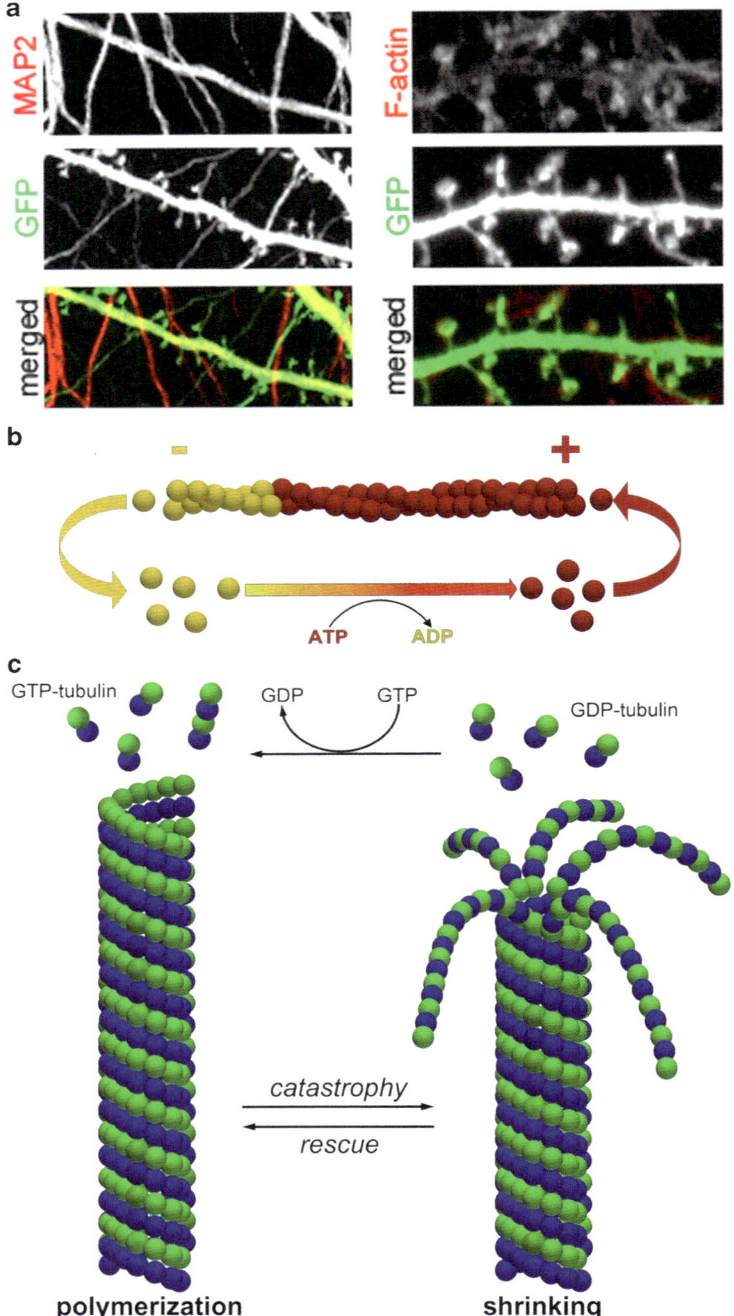

Fig. 12.3 *Spatial distribution and dynamics of dendritic compartment cytoskeleton.* (**a**) *Microfilaments and microtubules constitute major cytoskeletal components of dendritic cytoskeleton.* Stable microtubules dominate in dendritic shaft as shown by staining for MAP2 (*red*), a protein bound exclusively to microtubules (*left panel*), but are not present in dendritic spines.

provided by bound proteins (Fig. 12.3). Whether the microtubule grows, pauses, or undergoes rapid shrinkage strongly depends on the structure of the microtubule itself and activity of microtubule-binding proteins (MAPs). Several classes of MAPs have been described, and members of classical MAPs, microtubule plus-end tracking proteins (+TIPs), microtubule polymerizing and severing proteins, and tubulin regulating proteins are present in dendrites (Poulain and Sobel 2010). The presence of microtubules in spines has been controversial, but transient excursions of dynamic microtubules into spines, often reflected by the presence of +TIP protein EB3, have been recently described (Hu et al. 2008; Jaworski et al. 2009).

Microfilaments consist of polymerized actin (F-actin). Actin filaments are also polarized, indicating that they grow faster on one end (barbed end) than on the opposite end (pointed end) and resulting in the treadmilling of actin subunits from the barbed end to pointed end (Fig. 12.3). Unlike the microtubules, microfilaments, in addition to bundles, can also form branched networks. Growth dynamics and the spatial arrangement of microfilaments depend on the availability of monomeric actin and activity of a variety of actin-binding proteins. Several of these proteins have been detected in dendrites, similar to actin enriched in growth cones, filopodia, and dendritic spines (Hotulainen and Hoogenraad 2010; Schubert and Dotti 2007; Sekino et al. 2007; Zhang and Benson 2000). This includes actin-nucleating, depolymerizing, capping, and cross-linking proteins and their regulators, including members of the Rho GTPase family—RhoA, Rac1, and Cdc42 (Hotulainen and Hoogenraad 2010; Schubert and Dotti 2007; Sekino et al. 2007; Zhang and Benson 2000).

12.5 Role of the Cytoskeleton in Dendritic Arbor Morphogenesis

In different organisms, proper dendritic arbor morphology depends on MAPs (for review, see Conde and Caceres 2009; Georges et al. 2008; Poulain and Sobel 2010; Urbanska et al. 2008). Recently, +TIPs have begun to attract attention because they have been shown to control the regulation of intracellular transport and cross-talk between microtubules and the actin cytoskeleton, in addition to microtubule dynamics (Lansbergen and Akhmanova 2006; Siegrist and Doe 2007). Numerous reports describe the contribution of +TIPs to neuronal polarization and axonal growth (Grabham et al. 2007; Jimenez-Mateos et al. 2005; Lee et al. 2004;

Fig. 12.3 (continued) On the other hand, F-actin (*red*) is enriched in dendritic protrusions (filopodia and spines) (*right panel*). (**b**) *Microfilament dynamics.* Microfilaments consist of polymerized actin and are polarized, indicating that they grow faster on one end (barbed end) than on the opposite end (pointed end) and resulting in the treadmilling of actin subunits from the barbed end to pointed end. Treadmilling is an important mechanism responsible for microfilament dynamics in vivo. (**c**) *Dynamic instability of microtubules.* This dynamic instability is regulated by availability and posttranslational modifications of tubulin and is major cause of microtubule dynamics

Zhou et al. 2004), but very little is known about +TIP function in developing dendrites. The primary evidence of their importance derives from studies in *Drosophila*. At least three proteins, DLIS-1, Dhc64, and shortstop (also known as kakapo), which are respective homologs of Lis1, dynein heavy chain, and ACF7, were shown to regulate dendritic arborization in fruit flies (Gao et al. 1999; Liu et al. 2000; Prokop et al. 1998; Satoh et al. 2008). Recently, Kapitein et al. (2010) confirmed the importance of dynein for the dendrite morphology of mammalian cells. Overexpression of p50, which is known to block dynein function, resulted in simplification of the dendritic tree. Still unclear, however, is how +TIPs contribute to dendritic growth. Satoh et al. (2008) postulated that the dynein-dependent transport of "branching machinery" from the cell soma to distal dendrite regions is crucial for the proper patterning of *Drosophila dendritic arborization* (da) neurons. +TIPs can also contribute to dendrite growth in other ways. All three +TIPs—Lis1, dynein, and ACF7—can directly or indirectly link microtubules to the actin cytoskeleton (Kodama et al. 2003; Lansbergen and Akhmanova 2006), which is another important factor for dendritic patterning.

How do extracellular signals engage MAPs in plastic changes that occur during dendritic growth? MAP2, a microtubule cross-bridging protein, is likely the best-studied example. MAP2 is a substrate for different kinases, and its phosphorylation has pronounced effects on dendritic arborization (Bjorkblom et al. 2005; Podkowa et al. 2010; Terabayashi et al. 2007). For example, MAP2 phosphorylation by c-*jun* *N*-terminal kinase 1 (JNK1) contributes positively to dendrite elongation (Bjorkblom et al. 2005). The lack of JNK1 in cerebellar granular neurons results in MAP2 dephosphorylation, dendrite shortening, and increased dendritic branching. JNK1 activity can be induced by factors known to control dendritic branching (e.g., BMP7) (Podkowa et al. 2010).

Recent work in our laboratory also indicated that phosphorylation is an important regulator of microtubule dynamics during dendritic growth. mTOR is a known regulator of protein synthesis needed for BDNF-PI3K-Akt-induced dendritic development (Jaworski et al. 2005; Kumar et al. 2005). mTOR, however, is also known as CLIP-170 kinase in nonneuronal cells (Choi et al. 2002). CLIP-170 is a +TIP involved in the regulation of microtubule dynamics and microtubule minus-end-directed transport. We have shown that this protein is necessary for PI3K-induced dendritic arborization (Swiech et al. 2011). Whether the phosphorylation of other +TIPs might be important for dendritic growth needs to be established, but studies focusing on axonal growth cone navigation support this hypothesis (Jaworski et al. 2008).

In addition to activity regulation, increasing evidence suggests that the control of MAPs expression levels is important for dendritic development. For example, Chen and Firestein (2007) provided evidence that RhoA decreases the translation of cypin, a guanine deaminase that promotes microtubule assembly and increases dendrite number (Akum et al. 2004; Firestein et al. 1999). Furthermore, experiments by Wu et al. (2007) suggest that the transcription of neuronal navigator 3 (NAV3), another +TIP family member, depends on the presence of a chromatin remodeling complex that contains BAF35b, which is crucial for depolarization-induced dendritic growth (Wu et al. 2007). Although the function of NAV proteins has not been studied yet in the context of dendritic arborization, the importance of NAV2 transcription was demonstrated for neurite outgrowth in SH-SY5Y cells treated with retinoic acid (Muley et al. 2008).

Our understanding of how microfilaments are influenced by extracellular stimuli during dendritogenesis comes mostly from studies on RhoA, Rac1, and Cdc42. These proteins can act on actin dynamics and spatial microfilament arrangement via several pathways (Etienne-Manneville and Hall 2002; Jaffe and Hall 2005). The activation of Rac1 induces PAK1 and Arp2/3. The activation of PAK leads to increased actomyosin contractility and the inhibition of cofilin-dependent actin depolymerization. Arp2/3 activated by Rac1 induces the polymerization of branched F-actin networks. Active RhoA affects the cytoskeleton mostly via ROCK activation. The increased activity of RhoA and inhibition of Rac1 or Cdc42 result in significant simplification of dendritic trees in many types of neurons (Hayashi et al. 2002; Nakayama et al. 2000; Threadgill et al. 1997). In contrast, the activation of Rac1 or Cdc42 increased the number of dendritic branches (Hayashi et al. 2002; Nakayama et al. 2000; Threadgill et al. 1997). The same is true for downstream effectors of Rac1 and RhoA (Hayashi et al. 2002; Jacobs et al. 2007; Nakayama et al. 2000; Sin et al. 2002). Various extracellular signals regulate the activity of Rho GTPases during dendritic growth. Using a model of developing *Xenopus* tectal neurons, Cline and coworkers demonstrated that dendritic growth, which depends on neuronal activity, requires parallel Rac1 activation and RhoA inactivation (Li et al. 2000, 2002; Sin et al. 2002). Evidence derived from additional models clearly shows that other extracellular signals, such as BDNF, Wnt-7, and ephrinB–EphB interaction, control dendritic growth by recruiting Rho GTPases as mediators (Penzes et al. 2003; Rosso et al. 2005; Takemoto-Kimura et al. 2007). How can such diverse stimuli converge on Rho family members? A diversity of proteins that regulate the activity of Rho GTPases provides an answer to this question. Small GTPases are molecular switches that cycle between active (GTP-bound) and inactive (GDP-bound) states (Etienne-Manneville and Hall 2002). Thus, the degree of their activation is an outcome of the balance between guanine nucleotide exchange factors (GEFs) and GTPase-activating proteins (GAPs). GEFs and GAPs of Rho GTPases are recruited to act downstream of inducers of dendritogenesis in different ways. For example, the Rac 1 GEF Tiam1 interacts directly with the NR1 NMDA receptor subunit (Tolias et al. 2005), and this interaction is needed for NMDA-dependent activation of Rac1. At the same time, Tiam1 knockdown results in decreased dendritic complexity. Moreover, BDNF-induced dendritic growth requires the activity of a Rac1 GEF that is closely related to Tiam1, named SIF, and Tiam1-like exchange factor (STEF) (Takemoto-Kimura et al. 2007). However, extracellular stimuli not only regulate the activity of GEFs and GAPs of Rho GTPases but also control their expression levels. For example, Wu et al. (2007) reported that the expression of the putative Rac GEF ephexin1 decreased and RacGAP increased in neurons lacking *BAF53b*. Furthermore, KCl-induced dendritic growth was restored in *BAF53b*$^{-/-}$ neurons that overexpress ephexin1. Interestingly, local translation of Rac1 in dendrites was also shown to contribute to dendritic arborization in *Drosophila* (Lee et al. 2003).

While the role of Rho GTPases and their direct downstream effectors in dendritogenesis was the subject of very intensive studies, other actin binding proteins were not investigated so extensively in this context. Recently, however, proteins involved in actin nucleation have been studied more carefully. Wiskott-Aldrich syndrome (WASP) family protein N-WASP and its interacting protein complex Arp2/3, the activity of which is responsible for the formation of a branched filament network (Burridge and Wennerberg 2004; Etienne-Manneville and Hall 2002; Jaffe and Hall 2005), can serve as a first example. Overexpression of N-WASP increased the number of neurites and branch points in developing hippocampal neurons in vitro (Pinyol et al. 2007). Rocca et al. (2008) showed that Arp2/3 actin nucleation activity is important for proper dendritic branching patterns in developing hippocampal neurons. Factors involved in polymerization of unbranched actin filaments also appear to be important for proper dendritic arborization. Cordon-bleu (Cobl) is an actin nucleation factor enriched in the brain, and its activity is needed for dendritic growth and branching (Ahuja et al. 2007). In addition to the above examples, profilin 2a, a protein that enhances the actin treadmilling rate, has been shown to act downstream of pan-neurotrophin receptor p75 (NTR) to regulate dendritic arbor morphology (Michaelsen et al. 2010).

12.6 Actin and Microtubule Contribution to Dendritic Spine Structural Plasticity

Microfilaments are the most prominent components of dendritic filopodia and the dendritic spine cytoskeleton, and their dynamics play a role in the morphological changes of dendritic spines in developing and adult neurons. Pharmacological manipulations of the dynamics of polymerization and stability of microfilaments change the ratio of spines to filopodia in mature neurons and interfere with structural plasticity induced by synaptic transmission (Honkura et al. 2008; Jaworski et al. 2009; Okamoto et al. 2004). Pharmacological evidence to support the role of microfilament dynamics for the developmental transition of dendritic filopodia to spines is missing, but some supporting evidence exists. First, the inhibition of actin polymerization with cytochalasin D in acute slices from the developing cortex led to a substantial decrease in the number and motility and significant increase in the life of dendritic filopodia (Portera-Cailliau et al. 2003). The strongest proof, however, comes from numerous studies of the role of proteins that regulate F-actin in the morphological changes of dendritic filopodia and spines (Hotulainen and Hoogenraad 2010; Pontrello and Ethell 2009; Schubert and Dotti 2007; Sekino et al. 2007; Tada and Sheng 2006). One must remember, however, that our interpretation of such studies is deeply biased by our assumptions regarding the organization and dynamics of actin in filopodia and spines.

For many years, microfilament organization of dendritic filopodia and the spine neck was believed to resemble stiff, long bundles described for conventional filopodia of nonneuronal cells and axonal growth cones. Just recently, however, the detailed spatial arrangement and dynamics of actin filaments in dendritic filopodia and dendritic spines has been resolved, which proved that our assumptions were incorrect (Honkura et al. 2008; Hotulainen et al. 2009; Korobova and Svitkina 2010). Snapshots of the cytoskeleton of dendritic filopodia obtained by platinum replica electron microscopy clearly show striking differences between the microfilament arrangements in conventional and dendritic filopodia (Korobova and Svitkina 2010). In conventional filopodia, microfilaments are organized as unipolar bundles of parallel filaments. Dendritic filopodia contain stretched networks of branched and linear filaments of mixed polarization (Korobova and Svitkina 2010). The molecular organization also appears to be quite different. For example, Korobova and Svitkina (2010) and Hotulainen et al. (2009) reported the presence of the Arp2/3 complex in dendritic protrusion, which is not present in conventional filopodia. Finally, unlike in conventional filopodia in dendritic ones, actin polymerization occurs at their tips and bases (Hotulainen et al. 2009).

The described "stretched" network organization of the actin cytoskeleton has also been observed in the neck of dendritic spines of different shapes (Korobova and Svitkina 2010). The Arp2/3 complex and capping protein were also present in the spine neck, similar to dendritic spine filopodia (Korobova and Svitkina 2010). Microfilaments in the spine head form a meshwork of branched filaments, similar to those observed for lamellipodia (Korobova and Svitkina 2010). The major difference between F-actin organization in a spine head and the "classical" lamellipodium is the length of the individual microfilaments (Korobova and Svitkina 2010). Electron microscopy and single actin molecule tracking using photo-activation localization microscopy (PALM) suggest that filaments in spine heads are rarely longer than 200 nm (Frost et al. 2010b; Korobova and Svitkina 2010; Tatavarty et al. 2009). Based on these observations and the tight regulation of spine head actin dynamics by neuronal activity (Okamoto et al. 2004), Frost and colleagues (Frost et al. 2010a, b) proposed that a network of short, tightly controlled filaments is advantageous because it ensures the optimal spatial and temporal response of the actin cytoskeleton. Recently, several studies addressed the issue of actin dynamics within spines. Several groups revealed within spine heads the existence of separate pools of actin with different stability, most likely contributing to either spine volume stabilization or expansion (Frost et al. 2010a, b; Honkura et al. 2008).

Several molecular mechanisms have been proposed to explain spine shape changes, but most of them lacked complete information about the structure and dynamics of actin in filopodia and spines. However, Hotulainen and Hoogenraad (2010) recently proposed a new model, the most intriguing step of which is the initiation of dendritic filopodia. Hotulainen and Hoogenraad (2010) listed six potential scenarios for this event. For example, formin-dependent actin polymerization could initiate the growth of dendritic filopodia. This scenario is supported by a decrease in the number of dendritic filopodia caused by knockdown of mDia2, which is one of the formins. The next steps (i.e., the transformation of dendritic filopodia into immature dendritic spines and subsequent formation of mushroom spines) require the expansion of a spine head and its neck shrinkage. Several studies showed that Arp2/3 complex-dependent expansion of the branched microfilament network is needed for spine head enlargement (Grove et al. 2004; Hotulainen et al. 2009; Soderling et al. 2007; Wegner et al. 2008). One recent study suggested a surprising scenario for Arp2/3 activation during spine head expansion that involves dynamic microtubule entrance into spine (Hoogenraad and Bradke 2009; Jaworski et al. 2009, see also Chap. 9). As mentioned above, actin dynamics in dendritic spines is under very tight control. Therefore, the model of spine maturation proposed by Hotulainen and Hoogenraad (2010) also includes the activity of capping and severing proteins, such Eps8 and cofilin 1, respectively. Indeed, cofilin 1 is required for the proper formation of dendritic spines (Hotulainen et al. 2009). Several other actin-regulating proteins are required for proper spine morphology (Hotulainen and Hoogenraad 2010; Schubert and Dotti 2007; Sekino et al. 2007), and most of them fit nicely into the model proposed by Hotulainen and Hoogenraad (2010). But how exactly the inducers of developmental structural plasticity link to actin or microtubule binding proteins responsible for changes in spine morphology is unclear.

Like in dendritogenesis, also during spine morphogenesis, extracellular stimuli influence actin cytoskeleton dynamics via diverse signal transduction pathways, among which small GTPases play a prominent role (Saneyoshi et al. 2010, see also Chap. 5). In principle, the activation of Rac1 induces spine maturation, whereas RhoA activation leads to opposite effects (Tashiro et al. 2000; Yuste 2010). However, several examples show that growth factors, neurotransmitters, and cell adhesion molecules utilize also Rho GTPases-independent pathways to signal to the actin cytoskeleton and regulate spine morphology (Ethell and Pasquale 2005; Saneyoshi et al. 2010; Schubert and Dotti 2007). Good examples are NMDA receptor activation-dependent recruitment of profilin to and disappearance of cortactin and drebrin from dendritic spines (Ackermann and Matus 2003; Hering and Sheng 2003; Sekino et al. 2006). Activation of Trk-B by BDNF results in MAPK-dependent phosphorylation of cortactin and its recruitment to synapses (Iki et al. 2005). Finally, calcium ions can regulate actin dynamics either directly through calcium- and actin-binding proteins (e.g., gelsolin) or via CamK (Saneyoshi et al. 2010; Schubert and Dotti 2007).

12.7 Future Directions

We are aware that we have not described in this chapter all of the possibilities of cytoskeleton regulation by extracellular modulators during developmental structural plasticity, but we hope that we have covered the major topics. The research on this topic has been conducted for at least the past two decades, but several questions still remain unanswered. Some new, exciting directions are currently emerging, including the role of dynamic microtubules and cross-talk between microtubules and microfilaments. The contribution of cytoskeleton dynamics to other cellular process that may greatly affect structural plasticity, such as membrane trafficking

(Frost et al. 2010a) or local protein translation (Bramham 2008), requires more research. Understanding these mechanisms is greatly important because proper dendritic arborization and spine morphology are involved in brain plasticity and severely impaired in several neurodevelopmental disorders (Fiala et al. 2002; Kasai et al. 2010a; Kaufmann and Moser 2000; van Spronsen and Hoogenraad 2010).

Acknowledgments This work was supported by a Polish-Norwegian Research Fund grant (PNRF-96-A I-1/07), FP7 "NeuroGSK3" grant #223276, and FP7-HealthProt grant #229676.

References

Acebes, A., & Ferrus, A. (2000). Cellular and molecular features of axon collaterals and dendrites. *Trends in Neurosciences, 23*, 557–565.

Ackermann, M., & Matus, A. (2003). Activity-induced targeting of profilin and stabilization of dendritic spine morphology. *Nature Neuroscience, 6*, 1194–1200.

Ahuja, R., Pinyol, R., Reichenbach, N., Custer, L., Klingensmith, J., Kessels, M. M., & Qualmann, B. (2007). Cordon-bleu is an actin nucleation factor and controls neuronal morphology. *Cell, 131*, 337–350.

Akum, B. F., Chen, M., Gunderson, S. I., Riefler, G. M., Scerri-Hansen, M. M., & Firestein, B. L. (2004). Cypin regulates dendrite patterning in hippocampal neurons by promoting microtubule assembly. *Nature Neuroscience, 7*, 145–152.

Arellano, J. I., Benavides-Piccione, R., Defelipe, J., & Yuste, R. (2007). Ultrastructure of dendritic spines: Correlation between synaptic and spine morphologies. *Frontiers in Neuroscience, 1*, 131–143.

Bjorkblom, B., Ostman, N., Hongisto, V., Komarovski, V., Filen, J. J., Nyman, T. A., Kallunki, T., Courtney, M. J., & Coffey, E. T. (2005). Constitutively active cytoplasmic c-Jun N-terminal kinase 1 is a dominant regulator of dendritic architecture: Role of microtubule-associated protein 2 as an effector. *The Journal of Neuroscience, 25*, 6350–6361.

Bourne, J., & Harris, K. M. (2007). Do thin spines learn to be mushroom spines that remember? *Current Opinion in Neurobiology, 17*, 381–386.

Bradley, P., & Berry, M. (1976). The effects of reduced climbing and parallel fibre input on Purkinje cell dendritic growth. *Brain Research, 109*, 133–151.

Bramham, C. R. (2008). Local protein synthesis, actin dynamics, and LTP consolidation. *Current Opinion in Neurobiology, 18*, 524–531.

Brigman, J. L., Wright, T., Talani, G., Prasad-Mulcare, S., Jinde, S., Seabold, G. K., Mathur, P., Davis, M. I., Bock, R., Gustin, R. M., et al. (2010). Loss of GluN2B-containing NMDA receptors in CA1 hippocampus and cortex impairs long-term depression, reduces dendritic spine density, and disrupts learning. *The Journal of Neuroscience, 30*, 4590–4600.

Burridge, K., & Wennerberg, K. (2004). Rho and Rac take center stage. *Cell, 116*, 167–179.

Calabrese, B., Wilson, M. S., & Halpain, S. (2006). Development and regulation of dendritic spine synapses. *Physiology (Bethesda, MD), 21*, 38–47.

Chen, H., & Firestein, B. L. (2007). RhoA regulates dendrite branching in hippocampal neurons by decreasing cypin protein levels. *The Journal of Neuroscience, 27*, 8378–8386.

Chen, Y., Wang, P. Y., & Ghosh, A. (2005). Regulation of cortical dendrite development by Rap1 signaling. *Molecular and Cellular Neuroscience, 28*, 215–228.

Choi, J. H., Bertram, P. G., Drenan, R., Carvalho, J., Zhou, H. H., & Zheng, X. F. (2002). The FKBP12-rapamycin-associated protein (FRAP) is a CLIP-170 kinase. *EMBO Reports, 3*, 988–994.

Conde, C., & Caceres, A. (2009). Microtubule assembly, organization and dynamics in axons and dendrites. *Nature Reviews Neuroscience, 10*, 319–332.

Dotti, C. G., Sullivan, C. A., & Banker, G. A. (1988). The establishment of polarity by hippocampal neurons in culture. *The Journal of Neuroscience, 8*, 1454–1468.

Espinosa, J. S., Wheeler, D. G., Tsien, R. W., & Luo, L. (2009). Uncoupling dendrite growth and patterning: Single-cell knockout analysis of NMDA receptor 2B. *Neuron, 62*, 205–217.

Ethell, I. M., & Pasquale, E. B. (2005). Molecular mechanisms of dendritic spine development and remodeling. *Progress in Neurobiology, 75*, 161–205.

Etienne-Manneville, S., & Hall, A. (2002). Rho GTPases in cell biology. *Nature, 420*, 629–635.

Fiala, J. C., Spacek, J., & Harris, K. M. (2002). Dendritic spine pathology: Cause or consequence of neurological disorders? *Brain Research: Brain Research Reviews, 39*, 29–54.

Firestein, B. L., Brenman, J. E., Aoki, C., Sanchez-Perez, A. M., El-Husseini, A. E., & Bredt, D. S. (1999). Cypin: A cytosolic regulator of PSD-95 postsynaptic targeting. *Neuron, 24*, 659–672.

Frost, N. A., Kerr, J. M., Lu, H. E., & Blanpied, T. A. (2010a). A network of networks: Cytoskeletal control of compartmentalized function within dendritic spines. *Current Opinion in Neurobiology, 20*, 578–587.

Frost, N. A., Shroff, H., Kong, H., Betzig, E., & Blanpied, T. A. (2010b). Single-molecule discrimination of discrete perisynaptic and distributed sites of actin filament assembly within dendritic spines. *Neuron, 67*, 86–99.

Furutani, Y., Matsuno, H., Kawasaki, M., Sasaki, T., Mori, K., & Yoshihara, Y. (2007). Interaction between telencephalin and ERM family proteins mediates dendritic filopodia formation. *The Journal of Neuroscience, 27*, 8866–8876.

Gao, F. B., & Bogert, B. A. (2003). Genetic control of dendritic morphogenesis in Drosophila. *Trends in Neurosciences, 26*, 262–268.

Gao, F. B., Brenman, J. E., Jan, L. Y., & Jan, Y. N. (1999). Genes regulating dendritic outgrowth, branching, and routing in Drosophila. *Genes & Development, 13*, 2549–2561.

Gaudilliere, B., Konishi, Y., de la Iglesia, N., Yao, G., & Bonni, A. (2004). A CaMKII-NeuroD signaling pathway specifies dendritic morphogenesis. *Neuron, 41*, 229–241.

Georges, P. C., Hadzimichalis, N. M., Sweet, E. S., & Firestein, B. L. (2008). The yin-yang of dendrite morphology: Unity of actin and microtubules. *Molecular Neurobiology, 38*, 270–284.

Grabham, P. W., Seale, G. E., Bennecib, M., Goldberg, D. J., & Vallee, R. B. (2007). Cytoplasmic dynein and LIS1 are required for microtubule advance during growth cone remodeling and fast axonal outgrowth. *The Journal of Neuroscience, 27*, 5823–5834.

Grove, M., Demyanenko, G., Echarri, A., Zipfel, P. A., Quiroz, M. E., Rodriguiz, R. M., Playford, M., Martensen, S. A., Robinson, M. R., Wetsel, W. C., et al. (2004). ABI2-deficient mice exhibit defective cell migration, aberrant dendritic spine morphogenesis, and deficits in learning and memory. *Molecular and Cellular Biology, 24*, 10905–10922.

Harris, K. M. (1999). Structure, development, and plasticity of dendritic spines. *Current Opinion in Neurobiology, 9*, 343–348.

Hayashi, Y., & Majewska, A. K. (2005). Dendritic spine geometry: Functional implication and regulation. *Neuron, 46*, 529–532.

Hayashi, K., Ohshima, T., & Mikoshiba, K. (2002). Pak1 is involved in dendrite initiation as a downstream effector of Rac1 in cortical neurons. *Molecular and Cellular Neuroscience, 20*, 579–594.

Hering, H., & Sheng, M. (2001). Dendritic spines: Structure, dynamics and regulation. *Nature Reviews Neuroscience, 2*, 880–888.

Hering, H., & Sheng, M. (2003). Activity-dependent redistribution and essential role of cortactin in dendritic spine morphogenesis. *The Journal of Neuroscience, 23*, 11759–11769.

Holtmaat, A., Wilbrecht, L., Knott, G. W., Welker, E., & Svoboda, K. (2006). Experience-dependent and cell-type-specific spine growth in the neocortex. *Nature, 441*, 979–983.

Honkura, N., Matsuzaki, M., Noguchi, J., Ellis-Davies, G. C., & Kasai, H. (2008). The subspine organization of actin fibers regulates the structure and plasticity of dendritic spines. *Neuron, 57*, 719–729.

Hoogenraad, C. C., & Bradke, F. (2009). Control of neuronal polarity and plasticity – A renaissance for microtubules? *Trends in Cell Biology, 19*, 669–676.

Hotulainen, P., & Hoogenraad, C. C. (2010). Actin in dendritic spines: Connecting dynamics to function. *The Journal of Cell Biology, 189*, 619–629.

Hotulainen, P., Llano, O., Smirnov, S., Tanhuanpaa, K., Faix, J., Rivera, C., & Lappalainen, P. (2009). Defining mechanisms of actin polymerization and depolymerization during dendritic spine morphogenesis. *The Journal of Cell Biology, 185*, 323–339.

Hu, X., Viesselmann, C., Nam, S., Merriam, E., & Dent, E. W. (2008). Activity-dependent dynamic microtubule invasion of dendritic spines. *The Journal of Neuroscience, 28*, 13094–13105.

Iki, J., Inoue, A., Bito, H., & Okabe, S. (2005). Bi-directional regulation of postsynaptic cortactin distribution by BDNF and NMDA receptor activity. *European Journal of Neuroscience, 22*, 2985–2994.

Jacobs, T., Causeret, F., Nishimura, Y. V., Terao, M., Norman, A., Hoshino, M., & Nikolic, M. (2007). Localized activation of p21-activated kinase controls neuronal polarity and morphology. *The Journal of Neuroscience, 27*, 8604–8615.

Jaffe, A. B., & Hall, A. (2005). Rho GTPases: Biochemistry and biology. *Annual Review of Cell and Developmental Biology, 21*, 247–269.

Jan, Y. N., & Jan, L. Y. (2010). Branching out: Mechanisms of dendritic arborization. *Nature Reviews Neuroscience, 11*, 316–328.

Jaworski, J., Hoogenraad, C. C., & Akhmanova, A. (2008). Microtubule plus-end tracking proteins in differentiated mammalian cells. *The International Journal of Biochemistry & Cell Biology, 40*, 619–637.

Jaworski, J., Kapitein, L. C., Gouveia, S. M., Dortland, B. R., Wulf, P. S., Grigoriev, I., Camera, P., Spangler, S. A., Di Stefano, P., Demmers, J., et al. (2009). Dynamic microtubules regulate dendritic spine morphology and synaptic plasticity. *Neuron, 61*, 85–100.

Jaworski, J., Spangler, S., Seeburg, D. P., Hoogenraad, C. C., & Sheng, M. (2005). Control of dendritic arborization by the phosphoinositide-3'-kinase-Akt-mammalian target of rapamycin pathway. *The Journal of Neuroscience, 25*, 11300–11312.

Ji, Y., Lu, Y., Yang, F., Shen, W., Tang, T. T., Feng, L., Duan, S., & Lu, B. (2010). Acute and gradual increases in BDNF concentration elicit distinct signaling and functions in neurons. *Nature Neuroscience, 13*, 302–309.

Jimenez-Mateos, E. M., Paglini, G., Gonzalez-Billault, C., Caceres, A., & Avila, J. (2005). End binding protein-1 (EB1) complements microtubule-associated protein-1B during axonogenesis. *Journal of Neuroscience Research, 80*, 350–359.

Kapitein, L. C., Schlager, M. A., Kuijpers, M., Wulf, P. S., van Spronsen, M., Mackintosh, F. C., & Hoogenraad, C. C. (2010). Mixed microtubules steer dynein-driven cargo transport into dendrites. *Current Biology, 20*, 290–299.

Kasai, H., Fukuda, M., Watanabe, S., Hayashi-Takagi, A., & Noguchi, J. (2010a). Structural dynamics of dendritic spines in memory and cognition. *Trends in Neurosciences, 33*, 121–129.

Kasai, H., Hayama, T., Ishikawa, M., Watanabe, S., Yagishita, S., & Noguchi, J. (2010b). Learning rules and persistence of dendritic spines. *European Journal of Neuroscience, 32*, 241–249.

Kaufmann, W. E., & Moser, H. W. (2000). Dendritic anomalies in disorders associated with mental retardation. *Cerebral Cortex, 10*, 981–991.

Knott, G., & Holtmaat, A. (2008). Dendritic spine plasticity–current understanding from in vivo studies. *Brain Research Reviews, 58*, 282–289.

Kodama, A., Karakesisoglou, I., Wong, E., Vaezi, A., & Fuchs, E. (2003). ACF7: An essential integrator of microtubule dynamics. *Cell, 115*, 343–354.

Konur, S., & Ghosh, A. (2005). Calcium signaling and the control of dendritic development. *Neuron, 46*, 401–405.

Korobova, F., & Svitkina, T. (2010). Molecular architecture of synaptic actin cytoskeleton in hippocampal neurons reveals a mechanism of dendritic spine morphogenesis. *Molecular Biology of the Cell, 21*, 165–176.

Kumar, V., Zhang, M. X., Swank, M. W., Kunz, J., & Wu, G. Y. (2005). Regulation of dendritic morphogenesis by Ras-PI3K-Akt-mTOR and Ras-MAPK signaling pathways. *The Journal of Neuroscience, 25*, 11288–11299.

Lansbergen, G., & Akhmanova, A. (2006). Microtubule plus end: A hub of cellular activities. *Traffic, 7*, 499–507.

Laxson, L. C., & King, J. S. (1983). The development of the Purkinje cell in the cerebellar cortex of the opossum. *The Journal of Comparative Neurology, 214*, 290–308.

Lee, H., Engel, U., Rusch, J., Scherrer, S., Sheard, K., & Van Vactor, D. (2004). The microtubule plus end tracking protein Orbit/MAST/CLASP acts downstream of the tyrosine kinase Abl in mediating axon guidance. *Neuron, 42*, 913–926.

Lee, A., Li, W., Xu, K., Bogert, B. A., Su, K., & Gao, F. B. (2003). Control of dendritic development by the Drosophila fragile X-related gene involves the small GTPase Rac1. *Development, 130*, 5543–5552.

Li, Z., Aizenman, C. D., & Cline, H. T. (2002). Regulation of rho GTPases by crosstalk and neuronal activity in vivo. *Neuron, 33*, 741–750.

Li, Z., Van Aelst, L., & Cline, H. T. (2000). Rho GTPases regulate distinct aspects of dendritic arbor growth in Xenopus central neurons in vivo. *Nature Neuroscience, 3*, 217–225.

Liu, Z., Steward, R., & Luo, L. (2000). Drosophila Lis1 is required for neuroblast proliferation, dendritic elaboration and axonal transport. *Nature Cell Biology, 2*, 776–783.

Lohmann, C., Myhr, K. L., & Wong, R. O. (2002). Transmitter-evoked local calcium release stabilizes developing dendrites. *Nature, 418*, 177–181.

Majewska, A., & Sur, M. (2003). Motility of dendritic spines in visual cortex in vivo: Changes during the critical period and effects of visual deprivation. *Proceedings of the National Academy of Sciences of the United States of America, 100*, 16024–16029.

Marrs, G. S., Green, S. H., & Dailey, M. E. (2001). Rapid formation and remodeling of postsynaptic densities in developing dendrites. *Nature Neuroscience, 4*, 1006–1013.

Matsuzaki, M., Honkura, N., Ellis-Davies, G. C., & Kasai, H. (2004). Structural basis of long-term potentiation in single dendritic spines. *Nature, 429*, 761–766.

McAllister, A. K. (2000). Cellular and molecular mechanisms of dendrite growth. *Cerebral Cortex, 10*, 963–973.

Michaelsen, K., Murk, K., Zagrebelsky, M., Dreznjak, A., Jockusch, B. M., Rothkegel, M., & Korte, M. (2010). Fine-tuning of neuronal architecture requires two profilin isoforms. *Proceedings of the National Academy of Sciences of the United States of America, 107*, 15780–15785.

Mizrahi, A. (2007). Dendritic development and plasticity of adult-born neurons in the mouse olfactory bulb. *Nature Neuroscience, 10*, 444–452.

Mizrahi, A., & Katz, L. C. (2003). Dendritic stability in the adult olfactory bulb. *Nature Neuroscience, 6*, 1201–1207.

Muley, P. D., McNeill, E. M., Marzinke, M. A., Knobel, K. M., Barr, M. M., & Clagett-Dame, M. (2008). The atRA-responsive gene neuron navigator 2 functions in neurite outgrowth and axonal elongation. *Developmental Neurobiology, 68*, 1441–1453.

Nakayama, A. Y., Harms, M. B., & Luo, L. (2000). Small GTPases Rac and Rho in the maintenance of dendritic spines and branches in hippocampal pyramidal neurons. *The Journal of Neuroscience, 20*, 5329–5338.

Okamoto, K., Nagai, T., Miyawaki, A., & Hayashi, Y. (2004). Rapid and persistent modulation of actin dynamics regulates postsynaptic reorganization underlying bidirectional plasticity. *Nature Neuroscience, 7*, 1104–1112.

Parrish, J. Z., Emoto, K., Kim, M. D., & Jan, Y. N. (2007). Mechanisms that regulate establishment, maintenance, and remodeling of dendritic fields. *Annual Review of Neuroscience, 30*, 399–423.

Penzes, P., Beeser, A., Chernoff, J., Schiller, M. R., Eipper, B. A., Mains, R. E., & Huganir, R. L. (2003). Rapid induction of dendritic spine morphogenesis by trans-synaptic ephrinB-EphB receptor activation of the Rho-GEF kalirin. *Neuron, 37*, 263–274.

Pinyol, R., Haeckel, A., Ritter, A., Qualmann, B., & Kessels, M. M. (2007). Regulation of N-WASP and the Arp2/3 complex by Abp1 controls neuronal morphology. *PloS One, 2*, e400.

Podkowa, M., Zhao, X., Chow, C. W., Coffey, E. T., Davis, R. J., & Attisano, L. (2010). Microtubule stabilization by bone morphogenetic protein receptor-mediated scaffolding of c-Jun N-terminal kinase promotes dendrite formation. *Molecular and Cellular Biology, 30*, 2241–2250.

Pontrello, C. G., & Ethell, I. M. (2009). Accelerators, brakes, and gears of actin dynamics in dendritic spines. *The Open Neuroscience Journal, 3*, 67–86.

Portera-Cailliau, C., Pan, D. T., & Yuste, R. (2003). Activity-regulated dynamic behavior of early dendritic protrusions: evidence for different types of dendritic filopodia. *The Journal of Neuroscience, 23*, 7129–7142.

Poulain, F. E., & Sobel, A. (2010). The microtubule network and neuronal morphogenesis: Dynamic and coordinated orchestration through multiple players. *Molecular and Cellular Neuroscience, 43*, 15–32.

Prokop, A., Uhler, J., Roote, J., & Bate, M. (1998). The kakapo mutation affects terminal arborization and central dendritic sprouting of Drosophila motorneurons. *The Journal of Cell Biology, 143*, 1283–1294.

Redmond, L., & Ghosh, A. (2005). Regulation of dendritic development by calcium signaling. *Cell Calcium, 37*, 411–416.

Redmond, L., Kashani, A. H., & Ghosh, A. (2002). Calcium regulation of dendritic growth via CaM kinase IV and CREB-mediated transcription. *Neuron, 34*, 999–1010.

Rocca, D. L., Martin, S., Jenkins, E. L., & Hanley, J. G. (2008). Inhibition of Arp2/3-mediated actin polymerization by PICK1 regulates neuronal morphology and AMPA receptor endocytosis. *Nature Cell Biology, 10*, 259–271.

Rosso, S. B., Sussman, D., Wynshaw-Boris, A., & Salinas, P. C. (2005). Wnt signaling through Dishevelled, Rac and JNK regulates dendritic development. *Nature Neuroscience, 8*, 34–42.

Sala, C., Cambianica, I., & Rossi, F. (2008). Molecular mechanisms of dendritic spine development and maintenance. *Acta Neurobiologiae Experimentalis (Wars), 68*, 289–304.

Saneyoshi, T., Fortin, D. A., & Soderling, T. R. (2010). Regulation of spine and synapse formation by activity-dependent intracellular signaling pathways. *Current Opinion in Neurobiology, 20*, 108–115.

Satoh, D., Sato, D., Tsuyama, T., Saito, M., Ohkura, H., Rolls, M. M., Ishikawa, F., & Uemura, T. (2008). Spatial control of branching within dendritic arbors by dynein-dependent transport of Rab5-endosomes. *Nature Cell Biology, 10*, 1164–1171.

Schubert, V., & Dotti, C. G. (2007). Transmitting on actin: Synaptic control of dendritic architecture. *Journal of Cell Science, 120*, 205–212.

Segal, M. (2001). Rapid plasticity of dendritic spine: Hints to possible functions? *Progress in Neurobiology, 63*, 61–70.

Sekino, Y., Kojima, N., & Shirao, T. (2007). Role of actin cytoskeleton in dendritic spine morphogenesis. *Neurochemistry International, 51*, 92–104.

Sekino, Y., Tanaka, S., Hanamura, K., Yamazaki, H., Sasagawa, Y., Xue, Y., Hayashi, K., & Shirao, T. (2006). Activation of N-methyl-D-aspartate receptor induces a shift of drebrin distribution: Disappearance from dendritic spines and appearance in dendritic shafts. *Molecular and Cellular Neuroscience, 31*, 493–504.

Siegrist, S. E., & Doe, C. Q. (2007). Microtubule-induced cortical cell polarity. *Genes & Development, 21*, 483–496.

Sin, W. C., Haas, K., Ruthazer, E. S., & Cline, H. T. (2002). Dendrite growth increased by visual activity requires NMDA receptor and Rho GTPases. *Nature, 419*, 475–480.

Soderling, S. H., Guire, E. S., Kaech, S., White, J., Zhang, F., Schutz, K., Langeberg, L. K., Banker, G., Raber, J., & Scott, J. D. (2007). A WAVE-1 and WRP signaling complex regulates spine density, synaptic plasticity, and memory. *The Journal of Neuroscience, 27*, 355–365.

Stuart, G., Spruston, N., & Häusser, M. (2007). *Dendrites* (2nd ed.). Oxford: Oxford University Press.

Swiech, L., Blazejczyk, M., Urbanska, M., Pietruszka, P., Dortland ,B.R., Malik, A.R., Wulf, P.S., Hoogenraad, C.C., & Jaworski, J. (2011). CLIP-170 and IQGAP1 Cooperatively Regulate Dendrite Morphology. *J Neurosci., 31*, 4555–4568.

Tada, T., & Sheng, M. (2006). Molecular mechanisms of dendritic spine morphogenesis. *Current Opinion in Neurobiology, 16*, 95–101.

Takemoto-Kimura, S., Ageta-Ishihara, N., Nonaka, M., Adachi-Morishima, A., Mano, T., Okamura, M., Fujii, H., Fuse, T., Hoshino, M., Suzuki, S., et al. (2007). Regulation of dendritogenesis via a lipid-raft-associated Ca^{2+}/calmodulin-dependent protein kinase CLICK-III/CaMKIgamma. *Neuron, 54*, 755–770.

Tashiro, A., Minden, A., & Yuste, R. (2000). Regulation of dendritic spine morphology by the rho family of small GTPases: Antagonistic roles of Rac and Rho. *Cerebral Cortex, 10*, 927–938.

Tatavarty, V., Kim, E. J., Rodionov, V., & Yu, J. (2009). Investigating sub-spine actin dynamics in rat hippocampal neurons with super-resolution optical imaging. *PloS One, 4*, e7724.

Terabayashi, T., Itoh, T. J., Yamaguchi, H., Yoshimura, Y., Funato, Y., Ohno, S., & Miki, H. (2007). Polarity-regulating kinase partitioning-defective 1/microtubule affinity-regulating kinase 2 negatively regulates development of dendrites on hippocampal neurons. *The Journal of Neuroscience, 27*, 13098–13107.

Threadgill, R., Bobb, K., & Ghosh, A. (1997). Regulation of dendritic growth and remodeling by Rho, Rac, and Cdc42. *Neuron, 19*, 625–634.

Tolias, K. F., Bikoff, J. B., Burette, A., Paradis, S., Harrar, D., Tavazoie, S., Weinberg, R. J., & Greenberg, M. E. (2005). The Rac1-GEF Tiam1 couples the NMDA receptor to the activity-dependent development of dendritic arbors and spines. *Neuron, 45*, 525–538.

Tran, T. S., Rubio, M. E., Clem, R. L., Johnson, D., Case, L., Tessier-Lavigne, M., Huganir, R. L., Ginty, D. D., & Kolodkin, A. L. (2009). Secreted semaphorins control spine distribution and morphogenesis in the postnatal CNS. *Nature, 462*, 1065–1069.

Ultanir, S. K., Kim, J. E., Hall, B. J., Deerinck, T., Ellisman, M., & Ghosh, A. (2007). Regulation of spine morphology and spine density by NMDA receptor signaling in vivo. *Proceedings of the National Academy of Sciences of the United States of America, 104*, 19553–19558.

Urbanska, M., Blazejczyk, M., & Jaworski, J. (2008). Molecular basis of dendritic arborization. *Acta Neurobiologiae Experimentalis (Wars), 68*, 264–288.

van Spronsen, M., & Hoogenraad, C. C. (2010). Synapse pathology in psychiatric and neurologic disease. *Current Neurology and Neuroscience Reports, 10*, 207–214.

Wayman, G. A., Impey, S., Marks, D., Saneyoshi, T., Grant, W. F., Derkach, V., & Soderling, T. R. (2006). Activity-dependent dendritic arborization mediated by CaM-kinase I activation and enhanced CREB-dependent transcription of Wnt-2. *Neuron, 50*, 897–909.

Wegner, A. M., Nebhan, C. A., Hu, L., Majumdar, D., Meier, K. M., Weaver, A. M., & Webb, D. J. (2008). N-wasp and the arp2/3 complex are critical regulators of actin in the development of dendritic spines and synapses. *Journal of Biological Chemistry, 283*, 15912–15920.

Wu, J. I., Lessard, J., Olave, I. A., Qiu, Z., Ghosh, A., Graef, I. A., & Crabtree, G. R. (2007). Regulation of dendritic development by neuron-specific chromatin remodeling complexes. *Neuron, 56*, 94–108.

Wu, G. Y., Zou, D. J., Rajan, I., & Cline, H. (1999). Dendritic dynamics in vivo change during neuronal maturation. *The Journal of Neuroscience, 19*, 4472–4483.

Yasumatsu, N., Matsuzaki, M., Miyazaki, T., Noguchi, J., & Kasai, H. (2008). Principles of long-term dynamics of dendritic spines. *The Journal of Neuroscience, 28*, 13592–13608.

Yoshihara, Y., De Roo, M., & Muller, D. (2009). Dendritic spine formation and stabilization. *Current Opinion in Neurobiology, 19*, 146–153.

Yu, X., & Malenka, R. C. (2003). Beta-catenin is critical for dendritic morphogenesis. *Nature Neuroscience, 6*, 1169–1177.

Yuste, R. (2010). *Dendritic spines*. Cambridge: MIT Press.

Yuste, R., & Bonhoeffer, T. (2004). Genesis of dendritic spines: Insights from ultrastructural and imaging studies. *Nature Reviews Neuroscience, 5*, 24–34.

Zhang, W., & Benson, D. L. (2000). Development and molecular organization of dendritic spines and their synapses. *Hippocampus, 10*, 512–526.

Zhou, F. Q., Zhou, J., Dedhar, S., Wu, Y. H., & Snider, W. D. (2004). NGF-induced axon growth is mediated by localized inactivation of GSK-3beta and functions of the microtubule plus end binding protein APC. *Neuron, 42*, 897–912.

Chapter 13
Dendritic mRNA Targeting and Translation

Stefan Kindler and Hans-Jürgen Kreienkamp

Abstract Selective targeting of specific mRNAs into neuronal dendrites and their locally regulated translation at particular cell contact sites contribute to input-specific synaptic plasticity. Thus, individual synapses become decision-making units, which control gene expression in a spatially restricted and nucleus-independent manner. Dendritic targeting of mRNAs is achieved by active, microtubule-dependent transport. For this purpose, mRNAs are packaged into large ribonucleoprotein (RNP) particles containing an array of *trans*-acting RNA-binding proteins. These are attached to molecular motors, which move their RNP cargo into dendrites. A variety of proteins may be synthesized in dendrites, including signalling and scaffold proteins of the synapse and neurotransmitter receptors. In some cases, such as the alpha subunit of the calcium/calmodulin-dependent protein kinase II (αCaMKII) and the activity-regulated gene of 3.1 kb (Arg3.1, also referred to as activity-regulated cDNA, Arc), their local synthesis at synapses can modulate long-term changes in synaptic efficiency. Local dendritic translation is regulated by several signalling cascades including Akt/mTOR and Erk/MAP kinase pathways, which are triggered by synaptic activity. More recent findings show that miRNAs also play an important role in protein synthesis at synapses. Disruption of local translation control at synapses, as observed in the fragile X syndrome (FXS) and its mouse models and possibly also in autism spectrum disorders, interferes with cognitive abilities in mice and men.

Keywords Activity-dependent translation • Dendritic targeting element • Molecular motor • Synaptic plasticity • *Trans*-acting factor

S. Kindler (✉) • H.-J. Kreienkamp
University Medical Center Hamburg-Eppendorf, Institute for Human Genetics, Martinistr. 52, 20246 Hamburg, Germany
e-mail: kindler@uke.de; kreienkamp@uke.de

M.R. Kreutz and C. Sala (eds.), *Synaptic Plasticity*,
Advances in Experimental Medicine and Biology 970,
DOI 10.1007/978-3-7091-0932-8_13, © Springer-Verlag/Wien 2012

13.1 Introduction

The role of protein synthesis in memory formation was appreciated as early as 1963, when Flexner et al. reported that injection of the protein synthesis inhibitor puromycin into the temporal lobe of mice impairs the acquisition of memory (Flexner et al. 1963). Later it became evident that long-term modification of synapses requires the synthesis of new proteins. In electrophysiological paradigms of synaptic plasticity, the late phase of both long-term potentiation (LTP) and long-term depression (LTD) depends on a supply of newly synthesized proteins (Pfeiffer and Huber 2006). However, the input specificity of synaptic plasticity presents a considerable cell biological problem, as out of potentially thousands of synapses only a few need to be modified. How can proteins produced in neuronal cell bodies be addressed only to those postsynaptic sites, which have experienced an appropriate stimulus and are about to be modified? In this respect, the discovery of polysomes in dendrites (Steward and Levy 1982) was welcome news as it became obvious that some proteins might be produced locally in the vicinity of a synapse. More recently, Ostroff et al. (2010) observed an accumulation of polysomes and multivesicular bodies in dendritic shafts during fear conditioning training, indicating increased local protein synthesis and degradation during learning. By *in situ* hybridization in the rodent hippocampus, several mRNAs were identified, which were present not only in neuronal cell bodies but also in dendritic fields, including messages coding for microtubule-associated protein 2 (MAP2), αCaMKII, Arg3.1/Arc, the SH3 domain and ankyrin repeat protein (Shank1–3), the SAP and PSD-95-associated protein 3 (SAPAP3), dendrin and Jacob (Böckers et al. 2004; Burgin et al. 1990; Garner et al. 1988; Herb et al. 1997; Kindler et al. 2004, 2009; Link et al. 1995; Lyford et al. 1995; Welch et al. 2004). Based on these findings, stimulus-dependent local translation of mRNAs in dendrites was firmly established as a cellular mechanism contributing to long-lasting modifications of synapses. The relevance of this phenomenon for cognition in humans is underlined by cognitive deficits observed in patients suffering from fragile X syndrome (FXS). This disease is characterized by the functional loss of the fragile X mental retardation protein (FMRP), an RNA-binding protein (RBP) acting as a translational repressor in dendrites (Bassell and Warren 2008). Also, deficits in dendritic translational control have been implicated in other neuronal diseases, in particular autism (Kelleher and Bear 2008).

Here, we will address the following questions concerning dendritic transport and synaptic translation of mRNAs:

- How are specific mRNAs selected for dendritic transport, and how is this transport achieved by neuronal motor proteins?
- Which are the main dendritic mRNAs that may substantially contribute to alterations of the molecular composition and signalling capacity of individual postsynaptic sites?
- Which signalling pathways trigger dendritic translation?
- How do these phenomena relate to human disease?

13.2   Factors Involved in Dendritic mRNA Trafficking

It is understood that specific sequences in dendritically localized mRNAs should be responsible for trafficking into dendrites (Bramham and Wells 2007; Kindler et al. 2005). These dendritic targeting elements (DTEs) have been identified in cultured neurons using recombinant reporter RNAs. DTEs, identified so far, vary widely in size (from about 10 bases to 1.2 kb) and sequence, and it has not been possible to determine a commonly used consensus motif. Unfortunately, some results are also quite conflicting. Thus, five different studies identified four distinct, non-overlapping sequence elements, which promote dendritic targeting of αCaMKII mRNAs (Blichenberg et al. 2001; Gao et al. 2008; Huang et al. 2003; Mori et al. 2000; Tübing et al. 2010). Similarly, two different non-overlapping DTEs are described in Arg3.1/Arc mRNAs (Gao et al. 2008; Kobayashi et al. 2005). At present, we can only assume that some or all of the experimental approaches used are not close enough to the *in vivo* situation. As is true for most mRNAs, dendritic messages derive from precursors, which are spliced and processed in the nucleus, packaged into messenger RNPs and exported through nuclear pores, possibly followed by restructuring of the RNP complexes in the cytosol. Recombinant RNAs do not go through most of these processing steps. This may be highly relevant for transport efficiency, as it is likely that RNA sorting begins already in the nucleus. The large discrepancy between the size of most DTEs and the size of individual motifs recognized by RBPs suggests that a complement of several RBPs assembles on any functional DTE to enable dendritic trafficking.

Several RBPs specifically recognize individual DTEs and are therefore considered as potential *trans*-acting factors involved in mRNA targeting. Thus, MARTA1 associates with the MAP2-DTE (Rehbein et al. 2002), while the related zipcode binding protein 1 binds to the DTE (or zipcode) of β-actin mRNAs (Tiruchinapalli et al. 2003). Similar to hnRNPA2 that has been implicated in dendritic targeting of αCaMKII, Arg3.1/Arc and neurogranin C mRNAs (Gao et al. 2008), the cytoplasmic polyadenylation element (CPE) binding protein (CPEB) appears to play a role in dendritic targeting of several distinct transcripts (Huang et al. 2003). Also, the hematopoietic zinc finger (Hzf) binds to the inositol trisphosphate (IP3) receptor mRNA and is involved in its dendritic targeting in cerebellar Purkinje cells (Iijima et al. 2005). Finally, the two mammalian homologs of the *Drosophila* RBP Staufen interact with mRNAs in a sequence-independent manner and have been shown to be required for dendritic mRNA trafficking (Falley et al. 2009; Kanai et al. 2004; Monshausen et al. 2001; Tang et al. 2001).

Dendritic localization of RNAs requires motor proteins, which translocate RNP particles along cytoskeletal filaments. Kanai et al. (2004) identified one such motor showing that the cargo-binding domains of KIF5 family kinesins associate with large RNP complexes containing dendritic transcripts such as αCaMKII and Arg3.1/Arc mRNAs. Proteomic analysis of these granules revealed more than 30 distinct proteins, which associate with KIF5 in an RNA-dependent manner. The finding that an RNAi-mediated knockdown of some of these components interfered with dendritic localization of reporter transcripts suggests that KIF5-associated RNP complexes are bona fide mRNA transport granules, which are propelled into dendrites by KIF5. The fact that several predominantly nuclear RBPs are present in these RNP complexes further supports the idea that the assembly of dendritic RNP particles starts in the nucleus.

13.3 Contribution of Individual Dendritically Localized mRNAs to Synaptic Plasticity: Which mRNAs, and Why?

The question, which mRNAs are present in dendrites, and which of these are used to produce protein relevant for changes in synaptic function, is central for appreciating the functional relevance of dendritic mRNA transport. This issue has also been somewhat controversial, because the number of distinct dendritic mRNAs, and the extent to which each of these transcripts is transported into dendrites, remains unclear. Two extreme views of this issue have been published. Thus, the gold standard for dendritic localization of an mRNA is a corresponding strong *in situ* hybridization signal over molecular layers in the hippocampus or cerebellum. In this type of experiment, only few mRNAs have been unequivocally identified in dendrites, including MAP2, αCaMKII, Arg3.1/Arc, Shank, SAPAP3, dendrin and Jacob transcripts. In contrast, PCR-based techniques used to determine the full complement of dendritic mRNAs identified more than 400 putative dendritic mRNAs (Eberwine and Crino 2001). As both techniques differ in detection sensitivity, it appears likely that mRNAs identified by PCR techniques only are present in dendrites at very low levels as compared to the concentration in neuronal somata. Here, we will discuss the possible relevance of some major dendritic mRNAs for synaptic plasticity.

13.3.1 αCaMKII: A Chief Postsynaptic Signalling Molecule

αCaMKII is a key molecule involved in postsynaptic signal transduction. It is highly abundant in postsynaptic densities (PSDs) of excitatory synapses. Ouyang et al. (1999) provided evidence for a local synthesis of αCaMKII in hippocampal dendrites after tetanic stimulation. By replacing the DTE containing 3′-untranslated region (UTR) of endogenous αCaMKII mRNAs with a corresponding region of somatically restricted transcripts, Miller et al. (2002) were the first to analyse the physiological relevance of dendritic mRNA targeting *in vivo*. This manipulation results in a complete loss of dendritic αCaMKII mRNA targeting, accompanied with a 50% reduction in overall αCaMKII levels. Importantly, the protein is almost completely lost from PSDs, clearly indicating that its dendritic synthesis strongly contributes to postsynaptic targeting of αCaMKII. Whereas the early phase of LTP is unchanged in these mice, the late protein synthesis–dependent phase is reduced. These changes are accompanied by profound behavioural deficits in learning paradigms. As pointed out by Steward (2002), these results allow two interpretations: First, αCaMKII mRNAs need to be translated in dendrites for efficient postsynaptic targeting of the protein, and all changes observed with respect to LTP and behaviour result from the lack of postsynaptic αCaMKII. Second, local synthesis of αCaMKII at synapses is in itself a key element of synaptic plasticity, in particular the late phase of LTP.

13.3.2 AMPA Receptor Subunits

Most forms of synaptic plasticity are associated with local variations in the abundance of functional α-amino-3-hydroxy-5-methyl-4-isoxazolepropionic acid-type glutamate receptors (AMPA-Rs). Thus, local synthesis of any of the AMPA-R subunits (GluR) seems at first glance a direct and efficient way of changing the strength of particular synapses. However, *in situ* hybridization shows that in the rodent brain, dendritic layers of the hippocampus are rather devoid of GluR mRNAs (see, e.g. http://www. alleninstitute.org/science/public_resources/atlases/mouse_atlas.html). Yet, using sophisticated micromanipulation and imaging techniques, it was shown that GluR transcripts reside in dendrites of primary neurons, where the respective proteins are locally synthesized upon synaptic stimulation and are inserted into the synaptic plasma membrane as functional receptors (Grooms et al. 2006; Ju et al. 2004; Kacharmina et al. 2000).

13.3.3 Arg3.1/Arc: An Immediate Early Gene Product Involved in Glutamate Receptor Internalization

Immediately after strong stimulation, expression of the Arg3.1/Arc gene is induced, and the mRNA is rapidly transported into dendrites (Link et al. 1995; Lyford et al. 1995), where it appears to be locally translated under control of stimulating axons (Steward and Worley 2001). Arg3.1/Arc then interacts with proteins of the endocytosis machinery, in particular endophilin 2/3 and dynamin-2, and thus facilitates endocytosis of AMPA-Rs (Chowdhury et al. 2006). This is particularly relevant for metabotropic glutamate receptor (mGluR)-dependent LTD. Here, basal levels of Arg3.1/Arc are required for the reduction of synaptic AMPA-Rs by rapid endocytosis during the early phase of LTD. For the maintenance of LTD, however, new Arg3.1/Arc needs to be synthesized locally to maintain a higher rate of AMPA-R endocytosis (Waung et al. 2008). The physiological significance of this mechanism becomes obvious in gene knockout studies where the loss of Arg3.1/Arc leads to impaired memory formation in mice (Plath et al. 2006). Translation of Arg3.1/Arc mRNAs is controlled by FMRP, and loss of this regulation in FMRP-deficient mice leads to excessive LTD, which is independent of new protein synthesis.

Thus, excess Arg3.1/Arc synthesis may be one major factor contributing to deficits in synaptic plasticity associated with the FXS (Huber et al. 2002; Park et al. 2008; Waung et al. 2008).

13.3.4 Shank and SAPAP3, Main Scaffold Proteins of the PSD

Members of the Shank (Shank1–3) and SAPAP (SAPAP1–4) families are multidomain proteins with a strong and almost exclusive localization to PSDs of excitatory synapses (Kim et al. 1997; Naisbitt et al. 1999; Takeuchi et al. 1997; Zitzer et al. 1999). SAPAP1–4 reside in an intermediate layer of the PSD, linking PSD-95 family members (that are directly associated with synaptic glutamate receptors) with Shank1–3, which are more remote from the synaptic plasma membrane (Valtschanoff and Weinberg 2001). Genetic studies clearly link these proteins to mental diseases such as mental retardation, autism or obsessive-compulsive disorder (Berkel et al. 2010; Bonaglia et al. 2001; Durand et al. 2007; Welch et al. 2007).

mRNAs encoding SAPAP3 reside in dendritic layers of the hippocampus, whereas other SAPAP transcripts remain restricted to neuronal cell bodies (Kindler et al. 2004; Welch et al. 2004). SAPAP3 mRNAs are translationally repressed by FMRP, and translation may be induced by stimulation of mGluRs or dopamine D1 receptors (Narayanan et al. 2008; Wang et al. 2010). Shank1–3 mRNAs exhibit extensive dendritic localization, as both Shank1 and Shank3 transcripts are present in hippocampal dendrites, while Shank1 and Shank2 mRNAs are found in dendrites of cerebellar Purkinje cells (Böckers et al. 2004; Falley et al. 2009). Similar to SAPAP3 transcripts, Shank1 mRNAs are translationally repressed by FMRP, and translation is induced through mGluR signalling pathways (Schütt et al. 2009; Wang et al. 2010).

So far, the physiological relevance of a local dendritic synthesis of both SAPAP3 and Shank1–3 is unknown. Shank1–3 contribute to the functional and morphological maturation of postsynaptic specializations, as they help to recruit components of dendritic spines and PSDs to nascent synaptic sites (Sala et al. 2001). Thus, local synaptic synthesis of Shank1–3 may contribute to morphological changes observed in dendritic spines during and after LTP. Interestingly, Shank1–3 exhibit a strong ability to self-associate and form polymers, which have been suggested to provide a nucleus for the formation of PSDs (Baron et al. 2006; Gundelfinger et al. 2006). In this respect, it might be necessary that Shank1–3 are produced only in the vicinity of newly forming synaptic sites, as, for example in the cell body, these proteins might lead to the formation of aggregates (as observed by Romorini et al. (2004) upon Shank1 overexpression). This also points to the need for a tight regulation of synaptic Shank1 synthesis. Our observation that postsynaptic levels of Shank1 are significantly increased in FMRP-deficient mice may therefore provide an explanation for the aberrant spine formation in these animals. Overproduction of Shank1 without appropriate spatial and temporal control by FMRP is likely to lead to the stabilization of immature spines in the absence of an appropriate presynaptic input (Schütt et al. 2009).

13.3.5 *Jacob: A New Kid on the Block*

mRNAs encoding the plasticity-related protein Jacob are a relatively new addition to the list of dendritic mRNAs. Jacob associates with the postsynaptic Ca^{2+}-binding protein caldendrin. Activation of extrasynaptic N-methyl-D-aspartate receptors (NMDA-Rs), coupled with Ca^{2+}-influx, leads to cleavage of the N-terminal myristoyl membrane anchor of Jacob by the Ca^{2+}-dependent protease calpain (Kindler et al. 2009). This allows for association of Jacob with the nuclear import factor importin-α, followed by translocation of Jacob from dendrites to the nucleus. Here, Jacob affects transcription events, which eventually lead to a reduction of synaptic contacts, suggesting that Jacob plays a role in homeostatic synaptic plasticity (Dieterich et al. 2008). The suggested local translation of Jacob at synapses may be necessary for two obvious reasons: (1) Jacob levels decrease in the vicinity of activated NMDA-Rs due to calpain activity. Locally controlled protein synthesis would allow for an effective way to replenish local stores of Jacob without altering Jacob levels in other dendritic areas, where the protein has not been subject to activity-dependent degradation. (2) The nuclear localization signal of Jacob is occluded by the dendritic Ca^{2+}-binding protein caldendrin (Dieterich et al. 2008). If Jacob would be synthesized in neuronal cell bodies, insufficient amounts of caldendrin would allow for the nuclear translocation of Jacob, thus altering transcriptional events in the absence of an appropriate stimulus. Therefore, dendritic synthesis of Jacob appears to be required for the complex cellular trafficking pattern of the mature protein.

13.4 Regulation of Translation

In order to contribute to the regulation of input-specific synaptic plasticity, the translation of dendritically localized mRNAs must be tightly controlled. In particular, while en route dendritic transcripts must be translationally repressed. However, specific local stimuli at synapses should be capable to trigger *de novo* protein synthesis.

13.4.1 *General Mechanisms of Translational Control*

Translation of an mRNA advances in three sequential steps, namely, initiation, elongation and termination (Costa-Mattioli et al. 2009b; Gkogkas et al. 2010). While each step can be regulated, initiation is usually the rate-limiting event and therefore represents the primary target for control (Sonenberg and Hinnebusch 2009). There are two major possibilities for initiation: (1) $5'$cap-independent initiation that requires an internal ribosomal entry site (IRES). While several dendritically localized mRNAs seem to contain functional IRES elements (Pinkstaff et al. 2001), the relevance of this regulatory mechanism remains completely unclear. (2) 5′cap-dependent initiation is controlled by elements in both 5′- and 3′-UTRs. 3′-UTRs frequently contain binding sites for *trans*-acting factors such as poly-A-binding protein (PABP) or CPEB. Some dendritic mRNAs contain long and GC-rich 5′-UTRs, which mostly seem to act as obstacles for the scanning of small ribosomal subunits and thereby inhibit initiation. For example, a strong secondary structure and numerous upstream open reading frames inhibit translation initiation of Shank1 mRNAs, thus potentially providing a silencing mechanism during mRNA transport (Falley et al. 2009).

During initiation of 5′cap-dependent translation, binding of a ternary complex consisting of eukaryotic initiation factor 2, GTP and methionyl-tRNA (eIF2•GTP•Met-tRNA$_i$) to the 40S ribosomal subunit leads to the formation of the so-called 43S preinitiation complex (Fig. 13.1). With the help of eIF4F, this complex associates with the 5′-end of an mRNA. eIF4F consists of three different proteins, namely, eIF4E, eIF4G and eIF4A. eIF4E specifically recognizes the 5′cap, and this interaction is a critical step in the initiation process (Jackson et al. 2010; Sonenberg and Hinnebusch 2009). Different mechanisms regulate eIF4E activity. Three binding proteins (4E-BPs) can sequester eIF4E and occupy the eIF4G

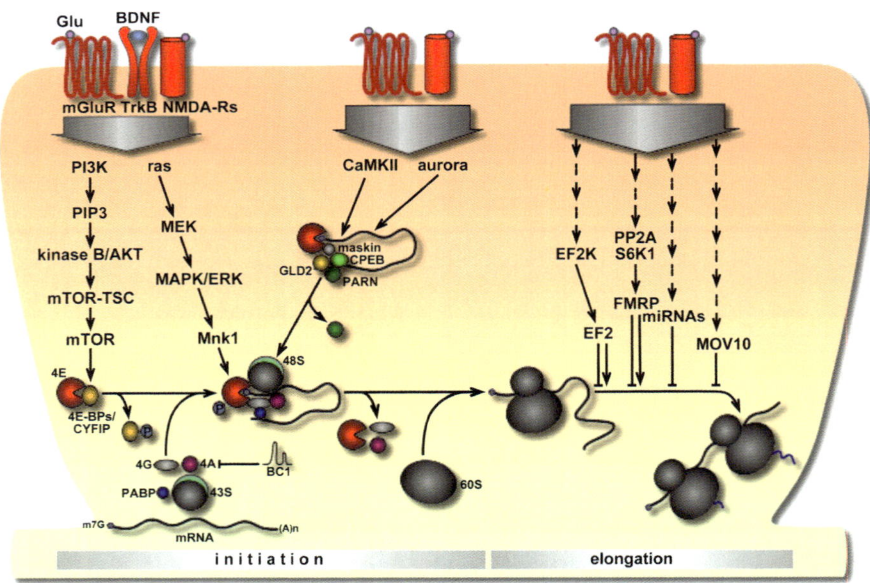

Fig. 13.1 Schematic representation of major pathways regulating protein synthesis at synapses. Activation of various receptor systems (*upper part*) stimulates distinct signaling pathways (*middle*) and thus regulates both 5′cap-dependent initiation (*lower left*) and elongation (*lower right*) of mRNA translation. Factors are not drawn to scale. 4A (eIF4A), 4E (eIF4E), 4G (eIF4G); BDNF receptor (TrkB); eEF2 (EF2); glutamate (Glu); 7-methylguanosine cap structure (m7G); rat sarcoma (ras). See text for detailed explanations and further abbreviations

binding site on eIF4E. Thereby, they hinder eIF4F assembly and translation initiation. Hyperphosphorylation of 4E-BPs disrupts their interaction with eIF4E and leads to the formation of eIF4F. Moreover, recruitment of eIF4E into the eIF4F complex stimulates phosphorylation of eIF4E, thereby decreases its affinity for the 5'cap and thus promotes binding of the 43S preinitiation complex to the initiator codon to form a 48S complex. Subsequently, eIF4 proteins dissociate from the complex, while joining of the 60S ribosomal subunit leads to the formation of an 80S ribosomal complex and the start of elongation (Jackson et al. 2010). During this phase, eukaryotic elongation factor 1A (eEF1A) recruits aminoacyl-tRNAs to the ribosomal A site. After peptide bond formation, elongation factor eEF2 catalyzes the downstream movement of the ribosome along the mRNA by one codon. Finally, various release factors mediate the termination of translation at the stop codon.

13.4.2 *Local Control of Protein Synthesis at Synapses*

Local translation at particular synapses can be triggered by activation of different transmembrane proteins, including receptors for neurotransmitters, hormones, neurotrophins and extracellular matrix molecules (Bramham and Wells 2007; Cajigas et al. 2010; Kindler et al. 2005; Sutton and Schuman 2006). For example, activation of NMDA-Rs and mGluRs, but not AMPA-Rs, stimulates translation of a dendritic reporter mRNA (Gong et al. 2006). Also, application of dihydroxyphenylglycine (DHPG), an agonist of group I mGluRs, induces LTD (DHPG-LTD), a process that depends on dendritic protein synthesis (Bramham and Wells 2007). Although application of brain-derived neurotrophic factor (BDNF) induces the opposite effect, namely, LTP (BDNF-LTP), it also requires translation of dendritic mRNAs (Kang and Schuman 1996). Consistent with the finding that β-adrenergic receptors (β-ARs) can influence the formation of long-lasting memories, their activation stimulates protein synthesis during long-term synaptic potentiation (Gelinas et al. 2007). In addition, a spatially restricted translation along dendrites appears to be established by the physical interaction of transmembrane receptors with components of the protein synthesis machinery. DCC, a receptor for the extracellular factor netrin, is found along dendrites where it assembles into complexes containing components of the translational machinery (Tcherkezian et al. 2010). Upon netrin binding, translation components are released from the complex, and local protein synthesis is stimulated.

During initiation, eIF4E is the main target of control by these transmembrane receptors (Bramham and Wells 2007). Two receptor-coupled kinase pathways are involved, the mammalian target of rapamycin (mTOR) and the mitogen-activated protein kinase (MAPK)/extracellular signal–regulated kinase (ERK) signalling cascade. Binding of growth factors such as BDNF to their respective receptors activates phosphatidylinositol 3-kinase (PI3K) and enhances production of phosphatidylinositol-3,4,5-triphosphate (PIP3). The latter serves as a second messenger that induces membrane recruitment and activation of kinases, including protein kinase B/Akt. Akt phosphorylates the tuberous sclerosis complex (TSC) and thereby abolishes its inhibitory effect on the kinase mTOR. The finding that rapamycin, a potent inhibitor of mTOR activity, disrupts the protein synthesis–dependent late phase of LTP induced by either high-frequency stimulation or BDNF treatment provided a first clue that the mTOR pathway plays a crucial role in controlling dendritic translation (Tang et al. 2002). Interestingly, despite an opposing change in synaptic efficacy, DHPG-LTD also involves activation of the mTOR signalling cascade (Bramham and Wells 2007). Finally, NMDA-R activity can also trigger mTOR-mediated dendritic protein synthesis (Gong et al. 2006). In dendrites, activated mTOR phosphorylates 4E-BPs resulting in the dissociation of eIF4E and 4E-BPs, an enhanced formation of eIF4F and an increased translation rate. In agreement with this finding, key components of the mTOR signalling cascade, including mTOR, eIF4E and 4E-BPs, are present at postsynaptic sites. Also, in primary neurons and isolated dendrites, BDNF induces phosphorylation of mTOR and 4E-BPs and stimulates the local dendritic translation of reporter mRNAs (Aakalu et al. 2001; Takei et al. 2001, 2004; Tang et al. 2002). A number of dendritic transcripts are targets of the mTOR pathway, including αCaMKII, MAP2 and GluR1 mRNAs (Gong et al. 2006; Schratt et al. 2004; Slipczuk et al. 2009). The BDNF-dependent translation of GluR1 mRNAs appears to be required for particular forms of memory consolidation (Slipczuk et al. 2009). Interestingly, suppression of mTOR pathway activity may also stimulate dendritic protein synthesis (Raab-Graham et al. 2006). Thus, in mammalian dendrites, the mTOR signalling cascade appears to serve as a cellular switch that differentially regulates local translation of distinct sets of mRNAs.

The MAPK/ERK signalling pathway also stimulates dendritic protein synthesis (Kelleher et al. 2004). Activated MAPK/ERK phosphorylates and thereby activates MAP kinase-interacting kinase 1 (Mnk1) (Fukunaga and Hunter 1997; Waskiewicz et al. 1999). Mnk1 subsequently phosphorylates eIF4E residing in eIF4F complexes, thus reducing eIF4E affinity for the $5'$cap, promoting binding of the 43S complex to the initiator codon and increasing the translation initiation rate (Pyronnet et al. 1999; Waskiewicz et al. 1999). For example, in hippocampal slices, activation of NMDA-Rs induces ERK- and PKA-dependent Mnk1 activation and increased eIF4E phosphorylation (Banko et al. 2004). In addition, BDNF-LTP involves MEK-MAPK/ERK-dependent phosphorylation of eIF4E and triggers αCaMKII synthesis (Kanhema et al. 2006).

Taken together, these data indicate that both MAPK/ERK and mTOR signalling pathways co-regulate $5'$cap-dependent translation in dendrites by targeting eIF4E activity. In rats, both the formation and retention of long-term fear conditioning memory require activation of mTOR pathways (Parsons et al. 2006). Also, MAPK/ERK and mTOR pathways interact in LTP formation (Tsokas et al. 2007). In particular, coincident activity of phosphatidylinositide 3-kinase and MAPK/ERK pathways are required for mTOR-mediated dendritic translation of mRNAs containing a $5'$ terminal oligopyrimidine tract (TOP). As $5'$ TOP transcripts encode distinct components of the protein synthesis machinery, this process appears to increase the general translation capacity of dendrites. Moreover, co-activation of both MAPK/ERK and mTOR pathways is required for both mGluR-mediated LTD and β-AR-induced LTP (Banko et al. 2006; Gelinas et al. 2007).

Together with the scaffolding protein eIF4G and the ATP-dependent RNA helicase eIF4A, eIF4E constitutes heterotrimeric eIF4F, which mediates the recruitment of the 43S complex to the 5′cap (Sonenberg and Hinnebusch 2009). eIF4F activity is enhanced by its interaction with PABP associated with the poly(A) tails of mRNAs. This interaction is controlled by the dendritically localized, untranslated BC1 RNA. It represses translation initiation by preventing the assembly of the 48S preinitiation complex (Wang et al. 2002, 2005). In particular, BC1 blocks the RNA duplex unwinding activity of eIF4A while stimulating its ATPase activity (Lin et al. 2008). Uncoupling of both processes prevents the recruitment of the small ribosomal subunit to complex 5′-UTR structures. BC1 loss results in an mGluR-stimulated and translation-dependent hyperexcitability of neurons (Zhong et al. 2009).

Another mechanism controlling translation initiation involves CPEs in 3′-UTRs. Several dendritic mRNAs appear to remain translationally dormant until their short poly(A) tails are extended (Bramham and Wells 2007; Costa-Mattioli et al. 2009a; Wells 2006). Upon phosphorylation of the *trans*-acting CPEB by CaMKII or aurora, a deadenylase (PARN) is released from the 3′-UTR, while the poly(A) polymerase GLD2 polyadenylates the mRNA. This results in the dissociation of the 4E-BP maskin from eIF4E and the activation of translation (Atkins et al. 2004; Barnard et al. 2004; Costa-Mattioli et al. 2009b). αCaMKII transcripts contain two CPE-like sequences (Wu et al. 1998). In neurons, NMDA-R and mGluR activations stimulate the CPE/CPEB pathway and trigger dendritic polyadenylation and translation of mRNAs encoding αCaMKII (Huang et al. 2002; Wells et al. 2001) and plasminogen activator, respectively (Shin et al. 2004).

Elongation appears to represent another rather rare target for local dendritic translation control. This mechanism often involves modulation of eEF2 activity. While phosphorylation of eEF2 by eEF2 kinase (eEF2K) slows elongation of translation, thus reducing overall protein synthesis rate (Nairn et al. 2001), it enhances translation of select mRNAs in neurons (Scheetz et al. 2000). For example, during mGluR-activated LTD, eEF2 phosphorylation by eEF2K inhibits general protein synthesis while simultaneously increasing Arg3.1/Arc mRNA translation (Park et al. 2008) and AMPA-R endocytosis rates (Waung et al. 2008). As translational regulation of Arg3.1/Arc mRNAs is disrupted in FMRP-deficient mice, both eEF2K-eEF2 and FMRP appear to co-ordinately regulate local Arg3.1/Arc synthesis in dendrites (Park et al. 2008). Similarly, mGluR-mediated phosphorylation of eEF2 stimulates dendritic synthesis of BDNF, which stabilizes dendritic spines (Verpelli et al. 2010).

13.4.3   FMRP and miRNAs: A Role in Cognitive Functions in Mice and Men

Although FMRP has since long been implicated in dendritic translation control, its exact mode of action is still unclear (Bassell and Warren 2008). It has been proposed that FMRP associates with BC1 to repress translation of particular dendritic transcripts, which base pair with BC1 (Zalfa et al. 2003). However, whether FMRP can indeed specifically associate with BC1 remains a matter of debate (Iacoangeli et al. 2008a, b). Also, the finding that FMRP primarily associates with polysomes (see review by Kindler et al. 2005 and publications cited therein), whereas BC1 resides in lighter RNP fractions (Krichevsky and Kosik 2001), supports an independent action of both molecules. In addition, it is not yet clear which phase of translation is targeted by FMRP. While the preferential association of FMRP with polysomes implicates a function after initiation, sequestration of eIF4E by the cytoplasmic FMRP interacting protein 1 (CYFIP1) may control initiation by inhibiting eIF4E function (Napoli et al. 2008).

Which signalling events act on FMRP? Upon activation of BDNF receptors or mGluRs, CYFIP1 dissociates from eIF4E, thereby stimulating synaptic protein synthesis (Napoli et al. 2008). Activation of mGluRs on primary neurons abolishes the FMRP-mediated translation block of reporter mRNAs carrying the Shank1-DTE and of endogenous transcripts encoding the catalytic subunit of phosphoinositide 3-kinase (PI3K) (Gross et al. 2010; Schütt et al. 2009). Consistently, neuronal loss of FMRP leads to increased postsynaptic levels of Shank1 and an enlarged PI3K activity. S6 kinase 1 (S6K1) phosphorylates FMRP at Ser499. Non-phosphorylated FMRP primarily associates with actively translating polysomes, while phospho-FMRP tends to associate with apparently stalled polysomes (Ceman et al. 2003). Therefore, the phosphorylated form of FMRP inhibits translation (Ceman et al. 2003; Narayanan et al. 2008). mGluR activity induces a brief dephosphorylation of phospho-FMRP by PP2A. This abolishes the translational block by FMRP (as shown for dendritic SAPAP3 mRNAs). Shortly thereafter, mGluR activity induces the re-phosphorylation of FMRP by S6K1. Thus, loss of this kinase activity results in the absence of phospho-FMRP and increased levels of postsynaptic SAPAP3 (as observed in S6K1- and FMRP-deficient mice) (Narayanan et al. 2008; Schütt et al. 2009). These data suggest that the mGluR/FMRP pathway allows for a brief phase of synthesis from otherwise translationally repressed messages, with the capacity to shut down translation after an appropriate period of time.

Another proposed mechanism links FMRP with microRNAs (miRNAs) (Cheever and Ceman 2009; Jin et al. 2004). miRNAs are small RNAs of 21–24 nucleotides in length that serve as translation repressors through partial base pairing with their target mRNAs (Slezak-Prochazka et al. 2010). FMRP associates with distinct miRNAs, including miR-125b and miR-132 (Edbauer et al. 2010). miR-125b-associated FMRP suppresses translation of NMDA-R mRNAs. Several recent findings support the significance of miRNAs for translation control at synapses (Schratt 2009). For example, miR-134 inhibits dendritic translation of mRNAs encoding the protein kinase Limk1 (Schratt et al. 2006). BDNF treatment abolishes this translational block and thus contributes to dendritic spine maturation and synaptic plasticity. Moreover, an interaction of the DExD-box RNA helicase MOV10 with the respective 3′-UTRs appears to suppress translation of mRNAs encoding αCaMKII, Limk1 and the depalmitoylating enzyme lysophospholipase1 (Lypla1) at synapses (Banerjee et al. 2009). Translational repression of Limk1 transcripts seems to involve miR-138. In addition, activity-dependent degradation of MOV10 by the proteasome removes the translational block on αCaMKII and Lypla1 mRNAs.

13.5 Perspectives

Dendritic mRNA targeting and local translation have now been firmly established as cellular mechanisms to provide proteins to specific dendritic domains under local translational control. The relevance of this phenomenon for synaptic plasticity and human cognition is underlined by several mouse models and human genetic diseases. Most importantly, deficits in translational control as observed in FXS and possibly also autism are associated with aberrant neuronal morphology and synaptic plasticity in mice and men. Nevertheless, numerous open questions remain, which will require further research. Further quantitative work is needed to determine which mRNAs are present in dendrites in physiologically significant amounts. Possible approaches here should include microarray hybridization approaches (Zhong et al. 2006) and quantitative real-time RT-PCR on biochemically isolated tissue fractions such as synaptosomes (compared to total brain homogenates).

Similarly, cellular mechanisms of dendritic mRNA targeting require further clarification. Though the molecular machinery of dendritic mRNA transport is now being approached from at least two sides (motor proteins and identified DTEs), it is still unclear how dendritic RNPs are attached to cargo-binding domains of motor proteins, which interactions of DTEs with RBPs are relevant to select individual RNAs for transport and which processing events in the nucleus determine extrasomatic RNA trafficking. Importantly, the field suffers from an almost complete lack of *in vivo* studies. Although different RBP genes have been deleted in mice, in most cases, these animals have apparently not been analysed with respect to dendritic targeting of some major dendritic mRNAs such as αCaMKII, Arg3.1/Arc or MAP2 messages. One notable exception is the Hzf knockout mouse, in which loss of the RBP leads to reduced dendritic localization of IP3 receptor mRNAs in Purkinje cells (Iijima et al. 2005). Also, RBP knockdown studies in primary neurons have been limited to the analysis of overexpressed dendritic reporter mRNAs.

Finally, it is unknown in most cases why a given protein needs to be synthesized in dendrites. This uncertainty may be best illustrated in the case of AMPA-Rs, or membrane proteins in general. Local translation of, for example, GluR1 and GluR2 mRNAs (Grooms et al. 2006; Ju et al. 2004; Kacharmina et al. 2000) produces membrane proteins, which need to be processed by the Golgi apparatus. However, most dendrites do not contain Golgi cisternae, and usually, only one dendrite per neuron harbours a limited number of isolated Golgi stacks (termed Golgi outposts by Hanus and Ehlers (2008)). Thus, for AMPA-R subunits as for most membrane proteins, the advantage of being synthesized close to a postsynaptic site might be taken away by the need for retrograde transport to the Golgi apparatus, followed again by anterograde transport into dendrites and eventual insertion into the postsynaptic membrane. It is difficult to envision how this process contributes to the rapid and synapse-specific turnover of AMPA-Rs, which is observed in LTP and LTD. In these and other cases, it will be helpful to genetically interfere with dendritic mRNA targeting *in vivo*, for example by deleting DTEs or by knockout/knockdown of proteins involved in mRNA transport. In this respect, deletion of the DTE in αCaMKII mRNAs in mice (Miller et al. 2002) represents a lone example where deficits in dendritic mRNA targeting have been linked to significant changes in synaptic protein content and function *in vivo*. More experiments like this will be required to fully understand how synaptic activity may be modified by local protein synthesis.

Acknowledgements Work in the authors' laboratories is supported by grants from FRAXA, Deutsche Forschungsgemeinschaft and Fritz-Thyssen-Stiftung.

References

Aakalu, G., Smith, W. B., Nguyen, N., Jiang, C., & Schuman, E. M. (2001). Dynamic visualization of local protein synthesis in hippocampal neurons. *Neuron, 2*, 489–502.

Atkins, C. M., Nozaki, N., Shigeri, Y., & Soderling, T. R. (2004). Cytoplasmic polyadenylation element binding protein-dependent protein synthesis is regulated by calcium/calmodulin-dependent protein kinase II. *The Journal of Neuroscience, 22*, 5193–5201.

Banerjee, S., Neveu, P., & Kosik, K. S. (2009). A coordinated local translational control point at the synapse involving relief from silencing and MOV10 degradation. *Neuron, 6*, 871–884.

Banko, J. L., Hou, L., & Klann, E. (2004). NMDA receptor activation results in PKA- and ERK-dependent Mnk1 activation and increased eIF4E phosphorylation in hippocampal area CA1. *Journal of Neurochemistry, 2*, 462–470.

Banko, J. L., Hou, L., Poulin, F., Sonenberg, N., & Klann, E. (2006). Regulation of eukaryotic initiation factor 4E by converging signaling pathways during metabotropic glutamate receptor-dependent long-term depression. *The Journal of Neuroscience, 8*, 2167–2173.

Barnard, D. C., Ryan, K., Manley, J. L., & Richter, J. D. (2004). Symplekin and xGLD-2 are required for CPEB-mediated cytoplasmic polyadenylation. *Cell, 5*, 641–651.

Baron, M. K., Boeckers, T. M., Vaida, B., Faham, S., Gingery, M., Sawaya, M. R., Salyer, D., Gundelfinger, E. D., & Bowie, J. U. (2006). An architectural framework that may lie at the core of the postsynaptic density. *Science, 5760*, 531–535.

Bassell, G. J., & Warren, S. T. (2008). Fragile X syndrome: Loss of local mRNA regulation alters synaptic development and function. *Neuron, 2*, 201–214.

Berkel, S., Marshall, C. R., Weiss, B., Howe, J., Roeth, R., Moog, U., Endris, V., Roberts, W., Szatmari, P., Pinto, D., Bonin, M., Riess, A., Engels, H., Sprengel, R., Scherer, S. W., & Rappold, G. A. (2010). Mutations in the SHANK2 synaptic scaffolding gene in autism spectrum disorder and mental retardation. *Nature Genetics, 6*, 489–491.

Blichenberg, A., Rehbein, M., Muller, R., Garner, C. C., Richter, D., & Kindler, S. (2001). Identification of a *cis*-acting dendritic targeting element in the mRNA encoding the alpha subunit of Ca^{2+}/calmodulin-dependent protein kinase II. *European Journal of Neuroscience, 10*, 1881–1888.

Böckers, T. M., Segger-Junius, M., Iglauer, P., Bockmann, J., Gundelfinger, E. D., Kreutz, M. R., Richter, D., Kindler, S., & Kreienkamp, H. J. (2004). Differential expression and dendritic transcript localization of Shank family members: Identification of a dendritic targeting element in the 3′ untranslated region of Shank1 mRNA. *Molecular and Cellular Neuroscience, 1*, 182–190.

Bonaglia, M. C., Giorda, R., Borgatti, R., Felisari, G., Gagliardi, C., Selicorni, A., & Zuffardi, O. (2001). Disruption of the ProSAP2 gene in a t(12;22)(q24.1;q13.3) is associated with the 22q13.3 deletion syndrome. *American Journal of Human Genetics, 2*, 261–268.

Bramham, C. R., & Wells, D. G. (2007). Dendritic mRNA: Transport, translation and function. *Nature Reviews Neuroscience, 10*, 776–789.

Burgin, K. E., Waxham, M. N., Rickling, S., Westgate, S. A., Mobley, W. C., & Kelly, P. T. (1990). In situ hybridization histochemistry of Ca^{2+}/calmodulin-dependent protein kinase in developing rat brain. *The Journal of Neuroscience, 6*, 1788–1798.

Cajigas, I. J., Will, T., & Schuman, E. M. (2010). Protein homeostasis and synaptic plasticity. *The EMBO Journal, 16*, 2746–2752.

Ceman, S., O'Donnell, W. T., Reed, M., Patton, S., Pohl, J., & Warren, S. T. (2003). Phosphorylation influences the translation state of FMRP-associated polyribosomes. *Human Molecular Genetics, 24*, 3295–3305.

Cheever, A., & Ceman, S. (2009). Translation regulation of mRNAs by the fragile X family of proteins through the microRNA pathway. *RNA Biology, 2*, 175–178.

Chowdhury, S., Shepherd, J. D., Okuno, H., Lyford, G., Petralia, R. S., Plath, N., Kuhl, D., Huganir, R. L., & Worley, P. F. (2006). Arc/Arg3.1 interacts with the endocytic machinery to regulate AMPA receptor trafficking. *Neuron, 3*, 445–459.

Costa-Mattioli, M., Sonenberg, N., & Richter, J. D. (2009a). Chapter 8 translational regulatory mechanisms in synaptic plasticity and memory storage. *Progress in Molecular Biology and Translational Science, 90*, 293–311.

Costa-Mattioli, M., Sossin, W. S., Klann, E., & Sonenberg, N. (2009b). Translational control of long-lasting synaptic plasticity and memory. *Neuron, 1*, 10–26.

Dieterich, D. C., Karpova, A., Mikhaylova, M., Zdobnova, I., Konig, I., Landwehr, M., Kreutz, M., Smalla, K. H., Richter, K., Landgraf, P., Reissner, C., Boeckers, T. M., Zuschratter, W., Spilker, C., Seidenbecher, C. I., Garner, C. C., Gundelfinger, E. D., & Kreutz, M. R. (2008). Caldendrin-Jacob: A protein liaison that couples NMDA receptor signalling to the nucleus. *PLoS Biology, 2*, e34.

Durand, C. M., Betancur, C., Boeckers, T. M., Bockmann, J., Chaste, P., Fauchereau, F., Nygren, G., Rastam, M., Gillberg, I. C., Anckarsater, H., Sponheim, E., Goubran-Botros, H., Delorme, R., Chabane, N., Mouren-Simeoni, M. C., de Mas, P., Bieth, E., Roge, B., Heron, D., Burglen, L., Gillberg, C., Leboyer, M., & Bourgeron, T. (2007). Mutations in the gene encoding the synaptic scaffolding protein SHANK3 are associated with autism spectrum disorders. *Nature Genetics, 1*, 25–27.

Eberwine, J., & Crino, P. (2001). Analysis of mRNA populations from single live and fixed cells of the central nervous system. *Current Protocols in Neuroscience*, Unit 5.3.

Edbauer, D., Neilson, J. R., Foster, K. A., Wang, C. F., Seeburg, D. P., Batterton, M. N., Tada, T., Dolan, B. M., Sharp, P. A., & Sheng, M. (2010). Regulation of synaptic structure and function by FMRP-associated microRNAs miR-125b and miR-132. *Neuron, 3*, 373–384.

Falley, K., Schütt, J., Iglauer, P., Menke, K., Maas, C., Kneussel, M., Kindler, S., Wouters, F. S., Richter, D., & Kreienkamp, H. J. (2009). Shank1 mRNA: Dendritic transport by kinesin and translational control by the 5′untranslated region. *Traffic, 7*, 844–857.

Flexner, J. B., Flexner, L. B., & Stellar, E. (1963). Memory in mice as affected by intracerebral puromycin. *Science, 141*, 57–59.

Fukunaga, R., & Hunter, T. (1997). MNK1, a new MAP kinase-activated protein kinase, isolated by a novel expression screening method for identifying protein kinase substrates. *The EMBO Journal, 8*, 1921–1933.

Gao, Y., Tatavarty, V., Korza, G., Levin, M. K., & Carson, J. H. (2008). Multiplexed dendritic targeting of alpha calcium calmodulin-dependent protein kinase II, neurogranin, and activity-regulated cytoskeleton-associated protein RNAs by the A2 pathway. *Molecular Biology of the Cell, 5*, 2311–2327.

Garner, C. C., Tucker, R. P., & Matus, A. (1988). Selective localization of messenger RNA for cytoskeletal protein MAP2 in dendrites. *Nature, 6200*, 674–677.

Gelinas, J. N., Banko, J. L., Hou, L., Sonenberg, N., Weeber, E. J., Klann, E., & Nguyen, P. V. (2007). ERK and mTOR signaling couple beta-adrenergic receptors to translation initiation machinery to gate induction of protein synthesis-dependent long-term potentiation. *Journal of Biological Chemistry, 37*, 27527–27535.

Gkogkas, C., Sonenberg, N., & Costa-Mattioli, M. (2010). Translational control mechanisms in long-lasting synaptic plasticity and memory. *Journal of Biological Chemistry, 42*, 31913–31917.

Gong, R., Park, C. S., Abbassi, N. R., & Tang, S. J. (2006). Roles of glutamate receptors and the mammalian target of rapamycin (mTOR) signaling pathway in activity-dependent dendritic protein synthesis in hippocampal neurons. *Journal of Biological Chemistry, 27*, 18802–18815.

Grooms, S. Y., Noh, K. M., Regis, R., Bassell, G. J., Bryan, M. K., Carroll, R. C., & Zukin, R. S. (2006). Activity bidirectionally regulates AMPA receptor mRNA abundance in dendrites of hippocampal neurons. *The Journal of Neuroscience, 32*, 8339–8351.

Gross, C., Nakamoto, M., Yao, X., Chan, C. B., Yim, S. Y., Ye, K., Warren, S. T., & Bassell, G. J. (2010). Excess phosphoinositide 3-kinase subunit synthesis and activity as a novel therapeutic target in fragile X syndrome. *The Journal of Neuroscience, 32*, 10624–10638.

Gundelfinger, E. D., Boeckers, T. M., Baron, M. K., & Bowie, J. U. (2006). A role for zinc in postsynaptic density asSAMbly and plasticity? *Trends in Biochemical Sciences, 7*, 366–373.

Hanus, C., & Ehlers, M. D. (2008). Secretory outposts for the local processing of membrane cargo in neuronal dendrites. *Traffic, 9*, 1437–1445.

Herb, A., Wisden, W., Catania, M., Marechal, D., Dresse, A., & Seeburg, P. (1997). Prominent dendritic localization in forebrain neurons of a novel mRNA and its product, dendrin. *Molecular and Cellular Neuroscience, 8*, 367–374.

Huang, Y. S., Carson, J. H., Barbarese, E., & Richter, J. D. (2003). Facilitation of dendritic mRNA transport by CPEB. *Genes & Development, 5*, 638–653.

Huang, Y. S., Jung, M. Y., Sarkissian, M., & Richter, J. D. (2002). N-methyl-D-aspartate receptor signaling results in Aurora kinase-catalyzed CPEB phosphorylation and alpha CaMKII mRNA polyadenylation at synapses. *The EMBO Journal, 9*, 2139–2148.

Huber, K. M., Gallagher, S. M., Warren, S. T., & Bear, M. F. (2002). Altered synaptic plasticity in a mouse model of fragile X mental retardation. *Proceedings of the National Academy of Sciences of the United States of America, 11*, 7746–7750.

Iacoangeli, A., Rozhdestvensky, T. S., Dolzhanskaya, N., Tournier, B., Schütt, J., Brosius, J., Denman, R. B., Khandjian, E. W., Kindler, S., & Tiedge, H. (2008a). On BC1 RNA and the fragile X mental retardation protein. *Proceedings of the National Academy of Sciences of the United States of America, 2*, 734–739.

Iacoangeli, A., Rozhdestvensky, T. S., Dolzhanskaya, N., Tournier, B., Schütt, J., Brosius, J., Denman, R. B., Khandjian, E. W., Kindler, S., & Tiedge, H. (2008b). Reply to Bagni: On BC1 RNA and the fragile X mental retardation protein. *Proceedings of the National Academy of Sciences of the United States of America, 22*, E29.

Iijima, T., Imai, T., Kimura, Y., Bernstein, A., Okano, H. J., Yuzaki, M., & Okano, H. (2005). Hzf protein regulates dendritic localization and BDNF-induced translation of type 1 inositol 1,4,5-trisphosphate receptor mRNA. *Proceedings of the National Academy of Sciences of the United States of America, 47*, 17190–17195.

Jackson, R. J., Hellen, C. U., & Pestova, T. V. (2010). The mechanism of eukaryotic translation initiation and principles of its regulation. *Nature Reviews: Molecular Cell Biology, 2*, 113–127.

Jin, P., Alisch, R. S., & Warren, S. T. (2004). RNA and microRNAs in fragile X mental retardation. *Nature Cell Biology, 11*, 1048–1053.

Ju, W., Morishita, W., Tsui, J., Gaietta, G., Deerinck, T. J., Adams, S. R., Garner, C. C., Tsien, R. Y., Ellisman, M. H., & Malenka, R. C. (2004). Activity-dependent regulation of dendritic synthesis and trafficking of AMPA receptors. *Nature Neuroscience, 3*, 244–253.

Kacharmina, J. E., Job, C., Crino, P., & Eberwine, J. (2000). Stimulation of glutamate receptor protein synthesis and membrane insertion within isolated neuronal dendrites. *Proceedings of the National Academy of Sciences of the United States of America, 21*, 11545–11550.

Kanai, Y., Dohmae, N., & Hirokawa, N. (2004). Kinesin transports RNA: Isolation and characterization of an RNA-transporting granule. *Neuron, 4*, 513–525.

Kang, H., & Schuman, E. M. (1996). A requirement for local protein synthesis in neurotrophin-induced hippocampal synaptic plasticity. *Science, 5280*, 1402–1406.

Kanhema, T., Dagestad, G., Panja, D., Tiron, A., Messaoudi, E., Havik, B., Ying, S. W., Nairn, A. C., Sonenberg, N., & Bramham, C. R. (2006). Dual regulation of translation initiation and peptide chain elongation during BDNF-induced LTP in vivo: Evidence for compartment-specific translation control. *Journal of Neurochemistry, 5*, 1328–1337.

Kelleher, R. J., III, & Bear, M. F. (2008). The autistic neuron: Troubled translation? *Cell, 3*, 401–406.

Kelleher, R. J., III, Govindarajan, A., Jung, H. Y., Kang, H., & Tonegawa, S. (2004). Translational control by MAPK signaling in long-term synaptic plasticity and memory. *Cell, 3*, 467–479.

Kim, E., Naisbitt, S., Hsueh, Y. P., Rao, A., Rothschild, A., Craig, A. M., & Sheng, M. (1997). GKAP, a novel synaptic protein that interacts with the guanylate kinase-like domain of the PSD-95/SAP90 family of channel clustering molecules. *The Journal of Cell Biology, 3*, 669–678.

Kindler, S., Dieterich, D. C., Schutt, J., Sahin, J., Karpova, A., Mikhaylova, M., Schob, C., Gundelfinger, E. D., Kreienkamp, H. J., & Kreutz, M. R. (2009). Dendritic mRNA targeting of Jacob and N-methyl-d-aspartate-induced nuclear translocation after calpain-mediated proteolysis. *Journal of Biological Chemistry, 37*, 25431–25440.

Kindler, S., Rehbein, M., Classen, B., Richter, D., & Böckers, T. M. (2004). Distinct spatiotemporal expression of SAPAP transcripts in the developing rat brain: A novel dendritically localized mRNA. *Brain Research: Molecular Brain Research, 1*, 14–21.

Kindler, S., Wang, H., Richter, D., & Tiedge, H. (2005). RNA transport and local control of translation. *Annual Review of Cell and Developmental Biology, 21*, 223–245.

Kobayashi, H., Yamamoto, S., Maruo, T., & Murakami, F. (2005). Identification of a cis-acting element required for dendritic targeting of activity-regulated cytoskeleton-associated protein mRNA. *European Journal of Neuroscience, 12*, 2977–2984.

Krichevsky, A. M., & Kosik, K. S. (2001). Neuronal RNA granules: A link between RNA localization and stimulation-dependent translation. *Neuron, 4*, 683–696.

Lin, D., Pestova, T. V., Hellen, C. U., & Tiedge, H. (2008). Translational control by a small RNA: Dendritic BC1 RNA targets the eukaryotic initiation factor 4A helicase mechanism. *Molecular and Cellular Biology, 9*, 3008–3019.

Link, W., Konietzko, U., Kauselmann, G., Krug, M., Schwanke, B., Frey, U., & Kuhl, D. (1995). Somatodendritic expression of an immediate early gene is regulated by synaptic activity. *Proceedings of the National Academy of Sciences of the United States of America, 12*, 5734–5738.

Lyford, G. L., Yamagata, K., Kaufmann, W. E., Barnes, C. A., Sanders, L. K., Copeland, N. G., Gilbert, D. J., Jenkins, N. A., Lanahan, A. A., & Worley, P. F. (1995). Arc, a growth factor and activity-regulated gene, encodes a novel cytoskeleton-associated protein that is enriched in neuronal dendrites. *Neuron, 2*, 433–445.

Miller, S., Yasuda, M., Coats, J. K., Jones, Y., Martone, M. E., & Mayford, M. (2002). Disruption of dendritic translation of CaMKIIalpha impairs stabilization of synaptic plasticity and memory consolidation. *Neuron, 3*, 507–519.

Monshausen, M., Putz, U., Rehbein, M., Schweizer, M., DesGroseillers, L., Kuhl, D., Richter, D., & Kindler, S. (2001). Two rat brain Staufen isoforms differentially bind RNA. *Journal of Neurochemistry, 1*, 155–165.

Mori, Y., Imaizumi, K., Katayama, T., Yoneda, T., & Tohyama, M. (2000). Two cis-acting elements in the 3$'$ untranslated region of alpha-CaMKII regulate its dendritic targeting. *Nature Neuroscience, 11*, 1079–1084.

Nairn, A. C., Matsushita, M., Nastiuk, K., Horiuchi, A., Mitsui, K., Shimizu, Y., & Palfrey, H. C. (2001). Elongation factor-2 phosphorylation and the regulation of protein synthesis by calcium. *Progress in Molecular and Subcellular Biology, 27*, 91–129.

Naisbitt, S., Kim, E., Tu, J. C., Xiao, B., Sala, C., Valtschanoff, J., Weinberg, R. J., Worley, P. F., & Sheng, M. (1999). Shank, a novel family of postsynaptic density proteins that binds to the NMDA receptor/PSD-95/GKAP complex and cortactin. *Neuron, 3*, 569–582.

Napoli, I., Mercaldo, V., Boyl, P. P., Eleuteri, B., Zalfa, F., De Rubeis, S., Di Marino, D., Mohr, E., Massimi, M., Falconi, M., Witke, W., Costa-Mattioli, M., Sonenberg, N., Achsel, T., & Bagni, C. (2008). The fragile X syndrome protein represses activity-dependent translation through CYFIP1, a new 4E-BP. *Cell, 6*, 1042–1054.

Narayanan, U., Nalavadi, V., Nakamoto, M., Thomas, G., Ceman, S., Bassell, G. J., & Warren, S. T. (2008). S6K1 phosphorylates and regulates fragile X mental retardation protein (FMRP) with the neuronal protein synthesis-dependent mammalian target of rapamycin (mTOR) signaling cascade. *Journal of Biological Chemistry, 27*, 18478–18482.

Ostroff, L. E., Cain, C. K., Bedont, J., Monfils, M. H., & Ledoux, J. E. (2010). Fear and safety learning differentially affect synapse size and dendritic translation in the lateral amygdala. *Proceedings of the National Academy of Sciences of the United States of America, 20*, 9418–9423.

Ouyang, Y., Rosenstein, A., Kreiman, G., Schuman, E. M., & Kennedy, M. B. (1999). Tetanic stimulation leads to increased accumulation of Ca(2+)/calmodulin-dependent protein kinase II via dendritic protein synthesis in hippocampal neurons. *The Journal of Neuroscience, 18*, 7823–7833.

Park, S., Park, J. M., Kim, S., Kim, J. A., Shepherd, J. D., Smith-Hicks, C. L., Chowdhury, S., Kaufmann, W., Kuhl, D., Ryazanov, A. G., Huganir, R. L., Linden, D. J., & Worley, P. F. (2008). Elongation factor 2 and fragile X mental retardation protein control the dynamic translation of Arc/Arg3.1 essential for mGluR-LTD. *Neuron, 1*, 70–83.

Parsons, R. G., Gafford, G. M., & Helmstetter, F. J. (2006). Translational control via the mammalian target of rapamycin pathway is critical for the formation and stability of long-term fear memory in amygdala neurons. *The Journal of Neuroscience, 50*, 12977–12983.

Pfeiffer, B. E., & Huber, K. M. (2006). Current advances in local protein synthesis and synaptic plasticity. *The Journal of Neuroscience, 27*, 7147–7150.

Pinkstaff, J. K., Chappell, S. A., Mauro, V. P., Edelman, G. M., & Krushel, L. A. (2001). Internal initiation of translation of five dendritically localized neuronal mRNAs. *Proceedings of the National Academy of Sciences of the United States of America, 5*, 2770–2775.

Plath, N., Ohana, O., Dammermann, B., Errington, M. L., Schmitz, D., Gross, C., Mao, X., Engelsberg, A., Mahlke, C., Welzl, H., Kobalz, U., Stawrakakis, A., Fernandez, E., Waltereit, R., Bick-Sander, A., Therstappen, E., Cooke, S. F., Blanquet, V., Wurst, W., Salmen, B., Bosl, M. R., Lipp, H. P., Grant, S. G., Bliss, T. V., Wolfer, D. P., & Kuhl, D. (2006). Arc/Arg3.1 is essential for the consolidation of synaptic plasticity and memories. *Neuron, 3*, 437–444.

Pyronnet, S., Imataka, H., Gingras, A. C., Fukunaga, R., Hunter, T., & Sonenberg, N. (1999). Human eukaryotic translation initiation factor 4G (eIF4G) recruits mnk1 to phosphorylate eIF4E. *The EMBO Journal, 1*, 270–279.

Raab-Graham, K. F., Haddick, P. C., Jan, Y. N., & Jan, L. Y. (2006). Activity- and mTOR-dependent suppression of Kv1.1 channel mRNA translation in dendrites. *Science, 5796*, 144–148.

Rehbein, M., Wege, K., Buck, F., Schweizer, M., Richter, D., & Kindler, S. (2002). Molecular characterization of MARTA1, a protein interacting with the dendritic targeting element of MAP2 mRNAs. *Journal of Neurochemistry, 5*, 1039–1046.

Romorini, S., Piccoli, G., Jiang, M., Grossano, P., Tonna, N., Passafaro, M., Zhang, M., & Sala, C. (2004). A functional role of postsynaptic density-95-guanylate kinase-associated protein complex in regulating Shank assembly and stability to synapses. *The Journal of Neuroscience, 42*, 9391–9404.

Sala, C., Piech, V., Wilson, N. R., Passafaro, M., Liu, G., & Sheng, M. (2001). Regulation of dendritic spine morphology and synaptic function by Shank and Homer. *Neuron, 1*, 115–130.

Scheetz, A. J., Nairn, A. C., & Constantine-Paton, M. (2000). NMDA receptor-mediated control of protein synthesis at developing synapses. *Nature Neuroscience, 3*, 211–216.

Schratt, G. (2009). microRNAs at the synapse. *Nature Reviews Neuroscience, 12*, 842–849.

Schratt, G. M., Nigh, E. A., Chen, W. G., Hu, L., & Greenberg, M. E. (2004). BDNF regulates the translation of a select group of mRNAs by a mammalian target of rapamycin-phosphatidylinositol 3-kinase-dependent pathway during neuronal development. *The Journal of Neuroscience, 33*, 7366–7377.

Schratt, G. M., Tuebing, F., Nigh, E. A., Kane, C. G., Sabatini, M. E., Kiebler, M., & Greenberg, M. E. (2006). A brain-specific microRNA regulates dendritic spine development. *Nature, 7074*, 283–289.

Schütt, J., Falley, K., Richter, D., Kreienkamp, H. J., & Kindler, S. (2009). Fragile X mental retardation protein regulates the levels of scaffold proteins and glutamate receptors in postsynaptic densities. *Journal of Biological Chemistry, 38*, 25479–25487.

Shin, C. Y., Kundel, M., & Wells, D. G. (2004). Rapid, activity-induced increase in tissue plasminogen activator is mediated by metabotropic glutamate receptor-dependent mRNA translation. *The Journal of Neuroscience, 42*, 9425–9433.

Slezak-Prochazka, I., Durmus, S., Kroesen, B. J., & van den Berg, A. (2010). MicroRNAs, macrocontrol: Regulation of miRNA processing. *RNA, 6*, 1087–1095.

Slipczuk, L., Bekinschtein, P., Katche, C., Cammarota, M., Izquierdo, I., & Medina, J. H. (2009). BDNF activates mTOR to regulate GluR1 expression required for memory formation. *PloS One, 6*, e6007.

Sonenberg, N., & Hinnebusch, A. G. (2009). Regulation of translation initiation in eukaryotes: Mechanisms and biological targets. *Cell, 4*, 731–745.

Steward, O. (2002). mRNA at synapses, synaptic plasticity, and memory consolidation. *Neuron, 3*, 338–340.

Steward, O., & Levy, W. B. (1982). Preferential localization of polyribosomes under the base of dendritic spines in granule cells of the dentate gyrus. *The Journal of Neuroscience, 3*, 284–291.

Steward, O., & Worley, P. F. (2001). Selective targeting of newly synthesized Arc mRNA to active synapses requires NMDA receptor activation. *Neuron, 1*, 227–240.

Sutton, M. A., & Schuman, E. M. (2006). Dendritic protein synthesis, synaptic plasticity, and memory. *Cell, 1*, 49–58.

Takei, N., Inamura, N., Kawamura, M., Namba, H., Hara, K., Yonezawa, K., & Nawa, H. (2004). Brain-derived neurotrophic factor induces mammalian target of rapamycin-dependent local activation of translation machinery and protein synthesis in neuronal dendrites. *The Journal of Neuroscience, 44*, 9760–9769.

Takei, N., Kawamura, M., Hara, K., Yonezawa, K., & Nawa, H. (2001). Brain-derived neurotrophic factor enhances neuronal translation by activating multiple initiation processes: Comparison with the effects of insulin. *Journal of Biological Chemistry, 46*, 42818–42825.

Takeuchi, M., Hata, Y., Hirao, K., Toyoda, A., Irie, M., & Takai, Y. (1997). SAPAPs: A family of PSD-95/SAP90-associated proteins localized at postsynaptic density. *Journal of Biological Chemistry, 18*, 11943–11951.

Tang, S. J., Meulemans, D., Vazquez, L., Colaco, N., & Schuman, E. (2001). A role for a rat homolog of staufen in the transport of RNA to neuronal dendrites. *Neuron, 3*, 463–475.

Tang, S. J., Reis, G., Kang, H., Gingras, A. C., Sonenberg, N., & Schuman, E. M. (2002). A rapamycin-sensitive signaling pathway contributes to long-term synaptic plasticity in the hippocampus. *Proceedings of the National Academy of Sciences of the United States of America, 1*, 467–472.

Tcherkezian, J., Brittis, P. A., Thomas, F., Roux, P. P., & Flanagan, J. G. (2010). Transmembrane receptor DCC associates with protein synthesis machinery and regulates translation. *Cell, 4*, 632–644.

Tiruchinapalli, D. M., Oleynikov, Y., Kelic, S., Shenoy, S. M., Hartley, A., Stanton, P. K., Singer, R. H., & Bassell, G. J. (2003). Activity-dependent trafficking and dynamic localization of zipcode binding protein 1 and beta-actin mRNA in dendrites and spines of hippocampal neurons. *The Journal of Neuroscience, 8*, 3251–3261.

Tsokas, P., Ma, T., Iyengar, R., Landau, E. M., & Blitzer, R. D. (2007). Mitogen-activated protein kinase upregulates the dendritic translation machinery in long-term potentiation by controlling the mammalian target of rapamycin pathway. *The Journal of Neuroscience, 22*, 5885–5894.

Tübing, F., Vendra, G., Mikl, M., Macchi, P., Thomas, S., & Kiebler, M. A. (2010). Dendritically localized transcripts are sorted into distinct ribonucleoprotein particles that display fast directional motility along dendrites of hippocampal neurons. *The Journal of Neuroscience, 11*, 4160–4170.

Valtschanoff, J. G., & Weinberg, R. J. (2001). Laminar organization of the NMDA receptor complex within the postsynaptic density. *The Journal of Neuroscience, 4*, 1211–1217.

Verpelli, C., Piccoli, G., Zanchi, A., Gardoni, F., Huang, K., Brambilla, D., Di Luca, M., Battaglioli, E., & Sala, C. (2010). Synaptic activity controls dendritic spine morphology by modulating eEF2-dependent BDNF synthesis. *The Journal of Neuroscience, 17*, 5830–5842.

Wang, H., Iacoangeli, A., Lin, D., Williams, K., Denman, R. B., Hellen, C. U., & Tiedge, H. (2005). Dendritic BC1 RNA in translational control mechanisms. *The Journal of Cell Biology, 5*, 811–821.

Wang, H., Iacoangeli, A., Popp, S., Muslimov, I. A., Imataka, H., Sonenberg, N., Lomakin, I. B., & Tiedge, H. (2002). Dendritic BC1 RNA: Functional role in regulation of translation initiation. *The Journal of Neuroscience, 23*, 10232–10241.

Wang, H., Kim, S. S., & Zhuo, M. (2010). Roles of fragile X mental retardation protein in dopaminergic stimulation-induced synapse-associated protein synthesis and subsequent alpha-amino-3-hydroxyl-5-methyl-4-isoxazole-4-propionate (AMPA) receptor internalization. *Journal of Biological Chemistry, 28*, 21888–21901.

Waskiewicz, A. J., Johnson, J. C., Penn, B., Mahalingam, M., Kimball, S. R., & Cooper, J. A. (1999). Phosphorylation of the cap-binding protein eukaryotic translation initiation factor 4E by protein kinase Mnk1 in vivo. *Molecular and Cellular Biology, 3*, 1871–1880.

Waung, M. W., Pfeiffer, B. E., Nosyreva, E. D., Ronesi, J. A., & Huber, K. M. (2008). Rapid translation of Arc/Arg3.1 selectively mediates mGluR-dependent LTD through persistent increases in AMPAR endocytosis rate. *Neuron, 1*, 84–97.

Welch, J. M., Lu, J., Rodriguiz, R. M., Trotta, N. C., Peca, J., Ding, J. D., Feliciano, C., Chen, M., Adams, J. P., Luo, J., Dudek, S. M., Weinberg, R. J., Calakos, N., Wetsel, W. C., & Feng, G. (2007). Cortico-striatal synaptic defects and OCD-like behaviours in Sapap3-mutant mice. *Nature, 7156*, 894–900.

Welch, J. M., Wang, D., & Feng, G. (2004). Differential mRNA expression and protein localization of the SAP90/PSD-95-associated proteins (SAPAPs) in the nervous system of the mouse. *The Journal of Comparative Neurology, 1*, 24–39.

Wells, D. G. (2006). RNA-binding proteins: A lesson in repression. *The Journal of Neuroscience, 27*, 7135–7138.

Wells, D. G., Dong, X., Quinlan, E. M., Huang, Y. S., Bear, M. F., Richter, J. D., & Fallon, J. R. (2001). A role for the cytoplasmic polyadenylation element in NMDA receptor-regulated mRNA translation in neurons. *The Journal of Neuroscience, 24*, 9541–9548.

Wu, L., Wells, D., Tay, J., Mendis, D., Abbott, M. A., Barnitt, A., Quinlan, E., Heynen, A., Fallon, J. R., & Richter, J. D. (1998). CPEB-mediated cytoplasmic polyadenylation and the regulation of experience-dependent translation of alpha-CaMKII mRNA at synapses. *Neuron, 5*, 1129–1139.

Zalfa, F., Giorgi, M., Primerano, B., Moro, A., Di Penta, A., Reis, S., Oostra, B., & Bagni, C. (2003). The fragile X syndrome protein FMRP associates with BC1 RNA and regulates the translation of specific mRNAs at synapses. *Cell, 3*, 317–327.

Zhong, J., Chuang, S. C., Bianchi, R., Zhao, W., Lee, H., Fenton, A. A., Wong, R. K., & Tiedge, H. (2009). BC1 regulation of metabotropic glutamate receptor-mediated neuronal excitability. *The Journal of Neuroscience, 32*, 9977–9986.

Zhong, J., Zhang, T., & Bloch, L. M. (2006). Dendritic mRNAs encode diversified functionalities in hippocampal pyramidal neurons. *BMC Neuroscience, 7*, 17.

Zitzer, H., Hönck, H. H., Bächner, D., Richter, D., & Kreienkamp, H. J. (1999). Somatostatin receptor interacting protein defines a novel family of multidomain proteins present in human and rodent brain. *Journal of Biological Chemistry, 46*, 32997–33001.

Chapter 14
Gliotransmission and the Tripartite Synapse

Mirko Santello, Corrado Calì, and Paola Bezzi

Please note the erratum to this chapter at the end of the book.

Abstract In the last years, the classical view of glial cells (in particular of astrocytes) as a simple supportive cell for neurons has been replaced by a new vision in which glial cells are active elements of the brain. Such a new vision is based on the existence of a bidirectional communication between astrocytes and neurons at synaptic level. Indeed, perisynaptic processes of astrocytes express active G-protein-coupled receptors that are able (1) to sense neurotransmitters released from the synapse during synaptic activity, (2) to increase cytosolic levels of calcium, and (3) to stimulate the release of gliotransmitters that in turn can interact with the synaptic elements. The mechanism(s) by which astrocytes can release gliotransmitter has been extensively studied during the last years. Many evidences have suggested that a fraction of astrocytes in situ release neuroactive substances both with calcium-dependent and calcium-independent mechanism(s); whether these mechanisms coexist and under what physiological or pathological conditions they occur, it remains unclear. However, the calcium-dependent exocytotic vesicular release has received considerable attention due to its potential to occur under physiological conditions via a finely regulated way. By releasing gliotransmitters in millisecond time scale with a specific vesicular apparatus, astrocytes can integrate and process synaptic information and control or modulate synaptic transmission and plasticity.

Keywords Astrocytes • D-serine • Exocytosis • Extrasynaptic NMDA-receptors • Gliotransmitters

M. Santello • C. Calì • P. Bezzi (✉)
DBCM, Department of Physiology, University of Bern, Bühlplatz 5, 3012 Bern, Switzerland
e-mail: paola.bezzi@unil.ch

M.R. Kreutz and C. Sala (eds.), *Synaptic Plasticity*,
Advances in Experimental Medicine and Biology 970,
DOI 10.1007/978-3-7091-0932-8_14, © Springer-Verlag/Wien 2012

14.1   Introduction

At the beginning of the twentieth century, Camillo Golgi (1843–1926) and Santiago Ramón J Cajal (1852–1934), using various ingenious staining and microscopic techniques, discovered a huge diversity of glial cells in the brain and found the contacts formed between glial cells and blood vessels (Ramon and Cajal 1899). Further enhancements in morphological characterization of astrocytes, thanks to the improvements in cellular labeling and imaging technologies, showed that astrocytic morphology is far more complicated than previously thought (Fig. 14.1). Ultrastructural examination of the nervous system, for instance, revealed that astrocytes can be intimately associated with synapses, literally enwrapping many pre- and postsynaptic terminals (Ventura and Harris 1999). Nonetheless, for the following decades, glial cells were still considered passive elements in the central nervous system (CNS), bearing mostly supportive and nutritional roles. The fundamental difference between neurons and astrocytes lies in their electrical excitability – neurons are electrically excitable cells whereas astrocytes (as other glial cells) are nonexcitable neural cells. Neurons are able to respond to external stimuli by generation of a plasmalemmal "all-or-none" action potential, capable of propagating through the neuronal network, although not all neurons generate action potentials. Glial cells are

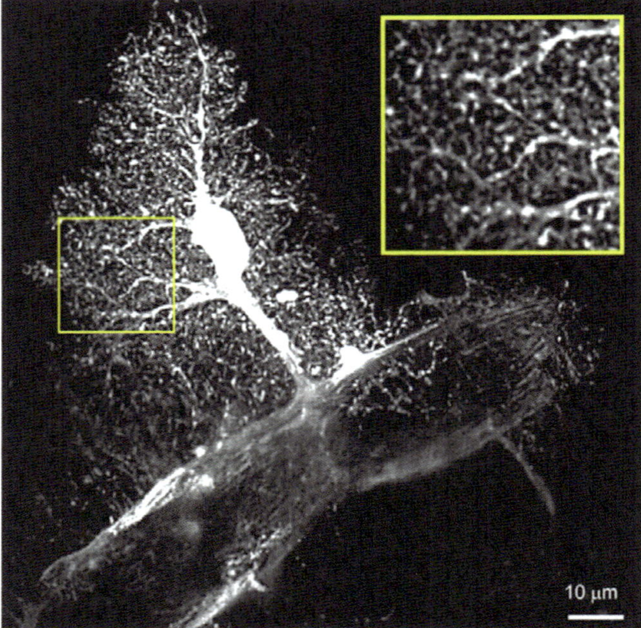

Fig. 14.1 The structural complexity of a protoplasmic astrocyte. A single astrocyte labeled with enhanced green fluorescent protein (eGFP) contacting a large blood vessel. Insert shows astrocytic processes at higher magnification (Adapted from Nedergaard et al. (2010))

unable to generate action potential in their plasma membrane (although they are able to express some voltage-gated channels). The advent of modern physiological techniques, most notably the patch-clamp and fluorescent calcium dyes, has dramatically changed this image of glia as "silent" brain cells.

Recently, by filling single astrocytes with fluorescent dyes, it has been found that astrocytes occupy nonoverlapping spatial territories in which a single astrocyte contacts hundreds of neuronal processes and multiple neuronal cell bodies (Volterra and Meldolesi 2005; Halassa et al. 2007; Bushong et al. 2005; Fig. 14.2). The processes of one astrocyte contact tens of thousands of synapses, with more than 50% of hippocampal excitatory synapses being closely opposed to an astrocytic process (Ventura and Harris 1999) at a structure called the "tripartite synapse" to recognize the structural and functional relationship between the astrocytes and the pre- and postsynaptic elements (Perea et al. 2009).

At perisynaptic sites, astrocytes exchange information with the synaptic neuronal elements, both responding to the neuron and regulating synaptic transmission. The concept of "tripartite synapse" began with a series of evidences obtained by many laboratories during the 1990s that revealed the existence of bidirectional communication between neurons and astrocytes (Bezzi and Volterra 2001; Perea et al. 2009). The signaling pathway between neurons and astrocytes at the "tripartite synapse" is reciprocal; astrocytes sense neuronal activity by increasing intracellular levels of calcium (Ca^{2+}) and respond by releasing a variety of different molecules (the so-called gliotransmitters; Bezzi and Volterra 2001). However, tripartite synapse has been debated because not all astrocytic calcium increases cause gliotransmitter release (Fiacco et al. 2007; Petravicz et al. 2008). The possibility

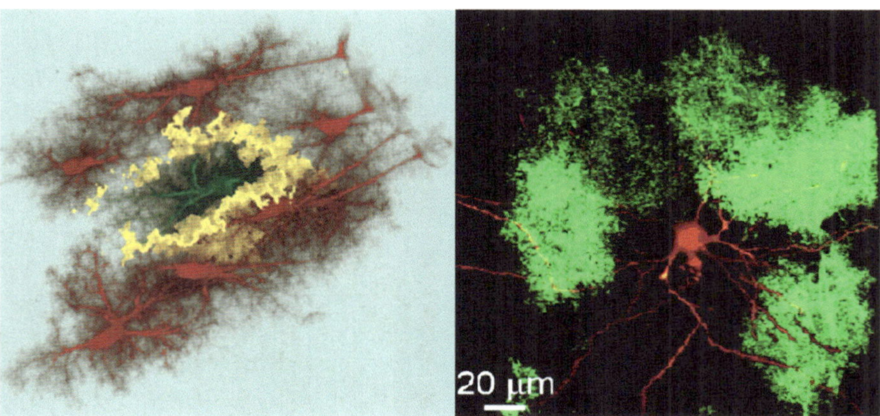

Fig. 14.2 (*Left*) Astrocytes of the hippocampal CA1 region filled with different florescent dyes by microinjection. *Yellow* represents limited overlapping regions between adjacent astrocytes. (*Right*) Top-view reconstruction showing a biocytin-filled layer 2/3 cortical neuron in a slice from GFAP–EGFP animals. Note that a single astrocyte enwraps different dendrites of the same neuron. *Left image*, courtesy of E. Bushong and M. Ellisman, The National Center for Microscopy and Imaging Research, University of California, San Diego, USA; *right image* (Adapted from Halassa et al. (2007))

for perisynaptic astrocytic processes to communicate with neurons is indeed a new concept in synaptic physiology wherein, in addition to the information flow between the pre- and postsynaptic neurons, astrocytes exchange information with synaptic elements by responding to synaptic activity and possibly by modulating synaptic transmission.

14.2 Astrocytes Exhibit Ca^{2+} Excitability

The morphology and the location of astrocytes place them in a unique position to be able to listen and respond to neuronal activity. Although nonexcitable cells and thus not equipped with the cellular machinery necessary for generating action potentials, astrocytes express a wide variety of functional neurotransmitter receptors essential for sensing neuronal activity. Many of these receptors are G-protein-coupled receptors (GPCRs) that, upon activation, stimulate phospholipase C and formation of inositol (1,4,5)-triphosphate (IP3) which increases the intracellular Ca^{2+} concentration through the release of Ca^{2+} from intracellular Ca^{2+} stores. In the mid-1990s, the discovery of new fluorescent tools for studying intracellular ions in living cells (in particular the Ca^{2+} sensors), together with the advancements of microscopy imaging technologies, provided the technical background to make breakthroughs in our understanding of astrocytes functions. Indeed, the Ca^{2+}-imaging technique allowed many laboratories to demonstrate that in vitro and in situ astrocytes can respond to neurotransmitters released from synaptic terminals during neuronal activity with GPCR-mediated intracellular Ca^{2+} increases (Pasti et al. 1997; Porter and McCarthy 1996; Kang et al. 1998; Araque et al. 2002; Perea and Araque 2005; Navarrete and Araque 2008; Honsek et al. 2010). More recently, activation of this signaling pathway (s) has been demonstrated to occur in vivo in response to sensory stimulation (Wang et al. 2006; Winship et al. 2007; Petzold et al. 2008; Schummers et al. 2008; Nimmerjahn et al. 2009), suggesting that astrocytes are indeed activated during physiological running of the brain circuitry. These findings are of particular relevance because they demonstrate the existence of neuron-to-astrocyte communication and that the GPCRs appear to be the first link between neuronal activity and perisynaptic astrocytes. Thus, stimulation of GPCRs and the subsequent intracellular Ca^{2+} rises is now widely considered a form of glial excitability, the so-called Ca^{2+} excitability.

Astrocytic Ca^{2+} signaling in vivo (both in anesthetized and in freely moving animals) is mediated by activation of glutamate or purinergic metabotropic receptors. However, depending on the brain areas and layers, and similarly to neurons, astrocytes display considerable heterogeneity, including a markedly different pattern of Ca^{2+} excitation (Fiacco and McCarthy 2006; Takata and Hirase 2008). For instance, in the cerebellum, an area implicated in locomotor coordination, particular astrocytes called Bergmann glia cells exhibit three forms of Ca^{2+} transients in awake mice (Nimmerjahn et al. 2009). One of these subtypes (named "Ca^{2+} flares") was triggered during locomotion and extended over a large network of astrocytes (at least 100 microns extension). The other two forms of Ca^{2+} excitation (called "sparkles" and "bursts") appear to be restricted to individual fibers or to a maximum of 40 cells, respectively. Interestingly, the dependence of flares on locomotion and the absence of flares and near abolition of sparkles under anesthesia emphasize that studies in awake animals are essential to understand the real physiological role of intracellular Ca^{2+} rises in astrocytes.

Intracellular Ca^{2+} rises in astrocytes are not stereotyped signals; there are actually multiple and varied spatiotemporal patterns of Ca^{2+} elevations, which probably underlie different types of function, including generation of diverse output signals. The significance of the different features of intracellular Ca^{2+} events generated in astrocytes by neuronal activity, e.g., amplitude, frequency, and extent of propagation, remains, however, largely unknown. The spatiotemporal characteristics of Ca^{2+} events in astrocytes might be heterogeneous and dependent on the location of astrocytes. For instance, in the hippocampal CA1 region, the amplitude of astrocytic Ca^{2+} elevations is correlated with the number of simultaneously activated synapses (Honsek et al. 2010); in the cerebellum, instead, the extent of intracellular Ca^{2+} propagation, rather than the amplitude, depends on the characteristics of the neuronal input. Indeed, axonal firing at low frequency generates in Bergmann glial cells' intracellular Ca^{2+} rises that are spatially restricted to the cell periphery and might represent functional independent localized microdomains (Grosche et al. 1999). The existence of localized functional domain of calcium in restricted regions of astrocytes is an intriguing observation. In fact, while most of the present literature supports the general view that intracellular Ca^{2+} events in astrocytes are slow and long lasting (therefore, well distinct from the neuronal ones), recent evidences reports in vitro (Marchaland et al. 2008), in situ (Santello et al. 2011; Di Castro et al. 2011), and in vivo (Winship et al. 2007) the existence of astrocytic Ca^{2+} responses that are as fast as in neurons (time to peak within 500 ms from stimulation). The localized and fast Ca^{2+} events in astrocytes are evoked by endogenous synaptic activity (Chuquet et al. 2010) or by activation of metabotropic receptors (Santello et al. 2011; Marchaland et al. 2008), occur in millisecond time scale and are therefore compatible with a physiological role in fast, activity-dependent synaptic modulation (Santello and Volterra 2009; Henneberger and Rusakov 2010).

The localized variations of Ca^{2+} may represent a sophisticated signaling mechanism controlling a large variety of cellular process, including the release of gliotransmitters. Our group has established that such localized Ca^{2+} rises are implicated in the regulated exocytosis processes of glutamatergic vesicles in vitro (Marchaland et al. 2008). By mean of total internal reflection fluorescence (TIRF) microscopy, Marchaland and colleagues observed that endoplasmic reticulum (ER) tubules come in tight apposition with the plasma membrane, forming a complex structural microdomain of submicrometer volume that most likely limits diffusion of signaling molecules, including Ca^{2+}. In this microdomain, glutamatergic vesicles lie in the in tight spatial proximity with ER structures. Moreover, like at glutamatergic synapses, the cytoskeleton could provide a structural organization of the astrocytic microdomain. Indeed, scaffold proteins like Homer(s) could provide a molecular link between IP3 receptors located at the tip of the ER tubules and the metabotropic glutamate receptors (mGluRs) on the plasma membrane (Sala et al. 2005). The activation glutamatergic GPCRs generates, in cultured astrocytes, store-dependent submicrometer Ca^{2+} events characterized by fast kinetics and spatial segregation.

Interestingly, most of the Ca^{2+} events occurred at or near sites in which glutamatergic vesicles underwent exocytosis and were in strict temporal and spatial correlation with fusion events. How the subplasma membrane Ca^{2+} events regulate exocytosis of glutamatergic vesicles and whether this is similar to synaptic release is not clear yet. These first findings can be relevant to understand the functional role of transmitter release from astrocytes in the brain function; indeed, the fact that fast astrocytic Ca^{2+} events occur in intact tissue (Winship et al. 2007; Santello et al. 2011) suggests that such Ca^{2+} events are not peculiarity of astrocytes in cell culture but may correspond to events taking place in astrocytes of the living brain.

14.3 Astrocytes Release Chemical Transmitters (Gliotransmitters)

The intracellular cascade resulting in Ca^{2+} rise in astrocytes is the main mechanism these cells use to transduce synaptic activity. What is the functional meaning of the intracellular Ca^{2+} rises in astrocytes? It is now well established that GPCR-mediated Ca^{2+} variations in astrocytes can trigger the release of chemical substances (the so-called gliotransmitters; Bezzi and Volterra 2001; Volterra and Meldolesi 2005). The existence of communication systems based on the release of chemical transmitters from astrocytes to other brain cells was first hypothesized at the end of the 1980s based on the observation that glial cells contain, synthesize, and release a variety of compounds (Martin 1992). At that time, however, the common view of astrocytes as passive elements was dogmatic, thus the whole concept remained largely hypothetical. Nevertheless, over the last 15 years, a number of evidences have shown that astrocytes are indeed highly secretive cells, able to release different chemical transmitters in response to an active stimulus (such as the activation of GPCRs). The term "gliotransmitters" is broad and includes a huge number of neuroactive molecules, such as (1) excitatory and inhibitory amino acids (D-serine, glutamate, aspartate, GABA, glycine, and taurine), (2) ATP and related nucleotides and nucleosides (purine nucleotides ATP), (3) eicosanoids and other lipid mediators (prostaglandins), (4) neuropeptides (proenkephalin, angiotensinogen, endothelins), (5) neurotrophins (nerve growth factor, neurotrophin-3, brain-derived neurotrophic factor), (6) cytokines (interleukins (IL), interferons (IFN), tumor necrosis factors alpha (TNFα)), (7) structurally associated chemokines, and (8) growth factors (Bergami et al. 2008; Bezzi and Volterra 2001; Blum et al. 2008; Fields and Stevens 2000; Fujita et al. 2009; Hussy et al. 2000; Kang et al. 2008; Liu et al. 2008; Medhora 2000; Sanzgiri et al. 1999; Snyder and Kim 2000). Among them, compelling evidence supporting Ca^{2+}-dependent gliotransmission has been provided for glutamate, D-serine, and ATP.

During the last 15 years, numbers of laboratories focused their studies on mechanisms of amino acid release from astrocytes (Malarkey and Parpura 2008); several different mechanisms of release from cultured astroglia have been documented, including (1) volume-sensitive organic anion channels (Haskew-Layton et al. 2008;

Kimelberg et al. 1990; Mongin and Kimelberg 2002), (2) hemichannels (Cotrina et al. 1998; Stout et al. 2002; Ye et al. 2003), (3) P2X7 receptor channels (Duan et al. 2003; Kukley et al. 2001), (4) reversed operation of reuptake carriers (Attwell et al. 1993; Longuemare and Swanson 1997; Re et al. 2006; Rossi et al. 2000; Szatkowski et al. 1990; Volterra et al. 1996), or (5) exchange via the cystine–glutamate antiporter (Allen et al. 2001; Baker et al. 2002; Bender et al. 2000; Moran et al. 2003, 2005; Shanker and Aschner 2001; Tang and Kalivas 2003), occurring via Ca^{2+}-independent processes and pimarily under pathological conditions. Recent data, however, have suggested that a fraction of astrocytes in situ release neuroactive substances with Ca^{2+}-dependent mechanism(s) (Bezzi et al. 1998; Fiacco and McCarthy 2004; Kang et al. 1998; Lee et al. 2007; Mothet et al. 2005; Navarrete and Araque 2008; Pascual et al. 2005; Pasti et al. 1997; Serrano et al. 2006; Jourdain et al. 2007; Yang et al. 2003; Santello et al. 2011). Whether Ca^{2+}-dependent and independent mechanisms coexist and under what physiological or pathological conditions they occur remains unclear. However, the Ca^{2+}-dependent exocytotic vesicular release has received considerable attention due to its potential to occur under physiological conditions via a finely regulated way.

The regulated exocytosis is a process by which secretory vesicles, via formation of the soluble N-ethylmaleimide-sensitive fusion protein attachment protein receptor (SNARE) complex, fuse with the plasma membrane and release their content into the extracellular space. Until few years ago, the most direct evidence that Ca^{2+}-dependent release of chemical transmitters (most specifically glutamate) occurs via exocytosis in astrocytes came from pharmacological experiments using clostridial neurotoxins and other agents that selectively interfere with neuronal exocytosis (Bezzi et al. 1998; Pascual et al. 2001; Pasti et al. 2001). If clostridial toxins blocked release of glutamate, then astrocytes must express proteins that are substrate for these toxins. Indeed, astrocytes express the core machinery proteins involved in forming the SNARE complex, such as synaptobrevin II and its homologue cellubrevin (Bezzi et al. 2004; Jourdain et al. 2007) and SNAP-23 (Hamilton and Attwell 2010). Similar to clostridial toxins, astrocytic glutamate release is inhibited by bafilomycin A1 (Baf A1). This suggests that these cells must possess organelles expressing proton-dependent vesicular glutamate transporter (VGLUT). Baf A1, in fact, interferes with H+-ATPase, leading to alkalinization of vesicular lumen and collapsing the proton gradient necessary for VGLUT to transport glutamate into glutamatergic vesicles. This hypothesis was confirmed by studies on cultured astrocytes, where SNARE proteins colocalize with a number of vesicular organelles, including small vesicles positive for VGLUT1–3 (Bezzi et al. 2004; Kreft et al. 2004; Montana et al. 2004; Zhang et al. 2004), ATP-storing vesicles (Coco et al. 2003), neuropeptide-storing granules (Ramamoorthy and Whim 2008; Prada et al. 2011), and D-serine-containing vesicles (Martineau et al. 2008), suggesting the involvement of vesicular mechanisms in the release of these gliotransmitters. Despite these indications, a conclusive demonstration of the existence of a secretory compartment in astrocytes was unresolved until few years ago when our laboratory, in collaboration with Vidar Gundersen (University of Oslo), identified in hippocampal dentate gyrus a glutamate-storing vesicular compartment with properties similar to those of synaptic vesicles in glutamatergic terminals

(Bezzi et al. 2004; Jourdain et al. 2007). In intact tissue, vesicles containing glutamate in astrocytes (1) are grouped very close to plasma membrane (about 100 nm), (2) have a clear appearance (they are not electrondense), and (3) have a small diameter (about 30–50 nm of diameter).

14.4 Astrocytes Possess Different Exocytotic Organelles and Contain Different Gliotransmitters

Astrocytes, like specialized professional secretory cells (exocrine, endocrine, and neurons), contain at least the three major classes of secretory organelles: the small synaptic-like microvesicles (SLMV; Bezzi et al. 2004; Jourdain et al. 2007; Bergersen and Gundersen 2009); the large dense-core granules (LDCGs ; Coco et al. 2003; Ramamoorthy and Whim 2008; Prada et al. 2011), which store and release distinct cargo; and lysosomes (Jaiswal et al. 2007; Li et al. 2008; Zhang et al. 2007). In neurons and specialized secretory cells, classical neurotransmitters and peptides are located in small SLMVs and LDCGs, respectively (Kupfermann 1991). These organelles have specialized physiological functions and are typically found in different regions of the cell. For instance, in neurons, synaptic vesicles responsible for the fast release of neurotransmitters during synaptic activity are clustered at the active zones, while peptide-containing LDCGs, typical exocytic organelles involved in maintaining the tonic level of hormones and neuropeptides in endocrine cells and neurons, are diffusively distributed in axons or dendrites (Meldolesi et al. 2004; Pickel et al. 1995). In contrast to other secretory cells, morphological, molecular, and physiological properties of the two secretory processes in astrocytes are still largely unknown. SLMVs represent the best characterized secretory organelles in astrocytes. Morphologically, they strongly resemble of synaptic vesicles of nerve terminals (Bergersen and Gundersen 2009; Bezzi et al. 2004; Crippa et al. 2006; Jourdain et al. 2007), are equipped with transport proteins for uptake of transmitters (VGLUTs), and contain glutamate and possibly D-serine (Calì et al. 2009; Marchaland et al. 2008; Mothet et al. 2006). D-Serine is a small amino acid synthesized by the enzyme serine racemase that in many brain areas represents an endogenous ligand for NMDA glutamate receptor. Fusion of D-serine-containing vesicles in astrocytes is a Ca^{2+}-dependent process mechanically similar to those of glutamatergic SLMVs. Whether glutamate and D-serine are coreleased in the same brain areas and by the same pool of vesicles is still largely unknown. SLMVs-containing glutamate have been extensively studied during the last years; in a recent work from our laboratory, we took advantage of chimerical protein VGLUT1-pHluorin (Voglmaier et al. 2006) to study in detail the characteristics of exoendocytosis and recycling processes at the single-vesicle level with TIRF illumination (Marchaland et al. 2008). The fast imaging protocol (40 Hz) applied in these sets of experiments provided some important information on both the kinetics and modalities of exocytosis and recycling. Upon GPCR stimulation, glutamatergic SLMVs undergo Ca^{2+}-dependent-regulated exocytosis in a burst that displays a bimodal distribution (Marchaland et al. 2008): a rapid phase sustained almost exclusively by "resident" vesicles (vesicles already docked to the plasma membrane before the stimulus) mostly undergoing kiss-and-run fusion, and a relatively slower phase sustained mainly by "newcomers" vesicles (vesicles that approach the plasma membrane after the stimulus), mostly undergoing full-collapse fusion. This duality of fusion events is reminiscent of observation in neurons where only readily releasable vesicles are rapidly recycled and reused (Harata et al. 2006). Indeed, recent observations show that this bimodal fusion is essential for activation of neuronal receptors by astrocytic glutamate (Santello et al. 2011).

There is significantly less information concerning LDCGs in astrocytes. Proteins belonging to the family of granins (such as chromogranins and secretogranins) are known to be stored in LDCGs of neuroendocrine cells together with neuropeptides and hormones (Malosio et al. 2004; Meldolesi et al. 2004; Rosa and Gerdes 1994). Therefore, granins are the most useful markers to investigate the presence of LDCGs in neurons in different areas of the mammalian brain (Meldolesi et al. 2004). In 1999, Calegari and colleagues showed for the first time that secretogranin II (SgII) is also expressed in cultured hippocampal astrocytes (Calegari et al. 1999). At the ultrastructural level, SgII appeared to be packaged in LDCGs (diameter > 100 nm) located in the Golgi apparatus and near the tubular structure of the trans-Golgi networks, where biogenesis of secretory granules is known to take place. Release of intracellularly stored SgII was evoked by treatment with various secretagogues (e.g., ionomycin, dibutyryl-cAMP, and bradykinin) in a Ca^{2+}-dependent manner. Later on, it was found that LDCGs contain ATP and does not colocalize with the SNARE synaptobrevin/VAMP2, thus representing another, distinct population of organelles (Coco et al. 2003). Recently, Prada and colleagues (2011), moreover, found that the expression LDCVs and their regulated discharge are governed by REST (otherwise called NRSF), the transcription repressor encoded by the master gene that orchestrates differentiation of nerve cells (Ballas and Mandel 2005; D'Alessandro et al. 2008). These findings suggested the possible existence of two distinct classes of secretory vesicles in astrocytes: SLMVs and LDCGs. Therefore, astrocytes, like neurons, might have a regulated secretory pathway that is responsible for the release of multiple classes of chemical transmitters. Both of these processes may be involved in the regulation of synaptic transmission by astrocyte-released molecules (see below).

Lysosomes have been considered to be a major storage site of immune-signaling substances, such as proinflammatory cytokines (Andrei et al. 2004) and adenosine (Lukashev et al. 2004; Pisoni and Thoene 1989), and have been shown to be implicated in intercellular communication at the immunological synapse (McNeil and Kirchhausen 2005). Secretory lysosomes are also enriched in certain types of glial cells like oligodendrocytes, where they are employed for myelin proteins secretion and therefore likely play a critical role in myelination (Trajkovic et al. 2006). Additionally, recent studies have revealed that elevated calcium in astrocytes does induce a special kind of regulated secretion from secretory lysosomes (Jaiswal et al. 2007; Li et al. 2008; Zhang et al. 2007). Indeed, in astrocytes, secretory lysosomes release ATP, and blockade of this release prevents the propagation of calcium waves between neighboring astrocytes (Bowser and Khakh 2007). Although these studies have been so far focused on astrocytes in culture, a similar mechanism of release is likely to occur in vivo, as acutely isolated astrocytes express some mRNAs of proteins involved in lysosome secretion (Cahoy et al. 2008). However, activated at physiological cytosolic Ca^{2+} concentrations, the observed Ca^{2+}-dependent release of astrocytic secretory lysosomes operates on timescale orders of magnitude slower than neurotransmission.

14.5 Gliotransmitters Modulate Synaptic Transmission (Focus on Glutamate)

For a long time, intercellular signaling underlying information transfer and processing in the brain was considered to occur exclusively between neurons. Numerous studies performed during the past few years have instead established the existence of bidirectional signaling between neurons and astrocytes. Indeed, gliotransmitters released upon synaptic stimulation such as glutamate, ATP, and D-serine are able to regulate neuronal excitability (Sasaki et al. 2011) and synaptic transmission (Araque et al. 2001; Bezzi et al. 2001a; Perea et al. 2009). These findings led to the establishment of a new concept in synaptic physiology, "the tripartite synapse," in which astrocytes exchange information with the neuronal synaptic elements (Araque et al. 1999). Consequently, astrocytes can be considered an integral part of the synapses, being involved not only in maintaining passively the homeostatic conditions for proper synaptic transmission but also participating actively in synaptic function (Santello and Volterra 2010).

Glutamate is probably the best characterized gliotransmitter, able to modulate synaptic transmission. In the hippocampal CA1 region, astrocytes of the stratum radiatum sense the activity of Schaffer collateral afferents and respond to it with $Ca^{2+}i$ elevations and release of glutamate. The astrocytic glutamate acts on extrasynaptic NR2B-containing NMDA receptors located on the dendrites of CA1 pyramidal cells. Activation of such receptors results in large, slow inward currents (SICs) in the pyramidal cells able to significantly depolarize the cells and even to trigger their firing (Angulo et al. 2004; Fellin et al. 2004; Perea and Araque 2005; Navarrete and Araque 2008). Astrocyte-evoked SICs have been found to occur in two or more neighboring pyramidal cells in strict temporal correlation, which has been proposed to induce their synchronous firing. In addition, astrocytic glutamate might also activate receptors localized at presynaptic level at the same synapses. Through activation of group I metabotropic glutamate receptors (mGluRs) (Perea and Araque 2007; Navarrete and Araque 2010), astrocytes enhance the frequency of spontaneous and evoked excitatory synaptic currents. Alternatively, astrocytes induce the potentiation or depression of inhibitory synaptic transmission by activation of presynaptic kainate or II/III mGlu receptors, respectively (Liu et al. 2004a, b). In addition, it has been recently shown that glutamate release from cortical astrocytes is also able to broaden action potentials and therefore to facilitate ensuing synaptic transmission (Sasaki et al. 2011). Therefore, a single gliotransmitter can exert multiple effects depending on the sites of action and the activated receptor subtypes, which provides a high degree of complexity to astrocyte–neuron communication. This complexity becomes even higher when considering that other gliotransmitters, such as GABA, ATP, adenosine (a metabolic product of ATP), or D-serine, could have converging actions on the same neuron or, on the contrary, divergently act on several cells (both neurons and glia), thus evoking distinctive responses (Perea et al. 2009).

In our lab, we demonstrated that, at perforant path–granule cell (PP–GC) synapses in the hippocampal dentate gyrus, astrocytes of the outer molecular layer sense synaptic activity, elevate their intracellular Ca^{2+} elevations, and release glutamate via exocytosis of SLMVs. Indeed, by using dual patch-clamp experiments on pairs of dentate GC and molecular layer astrocytes, we have shown that direct electrical stimulation of astrocytes induces strengthening of synaptic transmission. The same effect was observed when applying an agonist of the astrocytic purinergic P2Y1 receptors (P2Y1Rs). Astrocytic glutamate is released at presynaptic level in close proximity of NR2B-containing NMDA receptors: activation of these receptors results in an increased synaptic transmitter release and in the strengthening of synaptic transmission (Jourdain et al. 2007).

This is the only neuromodulatory action of astrocytes for which a precise ultrastructural correlate has been established. Thus, we found that excitatory nerve terminals in the dentate outer molecular layer express NR2B subunits and that the distribution of such NR2B subunits is particularly abundant in the extrasynaptic terminal membrane opposed to astrocytic processes containing SLMVs (Fig. 14.3). Moreover, the astrocyte input to the synapse is blocked by introducing the tetanus

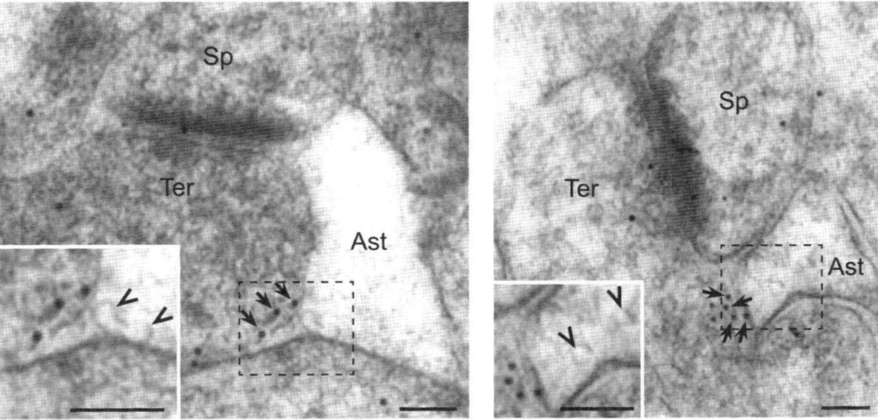

Fig. 14.3 Electron micrographs showing NR2B (*gold particles*) in extrasynaptic membranes (*arrows*) of nerve terminals, (*Ter*) making asymmetric synapses with dendritic spines (*Sp*) in the hippocampal dentate molecular layer. NR2B particles face astrocytic processes (*Ast*) containing SLMVs. There is a close proximity of NR2B to astrocytic SLMVs. Insets: higher magnification showing NR2B gold particles and astrocytic SLMVs (*arrowheads*). Scale bars, 100 nm (Adapted from Jourdain et al. (2007), courtesy of Nature Neuroscience)

toxin light chain (a specific exocytosis inhibitor) through the patch pipette into the stimulated astrocyte, indicative of an obligatory role of exocytosis for the synaptic modulation. The distance separating NR2B subunits in nerve terminals from SLMVs in surrounding astrocytic processes was found to be in the majority of the cases similar to the one separating postsynaptic receptors from "readily releasable" synaptic vesicles at the active zone of nerve terminals (Gitler et al. 2004).

In contrast, SICs are not observed in response to the activity of PP–GC synapses. What is the reason for this discrepancy? A possible explanation could reside in the existence of structural–functional differences between excitatory synapses in the CA1 region and in the dentate gyrus. For instance, at PP–GC synapses, activation of presynaptic ifenprodil-sensitive NMDARs seems to predominate with respect to activation of extrasynaptic (dendritic) ifenprodil-sensitive NMDARs (Dalby and Mody 2003; Jourdain et al. 2007).

Independent from the structural–functional differences, it is interesting to note that at both CA1 and dentate synapses astrocytic glutamate is able to directly activate NMDARs. This is probably because NMDARs have much higher affinity for glutamate than all other glutamate receptors; therefore, they could be particularly suited for nonsynaptic communication that implies wider diffusion and lower local accumulation of glutamate (in the synapse, glutamate reaches mM concentrations once being released).

NMDARs should open only upon membrane depolarization: this raise an apparent paradox on how astrocyte glutamate might activate these receptors. One possible explanation might be that NMDARs targeted by astrocyte-released glutamate could have a peculiar subunit composition conferring them low sensitivity to Mg^{2+} block, like those present on oligodendrocytes (Burzomato et al. 2010) and in presynaptic terminals in the cortex (Larsen et al. 2011), or such receptors could be located in small volume structures that might be relatively depolarized or have high input resistance: in this case, no or very small inward currents would be sufficient to induce significant depolarization and relieve the Mg^{2+} block. Alternatively, other reasons might explain why NMDA receptors are often particular targets for astrocyte glutamate. For instance, together with glutamate, astrocytes might corelease depolarizing agents and/or facilitatory factors with specificity for NMDARs, notably D-serine (Mothet et al. 2005; Bergersen and Gundersen 2009; Henneberger et al. 2010). D-Serine interacts with the so-called glycine-binding site of the NMDAR, allowing its transmembrane channel to open when glutamate binds. Henneberger and colleagues (2010) showed that high frequency stimulation of Schaffer collaterals in the hippocampus gives rise to astrocyte intracellular Ca^{2+} increase that controls the induction of long-term potentiation by releasing D-serine. D-Serine binds to NMDA receptors to promote LTP establishment when glutamate is released from the presynaptic terminal. It is therefore theoretically possible that also opening of extrasynaptic NMDARs is facilitated by D-serine release from astrocytes, although it has been reported that glycine-binding site on presynaptic NMDARs, contrarily to the synaptic one, might be already saturated (Li and Han 2007; Li et al. 2009).

14.6 Gliotransmission Is Controversial

Albeit multiple experimental evidences have been accumulating during the last 15 years in favor of an active role of astrocytes in some forms of synaptic plasticity, some recent studies challenged these findings. Concerns are mainly focused on actual possibility that astrocytes may not contain the machinery to exocytose glutamate and/or that glutamate in the astrocyte cytoplasm could not be sufficient for efficient vesicular loading of the transmitter (Hamilton and Attwell 2010). Moreover, regardless of the release machinery, activation of exogenous GPCRs or knock-out of IP3R in astrocytes failed to modify synaptic transmission at hippocampal CA1 pyramidal cells (Fiacco et al. 2007; Agulhon et al. 2010).

14.6.1 Astrocyte Glutamate Cytosolic Levels and Vesicular Filling

Glutamate that is taken up by astrocytes is converted to glutamine by the enzyme glutamine synthetase (GS), before being passed back to synaptic terminals, in which it is converted back to glutamate. Because of the high activity of GS in astrocytes, the cytoplasmic level of glutamate is substantially lower than in neurons. This raises the question of whether a sufficiently high concentration of glutamate could be accumulated in astrocytic vesicles to activate neuronal receptors when released (Barres 2008). Nevertheless, it has to be noted that the Km of vesicular glutamate transporter is lower than that of GS, proving the opportunity for glutamate transport into vesicles (Halassa and Haydon 2010). Moreover, even assuming a partial filling of the vesicles, theoretical calculations show that the organelles would contain enough glutamate to activate extrasynaptic NMDARs (Hamilton and Attwell 2010).

Studies performed in cell culture, brain slices, acutely isolated astrocytes, and tissue sections provide compelling support for the presence of vesicular machinery for glutamatergic gliotransmission (Halassa and Haydon 2010). However, in contrast, two studies, using microarrays, have not detected the message for vesicular glutamate transporters in the astrocyte transcriptome (Lovatt et al. 2007; Cahoy et al. 2008). Further work is required to determine the reasons of such discrepancy, but there is lack of consensus in data acquired by commercially available microarray platforms given the high variability in gene expression profiles obtained from different laboratories performing seemingly identical experiments (Shi et al. 2008).

14.6.2 Astrocyte Involvement in Synaptic Plasticity

Molecular genetic techniques have been used to express novel GPCRs in astrocytes to ask whether their selective activation triggers glutamatergic gliotransmission (Fiacco et al. 2007; Agulhon et al. 2010). One of these receptors, MrgA1, is normally expressed by dorsal root ganglion neurons but not in the central nervous system (CNS). Expression of this receptor in astrocytes results in Ca^{2+} transients in astrocytes in response to the peptide ligand FLRFa. However, despite a robust volume-averaged Ca^{2+} signal, no SIC, modulation of basal synaptic transmission or LTP gating has been detected. Moreover, knock-out of IP3R2, believed to be astrocyte specific, does not enhance or lower baseline of CA1 pyramidal neuron synaptic activity or affects LTP (Petravicz et al. 2008; Agulhon et al. 2010). Opposite results have been instead reported at the same synapses (Fellin et al. 2004; Perea and Araque 2007; Henneberger et al. 2010) (Fig. 14.4), but some of them have been criticized as obtained with nonphysiological astrocyte manipulations (such as IP3 or Ca^{2+} uncaging). Indeed, achieving conditions compatible with brain physiology is an issue of fundamental importance in experimental studies in vitro (Agulhon et al. 2008). Ideally, one would aim to identify and activate selectively the molecular trigger(s) of Ca^{2+}-dependent release in astrocytes.

One conceptual difficulty is that the current experimental techniques employed to evoke intra-astrocytic Ca^{2+} elevations do not reproduce the spatiotemporal

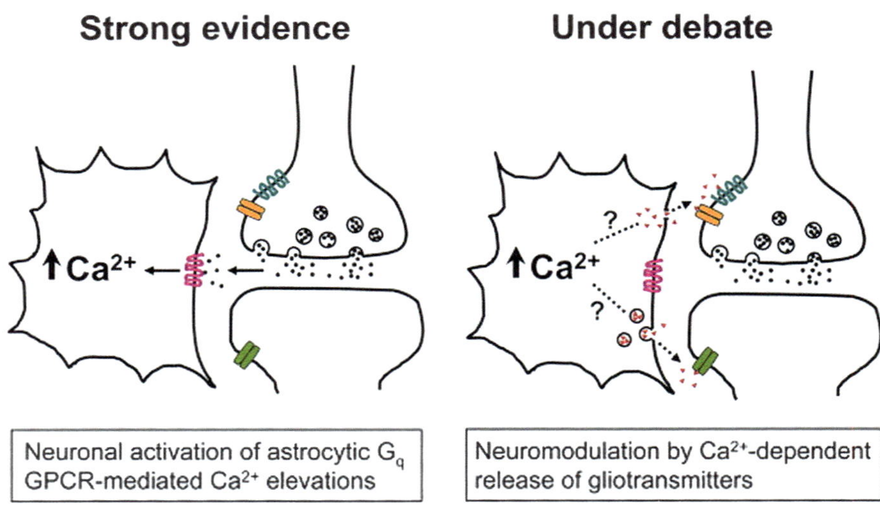

Fig. 14.4 Schematic showing the current understanding of astrocytic Ca^{2+}-signaling involvement at the synapse. (*Left panel*) Both in situ and in vivo studies strongly support the conclusion that synaptic release of neurotransmitters, under basal and heightened levels of stimulation, elicits Ca^{2+} increases in astrocytes mostly via the activation of Gq GPCRs. These astrocytic Ca^{2+} elevations can remain localized within small territories (microdomains) within the cell or propagate as intracellular waves into more distant compartments, depending on the level of neuronal activity. (*Right panel*) Whether or not astrocytic Ca^{2+} increases evoke the release of gliotransmitters to modulate pre- or postsynaptic metabotropic or ionotropic neuronal receptors is still under debate. To date, there are no in vivo data available, and data in situ argue both for and against the concept of gliotransmission. The potential significance of gliotransmission in neurophysiology and neuropathophysiology remains an open issue (Adapted from Agulhon et al. (2008))

aspects of physiological Ca^{2+} signaling. Anyway, Ca^{2+}-dependent glutamate release from astrocytes has been associated with an action of endogenous endocannabinoids or ATP released by neurons able to modify synaptic transmission (Jourdain et al. 2007; Navarrete and Araque 2008, 2010). This disparity suggests that the origin and propagation of a physiological Ca^{2+} signal could depend critically on GPCR localization or, more generally, on the spatial relationships among the intracellular players involved in Ca^{2+} signaling in astrocytes: GPCRs, IP3Rs, Ca^{2+} stores, and the Ca^{2+}-sensing molecular targets such as the trigger of transmitter release. Little is known about the intracellular distribution of such players, either on the scale of the entire astrocytic arbor or within the fine astrocyte processes that approach synaptic structures. Therefore, as a large number of information on the way astrocytes might influence neuronal functions is still not available, no definitive conclusion can be made on negative results.

14.7 Glutamatergic Gliotransmission and TNFα

One of the common effects of immune activation is the production of cytokines. In the CNS, cytokines are primarily produced by activated microglia, but are also generated by astrocytes and infiltrating immune cells upon brain injury (Bailey et al. 2006). They are the secreted molecules that mediate communication between immune cells and between immune system and host. Cytokines encompass a broad class of signaling molecules that have the potential to influence an immense variety of signals that regulate CNS function, including growth factor production, electrical activity, synaptic function, and axonal path finding (Carpentier and Palmer 2009).

Among the cytokine family, TNFα is well known for its proinflammatory functions in the immune system, where it is produced by a variety of cells including T cells and macrophages. In the brain, TNFα has the paradoxical ability to both protect and destroy neurons depending on a number of factors (McCoy and Tansey 2008). TNFα signals through two distinct receptors: TNF receptor 1 (TNFR1 or p55TNFR), the major mediator of proinflammatory and proapoptotic functions of TNFα, and TNFR2 (or p75TNFR), which activates more progrowth and survival pathways.

TNFα and its receptors, TNFR1 and TNFR2, are constitutively expressed in healthy brain, both in neurons and glial cells; this means that cells in the brain must be able to respond to a signaling mediated by TNFα and its receptors. For example, blocking IL-1β or TNFα by several independent means alters regulation of sleep (Imeri and Opp 2009). Other possible roles in synaptic physiology have been investigated by the labs of Malenka and Turrigiano, showing a major involvement of TNFα in synaptic plasticity and synaptic scaling (Beattie et al. 2002; Stellwagen and Malenka 2006; Kaneko et al. 2008; Steinmetz and Turrigiano 2010).

In astrocytes, recent papers reported that TNFα and its cognate receptor TNFR1 play an important role in the modulation of the regulated secretion of glutamate (Bezzi et al. 2001b; Rossi et al. 2005; Domercq et al. 2006). TNFα could directly influence glial cells potentially resulting in complex changes in the brain network. Thus, when a local inflammatory reaction is triggered in the brain, microglial cells that rapidly migrate to the injury site (Davalos et al. 2005; Nimmerjahn et al. 2005) become activated and start releasing a number of mediators such as TNFα, deeply altering the properties of glial networks (Bezzi and Volterra 2001). Indeed, TNFα at pathological concentrations appears to exert a potent control on Ca^{2+}-dependent glutamate release from astrocytes.

The first evidence for this was reported in 2001 when Bezzi and colleagues reported two seminal observations: (1) glutamate release from astrocytes induced by the CXCR4 receptor agonist SDF1α was hampered in TNF$\alpha^{-/-}$ preparations and (2) microglial TNFα release induces a massive glutamate release from astrocytes (threefold more than the one induced by other agonists) via prostaglandin PGE_2 production, amplifying CXCR4-induced glutamate release (Bezzi et al. 2001b). This massive glutamate release can cause neuronal excitotoxicity both in culture and in vivo.

However, TNFα is expressed also in the normal brain, albeit at much lower levels than during inflammatory reactions and participates in homeostatic brain functions (Boulanger 2009; Vitkovic et al. 2000). In particular, constitutive TNFα has recently been implicated in control of the stability of neuronal networks in response to prolonged changes in activity via the phenomenon of synaptic scaling (Stellwagen and Malenka 2006; Turrigiano 2008) and plays a specific role in ocular dominance plasticity upon monocular visual deprivation (Kaneko et al. 2008). The TNFα released from astrocytes was able to strengthen excitatory synaptic transmission by promoting insertion of AMPA receptor subunits at the surface (Bains and Oliet 2007; Beattie et al. 2002; Stellwagen et al. 2005). The involvement of TNFα in regulating glutamate release from astrocytes during physiological conditions have been found recently by Domercq et al.; the authors showed that activation of another GPCR, the purinergic P2Y1 receptor (P2Y1R), evoked glutamate release from astrocytes via exocytosis of SLMVs (Domercq et al. 2006). Interestingly, glutamate release is impaired in TNFα knock-out and TNFR1 knock-out slices and cultures, pointing to a permissive role of the cytokine in the exocytosis of glutamate from astrocytes induced by purinergic GPCR activation (Domercq et al. 2006).

A recent paper sheds light on the way TNFα can modulate glutamate release from astrocytes and how this impinges on synaptic activity. Indeed, Santello and colleagues found that the cytokine is an essential factor for functional glutamatergic gliotransmission (Santello et al. 2011). In the hippocampal dentate gyrus, astrocytes exert a modulatory action of GC synapses via Ca^{2+}-dependent glutamate release from astrocytes that is controlled by TNFα. In other words, the constitutive levels of the cytokine (estimated to be in the low picomolar range) need to be present for the neuromodulation to occur. Which is the mechanism(s) by which TNFα control glutamatergic gliotransmission? In astrocytes, the cytokine controls some steps of the stimulus-secretion coupling downstream of GPCR-evoked intracellular Ca^{2+} elevations. In particular, by using TIRF illumination, Santello and colleagues identified in cultured astrocytes obtained from TNF$^{-/-}$ mice, a defect in the functional docking of glutamatergic SLMVs. It appears that a constitutive TNFα level is necessary for SLMVs docking and thus of the subsequent synchronous fusion upon stimulation of metabotropic purinergic receptor (P2Y1R). Thus, in absence of the cytokine, the majority of the vesicles are far from the docking places and thus are not ready to fuse; as a consequence, the kinetics of evoked exocytosis dramatically slow down (vesicles are released over several seconds after stimulus). The asynchronous and slowly release of glutamate from astrocytes then is efficiently scavenged by the glutamate transporters (present on the astrocytic membrane) that therefore prevent presynaptic NMDARs activation and synaptic strengthening. Interestingly, TNFα absence did not alter localized Ca^{2+} increases in the astrocyte process, but specifically dampers vesicular fusion (Santello et al. 2011) (Fig. 14.5).

Which could be the downstream molecule/s responsible for TNFα regulation of astrocyte exocytosis? A link could potentially exist between Rab proteins and TNFα through the protein DENN/MADD. Rab proteins, a large family of small GTPase, are central in ensuring that vesicles are delivered to their correct destinations and play an important role in the regulation of vesicle traffic and fusion in eukaryotic cells (Ferro-Novick and Novick 1993; Geppert and Südhof 1998; Stenmark 2009). Among Rab proteins, Rab3 proteins are associated with secretory vesicles of exocrine, endocrine cells (Zerial and McBride 2001), and neuronal cells (Schlüter et al. 2002) and are part of the machinery controlling exocytosis (Takai et al. 1996; Geppert and Südhof 1998; Lang and Jahn 2008). Rab3 can switch between two functionally distinct conformational states: GDP (inactive) and GTP

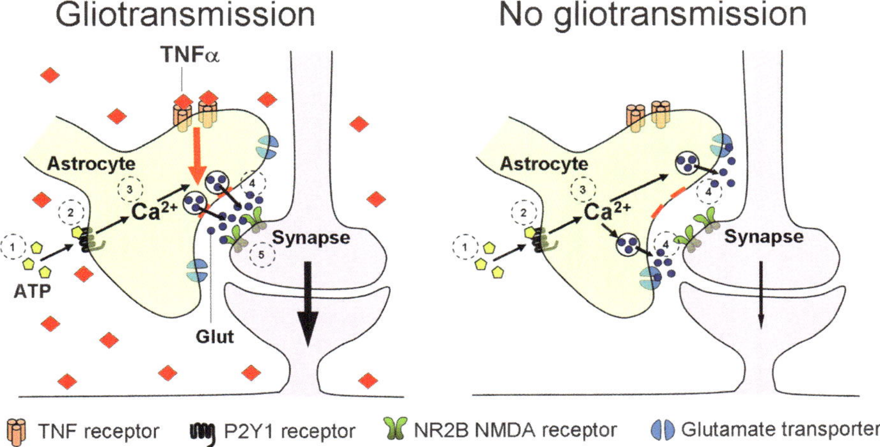

Fig. 14.5 Schematic summary of the TNFα control on gliotransmission at PP–GC synapses in the hippocampal dentate gyrus. (*Left*) In the presence of constitutive TNFα (*red diamonds*), astrocyte vesicles containing glutamate (*Glut, blue dots*) are functionally docked at putative active zones on the plasma membrane of a perisynaptic astrocytic process. When ATP (*yellow pentangles*) is released (*1*) from GC synapses or the astrocytes (Jourdain et al. 2007), it activates P2Y1 receptors (*2*) and causes Ca^{2+} release from the internal stores in the astrocyte microcompartment (*3*). This in turn triggers fusion of the astrocyte vesicles in proximity of presynaptic NR2B-containing NMDARs (*4*), eventually causing an increase in excitatory synaptic activity (*5*). (*Right*) In the absence of TNFα, astrocytic glutamatergic vesicles are not correctly docked and ready to fuse. Therefore, when ATP triggers the usual signal transduction in astrocytes, glutamate release occurs slowly and asynchronously and is scavenged by glutamate transporters before reaching pre-NMDA receptors to induce synaptic modulation (From Santello et al. (2011))

bound (active), respectively. Rab3 GDP/GTP exchange protein (Rab3GEP also known as DENN/MADD) catalyzes the replacement of GDP by GTP and, consequently, favors the activation of Rab3.

Interestingly, it is of recent acquisition that DENN/MADD can also bind the TNFR1 through a death domain that is located at the C-terminus of the protein (Miyoshi and Takai 2004). Given that (a) when TNFα is not bound to the receptor, TNFR1–DENN/MADD interaction is extremely strong; (b) Ca^{2+}-dependent release of glutamate from astrocytes is impaired when DENN/MADD levels are decreased (Rossi et al. 2005), an intriguing hypothesis could be that, in the absence of TNFα, DENN/MADD would be mainly bound to TNFR1 and therefore unable to function as RabGEP. Conversely, basal levels of TNFα would allow dissociation of DENN/MADD from the receptor which would promote efficient docking and fusion of SLMV in astrocytes.

References

Agulhon, C., Fiacco, T. A., & McCarthy, K. D. (2010). Hippocampal short- and long-term plasticity are not modulated by astrocyte Ca^{2+} signaling. *Science, 327*, 1250–1254.

Agulhon, C., Petravicz, J., McMullen, A. B., Sweger, E. J., Minton, S. K., Taves, S. R., Casper, K. B., Fiacco, T. A., & McCarthy, K. D. (2008). What is the role of astrocyte calcium in neurophysiology? *Neuron, 59*, 932–946.

Allen, J. W., Shanker, G., & Aschner, M. (2001). Methylmercury inhibits the in vitro uptake of the glutathione precursor, cystine, in astrocytes, but not in neurons. *Brain Research, 894*, 131–140.

Andrei, C., Margiocco, P., Poggi, A., Lotti, L. V., Torrisi, M. R., & Rubartelli, A. (2004). Phospholipases C and A2 control lysosome-mediated IL-1 beta secretion: Implications for inflammatory processes. *Proceedings of the National Academy of Sciences of the United States of America, 101*, 9745–9750.

Angulo, M. C., Kozlov, A. S., Charpak, S., & Audinat, E. (2004). Glutamate released from glial cells synchronizes neuronal activity in the hippocampus. *Journal of Neuroscience, 24*, 6920–6927.

Araque, A., Carmignoto, G., & Haydon, P. G. (2001). Dynamic signaling between astrocytes and neurons. *Annual Review of Physiology, 63*, 795–813.

Araque, A., Martín, E. D., Perea, G., Arellano, J. I., & Buño, W. (2002). Synaptically released acetylcholine evokes Ca^{2+} elevations in astrocytes in hippocampal slices. *Journal of Neuroscience, 22*, 2443–2450.

Araque, A., Parpura, V., Sanzgiri, R. P., & Haydon, P. G. (1999). Tripartite synapses: Glia, the unacknowledged partner. *Trends in Neurosciences, 22*, 208–215.

Attwell, D., Barbour, B., & Szatkowski, M. (1993). Nonvesicular release of neurotransmitter. *Neuron, 11*, 401–407.

Bailey, S. L., Carpentier, P. A., McMahon, E. J., Begolka, W. S., & Miller, S. D. (2006). Innate and adaptive immune responses of the central nervous system. *Critical Reviews in Immunology, 26*, 149–188.

Bains, J. S., & Oliet, S. H. (2007). Glia: They make your memories stick! *Trends in Neurosciences, 30*, 417–424.

Baker, D. A., Xi, Z. X., Shen, H., Swanson, C. J., & Kalivas, P. W. (2002). The origin and neuronal function of in vivo nonsynaptic glutamate. *The Journal of Neuroscience, 22*, 9134–9141.

Ballas, N., & Mandel, G. (2005). The many faces of REST oversee epigenetic programming of neuronal genes. *Current Opinion in Neurobiology, 15*, 500–506.

Barres, B. A. (2008). The mystery and magic of glia: A perspective on their roles in health and disease. *Neuron, 60*, 430–440.

Beattie, E. C., Stellwagen, D., Morishita, W., Bresnahan, J. C., Ha, B. K., Von Zastrow, M., Beattie, M. S., & Malenka, R. C. (2002). Control of synaptic strength by glial TNFalpha. *Science, 295*, 2282–2285.

Bender, A. S., Reichelt, W., & Norenberg, M. D. (2000). Characterization of cystine uptake in cultured astrocytes. *Neurochemistry International, 37*, 269–276.

Bergami, M., Santi, S., Formaggio, E., Cagnoli, C., Verderio, C., Blum, R., Berninger, B., Matteoli, M., & Canossa, M. (2008). Uptake and recycling of pro-BDNF for transmitter-induced secretion by cortical astrocytes. *The Journal of Cell Biology, 183*, 213–221.

Bergersen, L. H., & Gundersen, V. (2009). Morphological evidence for vesicular glutamate release from astrocytes. *Neuroscience, 158*, 260–265.

Bezzi, P., Carmignoto, G., Pasti, L., Vesce, S., Rossi, D., Rizzini, B. L., Pozzan, T., & Volterra, A. (1998). Prostaglandins stimulate calcium-dependent glutamate release in astrocytes. *Nature, 391*, 281–285.

Bezzi, P., Domercq, M., Brambilla, L., Galli, R., Schols, D., De Clercq, E., Vescovi, A., Bagetta, G., Kollias, G., Meldolesi, J., & Volterra, A. (2001a). CXCR4-activated astrocyte glutamate release via TNFalpha: Amplification by microglia triggers neurotoxicity. *Nature Neuroscience, 4*, 702–710.

Bezzi, P., Domercq, M., Vesce, S., & Volterra, A. (2001b). Neuron-astrocyte cross-talk during synaptic transmission: Physiological and neuropathological implications. *Progress in Brain Research, 132*, 255–265.

Bezzi, P., Gundersen, V., Galbete, J. L., Seifert, G., Steinhauser, C., Pilati, E., & Volterra, A. (2004). Astrocytes contain a vesicular compartment that is competent for regulated exocytosis of glutamate. *Nature Neuroscience, 7*, 613–620.

Bezzi, P., & Volterra, A. (2001). A neuron-glia signalling network in the active brain. *Current Opinion in Neurobiology, 11*, 387–394.

Blum, A. E., Joseph, S. M., Przybylski, R. J., & Dubyak, G. R. (2008). Rho-family GTPases modulate Ca(2+)-dependent ATP release from astrocytes. *American Journal of Physiology: Cell Physiology, 295*, C231–C241.

Boulanger, L. M. (2009). Immune proteins in brain development and synaptic plasticity. *Neuron, 64*, 93–109.

Bowser, D. N., & Khakh, B. S. (2007). Vesicular ATP is the predominant cause of intercellular calcium waves in astrocytes. *Journal of General Physiology, 129*, 485–491.

Burzomato, V., Frugier, G., Pérez-Otaño, I., Kittler, J. T., & Attwell, D. (2010). The receptor subunits generating NMDA receptor mediated currents in oligodendrocytes. *Journal of Physiology, 588*, 3403–3414.

Bushong, E. A., Martone, M. E., Jones, Y. Z., & Ellisman, M. H. (2005). Protoplasmic astrocytes in CA1 stratum radiatum occupy separate anatomical domains. *The Journal of Neuroscience, 22*, 183–192.

Cahoy, J. D., Emery, B., Kaushal, A., Foo, L. C., Zamanian, J. L., Christopherson, K. S., Xing, Y., Lubischer, J. L., Krieg, P. A., Krupenko, S. A., Thompson, W. J., & Barres, B. A. (2008). A transcriptome database for astrocytes, neurons, and oligodendrocytes: A new resource for understanding brain development and function. *The Journal of Neuroscience, 28*, 264–278.

Calegari, F., Coco, S., Taverna, E., Bassetti, M., Verderio, C., Corradi, N., Matteoli, M., & Rosa, P. (1999). A regulated secretory pathway in cultured hippocampal astrocytes. *The Journal of Biological Chemistry, 274*, 22539–22547.

Calì, C., Marchaland, J., Spagnuolo, P., Gremion, J., & Bezzi, P. (2009). Regulated exocytosis from astrocytes physiological and pathological related aspects. *International Review of Neurobiology, 85*, 261–293.

Carpentier, P. A., & Palmer, T. D. (2009). Immune influence on adult neural stem cell regulation and function. *Neuron, 64*, 79–92.

Coco, S., Calegari, F., Pravettoni, E., Pozzi, D., Taverna, E., Rosa, P., Matteoli, M., & Verderio, C. (2003). Storage and release of ATP from astrocytes in culture. *The Journal of Biological Chemistry, 278*, 1354–1362.

Cotrina, M. L., Lin, J. H., Alves-Rodrigues, A., Liu, S., Li, J., Azmi-Ghadimi, H., Kang, J., Naus, C. C., & Nedergaard, M. (1998). Connexins regulate calcium signaling by controlling ATP release. *Proceedings of the National Academy of Sciences of the United States of America, 95,* 15735–15740.

Crippa, D., Schenk, U., Francolini, M., Rosa, P., Verderio, C., Zonta, M., Pozzan, T., Matteoli, M., & Carmignoto, G. (2006). Synaptobrevin2-expressing vesicles in rat astrocytes: Insights into molecular characterization, dynamics and exocytosis. *The Journal of Physiology, 570,* 567–582.

D'Alessandro, R., Klajn, A., Stucchi, L., Podini, P., Malosio, M. L., & Meldolesi, J. (2008). Expression of the neurosecretory process in PC12 cells is governed by REST. *Journal of Neurochemistry, 105,* 1369–1383.

Dalby, N. O., & Mody, I. (2003). Activation of NMDA receptors in rat dentate gyrus granule cells by spontaneous and evoked transmitter release. *Journal of Neurophysiology, 90,* 786–797.

Davalos, D., Grutzendler, J., Yang, G., Kim, J. V., Zuo, Y., Jung, S., Littman, D. R., Dustin, M. L., & Gan, W. B. (2005). ATP mediates rapid microglial response to local brain injury in vivo. *Nature Neuroscience, 8,* 752–758.

Di Castro MA., Chuquet J., Liaudet N., Bhaukaurally K., Santello M., Bouvier D., Tiret P., Volterra (2011). Local Ca2$^+$ detection and modulation of synaptic release by astrocytes. *Nature Neuroscience, 14,* 1276–84.

Domercq, M., Brambilla, L., Pilati, E., Marchaland, J., Volterra, A., & Bezzi, P. (2006). P2Y1 receptor-evoked glutamate exocytosis from astrocytes: Control by tumor necrosis factor-alpha and prostaglandins. *The Journal of Biological Chemistry, 281,* 30684–30696.

Duan, S., Anderson, C. M., Keung, E. C., Chen, Y., Chen, Y., & Swanson, R. A. (2003). P2X7 receptor-mediated release of excitatory amino acids from astrocytes. *The Journal of Neuroscience, 23,* 1320–1328.

Fellin, T., Pascual, O., Gobbo, S., Pozzan, T., Haydon, P. G., & Carmignoto, G. (2004). Neuronal synchrony mediated by astrocytic glutamate through activation of extrasynaptic NMDA receptors. *Neuron, 43,* 729–743.

Ferro-Novick, S., & Novick, P. (1993). The role of GTP-binding proteins in transport along the exocytic pathway. *Annual Review of Cell Biology, 9,* 575–599.

Fiacco, T. A., Agulhon, C., Taves, S. R., Petravicz, J., Casper, K. B., Dong, X., Chen, J., & McCarthy, K. D. (2007). Selective stimulation of astrocyte calcium in situ does not affect neuronal excitatory synaptic activity. *Neuron, 54,* 611–626.

Fiacco, T. A., & McCarthy, K. D. (2004). Intracellular astrocyte calcium waves in situ increase the frequency of spontaneous AMPA receptor currents in CA1 pyramidal neurons. *The Journal of Neuroscience, 24,* 722–732.

Fiacco, T. A., & McCarthy, K. D. (2006). Astrocyte calcium elevations: Properties, propagation, and effects on brain signaling. *Glia, 54,* 676–690.

Fields, R. D., & Stevens, B. (2000). ATP: An extracellular signaling molecule between neurons and glia. *Trends in Neurosciences, 23,* 625–633.

Fujita, T., Tozaki-Saitoh, H., & Inoue, K. (2009). P2Y1 receptor signaling enhances neuroprotection by astrocytes against oxidative stress via IL-6 release in hippocampal cultures. *Glia, 57,* 244–257.

Geppert, M., & Südhof, T. C. (1998). RAB3 and synaptotagmin: The yin and yang of synaptic membrane fusion. *Annual Review of Neuroscience, 21,* 75–95.

Gitler, D., Takagishi, Y., Feng, J., Ren, Y., Rodriguiz, R. M., Wetsel, W. C., Greengard, P., & Augustine, G. J. (2004). Different presynaptic roles of synapsins at excitatory and inhibitory synapses. *Journal of Neuroscience, 24,* 11368–11380.

Grosche, J., Matyash, V., Möller, T., Verkhratsky, A., Reichenbach, A., & Kettenmann, H. (1999). Microdomains for neuron-glia interaction: Parallel fiber signaling to Bergmann glial cells. *Nature Neuroscience, 2,* 139–143.

Halassa, M. M., Fellin, T., Takano, H., Dong, J. H., & Haydon, P. G. (2007). Synaptic islands defined by the territory of a single astrocyte. *The Journal of Neuroscience, 27,* 6473–6477.

Halassa, M. M., & Haydon, P. G. (2010). Integrated brain circuits: Astrocytic networks modulate neuronal activity and behavior. *Annual Review of Physiology, 72*, 335–355.

Hamilton, N. B., & Attwell, D. (2010). Do astrocytes really exocytose neurotransmitters? *Nature Reviews Neuroscience, 11*, 227–238.

Harata, N. C., Aravanis, A. M., & Tsien, R. W. (2006). Kiss-and-run and full-collapse fusion as modes of exo-endocytosis in neurosecretion. *Journal of Neurochemistry, 97*, 1546–1570.

Haskew-Layton, R. E., Rudkouskaya, A., Jin, Y., Feustel, P. J., Kimelberg, H. K., & Mongin, A. A. (2008). Two distinct modes of hypoosmotic medium-induced release of excitatory amino acids and taurine in the rat brain in vivo. *PLoS One, 3*, e3543.

Henneberger, C., Papouin, T., Oliet, S. H., & Rusakov, D. A. (2010). Long-term potentiation depends on release of D-serine from astrocytes. *Nature, 463*, 232–236.

Henneberger, C., & Rusakov, D. A. (2010). Synaptic plasticity and Ca^{2+} signalling in astrocytes. *Neuron Glia Biology, 13*, 1–6.

Honsek, S. D., Walz, C., Kafitz, K. W., & Rose, C. R. (2010). Astrocyte calcium signals at Schaffer collateral to CA1 pyramidal cell synapses correlate with the number of activated synapses but not with synaptic strength. *Hippocampus*. doi:10.1002/hipo.20843.

Hussy, N., Deleuze, C., Desarménien, M. G., & Moos, F. C. (2000). Osmotic regulation of neuronal activity: A new role for taurine and glial cells in a hypothalamic neuroendocrine structure. *Progress in Neurobiology, 62*, 113–134.

Imeri, L., & Opp, M. R. (2009). How (and why) the immune system makes us sleep. *Nature Reviews Neuroscience, 10*, 199–210.

Jaiswal, J. K., Fix, M., Takano, T., Nedergaard, M., & Simon, S. M. (2007). Resolving vesicle fusion from lysis to monitor calcium-triggered lysosomal exocytosis in astrocytes. *Proceedings of the National Academy of Sciences of the United States of America, 104*, 14151–14156.

Jourdain, P., Bergersen, L. H., Bhaukaurally, K., Bezzi, P., Santello, M., Domercq, M., Matute, C., Tonello, F., Gundersen, V., & Volterra, A. (2007). Glutamate exocytosis from astrocytes controls synaptic strength. *Nature Neuroscience, 10*, 331–339.

Kaneko, M., Stellwagen, D., Malenka, R. C., & Stryker, M. P. (2008). Tumor necrosis factor-alpha mediates one component of competitive, experience-dependent plasticity in developing visual cortex. *Neuron, 58*, 673–680.

Kang, J., Jiang, L., Goldman, S. A., & Nedergaard, M. (1998). Astrocyte-mediated potentiation of inhibitory synaptic transmission. *Nature Neuroscience, 1*, 683–692.

Kang, J., Kang, N., Lovatt, D., Torres, A., Zhao, Z., Lin, J., & Nedergaard, M. (2008). Connexin 43 hemichannels are permeable to ATP. *Journal of Neuroscience, 28*, 4702–4711.

Kimelberg, H. K., Goderie, S. K., Higman, S., Pang, S., & Waniewski, R. A. (1990). Swelling-induced release of glutamate, aspartate, and taurine from astrocyte cultures. *The Journal of Neuroscience, 10*, 1583–1591.

Kreft, M., Stenovec, M., Rupnik, M., Grilc, S., Krzan, M., Potokar, M., Pangrsic, T., Haydon, P. G., & Zorec, R. (2004). Properties of Ca(2+)-dependent exocytosis in cultured astrocytes. *Glia, 46*, 437–445.

Kukley, M., Barden, J. A., Steinhäuser, C., & Jabs, R. (2001). Distribution of P2X receptors on astrocytes in juvenile rat hippocampus. *Glia, 36*, 11–21.

Kupfermann, I. (1991). Functional studies of cotransmission. *Physiological Reviews, 71*, 683–732.

Lang, T., & Jahn, R. (2008). Core proteins of the secretory machinery. *Handbook of Experimental Pharmacology, 184*, 107–127.

Larsen, R. S., Corlew, R. J., Henson, M. A., Roberts, A. C., Mishina, M., Watanabe, M., Lipton, S. A., Nakanishi, N., Pérez-Otaño, I., Weinberg, R. J., & Philpot, B. D. (2011). NR3A-containing NMDARs promote neurotransmitter release and spike timing-dependent plasticity. *Nature Neuroscience, 14*, 338–344.

Lee, C. J., Mannaioni, G., Yuan, H., Woo, D. H., Gingrich, M. B., & Traynelis, S. F. (2007). Astrocytic control of synaptic NMDA receptors. *The Journal of Physiology, 581*, 1057–1081.

Li, Y. H., & Han, T. Z. (2007). Glycine binding sites of presynaptic NMDA receptors may tonically regulate glutamate release in the rat visual cortex. *Journal of Neurophysiology, 97*, 817–823.

Li, P., Li, Y. H., & Han, T. Z. (2009). NR2A-containing NMDA receptors are required for LTP induction in rat dorsolateral striatum in vitro. *Brain Research, 1274*, 40–46.

Li, D., Ropert, N., Koulakoff, A., Giaume, C., & Oheim, M. (2008). Lysosomes are the major vesicular compartment undergoing Ca^{2+}-regulated exocytosis from cortical astrocytes. *Journal of Neuroscience, 28*, 7648–7658.

Liu, H. T., Sabirov, R. Z., & Okada, Y. (2008). Oxygen-glucose deprivation induces ATP release via maxi-anion channels in astrocytes. *Purinergic Signal, 4*, 147–154.

Liu, Q. S., Xu, Q., Arcuino, G., Kang, J., & Nedergaard, M. (2004a). Astrocyte-mediated activation of neuronal kainate receptors. *Proceedings of the National Academy of Sciences of the United States of America, 101*, 3172–3177.

Liu, Q. S., Xu, Q., Kang, J., & Nedergaard, M. (2004b). Astrocyte activation of presynaptic metabotropic glutamate receptors modulates hippocampal inhibitory synaptic transmission. *Neuron Glia Biology, 1*, 307–316.

Longuemare, M. C., & Swanson, R. A. (1997). Net glutamate release from astrocytes is not induced by extracellular potassium concentrations attainable in brain. *Journal of Neurochemistry, 69*, 879–882.

Lovatt, D., Sonnewald, U., Waagepetersen, H. S., Schousboe, A., He, W., Lin, J. H., Han, X., Takano, T., Wang, S., Sim, F. J., Goldman, S. A., & Nedergaard, M. (2007). The transcriptome and metabolic gene signature of protoplasmic astrocytes in the adult murine cortex. *The Journal of Neuroscience, 27*, 12255–12266.

Lukashev, D., Ohta, A., Apasov, S., Chen, J. F., & Sitkovsky, M. (2004). Cutting edge: Physiologic attenuation of proinflammatory transcription by the Gs protein-coupled A2A adenosine receptor in vivo. *Journal of Immunology, 173*, 21–24.

Malarkey, E. B., & Parpura, V. (2008). Mechanisms of glutamate release from astrocytes. *Neurochemistry International, 52*, 142–154.

Malosio, M. L., Giordano, T., Laslop, A., & Meldolesi, J. (2004). Dense-core granules: A specific hallmark of the neuronal/neurosecretory cell phenotype. *Journal of Cell Science, 117*, 743–749.

Marchaland, J., Cali, C., Voglmaier, S. M., Li, H., Regazzi, R., Edwards, R. H., & Bezzi, P. (2008). Fast sub-plasma membrane Ca^{2+} transients control exo-endocytosis of SLMVs in astrocytes. *Journal of Neuroscience, 28*, 9122–9132.

Martin, D. L. (1992). Synthesis and release of neuroactive substances by glial cells. *Glia, 5*, 81–94.

Martineau, M., Galli, T., Baux, G., & Mothet, J. P. (2008). Confocal imaging and tracking of the exocytotic routes for D-serine-mediated gliotransmission. *Glia, 56*, 1271–1284.

McCoy, M. K., & Tansey, M. G. (2008). TNF signaling inhibition in the CNS: Implications for normal brain function and neurodegenerative disease. *Journal of Neuroinflammation, 5*, 45.

McNeil, P. L., & Kirchhausen, T. (2005). An emergency response team for membrane repair. *Nature Reviews: Molecular Cell Biology, 6*, 499–505.

Medhora, M. M. (2000). Retinoic acid upregulates beta(1)-integrin in vascular smooth muscle cells and alters adhesion to fibronectin. *American Journal of Physiology: Heart and Circulatory Physiology, 279*, H382–H387.

Meldolesi, J., Chieregatti, E., & Luisa, M. M. (2004). Requirements for the identification of dense-core granules. *Trends in Cell Biology, 14*, 13–19.

Miyoshi, J., & Takai, Y. (2004). Dual role of DENN/MADD (Rab3GEP) in neurotransmission and neuroprotection. *Trends in Molecular Medicine, 10*, 476–480.

Mongin, A. A., & Kimelberg, H. K. (2002). ATP potently modulates anion channel-mediated excitatory amino acid release from cultured astrocytes. *American Journal of Physiology: Cell Physiology, 283*, C569–C578.

Montana, V., Ni, Y., Sunjara, V., Hua, X., & Parpura, V. (2004). Vesicular glutamate transporter-dependent glutamate release from astrocytes. *Journal of Neuroscience, 24*, 2633–2642.

Moran, M. M., McFarland, K., Melendez, R. I., Kalivas, P. W., & Seamans, J. K. (2005). Cystine/glutamate exchange regulates metabotropic glutamate receptor presynaptic inhibition of excitatory transmission and vulnerability to cocaine seeking. *Journal of Neuroscience, 25*, 6389–6393.

Moran, M. M., Melendez, R., Baker, D., Kalivas, P. W., & Seamans, J. K. (2003). Cystine/glutamate antiporter regulation of vesicular glutamate release. *Annals of the New York Academy of Sciences, 1003*, 445–447.

Mothet, J. P., Pollegioni, L., Ouanounou, G., Martineau, M., Fossier, P., & Baux, G. (2005). Glutamate receptor activation triggers a calcium-dependent and SNARE protein-dependent release of the gliotransmitter D-serine. *Proceedings of the National Academy of Sciences of the United States of America, 102*, 5606–5611.

Mothet, J. P., Rouaud, E., Sinet, P. M., Potier, B., Jouvenceau, A., Dutar, P., Videau, C., Epelbaum, J., & Billard, J. M. (2006). A critical role for the glial-derived neuromodulator D-serine in the age-related deficits of cellular mechanisms of learning and memory. *Aging Cell, 5*, 267–274.

Navarrete, M., & Araque, A. (2008). Endocannabinoids mediate neuron-astrocyte communication. *Neuron, 57*, 883–893.

Navarrete, M., & Araque, A. (2010). Endocannabinoids potentiate synaptic transmission through stimulation of astrocytes. *Neuron, 68*, 113–126.

Nedergaard, M., Rodríguez, J. J., & Verkhratsky, A. (2010). Glial calcium and diseases of the nervous system. *Cell Calcium, 47*, 140–149.

Nimmerjahn, A., Kirchhoff, F., & Helmchen, F. (2005). Resting microglial cells are highly dynamic surveillants of brain parenchyma in vivo. *Science, 308*, 1314–1318.

Nimmerjahn, A., Mukamel, E. A., & Schnitzer, M. J. (2009). Motor behavior activates Bergmann glial networks. *Neuron, 62*, 400–412.

Pascual, O., Casper, K. B., Kubera, C., Zhang, J., Revilla-Sanchez, R., Sul, J. Y., Takano, H., Moss, S. J., McCarthy, K., & Haydon, P. G. (2005). Astrocytic purinergic signaling coordinates synaptic networks. *Science, 310*, 113–116.

Pascual, M., Climent, E., & Guerri, C. (2001). BDNF induces glutamate release in cerebrocortical nerve terminals and in cortical astrocytes. *Neuroreport, 12*, 2673–2677.

Pasti, L., Volterra, A., Pozzan, T., & Carmignoto, G. (1997). Intracellular calcium oscillations in astrocytes: A highly plastic, bidirectional form of communication between neurons and astrocytes in situ. *Journal of Neuroscience, 17*, 7817–7830.

Pasti, L., Zonta, M., Pozzan, T., Vicini, S., & Carmignoto, G. (2001). Cytosolic calcium oscillations in astrocytes may regulate exocytotic release of glutamate. *Journal of Neuroscience, 21*, 477–484.

Perea, G., & Araque, A. (2005). Properties of synaptically evoked astrocyte calcium signal reveal synaptic information processing by astrocytes. *Journal of Neuroscience, 25*, 2192–2203.

Perea, G., & Araque, A. (2007). Astrocytes potentiate transmitter release at single hippocampal synapses. *Science, 317*, 1083–1086.

Perea, G., Navarrete, M., & Araque, A. (2009). Tripartite synapses: Astrocytes process and control synaptic information. *Trends in Neurosciences, 32*, 421–431.

Petravicz, J., Fiacco, T. A., & McCarthy, K. D. (2008). Loss of IP3 receptor-dependent Ca^{2+} increases in hippocampal astrocytes does not affect baseline CA1 pyramidal neuron synaptic activity. *Journal of Neuroscience, 28*, 4967–4973.

Petzold, G. C., Albeanu, D. F., Sato, T. F., & Murthy, V. N. (2008). Coupling of neural activity to blood flow in olfactory glomeruli is mediated by astrocytic pathways. *Neuron, 58*, 897–910.

Pickel, V. M., Chan, J., Veznedaroglu, E., & Milner, T. A. (1995). Neuropeptide Y and dynorphin-immunoreactive large dense-core vesicles are strategically localized for presynaptic modulation in the hippocampal formation and substantia nigra. *Synapse, 19*, 160–169.

Pisoni, R. L., & Thoene, J. G. (1989). Detection and characterization of a nucleoside transport system in human fibroblast lysosomes. *The Journal of Biological Chemistry, 264*, 4850–4856.

Porter, J. T., & McCarthy, K. D. (1996). Hippocampal astrocytes in situ respond to glutamate released from synaptic terminals. *Journal of Neuroscience, 16*, 5073–5081.

Prada, I., Marchaland, M., Podini, P., Magrassi, L., D'Alessandro, R., Bezzi, P., & Meldolesi, J. (2011). REST/NRSF governs the expression of dense-core vesicle gliosecretion in astrocytes. *Journal of Cell Biology, 193*, 537–549.

Ramamoorthy, P., & Whim, M. D. (2008). Trafficking and fusion of neuropeptide Y-containing dense-core granules in astrocytes. *Journal of Neuroscience, 28*, 13815–13827.

Ramon y Cajal, S. (1995). Histology of the nervous system, translated by N.Swanson and L.Swanson, Oxford University Press

Re, D. B., Nafia, I., Melon, C., Shimamoto, K., Kerkerian-Le Goff, L., & Had-Aissouni, L. (2006). Glutamate leakage from a compartmentalized intracellular metabolic pool and activation of the lipoxygenase pathway mediate oxidative astrocyte death by reversed glutamate transport. *Glia, 54*, 47–57.

Rosa, P., & Gerdes, H. H. (1994). The granin protein family: Markers for neuroendocrine cells and tools for the diagnosis of neuroendocrine tumors. *Journal of Endocrinological Investigation, 17*, 207–225.

Rossi, D., Brambilla, L., Valori, C. F., Crugnola, A., Giaccone, G., Capobianco, R., Mangieri, M., Kingston, A. E., Bloc, A., Bezzi, P., & Volterra, A. (2005). Defective tumor necrosis factor-alpha-dependent control of astrocyte glutamate release in a transgenic mouse model of Alzheimer disease. *The Journal of Biological Chemistry, 280*, 42088–42096.

Rossi, D. J., Oshima, T., & Attwell, D. (2000). Glutamate release in severe brain ischaemia is mainly by reversed uptake. *Nature, 403*, 316–321.

Sala, C., Roussignol, G., Meldolesi, J., & Fagni, L. (2005). Key role of the postsynaptic density scaffold proteins Shank and Homer in the functional architecture of Ca^{2+} homeostasis at dendritic spines in hippocampal neurons. *Journal of Neuroscience, 25*, 4587–4592.

Santello, M., Bezzi, P., & Volterra, A. (2011). TNFa controls glutamatergic gliotransmission in the hippocampal dentate gyrus. *Neuron, 69*, 988–1001.

Santello, M., & Volterra, A. (2009). Synaptic modulation by astrocytes via Ca^{2+} -dependent glutamate release. *Neuroscience, 158*, 253–259.

Santello, M., & Volterra, A. (2010). Neuroscience: Astrocytes as aide-mémoires. *Nature, 463*, 169–170.

Sanzgiri, R. P., Araque, A., & Haydon, P. G. (1999). Prostaglandin E(2) stimulates glutamate receptor-dependent astrocyte neuromodulation in cultured hippocampal cells. *Journal of Neurobiology, 41*, 221–229.

Sasaki, T., Kuga, N., Namiki, S., Matsuki, N., & Ikegaya, Y. (2011). Locally synchronized astrocytes. *Cerebral Cortex*. [Epub ahead of print].

Schlüter, O. M., Khvotchev, M., Jahn, R., & Südhof, T. C. (2002). Localization versus function of Rab3 proteins. Evidence for a common regulatory role in controlling fusion. *The Journal of Biological Chemistry, 277*, 40919–40929.

Schummers, J., Yu, H., & Sur, M. (2008). Tuned responses of astrocytes and their influence on hemodynamic signals in the visual cortex. *Science, 320*, 1638–1643.

Serrano, A., Haddjeri, N., Lacaille, J. C., & Robitaille, R. (2006). GABAergic network activation of glial cells underlies hippocampal heterosynaptic depression. *Journal of Neuroscience, 26*, 5370–5382.

Shanker, G., & Aschner, M. (2001). Identification and characterization of uptake systems for cystine and cysteine in cultured astrocytes and neurons: Evidence for methylmercury-targeted disruption of astrocyte transport. *Journal of Neuroscience Research, 66*, 998–1002.

Shi, Y., Liu, X., Gebremedhin, D., Falck, J. R., Harder, D. R., & Koehler, R. C. (2008). Interaction of mechanisms involving epoxyeicosatrienoic acids, adenosine receptors, and metabotropic glutamate receptors in neurovascular coupling in rat whisker barrel cortex. *Journal of Cerebral Blood Flow and Metabolism, 28*, 111–125.

Snyder, S. H., & Kim, P. M. (2000). D-amino acids as putative neurotransmitters: Focus on D-serine. *Neurochemical Research, 25*, 553–560.

Steinmetz, C. C., & Turrigiano, G. G. (2010). Tumor necrosis factor-α signaling maintains the ability of cortical synapses to express synaptic scaling. *The Journal of Neuroscience, 30*, 14685–14690.

Stellwagen, D., Beattie, E. C., Seo, J. Y., & Malenka, R. C. (2005). Differential regulation of AMPA receptor and GABA receptor trafficking by tumor necrosis factor-alpha. *The Journal of Neuroscience, 25*, 3219–3228.

Stellwagen, D., & Malenka, R. C. (2006). Synaptic scaling mediated by glial TNF-alpha. *Nature, 440*, 1054–1059.

Stenmark, H. (2009). Rab GTPases as coordinators of vesicle traffic. *Nature Reviews Molecular Cell Biology, 10*, 513–525.

Stout, C. E., Costantin, J. L., Naus, C. C., & Charles, A. C. (2002). Intercellular calcium signaling in astrocytes via ATP release through connexin hemichannels. *The Journal of Biological Chemistry, 277*, 10482–10488.

Szatkowski, M., Barbour, B., & Attwell, D. (1990). Non-vesicular release of glutamate from glial cells by reversed electrogenic glutamate uptake. *Nature, 348*, 443–446.

Takai, Y., Sasaki, T., Shirataki, H., & Nakanishi, H. (1996). Rab3A small GTP-binding protein in Ca(2+)-dependent exocytosis. *Genes to Cells, 1*, 615–632.

Takata, N., & Hirase, H. (2008). Cortical layer 1 and layer 2/3 astrocytes exhibit distinct calcium dynamics in vivo. *PLoS One, 3*, e2525.

Tang, X. C., & Kalivas, P. W. (2003). Bidirectional modulation of cystine/glutamate exchanger activity in cultured cortical astrocytes. *Annals of the New York Academy of Sciences, 1003*, 472–475.

Trajkovic, K., Dhaunchak, A. S., Goncalves, J. T., Wenzel, D., Schneider, A., Bunt, G., Nave, K.-A., et al. (2006). Neuron to glia signaling triggers myelin membrane exocytosis from endosomal storage sites. *The Journal of cell biology, 172(6)*, 937–948. doi:10.1083/jcb.200509022.

Turrigiano, G. G. (2008). The self-tuning neuron: Synaptic scaling of excitatory synapses. *Cell, 135*, 422–435.

Ventura, R., & Harris, K. M. (1999). Three-dimensional relationships between hippocampal synapses and astrocytes. *The Journal of Neuroscience, 19*, 6897–6906.

Vitkovic, L., Bockaert, J., & Jacque, C. (2000). "Inflammatory" cytokines: Neuromodulators in normal brain? *Journal of Neurochemistry, 74*, 457–471.

Voglmaier, S. M., Kam, K., Yang, H., Fortin, D. L., Hua, Z., Nicoll, R. A., & Edwards, R. H. (2006). Distinct endocytic pathways control the rate and extent of synaptic vesicle protein recycling. *Neuron, 51*, 71–84.

Volterra, A., Bezzi, P., Rizzini, B. L., Trotti, D., Ullensvang, K., Danbolt, N. C., & Racagni, G. (1996). The competitive transport inhibitor L-trans-pyrrolidine-2, 4-dicarboxylate triggers excitotoxicity in rat cortical neuron-astrocyte co-cultures via glutamate release rather than uptake inhibition. *European Journal of Neuroscience, 8*, 2019–2028.

Volterra, A., & Meldolesi, J. (2005). Astrocytes, from brain glue to communication elements: The revolution continues. *Nature Reviews Neuroscience, 6*, 626–640.

Wang, X., Lou, N., Xu, Q., Tian, G. F., Peng, W. G., Han, X., Kang, J., Takano, T., & Nedergaard, M. (2006). Astrocytic Ca^{2+} signaling evoked by sensory stimulation in vivo. *Nature Neuroscience, 9*, 816–823.

Winship, I. R., Plaa, N., & Murphy, T. H. (2007). Rapid astrocyte calcium signals correlate with neuronal activity and onset of the hemodynamic response in vivo. *Journal of Neuroscience, 27*, 6268–6272.

Yang, Y., Ge, W., Chen, Y., Zhang, Z., Shen, W., Wu, C., Poo, M., & Duan, S. (2003). Contribution of astrocytes to hippocampal long-term potentiation through release of D-serine. *Proceedings of the National Academy of Sciences of the United States of America, 100*, 15194–15199.

Ye, Z. C., Wyeth, M. S., Baltan-Tekkok, S., & Ransom, B. R. (2003). Functional hemichannels in astrocytes: A novel mechanism of glutamate release. *The Journal of Neuroscience, 23*, 3588–3596.

Zerial, M., & McBride, H. (2001). Rab proteins as membrane organizers. *Nature Reviews: Molecular Cell Biology, 2*, 107–117.

Zhang, Z., Chen, G., Zhou, W., Song, A., Xu, T., Luo, Q., Wang, W., Gu, X. S., & Duan, S. (2007). Regulated ATP release from astrocytes through lysosome exocytosis. *Nature Cell Biology, 9*, 945–953.

Zhang, Q., Fukuda, M., Van Bockstaele, E., Pascual, O., & Haydon, P. G. (2004). Synaptotagmin IV regulates glial glutamate release. *Proceedings of the National Academy of Sciences of the United States of America, 101*, 9441–9446.

Part III
Synapse-to-Nucleus Communication

Chapter 15
Roles of Neuronal Activity-Induced Gene Products in Hebbian and Homeostatic Synaptic Plasticity, Tagging, and Capture

Yasunori Hayashi, Ken-ichi Okamoto, Miquel Bosch, and Kensuke Futai

Abstract The efficiency of synaptic transmission undergoes plastic modification in response to changes in input activity. This phenomenon is most commonly referred to as synaptic plasticity and can involve different cellular mechanisms over time. In the short term, typically in the order of minutes to 1 h, synaptic plasticity is mediated by the actions of locally existing proteins. In the longer term, the synthesis of new proteins from existing or newly synthesized mRNAs is required to maintain the changes in synaptic transmission. Many studies have attempted to identify genes induced by neuronal activity and to elucidate the functions of the encoded proteins. In this chapter, we describe our current understanding of how activity can regulate the synthesis of new proteins, how the distribution of the newly synthesized protein is regulated in relation to the synapses undergoing plasticity and the function of these proteins in both Hebbian and homeostatic synaptic plasticity.

Keywords Arc/Arg3.1 • Homeostatic plasticity • Homer/vesl • Synapse tagging and capture • Synaptic plasticity

Y. Hayashi (✉)
Brain Science Institute, Central Building S506, RIKEN, 2-1 Hirosawa, Wako, Saitama 351-0198, Japan
e-mail: yhayashi@brain.riken.jp

K.-i. Okamoto
Samuel Lunenfeld Research Institute, Mount Sinai Hospital, 600 University Avenue, Toronto, ON, M5G 1X5, Canada

M. Bosch
Department of Brain and Cognitive Sciences, The Picower Institute for Learning and Memory, Massachusetts Institute of Technology, Cambridge, MA 02139, USA

K. Futai
Department of Psychiatry, Brudnick Neuropsychiatric Research Institute, University of Massachusetts Medical School, Worcester, MA 01604-1676, USA

M.R. Kreutz and C. Sala (eds.), *Synaptic Plasticity*,
Advances in Experimental Medicine and Biology 970,
DOI 10.1007/978-3-7091-0932-8_15, © Springer-Verlag/Wien 2012

15.1 Introduction: Memory, Synaptic Plasticity, and New Protein Synthesis

The synthesis of the right species of protein in the right cells at the right time is crucial for every aspect of cellular function. Therefore, protein synthesis is tightly genetically defined depending on cell type and developmental status. In addition, gene expression is also regulated by external factors. Various cellular environments can also affect the timing, species, and the amount of synthesized proteins.

Protein synthesis in neurons is not an exception from such regulation (for review, see Loebrich and Nedivi 2009; Okuno 2011). The first evidence to link gene expression to the functional modification of neurons comes from the observation that pharmacological stimulation of cultured neuronal cells induces expression of a specific set of genes (Greenberg et al. 1986). Similarly, induction of kindling in animals induces gene transcription (Dragunow and Robertson 1987; Saffen et al. 1988). These observations led to a seminal study by Cole et al. (1989) that linked gene expression to synaptic plasticity for the first time. They induced synaptic plasticity in the hippocampus by delivering a high-frequency stimulation to the input fibers, which leads to long-term potentiation (LTP) of synaptic transmission (Bliss and Lømo 1973; Bliss and Collingridge 1993), and found a correlated increase of specific genes. This induction of gene expression required activation of N-methyl-D-aspartate type glutamate receptor (NMDAR), which is also necessary for hippocampal LTP indicating that the electrophysiologically measured LTP and the induction of new gene transcription share at least a common part of cellular signaling (Herron et al. 1986). Arguably, LTP is a cellular counterpart of learning and memory (Bliss and Collingridge 1993), and this work was the first to demonstrate the expression of specific genes in the context of learning and memory. Subsequently, many paradigms of learning such as exposure to a novel environment, fear, and pheromones have been shown to induce expression of genes in specific neurons (Brennan et al. 1999; Guzowski et al. 2001; Hall et al. 2001). Furthermore, because the induction is so robust and reproducible, the induction of those genes could be even used to identify neurons that are activated during learning (Frankland et al. 2004; Reijmers et al. 2007; Kitanishi et al. 2009; Sacco and Sacchetti 2010).

Are gene expression and protein synthesis required for learning and memory? Earlier studies using pharmacological inhibitors of protein synthesis in animals resulted in an inhibition of memory formation (Flexner et al. 1963; Barondes and Cohen 1968). This is followed by multiple studies that tested the effect of protein synthesis inhibitors in various learning paradigms in different animals (Davis and Squire 1984, for review). Overall, the results consistently showed that animals exhibited normal memory performance for a short period of time after the initial training but when tested several days later, their memory was impaired. These observations indicate that the initial formation of memory does not require *de novo* protein synthesis whereas the retention of the memory over extended periods of time requires newly synthesized proteins. Interestingly, when protein synthesis inhibitors were administered at later time points after the initial memory formation, the memory became resistant to the treatment, indicating that protein synthesis is not required for retention or recall once memory becomes consolidated (Davis and Squire 1984).

Consistent with the effect of protein synthesis inhibitors on memory formation, these reagents also blocked LTP consolidation. Inhibition of either protein translation (Krug et al. 1984; Otani et al. 1989) or transcription (Frey et al. 1996) did not affect the early phase of LTP which occurred within 1 h after stimulation (early LTP or E-LTP), suggesting that this initial phase is mediated by existing proteins. However, these treatments did block the maintenance of LTP, especially during the late phase which typically occurs 2 h after the induction (late LTP or L-LTP), which indicates that *de novo* protein synthesis is required for the retention of LTP. Similarly, if transport of newly synthesized mRNA and protein was prevented by separating dendrites and cell bodies of hippocampal neurons, L-LTP was blocked while E-LTP was not affected, indicating a requirement for transport of newly synthesized molecules from the cell body for maintenance of LTP at the late phase (Frey et al. 1989).

15.2   Search for Activity-Regulated Genes

These observations triggered many investigations into the involvement of transcription and translation in synaptic plasticity and learning. A number of laboratories set forth to systematically identify genes induced by neuronal activity in an unbiased way. One of the earlier screening efforts was performed by Nedivi and colleagues (1993). They carried out differential cDNA screening between mRNAs from hippocampal dentate gyrus treated with or without an excitatory amino acid analog kainate. They isolated 52 activity-induced genes, of which 35 were novel at the time of report and 17 were known genes such as c-*fos*, c-*jun*, and *zif/268*, which were already reported to be neuronal activity-induced genes, indicating the validity of their approach.

Similar approaches were also undertaken independently around the same time by several other laboratories including those of Inokuchi (e.g., Kato et al. 1997; Matsuo et al. 1998), Worley (Yamagata et al. 1993; Lanahan and Worley 1998), Bliss (Fazeli et al. 1993), Kuhl (Link et al. 1995), and Kandel (Qian et al. 1993). Over the last decade, further analyses furnished with new information and technologies such as entire genome sequence and microarray have led to the identification of many other neuronal activity-induced genes (French et al. 2001; Elliott et al. 2003; Newton et al. 2003; Altar et al. 2004; Hong et al. 2004; Park et al. 2006; Kitamura et al. 2007; Xiang et al. 2007). Currently, it is estimated that around 500–1,000 different neuronal activity-induced genes exist (Nedivi et al. 1993; Loebrich and Nedivi 2009). The identified genes encode proteins ranging from transcription factors, enzymes involved in metabolism, intracellular and extracellular signaling molecules, to cytoskeletal proteins (for review, see Loebrich and Nedivi 2009; Okuno 2011).

Among the various mRNA species detected in such studies, those which increased their amount in the presence of protein synthesis inhibitors were named *immediate early genes* (Cochran et al. 1983). These mRNA species increase their content utilizing only the transcription machineries already present under basal cellular activity and therefore are the primary responder to the external stimulation. Interestingly, some of them encode proto-oncogene transcription factors such as *c-fos, c-jun, jun-B, c-myc,* and *zif/268* (also named *Egr1, NGFI-A, Krox 24*), which have been identified in multiple studies to isolate activity-induced genes both from neuronal and nonneuronal cells (Greenberg and Ziff 1984; Greenberg et al. 1986; Franza et al. 1988; Rauscher et al. 1988; Ryder et al. 1988). Once they are induced, they in turn induce a second surge of gene expression. For example, *c-fos* and *c-jun* gene products associate with each other to form activator protein 1 (AP-1) transcription factor (Raivich and Behrens 2006). The AP-1-responsive consensus sequence is found in genes implicated in the synaptic functions such as α-amino-3-hydroxyl-5-methyl-4-isoxazole-propionate-type glutamate receptor (AMPAR) subunit GluR2, growth-associated protein 43 (GAP-43), and the cyclin-dependent protein kinase Cdk5 (Rylski and Kaczmarek 2004; Raivich and Behrens 2006).

In addition to regulation at the transcriptional level, neuronal activity can trigger translation of new proteins from existing mRNAs. Such translation is regulated in several phases (see Wang et al. 2010 for review). While neurons are at basal activity, the translation from these mRNA must be repressed. When the activity level goes up beyond a certain level, this repression must be removed and the translational machinery recruited. Several proteins and noncoding RNAs are involved in the regulation (Feng et al. 1995; Costa-Mattioli et al. 2009; Wang et al. 2010). Of note, abnormality in the translational regulation is suspected to be a major cause of a hereditary mental retardation syndrome, fragile X disease. The disease is caused by a mutation in fragile X mental retardation protein (FMRP), a mRNA binding protein that represses the translation of its target (Feng et al. 1995; Kindler and Monshausen 2002; Bassell and Warren 2008; De Rubeis and Bagni 2010). Autopsy examination of fragile X patients revealed abnormal dendritic spine morphology (Irwin et al. 2001). This is recapitulated in mice lacking FMRP, which show impairment in both synaptic function and morphology, highlighting the importance of the translational regulation for proper synaptic functions (Comery et al. 1997; Nimchinsky et al. 2001; Huber et al. 2002; Li et al. 2002; Schütt et al. 2009). There is also evidence that FMRP can regulate the translation of NMDA receptor subunit NR2A through a specific micro RNA (miRNA), miR-125b (Edbauer et al. 2010).

15.3 Synaptic Tagging and Capture Hypothesis

In the aforementioned screening studies, multiple molecules implicated in synaptic function have been isolated. This includes synaptic scaffolding proteins, e.g., Homer1a/vesl-1s (Brakeman et al. 1997; Kato et al. 1998), Arc/Arg3.1 (Link et al. 1995; Lyford et al. 1995), and candidate plasticity gene (CPG) 2 (Cottrell et al. 2004);

intracellular signaling molecules such as kinases, e.g., SNK/polo-like kinase 2 (Kauselmann et al. 1999), protein kinase Mζ (Osten et al. 1996), and GTP-binding proteins or their associated proteins, e.g., RGS2 (Ingi et al. 1998) and Rheb (Yamagata et al. 1994); extracellular signaling molecules, e.g., brain-derived neurotrophic factor (BDNF) (Hughes et al. 1993) and CPG15 (Nedivi et al. 1998); and cell surface adhesion molecules involved in mediating cell–cell communication, e.g., Arcadin (Yamagata et al. 1999) (for review, see Loebrich and Nedivi 2009; Okuno 2011).

One basic feature of LTP is "input specificity," where only stimulated synapses become potentiated and unstimulated synapses are not affected (Bliss and Collingridge 1993). This feature applies not only to the early phase of LTP, which is independent of protein synthesis, but also to the protein synthesis–dependent late phase of LTP. So there may be a mechanism that ensures newly synthesized proteins to function only at potentiated synapses. The presence of such a mechanism was experimentally demonstrated by Frey and Morris (1997). In two-pathway hippocampal recordings, when a high-frequency stimulation was given to one pathway in the presence of a protein synthesis inhibitor, the slice exhibited E-LTP but not L-LTP, consistent with the requirement of protein synthesis in L-LTP. But if a high-frequency stimulation was given to the second pathway before the application of a protein synthesis inhibitor, both pathways could elicit L-LTP. This is interpreted as follows: when the tetanus is given, it generates a "tag" specifically at the stimulated synapse independent of protein synthesis. This will serve as a synapse-specific binding site where newly synthesized proteins required for L-LTP are "captured." The L-LTP induced in the second pathway supplied the necessary protein not only for itself but also for the first pathway stimulated in the presence of protein synthesis inhibitors. The first pathway by itself could not induce protein synthesis because it received tetanic stimulation in the presence of the inhibitor but could still form a "tag," which can "capture" the newly synthesized protein induced by the second pathway and therefore, can induce L-LTP if the second pathway is stimulated in the absence of protein synthesis inhibitor.

This "synapse tag and capture" hypothesis was an attractive model to explain the input specificity of synaptic plasticity and multiple studies have been conducted ever since the proposal (Redondo and Morris 2011 for review). In particular, the identity of the tag has been the source of major research interest. The tag has to fulfill at least four criteria: (1) The tag is formed specifically at potentiated synapses, (2) it does not require synthesis of new proteins for formation, (3) it must stay at the synapse for at least 1–2 h, and (4) it should have a structure that is capable of recruiting newly synthesized plasticity-related proteins (Okamoto et al. 2009).

Okamoto et al. (2004) found that LTP induction induced a rapid formation of filamentous actin (F-actin) in dendritic spines (Fig. 15.1a). F-actin is the major cytoskeletal protein in dendritic spines that serves both as a structural framework of the synapse and as a binding site for other postsynaptic proteins. Consistent with the increase in F-actin, LTP induction also caused an enlargement of dendritic spines that persisted for up to 1 h (Matsuzaki et al. 2004; Okamoto et al. 2004; Honkura et al. 2008). A pharmacological formation of more F-actin at the dendritic spine was sufficient to deliver postsynaptic proteins to the synapse (Okamoto et al. 2004)

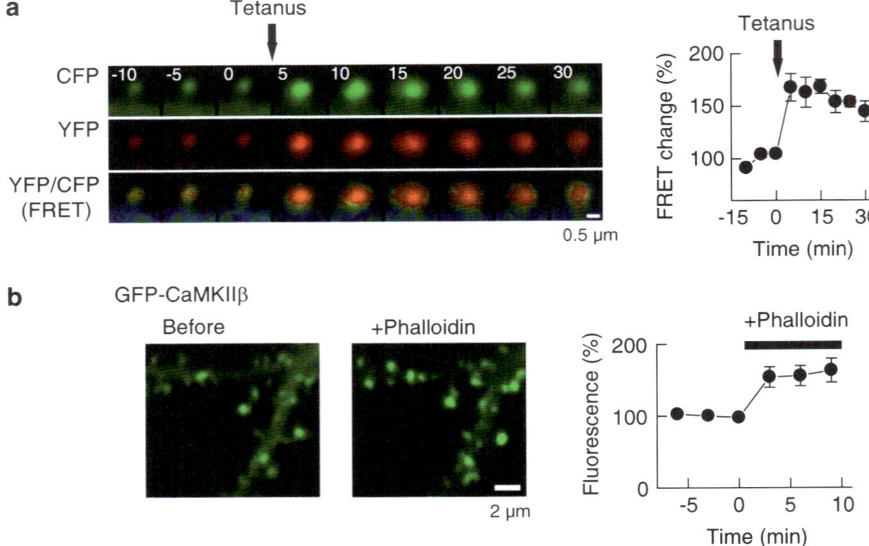

Fig. 15.1 *Polymerization of F-actin induced by local tetanic stimulation and its effect on synaptic protein distribution.* (**a**) F-actin formation was visualized by Förster resonance energy transfer (FRET) between donor- and acceptor-tagged actin molecules. Warmer hue indicates higher FRET. Time stamp in minutes. (**b**) Pharmacological polymerization of actin is sufficient to deliver CaMKII into dendritic spine. A neuron was transfected with GFP-tagged CaMKIIβ and injected with phalloidin, which stabilizes F-actin. The synaptic distribution of GFP-CaMKIIβ was monitored before and after the injection (Modified from Okamoto et al. 2004)

(Fig. 15.1b). This increase in F-actin levels along with the resultant increase in spine volume and the enhanced capacity of binding of the synapse perfectly fulfill all four criteria listed above. In fact, it has been recently shown that a pharmacological disruption of F-actin prevented synapse tagging (Ramachandran and Frey 2009). Therefore, formation of new F-actin together with the resulting structural changes is a prime candidate for the synapse tag.

15.4   Synaptic Capture of Newly Synthesized Protein

Another feature of the synapse tag and capture hypothesis is the selective capture of newly synthesized proteins at potentiated synapses. There can be several ways to deliver proteins to the tagged synapses. One is to deliver mRNA to the tagged synapse and have the proteins locally translated. The dendrite contains hundreds of mRNA species, ribosomes and intracellular organelles such as endoplasmic reticulum or Golgi apparatus required for protein synthesis (Steward and Levy 1982; Eberwine et al. 2002; Kindler and Monshausen 2002; Ostroff et al. 2002; Horton and Ehlers 2003; Moccia et al. 2003; Poon et al. 2006).

One representative mRNA of a dendritically localized immediate early gene is Arc/Arg3.1. It was initially identified as an immediate early gene induced after neuronal activation (Link et al. 1995; Lyford et al. 1995; Hevroni et al. 1998; Steward et al. 1998; Steward and Worley 2001). In dendrites, the Arc/Arg3.1 mRNA is specifically localized to the activated dendritic region thereby confining the protein product within the vicinity of the activated synapse (Steward et al. 1998; Steward and Worley 2001).

The mRNA of Ca^{2+}/calmodulin-dependent protein kinase IIα (CaMKIIα) is another abundant dendritic mRNA. It has a dendritic localization element in the $3'$-untranslated region (UTR) (Mayford et al. 1996). LTP induction induced a rapid increase in local translation of CaMKIIα (Ouyang et al. 1999; Bagni et al. 2000). Genetic elimination of the $3'$-UTR in mice not only disrupted the dendritic targeting of CaMKIIα but also impaired the stabilization of synaptic plasticity and memory consolidation (Miller et al. 2002).

However, these studies still have not demonstrated whether the newly synthesized proteins are specifically captured specifically at the potentiated synapse or not. It has been difficult to address this issue in the mammalian central nervous system, primarily due to the small size of individual synapses, the high density of synapses, and the lack of appropriate methods to induce synaptic plasticity specifically in the synapse under observation. To overcome these problems, Wang et al. (2009) used the *Aplysia* sensory-motor neuron co-culture system that mimics a simple neuronal circuit which underlies sensitization and habituation of gill-withdrawal reflex. In this preparation, focal application of neurotransmitter serotonin can trigger protein synthesis-dependent synapse-specific plasticity (Martin et al. 1997). By time-lapse imaging of photoconvertible translational reporters introduced to neurons in this system, they demonstrated that the translation was spatially restricted to the activated synapse.

Recently, Inokuchi's group tested whether the somatically synthesized immediate early gene product, Homer1a, can be trapped at the activated synapse in cultured hippocampal neurons (Okada et al. 2009). Homer (also called cupidin/Vesl/PSD-Zip45/Ania3) is a family of synaptic scaffolding protein identified in various studies as a gene product induced by neuronal activity (Nedivi et al. 1993; Brakeman et al. 1997; Kato et al. 1997; Xiao et al. 2000; Shiraishi-Yamaguchi and Furuichi 2007). Okada et al. (2009) expressed one of the subtypes, Homer1a fused with green fluorescent protein (GFP) and monitored its translocation in neurons. When synapses were locally stimulated with NMDA and glycine, more GFP-tagged Homer1a became localized at the stimulated synapse. To test if the somatically synthesized Homer1a could be captured by the stimulated synapse, they used photoactivatable GFP (PA-GFP)-tagged Homer1a and selectively photoactivated PA-GFP-Homer1a in the cell body. The authors demonstrated that the soma-derived PA-GFP-tagged Homer1a protein was captured by activated synapses.

Is the capture of protein to activated synapses a unique property of Homer1a, or is it a more general property shared by various synaptic proteins (Okabe 2007; Sheng and Hoogenraad 2007)? To address this question, Bosch et al. (2009) investigated the time course of translocation of multiple postsynaptic proteins including receptors, enzymes, and scaffolding components of the postsynaptic density (PSD), during the selective induction of LTP onto single dendritic spines using two-photon glutamate uncaging. They cotransfected the GFP-fused protein of interest with a red fluorescent protein (RFP) to simultaneously observe the localization of the protein and spine volume changes upon synaptic stimulation. They found that most of the proteins were translocated to the synapse following LTP induction. For the majority of them, including Homer1a, AMPAR GluR1, actin, and CaMKIIα, the total amount of protein that accumulated at the spine was comparable to the changes in spine volume, making their concentration before and after LTP induction equivalent (Okamoto et al. 2004; Kuriu et al. 2006; Bosch et al. 2009). This supports the idea that the change in the volume of the spines induced by F-actin polymerization provides an enhanced binding capacity to that synapse and acts as the tag that can quantitatively capture the new proteins needed to maintain the potentiated state (Okamoto et al. 2009; Redondo and Morris 2011).

15.5   The Functions of the Neuronal Activity-Induced Gene Products in Hebbian and Homeostatic Plasticity

As we have seen above, synaptic activity induces the expression of many neuronal genes. How then do these neuronal activity-induced gene products affect synaptic transmission in turn? We overexpressed Homer1a in hippocampal pyramidal neurons and compared the response amplitude of glutamatergic synaptic transmission between control and overexpressing neurons (Sala et al. 2003) (Fig. 15.2a). Surprisingly, overexpression of Homer1a reduced the postsynaptic current amplitude instead of increasing it. The reduction in transmission occurred in both AMPARs and NMDAR-mediated synaptic currents. Sala et al. (2003) carried out morphological analyses of Homer1a overexpressing neurons and found that the neurons not only had a reduced surface glutamate receptors but also a decrease in various postsynaptic proteins as well as the density and size of dendritic spines (Fig. 15.2b). So the effect of Homer1a is the global suppression of both synaptic structure and function, which is seemingly counterintuitive considering that Homer1a is induced by neuronal activity.

To elucidate the molecular mechanisms of this effect, a better understanding of the role of Homer1b, the long form of Homer is critical (Shiraishi-Yamaguchi and Furuichi 2007). Homer1b is generated from the same gene as Homer1a but from a longer transcript (Fig. 15.3a). Homer1a is a short, monomeric form that contains only an EVH1 domain and is expressed in an activity-dependent manner. In contrast, the long forms, Homer1b and c, have both the EVH1 and the coiled-coil domains that form a tetramer and are constitutively expressed (Hayashi et al. 2006, 2009). Through the EVH1 domain, both short and long Homers can bind to various other scaffolding and signal transduction molecules, which include group I metabotropic glutamate receptors (mGluR) (Brakeman et al. 1997), inositol-1,4,5-trisphosphate receptors

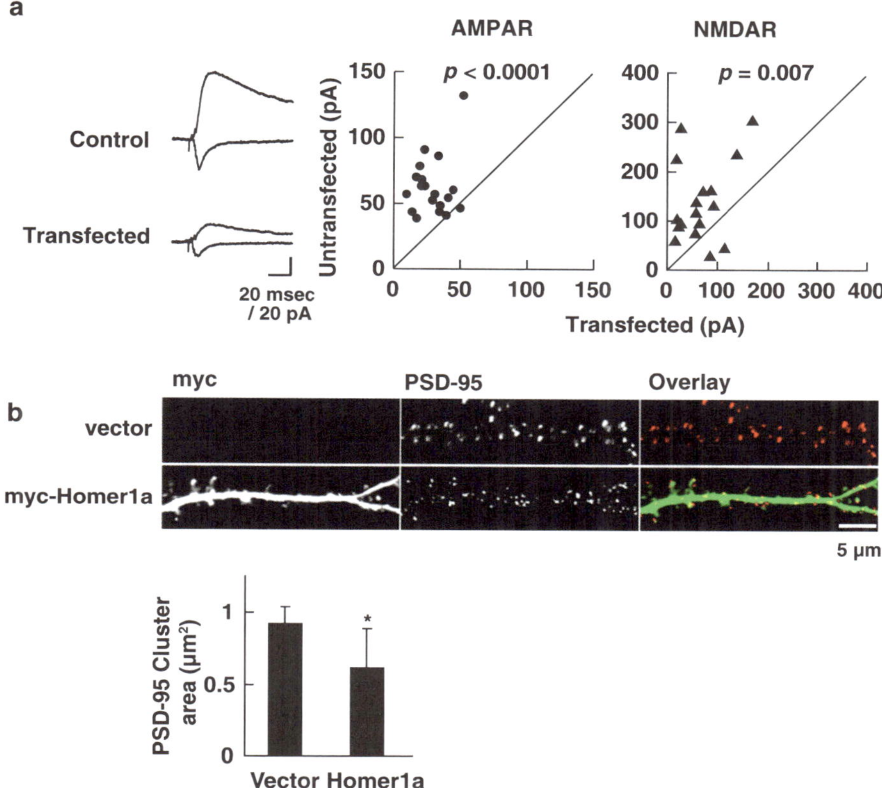

Fig. 15.2 *Reduction of synaptic current and components in neurons expressing Homer1a.*
(**a**) A paired recording from a hippocampal pyramidal CA1 pyramidal neuron expressing Homer1a
or adjacent untransfected neuron. *Downward trace*: AMPAR EPSC recorded at −60 mV; *upward
trace*: NMDAR-EPSC recorded at +40 mV in the presence of AMPAR blocker. Results of recording
from multiple cells. Note that both AMPAR and NMDAR components are similarly reduced.
(**b**) Reduction of PDS-95 synaptic clusters, a representative postsynaptic protein, in neurons
overexpressing Homer1a. $^*p < 0.01$ (Modified from Sala et al. 2003)

(IP$_3$R) (Tu et al. 1998), Shank (Tu et al. 1999), TRPC family channels (Yuan et al.
2003), and PI3 kinase enhancer (Rong et al. 2003).

Sala et al. (2001) found that when Homer1b is coexpressed with its binding
partner Shank in neurons, an enlargement in the size of dendritic spines together
with the recruitment of multiple postsynaptic proteins was observed. Shank itself can
also form oligomers through homomeric association (Im et al. 2003; Romorini et al.
2004; Baron et al. 2006). Hayashi et al. (2009) found that when Homer1b and Shank
were mixed together, they formed a high-order mesh-like complex (Fig. 15.3b). This
complex can carry postsynaptic adapter protein GKAP, which is further linked to
synaptic surface glutamate receptor proteins. When tetramer formation of Homer1b
or interaction with Shank was prevented by point mutations, Homer and Shank could

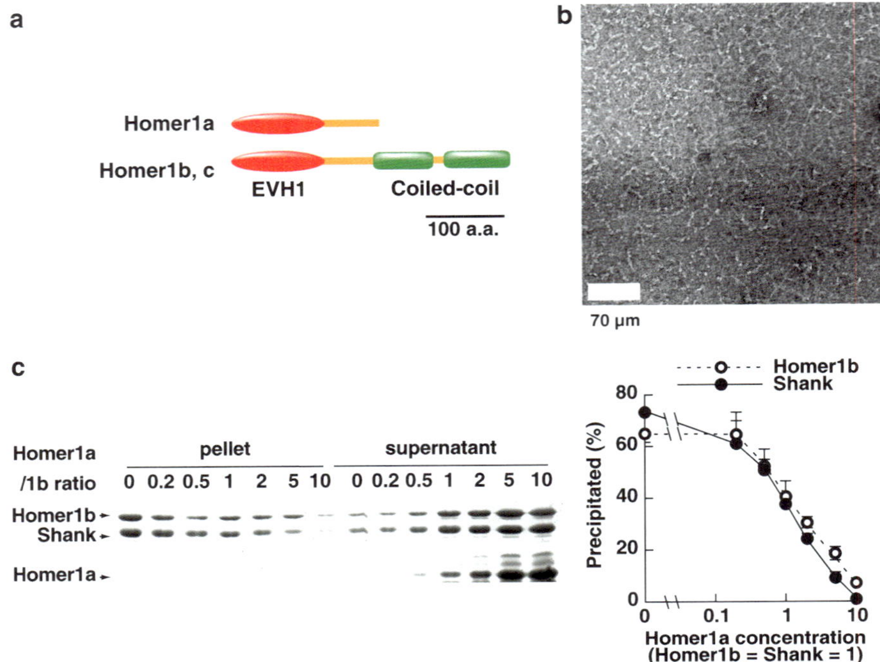

Fig. 15.3 *Interaction between Homer1a and Shank-Homer1b complex*. (**a**) A schematic drawing of Homer1a and Homer1b. (**b**) Electron microscopic picture of complex between Homer1b and Shank. The proteins were expressed separately, reconstituted in vitro, and observed with electron microscopy. Note a mesh-like high-order structure. (**c**) Inhibition of Homer1b-Shank by the addiction of Homer1a. To the fixed amount of Homer1b and Shank, an increasing amount of Homer1a was added. The mixture was centrifuged to precipitate high order complex and the amount of Homer1b and Shank in pellet fraction was measured (Modified from Hayashi et al. 2009)

not form the high order complex. When the mutant that abolish tetramer formation was introduced to a neuron, the number of dendritic spines decreased and the spines became longer and thinner similarly to those which overexpress Homer1a (Sala et al. 2003; Hayashi et al. 2009). These results indicate that the mesh formation between Homer1b and Shank is required for the maturation of synapses, likely by forming a two-dimensional lattice where other postsynaptic proteins rest. Interestingly, addition of Homer1a prevented the Homer1b-Shank mesh formation in a dose-dependent fashion. This observation reasonably explains the general suppressive function of Homer1a on postsynaptic structure and function (Fig. 15.3c).

Hu et al. (2010) also found that Homer1a reduces synaptic AMPAR currents similarly to Sala et al. (2003). But they considered this effect to be mediated by the activation of group I mGluR. Binding of Homer1a with the intracellular carboxyl tail of group I mGluR activates the receptor without glutamate (Ango et al. 2001), which leads to the reduction of tyrosine phosphorylation of AMPAR subunit GluR2 and then to the reduction of surface AMPAR (Hayashi et al. 1999; Hayashi and Huganir 2004).

However, this mechanism does not fully explain the reduction of other postsynaptic proteins such as NMDA receptor, Homer1b, actin, and Shank or the shrinkage of the overall structure of dendritic spines (Sala et al. 2003). Therefore, it still remains to be determined what is the exact role of Homer1a at the synapse.

Overexpression of Arc/Arg3.1 also reduced the level of AMPAR-mediated synaptic transmission without changing NMDAR-mediated synaptic transmission unlike Homer1a, indicative of different mechanism of action between these two proteins (Chowdhury et al. 2006; Rial Verde et al. 2006; Shepherd et al. 2006). Arc/Arg3.1 interacts with endophilin 2 and 3 and dynamin, which are components of the clathrin-mediated endocytotic machinery. The interaction stimulates the clathrin-mediated endocytosis of synaptic AMPAR, thereby leading to a specific reduction of AMPAR-mediated synaptic transmission. This mechanism works even at a single synapse level (Béïque et al. 2011). Mice with Arc/Arg3.1 gene disruption exhibited an enhanced E-LTP but diminished L-LTP and impaired memory (Plath et al. 2006). However, it remains to be determined whether Arc/Arg3.1 protein is captured by an activated synapse to the same extent as that reported for Homer1a. Further study is needed to clarify the role of Arc/Arg3.1 at specific synapses during memory formation.

There is a special isoform of atypical protein kinase C (PKC) called PKMζ. Unlike full-length PKC, which requires diacylglycerol and Ca^{2+} for activation, PKMζ is independent of Ca^{2+}. Instead, the activity of PKMζ is regulated through a unique translational machanism. Under basal neuronal activity, translation of PKMζ is repressed. But by LTP induction and resultant activation of intracellular signaling cascade, this repression is unmasked and translation of PKMζ protein is initiated (Hernandez et al. 2003). Once active PKMζ is formed, it induces its own translation, thereby maintaining its own protein levels (Westmark et al. 2010). The activated PKMζ reduces internalization of AMPAR subunit GluR2 by increasing the interaction between GluR2 and NSF, thereby increasing the surface amount of AMPAR (Yao et al. 2008). Interestingly, once PKMζ increases the synaptic GluR2, it is kept within the vicinity of GluR2 through a PDZ containing protein Pick1 and this is proposed to be a self-perpetuating mechanism to maintain the increased transmission seen in LTP (Yao et al. 2008). Consistently, a peptide inhibitor of PKMζ blocks the maintenance of LTP as well as retention of memory (Sajikumar et al. 2005; Shema et al. 2007). However, again, a detailed intracellular distribution of PKMζ has not been visualized at the resolution carried out for Homer1a. Therefore, it is still an open question whether the proposed mechanism works specifically at the potentiated synapse and, if it is the case, what is the mechanism for the selective action of the protein. Also, the fact that L-LTP can be still induced in GluR2 knockout animals contradicts the proposed self-perpetuating mechanism (Asrar et al. 2009).

CPG2 was isolated and functionally characterized by Nedivi and colleagues (Nedivi et al. 1993; Hevroni et al. 1998; Cottrell et al. 2004; Loebrich and Nedivi 2009). It encodes a protein with homology to dystrophin and also contains several structural domains, such as spectrin repeats and coiled-coil domain. Interestingly, this protein is localized to the postsynaptic endocytotic zone and is also involved in AMPAR internalization (Cottrell et al. 2004). RNAi-mediated suppression of

CPG2 resulted in a decreased internalization of AMPAR and an increase in the size of dendritic spines. Overexpression of CPG2 reduces the size of dendritic spines.

Another activity-induced gene product with a known synaptic role is SNK/polo-like kinase 2, isolated by Kuhl's group (Kauselmann et al. 1999). Seeburg et al. found that it phosphorylates spine-associated RapGAP (SPAR) and destines it to the degradation pathway, which ultimately leads to a reduction of synaptic transmission (Seeburg et al. 2008; Seeburg and Sheng 2008). Arcadin is an activity-induced cadherin-like trans-membrane molecule (Yamagata et al. 1999). Overexpression of this molecule in hippocampal neuron makes dendritic spines smaller (Yasuda et al. 2007).

From these studies, a feature shared by many, though not all, of the neuronal activity-induced gene products emerges. They are consistent in reducing excitatory synaptic transmission, rather than potentiating the transmission. Considering how most of these genes were isolated, it is actually a logical consequence. Most of the studies used massive neuronal stimulation typically by pharmacological reagents or electrical stimulation above the physiological range of neuronal activity, such as seizure and kindling. This is an understandable experimental choice to obtain sufficient sample materials for biochemical or molecular biological identification of the genes. But as a result, most of the identified activity-induced gene products, instead of potentiating the excitatory synaptic response, downregulate the synaptic response. These genes are most likely involved in homeostatic plasticity, the neurons ability to reduce the input activity when their excitability is too high (Turrigiano et al. 1998; Turrigiano 1999). This indicates that while these studies show that neuronal activity-induced genes shapes our initial understanding of the biology of synapse, there is still a lot to be investigated in order to fully comprehend the roles of neuronal activity-induced gene in regulating synaptic plasticity.

15.6 Concluding Remarks

Experimental efforts from a number of laboratories over the last two decades have elucidated the roles of various neuronal activity-induced genes. Among the estimated 500–1,000 neuronal activity-induced genes (Loebrich and Nedivi 2009), only a handful of them have been characterized in any great detail. Nevertheless, these studies have already illustrated the diverse mechanisms by which neuronal activity-induced gene products regulate synaptic transmission. Interestingly, some of these genes have been found to be negative homeostatic regulators of neuronal function. Critical information still largely lacking is the precise intracellular distri-bution of the neuronal activity-induced gene products in relation to the synapse underwent plasticity. It is not clearly known whether these proteins specifically act on potentiated synapses or they act globally on all synapses. Proteins in neurons are diffusible and can even be shared between neighboring synapses (Gray et al. 2006; Kuriu et al. 2006; Dieterich et al. 2010). For neuronal activity-induced genes to function at a synapse which has undergone Hebbian-fashion potentiation, it has to be captured specifically at the activated synapse but not at others. Further examination of the precise intracellular dynamics of these activity-induced gene products using more advanced imaging techniques will be necessary to fully understand the role of neuronal activity-induced genes.

Acknowledgments We would like to express our sincere gratitude to collaborators of our works introduced here, especially Drs. Mariko Kato-Hayashi, Carlo Sala, Morgan Sheng, Atsushi Miyawaki, Rui-ming Xu, Huilin Li, Mriganka Sur, and members of their laboratories. We thank Drs. Dan Ohtan Wang and Lily Yu for comments on the manuscript. This work was supported by RIKEN, NIH grant R01DA17310, Grant-in-Aid for Scientific Research (A) and Grant-in-Aid for Scientific Research on Innovative Area "Foundation of Synapse and Neurocircuit Pathology" from the Ministry of Education, Science, and Culture of Japan to YH. MB is a recipient of a Beatriu de Pinós fellowship from the Generalitat de Catalunya.

References

Altar, C. A., Laeng, P., Jurata, L. W., Brockman, J. A., Lemire, A., Bullard, J., Bukhman, Y. V., Young, T. A., Charles, V., & Palfreyman, M. G. (2004). Electroconvulsive seizures regulate gene expression of distinct neurotrophic signaling pathways. *Journal of Neuroscience, 24,* 2667–2677.

Ango, F., Prezeau, L., Muller, T., Tu, J. C., Xiao, B., Worley, P. F., Pin, J. P., Bockaert, J., & Fagni, L. (2001). Agonist-independent activation of metabotropic glutamate receptors by the intracellular protein Homer. *Nature, 411,* 962–965.

Asrar, S., Zhou, Z., Ren, W., & Jia, Z. (2009). Ca^{2+} permeable AMPA receptor induced long-term potentiation requires PI3/MAP kinases but not Ca/CaM-dependent kinase II. *PLoS One, 4,* e4339.

Bagni, C., Mannucci, L., Dotti, C. G., & Amaldi, F. (2000). Chemical stimulation of synaptosomes modulates α-Ca^{2+}/calmodulin-dependent protein kinase II mRNA association to polysomes. *Journal of Neuroscience, 20,* RC76.

Baron, M. K., Boeckers, T. M., Vaida, B., Faham, S., Gingery, M., Sawaya, M. R., Salyer, D., Gundelfinger, E. D., & Bowie, J. U. (2006). An architectural framework that may lie at the core of the postsynaptic density. *Science, 311,* 531–535.

Barondes, S. H., & Cohen, H. D. (1968). Memory impairment after subcutaneous injection of acetoxycycloheximide. *Science, 160,* 556–557.

Bassell, G. J., & Warren, S. T. (2008). Fragile X syndrome: Loss of local mRNA regulation alters synaptic development and function. *Neuron, 60,* 201–214.

Béïque, J. C., Na, Y., Kuhl, D., Worley, P. F., & Huganir, R. L. (2011). Arc-dependent synapse-specific homeostatic plasticity. *Proceedings of the National Academy of Sciences of the United States of America, 108,* 816–821.

Bliss, T. V., & Collingridge, G. L. (1993). A synaptic model of memory: Long-term potentiation in the hippocampus. *Nature, 361,* 31–39.

Bliss, T. V., & Lømo, T. (1973). Long-lasting potentiation of synaptic transmission in the dentate area of the anaesthetized rabbit following stimulation of the perforant path. *The Journal of Physiology (London), 232,* 331–356.

Bosch, M., Castro, J., Narayanan, R., Sur, M., & Hayashi, Y. (2009). *Structural and molecular reorganization of a single dendritic spine during long-term potentiation.* In Neuroscience meeting, Society for Neuroscience, Chicago, IL.

Brakeman, P. R., Lanahan, A. A., O'Brien, R., Roche, K., Barnes, C. A., Huganir, R. L., & Worley, P. F. (1997). Homer: A protein that selectively binds metabotropic glutamate receptors. *Nature, 386,* 284–288.

Brennan, P. A., Schellinck, H. M., & Keverne, E. B. (1999). Patterns of expression of the immediate-early gene egr-1 in the accessory olfactory bulb of female mice exposed to pheromonal constituents of male urine. *Neuroscience, 90,* 1463–1470.

Chowdhury, S., Shepherd, J. D., Okuno, H., Lyford, G., Petralia, R. S., Plath, N., Kuhl, D., Huganir, R. L., & Worley, P. F. (2006). Arc/Arg3.1 interacts with the endocytic machinery to regulate AMPA receptor trafficking. *Neuron, 52*, 445–459.

Cochran, B. H., Reffel, A. C., & Stiles, C. D. (1983). Molecular cloning of gene sequences regulated by platelet-derived growth factor. *Cell, 33*, 939–947.

Cole, A. J., Saffen, D. W., Baraban, J. M., & Worley, P. F. (1989). Rapid increase of an immediate early gene messenger RNA in hippocampal neurons by synaptic NMDA receptor activation. *Nature, 340*, 474–476.

Comery, T. A., Harris, J. B., Willems, P. J., Oostra, B. A., Irwin, S. A., Weiler, I. J., & Greenough, W. T. (1997). Abnormal dendritic spines in fragile X knockout mice: Maturation and pruning deficits. *Proceedings of the National Academy of Sciences of the United States of America, 94*, 5401–5404.

Costa-Mattioli, M., Sossin, W. S., Klann, E., & Sonenberg, N. (2009). Translational control of long-lasting synaptic plasticity and memory. *Neuron, 61*, 10–26.

Cottrell, J. R., Borok, E., Horvath, T. L., & Nedivi, E. (2004). CPG2: A brain- and synapse-specific protein that regulates the endocytosis of glutamate receptors. *Neuron, 44*, 677–690.

Davis, H. P., & Squire, L. R. (1984). Protein synthesis and memory: A review. *Psychological Bulletin, 96*, 518–559.

De Rubeis, S., & Bagni, C. (2010). Fragile X mental retardation protein control of neuronal mRNA metabolism: Insights into mRNA stability. *Molecular and Cellular Neuroscience, 43*, 43–50.

Dieterich, D. C., Hodas, J. J., Gouzer, G., Shadrin, I. Y., Ngo, J. T., Triller, A., Tirrell, D. A., & Schuman, E. M. (2010). In situ visualization and dynamics of newly synthesized proteins in rat hippocampal neurons. *Nature Neuroscience, 13*, 897–905.

Dragunow, M., & Robertson, H. A. (1987). Kindling stimulation induces c-*fos* protein(s) in granule cells of the rat dentate gyrus. *Nature, 329*, 441–442.

Eberwine, J., Belt, B., Kacharmina, J. E., & Miyashiro, K. (2002). Analysis of subcellularly localized mRNAs using in situ hybridization, mRNA amplification, and expression profiling. *Neurochemical Research, 27*, 1065–1077.

Edbauer, D., Neilson, J. R., Foster, K. A., Wang, C. F., Seeburg, D. P., Batterton, M. N., Tada, T., Dolan, B. M., Sharp, P. A., & Sheng, M. (2010). Regulation of synaptic structure and function by FMRP-associated microRNAs miR-125b and miR-132. *Neuron, 65*, 373–384.

Elliott, R. C., Miles, M. F., & Lowenstein, D. H. (2003). Overlapping microarray profiles of dentate gyrus gene expression during development- and epilepsy-associated neurogenesis and axon outgrowth. *Journal of Neuroscience, 23*, 2218–2227.

Fazeli, M. S., Corbet, J., Dunn, M. J., Dolphin, A. C., & Bliss, T. V. (1993). Changes in protein synthesis accompanying long-term potentiation in the dentate gyrus in vivo. *Journal of Neuroscience, 13*, 1346–1353.

Feng, Y., Zhang, F., Lokey, L. K., Chastain, J. L., Lakkis, L., Eberhart, D., & Warren, S. T. (1995). Translational suppression by trinucleotide repeat expansion at FMR1. *Science, 268*, 731–734.

Flexner, J. B., Flexner, L. B., & Stellar, E. (1963). Memory in mice as affected by intracerebral puromycin. *Science, 141*, 57–59.

Frankland, P. W., Bontempi, B., Talton, L. E., Kaczmarek, L., & Silva, A. J. (2004). The involvement of the anterior cingulate cortex in remote contextual fear memory. *Science, 304*, 881–883.

Franza, B. R., Jr., Rauscher, F. J., 3rd, Josephs, S. F., & Curran, T. (1988). The Fos complex and Fos-related antigens recognize sequence elements that contain AP-1 binding sites. *Science, 239*, 1150–1153.

French, P. J., O'Connor, V., Voss, K., Stean, T., Hunt, S. P., & Bliss, T. V. (2001). Seizure-induced gene expression in area CA1 of the mouse hippocampus. *European Journal of Neuroscience, 14*, 2037–2041.

Frey, U., & Morris, R. G. (1997). Synaptic tagging and long-term potentiation. *Nature, 385*, 533–536.

Frey, U., Krug, M., Brodemann, R., Reymann, K., & Matthies, H. (1989). Long-term potentiation induced in dendrites separated from rat's CA1 pyramidal somata does not establish a late phase. *Neuroscience Letters, 97*, 135–139.

Frey, U., Frey, S., Schollmeier, F., & Krug, M. (1996). Influence of actinomycin D, a RNA synthesis inhibitor, on long-term potentiation in rat hippocampal neurons in vivo and in vitro. *The Journal of Physiology, 490*, 703–711.

Gray, N. W., Weimer, R. M., Bureau, I., & Svoboda, K. (2006). Rapid redistribution of synaptic PSD-95 in the neocortex in vivo. *PLoS Biology, 4*, e370.

Greenberg, M. E., & Ziff, E. B. (1984). Stimulation of 3 T3 cells induces transcription of the c-*fos* proto-oncogene. *Nature, 311*, 433–438.

Greenberg, M. E., Ziff, E. B., & Greene, L. A. (1986). Stimulation of neuronal acetylcholine receptors induces rapid gene transcription. *Science, 234*, 80–83.

Guzowski, J. F., Setlow, B., Wagner, E. K., & McGaugh, J. L. (2001). Experience-dependent gene expression in the rat hippocampus after spatial learning: A comparison of the immediate-early genes Arc, c-fos, and zif268. *Journal of Neuroscience, 21*, 5089–5098.

Hall, J., Thomas, K. L., & Everitt, B. J. (2001). Fear memory retrieval induces CREB phosphorylation and Fos expression within the amygdala. *European Journal of Neuroscience, 13*, 1453–1458.

Hayashi, T., & Huganir, R. L. (2004). Tyrosine phosphorylation and regulation of the AMPA receptor by Src family tyrosine kinases. *Journal of Neuroscience, 24*, 6152–6160.

Hayashi, T., Umemori, H., Mishina, M., & Yamamoto, T. (1999). The AMPA receptor interacts with and signals through the protein tyrosine kinase Lyn. *Nature, 397*, 72–76.

Hayashi, M. K., Ames, H. M., & Hayashi, Y. (2006). Tetrameric hub structure of postsynaptic scaffolding protein homer. *Journal of Neuroscience, 26*, 8492–8501.

Hayashi, M. K., Tang, C., Verpelli, C., Narayanan, R., Stearns, M. H., Xu, R. M., Li, H., Sala, C., & Hayashi, Y. (2009). The postsynaptic density proteins Homer and Shank form a polymeric network structure. *Cell, 137*, 159–171.

Hernandez, A. I., Blace, N., Crary, J. F., Serrano, P. A., Leitges, M., Libien, J. M., Weinstein, G., Tcherapanov, A., & Sacktor, T. C. (2003). Protein kinase M zeta synthesis from a brain mRNA encoding an independent protein kinase C zeta catalytic domain. Implications for the molecular mechanism of memory. *Journal of Biological Chemistry, 278*, 40305–40316.

Herron, C. E., Lester, R. A., Coan, E. J., & Collingridge, G. L. (1986). Frequency-dependent involvement of NMDA receptors in the hippocampus: A novel synaptic mechanism. *Nature, 322*, 265–268.

Hevroni, D., Rattner, A., Bundman, M., Lederfein, D., Gabarah, A., Mangelus, M., Silverman, M. A., Kedar, H., Naor, C., Kornuc, M., Hanoch, T., Seger, R., Theill, L. E., Nedivi, E., Richter-Levin, G., & Citri, Y. (1998). Hippocampal plasticity involves extensive gene induction and multiple cellular mechanisms. *Journal of Molecular Neuroscience, 10*, 75–98.

Hong, S. J., Li, H., Becker, K. G., Dawson, V. L., & Dawson, T. M. (2004). Identification and analysis of plasticity-induced late-response genes. *Proceedings of the National Academy of Sciences of the United States of America, 101*, 2145–2150.

Honkura, N., Matsuzaki, M., Noguchi, J., Ellis-Davies, G. C., & Kasai, H. (2008). The subspine organization of actin fibers regulates the structure and plasticity of dendritic spines. *Neuron, 57*, 719–729.

Horton, A. C., & Ehlers, M. D. (2003). Dual modes of endoplasmic reticulum-to-Golgi transport in dendrites revealed by live-cell imaging. *Journal of Neuroscience, 23*, 6188–6199.

Hu, J. H., Park, J. M., Park, S., Xiao, B., Dehoff, M. H., Kim, S., Hayashi, T., Schwarz, M. K., Huganir, R. L., Seeburg, P. H., Linden, D. J., & Worley, P. F. (2010). Homeostatic scaling requires group I mGluR activation mediated by Homer1a. *Neuron, 68*, 1128–1142.

Huber, K. M., Gallagher, S. M., Warren, S. T., & Bear, M. F. (2002). Altered synaptic plasticity in a mouse model of fragile X mental retardation. *Proceedings of the National Academy of Sciences of the United States of America, 99*, 7746–7750.

Hughes, P., Beilharz, E., Gluckman, P., & Dragunow, M. (1993). Brain-derived neurotrophic factor is induced as an immediate early gene following *N*-methyl-D-aspartate receptor activation. *Neuroscience, 57*, 319–328.

Im, Y. J., Lee, J. H., Park, S. H., Park, S. J., Rho, S. H., Kang, G. B., Kim, E., & Eom, S. H. (2003). Crystal structure of the Shank PDZ-ligand complex reveals a class I PDZ interaction and a novel PDZ-PDZ dimerization. *Journal of Biological Chemistry, 278*, 48099–48104.

Ingi, T., Krumins, A. M., Chidiac, P., Brothers, G. M., Chung, S., Snow, B. E., Barnes, C. A., Lanahan, A. A., Siderovski, D. P., Ross, E. M., Gilman, A. G., & Worley, P. F. (1998). Dynamic regulation of RGS2 suggests a novel mechanism in G-protein signaling and neuronal plasticity. *Journal of Neuroscience, 18*, 7178–7188.

Irwin, S. A., Patel, B., Idupulapati, M., Harris, J. B., Crisostomo, R. A., Larsen, B. P., Kooy, F., Willems, P. J., Cras, P., Kozlowski, P. B., Swain, R. A., Weiler, I. J., & Greenough, W. T. (2001). Abnormal dendritic spine characteristics in the temporal and visual cortices of patients with fragile-X syndrome: A quantitative examination. *American Journal of Medical Genetics, 98*, 161–167.

Kato, A., Ozawa, F., Saitoh, Y., Hirai, K., & Inokuchi, K. (1997). vesl, a gene encoding VASP/Ena family related protein, is upregulated during seizure, long-term potentiation and synaptogenesis. *FEBS Letters, 412*, 183–189.

Kato, A., Ozawa, F., Saitoh, Y., Fukazawa, Y., Sugiyama, H., & Inokuchi, K. (1998). Novel members of the Vesl/Homer family of PDZ proteins that bind metabotropic glutamate receptors. *Journal of Biological Chemistry, 273*, 23969–23975.

Kauselmann, G., Weiler, M., Wulff, P., Jessberger, S., Konietzko, U., Scafidi, J., Staubli, U., Bereiter-Hahn, J., Strebhardt, K., & Kuhl, D. (1999). The polo-like protein kinases Fnk and Snk associate with a Ca^{2+}- and integrin-binding protein and are regulated dynamically with synaptic plasticity. *EMBO Journal, 18*, 5528–5539.

Kindler, S., & Monshausen, M. (2002). Candidate RNA-binding proteins regulating extrasomatic mRNA targeting and translation in mammalian neurons. *Molecular Neurobiology, 25*, 149–165.

Kitamura, C., Takahashi, M., Kondoh, Y., Tashiro, H., & Tashiro, T. (2007). Identification of synaptic activity-dependent genes by exposure of cultured cortical neurons to tetrodotoxin followed by its withdrawal. *Journal of Neuroscience Research, 85*, 2385–2399.

Kitanishi, T., Ikegaya, Y., Matsuki, N., & Yamada, M. K. (2009). Experience-dependent, rapid structural changes in hippocampal pyramidal cell spines. *Cerebral Cortex, 19*, 2572–2578.

Krug, M., Lossner, B., & Ott, T. (1984). Anisomycin blocks the late phase of long-term potentiation in the dentate gyrus of freely moving rats. *Brain Research Bulletin, 13*, 39–42.

Kuriu, T., Inoue, A., Bito, H., Sobue, K., & Okabe, S. (2006). Differential control of postsynaptic density scaffolds via actin-dependent and -independent mechanisms. *Journal of Neuroscience, 26*, 7693–7706.

Lanahan, A., & Worley, P. (1998). Immediate-early genes and synaptic function. *Neurobiology of Learning and Memory, 70*, 37–43.

Li, J., Pelletier, M. R., Perez Velazquez, J. L., & Carlen, P. L. (2002). Reduced cortical synaptic plasticity and GluR1 expression associated with fragile X mental retardation protein deficiency. *Molecular and Cellular Neuroscience, 19*, 138–151.

Link, W., Konietzko, U., Kauselmann, G., Krug, M., Schwanke, B., Frey, U., & Kuhl, D. (1995). Somatodendritic expression of an immediate early gene is regulated by synaptic activity. *Proceedings of the National Academy of Sciences of the United States of America, 92*, 5734–5738.

Loebrich, S., & Nedivi, E. (2009). The function of activity-regulated genes in the nervous system. *Physiological Reviews, 89*, 1079–1103.

Lyford, G. L., Yamagata, K., Kaufmann, W. E., Barnes, C. A., Sanders, L. K., Copeland, N. G., Gilbert, D. J., Jenkins, N. A., Lanahan, A. A., & Worley, P. F. (1995). Arc, a growth factor and activity-regulated gene, encodes a novel cytoskeleton-associated protein that is enriched in neuronal dendrites. *Neuron, 14*, 433–445.

Martin, K. C., Casadio, A., Zhu, H., Yaping, E., Rose, J. C., Chen, M., Bailey, C. H., & Kandel, E. R. (1997). Synapse-specific, long-term facilitation of aplysia sensory to motor synapses: A function for local protein synthesis in memory storage. *Cell, 91*, 927–938.

Matsuo, R., Kato, A., Sakaki, Y., & Inokuchi, K. (1998). Cataloging altered gene expression during rat hippocampal long-term potentiation by means of differential display. *Neuroscience Letters, 244*, 173–176.

Matsuzaki, M., Honkura, N., Ellis-Davies, G. C., & Kasai, H. (2004). Structural basis of long-term potentiation in single dendritic spines. *Nature, 429*, 761–766.

Mayford, M., Baranes, D., Podsypanina, K., & Kandel, E. R. (1996). The 3′-untranslated region of CaMKIIα is a cis-acting signal for the localization and translation of mRNA in dendrites. *Proceedings of the National Academy of Sciences of the United States of America, 93*, 13250–13255.

Miller, S., Yasuda, M., Coats, J. K., Jones, Y., Martone, M. E., & Mayford, M. (2002). Disruption of dendritic translation of CaMKIIα impairs stabilization of synaptic plasticity and memory consolidation. *Neuron, 36*, 507–519.

Moccia, R., Chen, D., Lyles, V., Kapuya, E., E, Y., Kalachikov, S., Spahn, C. M., Frank, J., Kandel, E. R., Barad, M., & Martin, K. C. (2003). An unbiased cDNA library prepared from isolated *Aplysia* sensory neuron processes is enriched for cytoskeletal and translational mRNAs. *Journal of Neuroscience, 23*, 9409–9417.

Nedivi, E., Hevroni, D., Naot, D., Israeli, D., & Citri, Y. (1993). Numerous candidate plasticity-related genes revealed by differential cDNA cloning. *Nature, 363*, 718–722.

Nedivi, E., Wu, G. Y., & Cline, H. T. (1998). Promotion of dendritic growth by CPG15, an activity-induced signaling molecule. *Science, 281*, 1863–1866.

Newton, S. S., Collier, E. F., Hunsberger, J., Adams, D., Terwilliger, R., Selvanayagam, E., & Duman, R. S. (2003). Gene profile of electroconvulsive seizures: Induction of neurotrophic and angiogenic factors. *Journal of Neuroscience, 23*, 10841–10851.

Nimchinsky, E. A., Oberlander, A. M., & Svoboda, K. (2001). Abnormal development of dendritic spines in FMR1 knock-out mice. *Journal of Neuroscience, 21*, 5139–5146.

Okabe, S. (2007). Molecular anatomy of the postsynaptic density. *Molecular and Cellular Neuroscience, 34*, 503–518.

Okada, D., Ozawa, F., & Inokuchi, K. (2009). Input-specific spine entry of soma-derived Vesl-1 S protein conforms to synaptic tagging. *Science, 324*, 904–909.

Okamoto, K., Nagai, T., Miyawaki, A., & Hayashi, Y. (2004). Rapid and persistent modulation of actin dynamics regulates postsynaptic reorganization underlying bidirectional plasticity. *Nature Neuroscience, 7*, 1104–1112.

Okamoto, K., Bosch, M., & Hayashi, Y. (2009). The roles of CaMKII and F-actin in the structural plasticity of dendritic spines: A potential molecular identity of a synaptic tag? *Physiology (Bethesda, MD), 24*, 357–366.

Okuno, H. (2011). Regulation and function of immediate-early genes in the brain: Beyond neuronal activity markers. *Neuroscience Research, 69*, 175–186.

Osten, P., Valsamis, L., Harris, A., & Sacktor, T. C. (1996). Protein synthesis-dependent formation of protein kinase Mzeta in long-term potentiation. *Journal of Neuroscience, 16*, 2444–2451.

Ostroff, L. E., Fiala, J. C., Allwardt, B., & Harris, K. M. (2002). Polyribosomes redistribute from dendritic shafts into spines with enlarged synapses during LTP in developing rat hippocampal slices. *Neuron, 35*, 535–545.

Otani, S., Marshall, C. J., Tate, W. P., Goddard, G. V., & Abraham, W. C. (1989). Maintenance of long-term potentiation in rat dentate gyrus requires protein synthesis but not messenger RNA synthesis immediately post-tetanization. *Neuroscience, 28*, 519–526.

Ouyang, Y., Rosenstein, A., Kreiman, G., Schuman, E. M., & Kennedy, M. B. (1999). Tetanic stimulation leads to increased accumulation of Ca^{2+}/calmodulin-dependent protein kinase II via dendritic protein synthesis in hippocampal neurons. *Journal of Neuroscience, 19*, 7823–7833.

Park, C. S., Gong, R., Stuart, J., & Tang, S. J. (2006). Molecular network and chromosomal clustering of genes involved in synaptic plasticity in the hippocampus. *Journal of Biological Chemistry, 281*, 30195–30211.

Plath, N., Ohana, O., Dammermann, B., Errington, M. L., Schmitz, D., Gross, C., Mao, X., Engelsberg, A., Mahlke, C., Welzl, H., Kobalz, U., Stawrakakis, A., Fernandez, E., Waltereit, R., Bick-Sander, A., Therstappen, E., Cooke, S. F., Blanquet, V., Wurst, W., Salmen, B., Bosl, M. R., Lipp, H. P., Grant, S. G., Bliss, T. V., Wolfer, D. P., & Kuhl, D. (2006). Arc/Arg3.1 is essential for the consolidation of synaptic plasticity and memories. *Neuron, 52*, 437–444.

Poon, M. M., Choi, S. H., Jamieson, C. A., Geschwind, D. H., & Martin, K. C. (2006). Identification of process-localized mRNAs from cultured rodent hippocampal neurons. *Journal of Neuroscience, 26*, 13390–13399.

Qian, Z., Gilbert, M. E., Colicos, M. A., Kandel, E. R., & Kuhl, D. (1993). Tissue-plasminogen activator is induced as an immediate-early gene during seizure, kindling and long-term potentiation. *Nature, 361*, 453–457.

Raivich, G., & Behrens, A. (2006). Role of the AP-1 transcription factor c-Jun in developing, adult and injured brain. *Progress in Neurobiology, 78*, 347–363.

Ramachandran, B., & Frey, J. U. (2009). Interfering with the actin network and its effect on long-term potentiation and synaptic tagging in hippocampal CA1 neurons in slices in vitro. *Journal of Neuroscience, 29*, 12167–12173.

Rauscher, F. J., 3rd, Cohen, D. R., Curran, T., Bos, T. J., Vogt, P. K., Bohmann, D., Tjian, R., & Franza, B. R., Jr. (1988). Fos-associated protein p39 is the product of the *jun* proto-oncogene. *Science, 240*, 1010–1016.

Redondo, R. L., & Morris, R. G. (2011). Making memories last: The synaptic tagging and capture hypothesis. *Nature Reviews Neuroscience, 12*, 17–30.

Reijmers, L. G., Perkins, B. L., Matsuo, N., & Mayford, M. (2007). Localization of a stable neural correlate of associative memory. *Science, 317*, 1230–1233.

Rial Verde, E. M., Lee-Osbourne, J., Worley, P. F., Malinow, R., & Cline, H. T. (2006). Increased expression of the immediate-early gene *Arc/Arg3.1* reduces AMPA receptor-mediated synaptic transmission. *Neuron, 52*, 461–474.

Romorini, S., Piccoli, G., Jiang, M., Grossano, P., Tonna, N., Passafaro, M., Zhang, M., & Sala, C. (2004). A functional role of postsynaptic density-95-guanylate kinase-associated protein complex in regulating Shank assembly and stability to synapses. *Journal of Neuroscience, 24*, 9391–9404.

Rong, R., Ahn, J. Y., Huang, H., Nagata, E., Kalman, D., Kapp, J. A., Tu, J., Worley, P. F., Snyder, S. H., & Ye, K. (2003). PI3 kinase enhancer-Homer complex couples mGluRI to PI3 kinase, preventing neuronal apoptosis. *Nature Neuroscience, 6*, 1153–1161.

Ryder, K., Lau, L. F., & Nathans, D. (1988). A gene activated by growth factors is related to the oncogene v-*jun*. *Proceedings of the National Academy of Sciences of the United States of America, 85*, 1487–1491.

Rylski, M., & Kaczmarek, L. (2004). AP-1 targets in the brain. *Frontiers in Bioscience, 9*, 8–23.

Sacco, T., & Sacchetti, B. (2010). Role of secondary sensory cortices in emotional memory storage and retrieval in rats. *Science, 329*, 649–656.

Saffen, D. W., Cole, A. J., Worley, P. F., Christy, B. A., Ryder, K., & Baraban, J. M. (1988). Convulsant-induced increase in transcription factor messenger RNAs in rat brain. *Proceedings of the National Academy of Sciences of the United States of America, 85*, 7795–7799.

Sajikumar, S., Navakkode, S., Sacktor, T. C., & Frey, J. U. (2005). Synaptic tagging and cross-tagging: The role of protein kinase Mzeta in maintaining long-term potentiation but not long-term depression. *Journal of Neuroscience, 25*, 5750–5756.

Sala, C., Piech, V., Wilson, N. R., Passafaro, M., Liu, G., & Sheng, M. (2001). Regulation of dendritic spine morphology and synaptic function by Shank and Homer. *Neuron, 31*, 115–130.

Sala, C., Futai, K., Yamamoto, K., Worley, P. F., Hayashi, Y., & Sheng, M. (2003). Inhibition of dendritic spine morphogenesis and synaptic transmission by activity-inducible protein Homer1a. *Journal of Neuroscience, 23*, 6327–6337.

Schütt, J., Falley, K., Richter, D., Kreienkamp, H. J., & Kindler, S. (2009). Fragile X mental retardation protein regulates the levels of scaffold proteins and glutamate receptors in postsynaptic densities. *Journal of Biological Chemistry, 284*, 25479–25487.

Seeburg, D. P., & Sheng, M. (2008). Activity-induced Polo-like kinase 2 is required for homeostatic plasticity of hippocampal neurons during epileptiform activity. *Journal of Neuroscience, 28*, 6583–6591.

Seeburg, D. P., Feliu-Mojer, M., Gaiottino, J., Pak, D. T., & Sheng, M. (2008). Critical role of CDK5 and Polo-like kinase 2 in homeostatic synaptic plasticity during elevated activity. *Neuron, 58*, 571–583.

Shema, R., Sacktor, T. C., & Dudai, Y. (2007). Rapid erasure of long-term memory associations in the cortex by an inhibitor of PKM zeta. *Science, 317*, 951–953.

Sheng, M., & Hoogenraad, C. C. (2007). The postsynaptic architecture of excitatory synapses: A more quantitative view. *Annual Review of Biochemistry, 76*, 823–847.

Shepherd, J. D., Rumbaugh, G., Wu, J., Chowdhury, S., Plath, N., Kuhl, D., Huganir, R. L., & Worley, P. F. (2006). Arc/Arg3.1 mediates homeostatic synaptic scaling of AMPA receptors. *Neuron, 52*, 475–484.

Shiraishi-Yamaguchi, Y., & Furuichi, T. (2007). The Homer family proteins. *Genome Biology, 8*, 206.

Steward, O., & Levy, W. B. (1982). Preferential localization of polyribosomes under the base of dendritic spines in granule cells of the dentate gyrus. *Journal of Neuroscience, 2*, 284–291.

Steward, O., & Worley, P. (2001). Localization of mRNAs at synaptic sites on dendrites. *Results and Problems in Cell Differentiation, 34*, 1–26.

Steward, O., Wallace, C. S., Lyford, G. L., & Worley, P. F. (1998). Synaptic activation causes the mRNA for the IEG *Arc* to localize selectively near activated postsynaptic sites on dendrites. *Neuron, 21*, 741–751.

Tu, J. C., Xiao, B., Yuan, J. P., Lanahan, A. A., Leoffert, K., Li, M., Linden, D. J., & Worley, P. F. (1998). Homer binds a novel proline-rich motif and links group 1 metabotropic glutamate receptors with IP3 receptors. *Neuron, 21*, 717–726.

Tu, J. C., Xiao, B., Naisbitt, S., Yuan, J. P., Petralia, R. S., Brakeman, P., Doan, A., Aakalu, V. K., Lanahan, A. A., Sheng, M., & Worley, P. F. (1999). Coupling of mGluR/Homer and PSD-95 complexes by the Shank family of postsynaptic density proteins. *Neuron, 23*, 583–592.

Turrigiano, G. G. (1999). Homeostatic plasticity in neuronal networks: The more things change, the more they stay the same. *Trends in Neurosciences, 22*, 221–227.

Turrigiano, G. G., Leslie, K. R., Desai, N. S., Rutherford, L. C., & Nelson, S. B. (1998). Activity-dependent scaling of quantal amplitude in neocortical neurons. *Nature, 391*, 892–896.

Wang, D. O., Kim, S. M., Zhao, Y., Hwang, H., Miura, S. K., Sossin, W. S., & Martin, K. C. (2009). Synapse- and stimulus-specific local translation during long-term neuronal plasticity. *Science, 324*, 1536–1540.

Wang, D. O., Martin, K. C., & Zukin, R. S. (2010). Spatially restricting gene expression by local translation at synapses. *Trends in Neurosciences, 33*, 173–182.

Westmark, P. R., Westmark, C. J., Wang, S., Levenson, J., O'Riordan, K. J., Burger, C., & Malter, J. S. (2010). Pin1 and PKMzeta sequentially control dendritic protein synthesis. *Science Signaling, 3*, ra18.

Xiang, G., Pan, L., Xing, W., Zhang, L., Huang, L., Yu, J., Zhang, R., Wu, J., Cheng, J., & Zhou, Y. (2007). Identification of activity-dependent gene expression profiles reveals specific subsets of genes induced by different routes of Ca^{2+} entry in cultured rat cortical neurons. *Journal of Cellular Physiology, 212*, 126–136.

Xiao, B., Tu, J. C., & Worley, P. F. (2000). Homer: A link between neural activity and glutamate receptor function. *Current Opinion in Neurobiology, 10*, 370–374.

Yamagata, K., Andreasson, K. I., Kaufmann, W. E., Barnes, C. A., & Worley, P. F. (1993). Expression of a mitogen-inducible cyclooxygenase in brain neurons: Regulation by synaptic activity and glucocorticoids. *Neuron, 11*, 371–386.

Yamagata, K., Sanders, L. K., Kaufmann, W. E., Yee, W., Barnes, C. A., Nathans, D., & Worley, P. F. (1994). rheb, a growth factor- and synaptic activity-regulated gene, encodes a novel Ras-related protein. *Journal of Biological Chemistry, 269*, 16333–16339.

Yamagata, K., Andreasson, K. I., Sugiura, H., Maru, E., Dominique, M., Irie, Y., Miki, N., Hayashi, Y., Yoshioka, M., Kaneko, K., Kato, H., & Worley, P. F. (1999). Arcadlin is a neural activity-regulated cadherin involved in long term potentiation. *Journal of Biological Chemistry, 274*, 19473–11979.

Yao, Y., Kelly, M. T., Sajikumar, S., Serrano, P., Tian, D., Bergold, P. J., Frey, J. U., & Sacktor, T. C. (2008). PKM zeta maintains late long-term potentiation by N-ethylmaleimide-sensitive factor/

GluR2-dependent trafficking of postsynaptic AMPA receptors. *Journal of Neuroscience, 28,* 7820–7827.

Yasuda, S., Tanaka, H., Sugiura, H., Okamura, K., Sakaguchi, T., Tran, U., Takemiya, T., Mizoguchi, A., Yagita, Y., Sakurai, T., De Robertis, E. M., & Yamagata, K. (2007). Activity-induced protocadherin arcadlin regulates dendritic spine number by triggering N-cadherin endocytosis via TAO2β and p38 MAP kinases. *Neuron, 56,* 456–471.

Yuan, J. P., Kiselyov, K., Shin, D. M., Chen, J., Shcheynikov, N., Kang, S. H., Dehoff, M. H., Schwarz, M. K., Seeburg, P. H., Muallem, S., & Worley, P. F. (2003). Homer binds TRPC family channels and is required for gating of TRPC1 by IP_3 receptors. *Cell, 114,* 777–789.

Chapter 16
Long-Distance Signaling from Synapse to Nucleus via Protein Messengers

Anna Karpova, Julia Bär, and Michael R. Kreutz

Abstract The communication between synapses and the cell nucleus has attracted considerable interest for many years. This interest is largely fueled by the idea that synapse-to-nucleus signaling might specifically induce the expression of genes that make long-term memory "stick." However, despite many years of research, it is still essentially unclear how synaptic signals are conveyed to the nucleus, and it remains to a large degree enigmatic how activity-induced gene expression feeds back to synaptic function. In this chapter, we will focus on the activity-dependent synapto-nuclear trafficking of protein messengers and discuss the underlying mechanisms of their retrograde transport and their supposed functional role in neuronal plasticity.

Keywords Activity-dependent gene expression • Importins • Jacob/Nelf • Micro-tubule • NMDA-receptors

16.1   Introduction

Synapse-to-nucleus communication is a classical topic in neuroscience since multiple signaling pathways converge in the nucleus that drive gene expression associated with long-term structural changes of synapto-dendritic input and the formation of long-term memories (Greer and Greenberg 2008; Cohen and Greenberg 2008; Flavell and Greenberg 2008; Alberini 2009). Given that neurons regulate more genes than any other cell type (Deisseroth et al. 2003; Alberini 2009), it is unlikely that nuclear Ca^{2+} rises alone could generate the varied and complex responses to the diverse array of extracellular stimuli involved in neuronal signaling. It has been proposed that the shuttling of synaptic proteins into the nucleus may

A. Karpova • J. Bär • M.R. Kreutz (✉)
PG Neuroplasticity, Leibniz Institute for Neurobiology, Brenneckestr.6, 39118 Magdeburg, Germany
e-mail: kreutz@ifn-magdeburg.de

M.R. Kreutz and C. Sala (eds.), *Synaptic Plasticity*,
Advances in Experimental Medicine and Biology 970,
DOI 10.1007/978-3-7091-0932-8_16, © Springer-Verlag/Wien 2012

"

provide further specificity required for the integration of multiple signaling pathways to the nucleus (Jordan and Kreutz 2009).

A number of arguments speak in favor of the existence of synapto-nuclear protein messenger pathways to the nucleus. Synapses contain components of the nuclear import machinery like importin-α and importin-β (Thompson et al. 2004), and it has been shown that they translocate to the nucleus in an activity-dependent manner (Thompson et al. 2004; Dieterich et al. 2008). In addition, synapses also contain several nuclear localization signal (NLS) containing cargo proteins. Different proteomic studies aimed to elucidate the protein composition of the postsynaptic density (PSD) revealed that at least 166 of more than 1,100 proteins contain *bona fide* NLSs (Jordan et al. 2004; Jordan and Kreutz 2009). Many of these proteins exhibit a dual synaptic and nuclear localization with the latter being frequently overlooked. Together with evidence from proteins like AIDA-1d (Jordan et al. 2007), Jacob (Dieterich et al. 2008), Abi-1 (Proepper et al. 2007), CREB2/ATF4 (Lai et al. 2008), and Lapser1 (Schmeisser et al. 2009), this strongly supports that nucleocytoplasmic shuttling of proteins is an important component of synapse-to-nucleus signaling. Finally, the signaling-dependent nuclear translocation of proteins from cell-cell junctions is an established principle in many cell types, and since synaptic junctions resemble other cell junctions in many aspects, it is as such not a surprise that in recent years many proteins were identified that are able to transit from dendrites to the nucleus in response to various kinds of neuronal stimuli (Jordan and Kreutz 2009; Table 16.1).

At present, however, neither the mechanisms underlying activity-dependent nuclear signaling nor the forms of synaptic plasticity, which are controlled by it, are well understood. There are many remaining concerns, including the lack of demonstrative evidence that nuclear AIDA-1, Jacob, Abi-1, or other nucleocytoplasmic shuttling proteins have a synaptic origin. Moreover, synapses at distal dendrites can be several hundred microns away from the nucleus; it is therefore another principal question how proteins can translocate over long distances. Even though typical textbook illustrations show various signaling molecules traveling from synapses to the nucleus, long-distance travel along axons and/or dendrites can result in the rapid decay of signals. This may result from degradation of signaling molecules or reversal of posttranslational modifications required for nuclear import. Protein modifications like phosphorylation during their way along dendrites or other signaling decay would be more pronounced when transport is based on passive diffusion as it was suggested recently for ERK1/2 (Wiegert et al. 2007). Given the exponentially decaying strength of diffusible molecules, it was therefore concluded that successful nuclear signaling is dependent on the proximity of activated synapses to the nucleus. Accordingly, modeling studies of mechanisms associated with nuclear import have revealed that simple diffusion is inefficient when compared to active transport along microtubules (Howe and Mobley 2005). However, these models describe protein mobility in large and relatively unconfined spaces (unlike dendrites or axons) and did not incorporate directional gradient-dependent motion. Nonetheless, these motions are likely to be low for nuclear messengers and, therefore, might be negligible. Further questions regarding diffusion therefore also include the directionality of signals, such as how are signals directed toward the nucleus as opposed to elsewhere or even into adjacent spines? Active

Table 16.1 Potential synapto-nuclear messengers

Synaptic proteins which can be found in the nucleus	
Name	Putative nuclear function
Cytoskeletal and scaffolding proteins	
GRIP1	Trafficking (Ataman et al. 2006), regulate transcription (Yu et al. 2001; Nakata et al. 2004)
SAP97	?[a] (Kohu et al. 2002)
CASK	Regulate transcription (Tbr-1) (Hsueh et al. 2000)
α-actinin4	Antagonize HDAC7 activity (Chakraborty et al. 2006)
Band 4.1	Splicing (Lallena et al. 1998; Shen et al. 2000)
Ezrin	? (Kaul et al. 1999)
ZO-1	Regulate transcription (ZONAB) (Balda and Matter 2000; Kavanagh et al. 2006)
Catenins	
β-catenin	Regulate LEF-1 transcription (Behrens et al. 1996; Molenaar et al. 1996; Huber et al. 1996)
γ-catenin (plakoglobin)	Regulate LEF-1 transcription (Simcha et al. 1998)
δ-catenin (NPRAP)	Transcriptional regulation? (Rodova et al. 2004)
p120	Regulate KAISO-dependent repression (Kelly et al. 2004)
ARVCF	Regulate ZONAB transcription (?)
JAB-1 (sub 3)	c-Jun, JunD coactivator (Claret et al. 1996)
p0071	Transcriptional regulation?
Proteins that shuttle into the nucleus in response to neuronal activity	
CREB2	Transcriptional regulation (Lai et al. 2008)
AIDA-1	Nucleolar assembly (Jordan et al. 2007)
Jacob	Transcriptional regulation (CREB, Dieterich et al. 2008)
Abi-1	Transcriptional regulation (c-Myc, Proepper et al. 2007)
NF-κB	Transcription factor (Guerrini et al. 1995; Kaltschmidt et al. 1995; Meffert et al. 2003)
CAMAP	Transcriptional coactivator CREB1 (Lee et al. 2007)
NFATc4	Transcription factor (Graef et al. 1999)
HDAC4, HDAC5	Histone deacetylases (Chawla et al. 2003)
LAPSER1	Modulation of gene transcription (Schmeisser et al. 2009)
Transmembrane proteolytic fragments	
APP intracellular domain	Transcriptional regulation (Cao and Südhof 2001)
N-cadherin	Transcriptional regulation (Marambaud et al. 2003)
ErbB4	Transcriptional regulation (STAT5A, N-CoR, Sardi et al. 2006)
Frizzled2-C	? (Mathew et al. 2005; Ataman et al. 2006)
L-type calcium channel	Transcription factor (Gomez-Ospina et al. 2006)
Protocadherin-γ	Transcriptional regulation (Haas et al. 2005; Hambsch et al. 2005)
Neuregulin-1	Transcriptional regulation (Eos, Bao et al. 2004)
Notch	Transcriptional regulation, CBF1 binding (Lu and Lux 1996; Alberi et al. 2011)

[a]Denotes lacking or incomplete evidence

transport along microtubules is also often discussed in the context of nucleocytoplasmic shuttling of nonendosomal proteins, and the retrograde transport via importins attached to dynein motors along microtubules appears to be a plausible mechanism (Thompson et al. 2004; Hanz et al. 2003; Perlson et al. 2005, 2006. However, in a large-scale study to address this issue, Roth et al. (2007) explored the nuclear import of several proteins in the presence or absence of intact microtubules in nonneuronal cells. Surprisingly, seven out of ten proteins showed no significant reduction in nuclear accumulation in the absence of microtubules. Although the relevance of dynein-dependent transport along microtubules for nucleocytoplasmic shuttling of synaptic proteins has not been systemically addressed yet, it was reported that nuclear translocation of AIDA-1d does not require intact microtubules (Jordan et al. 2007). Therefore, questions remain how nuclear messengers arrive at the nuclear pore complex (NPC). The microtubule-depolymerizing drug nocodazole that was used in many studies only depolymerizes tyrosinated α-tubulin. However, microtubules that are rich in detyrosinated and acetylated α-tubulin are resistant to this treatment (Conde and Cáceres 2009). Detyrosinatinon is a reversible posttranslational modification of tubulin subunits. A yet unknown carboxypeptidase removes the C-terminal tyrosine of α-tubulin, and tubulin tyrosine ligase catalyzes tyrosination (Hammond et al. 2008). Therefore, nocodazole would prevent the movement of cargo only along tyrosinated microtubules, and the transport along detyrosinated microtubules will remain intact. Finally, an aspect that has not been stressed at all is the triggering of highly diverse events by the nuclear enrichment of synaptic proteins. Thus, the possible functions for learning and memory that have been attributed to synapse-to-nucleus communication in recent years with particular emphasis on synaptic proteins have not been addressed yet.

16.2   The Many Different Ways to the Nucleus

16.2.1   The Fast Ca^{2+} Track

The fast track from synapse to nucleus is initiated by local membrane depolarization that initiates backpropagating dendritic action potentials (Fig.16.1.I.1), as well as by regenerative calcium waves (Fig.16.1.I.2) that propagate along the endoplasmic reticulum (ER) toward the nucleus. This aspect of synapse-to-nucleus communication is extensively covered in the chapter of Bengtson and Bading in this book and is therefore only briefly summarized here.

The so-called rapid electrochemical model (Adams and Dudek 2005; Saha and Dudek 2008) suggests that synaptic depolarization triggers multiple action potentials which in turn result in an elevation of somatic Ca^{2+} concentration from different sources including influx through voltage-operated calcium channels (VOCC) and release from ER. Alternatively, the "regenerative calcium waves" model suggests that the activation of inositol-1,4,5-triphosphate receptors (IP3Rs) and ryanodine

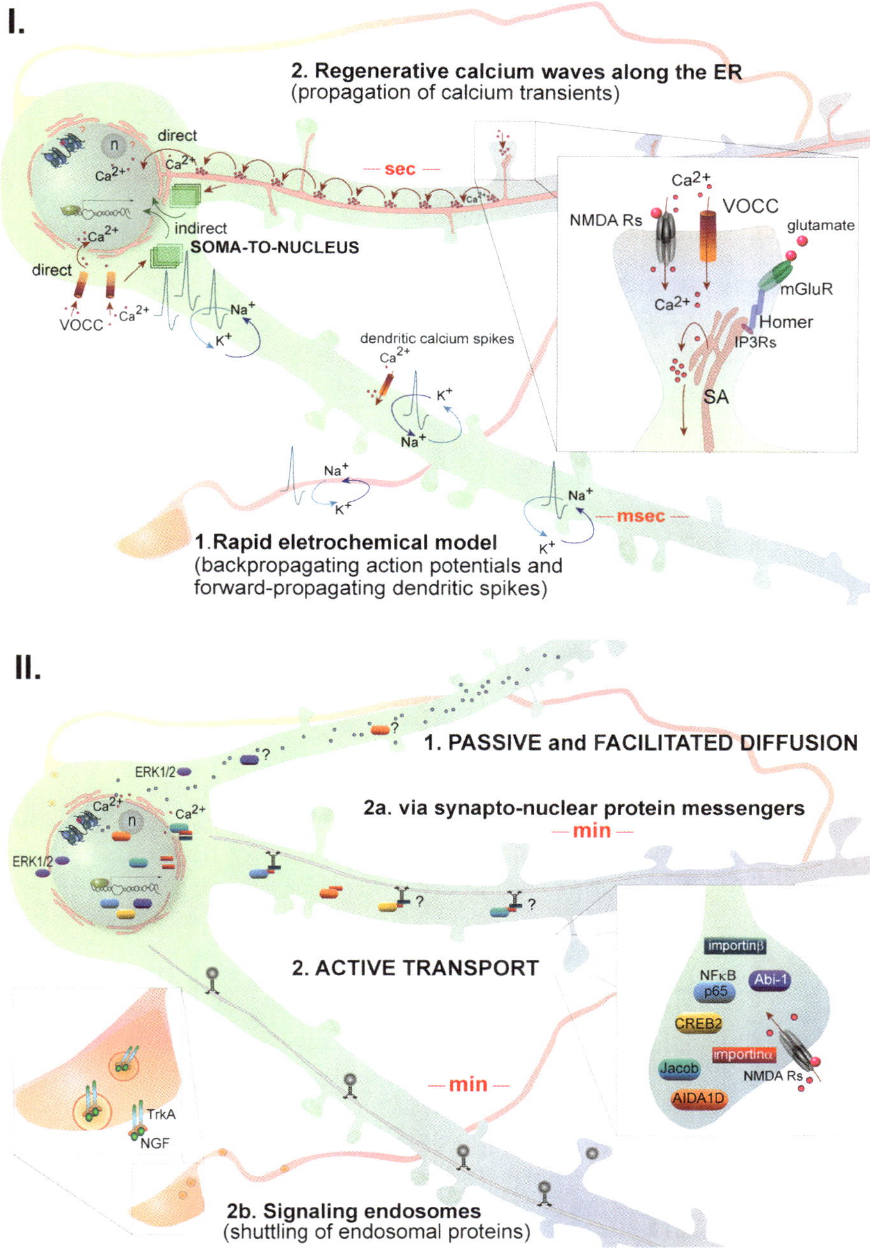

Fig. 16.1 The many ways from synapse to the nucleus. There are many possible ways for the signals to get transduced from a subset of activated synapses into the nucleus where they mediate transcriptional activity and nucleolar (n) assembly and where they might be involved in regulation of epigenetic DNA modification

receptors (RYRs) leads to Ca^{2+}- induced Ca^{2+} release from internal stores, and the resulting Ca^{2+} transients can thereafter rapidly propagate along the ER (Simpson et al. 1995; Berridge 1998; Berridge et al. 2003). A possible starting point for such waves would be the spine apparatus, which represents the specialized ER at the synapse. It continues along the dendrite and fuses with the outer and inner nuclear membrane. Ca^{2+} signals arising from synaptic VOCC and NMDARs can plausibly initiate such regenerative waves along the ER (Kapur et al. 2001; Nakamura et al. 2002; Ross et al. 2005). In addition, metabotropic glutamate receptors (mGluRs) provide a physical link to IP3Rs via Homer proteins and might contribute to the neurotransmitter-induced Ca^{2+} release from internal Ca^{2+} stores. In both scenarios, somatic Ca^{2+} might either directly enter the nucleus to regulate gene expression or initiate soma-to-nucleus signaling via the Ca^{2+}-activated nuclear import of messengers like NFAT (Fig. 16.1.I).

16.2.2 Passive and Facilitated Diffusion Across the Nuclear Pore

Transport of macromolecules across the nuclear border is either realized by passive (energy- and carrier-independent) and facilitating diffusion (Fig. 16.1.II.1) or active importin-α/importin-β-mediated Ran- and GTP-hydrolysis-dependent transport (Fig. 16.1.II.2). The limiting factor for passive diffusion through the pore is the size of the messenger. The diffusion limit for protein passage through the nuclear pore is in the range between 40 and 60 kDa (Paine et al. 1975), but diffusion becomes highly inefficient with increasing molecular weight (Görlich and Kutay 1999).

Facilitated passage (facilitated diffusion) through the nuclear pore is accomplished by direct binding of the substrate to the NPCs and can be mediated by other carriers, distinct from importin-β transport receptors, and therefore does neither require Ran nor its GTP hydrolysis (Görlich and Kutay 1999; Yokoya et al. 1999). It is widely believed that the MAPK-ERK pathway mediates synapse-to-nucleus signaling and is involved in the regulation of activity-dependent gene expression that is required for neuronal plasticity and long-term memory (Impey et al. 1998; Hardingham et al. 2001a, b; Sweatt 2004; Wiegert et al. 2007; Wiegert and Bading 2011). For nonneuronal cells, it has been shown that the nuclear entry of ERK can be accomplished by different mechanisms: direct facilitated diffusion via interaction with nucleoporins or an active importin-7 (importin-β-like transport receptor)-mediated nuclear transport (Chuderland et al. 2008; Jivan et al. 2010). In hippocampal neurons, ERK translocate into the nucleus upon NMDA and TrkB receptor activation, and the passage across the nuclear envelope is largely mediated by passive and facilitated diffusion mechanisms (Wiegert et al. 2007). This, however, is a highly inefficient mechanism for transport from distal dendrites and would limit the activity-dependent translocation of ERK pools into the soma and proximal dendrites. However, the possibility remains that active ERK can be actively transported from distal locations to the nucleus via its interaction with synaptic and dendritic proteins as it has been shown for long-distance transport in axons (Perlson et al. 2005). Such a mechanism would only concern a subfraction of synaptic or dendritic ERK, and the majority of kinase would remain stationary in distal dendrites.

16.2.3 Signaling Endosomes

A pervasive model of signal transmission from axonal synapses to the cell nucleus is the retrograde transport of signaling endosomes (Ginty and Segal 2002; Miaczynska et al. 2004; Howe and Mobley 2005; Cosker et al. 2008; Wu et al. 2009). Especially the transport of activated Trk receptors upon neurotrophin and here especially NGF binding to TrkA has been investigated. The endocytosis of these receptor-ligand complexes into early endosomes (Delcroix et al. 2003) has been shown to be clathrin dependent (Howe et al. 2001). Signaling endosomes are subsequently actively transported to the soma along microtubules (Watson et al. 1999) associated with dynein motors (Heerssen et al. 2004; Yano et al. 2001). Interestingly, endosome isolation (Howe et al. 2001) and studies using compartmentalized chambers with cultured neurons showed that receptor-ligand complexes are transported together with molecules of downstream signaling pathways (e.g., p-ERK of MAP/ERK pathway, Howe et al. 2001) and even the transcription factor CREB is associated with endosomes and can be activated during transport (Cox et al. 2008). Therefore, these self-regenerating organelles (Ye et al. 2003) provide a basis of specifically regulated transport of signals, e.g., supporting neuronal survival, over long distances without decay of signal integrity.

16.2.4 Active Retrograde Transport of Synapto-Nuclear Messenger Proteins

Neuronal processes extend several hundreds of micrometers away from the cell soma. Therefore, signals from activated synapses at distal dendrites and from axonal terminals have to travel a long route to and, eventually, into the nucleus to modulate neuronal function. In pyramidal neurons, the closest dendritic spiny synapses are located at least 40 μm away from the nucleus. This raises a number of important questions on how proteins can translocate from distal synaptic sites to the nucleus. One possible mechanism is the active retrograde transport mediated by the importin-α/importin-β complex. Both importin-α and importin-β are present in synaptic compartment where they are well positioned to mediate direct synapse-to-nucleus signaling. Importin-α family members (importin-α1 and importin-α2) directly associate with the postsynaptic density (Thompson et al. 2004), and particularly, importin-α1 might be docked at synaptic sites by interaction with certain splice isoforms of the NR1 NMDAR subunit (Jeffrey et al. 2009). The fact that importin-α and importin-β1 undergo an NMDA receptor–dependent nuclear translocation (Thompson et al. 2004; Dieterich et al. 2008) suggests that this transport mechanism might be involved in NMDA receptor–activated gene expression. In line with this notion, a number of potential synapto-nuclear protein messengers have been identified in recent years (Jordan and Kreutz 2009). Proteins like Abi-1, AIDA-1d, CREB2/ATF4, Jacob, and p65/RelA (NF-κB) that are localized at postsynaptic sites and translocate to the nucleus in response to NMDA receptor activity are of particular interest. All abovementioned cargo proteins with the exception of Abi-1 possess NLSs that are recognized by certain members of the importin-α nuclear transport adaptor protein family.

Conventionally, importin-α binds the cargo protein and subsequently forms a heterotrimeric nuclear pore-targeting complex with importin-β1 (Goldfarb et al. 2004). Previous reports have shown that the murine importin-α/karyopherin-α gene family of nuclear transport adaptor proteins comprises at least five members (Otis et al. 2006). Based on sequence homology, importin-α family members are classified into three subfamilies: α-P (Imp-α2/karyopherin-α2/*Gene ID:16647*), α-Q (Imp-α3/Q2/karyopherin-α3/*Gene ID:16648* and Imp-α4/Q1/karyopherin-α4/*Gene ID:16649*), and α-S (Imp-α1/S1/karyopherin-α1/*Gene ID:16646* and Imp-α 6/S2/karyopherin-α6/*Gene ID:16650*). They exhibit differential expression patterns in brain and other tissues (Kamei et al. 1999; Yoneda 2000; Jans et al. 2000; Lai et al. 2008; Hosokawa et al. 2008; Yasuhara et al. 2009). The vast majority of importin-α family members are highly expressed in hippocampal pyramidal neurons (Hosokawa et al. 2008). Recently, a novel member of the murine importin-α/karyopherin-α gene family was identified (Knap7/*Gene ID:381686*, Hu et al. 2010). Knap7 has been shown to interact with the importin-β1 transport receptor, but its expression in brain has not been investigated yet.

The expression of importin-α/karyopherin-α family members is regulated during neural differentiation of mouse embryonic stem cells, and it has been suggested that the switching of importin-α subtype expression might be important for neuronal differentiation (Yasuhara et al. 2007). The multiple family members of importin-α possess both distinct and overlapping cargo specificities (Jans et al. 2000; Yasuhara et al. 2007; Shmidt et al. 2007). It has been reported that a compensatory mechanism for importin-α/karyopherin-α family members might exist which indicates the overlapping cargo specificities. Particularly, importin-α1/karyopherin-α1 (reported as importin-α5) seems to be involved in neuronal differentiation (Yasuhara et al. 2007, 2009). Surprisingly, the homo- and heterozygous karyopherin-α1 knockout mice have no obvious defect in brain development (Shmidt et al. 2007). On the other side, the level of karyopherin-α3 (reported as importin-α4) in brain and other tissues of these animals is dramatically upregulated (Shmidt et al. 2007) suggesting a compensatory mechanism for the depletion of karyopherin-α1. Based on this observation, it has been suggested that some importin-α family members might, at least to a certain degree, functionally substitute each other. Another example of overlapping cargo specificities of importin-α family members in brain tissue has been reported for karyopherin-α1 and karyopherin-α6 regarding CREB2/ATF4 NLS recognition. The primary structure of CREB2/ATF4 harbors defined nuclear targeting sequences: a bipartite NLS (KKLKK motif, Cibelli et al. 1999) and a second putative NLS (RYRQKKR motif). Both are potential recognition sites for transport adaptors. Screening for CREB2/ATF4 interaction with all importin-α family members revealed that it binds exclusively to the importin-αS subclass (importin-α1/karyopherin-α1 and -α6, Lai et al. 2008). Conceivably, the transport of distinct cargos mediated by distinct importins upon particular neuronal stimuli might provide signal specificity for the nuclear response. It remains elusive whether different transport adaptors compete for the same cargo protein and/or vice versa.

Another synapto-nuclear protein messenger, AIDA-1d, harbors a monopartite NLS. Although its direct interaction with importin-α/karyopherin-α family members has not been reported, mutagenesis of the NLS prevents the accumulation of overexpressed protein in the nucleoli (Jordan et al. 2007) indicating that its translocation is indeed mediated by importins. Abi-1 is transported from synaptic sites into the nucleus upon NMDA application. This process could be abolished by the destruction of the microtubules and microfilaments (Proepper et al. 2007). Since there is no apparent NLS present in Abi-1 and binding to importin family members has not been reported, it remains unclear how Abi-1 is delivered to the nucleus.

Various studies have provided substantial evidences that both p50 and p65/RelA subunits of NF-κB are present at synaptic compartments of hippocampal pyramidal neurons and associate with PSDs (Kaltschmidt et al. 1993; Meberg et al. 1996; Meffert et al. 2003; Marcora and Kennedy 2010). Both p65/RelA and p50 might be targeted to the PSD95 via huntingtin protein (Htt, Takano and Gusella 2002; Marcora and Kennedy 2010). Importin-α1 (Nadler et al. 1997; Cunningham et al. 2003) and importin-α2 (Cunningham et al. 2003; Marcora and Kennedy 2010) recognize the NLS of p65/RelA (KRKR motif) and were proposed to modulate its nuclear transport. In addition, Marcora and Kennedy (2010) could demonstrate that Htt preferentially associates with activated p65/RelA at synapses and facilitates its transport from the PSD toward the nucleus by binding to a dynein/dynactin motor complex. This finding is consistent with previous observations that retrograde transport of p65/RelA along dendrites to the nucleus requires microtubules and is mediated by the dynein/dynactin complex (Mikenberg et al. 2007; Shrum et al. 2009). Jacob's primary structure harbors a bipartite NLS (aa 247–266), and the 247–252 amino acid stretch is necessary for interaction with importin-α1 since the binding was abolished when the RKRRKR motif was deleted (Dieterich et al. 2008). A colocalization study revealed that Jacob and importin-α1 interaction might occur at synaptic sites at distal dendrites. This supports the idea that this transport adaptor protein might interact with Jacob at synaptic sites and therefore mediates direct synapse-to-nucleus signaling (Fig. 16.2)

16.2.5 Control of Nuclear Transport by Regulation of Importin-Cargo Binding

The accessibility of both the NLS on cargo proteins and the NLS recognition site on importins is a prerequisite for the cellular nuclear import machinery. Conventionally, the N-terminal importin-β binding domain of importin-α is autoinhibitory (Kobe 1999; Goldfarb et al. 2004). In the absence of an NLS-containing cargo,

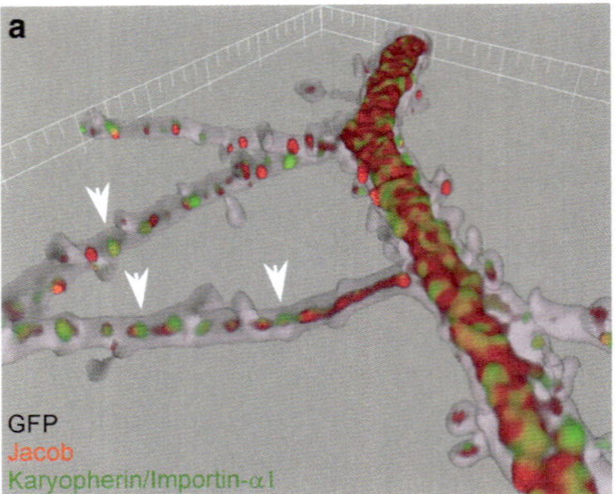

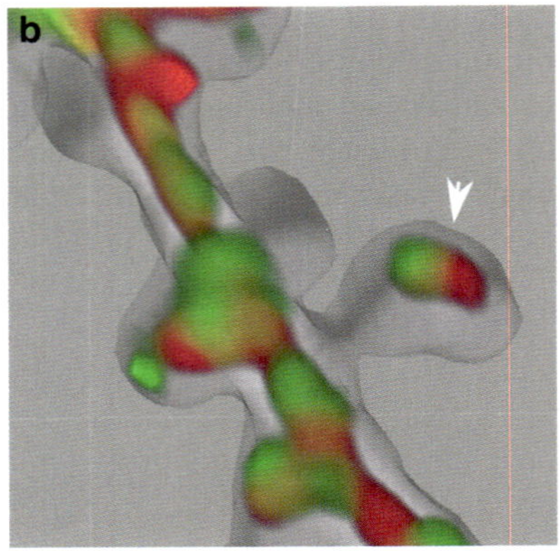

Fig. 16.2 Jacob/importin-α1 clusters are present at the distal dendrites and spines (**a**) and (**b**) 3D reconstructions (Imaris 6.2 software, Bitplane AG, Zürich, Switzerland) of GFP-filled dendrites (gray transparent isosurface) with spines of hippocampal 17DIV neurons immunolabeled with Jacob (red) and karyopherin-α1/importin-α1 (green, BD Biosciences) antibodies. Confocal Z-stacks were acquired using LAS AF (Leica Application Suite Advanced Fluorescence) imaging software and deconvoluted in three dimensions using AutoQuantX2.2, Media Cybernetics. For deconvolution, adaptive PSF (Blind), medium noise suppression, and two iterations were used. The punctated staining of Jacob/importin-α1 within the dendrites and spines was determined by masking these channels with an isosurface generated from GFP fluorescence. Box size is 5 μm. The fact that both proteins are detectable and found partially being colocalized in distal dendrites and spines strongly supports the idea that the transport adaptor importin-α1 might interact with Jacob at synaptic sites and mediate direct synapse-to-nucleus signaling pathway in vivo

the importin-β binding domain can form an intramolecular interaction with the cargo-NLS-binding pocket (Kobe 1999) and therefore masks the NLS recognition sequence (Fig. 16.3.1). However, it remains unclear how this is regulated at the synapse. A mechanism how importin-α might be docked at synapses in an activity-dependent manner was recently provided by Jeffrey et al. (2009). It is plausible to assume that synaptic proteins bearing the NLS sequence might target importins to the synapse. Importin-α1 binds to a bipartite NLS in the NR1-1a subunit of the NMDAR, and this interaction is regulated by phosphorylation of the NLS by PKC (Fig. 16.3.2). Upon activation of NMDAR, importin-α1 is released from the complex and becomes accessible for the interaction with cargo proteins (Jeffrey et al. 2009). Remarkably, the interaction between NR1-1a and importin-α1 is disrupted upon stimuli known to induce late long-term potentiation (late LTP) but not early LTP in CA1 Schaffer collateral synapses.

Synapse-to-nucleus signaling may also be regulated at the level of individual cargos at the synapse. An example for such a regulation is the Ca^{2+}-dependent binding of the IQ domain of caldendrin to the α-helical region of Jacob (Dieterich et al. 2008). Upon synaptic activity, importin-α1 competes with caldendrin for Jacob binding. Therefore, the amount of Jacob that is accessible for the interaction with importin-α1 is regulated by the amount of caldendrin at the synapse

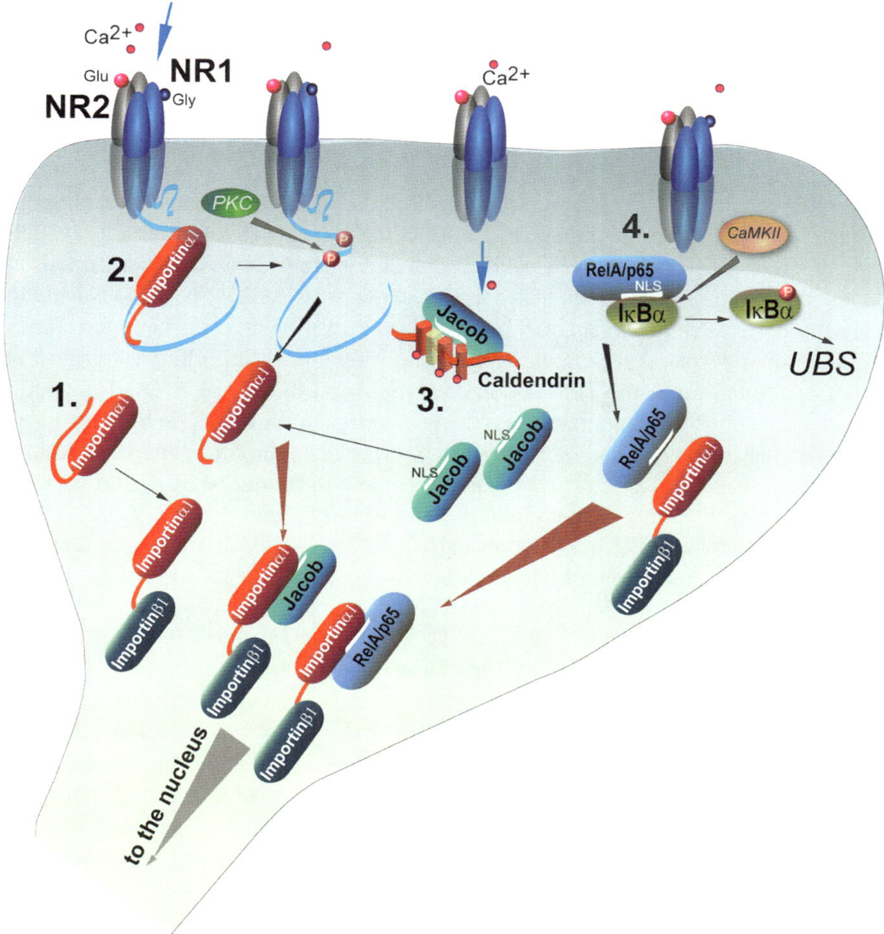

Fig. 16.3 Regulation of synapse-to-nucleus signaling by importin-α/karyopherin-α-cargo binding. *1* – Autoregulation via masking of the NLS recognition sequence in importins; *2* – Anchoring of importin-α at the synapse; *3* – Regulation via the amount of available individual cargoes; *4* – Masking the signal on the cargo by heterologous molecules

(Fig. 16.3.3 see also below). Targeting sequence masking through specific protein binding is best characterized for p65/RelA and IκBα. Phosphorylation of IκBα in an activity-dependent manner, which in neurons can be mediated by CaMKII (Marcora and Kennedy 2010) and subsequent proteolytic degradation, is required for unmasking the NLS of p65/RelA. Thus, unmasking the NLS on the cargo protein and the NLS recognition site of importin-α is conceivably important regulatory mechanisms at synapses to initiate synapto-nuclear trafficking (Fig. 16.3.4).

16.3 The Functional Role of Synapto-Nuclear Protein Messengers

16.3.1 Jacob

Jacob is a putative synapto-nuclear protein messenger that was originally identified as a binding partner of the neuronal calcium-binding protein caldendrin in a yeast two-hybrid screen (Dieterich et al. 2008). It is abundantly expressed in the limbic brain and cortex and prominently present in synapses and neuronal nuclei. The protein is highly conserved between mouse, rat, and human (95% identity) and other mammals. Database searches revealed no known invertebrate orthologue. Subcellular fractionation experiments confirmed that Jacob is enriched in synaptosomes and PSDs of excitatory synapses. Jacob RNA can undergo extensive alternative splicing. Out of 16 exons, at least 5 exons can be alternatively spliced, alone and in various combinations (Kindler et al. 2009). Jacob mRNA is prominently localized in dendrites, as it harbors a cis-acting dendritic targeting element in its $3'$-untranslated region (Kindler et al. 2009; see also the chapter of Kindler and Kreienkamp for a more detailed account). The dendritic mRNA of Jacob might replenish local pools after nuclear translocation of Jacob.

Jacob harbors a classical bipartite NLS that is a prerequisite for its nuclear localization as well as an N-myristoylation site, which anchors the protein to membranes and is required for its extranuclear localization. An N-terminal fragment of Jacob can be cleaved by the NMDA receptor– and Ca^{2+}-activated protease calpain (Kindler et al. 2009). Importantly, caldendrin binding to Jacob masks the NLS and competes with importin-$\alpha 1$ binding in a Ca^{2+}-dependent manner. In consequence, the importin-$\alpha 1$-dependent translocation of Jacob can take place only if (a) the myristoylation site is cleaved from Jacob's N-terminus and (b) Jacob is not bound to caldendrin either due to the lack of the latter at the corresponding subsynaptic site or calcium levels that do not allow for both proteins to interact.

In the nucleus, Jacob is associated with zones of active gene transcription (Dieterich et al. 2008). These findings suggest that it can directly or indirectly influence NMDA receptor–regulated gene transcription. Enhancing neuronal activity via bath application of glutamate and NMDA leads to increased Jacob levels in the nucleus (Dieterich et al. 2008). This increase can be blocked by addition of NMDAR antagonists. NMDARs are present at both synaptic and extrasynaptic sites. Differential activation of synaptic vs. extrasynaptic NMDARs showed that the latter is a much more efficient stimulus to drive Jacob into the nucleus. Moreover, the nuclear accumulation of Jacob can be blocked by the NR2B-specific NMDAR antagonist ifenprodil. Thus, Jacob translocates to the nucleus strictly after activation of NR2B containing NMDAR; depolarization alone (e.g., due to KCl) is not sufficient (Dieterich et al. 2008; Rönicke et al. 2011). This suggests that at least an indirect association of Jacob with NMDARs might exist.

Many studies have shown that the stimulation of extrasynaptic NMDARs leads to a long-lasting dephosphorylation of the transcription factor CREB at a serine at position 133, which renders CREB transcriptionally inactive, a phenomenon called CREB shut-off (Sala et al. 2000; Hardingham et al. 2002; Chandler et al. 2001; Kim et al. 2005; Hardingham and Bading 2002). It is now well established that extrasynaptic NMDA receptors as opposed to their synaptic counterparts trigger the CREB shut-off pathway and cell death. Signaling from extrasynaptic NMDA receptors to the nucleus has been linked to neurodegeneration in a variety of brain disease states including ischemia (Tu et al. 2010) and Huntington's disease (Milnerwood et al. 2010; Hardingham and Bading 2010). We found that nuclear knockdown of Jacob prevents CREB shut-off after extrasynaptic NMDA receptor activation while its nuclear overexpression induces CREB shut-off without NMDA receptor stimulation (Dieterich et al. 2008). Importantly, nuclear knockdown of Jacob attenuates NMDA-induced loss of synaptic contacts and neuronal degeneration (Dieterich et al. 2008). This defines a novel mechanism of synapse-to-nucleus communication via a synaptic Ca^{2+}-sensor protein, which links the activity of NMDA receptors to nuclear signaling events involved in modeling synapto-dendritic input and NMDA receptor–induced cellular degeneration.

However, we also observed a less prominent nuclear accumulation of the protein after triggering the activity of synaptic NMDA receptors (Dieterich et al. 2008). Since this pathway promotes cell survival and induces the expression of plasticity-related genes, we wondered whether Jacob might be also a messenger on this synaptic NMDA receptor pathway to the nucleus in the cellular models of synaptic plasticity. LTP and LTD are activity-dependent forms of synaptic plasticity that, in the cornu ammonis 1 (CA1) region of the hippocampus, require a calcium influx through NMDARs (Bliss and Lomo 1973; Morris and Frey 1997). The induction of LTP and LTD at these synapses correlates with learning processes in vivo and is thought to underlie memory formation (Nguyen et al. 1994; Reymann and Frey 2007). We recently found that LTP-inducing stimuli (strong tetanization consisting of three 1 s trains at 100 Hz; intertrain interval was 10 min) were sufficient to rise the Jacob nuclear level already during tetanization face. This might be a requirement for gene expression that stabilizes LTP type of synaptic plasticity and contributes to the LTP maintenance. Late-LTD-inducing stimuli (900 bursts at 1 Hz; one burst consists of three stimuli with 50 ms interstimulation interval) had no influence on Jacob nuclear import, suggesting that synapto-nuclear protein messengers might provide the input specificity that required for plasticity events (Behnisch et al. 2011). Interestingly, it was shown in one previous study that the transcription factor cyclic AMP-response element-binding protein 2 (CREB2) transits to the nucleus during LTD but not LTP of synaptic transmission in hippocampal primary neurons. Taken together, these findings suggest that the two major forms of NMDA receptor–dependent synaptic plasticity, LTP and LTD, elicit the transition of different synapto-nuclear protein messengers, albeit in both cases importin-mediated retrograde transport and NMDA receptor activation are required.

16.3.2 NF-κB

The transcription factor NF-κB is a homo- or heterodimer of the subunits RelA (also called p65), RelB, c-Rel, p50, and p52. The most common active dimer in neurons is p65:p50 (Kaltschmidt et al. 1993; Bakalkin et al. 1993; Schmidt-Ullrich et al. 1996; Meffert et al. 2003). Besides the constitutive NF-κB activity in neurons (Kaltschmidt et al. 1994), an inducible, IκB-bound pool also exists (Kaltschmidt and Kaltschmidt 2009). The presence of NF-κB in synaptosomes (Kaltschmidt et al. 1993; Meberg et al. 1996; Meffert et al. 2003; Marcora and Kennedy 2010) and the fact that it can be activated by glutamate stimulation (Guerrini et al. 1995) gave rise to the idea that it plays a role in synapse-to-nucleus communication. It has been shown that p65 translocates to the nucleus upon NMDAR stimulation in a CaMKII- (Meffert et al. 2003) and NLS-dependent manner (Wellmann et al. 2001). Additionally, NF-κB transcriptional activity is enhanced following neuronal depolarization via KCl and kainate stimulation in primary neuronal cultures (Kaltschmidt et al. 1995). Therefore, this transcription factor might directly transmit synapto-nuclear protein messenger.

16.3.3 Abi-1

Abelson interacting protein 1 (Abi-1) is a synaptic and nuclear protein functioning as a regulator of dendritic growth and synaptic contacts (Proepper et al. 2007; Ito et al. 2010). As a direct binding partner of ProSAP2, it is located at PSDs. Upon NMDAR stimulation, Abi-1 immunoreactivity is increased in the nucleus, whereas the staining is diminished in dendritic branches in the presence of protein synthesis inhibitors. This translocation is reversible and microtubule dependent. Inside the nucleus, Abi-1 associates with c-Myc/Max complex and thereby influences gene expression.

16.3.4 CREB2

The CREB repressor and transcription factor, CREB2, is expressed in synapses and distal dendrites, as well as in the nucleus of hippocampal pyramidal neurons (Lai et al. 2008). Its nucleocytoplasmic shuttling upon NMDAR activation is mediated by importins. While the LTD-inducing protocol of NMDA and glycine application leads to an intense nuclear accumulation of CREB2, nuclear levels remain unaffected upon LTP induction with glycine application. This association with plasticity and memory formation has been shown both in rodent models and in *aplysia* (Bartsch et al. 1995; Chen et al. 2003; Lai et al. 2008).

16.3.5 AIDA-1d

Amyloid precursor protein intracellular domain–associated protein-1 (AIDA-1d) is a synaptically localized protein carrying a functional bipartite nuclear localization signal (NLS) in its N-terminus (Jordan et al. 2004). Its binding to PSD-95 associates AIDA-1d to the NMDAR complex (Jordan et al. 2007). Upon NMDAR stimulation, AIDA-1d is proteolytically cleaved, and the N-terminus translocates to the nucleus (Jordan et al. 2007). The observed AIDA-1d dependent increase in the number of nucleoli, its enrichment in especially Cajal bodies (Jacob et al. 2010; Xu and Hebert 2005), and the increased global protein synthesis suggest a structural function of this synapto-nuclear protein messenger.

16.4   Conclusions and Future Directions

Apart from the lack of demonstrative evidence that synapto-nuclear protein messengers indeed translocate from synapse to nucleus, a number of other open questions have to be addressed in the forthcoming years. It is not clear at all whether synapse-to-nucleus communication via protein messengers is required for long-term memory formation, and if yes, why. It remains to be established which nuclear events are crucial in this respect and how regulation of gene expression feeds back to synaptic integrity and function. In addition, the transport mechanisms have not been analyzed in any detail yet. The association of the messengers to the direction-specific dynein/kinesin motors does not explain their translocation to the nucleus completely, as the microtubules with which these motors in turn associate show a mixed polarity in dendrites. The same ambiguity exists even for those messengers associated with importin-α. Another intriguing question concerns the mechanism how synapto-nuclear protein messengers leave the synapse. In many cases, they are supposedly tightly anchored to the postsynaptic scaffold, and the processes that lead to their release from synaptic sites are not understood at all. Binding to importin-α could be a requisite but does neither explain how the complex leaves the synapse nor how the cargo was set free for nuclear import.

Acknowledgement   The work in the lab of M.R.K. is supported by grants from the CBBS/EFRE, the Deutsche Forschungsgemeinschaft (DFG), the German-Israeli (DIP) grant, the DZNE, the Deutsche Akademische Austauschdienst (DAAD), the Bundesministerium für Bildung und Forschung (BMBF), the Leibniz Foundation (Pakt für Forschung), and the Schram Foundation.

References

Adams, J. P., & Dudek, S. M. (2005). Late-phase long-term potentiation: Getting to the nucleus. *Nature Reviews Neuroscience, 6*, 737–743.

Alberi, L., Liu, S., Wang, Y., Badie, R., Smith-Hicks, C., Wu, J., Pierfelice, T. J., Abazyan, B., Mattson, M. P., Kuhl, D., Pletnikov, M., Worley, P. F., & Gaiano, N. (2011). Activity-induced Notch signaling in neurons requires Arc/Arg3.1 and is essential for synaptic plasticity in hippocampal networks. *Neuron, 69*, 437–444.

Alberini, C. M. (2009). Transcription factors in long-term memory and synaptic plasticity. *Physiological Reviews, 89*, 121–145.

Ataman, B., Ashley, J., Gorczyca, D., Gorczyca, M., Mathew, D., Wichmann, C., Sigrist, S. J., & Budnik, V. (2006). Nuclear trafficking of Drosophila Frizzled-2 during synapse development requires the PDZ protein dGRIP. *Proceedings of the National Academy of Sciences of the United States of America, 103*, 7841–7846.

Bakalkin, G. Ya., Yakovleva, T., & Terenius, L. (1993). NF-kappa B-like factors in the murine brain. Developmentally-regulated and tissue-specific expression. *Brain Research. Molecular Brain Research, 20*, 137–146.

Balda, M. S., & Matter, K. (2000). The tight junction protein ZO-1 and an interacting transcription factor regulate ErbB-2 expression. *EMBO Journal, 19*, 2024–2033.

Bao, J., Lin, H., Ouyang, Y., Lei, D., Osman, A., Kim, T. W., Mei, L., Dai, P., Ohlemiller, K. K., & Ambron, R. T. (2004). Activity-dependent transcription regulation of PSD-95 by neuregulin-1 and Eos. *Nature Neuroscience, 7*, 1250–1258.

Bartsch, D., Ghirardi, M., Skehel, P. A., Karl, K. A., Herder, S. P., Chen, M., Bailey, C. H., & Kandel, E. R. (1995). Aplysia CREB2 represses long-term facilitation: Relief of repression converts transient facilitation into long-term functional and structural change. *Cell, 83*, 979–992.

Behnisch, T., Yuanxiang, P., Bethge, P., Parvez, S., Chen, Y., Yu, J., Karpova, A., Frey, J. U., Mikhaylova, M., & Kreutz, M. R. (2011). Nuclear translocation of Jacob in hippocampal neurons after stimuli inducing long-term potentiation but not long-term depression. *PLoS One, 6*, e17276.

Behrens, J., von Kries, J. P., Kühl, M., Bruhn, L., Wedlich, D., Grosschedl, R., & Birchmeier, W. (1996). Functional interaction of beta-catenin with the transcription factor LEF-1. *Nature, 382*, 638–642.

Berridge, M. J. (1998). Neuronal calcium signaling. *Neuron, 21*, 13–26.

Berridge, M. J., Bootman, M. D., & Roderick, H. L. (2003). Calcium signalling: Dynamics, homeostasis and remodelling. *Nature Reviews Molecular Cell Biology, 4*, 517–529.

Bliss, T. V., & Lomo, T. (1973). Long-lasting potentiation of synaptic transmission in the dentate area of the anaesthetized rabbit following stimulation of the perforant path. *The Journal of Physiology, 232*, 331–356.

Cao, X., & Südhof, T. C. (2001). A transcriptionally [correction of transcriptively] active complex of APP with Fe65 and histone acetyltransferase Tip60. *Science, 293*, 115–120.

Chakraborty, S., Reineke, E. L., Lam, M., Li, X., Liu, Y., Gao, C., Khurana, S., & Kao, H. Y. (2006). Alpha-actinin 4 potentiates myocyte enhancer factor-2 transcription activity by antagonizing histone deacetylase 7. *Journal of Biological Chemistry, 281*, 35070–35080.

Chandler, L. J., Sutton, G., Dorairaj, N. R., & Norwood, D. (2001). N-methyl D-aspartate receptor-mediated bidirectional control of extracellular signal-regulated kinase activity in cortical neuronal cultures. *Journal of Biological Chemistry, 276*, 2627–2636.

Chawla, S., Vanhoutte, P., Arnold, F. J., Huang, C. L., & Bading, H. (2003). Neuronal activity-dependent nucleocytoplasmic shuttling of HDAC4 and HDAC5. *Journal of Neurochemistry, 85*, 151–159.

Chen, A., Muzzio, I. A., Malleret, G., Bartsch, D., Verbitsky, M., Pavlidis, P., Yonan, A. L., Vronskaya, S., Grody, M. B., Cepeda, I., Gilliam, T. C., & Kandel, E. R. (2003). Inducible enhancement of memory storage and synaptic plasticity in transgenic mice expressing an inhibitor of ATF4 (CREB-2) and C/EBP proteins. *Neuron, 39*, 655–669.

Chuderland, D., Konson, A., & Seger, R. (2008). Identification and characterization of a general nuclear translocation signal in signaling proteins. *Molecular Cell, 31*, 850–861.

Cibelli, G., Schoch, S., & Thiel, G. (1999). Nuclear targeting of cAMP response element binding protein 2 (CREB2). *European Journal of Cell Biology, 78*, 642–649.

Claret, F. X., Hibi, M., Dhut, S., Toda, T., & Karin, M. (1996). A new group of conserved coactivators that increase the specificity of AP-1 transcription factors. *Nature, 383*, 453–457.

Cohen, S., & Greenberg, M. E. (2008). Communication between the synapse and the nucleus in neuronal development, plasticity, and disease. *Annual Review of Cell and Developmental Biology, 24*, 183–209.

Conde, C., & Cáceres, A. (2009). Microtubule assembly, organization and dynamics in axons and dendrites. *Nature Reviews Neuroscience, 10*, 319–332.

Cosker, K. E., Courchesne, S. L., & Segal, R. A. (2008). Action in the axon: Generation and transport of signaling endosomes. *Current Opinion in Neurobiology, 18*, 270–275.

Cox, L. J., Hengst, U., Gurskaya, N. G., Lukyanov, K. A., & Jaffrey, S. R. (2008). Intra-axonal translation and retrograde trafficking of CREB promotes neuronal survival. *Nature Cell Biology, 10*, 149–159.

Cunningham, M. D., Cleaveland, J., & Nadler, S. G. (2003). An intracellular targeted NLS peptide inhibitor of karyopherin alpha: NF-kappa B interactions. *Biochemical and Biophysical Research Communications, 300*, 403–407.

Deisseroth, K., Mermelstein, P. G., Xia, H., & Tsien, R. W. (2003). Signaling from synapse to nucleus: The logic behind the mechanisms. *Current Opinion in Neurobiology, 13*, 354–365.

Delcroix, J. D., Valletta, J. S., Wu, C., Hunt, S. J., Kowal, A. S., & Mobley, W. C. (2003). NGF signaling in sensory neurons: Evidence that early endosomes carry NGF retrograde signals. *Neuron, 39*, 69–84.

Dieterich, D. C., Karpova, A., Mikhaylova, M., Zdobnova, I., König, I., Landwehr, M., Kreutz, M., Smalla, K. H., Richter, K., Landgraf, P., Reissner, C., Boeckers, T. M., Zuschratter, W., Spilker, C., Seidenbecher, C. I., Garner, C. C., Gundelfinger, E. D., & Kreutz, M. R. (2008). Caldendrin-Jacob: A protein liaison that couples NMDA receptor signalling to the nucleus. *PLoS Biology, 6*, e34.

Flavell, S. W., & Greenberg, M. E. (2008). Signaling mechanisms linking neuronal activity to gene expression and plasticity of the nervous system. *Annual Review of Neuroscience, 31*, 563–590.

Ginty, D. D., & Segal, R. A. (2002). Retrograde neurotrophin signaling: Trk-ing along the axon. *Current Opinion in Neurobiology, 12*, 268–274.

Goldfarb, D. S., Corbett, A. H., Mason, D. A., Harreman, M. T., & Adam, S. A. (2004). Importin alpha: A multipurpose nuclear-transport receptor. *Trends in Cell Biology, 14*, 505–514.

Gomez-Ospina, N., Tsuruta, F., Barreto-Chang, O., Hu, L., & Dolmetsch, R. (2006). The C terminus of the L-type voltage-gated calcium channel Ca(V)1.2 encodes a transcription factor. *Cell, 127*, 591–606.

Görlich, D., & Kutay, U. (1999). Transport between the cell nucleus and the cytoplasm. *Annual Review of Cell and Developmental Biology, 15*, 607–660.

Graef, I. A., Mermelstein, P. G., Stankunas, K., Neilson, J. R., Deisseroth, K., Tsien, R. W., & Crabtree, G. R. (1999). L-type calcium channels and GSK-3 regulate the activity of NF-ATc4 in hippocampal neurons. *Nature, 401*, 703–708.

Greer, P. L., & Greenberg, M. E. (2008). From synapse to nucleus: Calcium-dependent gene transcription in the control of synapse development and function. *Neuron, 59*, 846–860.

Guerrini, L., Blasi, F., & Denis-Donini, S. (1995). Synaptic activation of NF-kappa B by glutamate in cerebellar granule neurons in vitro. *Proceedings of the National Academy of Sciences of the United States of America, 92*, 9077–9081.

Haas, I. G., Frank, M., Véron, N., & Kemler, R. (2005). Presenilin-dependent processing and nuclear function of gamma-protocadherins. *Journal of Biological Chemistry, 280*, 9313–9319.

Hambsch, B., Grinevich, V., Seeburg, P. H., & Schwarz, M. K. (2005). Gamma-Protocadherins, presenilin-mediated release of C-terminal fragment promotes locus expression. *Journal of Biological Chemistry, 280*, 15888–15897.

Hammond, J. W., Cai, D., & Verhey, K. J. (2008). Tubulin modifications and their cellular functions. *Current Opinion in Cell Biology, 20*, 71–76.

Hanz, S., Perlson, E., Willis, D., Zheng, J. Q., Massarwa, R., Huerta, J. J., Koltzenburg, M., Kohler, M., van-Minnen, J., Twiss, J. L., & Fainzilber, M. (2003). Axoplasmic importins enable retrograde injury signaling in lesioned nerve. *Neuron, 40*, 1095–1104.

Hardingham, G. E., & Bading, H. (2002). Coupling of extrasynaptic NMDA receptors to a CREB shut-off pathway is developmentally regulated. *Biochimica et Biophysica Acta, 1600*, 148–153.

Hardingham, G. E., & Bading, H. (2010). Synaptic versus extrasynaptic NMDA receptor signalling: Implications for neurodegenerative disorders. *Nature Reviews Neuroscience, 11*, 682–696.

Hardingham, G. E., Arnold, F. J., & Bading, H. (2001a). Nuclear calcium signaling controls CREB-mediated gene expression triggered by synaptic activity. *Nature Neuroscience, 4*, 261–267.

Hardingham, G. E., Arnold, F. J., & Bading, H. (2001b). A calcium microdomain near NMDA receptors: On switch for ERK-dependent synapse-to-nucleus communication. *Nature Neuroscience, 4*, 565–566.

Hardingham, G. E., Fukunaga, Y., & Bading, H. (2002). Extrasynaptic NMDARs oppose synaptic NMDARs by triggering CREB shut-off and cell death pathways. *Nature Neuroscience, 5*, 405–414.

Heerssen, H. M., Pazyra, M. F., & Segal, R. A. (2004). Dynein motors transport activated Trks to promote survival of target-dependent neurons. *Nature Neuroscience, 7*, 596–604.

Hosokawa, K., Nishi, M., Sakamoto, H., Tanaka, Y., & Kawata, M. (2008). Regional distribution of importin subtype mRNA expression in the nervous system: Study of early postnatal and adult mouse. *Neuroscience, 157*, 864–877.

Howe, C. L., & Mobley, W. C. (2005). Long-distance retrograde neurotrophic signaling. *Current Opinion in Neurobiology, 15*, 40–48.

Howe, C. L., Valletta, J. S., Rusnak, A. S., & Mobley, W. C. (2001). NGF signaling from clathrin-coated vesicles: Evidence that signaling endosomes serve as a platform for the Ras-MAPK pathway. *Neuron, 32*, 801–814.

Hsueh, Y. P., Wang, T. F., Yang, F. C., & Sheng, M. (2000). Nuclear translocation and transcription regulation by the membrane-associated guanylate kinase CASK/LIN-2. *Nature, 404*, 298–302.

Hu, J., Wang, F., Yuan, Y., Zhu, X., Wang, Y., Zhang, Y., Kou, Z., Wang, S., & Gao, S. (2010). Novel importin-alpha family member Kpna7 is required for normal fertility and fecundity in the mouse. *Journal of Biological Chemistry, 285*, 33113–33122.

Huber, O., Korn, R., McLaughlin, J., Ohsugi, M., Herrmann, B. G., & Kemler, R. (1996). Nuclear localization of beta-catenin by interaction with transcription factor LEF-1. *Mechanisms of Development, 59*, 3–10.

Impey, S., Obrietan, K., Wong, S. T., Poser, S., Yano, S., Wayman, G., Deloulme, J. C., Chan, G., & Storm, D. R. (1998). Cross talk between ERK and PKA is required for Ca^{2+} stimulation of CREB-dependent transcription and ERK nuclear translocation. *Neuron, 21*, 869–883.

Ito, H., Morishita, R., Shinoda, T., Iwamoto, I., Sudo, K., Okamoto, K., & Nagata, K. (2010). Dysbindin-1, WAVE2 and Abi-1 form a complex that regulates dendritic spine formation. *Molecular Psychiatry, 15*, 976–986.

Jacob, A. L., Jordan, B. A., & Weinberg, R. J. (2010). Organization of amyloid-beta protein precursor intracellular domain-associated protein-1 in the rat brain. *The Journal of Comparative Neurology, 518*, 3221–3236.

Jans, D. A., Xiao, C. Y., & Lam, M. H. (2000). Nuclear targeting signal recognition: A key control point in nuclear transport? *Bioessays, 22*, 532–544.

Jeffrey, R. A., Ch'ng, T. H., O'Dell, T. J., & Martin, K. C. (2009). Activity-dependent anchoring of importin alpha at the synapse involves regulated binding to the cytoplasmic tail of the NR1-1a subunit of the NMDA receptor. *Journal of Neuroscience, 29*, 15613–15620.

Jivan, A., Ranganathan, A., & Cobb, M. H. (2010). Reconstitution of the nuclear transport of the MAP kinase ERK2. *Methods in Molecular Biology, 661*, 273–285.

Jordan, B. A., & Kreutz, M. R. (2009). Nucleocytoplasmic protein shuttling: The direct route in synapse-to-nucleus signaling. *Trends in Neurosciences, 32*, 392–401.

Jordan, B. A., Fernholz, B. D., Boussac, M., Xu, C., Grigorean, G., Ziff, E. B., & Neubert, T. A. (2004). Identification and verification of novel rodent postsynaptic density proteins. *Molecular & Cellular Proteomics, 3*, 857–871.

Jordan, B. A., Fernholz, B. D., Khatri, L., & Ziff, E. B. (2007). Activity-dependent AIDA-1 nuclear signaling regulates nucleolar numbers and protein synthesis in neurons. *Nature Neuroscience, 10*, 427–435.

Kaltschmidt, B., & Kaltschmidt, C. (2009). NF-kappaB in the nervous system. *Cold Spring Harbor Perspectives in Biology, 1*, a001271.

Kaltschmidt, C., Kaltschmidt, B., & Baeuerle, P. A. (1993). Brain synapses contain inducible forms of the transcription factor NF-kappa B. *Mechanisms of Development, 43*, 135–147.

Kaltschmidt, C., Kaltschmidt, B., Neumann, H., Wekerle, H., & Baeuerle, P. A. (1994). Constitutive NF-kappa B activity in neurons. *Molecular and Cellular Biology, 14*, 3981–3992.

Kaltschmidt, C., Kaltschmidt, B., & Baeuerle, P. A. (1995). Stimulation of ionotropic glutamate receptors activates transcription factor NF-kappa B in primary neurons. *Proceedings of the National Academy of Sciences of the United States of America, 92*, 9618–9622.

Kamei, Y., Yuba, S., Nakayama, T., & Yoneda, Y. (1999). Three distinct classes of the alpha-subunit of the nuclear pore-targeting complex (importin-alpha) are differentially expressed in adult mouse tissues. *Journal of Histochemistry and Cytochemistry, 47*, 363–372.

Kapur, A., Yeckel, M., & Johnston, D. (2001). Hippocampal mossy fiber activity evokes Ca^{2+} release in CA3 pyramidal neurons via a metabotropic glutamate receptor pathway. *Neuroscience, 107*, 59–69.

Kaul, S. C., Kawai, R., Nomura, H., Mitsui, Y., Reddel, R. R., & Wadhwa, R. (1999). Identification of a 55-kDa ezrin-related protein that induces cytoskeletal changes and localizes to the nucleolus. *Experimental Cell Research, 250*, 51–61.

Kavanagh, E., Buchert, M., Tsapara, A., Choquet, A., Balda, M. S., Hollande, F., & Matter, K. (2006). Functional interaction between the ZO-1-interacting transcription factor ZONAB/DbpA and the RNA processing factor symplekin. *Journal of Cell Science, 119*, 5098–5105.

Kelly, K. F., Spring, C. M., Otchere, A. A., & Daniel, J. M. (2004). NLS-dependent nuclear localization of p120ctn is necessary to relieve Kaiso-mediated transcriptional repression. *Journal of Cell Science, 117*, 2675–2686.

Kim, M. J., Dunah, A. W., Wang, Y. T., & Sheng, M. (2005). Differential roles of NR2A- and NR2B-containing NMDA receptors in Ras-ERK signaling and AMPA receptor trafficking. *Neuron, 46*, 745–760.

Kindler, S., Dieterich, D. C., Schütt, J., Sahin, J., Karpova, A., Mikhaylova, M., Schob, C., Gundelfinger, E. D., Kreienkamp, H. J., & Kreutz, M. R. (2009). Dendritic mRNA targeting of Jacob and N-methyl-d-aspartate-induced nuclear translocation pool after calpain-mediated proteolysis. *Journal of Biological Chemistry, 284*, 25431–25440.

Kobe, B. (1999). Autoinhibition by an internal nuclear localization signal revealed by the crystal structure of mammalian importin alpha. *Nature Structural Biology, 6*, 388–397.

Kohu, K., Ogawa, F., & Akiyama, T. (2002). The SH3, HOOK and guanylate kinase-like domains of hDLG are important for its cytoplasmic localization. *Genes Cell, 7*, 707–715.

Lai, K. O., Zhao, Y., Ch'ng, T. H., & Martin, K. C. (2008). Importin-mediated retrograde transport of CREB2 from distal processes to the nucleus in neurons. *Proceedings of the National Academy of Sciences of the United States of America, 105*, 17175–17180.

Lallena, M. J., Martínez, C., Valcárcel, J., & Correas, I. (1998). Functional association of nuclear protein 4.1 with pre-mRNA splicing factors. *Journal of Cell Science, 111*, 1963–1971.

Lee, S. H., Lim, C. S., Park, H., Lee, J. A., Han, J. H., Kim, H., Cheang, Y. H., Lee, S. H., Lee, Y. S., Ko, H. G., Jang, D. H., Kim, H., Miniaci, M. C., Bartsch, D., Kim, E., Bailey, C. H., Kandel, E. R., & Kaang, B. K. (2007). Nuclear translocation of CAM-associated protein activates transcription for long-term facilitation in Aplysia. *Cell, 129*, 801–812.

Lu, F. M., & Lux, S. E. (1996). Constitutively active human Notch1 binds to the transcription factor CBF1 and stimulates transcription through a promoter containing a CBF1-responsive element. *Proceedings of the National Academy of Sciences of the United States of America, 93*, 5663–5667.

Marambaud, P., Wen, P. H., Dutt, A., Shioi, J., Takashima, A., Siman, R., & Robakis, N. K. (2003). A CBP binding transcriptional repressor produced by the PS1/epsilon-cleavage of N-cadherin is inhibited by PS1 FAD mutations. *Cell, 114*, 635–645.

Marcora, E., & Kennedy, M. B. (2010). The Huntington's disease mutation impairs Huntingtin's role in the transport of NF-κB from the synapse to the nucleus. *Human Molecular Genetics, 19*, 4373–4384.

Mathew, D., Ataman, B., Chen, J., Zhang, Y., Cumberledge, S., & Budnik, V. (2005). Wingless signaling at synapses is through cleavage and nuclear import of receptor DFrizzled2. *Science, 310*, 1344–1347.

Meberg, P. J., Kinney, W. R., Valcourt, E. G., & Routtenberg, A. (1996). Gene expression of the transcription factor NF-kappa B in hippocampus: Regulation by synaptic activity. *Brain Research. Molecular Brain Research, 38*, 179–190.

Meffert, M. K., Chang, J. M., Wiltgen, B. J., Fanselow, M. S., & Baltimore, D. (2003). NF-kappa B functions in synaptic signaling and behavior. *Nature Neuroscience, 6*, 1072–1078.

Miaczynska, M., Pelkmans, L., & Zerial, M. (2004). Not just a sink: Endosomes in control of signal transduction. *Current Opinion in Cell Biology, 16*, 400–406.

Mikenberg, I., Widera, D., Kaus, A., Kaltschmidt, B., & Kaltschmidt, C. (2007). Transcription factor NF-kappaB is transported to the nucleus via cytoplasmic dynein/dynactin motor complex in hippocampal neurons. *PLoS One, 2*, e589.

Milnerwood, A. J., Gladding, C. M., Pouladi, M. A., Kaufman, A. M., Hines, R. M., Boyd, J. D., Ko, R. W., Vasuta, O. C., Graham, R. K., Hayden, M. R., Murphy, T. H., & Raymond, L. A. (2010). Early increase in extrasynaptic NMDA receptor signaling and expression contributes to phenotype onset in Huntington's disease mice. *Neuron, 65*, 178–190.

Molenaar, M., van de Wetering, M., Oosterwegel, M., Peterson-Maduro, J., Godsave, S., Korinek, V., Roose, J., Destrée, O., & Clevers, H. (1996). XTcf-3 transcription factor mediates beta-catenin-induced axis formation in Xenopus embryos. *Cell, 86*, 391–399.

Morris, R. G., & Frey, U. (1997). Hippocampal synaptic plasticity: Role in spatial learning or the automatic recording of attended experience? *Philosophical Transactions of the Royal Society of London. Series B, Biological Sciences, 352*, 1489–1503.

Nadler, S. G., Tritschler, D., Haffar, O. K., Blake, J., Bruce, A. G., & Cleaveland, J. S. (1997). Differential expression and sequence-specific interaction of karyopherin alpha with nuclear localization sequences. *Journal of Biological Chemistry, 272*, 4310–4315.

Nakamura, T., Lasser-Ross, N., Nakamura, K., & Ross, W. N. (2002). Spatial segregation and interaction of calcium signalling mechanisms in rat hippocampal CA1 pyramidal neurons. *The Journal of Physiology, 543*, 465–480.

Nakata, A., Ito, T., Nagata, M., Hori, S., & Sekimizu, K. (2004). GRIP1tau, a novel PDZ domain-containing transcriptional activator, cooperates with the testis-specific transcription elongation factor SII-T1. *Genes to Cells, 9*, 1125–1135.

Nguyen, P. V., Abel, T., & Kandel, E. R. (1994). Requirement of a critical period of transcription for induction of late phase LTP. *Science, 265*, 1104–1107.

Otis, K. O., Thompson, K. R., & Martin, K. C. (2006). Importin-mediated nuclear transport in neurons. *Current Opinion in Neurobiology, 16*, 329–335.

Paine, P. L., Moore, L. C., & Horowitz, S. B. (1975). Nuclear envelope permeability. *Nature, 254*, 109–114.

Perlson, E., Hanz, S., Ben-Yaakov, K., Segal-Ruder, Y., Seger, R., & Fainzilber, M. (2005). Vimentin-dependent spatial translocation of an activated MAP kinase in injured nerve. *Neuron, 45*, 715–726.

Perlson, E., Michaelevski, I., Kowalsman, N., Ben-Yaakov, K., Shaked, M., Seger, R., Eisenstein, M., & Fainzilber, M. (2006). Vimentin binding to phosphorylated Erk sterically hinders enzymatic dephosphorylation of the kinase. *Journal of Molecular Biology, 364*, 938–944.

Proepper, C., Johannsen, S., Liebau, S., Dahl, J., Vaida, B., Bockmann, J., Kreutz, M. R., Gundelfinger, E. D., & Boeckers, T. M. (2007). Abelson interacting protein 1 (Abi-1) is essential for dendrite morphogenesis and synapse formation. *EMBO Journal, 26*, 1397–1409.

Reymann, K. G., & Frey, J. U. (2007). The late maintenance of hippocampal LTP: Requirements, phases, 'synaptic tagging', 'late-associativity' and implications. *Neuropharmacology, 52*, 24–40.

Rodova, M., Kelly, K. F., VanSaun, M., Daniel, J. M., & Werle, M. J. (2004). Regulation of the rapsyn promoter by kaiso and delta-catenin. *Molecular and Cellular Biology, 24*, 7188–7196.

Rönicke, R., Mikhaylova, M., Rönicke, S., Meinhardt, J., Schröder, U. H., Fändrich, M., Reiser, G., Kreutz, M. R., & Reymann, K. G. (2011). Early neuronal dysfunction by amyloid beta oligomers depends on activation of NR2B-containing NMDA receptors. *Neurobiology of Aging 32*, 2219–2228.

Ross, W. N., Nakamura, T., Watanabe, S., Larkum, M., & Lasser-Ross, N. (2005). Synaptically activated Ca^{2+} release from internal stores in CNS neurons. *Cellular and Molecular Neurobiology, 25*, 283–295.

Roth, D. M., Moseley, G. W., Glover, D., Pouton, C. W., & Jans, D. A. (2007). A microtubule-facilitated nuclear import pathway for cancer regulatory proteins. *Traffic, 8*, 673–686.

Saha, R. N., & Dudek, S. M. (2008). Action potentials: To the nucleus and beyond. *Experimental Biology and Medicine (Maywood, N.J.), 233*, 385–393.

Sala, C., Rudolph-Correia, S., & Sheng, M. (2000). Developmentally regulated NMDA receptor-dependent dephosphorylation of cAMP response element-binding protein (CREB) in hippocampal neurons. *Journal of Neuroscience, 20*, 3529–3536.

Sardi, S. P., Murtie, J., Koirala, S., Patten, B. A., & Corfas, G. (2006). Presenilin-dependent ErbB4 nuclear signaling regulates the timing of astrogenesis in the developing brain. *Cell, 127*, 185–197.

Schmeisser, M. J., Grabrucker, A. M., Bockmann, J., & Boeckers, T. M. (2009). Synaptic crosstalk between N-methyl-D-aspartate receptors and LAPSER1-beta-catenin at excitatory synapses. *Journal of Biological Chemistry, 284*, 29146–29157.

Schmidt-Ullrich, R., Mémet, S., Lilienbaum, A., Feuillard, J., Raphaël, M., & Israel, A. (1996). NF-kappaB activity in transgenic mice: Developmental regulation and tissue specificity. *Development (Cambridge, England), 122*, 2117–2128.

Shen, L., Liang, F., Walensky, L. D., & Huganir, R. L. (2000). Regulation of AMPA receptor GluR1 subunit surface expression by a 4. 1 N-linked actin cytoskeletal association. *Journal of Neuroscience, 20*, 7932–7940.

Shmidt, T., Hampich, F., Ridders, M., Schultrich, S., Hans, V. H., Tenner, K., Vilianovich, L., Qadri, F., Alenina, N., Hartmann, E., Köhler, M., & Bader, M. (2007). Normal brain development in importin-alpha5 deficient-mice. *Nature Cell Biology, 9*, 1337–1338.

Shrum, C. K., Defrancisco, D., & Meffert, M. K. (2009). Stimulated nuclear translocation of NF-kappaB and shuttling differentially depend on dynein and the dynactin complex. *Proceedings of the National Academy of Sciences of the United States of America, 106*, 2647–2652.

Simcha, I., Shtutman, M., Salomon, D., Zhurinsky, J., Sadot, E., Geiger, B., & Ben-Ze'ev, A. (1998). Differential nuclear translocation and transactivation potential of beta-catenin and plakoglobin. *The Journal of Cell Biology, 141*, 1433–1448.

Simpson, P. B., Challiss, R. A., & Nahorski, S. R. (1995). Neuronal Ca^{2+} stores: Activation and function. *Trends in Neurosciences, 18*, 299–306.

Sweatt, J. C. (2004). Mitogen-activated protein kinases in synaptic plasticity and memory. *Current Opinion in Neurobiology, 14*, 311–317.

Takano, H., & Gusella, J. F. (2002). The predominantly HEAT-like motif structure of huntingtin and its association and coincident nuclear entry with dorsal, an NF-kB/Rel/dorsal family transcription factor. *BMC Neuroscience, 3*, 15.

Thompson, K. R., Otis, K. O., Chen, D. Y., Zhao, Y., O'Dell, T. J., & Martin, K. C. (2004). Synapse to nucleus signaling during long-term synaptic plasticity: A role for the classical active nuclear import pathway. *Neuron, 44*, 997–1009.

Tu, W., Xu, X., Peng, L., Zhong, X., Zhang, W., Soundarapandian, M. M., Balel, C., Wang, M., Jia, N., Zhang, W., Lew, F., Chan, S. L., Chen, Y., & Lu, Y. (2010). DAPK1 interaction with NMDA receptor NR2B subunits mediates brain damage in stroke. *Cell, 140*, 222–234.

Watson, F. L., Heerssen, H. M., Moheban, D. B., Lin, M. Z., Sauvageot, C. M., Bhattacharyya, A., Pomeroy, S. L., & Segal, R. A. (1999). Rapid nuclear responses to target-derived neurotrophins require retrograde transport of ligand-receptor complex. *Journal of Neuroscience, 19*, 7889–7900.

Wellmann, H., Kaltschmidt, B., & Kaltschmidt, C. (2001). Retrograde transport of transcription factor NF-kappa B in living neurons. *Journal of Biological Chemistry, 276*, 11821–11829.

Wiegert, J. S., & Bading, H. (2011). Activity-dependent calcium signaling and ERK-MAP kinases in neurons: A link to structural plasticity of the nucleus and gene transcription regulation. *Cell Calcium 49*(5), 296–305.

Wiegert, J. S., Bengtson, C. P., & Bading, H. (2007). Diffusion and not active transport underlies and limits ERK1/2 synapse-to-nucleus signaling in hippocampal neurons. *Journal of Biological Chemistry, 282*, 29621–29633.

Wu, C., Cui, B., He, L., Chen, L., & Mobley, W. C. (2009). The coming of age of axonal neurotrophin signaling endosomes. *Journal of Proteomics, 72*, 46–55.

Xu, H., & Hebert, M. D. (2005). A novel EB-1/AIDA-1 isoform, AIDA-1c, interacts with the Cajal body protein coilin. *BMC Cell Biology, 6*, 23.

Yano, H., Lee, F. S., Kong, H., Chuang, J., Arevalo, J., Perez, P., Sung, C., & Chao, M. V. (2001). Association of Trk neurotrophin receptors with components of the cytoplasmic dynein motor. *Journal of Neuroscience, 21*, 125.

Yasuhara, N., Shibazaki, N., Tanaka, S., Nagai, M., Kamikawa, Y., Oe, S., Asally, M., Kamachi, Y., Kondoh, H., & Yoneda, Y. (2007). Triggering neural differentiation of ES cells by subtype switching of importin-alpha. *Nature Cell Biology, 9*, 72–79.

Yasuhara, N., Oka, M., & Yoneda, Y. (2009). The role of the nuclear transport system in cell differentiation. *Seminars in Cell & Developmental Biology, 20*, 590–599.

Ye, H., Kuruvilla, R., Zweifel, L. S., & Ginty, D. D. (2003). Evidence in support of signaling endosome-based retrograde survival of sympathetic neurons. *Neuron, 39*, 57–68.

Yokoya, F., Imamoto, N., Tachibana, T., & Yoneda, Y. (1999). beta-catenin can be transported into the nucleus in a Ran-unassisted manner. *Molecular Biology of the Cell, 10*, 1119–1131.

Yoneda, Y. (2000). Nucleocytoplasmic protein traffic and its significance to cell function. *Genes to Cells, 5*, 777–787.

Yu, G., Zerucha, T., Ekker, M., & Rubenstein, J. L. (2001). Evidence that GRIP, a PDZ-domain protein which is expressed in the embryonic forebrain, co-activates transcription with DLX homeodomain proteins. *Brain Research. Developmental Brain Research, 130*, 217–230.

Chapter 17
Nuclear Calcium Signaling

C. Peter Bengtson and Hilmar Bading

Abstract Calcium is the major intracellular messenger linking synaptic activity in neurons to gene expression to control diverse functions including adaptive responses to synaptic activity as well as survival and death (Bading et al. 1993; Hardingham et al. 1997; Chawla and Bading 2001; West et al. 2001; Zhang et al. 2007; Flavell and Greenberg 2008; Mellstrom et al. 2008; Redmond 2008; Wayman et al. 2008; Bootman et al. 2009; Zhang et al. 2009; Hardingham and Bading 2010). Calcium entry at the synapse acts locally to activate signaling cascades which regulate post-translational modifications essential for synaptic plasticity, such as the insertion of alpha-amino-3-hydroxy-5-methyl-4-isoxazolepropionic acid receptors (AMPARs) into the postsynaptic membrane (Soderling 2000; Malinow and Malenka 2002; Ehrlich and Malinow 2004). Synaptic activity can also evoke calcium signals in the nucleus which regulate gene pools largely through the phosphorylation of cAMP response element-binding protein (CREB) and its coactivator, CREB-binding protein (CBP) (Bading et al. 1993; Hardingham et al. 1997; Hardingham et al. 1999; Hu et al. 1999; Hardingham et al. 2001b; Impey et al. 2002; Zhang et al. 2009). Distinct mechanisms have been proposed to mediate synaptically generated calcium signals in subcompartments of pyramidal neurons; N-methyl-D-aspartate receptors (NMDARs) and ryanodine receptors have been implicated in the spine, inositol 3,4,5 triphosphate (IP3) receptors in the dendrites, and L-type voltage-gated calcium channels (VGCCs) at the soma and nucleus, although both NMDARs and IP3 receptors can also contribute to somatic and nuclear calcium signals under certain stimulation conditions (Nakamura et al. 1999; Bardo et al. 2006; Raymond and Redman 2006; Watanabe et al. 2006; Hong and Ross 2007; Hagenston et al. 2008; Bengtson et al. 2010). We review here the calcium signaling pathways underlying synaptically activated gene transcription leading to long-lasting changes in synaptic

C.P. Bengtson (✉) • H. Bading
Department of Neurobiology, Interdisciplinary Centre for Neurosciences (IZN), University of Heidelberg, INF 364, 69120 Heidelberg, Germany
e-mail: Bengtson@nbio.uni-heidelberg.de; Hilmar.Bading@uni-hd.de

M.R. Kreutz and C. Sala (eds.), *Synaptic Plasticity*,
Advances in Experimental Medicine and Biology 970,
DOI 10.1007/978-3-7091-0932-8_17, © Springer-Verlag/Wien 2012

efficacy and memory as well as the physiological mechanisms by which synaptic activity evokes nuclear calcium signals.

Keywords Activity-dependent gene expression • Nuclear calcium transients • Late phase long term depression • Late phase long term potentiation

17.1 Synaptic Activity Induces Nuclear Calcium Transients

Trains of synaptic stimulation evoke interacting and compartmentally specific calcium signals in neurons whose regulation of second messenger cascades and transcriptional responses is determinant for the induction and maintenance of synaptic plasticity. The generation of a nuclear calcium signal in response to synaptic activity determines the transcriptional output critical for late-phase plasticity and survival (see Sects. 17.4, 17.5). The nature of synaptic stimulation necessary for transcription-dependent plasticity is best understood in terms of late-phase long-term potentiation (L-LTP) (Huang 1998); however, the relationship between synaptic input and nuclear calcium output in the context of L-LTP is less understood.

Synaptic activity even at synapses over 150 μm from the soma can induce a somatic calcium signal under the right conditions. Presynaptic activity can evoke postsynaptic calcium signals which reach the soma and nucleus when a train of stimuli rather than a few stimuli are given (Fig. 17.1). Bursts of excitatory synaptic input sufficient to induce a burst of action potentials will evoke measurable calcium signals throughout the neuron including the soma and nucleus largely due to the activation of VGCCs (Miyakawa et al. 1992; Regehr and Tank 1992; Bengtson et al. 2010). Repetition of such stimulation trains is needed to induce L-LTP, and such repetition evokes much larger nuclear calcium signals and more postsynaptic action potentials (Fig. 17.2) (Johenning and Holthoff 2007; Bengtson et al. 2010). The mechanism of this parallel increase in spike numbers and nuclear calcium responses likely involves posttetanic potentiation which persists during the interburst interval (typically 30 s to 10 min for L-LTP induction). This makes nuclear calcium signals a reporter for recent bursts of synaptic input sufficient to activate posttetanic potentiation. The increase in nuclear calcium signals caused by repetition of stimulation trains may be critical to boost calcium signals to levels sufficient to activate transcriptional responses necessary for L-LTP (see Sect. 17.4). The physiological mechanisms and calcium sources which generate nuclear calcium signals are discussed in Sect. 17.3.

Several differences between nuclear and cytoplasmic calcium signals have been noted in nonneuronal cell types, whereas neurons show qualitatively similar somatic and nuclear signals apart from a slower rise and decay time of the nuclear signal (Bootman et al. 2009). Calcium can freely diffuse from the cytoplasm into the nucleus through the nuclear pore complex (NPC) in several cell types including neurons (Allbritton et al. 1994; Brini et al. 1994; Eder and Bading 2007). Although NPCs allow calcium diffusion to the nucleus, their limited number imposes a small delay on calcium entry responsible for the slower rise time in the nucleus than in the cytoplasm. The slower decay time of calcium signals in the nucleoplasm than in the

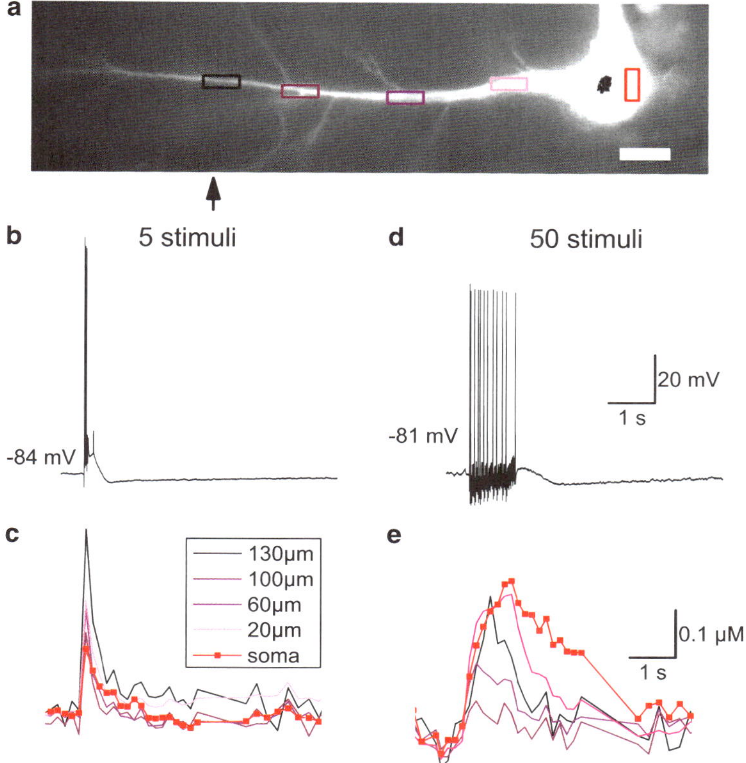

Fig. 17.1 Recordings from a CA1 hippocampal pyramidal cell filled with bis-fura2 (**a**) showing membrane potential (**b, d**) and changes in calcium concentrations (**c, e**) at the regions shown in (**a**). Responses are shown to five stimuli (**b, c**) and 50 stimuli (**d, e**) of high-frequency stimulation (100 Hz) from a field stimulator (glass theta pipette) placed in the stratum radiatum 130 µm from the soma (see *arrow* in **a**)

cytoplasm may be due to the absence of plasma membrane exchangers which act to extrude calcium from the cytoplasm and a lack of reuptake via sarcoplasmic and endoplasmic reticulum calcium ATPases (SERCA) pumps which may be absent on the inner nuclear membrane (Bootman et al. 2009).

Synaptic activity can also affect nuclear geometry in such a way to facilitate nuclear calcium signals. The nuclear envelope is a bilayer which can form infoldings which deeply invade some nuclei in dissociated and organotypic hippocampal cultures (Queisser et al. 2008; Wittmann et al. 2009). Moreover, the degree of infoldings and the percentage of nuclei showing infoldings increase dramatically over 60 min of synchronous network bursting activity. Infolded nuclei show a higher surface to volume ratio and more NPCs which are also present in the infoldings. Thus, infoldings improve the transfer of cytoplasmic calcium signals to deep nuclear regions, resulting in faster kinetics and higher amplitudes of nuclear calcium signals. The presence of infoldings in nuclei correlates with increased histone-3 phosphorylation, a marker of chromatin remodeling associated with the induction of transcription.

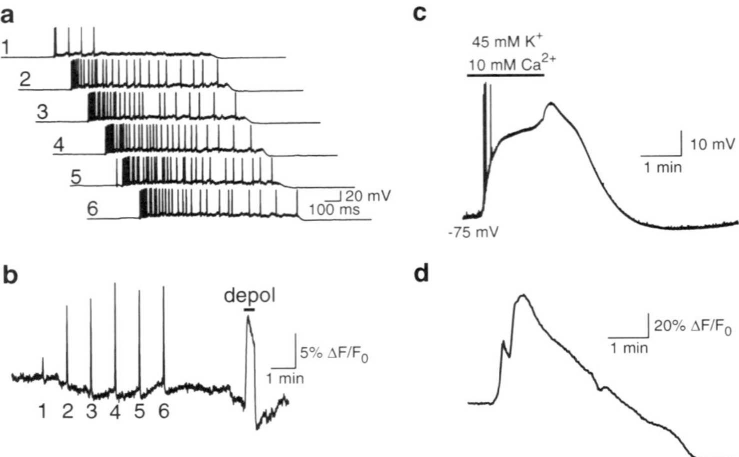

Fig. 17.2 Recordings from a CA1 hippocampal pyramidal cell showing membrane potential (**a, c**) and nuclear calcium signals measured with GCaMP2-NLS (**b, d**). (**a, b**) Responses to six repetitions of high-frequency stimulation (15 μA, 100 Hz, 1 s) of the Schaffer collaterals with a single-barrel glass pipette placed in the stratum radiatum 125 μm from the soma. Also shown is a calcium response to current injection through the patch pipette to depolarize (depol) the cell to approximately 0 mV (see *bar*). (**c, d**) Response to depolarization of the whole slice with elevated K^+ and Ca^{2+}

17.2 Measuring Nuclear Calcium Signals

Nuclear calcium signals can be directly visualized in living neurons; however, progress in this field has been impeded by the technical difficulty of unambiguously measuring calcium in the nucleus. The nuclear boundaries can be roughly distinguished using small molecule calcium indicator dyes such as fura2 or fluo3 which accumulate within intracellular compartments such as the nucleus. Optical sectioning (two-photon or confocal microscopy) is necessary to exclude cytoplasmic calcium signals above and below the nucleus in the z-axis.

Recombinant calcium indicators can be targeted to specific cell types and intracellular compartments using the appropriate promoter and localization sequences. Nuclear-targeted indicators can unambiguously measure nuclear signals without cytoplasmic calcium buffering or optical sectioning. A nuclear localization signal (NLS), which binds to transporter proteins at the NPC, has been used to target small molecule or recombinant calcium indicators to the nucleus in nonneuronal cell lines (Brini et al. 1993; Allbritton et al. 1994; Miyawaki et al. 1997). We have recently adapted nuclear-targeted indicators to neurons by expression of an NLS fused to the recombinant calcium indicator based on calmodulin fused to a circularly permutated green fluorescent protein (cpGFP), GCaMP2 (Bengtson et al. 2010). We used GCaMP2-NLS to study nuclear calcium responses to L-LTP induction protocols in brain slices (Fig. 17.2) and are currently applying similar technology to in vivo measurements of nuclear calcium signaling during olfactory learning paradigms (Weislogel et al., unpublished work).

Despite the existence of tools to visualize nuclear calcium signals, quantitative measurements of nuclear calcium concentration especially with simultaneous cytoplasmic measurements remain controversial. This is largely due to difficulties calibrating calcium indicators in the nucleus. Calibrations of cytoplasmic calcium indicators cannot be applied to nuclear indicators due to differences between the cytoplasm and the nucleoplasm in the capacity and speed of endogenous calcium buffers or other calcium-binding partners as well as differences in viscosity and indicator concentration (Bootman et al. 2009). The bleaching characteristics of recombinant indicators pose further challenges for their calibration such as the reversible bleaching (photoisomerization) of cpGFPs and distinct bleaching rates of fluorescent protein pairs used for fluorescent resonance energy transfer (FRET) (Pologruto et al. 2004; Zal and Gascoigne 2004).

17.3 The Cellular Mechanisms Mediating Nuclear Calcium Signals in Response to Synaptic Activity

The mechanism by which synaptically activated calcium signals reach the nucleus is only partly understood. Elevations in cytosolic calcium levels are counteracted by calcium uptake into mitochondria (mitochondrial uniporter or calcium channel) and the SERCA as well as extrusion through the plasmalemma by ion exchangers (Na^+/Ca^{2+} exchanger (NCX); $Na^+/Ca^{2+}/K^+$ exchanger (NCKX)) and pumps (plasma membrane calcium ATPase [PMCA]) which maintain cytosolic calcium at low concentrations in a homeostatic fashion. Cytosolic calcium signals are buffered by calcium-binding proteins which rapidly lower the concentration of free calcium and slow its diffusion. This effectively limits to a few micrometers the reach of calcium signals traveling from the synapse by diffusion alone (Neher 1986; Allbritton et al. 1992; Faas et al. 2011). However, other mechanisms including membrane depolarization and release from internal stores amplify and help propagate synaptically activated calcium signals along dendrites into the soma. In hippocampal CA1 pyramidal neurons, these mechanisms involve complex interactions between dendritic geometry, IP_3 receptors, backpropagating action potentials, NMDA receptors, and VGCCs (see below) (Nakazawa and Murphy 1999; Nakamura et al. 2002; Raymond and Redman 2006; Bengtson et al. 2010).

The prime player responsible for somatic calcium signals is the L-type VGCC. Blockade of L-type VGCCs blocks somatic and nuclear calcium responses and late-phase plasticity induced by repeated bursts of presynaptic activity (Raymond and Redman 2006; Johenning and Holthoff 2007; Bengtson et al. 2010). Excitatory synaptic input sufficient to induce postsynaptic action potentials at the soma will activate L-type VGCCs which, being selectively enriched at the base of the apical dendrite in CA1 pyramidal neurons (Westenbroek et al. 1990), are optimally localized for inducing a nuclear calcium signal.

Despite their distal location at the synapse, NMDA receptor activation in response to high-frequency stimulation of presynaptic input contributes to somatic depolarization and is necessary for the generation of bursts of somatic action potentials and nuclear calcium signals (Zhao et al. 2005; Bengtson et al. 2010). This positive feedback loop between action potential generation and NMDA receptor activation not only triggers local synaptic signaling cascades essential for AMPA receptor modification and insertion at the synapse but also underlies the nuclear calcium signal generated by excitatory synaptic input. When the arrival of backpropagating action potentials at the synapse coincides with the slow decay component of the EPSC, depolarization is sufficient to relieve the magnesium block of synaptic NMDA receptors, causing local calcium influx and spike-timing-dependent plasticity (Koester and Sakmann 1998; Kampa et al. 2004; Canepari et al. 2007). During burst activity, both NMDA receptors and VGCCs contribute to membrane depolarization and the generation of calcium spikes in distal dendrites, known to be important for the early phase of LTP (E-LTP) (Fig. 17.3) (Koester and Sakmann 1998; Takahashi and Magee 2009; Fuenzalida et al. 2010).

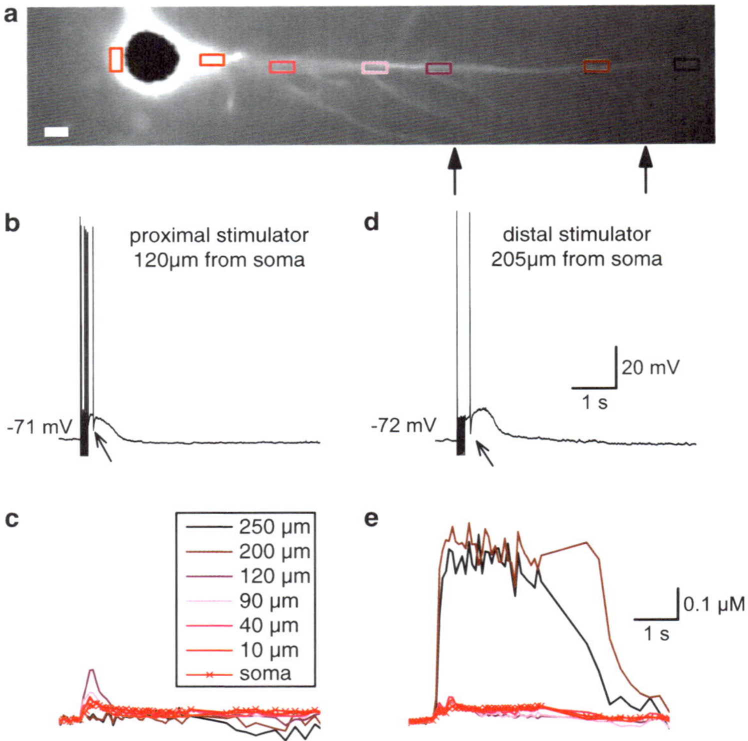

Fig. 17.3 Recordings from a CA1 hippocampal pyramidal cell filled with bis-fura2 (**a**) showing membrane potential (**b, d**) and change in calcium concentrations (**c, e**) at the regions shown in (**a**) in response to proximal (**b, c**) and distal (**d, e**) stimulation with a field stimulator (glass theta pipette, positions indicated by *arrows* in **a**) pulsed with high-frequency stimulation (100 Hz, 0.1 s) followed by current injection to evoke a backpropagating action potential (*arrow*). A calcium spike occurs in distal compartments when the stimulating electrode was placed distally

Bursts of synaptic activity can also activate a regenerative phenomenon involving IP$_3$ receptor-dependent release of calcium from internal stores. This can lead to a calcium wave which initiates in the proximal apical dendrite and propagates toward the soma and presumably the nucleus of pyramidal neurons in the CA1 and medial prefrontal cortex (Berridge 1998; Nakamura et al. 1999, 2000; Kapur et al. 2001; Power and Sah 2002; Larkum et al. 2003; Watanabe et al. 2006; Hagenston et al. 2008). While much of this work employs conditions, which pharmacologically enhance or isolate IP$_3$ signaling, synaptically evoked calcium waves can also occur in the absence of pharmacological manipulation (Nakamura et al. 1999; Watanabe et al. 2006; Hong and Ross 2007; Hagenston et al. 2008). However, IP$_3$ receptor function and release from internal stores do not contribute to nuclear calcium signals evoked by L-LTP induction protocols with repeated trains of TBS or HFS (Raymond and Redman 2006; Johenning and Holthoff 2007; Bengtson et al. 2010). This discrepancy may be due to differences in stimulation intensity used in these studies or penetration, stability, or specificity of IP$_3$ receptor antagonists or the incomplete emptying of intracellular stores with SERCA pump blockers. More effective techniques in this field may resolve such issues.

Evidence of IP$_3$ receptors mediating calcium release from the inner membrane of the nuclear envelope directly into the nucleus is mixed being shown with excised patch clamp recordings from cerebellar Purkinje neurons (Marchenko et al. 2005) but not with uncaging of IP$_3$ in rat basophilic leukemia cell lines (Allbritton et al. 1994) or immature hamster eggs (Shirakawa and Miyazaki 1996). If functional IP$_3$ receptors do exist on the inner nuclear membrane, it is unlikely that the IP$_3$ necessary to activate them comes from plasmalemmal receptors coupled to phospholipase C production, such as metabotropic glutamate or acetylcholine receptors, since such receptors are localized at synapses and not at the soma. Despite its relatively fast diffusion constant (Allbritton et al. 1992), IP$_3$ has a limited spatial range of action in neurons as it is rapidly inactivated by IP$_3$ phosphatase. Functional metabotropic glutamate receptors may be present, however, on the inner nuclear membrane of several neuron types, and the machinery necessary to produce IP$_3$ is present in the nucleoplasm (O'Malley et al. 2003; Jong et al. 2005, 2007; Visnjic and Banfic 2007; Kumar et al. 2008; Ye and Ahn 2008). This raises the intriguing possibility that intranuclear glutamate could activate IP$_3$ production and calcium release directly within the nucleus.

17.4 Late-Phase Plasticity and Long-Term Memory Require Transcription

Several lines of evidence indicate that gene expression is a requisite for memory consolidation and contributes to late-phase plasticity. Inhibitors of RNA synthesis reduce both L-LTP and late-phase long-term depression (L-LTD) (Nguyen et al. 1994; Frey et al. 1996; Linden 1996; Ahn et al. 1999; Huang et al. 2000). L-LTP is only moderately reduced, however, after physically separating dendrites from their somata, suggesting that ongoing translation also makes a major contribution to L-LTP (Kang and Schuman 1996; Vickers et al. 2005; Villers et al. 2010). Although their use in vivo has been problematic due to toxicity (reviewed elsewhere: Hernandez and Abel 2008; Alberini 2009), transcription blockers have also been shown to disrupt long-term memory (LTM) but not short-term memory (STM) (Agranoff et al. 1967; Squire and Barondes 1970; Thut and Lindell 1974; Montarolo et al. 1986; Pedreira et al. 1996; Igaz et al. 2002).

Robust transcription is activated by the induction of LTP in vitro and during learning in vivo. Indeed, recent studies using gene chip analysis to screen gene expression have identified hundreds of genes activated or suppressed by synaptic activity in hippocampal cultures and LTM consolidation in fear conditioning in vivo (Levenson et al. 2004; Keeley et al. 2006; Zhang et al. 2007, 2009; Burger et al. 2008; Wieczorek et al. 2010). While the identity and role of the many genes regulated by learning paradigms represent an enormous challenge to neuroscientists in the coming years, this chapter restricts itself to the role of calcium, which is the central signaling molecule activating the second messenger cascades and transcription factors mediating transcriptional regulation in response to learning paradigms and synaptic activity (Fig. 17.4).

Translation is also essential for the expression of late-phase plasticity and LTM (Linden 1996; Huang 1998; Miller et al. 2002; Villers et al. 2010). Translational

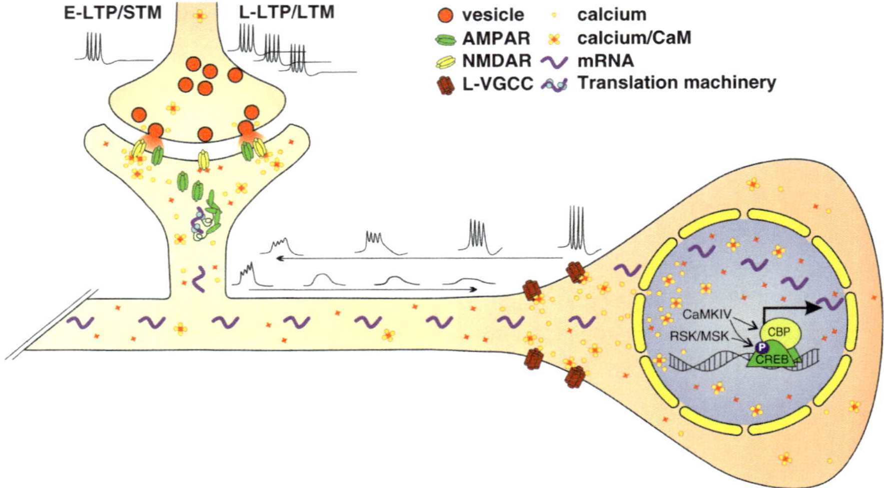

Fig. 17.4 Schematic neuron showing some key players in the switch from posttranslational modification and translation-dependent E-LTP/STM to translation- and transcription-dependent L-LTP/LTM. Postsynaptic potentials traveling toward the axon initial segment and action potentials backpropagating to the synapse facilitate L-type VGCC and NMDA receptor activation critical for nuclear calcium signals. A single train of presynaptic activity facilitates AMPA receptor insertion and phosphorylation. Repeated presynaptic trains activate sufficient nuclear calcium influx to activate transcription. Localization and density of all cellular components is not intended to accurately reflect reality but relates instead to their functional relevance to LTP (see text)

hotspots exist close to dendritic spines where RNA-anchoring proteins are localized which bind targeting elements within mRNA (Mayford et al. 1996; Mori et al. 2000; Aakalu et al. 2001). The capture of mRNA at such sites is believed to stabilize or prime synapses which have been tagged by previous activity (Frey and Morris 1997; Frey and Frey 2008; Redondo and Morris 2011). Thus, one role of transcription in L-LTP and LTM is the replenishment of dendritic mRNA required for ongoing translational activity which persists for hours after LTP induction to stabilize synaptic modifications. It is not surprising then that gene pools modulated by nuclear calcium and synaptic activity or by L-LTP induction or LTM paradigms include neurotrophic factors, cytoskeletal proteins, and transcription factors involved in survival, growth, and synaptogenesis (Thomas et al. 1994; Keeley et al. 2006; Havik et al. 2007; Zhang et al. 2007, 2009; Ploski et al. 2010).

17.5 Nuclear Calcium Signaling Is Required for Synaptic Activity–Induced Gene Expression

Experiments with calcium chelators have shown that intracellular calcium signals are essential for synaptic activity–induced LTP and nuclear calcium signals are necessary for the induction of cyclic AMP response element (CRE)–mediated transcription in response to L-type VGCC activation (Lynch et al. 1983; Hardingham et al. 1997). Evidence for the role of nuclear calcium signaling in synaptic activity–induced transcription, L-LTP, and LTM has come from experiments using a nuclear-targeted CaM-binding polypeptide 4 (CaMBP4) to prevent the activation of CaM kinases by calcium/CaM selectively in the nucleus. CaMBP4 is a nuclear localized protein composed of four copies of the M13 peptide which selectively binds and thus chelates calcium/CaM (Wang et al. 1995). CaMBP4 effectively blocks CRE-mediated gene expression induced by synaptic activity in hippocampal cultures (Papadia et al. 2005; Zhang et al. 2007, 2009), as well as reducing L-LTP in brain slices and impairing LTM but not STM in spatial learning, fear conditioning, and taste aversion tasks in mice (Limback-Stokin et al. 2004) and olfactory avoidance conditioning in adult *Drosophila melanogaster* (Weislogel et al., unpublished work).

The role of nuclear calcium as an activator of gene expression was first revealed by analysis of the promoter regions of the *c-fos gene* (Hardingham et al. 1997). Immediate early genes (IEGs) such as *c-fos*, *zif268*, *arc*, *bdnf*, and *homer* are activated to various extents by paradigms to induce neuronal activity such as depolarization, seizures, or developmental exposure to sensory stimulation as well as LTP induction protocols and various learning tasks (Flavell and Greenberg 2008; Alberini 2009). In line with the functional role of IEGs in plasticity and learning, the deletion of the *c-fos*, *zif268*, or *arc* genes in mice has been shown to cause deficits in hippocampal L-LTP and in consolidation of LTM in spatial and associative learning tasks (Jones et al. 2001; Fleischmann et al. 2003; Plath et al. 2006). Analysis of the promoter regions of the *c-fos* gene revealed that the CRE, which is also present in the promoter regions of many genes including *bdnf*, *zif268*, *somatostatin*, and *arc/arg3.1*, functions as a calcium-response element. The CREB family of transcription factors, which includes CREB, CREM, and ATF1, was revealed to bind to the CRE (Montminy et al. 1986; Montminy and Bilezikjian 1987; Gonzalez and Montminy 1989; Sheng and Greenberg 1990; Sheng et al. 1990, 1991). CREB forms homo- or heterodimers with other family members or splice variants containing the basic leucine zipper (bZIP) domain. This results in a large number of dimeric combinations of CREB isoforms which may show differences in their specificity for CREB target genes or differences in their activator or repressor activity (Mayr and Montminy 2001). One such bZIP-containing protein is the inducible cAMP early repressor (ICER), a product of the *cAMP response element modulator* (*CREM*) gene which is activity-induced and acts to repress CRE-mediated transcription by binding to CREB (Borlikova and Endo 2009). Further studies have shown that CRE-mediated gene expression occurs in L-LTP and L-LTD and that CREB overexpression enhances amygdala-dependent LTM while a CREB mutant which interferes with CREB binding to CRE impairs contextual fear conditioning and spatial memory (Impey et al. 1996; Ahn et al. 1999; Josselyn et al. 2001; Kida et al. 2002; Pittenger et al. 2002) as well as long-term facilitation in Aplysia and LTM in *Drosophila* (Yin et al. 1994; Bartsch et al. 1998). Conflicting results have come from memory testing in CREB$\alpha\delta$ knockout mice presumably due to differences in genetic background (C57Bl/6Jx129/SvEv exhibited deficits, but C57Bl/6xFVB/N mice did not) as well as the compensatory upregulation of CREBβ or CREM (see Olveira and Bading 2011).

The role of nuclear calcium in late-phase synaptic plasticity and LTM is associated with its ability to induce CRE-mediated transcription, which requires two events: the phosphorylation of CREB and the activation of CBP (Chawla et al. 1998). To activate transcription, CREB must become phosphorylated on its activator site Ser133 in the kinase-inducible domain (KID) which triggers recruitment of the transcriptional coactivator, CBP. In a second step that is required for transcription activation, CBP becomes activated by a nuclear calcium–/calmodulin-dependent protein kinase (CaMK) IV–mediated process that involves phosphorylation at Ser301 (Chawla et al. 1998; Hardingham et al. 1999; Hu et al. 1999; Mayr and Montminy 2001; Impey et al. 2002). CBP acts both as a platform for recruiting components of the transcription machinery and as a histone acetyltransferase to alter chromatin structure. Mutation or inhibition of CBP causes deficits in L-LTP and LTM consolidation in declarative and spatial memory but not fear conditioning (Alarcon et al. 2004; Korzus et al. 2004).

In order to initiate transcription in response to synaptic activity, both CREB and CBP need to be activated. While several calcium activated kinases can phosphorylate CREB at Ser133, nuclear calcium signaling through CaMKIV (see below) can activate both CREB and CBP, thus satisfying the requirements for initiating CRE-mediated transcription (Chrivia et al. 1993; Kwok et al. 1994; Chawla et al. 1998; Cruzalegui et al. 1999; Hu et al. 1999; Impey et al. 2002). Additional phosphorylation sites on CREB at Ser142/143 can be activated by other calcium signaling pathways (see below) and are thought to impair binding to the CREB-binding domain (KIX) of CBP and thus repress transcription (Sun et al. 1994; Kornhauser et al. 2002). The only other signal known to be sufficient to activate both CREB and CBP is an increase in the cAMP concentration (Montminy et al. 1990). The activation of the extracellular signal–regulated kinase (ERK)/mitogen-activated protein kinase (MAPK) pathway, while capable of causing CREB phosphorylation on Ser133, is not sufficient to stimulate CREB/CBP-mediated gene transcription, because this pathway does not activate CBP (Chawla et al. 1998) though can prolong CREB phosphorylation (see below).

17.6 Calcium Activates Multiple Interacting Signaling Pathways Mediating Transcription-Dependent Synaptic Plasticity

Synaptic activity can trigger calcium responses in neurons which signal both directly and indirectly to the nucleus to affect gene expression via multiple signaling pathways. Calcium enters the cytoplasm from several sources, but its reach is largely restricted to local microdomains due to the rapid buffering capacity of high-affinity calcium-binding proteins, the most ubiquitous of which in neurons is calmodulin (Burgoyne 2007). Calmodulin is also anchored to several plasma membrane proteins including the NR1 subunit of the NMDA receptor and the alpha subunit of L-type VGCCs, placing it in the immediate path of these calcium entry points (Ehlers et al. 1996; Dolmetsch et al. 2001; Kobayashi et al. 2007). Calmodulin when bound to calcium either facilitates or is a requisite for the activation of CaMKs, MAPKs, protein phosphatases, and adenylyl cyclase (AC) generation of cAMP and activation of protein kinase A (PKA). Many of these signaling molecules are anchored in the vicinity of specific calcium entry points, thereby linking specific calcium channels (i.e., VGCCs, NMDARs, ryanodine, IP$_3$ receptors, etc.) to distinct signaling pathways (Bading et al. 1993; Ghosh and Greenberg 1995; West et al. 2002; Krapivinsky et al. 2004). While CaMKs and MAPKs can induce CREB phosphorylation and may act to prolong CRE-mediated transcription, nuclear calcium signals can induce CBP phosphorylation necessary to initiate CRE-mediated transcription (Chawla et al. 1998; Hardingham et al. 2001a; Wu et al. 2001; Impey et al. 2002). Activators of AC or cAMP analogs are also sufficient to activate CRE-mediated transcription and L-LTP and inhibition of PKA blocks L-LTP, its associated CRE-mediated transcription, and LTM (Frey et al. 1993; Impey et al. 1996; Abel et al. 1997). However, tetanic stimulation of hippocampal slices induces only very moderate (less than twofold) increases in cAMP levels; moreover, in cultured hippocampal neurons, neuronal activity failed to measurably increase cAMP levels (Chetkovich et al. 1991; Frey et al. 1993; Pokorska et al. 2003). In light of the known inhibitory effect of PKA blockers on L-LTP and LTM formation, it remains possible that basal levels of PKA and cAMP may be required for synaptic activity–induced, nuclear calcium–mediated transcription, and thus gate late-phase plasticity and LTM.

Calcium signals activate the MAPK cascade leading to ERK 1 and 2 (ERK1/2) activation which has been shown to be important for LTP induction and maintenance, as well as LTM and survival (Rosen et al. 1994; English and Sweatt 1997; Orban et al. 1999; Sweatt 2004; Thomas and Huganir 2004). The MAPK pathway involves guanine nucleotide exchange factors (GEFs) which convert the small guanine nucleotide–binding protein, Ras, from a GDP-bound state into its GTP-bound state. Ras-GTP then leads to recruitment of Raf to the plasma membrane where it phosphorylates MAP and ERK 1 and 2 kinase (MEK1/2), which in turn phosphorylate ERK1/2. ERK1/2 then can dissociate from MEK1/2 which otherwise excludes ERK1/2 from the nucleus due to their nuclear export signal (NES). CREB can be phosphorylated at Ser133 by both downstream targets of pERK, mitogen- and stress-activated kinase 1/2 (MSK1/2) which is localized exclusively in the nucleoplasm, and the p90 ribosomal S6 kinase 2 (RSK2) which either translocates to the nucleus after its phosphorylation by pERK1/2 in the cytoplasm or is activated by ERK1/2 in the nucleus (Chen et al. 1992; Xing et al. 1996; Wiegert et al. 2007). ERK1-/2-mediated CREB phosphorylation seems to play an important role in prolonging the phosphorylation of CREB at Ser133 after the transient short-lived phosphorylation by CaMKIV has decayed (Impey et al. 1998; Hardingham et al. 2001a; Wu et al. 2001). The MAPK signaling pathway can be activated through a signaling cascade downstream of receptor tyrosine kinase (Trk) activation by neurotrophins such as brain-derived neurotrophic factor (BDNF). BDNF can act locally at the synapse to increase neurotransmitter release presynaptically and promote local translation from postsynaptic ribosomes, leading to synaptic potentiation (Kang and Schuman 1995, 1996). Several components of the MAPK pathway are coupled to NMDARs and the postsynaptic density (PSD) and can be positively modulated, independent of Trk activation, by calcium or calcium/CaM signals, resulting from either NMDAR or VGCC activation (Bading and Greenberg 1991; Rosen et al. 1994; Finkbeiner et al. 1997; Agell et al. 2002; Sweatt 2004; Thomas and Huganir 2004; Kim et al. 2005). Experiments using depolarizing stimuli suggest a strong link between L-type VGCCs and prolonged MAPK activity which may be critical for the expression of genes such as *bdnf* and for L-LTP (Impey et al. 1998; Tao et al. 1998; Dolmetsch et al. 2001). Both NMDA receptors, L-type VGCCs, and postsynaptic action potentials are required for ERK phosphorylation and somatic/nuclear calcium signals generated in response to high-frequency synaptic or antidromic stimulation in hippocampal slices (Dudek and Fields 2002; Zhao et al. 2005; Bengtson et al. 2010). CaMKI, CaMK kinase (CaMKK), Ras guanyl-nucleotide releasing factors (Ras-GRFs), synGAP, PKA, and PKC have all been proposed to activate or modulate the activation of Ras or ERK1/2 in response to calcium increases (Waltereit and Weller 2003; Sweatt 2004; Thomas and Huganir 2004; Wiegert and Bading 2011). Although the details of some of these signaling pathways are not fully resolved, cytoplasmic calcium signals induced by synaptic activity trigger MAPK-dependent dendritic protein synthesis as well as gene expression important for LTP and LTM.

The most extensively studied signaling mechanism of synaptic plasticity is the activation of CaMKIIα whose function is critical for LTP and memory (Giese et al. 1998; Lisman et al. 2002). CaMKII encodes four genes (α, β, γ, and δ) processed into a large family of 28 splice variants which form homo- and heteromultimers. CaMKIIα is the predominant neuronal form almost exclusively restricted to excitatory principal neurons where it is found tightly associated with the PSD and the GluN2B (also known as NR2B, product of the human *GRIN2B* gene) subunit of the NMDAR (Hudmon and Schulman 2002; Colbran 2004). CaMKIIα_B contains an NLS and shows nuclear localization; however, its expression may be limited to the midbrain and diencephalon (Brocke et al. 1995). Calcium entry through synaptic NMDARs forms calcium/CaM which binds Thr305/306 of CaMKII inducing autophosphorylation at Thr286/287 resulting in CaM independent activity, thus prolonging CaMKII activity beyond the duration of the calcium signal. CaMKII phosphorylation of the GluR1 subunit of AMPA receptors is required for hippocampal LTP and spatial learning (Barria et al. 1997; Lisman et al. 2002; Lee et al. 2003). GluR1 phosphorylation by CaMKII, as well as a likely involvement of transmembrane AMPA receptor regulatory proteins (TARPs) and cornichon proteins (CNIH-2/-3), increases the charge transfer of AMPA receptors and drives them into the synapse during LTP (Takahashi et al. 2003; Schwenk et al. 2009; Shi et al. 2010). Synaptic activity triggers the dendritic translation of CaMKIIα mRNA, and a mutation of the *cis*-acting dendritic targeting region impairs L-LTP and LTM but not E-LTP (Mayford et al. 1996; Wu et al. 1998; Mori et al. 2000; Miller et al. 2002). CaMKII protein constitutes a significant proportion of the PSD, and its presence correlates with synaptic strength of individual spines (Kennedy et al. 1983; Goldenring et al. 1984; Kelly et al. 1984; Asrican et al. 2007). The need for ongoing delivery of CaMKII mRNA to the spine to replenish and modulate the quantity of this major component of the PSD identifies one requirement of transcription during late-phase plasticity and learning (Havik et al. 2003; Murray et al. 2003).

A CaMK cascade has been described involving CaMKK, CaMKI, and CaMKIV which engage in sequential activation and interactive regulation (Colomer and Means 2007; Means 2008; Wayman et al. 2008). CaMKK is encoded by two independent genes, α and β, and is activated by calcium/CaM at concentrations of calcium close to basal levels in neurons (Edelman et al. 1996; Tokumitsu and Soderling 1996). In contrast to CaMKII, CaMKI (encoded by CaMKIα, γ, and δ) and CaMKIV (encoded by a single gene) are monomeric and require phosphorylation by CaMKK at a Thr residue within their activation loop (Corcoran and Means 2001; Means 2008). Following phosphorylation by CaMKK, CaMKIV but not CaMKI becomes independent of calcium/CaM and thus autonomously active (Haribabu et al. 1995; Chow et al. 2005).

Given that CaMKI, II, and IV are all activated by calcium/CaM and can all phosphorylate CREB (Sun et al. 1996), the localization of these kinases is determinant in their role in transcription-dependent plasticity. CaMKKα and β have wide and overlapping expression across many brain regions, and both show cytoplasmic localization (Sakagami et al. 2000). CaMKIβ2 is found in both cytoplasm and nucleus (Ueda et al. 1999; Rina et al. 2001). CaMKIα, γ, and δ are predominantly localized in the cytoplasm and have a nuclear export sequence (NES). However,

CaMKIα translocates to the nucleus in response to depolarization-induced calcium signals, and CaMKIδ contains an NLS but does not translocate with depolarization (Picciotto et al. 1995; Stedman et al. 2004; Sakagami et al. 2005). CaMKIV undergoes import into the nucleus and thus is mostly localized in the nucleus in its activated state (Jensen et al. 1991; Bito et al. 1996; Kasahara et al. 2001; Lemrow et al. 2004; Kotera et al. 2005). Several splice variants of CaMKII contain an NLS; however, this is inactive once phosphorylated by CaMKI or CaMKIV (Heist et al. 1998).

The CaMK cascade is crucially involved in synaptic plasticity and learning. Pharmacological inhibition of CaMKK or expression of a dominant-negative CaMKK or siRNA has shown that CaMKK activation is necessary for activity-induced increases in dendritic length as well as ERK-mediated L-LTP in hippocampal slices (Schmitt et al. 2005; Redmond 2008). CaMKKα-deficient mice show reduced fear conditioning along with reduced fear conditioning–induced CaMKIV and CREB activation but normal spatial learning and LTM, whereas CaMKKβ-deficient male mice show impaired spatial LTM and L-LTP but normal E-LTP (Peters et al. 2003; Blaeser et al. 2006; Mizuno et al. 2007). CaMKI activation is required for spinogenesis during development as well as activity-induced spine enlargement and the incorporation of calcium-permeable AMPA receptors into the PSD following glycine-induced LTP (Guire et al. 2008; Saneyoshi et al. 2008; Fortin et al. 2010). CaMKIV is phosphorylated by LTP-inducing synaptic stimulation in hippocampal slices (Kasahara et al. 2001). In the nucleus, CaMKIV is important for CBP recruitment by CREB in response to nuclear calcium/CaM signals (Chawla et al. 1998; Impey et al. 2002). Disruption of the CaMKIV gene or a dominant-negative form of CaMKIV impairs E-LTP, L-LTP, L-LTD (but not early phase-LTD), and LTM in spatial memory and fear conditioning (Ho et al. 2000; Ribar et al. 2000; Kang et al. 2001; Wei et al. 2002). Activated forms of CaMKIV induce dendritic growth; CaMKIV overexpression potentiates while kinase-dead or dominant-negative forms block activity-induced dendritic growth and/or complexity, although conflicting evidence comes from studies using nuclear-targeted variant of these constructs (Redmond 2008). Such results reveal limits to the interpretation of studies using recombinant variants of CaMKs where their localization is distinct from the endogenous protein, due to overexpression or the lack of the NLS, or where overexpression or constitutively active variants simply overpower the endogenous phosphatases (Wayman et al. 2008).

Extensive cross talk exists between MAPKs, CaMKs, PKA, and protein phosphatases, which is relevant to synaptic plasticity. For example, CaMKK activation of ERK through CaMKI partly mediates NMDAR-dependent hippocampal LTP (Schmitt et al. 2005). Also, CaMKIV forms complexes with protein phosphatase 2A which is likely to be responsible for its short-lived (approx. 10 min) activation by calcium/CaM (Anderson et al. 2004). While CaMKII can also phosphorylate CREB at Ser133, it also phosphorylates CREB at Ser142 which appears to block its transcriptional activity (Sun et al. 1994). CaMKI and CaMKIV minimize the phosphorylation of CREB Ser142 by phosphorylating CaMKII to inactivate its NLS, promoting its nuclear export.

The calcium-stimulated ACs, AC1 and AC8, are also components of long-term memory processing. Knockout of both isoforms impairs L-LTP and consolidation of LTM in fear conditioning and the Morris water maze (Wu et al. 1995; Wong et al. 1999; Wieczorek et al. 2010). Defects in the *Drosophila* AC gene in the mutants known as *rutabaga* and *dunce* cause defects in STM but not LTM induced by olfactory avoidance conditioning (Duerr and Quinn 1982; Tully and Quinn 1985; Davis 2005). A similar role for cAMP has also been revealed for LTM in the sea snail *Aplysia californica* (Kandel 2001) and the honey bee *Apis mellifera* (Menzel and Muller 1996). PKA activation downstream of AC activation is not necessary, however, for CREB phosphorylation following synaptic activity inducing either LTP or LTD in hippocampal cultures (Deisseroth et al. 1996) but is necessary for L-LTP in the hippocampus (Huang and Kandel 1994).

17.7 Other Calcium-Binding Proteins and Transcriptional Regulators

Several other transcription factors or regulators of CREB-mediated transcription may play a role in transcription-dependent synaptic plasticity in the adult brain. These include nuclear factor of activated T cells (NFAT), nuclear factor κ-light-chain enhancer of activated B cells (NF-κB), myocyte enhancer factor 2 (MEF2) family of transcription factors, Jacob, and calcineurin (Zhu and McKeon 2000; Lewis 2001; Meffert and Baltimore 2005; Flavell et al. 2006; Shalizi et al. 2006; Dieterich et al. 2008; Schwartz et al. 2009; Li et al. 2011). Here we review two nuclear calcium–regulated transcriptional repressors whose signaling mechanisms and importance for transcription-dependent plasticity are currently emerging.

Downstream regulatory element antagonist modulator (DREAM) is a member of the EF-hand superfamily of calcium-binding proteins which binds to the downstream regulatory element (DRE) of c-fos and dynorphin and blocks CRE-mediated transcription at basal calcium concentrations (Carrion et al. 1999; Osawa et al. 2001; Ledo et al. 2002). DREAM knockout mice have recently been shown to have enhanced LTP, STM, and LTM as well as increased levels of c-fos, bdnf, and c-jun mRNA (Alexander et al. 2009; Fontan-Lozano et al. 2009). Mice with a calcium-insensitive mutant of DREAM show impaired LTD (but not LTP) and contextual fear conditioning (Wu et al. 2010). Several cytoplasmic roles of DREAM may, however, mediate these phenotypes. DREAM in the cytoplasm has been found to modulate neuronal excitability by interacting with A-type potassium channels and was thus also named the potassium channel interacting protein-3 (KChIP3) (An et al. 2000). DREAM/KChIP3 has also been named calsenilin because it interacts with presenilin1 and 2 which has been suggested to regulate the calcium content of the endoplasmic reticulum (Fedrizzi et al. 2008). DREAM/KChIP3/calsenilin also colocalizes with NMDA receptors and affects their function (Wu et al. 2010; Zhang et al. 2010). Although DREAM/KChIP3/calsenilin functions as a calcium-regulated transcriptional repressor, its role in synaptic plasticity and LTM needs to be more clearly distinguished from its many cytoplasmic actions.

Methyl-CpG-binding protein 2 (MeCP2) is a nuclear localized transcriptional regulator involved in the stable repression of chromatin by recruiting a complex of chromatin remodeling enzymes including histone deacetylase (HDAC) 1 and histone methyltransferases that cause chromatin compaction and nucleosome clustering, thus silencing the DNA (Chahrour and Zoghbi 2007). Activity-dependent changes in chromatin structure are regulated by acetylation/deacetylation and phosphorylation of histone H3 and high-mobility group (HMG) protein 17 which is thought to increase acetylation and suppress methylation, thus stabilizing an open conformation of the DNA (Whitlock et al. 1983). MeCP2 is expressed in postmigrational neurons, and its mutation in Rett syndrome in humans leads to severe neurological symptoms after 1 year of age. Knockout of the MeCP2 gene results in synapse loss and overexpression of MeCP2 increases synapse number, although synaptic function in these mice mutants is otherwise generally unaffected (Nelson et al. 2006; Chao et al. 2007). MeCP2 mutant mice show selective loss of excitatory synapses, impaired spatial memory and contextual fear conditioning, and reduced hippocampal but not cortical LTP (Dani et al. 2005; Moretti et al. 2006; Dani and Nelson 2009). Synaptic activity promotes phosphorylation of MeCP2 at Ser421, a requisite for depolarization-induced transcription of genes including *bdnf* (Chen et al. 2003; Martinowich et al. 2003; Zhou et al. 2006). MeCP2 phosphorylation in the nucleus requires nuclear calcium signaling as well as CaMKII but not CaMKIV function (unpublished results from our lab, Zhou et al. 2006). This suggests the involvement of nuclear calcium and possibly a nuclear isoform of CaMKII in removing transcriptional repression by MeCP2 to permit gene expression following synaptic activity. Although MeCP2 function impacts largely on the development of the nervous system, these studies have illuminated the underlying importance of calcium in the regulation of DNA methylation for plasticity and memory also in the mature brain (Gupta et al. 2010).

Acknowledgments We wish to thank Simon Wiegert, David Lau, and Ana Oliveira for their comments on the manuscript.

References

Aakalu, G., Smith, W. B., Nguyen, N., Jiang, C., & Schuman, E. M. (2001). Dynamic visualization of local protein synthesis in hippocampal neurons. *Neuron, 30*, 489–502.

Abel, T., Nguyen, P. V., Barad, M., Deuel, T. A., Kandel, E. R., & Bourtchouladze, R. (1997). Genetic demonstration of a role for PKA in the late phase of LTP and in hippocampus-based long-term memory. *Cell, 88*, 615–626.

Agell, N., Bachs, O., Rocamora, N., & Villalonga, P. (2002). Modulation of the Ras/Raf/MEK/ERK pathway by Ca$^{(2+)}$, and calmodulin. *Cellular Signalling, 14*, 649–654.

Agranoff, B. W., Davis, R. E., Casola, L., & Lim, R. (1967). Actinomycin D blocks formation of memory of shock-avoidance in goldfish. *Science, 158*, 1600–1601.

Ahn, S., Ginty, D. D., & Linden, D. J. (1999). A late phase of cerebellar long-term depression requires activation of CaMKIV and CREB. *Neuron, 23*, 559–568.

Alarcon, J. M., Malleret, G., Touzani, K., Vronskaya, S., Ishii, S., Kandel, E. R., & Barco, A. (2004). Chromatin acetylation, memory, and LTP are impaired in CBP+/− mice: A model for the cognitive deficit in Rubinstein-Taybi syndrome and its amelioration. *Neuron, 42*, 947–959.

Alberini, C. M. (2009). Transcription factors in long-term memory and synaptic plasticity. *Physiological Reviews, 89*, 121–145.

Alexander, J. C., McDermott, C. M., Tunur, T., Rands, V., Stelly, C., Karhson, D., Bowlby, M. R., An, W. F., Sweatt, J. D., & Schrader, L. A. (2009). The role of calsenilin/DREAM/KChIP3 in contextual fear conditioning. *Learning and Memory, 16*, 167–177.

Allbritton, N. L., Meyer, T., & Stryer, L. (1992). Range of messenger action of calcium ion and inositol 1, 4, 5-trisphosphate. *Science, 258*, 1812–1815.

Allbritton, N. L., Oancea, E., Kuhn, M. A., & Meyer, T. (1994). Source of nuclear calcium signals. *Proceedings of the National Academy of Sciences of the United States of America, 91*, 12458–12462.

An, W. F., Bowlby, M. R., Betty, M., Cao, J., Ling, H. P., Mendoza, G., Hinson, J. W., Mattsson, K. I., Strassle, B. W., Trimmer, J. S., & Rhodes, K. J. (2000). Modulation of A-type potassium channels by a family of calcium sensors. *Nature, 403*, 553–556.

Anderson, K. A., Noeldner, P. K., Reece, K., Wadzinski, B. E., & Means, A. R. (2004). Regulation and function of the calcium/calmodulin-dependent protein kinase IV/protein serine/threonine phosphatase 2A signaling complex. *The Journal of Biological Chemistry, 279*, 31708–31716.

Asrican, B., Lisman, J., & Otmakhov, N. (2007). Synaptic strength of individual spines correlates with bound Ca^{2+}-calmodulin-dependent kinase II. *The Journal of Neuroscience, 27*, 14007–14011.

Bading, H., & Greenberg, M. E. (1991). Stimulation of protein tyrosine phosphorylation by NMDA receptor activation. *Science, 253*, 912–914.

Bading, H., Ginty, D. D., & Greenberg, M. E. (1993). Regulation of gene expression in hippocampal neurons by distinct calcium signaling pathways. *Science, 260*, 181–186.

Bardo, S., Cavazzini, M. G., & Emptage, N. (2006). The role of the endoplasmic reticulum Ca^{2+} store in the plasticity of central neurons. *Trends in Pharmacological Sciences, 27*, 78–84.

Barria, A., Muller, D., Derkach, V., Griffith, L. C., & Soderling, T. R. (1997). Regulatory phosphorylation of AMPA-type glutamate receptors by CaM-KII during long-term potentiation. *Science, 276*, 2042–2045.

Bartsch, D., Casadio, A., Karl, K. A., Serodio, P., & Kandel, E. R. (1998). CREB1 encodes a nuclear activator, a repressor, and a cytoplasmic modulator that form a regulatory unit critical for long-term facilitation. *Cell, 95*, 211–223.

Bengtson, C. P., Freitag, H. E., Weislogel, J. M., & Bading, H. (2010). Nuclear calcium sensors reveal that repetition of trains of synaptic stimuli boosts nuclear calcium signaling in CA1 pyramidal neurons. *Biophysical Journal, 99*, 4066–4077.

Berridge, M. J. (1998). Neuronal calcium signaling. *Neuron, 21*, 13–26.

Bito, H., Deisseroth, K., & Tsien, R. W. (1996). CREB phosphorylation and dephosphorylation: A $Ca^{(2+)}$- and stimulus duration-dependent switch for hippocampal gene expression. *Cell, 87*, 1203–1214.

Blaeser, F., Sanders, M. J., Truong, N., Ko, S., Wu, L. J., Wozniak, D. F., Fanselow, M. S., Zhuo, M., & Chatila, T. A. (2006). Long-term memory deficits in Pavlovian fear conditioning in Ca^{2+}/calmodulin kinase kinase alpha-deficient mice. *Molecular and Cellular Biology, 26*, 9105–9115.

Bootman, M. D., Fearnley, C., Smyrnias, I., MacDonald, F., & Roderick, H. L. (2009). An update on nuclear calcium signalling. *Journal of Cell Science, 122*, 2337–2350.

Borlikova, G., & Endo, S. (2009). Inducible cAMP early repressor (ICER) and brain functions. *Molecular Neurobiology, 40*, 73–86.

Brini, M., Murgia, M., Pasti, L., Picard, D., Pozzan, T., & Rizzuto, R. (1993). Nuclear Ca^{2+} concentration measured with specifically targeted recombinant aequorin. *The EMBO Journal, 12*, 4813–4819.

Brini, M., Marsault, R., Bastianutto, C., Pozzan, T., & Rizzuto, R. (1994). Nuclear targeting of aequorin. A new approach for measuring nuclear Ca^{2+} concentration in intact cells. *Cell Calcium, 16*, 259–268.

Brocke, L., Srinivasan, M., & Schulman, H. (1995). Developmental and regional expression of multifunctional Ca^{2+}/calmodulin-dependent protein kinase isoforms in rat brain. *The Journal of Neuroscience, 15*, 6797–6808.

Burger, C., Lopez, M. C., Baker, H. V., Mandel, R. J., & Muzyczka, N. (2008). Genome-wide analysis of aging and learning-related genes in the hippocampal dentate gyrus. *Neurobiology of Learning and Memory, 89*, 379–396.

Burgoyne, R. D. (2007). Neuronal calcium sensor proteins: Generating diversity in neuronal Ca^{2+} signalling. *Nature Reviews Neuroscience, 8*, 182–193.

Canepari, M., Djurisic, M., & Zecevic, D. (2007). Dendritic signals from rat hippocampal CA1 pyramidal neurons during coincident pre- and post-synaptic activity: A combined voltage- and calcium-imaging study. *The Journal of Physiology, 580*, 463–484.

Carrion, A. M., Link, W. A., Ledo, F., Mellstrom, B., & Naranjo, J. R. (1999). DREAM is a Ca2+-regulated transcriptional repressor. *Nature, 398*, 80–84.

Chahrour, M., & Zoghbi, H. Y. (2007). The story of Rett syndrome: From clinic to neurobiology. *Neuron, 56*, 422–437.

Chao, H. T., Zoghbi, H. Y., & Rosenmund, C. (2007). MeCP2 controls excitatory synaptic strength by regulating glutamatergic synapse number. *Neuron, 56*, 58–65.

Chawla, S., & Bading, H. (2001). CREB/CBP and SRE-interacting transcriptional regulators are fast on-off switches: Duration of calcium transients specifies the magnitude of transcriptional responses. *Journal of Neurochemistry, 79*, 849–858.

Chawla, S., Hardingham, G. E., Quinn, D. R., & Bading, H. (1998). CBP: A signal-regulated transcriptional coactivator controlled by nuclear calcium and CaM kinase IV. *Science, 281*, 1505–1509.

Chen, R. H., Sarnecki, C., & Blenis, J. (1992). Nuclear localization and regulation of erk- and rsk-encoded protein kinases. *Molecular and Cellular Biology, 12*, 915–927.

Chen, W. G., Chang, Q., Lin, Y., Meissner, A., West, A. E., Griffith, E. C., Jaenisch, R., & Greenberg, M. E. (2003). Derepression of BDNF transcription involves calcium-dependent phosphorylation of MeCP2. *Science, 302*, 885–889.

Chetkovich, D. M., Gray, R., Johnston, D., & Sweatt, J. D. (1991). N-methyl-D-aspartate receptor activation increases cAMP levels and voltage-gated Ca^{2+} channel activity in area CA1 of hippocampus. *Proceedings of the National Academy of Sciences of the United States of America, 88*, 6467–6471.

Chow, F. A., Anderson, K. A., Noeldner, P. K., & Means, A. R. (2005). The autonomous activity of calcium/calmodulin-dependent protein kinase IV is required for its role in transcription. *The Journal of Biological Chemistry, 280*, 20530–20538.

Chrivia, J. C., Kwok, R. P., Lamb, N., Hagiwara, M., Montminy, M. R., & Goodman, R. H. (1993). Phosphorylated CREB binds specifically to the nuclear protein CBP. *Nature, 365*, 855–859.

Colbran, R. J. (2004). Targeting of calcium/calmodulin-dependent protein kinase II. *The Biochemical Journal, 378*, 1–16.

Colomer, J., & Means, A. R. (2007). Physiological roles of the Ca^{2+}/CaM-dependent protein kinase cascade in health and disease. *Subcellular Biochemistry, 45*, 169–214.

Corcoran, E. E., & Means, A. R. (2001). Defining Ca^{2+}/calmodulin-dependent protein kinase cascades in transcriptional regulation. *The Journal of Biological Chemistry, 276*, 2975–2978.

Cruzalegui, F. H., Hardingham, G. E., & Bading, H. (1999). c-Jun functions as a calcium-regulated transcriptional activator in the absence of JNK/SAPK1 activation. *The EMBO Journal, 18*, 1335–1344.

Dani, V. S., & Nelson, S. B. (2009). Intact long-term potentiation but reduced connectivity between neocortical layer 5 pyramidal neurons in a mouse model of Rett syndrome. *The Journal of Neuroscience, 29*, 11263–11270.

Dani, V. S., Chang, Q., Maffei, A., Turrigiano, G. G., Jaenisch, R., & Nelson, S. B. (2005). Reduced cortical activity due to a shift in the balance between excitation and inhibition in a mouse model of Rett syndrome. *Proceedings of the National Academy of Sciences of the United States of America, 102*, 12560–12565.

Davis, R. L. (2005). Olfactory memory formation in Drosophila: From molecular to systems neuroscience. *Annual Review of Neuroscience, 28*, 275–302.

Deisseroth, K., Bito, H., & Tsien, R. W. (1996). Signaling from synapse to nucleus: Postsynaptic CREB phosphorylation during multiple forms of hippocampal synaptic plasticity. *Neuron, 16*, 89–101.

Dieterich, D. C., Karpova, A., Mikhaylova, M., Zdobnova, I., Konig, I., Landwehr, M., Kreutz, M., Smalla, K. H., Richter, K., Landgraf, P., Reissner, C., Boeckers, T. M., Zuschratter, W., Spilker, C., Seidenbecher, C. I., Garner, C. C., Gundelfinger, E. D., & Kreutz, M. R. (2008). Caldendrin-Jacob: A protein liaison that couples NMDA receptor signalling to the nucleus. *PLoS Biology, 6*, e34.

Dolmetsch, R. E., Pajvani, U., Fife, K., Spotts, J. M., & Greenberg, M. E. (2001). Signaling to the nucleus by an L-type calcium channel-calmodulin complex through the MAP kinase pathway. *Science, 294*, 333–339.

Dudek, S. M., & Fields, R. D. (2002). Somatic action potentials are sufficient for late-phase LTP-related cell signaling. *Proceedings of the National Academy of Sciences of the United States of America, 99*, 3962–3967.

Duerr, J. S., & Quinn, W. G. (1982). Three Drosophila mutations that block associative learning also affect habituation and sensitization. *Proceedings of the National Academy of Sciences of the United States of America, 79*, 3646–3650.

Edelman, A. M., Mitchelhill, K. I., Selbert, M. A., Anderson, K. A., Hook, S. S., Stapleton, D., Goldstein, E. G., Means, A. R., & Kemp, B. E. (1996). Multiple Ca^{2+}-calmodulin-dependent protein kinase kinases from rat brain. Purification, regulation by Ca^{2+}-calmodulin, and partial amino acid sequence. *The Journal of Biological Chemistry, 271*, 10806–10810.

Eder, A., & Bading, H. (2007). Calcium signals can freely cross the nuclear envelope in hippocampal neurons: Somatic calcium increases generate nuclear calcium transients. *BMC Neuroscience, 8*, 57–68.

Ehlers, M. D., Zhang, S., Bernhadt, J. P., & Huganir, R. L. (1996). Inactivation of NMDA receptors by direct interaction of calmodulin with the NR1 subunit. *Cell, 84*, 745–755.

Ehrlich, I., & Malinow, R. (2004). Postsynaptic density 95 controls AMPA receptor incorporation during long-term potentiation and experience-driven synaptic plasticity. *The Journal of Neuroscience, 24*, 916–927.

English, J. D., & Sweatt, J. D. (1997). A requirement for the mitogen-activated protein kinase cascade in hippocampal long term potentiation. *The Journal of Biological Chemistry, 272*, 19103–19106.

Faas, G. C., Raghavachari, S., Lisman, J. E., & Mody, I. (2011). Calmodulin as a direct detector of Ca^{2+} signals. *Nature Neuroscience, 14*, 301–304.

Fedrizzi, L., Lim, D., Carafoli, E., & Brini, M. (2008). Interplay of the Ca^{2+}-binding protein DREAM with presenilin in neuronal Ca^{2+} signaling. *The Journal of Biological Chemistry, 283*, 27494–27503.

Finkbeiner, S., Tavazoie, S. F., Maloratsky, A., Jacobs, K. M., Harris, K. M., & Greenberg, M. E. (1997). CREB: A major mediator of neuronal neurotrophin responses. *Neuron, 19*, 1031–1047.

Flavell, S. W., & Greenberg, M. E. (2008). Signaling mechanisms linking neuronal activity to gene expression and plasticity of the nervous system. *Annual Review of Neuroscience, 31*, 563–590.

Flavell, S. W., Cowan, C. W., Kim, T. K., Greer, P. L., Lin, Y., Paradis, S., Griffith, E. C., Hu, L. S., Chen, C., & Greenberg, M. E. (2006). Activity-dependent regulation of MEF2 transcription factors suppresses excitatory synapse number. *Science, 311*, 1008–1012.

Fleischmann, A., Hvalby, O., Jensen, V., Strekalova, T., Zacher, C., Layer, L. E., Kvello, A., Reschke, M., Spanagel, R., Sprengel, R., Wagner, E. F., & Gass, P. (2003). Impaired long-term memory and NR2A-type NMDA receptor-dependent synaptic plasticity in mice lacking c-Fos in the CNS. *The Journal of Neuroscience, 23*, 9116–9122.

Fontan-Lozano, A., Romero-Granados, R., del-Pozo-Martin, Y., Suarez-Pereira, I., Delgado-Garcia, J. M., Penninger, J. M., & Carrion, A. M. (2009). Lack of DREAM protein enhances learning and memory and slows brain aging. *Current Biology, 19*, 54–60.

Fortin, D. A., Davare, M. A., Srivastava, T., Brady, J. D., Nygaard, S., Derkach, V. A., & Soderling, T. R. (2010). Long-term potentiation-dependent spine enlargement requires synaptic Ca2+-permeable AMPA receptors recruited by CaM-kinase I. *The Journal of Neuroscience, 30*, 11565–11575.

Frey, S., & Frey, J. U. (2008). 'Synaptic tagging' and 'cross-tagging' and related associative reinforcement processes of functional plasticity as the cellular basis for memory formation. *Progress in Brain Research, 169*, 117–143.

Frey, U., & Morris, R. G. (1997). Synaptic tagging and long-term potentiation. *Nature, 385*, 533–536.

Frey, U., Huang, Y. Y., & Kandel, E. R. (1993). Effects of cAMP simulate a late stage of LTP in hippocampal CA1 neurons. *Science, 260*, 1661–1664.

Frey, U., Frey, S., Schollmeier, F., & Krug, M. (1996). Influence of actinomycin D, a RNA synthesis inhibitor, on long-term potentiation in rat hippocampal neurons in vivo and in vitro. *The Journal of Physiology, 490*(Pt 3), 703–711.

Fuenzalida, M., Fernandez de Sevilla, D., Couve, A., & Buno, W. (2010). Role of AMPA and NMDA receptors and back-propagating action potentials in spike timing-dependent plasticity. *Journal of Neurophysiology, 103*, 47–54.

Ghosh, A., & Greenberg, M. E. (1995). Calcium signaling in neurons: Molecular mechanisms and cellular consequences. *Science, 268*, 239–247.

Giese, K. P., Fedorov, N. B., Filipkowski, R. K., & Silva, A. J. (1998). Autophosphorylation at Thr286 of the alpha calcium-calmodulin kinase II in LTP and learning. *Science, 279*, 870–873.

Goldenring, J. R., McGuire, J. S., Jr., & DeLorenzo, R. J. (1984). Identification of the major postsynaptic density protein as homologous with the major calmodulin-binding subunit of a calmodulin-dependent protein kinase. *Journal of Neurochemistry, 42*, 1077–1084.

Gonzalez, G. A., & Montminy, M. R. (1989). Cyclic AMP stimulates somatostatin gene transcription by phosphorylation of CREB at serine 133. *Cell, 59*, 675–680.

Guire, E. S., Oh, M. C., Soderling, T. R., & Derkach, V. A. (2008). Recruitment of calcium-permeable AMPA receptors during synaptic potentiation is regulated by CaM-kinase I. *The Journal of Neuroscience, 28*, 6000–6009.

Gupta, S., Kim, S. Y., Artis, S., Molfese, D. L., Schumacher, A., Sweatt, J. D., Paylor, R. E., & Lubin, F. D. (2010). Histone methylation regulates memory formation. *The Journal of Neuroscience, 30*, 3589–3599.

Hagenston, A. M., Fitzpatrick, J. S., & Yeckel, M. F. (2008). MGluR-mediated calcium waves that invade the soma regulate firing in layer V medial prefrontal cortical pyramidal neurons. *Cerebral Cortex, 18*, 407–423.

Hardingham, G. E., & Bading, H. (2010). Synaptic versus extrasynaptic NMDA receptor signalling: Implications for neurodegenerative disorders. *Nature Reviews Neuroscience, 11*, 682–696.

Hardingham, G. E., Chawla, S., Johnson, C. M., & Bading, H. (1997). Distinct functions of nuclear and cytoplasmic calcium in the control of gene expression. *Nature, 385*, 260–265.

Hardingham, G. E., Chawla, S., Cruzalegui, F. H., & Bading, H. (1999). Control of recruitment and transcription-activating function of CBP determines gene regulation by NMDA receptors and L-type calcium channels. *Neuron, 22*, 789–798.

Hardingham, G. E., Arnold, F. J., & Bading, H. (2001a). A calcium microdomain near NMDA receptors: On switch for ERK-dependent synapse-to-nucleus communication. *Nature Neuroscience, 4*, 565–566.

Hardingham, G. E., Arnold, F. J., & Bading, H. (2001b). Nuclear calcium signaling controls CREB-mediated gene expression triggered by synaptic activity. *Nature Neuroscience, 4*, 261–267.

Haribabu, B., Hook, S. S., Selbert, M. A., Goldstein, E. G., Tomhave, E. D., Edelman, A. M., Snyderman, R., & Means, A. R. (1995). Human calcium-calmodulin dependent protein kinase I: cDNA cloning, domain structure and activation by phosphorylation at threonine-177 by calcium-calmodulin dependent protein kinase I kinase. *The EMBO Journal, 14*, 3679–3686.

Havik, B., Rokke, H., Bardsen, K., Davanger, S., & Bramham, C. R. (2003). Bursts of high-frequency stimulation trigger rapid delivery of pre-existing alpha-CaMKII mRNA to synapses: A mechanism in dendritic protein synthesis during long-term potentiation in adult awake rats. *The European Journal of Neuroscience, 17*, 2679–2689.

Havik, B., Rokke, H., Dagyte, G., Stavrum, A. K., Bramham, C. R., & Steen, V. M. (2007). Synaptic activity-induced global gene expression patterns in the dentate gyrus of adult behaving rats: Induction of immunity-linked genes. *Neuroscience, 148*, 925–936.

Heist, E. K., Srinivasan, M., & Schulman, H. (1998). Phosphorylation at the nuclear localization signal of Ca^{2+}/calmodulin-dependent protein kinase II blocks its nuclear targeting. *The Journal of Biological Chemistry, 273*, 19763–19771.

Hernandez, P. J., & Abel, T. (2008). The role of protein synthesis in memory consolidation: Progress amid decades of debate. *Neurobiology of Learning and Memory, 89*, 293–311.

Ho, N., Liauw, J. A., Blaeser, F., Wei, F., Hanissian, S., Muglia, L. M., Wozniak, D. F., Nardi, A., Arvin, K. L., Holtzman, D. M., Linden, D. J., Zhuo, M., Muglia, L. J., & Chatila, T. A. (2000). Impaired synaptic plasticity and cAMP response element-binding protein activation in Ca^{2+}/calmodulin-dependent protein kinase type IV/Gr-deficient mice. *The Journal of Neuroscience, 20*, 6459–6472.

Hong, M., & Ross, W. N. (2007). Priming of intracellular calcium stores in rat CA1 pyramidal neurons. *The Journal of Physiology, 584*, 75–87.

Hu, S. C., Chrivia, J., & Ghosh, A. (1999). Regulation of CBP-mediated transcription by neuronal calcium signaling. *Neuron, 22*, 799–808.

Huang, E. P. (1998). Synaptic plasticity: Going through phases with LTP. *Current Biology, 8*, R350–R352.

Huang, Y. Y., & Kandel, E. R. (1994). Recruitment of long-lasting and protein kinase A-dependent long-term potentiation in the CA1 region of hippocampus requires repeated tetanization. *Learning and Memory, 1*, 74–82.

Huang, Y. Y., Martin, K. C., & Kandel, E. R. (2000). Both protein kinase A and mitogen-activated protein kinase are required in the amygdala for the macromolecular synthesis-dependent late phase of long-term potentiation. *The Journal of Neuroscience, 20*, 6317–6325.

Hudmon, A., & Schulman, H. (2002). Neuronal CA^{2+}/calmodulin-dependent protein kinase II: The role of structure and autoregulation in cellular function. *Annual Review of Biochemistry, 71*, 473–510.

Igaz, L. M., Vianna, M. R., Medina, J. H., & Izquierdo, I. (2002). Two time periods of hippocampal mRNA synthesis are required for memory consolidation of fear-motivated learning. *The Journal of Neuroscience, 22*, 6781–6789.

Impey, S., Mark, M., Villacres, E. C., Poser, S., Chavkin, C., & Storm, D. R. (1996). Induction of CRE-mediated gene expression by stimuli that generate long-lasting LTP in area CA1 of the hippocampus. *Neuron, 16*, 973–982.

Impey, S., Obrietan, K., Wong, S. T., Poser, S., Yano, S., Wayman, G., Deloulme, J. C., Chan, G., & Storm, D. R. (1998). Cross talk between ERK and PKA is required for Ca^{2+} stimulation of CREB-dependent transcription and ERK nuclear translocation. *Neuron, 21*, 869–883.

Impey, S., Fong, A. L., Wang, Y., Cardinaux, J. R., Fass, D. M., Obrietan, K., Wayman, G. A., Storm, D. R., Soderling, T. R., & Goodman, R. H. (2002). Phosphorylation of CBP mediates transcriptional activation by neural activity and CaM kinase IV. *Neuron, 34*, 235–244.

Jensen, K. F., Ohmstede, C. A., Fisher, R. S., & Sahyoun, N. (1991). Nuclear and axonal localization of Ca^{2+}/calmodulin-dependent protein kinase type Gr in rat cerebellar cortex. *Proceedings of the National Academy of Sciences of the United States of America, 88*, 2850–2853.

Johenning, F. W., & Holthoff, K. (2007). Nuclear calcium signals during L-LTP induction do not predict the degree of synaptic potentiation. *Cell Calcium, 41*, 271–283.

Jones, M. W., Errington, M. L., French, P. J., Fine, A., Bliss, T. V., Garel, S., Charnay, P., Bozon, B., Laroche, S., & Davis, S. (2001). A requirement for the immediate early gene Zif268 in the expression of late LTP and long-term memories. *Nature Neuroscience, 4*, 289–296.

Jong, Y. J., Kumar, V., Kingston, A. E., Romano, C., & O'Malley, K. L. (2005). Functional metabotropic glutamate receptors on nuclei from brain and primary cultured striatal neurons. Role of transporters in delivering ligand. *The Journal of Biological Chemistry, 280*, 30469–30480.

Jong, Y. J., Schwetye, K. E., & O'Malley, K. L. (2007). Nuclear localization of functional metabotropic glutamate receptor mGlu1 in HEK293 cells and cortical neurons: Role in nuclear calcium mobilization and development. *Journal of Neurochemistry, 101*, 458–469.

Josselyn, S. A., Shi, C., Carlezon, W. A., Jr., Neve, R. L., Nestler, E. J., & Davis, M. (2001). Long-term memory is facilitated by cAMP response element-binding protein overexpression in the amygdala. *The Journal of Neuroscience, 21*, 2404–2412.

Kampa, B. M., Clements, J., Jonas, P., & Stuart, G. J. (2004). Kinetics of Mg^{2+} unblock of NMDA receptors: Implications for spike-timing dependent synaptic plasticity. *The Journal of Physiology, 556*, 337–345.

Kandel, E. R. (2001). The molecular biology of memory storage: A dialogue between genes and synapses. *Science, 294*, 1030–1038.

Kang, H., & Schuman, E. M. (1995). Long-lasting neurotrophin-induced enhancement of synaptic transmission in the adult hippocampus. *Science, 267*, 1658–1662.

Kang, H., & Schuman, E. M. (1996). A requirement for local protein synthesis in neurotrophin-induced hippocampal synaptic plasticity. *Science, 273*, 1402–1406.

Kang, H., Sun, L. D., Atkins, C. M., Soderling, T. R., Wilson, M. A., & Tonegawa, S. (2001). An important role of neural activity-dependent CaMKIV signaling in the consolidation of long-term memory. *Cell, 106*, 771–783.

Kapur, A., Yeckel, M., & Johnston, D. (2001). Hippocampal mossy fiber activity evokes Ca^{2+} release in CA3 pyramidal neurons via a metabotropic glutamate receptor pathway. *Neuroscience, 107*, 59–69.

Kasahara, J., Fukunaga, K., & Miyamoto, E. (2001). Activation of calcium/calmodulin-dependent protein kinase IV in long term potentiation in the rat hippocampal CA1 region. *The Journal of Biological Chemistry, 276*, 24044–24050.

Keeley, M. B., Wood, M. A., Isiegas, C., Stein, J., Hellman, K., Hannenhalli, S., & Abel, T. (2006). Differential transcriptional response to nonassociative and associative components of classical fear conditioning in the amygdala and hippocampus. *Learning and Memory, 13*, 135–142.

Kelly, P. T., McGuinness, T. L., & Greengard, P. (1984). Evidence that the major postsynaptic density protein is a component of a Ca^{2+}/calmodulin-dependent protein kinase. *Proceedings of the National Academy of Sciences of the United States of America, 81*, 945–949.

Kennedy, M. B., Bennett, M. K., & Erondu, N. E. (1983). Biochemical and immunochemical evidence that the "major postsynaptic density protein" is a subunit of a calmodulin-dependent protein kinase. *Proceedings of the National Academy of Sciences of the United States of America, 80*, 7357–7361.

Kida, S., Josselyn, S. A., Pena de Ortiz, S., Kogan, J. H., Chevere, I., Masushige, S., & Silva, A. J. (2002). CREB required for the stability of new and reactivated fear memories. *Nature Neuroscience, 5*, 348–355.

Kim, M. J., Dunah, A. W., Wang, Y. T., & Sheng, M. (2005). Differential roles of NR2A- and NR2B-containing NMDA receptors in Ras-ERK signaling and AMPA receptor trafficking. *Neuron, 46*, 745–760.

Kobayashi, T., Yamada, Y., Fukao, M., Tsutsuura, M., & Tohse, N. (2007). Regulation of Cav1.2 current: Interaction with intracellular molecules. *Journal of Pharmacological Sciences, 103*, 347–353.

Koester, H. J., & Sakmann, B. (1998). Calcium dynamics in single spines during coincident pre- and postsynaptic activity depend on relative timing of back-propagating action potentials and subthreshold excitatory postsynaptic potentials. *Proceedings of the National Academy of Sciences of the United States of America, 95*, 9596–9601.

Kornhauser, J. M., Cowan, C. W., Shaywitz, A. J., Dolmetsch, R. E., Griffith, E. C., Hu, L. S., Haddad, C., Xia, Z., & Greenberg, M. E. (2002). CREB transcriptional activity in neurons is regulated by multiple, calcium-specific phosphorylation events. *Neuron, 34*, 221–233.

Korzus, E., Rosenfeld, M. G., & Mayford, M. (2004). CBP histone acetyltransferase activity is a critical component of memory consolidation. *Neuron, 42*, 961–972.

Kotera, I., Sekimoto, T., Miyamoto, Y., Saiwaki, T., Nagoshi, E., Sakagami, H., Kondo, H., & Yoneda, Y. (2005). Importin alpha transports CaMKIV to the nucleus without utilizing importin beta. *The EMBO Journal, 24*, 942–951.

Krapivinsky, G., Medina, I., Krapivinsky, L., Gapon, S., & Clapham, D. E. (2004). SynGAP-MUPP1-CaMKII synaptic complexes regulate p38 MAP kinase activity and NMDA receptor-dependent synaptic AMPA receptor potentiation. *Neuron, 43*, 563–574.

Kumar, V., Jong, Y. J., & O'Malley, K. L. (2008). Activated nuclear metabotropic glutamate receptor mGlu5 couples to nuclear Gq/11 proteins to generate inositol 1,4,5-trisphosphate-mediated nuclear Ca^{2+} release. *The Journal of Biological Chemistry, 283*, 14072–14083.

Kwok, R. P., Lundblad, J. R., Chrivia, J. C., Richards, J. P., Bachinger, H. P., Brennan, R. G., Roberts, S. G., Green, M. R., & Goodman, R. H. (1994). Nuclear protein CBP is a coactivator for the transcription factor CREB. *Nature, 370*, 223–226.

Larkum, M. E., Watanabe, S., Nakamura, T., Lasser-Ross, N., & Ross, W. N. (2003). Synaptically activated Ca^{2+} waves in layer 2/3 and layer 5 rat neocortical pyramidal neurons. *The Journal of Physiology, 549*, 471–488.

Ledo, F., Kremer, L., Mellstrom, B., & Naranjo, J. R. (2002). Ca2+-dependent block of CREB-CBP transcription by repressor DREAM. *The EMBO Journal, 21*, 4583–4592.

Lee, H. K., Takamiya, K., Han, J. S., Man, H., Kim, C. H., Rumbaugh, G., Yu, S., Ding, L., He, C., Petralia, R. S., Wenthold, R. J., Gallagher, M., & Huganir, R. L. (2003). Phosphorylation of the AMPA receptor GluR1 subunit is required for synaptic plasticity and retention of spatial memory. *Cell, 112*, 631–643.

Lemrow, S. M., Anderson, K. A., Joseph, J. D., Ribar, T. J., Noeldner, P. K., & Means, A. R. (2004). Catalytic activity is required for calcium/calmodulin-dependent protein kinase IV to enter the nucleus. *The Journal of Biological Chemistry, 279*, 11664–11671.

Levenson, J. M., Choi, S., Lee, S. Y., Cao, Y. A., Ahn, H. J., Worley, K. C., Pizzi, M., Liou, H. C., & Sweatt, J. D. (2004). A bioinformatics analysis of memory consolidation reveals involvement of the transcription factor c-rel. *The Journal of Neuroscience, 24*, 3933–3943.

Lewis, R. S. (2001). Calcium signaling mechanisms in T lymphocytes. *Annual Review of Immunology, 19*, 497–521.

Li, H., Rao, A., & Hogan, P. G. (2011). Interaction of calcineurin with substrates and targeting proteins. *Trends in Cell Biology, 21*, 91–103.

Limback-Stokin, K., Korzus, E., Nagaoka-Yasuda, R., & Mayford, M. (2004). Nuclear calcium/calmodulin regulates memory consolidation. *The Journal of Neuroscience, 24*, 10858–10867.

Linden, D. J. (1996). A protein synthesis-dependent late phase of cerebellar long-term depression. *Neuron, 17*, 483–490.

Lisman, J., Schulman, H., & Cline, H. (2002). The molecular basis of CaMKII function in synaptic and behavioural memory. *Nature Reviews Neuroscience, 3*, 175–190.

Lynch, G., Larson, J., Kelso, S., Barrionuevo, G., & Schottler, F. (1983). Intracellular injections of EGTA block induction of hippocampal long-term potentiation. *Nature, 305*, 719–721.

Malinow, R., & Malenka, R. C. (2002). AMPA receptor trafficking and synaptic plasticity. *Annual Review of Neuroscience, 25*, 103–126.

Marchenko, S. M., Yarotskyy, V. V., Kovalenko, T. N., Kostyuk, P. G., & Thomas, R. C. (2005). Spontaneously active and InsP3-activated ion channels in cell nuclei from rat cerebellar Purkinje and granule neurones. *The Journal of Physiology, 565*, 897–910.

Martinowich, K., Hattori, D., Wu, H., Fouse, S., He, F., Hu, Y., Fan, G., & Sun, Y. E. (2003). DNA methylation-related chromatin remodeling in activity-dependent BDNF gene regulation. *Science, 302*, 890–893.

Mayford, M., Baranes, D., Podsypanina, K., & Kandel, E. R. (1996). The 3′-untranslated region of CaMKII alpha is a cis-acting signal for the localization and translation of mRNA in dendrites. *Proceedings of the National Academy of Sciences of the United States of America, 93*, 13250–13255.

Mayr, B., & Montminy, M. (2001). Transcriptional regulation by the phosphorylation-dependent factor CREB. *Nature Reviews Molecular Cell Biology, 2*, 599–609.

Means, A. R. (2008). The year in basic science: Calmodulin kinase cascades. *Molecular Endocrinology, 22*, 2759–2765.

Meffert, M. K., & Baltimore, D. (2005). Physiological functions for brain NF-kappaB. *Trends in Neurosciences, 28*, 37–43.

Mellstrom, B., Savignac, M., Gomez-Villafuertes, R., & Naranjo, J. R. (2008). Ca2+-operated transcriptional networks: Molecular mechanisms and in vivo models. *Physiological Reviews, 88*, 421–449.

Menzel, R., & Muller, U. (1996). Learning and memory in honeybees: From behavior to neural substrates. *Annual Review of Neuroscience, 19*, 379–404.

Miller, S., Yasuda, M., Coats, J. K., Jones, Y., Martone, M. E., & Mayford, M. (2002). Disruption of dendritic translation of CaMKIIalpha impairs stabilization of synaptic plasticity and memory consolidation. *Neuron, 36*, 507–519.

Miyakawa, H., Ross, W. N., Jaffe, D., Callaway, J. C., Lasser-Ross, N., Lisman, J. E., & Johnston, D. (1992). Synaptically activated increases in Ca^{2+} concentration in hippocampal CA1 pyramidal cells are primarily due to voltage-gated Ca^{2+} channels. *Neuron, 9*, 1163–1173.

Miyawaki, A., Llopis, J., Heim, R., McCaffery, J. M., Adams, J. A., Ikura, M., & Tsien, R. Y. (1997). Fluorescent indicators for Ca^{2+} based on green fluorescent proteins and calmodulin. *Nature, 388*, 882–887.

Mizuno, K., Antunes-Martins, A., Ris, L., Peters, M., Godaux, E., & Giese, K. P. (2007). Calcium/calmodulin kinase kinase beta has a male-specific role in memory formation. *Neuroscience, 145*, 393–402.

Montarolo, P. G., Goelet, P., Castellucci, V. F., Morgan, J., Kandel, E. R., & Schacher, S. (1986). A critical period for macromolecular synthesis in long-term heterosynaptic facilitation in Aplysia. *Science, 234*, 1249–1254.

Montminy, M. R., & Bilezikjian, L. M. (1987). Binding of a nuclear protein to the cyclic-AMP response element of the somatostatin gene. *Nature, 328*, 175–178.

Montminy, M. R., Sevarino, K. A., Wagner, J. A., Mandel, G., & Goodman, R. H. (1986). Identification of a cyclic-AMP-responsive element within the rat somatostatin gene. *Proceedings of the National Academy of Sciences of the United States of America, 83*, 6682–6686.

Montminy, M. R., Gonzalez, G. A., & Yamamoto, K. K. (1990). Regulation of cAMP-inducible genes by CREB. *Trends in Neurosciences, 13*, 184–188.

Moretti, P., Levenson, J. M., Battaglia, F., Atkinson, R., Teague, R., Antalffy, B., Armstrong, D., Arancio, O., Sweatt, J. D., & Zoghbi, H. Y. (2006). Learning and memory and synaptic plasticity are impaired in a mouse model of Rett syndrome. *The Journal of Neuroscience, 26*, 319–327.

Mori, Y., Imaizumi, K., Katayama, T., Yoneda, T., & Tohyama, M. (2000). Two cis-acting elements in the 3′ untranslated region of alpha-CaMKII regulate its dendritic targeting. *Nature Neuroscience, 3*, 1079–1084.

Murray, K. D., Isackson, P. J., & Jones, E. G. (2003). N-methyl-D-aspartate receptor dependent transcriptional regulation of two calcium/calmodulin-dependent protein kinase type II isoforms in rodent cerebral cortex. *Neuroscience, 122*, 407–420.

Nakamura, T., Barbara, J. G., Nakamura, K., & Ross, W. N. (1999). Synergistic release of Ca^{2+} from IP3-sensitive stores evoked by synaptic activation of mGluRs paired with backpropagating action potentials. *Neuron, 24*, 727–737.

Nakamura, T., Nakamura, K., Lasser-Ross, N., Barbara, J. G., Sandler, V. M., & Ross, W. N. (2000). Inositol 1,4,5-trisphosphate (IP3)-mediated Ca^{2+} release evoked by metabotropic agonists and backpropagating action potentials in hippocampal CA1 pyramidal neurons. *The Journal of Neuroscience, 20*, 8365–8376.

Nakamura, T., Lasser-Ross, N., Nakamura, K., & Ross, W. N. (2002). Spatial segregation and interaction of calcium signalling mechanisms in rat hippocampal CA1 pyramidal neurons. *The Journal of Physiology, 543*, 465–480.

Nakazawa, H., & Murphy, T. H. (1999). Activation of nuclear calcium dynamics by synaptic stimulation in cultured cortical neurons. *Journal of Neurochemistry, 73*, 1075–1083.

Neher, E. (Ed.). (1986). *Concentration profiles of intracellular calcium in the presence of a diffusible chelator* (Vol. 14). Berlin: Springer.

Nelson, E. D., Kavalali, E. T., & Monteggia, L. M. (2006). MeCP2-dependent transcriptional repression regulates excitatory neurotransmission. *Current Biology, 16*, 710–716.

Nguyen, P. V., Abel, T., & Kandel, E. R. (1994). Requirement of a critical period of transcription for induction of a late phase of LTP. *Science, 265*, 1104–1107.

Olveira, A. M. M., & Bading, H. (2011). Calcium signaling in cognition and aging-dependent cognitive decline. *Biofactors, 37*(3), 168–174.

O'Malley, K. L., Jong, Y. J., Gonchar, Y., Burkhalter, A., & Romano, C. (2003). Activation of metabotropic glutamate receptor mGlu5 on nuclear membranes mediates intranuclear Ca^{2+} changes in heterologous cell types and neurons. *The Journal of Biological Chemistry, 278*, 28210–28219.

Orban, P. C., Chapman, P. F., & Brambilla, R. (1999). Is the Ras-MAPK signalling pathway necessary for long-term memory formation? *Trends in Neurosciences, 22*, 38–44.

Osawa, M., Tong, K. I., Lilliehook, C., Wasco, W., Buxbaum, J. D., Cheng, H. Y., Penninger, J. M., Ikura, M., & Ames, J. B. (2001). Calcium-regulated DNA binding and oligomerization of the neuronal calcium-sensing protein, calsenilin/DREAM/KChIP3. *The Journal of Biological Chemistry, 276*, 41005–41013.

Papadia, S., Stevenson, P., Hardingham, N. R., Bading, H., & Hardingham, G. E. (2005). Nuclear Ca^{2+} and the cAMP response element-binding protein family mediate a late phase of activity-dependent neuroprotection. *The Journal of Neuroscience, 25*, 4279–4287.

Pedreira, M. E., Dimant, B., & Maldonado, H. (1996). Inhibitors of protein and RNA synthesis block context memory and long-term habituation in the crab Chasmagnathus. *Pharmacology Biochemistry and Behavior, 54*, 611–617.

Peters, M., Mizuno, K., Ris, L., Angelo, M., Godaux, E., & Giese, K. P. (2003). Loss of $Ca^{2+}/$ calmodulin kinase kinase beta affects the formation of some, but not all, types of hippocampus-dependent long-term memory. *The Journal of Neuroscience, 23*, 9752–9760.

Picciotto, M. R., Zoli, M., Bertuzzi, G., & Nairn, A. C. (1995). Immunochemical localization of calcium/calmodulin-dependent protein kinase I. *Synapse, 20*, 75–84.

Pittenger, C., Huang, Y. Y., Paletzki, R. F., Bourtchouladze, R., Scanlin, H., Vronskaya, S., & Kandel, E. R. (2002). Reversible inhibition of CREB/ATF transcription factors in region CA1 of the dorsal hippocampus disrupts hippocampus-dependent spatial memory. *Neuron, 34*, 447–462.

Plath, N., Ohana, O., Dammermann, B., Errington, M. L., Schmitz, D., Gross, C., Mao, X., Engelsberg, A., Mahlke, C., Welzl, H., Kobalz, U., Stawrakakis, A., Fernandez, E., Waltereit, R., Bick-Sander, A., Therstappen, E., Cooke, S. F., Blanquet, V., Wurst, W., Salmen, B., Bosl, M. R., Lipp, H. P., Grant, S. G., Bliss, T. V., Wolfer, D. P., & Kuhl, D. (2006). Arc/Arg3.1 is essential for the consolidation of synaptic plasticity and memories. *Neuron, 52*, 437–444.

Ploski, J. E., Park, K. W., Ping, J., Monsey, M. S., & Schafe, G. E. (2010). Identification of plasticity-associated genes regulated by Pavlovian fear conditioning in the lateral amygdala. *Journal of Neurochemistry, 112*, 636–650.

Pokorska, A., Vanhoutte, P., Arnold, F. J., Silvagno, F., Hardingham, G. E., & Bading, H. (2003). Synaptic activity induces signalling to CREB without increasing global levels of cAMP in hippocampal neurons. *Journal of Neurochemistry, 84*, 447–452.

Pologruto, T. A., Yasuda, R., & Svoboda, K. (2004). Monitoring neural activity and $[Ca^{2+}]$ with genetically encoded Ca^{2+} indicators. *The Journal of Neuroscience, 24*, 9572–9579.

Power, J. M., & Sah, P. (2002). Nuclear calcium signaling evoked by cholinergic stimulation in hippocampal CA1 pyramidal neurons. *The Journal of Neuroscience, 22*, 3454–3462.

Queisser, G., Wittmann, M., Bading, H., & Wittum, G. (2008). Filtering, reconstruction, and measurement of the geometry of nuclei from hippocampal neurons based on confocal microscopy data. *Journal of Biomedical Optics, 13*, 014009.

Raymond, C. R., & Redman, S. J. (2006). Spatial segregation of neuronal calcium signals encodes different forms of LTP in rat hippocampus. *The Journal of Physiology, 570*, 97–111.

Redmond, L. (2008). Translating neuronal activity into dendrite elaboration: Signaling to the nucleus. *Neurosignals, 16*, 194–208.

Redondo, R. L., & Morris, R. G. (2011). Making memories last: The synaptic tagging and capture hypothesis. *Nature Reviews Neuroscience, 12*, 17–30.

Regehr, W. G., & Tank, D. W. (1992). Calcium concentration dynamics produced by synaptic activation of CA1 hippocampal pyramidal cells. *The Journal of Neuroscience, 12*, 4202–4223.

Ribar, T. J., Rodriguiz, R. M., Khiroug, L., Wetsel, W. C., Augustine, G. J., & Means, A. R. (2000). Cerebellar defects in Ca^{2+}/calmodulin kinase IV-deficient mice. *The Journal of Neuroscience, 20*, RC107.

Rina, S., Jusuf, A. A., Sakagami, H., Kikkawa, S., Kondo, H., Minami, Y., & Terashima, T. (2001). Distribution of $Ca(^{2+})$/calmodulin-dependent protein kinase I beta 2 in the central nervous system of the rat. *Brain Research, 911*, 1–11.

Rosen, L. B., Ginty, D. D., Weber, M. J., & Greenberg, M. E. (1994). Membrane depolarization and calcium influx stimulate MEK and MAP kinase via activation of Ras. *Neuron, 12*, 1207–1221.

Sakagami, H., Umemiya, M., Saito, S., & Kondo, H. (2000). Distinct immunohistochemical localization of two isoforms of Ca^{2+}/calmodulin-dependent protein kinase kinases in the adult rat brain. *The European Journal of Neuroscience, 12*, 89–99.

Sakagami, H., Kamata, A., Nishimura, H., Kasahara, J., Owada, Y., Takeuchi, Y., Watanabe, M., Fukunaga, K., & Kondo, H. (2005). Prominent expression and activity-dependent nuclear translocation of Ca^{2+}/calmodulin-dependent protein kinase Idelta in hippocampal neurons. *The European Journal of Neuroscience, 22*, 2697–2707.

Saneyoshi, T., Wayman, G., Fortin, D., Davare, M., Hoshi, N., Nozaki, N., Natsume, T., & Soderling, T. R. (2008). Activity-dependent synaptogenesis: Regulation by a CaM-kinase kinase/CaM-kinase I/betaPIX signaling complex. *Neuron, 57*, 94–107.

Schmitt, J. M., Guire, E. S., Saneyoshi, T., & Soderling, T. R. (2005). Calmodulin-dependent kinase kinase/calmodulin kinase I activity gates extracellular-regulated kinase-dependent long-term potentiation. *The Journal of Neuroscience, 25*, 1281–1290.

Schwartz, N., Schohl, A., & Ruthazer, E. S. (2009). Neural activity regulates synaptic properties and dendritic structure in vivo through calcineurin/NFAT signaling. *Neuron, 62*, 655–669.

Schwenk, J., Harmel, N., Zolles, G., Bildl, W., Kulik, A., Heimrich, B., Chisaka, O., Jonas, P., Schulte, U., Fakler, B., & Klocker, N. (2009). Functional proteomics identify cornichon proteins as auxiliary subunits of AMPA receptors. *Science, 323*, 1313–1319.

Shalizi, A., Gaudilliere, B., Yuan, Z., Stegmuller, J., Shirogane, T., Ge, Q., Tan, Y., Schulman, B., Harper, J. W., & Bonni, A. (2006). A calcium-regulated MEF2 sumoylation switch controls postsynaptic differentiation. *Science, 311*, 1012–1017.

Sheng, M., & Greenberg, M. E. (1990). The regulation and function of c-fos and other immediate early genes in the nervous system. *Neuron, 4*, 477–485.

Sheng, M., McFadden, G., & Greenberg, M. E. (1990). Membrane depolarization and calcium induce c-fos transcription via phosphorylation of transcription factor CREB. *Neuron, 4*, 571–582.

Sheng, M., Thompson, M. A., & Greenberg, M. E. (1991). CREB: A $Ca(^{2+})$-regulated transcription factor phosphorylated by calmodulin-dependent kinases. *Science, 252*, 1427–1430.

Shi, Y., Suh, Y. H., Milstein, A. D., Isozaki, K., Schmid, S. M., Roche, K. W., & Nicoll, R. A. (2010). Functional comparison of the effects of TARPs and cornichons on AMPA receptor trafficking and gating. *Proceedings of the National Academy of Sciences of the United States of America, 107*, 16315–16319.

Shirakawa, H., & Miyazaki, S. (1996). Spatiotemporal analysis of calcium dynamics in the nucleus of hamster oocytes. *The Journal of Physiology, 494*, 29–40.

Soderling, T. R. (2000). CaM-kinases: Modulators of synaptic plasticity. *Current Opinion in Neurobiology, 10*, 375–380.

Squire, L. R., & Barondes, S. H. (1970). Actinomycin-D: Effects on memory at different times after training. *Nature, 225*, 649–650.

Stedman, D. R., Uboha, N. V., Stedman, T. T., Nairn, A. C., & Picciotto, M. R. (2004). Cytoplasmic localization of calcium/calmodulin-dependent protein kinase I-alpha depends on a nuclear export signal in its regulatory domain. *FEBS Letters, 566*, 275–280.

Sun, P., Enslen, H., Myung, P. S., & Maurer, R. A. (1994). Differential activation of CREB by Ca^{2+}/calmodulin-dependent protein kinases type II and type IV involves phosphorylation of a site that negatively regulates activity. *Genes & Development, 8*, 2527–2539.

Sun, P., Lou, L., & Maurer, R. A. (1996). Regulation of activating transcription factor-1 and the cAMP response element-binding protein by Ca^{2+}/calmodulin-dependent protein kinases type I, II, and IV. *The Journal of Biological Chemistry, 271*, 3066–3073.

Sweatt, J. D. (2004). Mitogen-activated protein kinases in synaptic plasticity and memory. *Current Opinion in Neurobiology, 14*, 311–317.

Takahashi, H., & Magee, J. C. (2009). Pathway interactions and synaptic plasticity in the dendritic tuft regions of CA1 pyramidal neurons. *Neuron, 62*, 102–111.

Takahashi, T., Svoboda, K., & Malinow, R. (2003). Experience strengthening transmission by driving AMPA receptors into synapses. *Science, 299*, 1585–1588.

Tao, X., Finkbeiner, S., Arnold, D. B., Shaywitz, A. J., & Greenberg, M. E. (1998). Ca^{2+} influx regulates BDNF transcription by a CREB family transcription factor-dependent mechanism. *Neuron, 20*, 709–726.

Thomas, G. M., & Huganir, R. L. (2004). MAPK cascade signalling and synaptic plasticity. *Nature Reviews Neuroscience, 5*, 173–183.

Thomas, K. L., Laroche, S., Errington, M. L., Bliss, T. V., & Hunt, S. P. (1994). Spatial and temporal changes in signal transduction pathways during LTP. *Neuron, 13*, 737–745.

Thut, P. D., & Lindell, T. J. (1974). Alpha-amanitin inhibition of mouse brain form II ribonucleic acid polymerase and passive avoidance retention. *Molecular Pharmacology, 10*, 146–154.

Tokumitsu, H., & Soderling, T. R. (1996). Requirements for calcium and calmodulin in the calmodulin kinase activation cascade. *The Journal of Biological Chemistry, 271*, 5617–5622.

Tully, T., & Quinn, W. G. (1985). Classical conditioning and retention in normal and mutant Drosophila melanogaster. *Journal of Comparative Physiology A, 157*, 263–277.

Ueda, T., Sakagami, H., Abe, K., Oishi, I., Maruo, A., Kondo, H., Terashima, T., Ichihashi, M., Yamamura, H., & Minami, Y. (1999). Distribution and intracellular localization of a mouse homologue of Ca^{2+}/calmodulin-dependent protein kinase Ibeta2 in the nervous system. *Journal of Neurochemistry, 73*, 2119–2129.

Vickers, C. A., Dickson, K. S., & Wyllie, D. J. (2005). Induction and maintenance of late-phase long-term potentiation in isolated dendrites of rat hippocampal CA1 pyramidal neurones. *The Journal of Physiology, 568*, 803–813.

Villers, A., Godaux, E., & Ris, L. (2010). Late phase of L-LTP elicited in isolated CA1 dendrites cannot be transferred by synaptic capture. *Neuroreport, 21*, 210–215.

Visnjic, D., & Banfic, H. (2007). Nuclear phospholipid signaling: Phosphatidylinositol-specific phospholipase C and phosphoinositide 3-kinase. *Pflügers Archiv, 455*, 19–30.

Waltereit, R., & Weller, M. (2003). Signaling from cAMP/PKA to MAPK and synaptic plasticity. *Molecular Neurobiology, 27*, 99–106.

Wang, J., Campos, B., Jamieson, G. A., Jr., Kaetzel, M. A., & Dedman, J. R. (1995). Functional elimination of calmodulin within the nucleus by targeted expression of an inhibitor peptide. *The Journal of Biological Chemistry, 270*, 30245–30248.

Watanabe, S., Hong, M., Lasser-Ross, N., & Ross, W. N. (2006). Modulation of calcium wave propagation in the dendrites and to the soma of rat hippocampal pyramidal neurons. *The Journal of Physiology, 575*, 455–468.

Wayman, G. A., Lee, Y. S., Tokumitsu, H., Silva, A. J., & Soderling, T. R. (2008). Calmodulin-kinases: Modulators of neuronal development and plasticity. *Neuron, 59*, 914–931.

Wei, F., Qiu, C. S., Liauw, J., Robinson, D. A., Ho, N., Chatila, T., & Zhuo, M. (2002). Calcium calmodulin-dependent protein kinase IV is required for fear memory. *Nature Neuroscience, 5*, 573–579.

Weislogel, J. M., Bengtson, C. P., Müller, M., Bujard, M., Schuster, C. M., & Bading H. (unpublished work). *Requirement for nuclear calcium signaling in Drosophila long-term memory.*

West, A. E., Chen, W. G., Dalva, M. B., Dolmetsch, R. E., Kornhauser, J. M., Shaywitz, A. J., Takasu, M. A., Tao, X., & Greenberg, M. E. (2001). Calcium regulation of neuronal gene expression. *Proceedings of the National Academy of Sciences of the United States of America, 98*, 11024–11031.

West, A. E., Griffith, E. C., & Greenberg, M. E. (2002). Regulation of transcription factors by neuronal activity. *Nature Reviews Neuroscience, 3*, 921–931.

Westenbroek, R. E., Ahlijanian, M. K., & Catterall, W. A. (1990). Clustering of L-type Ca^{2+} channels at the base of major dendrites in hippocampal pyramidal neurons. *Nature, 347*, 281–284.

Whitlock, J. P., Jr., Galeazzi, D., & Schulman, H. (1983). Acetylation and calcium-dependent phosphorylation of histone H3 in nuclei from butyrate-treated HeLa cells. *The Journal of Biological Chemistry, 258*, 1299–1304.

Wieczorek, L., Maas, J. W., Jr., Muglia, L. M., Vogt, S. K., & Muglia, L. J. (2010). Temporal and regional regulation of gene expression by calcium-stimulated adenylyl cyclase activity during fear memory. *PloS One, 5*, e13385.

Wiegert, J. S., & Bading, H. (2011). Activity-dependent calcium signaling and ERK-MAP kinases in neurons: A link to structural plasticity of the nucleus and gene transcription regulation. *Cell Calcium, 49*(5), 296–305.

Wiegert, J. S., Bengtson, C. P., & Bading, H. (2007). Diffusion and not active transport underlies and limits ERK1/2 synapse-to-nucleus signaling in hippocampal neurons. *The Journal of Biological Chemistry, 282*, 29621–29633.

Wittmann, M., Queisser, G., Eder, A., Wiegert, J. S., Bengtson, C. P., Hellwig, A., Wittum, G., & Bading, H. (2009). Synaptic activity induces dramatic changes in the geometry of the cell nucleus: Interplay between nuclear structure, histone H3 phosphorylation, and nuclear calcium signaling. *The Journal of Neuroscience, 29*, 14687–14700.

Wong, S. T., Athos, J., Figueroa, X. A., Pineda, V. V., Schaefer, M. L., Chavkin, C. C., Muglia, L. J., & Storm, D. R. (1999). Calcium-stimulated adenylyl cyclase activity is critical for hippocampus-dependent long-term memory and late phase LTP. *Neuron, 23*, 787–798.

Wu, Z. L., Thomas, S. A., Villacres, E. C., Xia, Z., Simmons, M. L., Chavkin, C., Palmiter, R. D., & Storm, D. R. (1995). Altered behavior and long-term potentiation in type I adenylyl cyclase mutant mice. *Proceedings of the National Academy of Sciences of the United States of America, 92*, 220–224.

Wu, L., Wells, D., Tay, J., Mendis, D., Abbott, M. A., Barnitt, A., Quinlan, E., Heynen, A., Fallon, J. R., & Richter, J. D. (1998). CPEB-mediated cytoplasmic polyadenylation and the regulation of experience-dependent translation of alpha-CaMKII mRNA at synapses. *Neuron, 21*, 1129–1139.

Wu, G. Y., Deisseroth, K., & Tsien, R. W. (2001). Activity-dependent CREB phosphorylation: Convergence of a fast, sensitive calmodulin kinase pathway and a slow, less sensitive mitogen-activated protein kinase pathway. *Proceedings of the National Academy of Sciences of the United States of America, 98*, 2808–2813.

Wu, L. J., Mellstrom, B., Wang, H., Ren, M., Domingo, S., Kim, S. S., Li, X. Y., Chen, T., Naranjo, J. R., & Zhuo, M. (2010). DREAM (downstream regulatory element antagonist modulator) contributes to synaptic depression and contextual fear memory. *Molecular Brain, 3*, 3–16.

Xing, J., Ginty, D. D., & Greenberg, M. E. (1996). Coupling of the RAS-MAPK pathway to gene activation by RSK2, a growth factor-regulated CREB kinase. *Science, 273*, 959–963.

Ye, K., & Ahn, J. Y. (2008). Nuclear phosphoinositide signaling. *Frontiers in Bioscience, 13*, 540–548.

Yin, J. C., Wallach, J. S., Del Vecchio, M., Wilder, E. L., Zhou, H., Quinn, W. G., & Tully, T. (1994). Induction of a dominant negative CREB transgene specifically blocks long-term memory in Drosophila. *Cell, 79*, 49–58.

Zal, T., & Gascoigne, N. R. (2004). Photobleaching-corrected FRET efficiency imaging of live cells. *Biophysical Journal, 86*, 3923–3939.

Zhang, S. J., Steijaert, M. N., Lau, D., Schutz, G., Delucinge-Vivier, C., Descombes, P., & Bading, H. (2007). Decoding NMDA receptor signaling: Identification of genomic programs specifying neuronal survival and death. *Neuron, 53*, 549–562.

Zhang, S. J., Zou, M., Lu, L., Lau, D., Ditzel, D. A., Delucinge-Vivier, C., Aso, Y., Descombes, P., & Bading, H. (2009). Nuclear calcium signaling controls expression of a large gene pool: Identification of a gene program for acquired neuroprotection induced by synaptic activity. *PLoS Genetics, 5*, e1000604.

Zhang, Y., Su, P., Liang, P., Liu, T., Liu, X., Liu, X. Y., Zhang, B., Han, T., Zhu, Y. B., Yin, D. M., Li, J., Zhou, Z., Wang, K. W., & Wang, Y. (2010). The DREAM protein negatively regulates the NMDA receptor through interaction with the NR1 subunit. *The Journal of Neuroscience, 30*, 7575–7586.

Zhao, M., Adams, J. P., & Dudek, S. M. (2005). Pattern-dependent role of NMDA receptors in action potential generation: Consequences on extracellular signal-regulated kinase activation. *The Journal of Neuroscience, 25*, 7032–7039.

Zhou, Z., Hong, E. J., Cohen, S., Zhao, W. N., Ho, H. Y., Schmidt, L., Chen, W. G., Lin, Y., Savner, E., Griffith, E. C., Hu, L., Steen, J. A., Weitz, C. J., & Greenberg, M. E. (2006). Brain-specific phosphorylation of MeCP2 regulates activity-dependent Bdnf transcription, dendritic growth, and spine maturation. *Neuron, 52*, 255–269.

Zhu, J., & McKeon, F. (2000). Nucleocytoplasmic shuttling and the control of NF-AT signaling. *Cellular and Molecular Life Sciences, 57*, 411–420.

Chapter 18
Integrating Neurotransmission in Striatal Medium Spiny Neurons

Jean-Antoine Girault

Abstract The striatum is a major entry structure of the basal ganglia. Its role in information processing in close interaction with the cerebral cortex and thalamus has various behavioral consequences depending on the regions concerned, including control of body movements and motivation. A general feature of striatal information processing is the control by reward-related dopamine signals of glutamatergic striatal inputs and of their plasticity. This relies on specific sets of receptors and signaling proteins in medium-sized spiny neurons which belong to two groups, striatonigral and striatopallidal neurons. Some signaling pathways are activated only by dopamine or glutamate, but many provide multiple levels of interactions. For example, the cAMP pathway is mostly regulated by dopamine D1 receptors in striatonigral neurons, whereas the ERK pathway detects a combination of glutamate and dopamine signals and is essential for long-lasting modifications. These adaptations require changes in gene expression, and the signaling pathways linking synaptic activity to nuclear function and epigenetic changes are beginning to be deciphered. Their alteration underlies many aspects of striatal dysfunction in pathological conditions which include a decrease or an increase in dopamine transmission, as encountered in Parkinson's disease or exposure to addictive drugs, respectively.

Keywords Cytonuclear signaling • DARPP-32 • Epigenetics • Extracellular signal-regulated kinase • Striatal medium sized spiny neurons

J.-A. Girault (✉)
Institut du Fer à Moulin, UMR-S 839, Inserm and Université Pierre et Marie Curie (UPMC),
17 rue du Fer à Moulin, 75005 Paris, France
e-mail: jean-antoine.girault@inserm.fr

M.R. Kreutz and C.Sala (eds.), *Synaptic Plasticity*,
Advances in Experimental Medicine and Biology 970,
DOI 10.1007/978-3-7091-0932-8_18, © Springer-Verlag/Wien 2012

18.1   Introduction: An Overview of Striatal Functions and Dysfunctions

The striatum is the largest component of the basal ganglia, a set of interconnected gray matter nuclei located deep in the forebrain, to which it provides a major site of entry. Its name comes from the striate aspect provided by radial crossing of white matter bundles. Although the whole striatum has some common principles of anatomical organization, it is in reality a heterogeneous structure with several types of divisions. A first major division corresponds to its dorsal and ventral parts. In many mammalian species, including carnivores and primates, the dorsal striatum is further separated by the internal capsule into two components, the putamen and the caudate nucleus. In mice or rats, this separation does not exist, and the dorsal striatum is often referred to as the caudate-putamen (CP). The ventral part of the striatum is known as the *nucleus accumbens septi* (the "nucleus of the septum which is lying down," although it is not directly connected to the septum) or nucleus accumbens (NAc). The NAc is comprised of a central core, similar to the CP, and a shell thought to be related to a set of scattered nuclei at the inferior part of the forebrain, the extended amygdala (Alheid and Heimer 1988; Voorn et al. 2004). There are no sharp anatomical borders between these striatal regions, and they have a similar histological appearance. However, important functional differences arise from the diversity of the brain regions to which they are connected.

The striatum receives excitatory glutamatergic inputs from the cerebral cortex and the thalamus. It is also the major target of the dopamine (DA) neurons whose cell bodies are located in the upper part of the brain stem, the mesencephalon. The substantia nigra (so called because of the black neuromelanin pigment accumulated in DA neurons in some species including primates) provides a dense innervation to the CP, whereas the ventral tegmental area mostly innervates the NAc (Fig. 18.1a–c). Globally, although they have relatively widespread terminals, the glutamatergic and dopaminergic fibers arriving to the striatum are topographically organized, providing the basis for its dorsoventral and lateromedial functional organization. As a unifying lead to understand the function of the striatum, it has been proposed that it is mostly involved in the selection of action (Mink 1996; Redgrave et al. 1999). The motor resources available to the brain are limited, and to drive efficient behavior, it must devote them to a single aim at a time. The striatum and basal ganglia are thought to be essential to the processing of this complex computing problem. The dorsal striatum in close association with the motor cortex is involved in selection of elaborated motor patterns and automated sets of behaviors (so-called extrapyramidal control of motor function), whereas the ventral striatum is implicated in the choice of behavioral orientation, interpreted in psychological terms as desire or motivation. The dopamine innervation is essential for both the efficient acute function of the striatum and its plasticity (Wickens et al. 2007), hence for acquiring elaborate motor abilities (a component of procedural memory) or motivated responses. Importantly, DA neurons firing is increased in relation with rewards. However, DA neurons do not code a simple reward, but

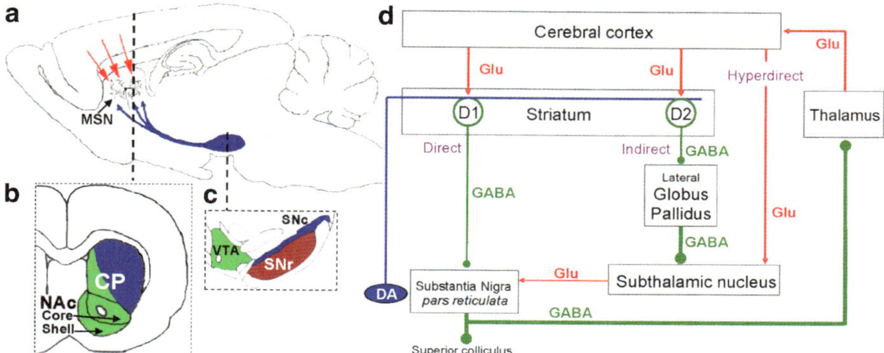

Fig. 18.1 *Anatomical organization of the striatum.* (**a**) Sagittal mouse brain section showing the striatal glutamate inputs from the cerebral cortex and the dopamine innervation from the mesencephalon. The thalamic input is not shown. (**b**) Coronal section at the level indicated by the *dashed line*, showing the dorsal region or caudate-putament (CP) and the ventral region, or nucleus accumbens (NAc) with its core and shell subdivisions. (**c**) Coronal section of the ventral mesencephalon at the level indicated by the *dashed line*, showing the localization of the *substantia nigra reticulata* (SNr) the main output structure of the basal ganglia, of the *substantia nigra compacta* (SNc) and the ventral tegmental area (VTA) which both contain dopamine neurons projecting to the striatum. The areas of projection of VTA (*green*) and SNc (*blue*) are shown in (**b**). (**d**) Schematic wiring diagram of the basal ganglia showing the opposite role of the two striatal efferent neurons. The medium-sized spiny neurons (MSN) of the direct pathway inhibit the substantia nigra reticulata neurons. The MSNs of the indirect pathway have an opposite effect because of the inhibitory link formed by the neuron of the lateral globus pallidus. SNr neurons are GABA neurons with a high basal firing rate. Activation of the direct striatonigral pathway disinhibits thalamocortical neurons. The indirect pathway has the opposite effect. DA reinforces the direct pathway and inhibits the indirect pathway, allowing a harmonious function of the basal ganglia circuits. In the absence of DA, the inhibitory effect of the SNr is unchecked and results in a lack of movement and rigidity

rather a reward error prediction signal, which has a great interest for the computing properties of the basal ganglia and the control of their plasticity (Schultz 2010). DA neurons are also augmented by alerting signals involved in rapid detection of potentially important sensory cues (Bromberg-Martin et al. 2010). This simple view allows understanding the two major pathological disturbances of the striatum and of its DA innervation (detailed in two other chapters of this book). In Parkinson's disease, the degeneration of DA neurons innervating the CP results in a decrease or absence of spontaneous movements (akinesia) and rigidity. Conversely, addictive drugs share the property to increase extracellular dopamine in the NAc, artificially mimicking a reward-related learning signal. This results in abnormal, chemically driven "learning" of drug consumption, which can escape any control however deleterious the consequences may be, thus corresponding to the definition of addiction. Hence, elucidating the cellular and molecular basis of neurotransmission and signaling in the striatum is of great interest to understand basic aspects of brain functions and major human diseases including Parkinson's and addiction.

18.2   The Striatal Medium-Sized Spiny Neurons

In rodents, the striatal neurons are made up of about 95% medium-sized spiny neurons (MSNs) and 5% interneurons (Kreitzer 2009; Tepper et al. 2010). The dendrites of MSNs are covered with spines which receive corticostriatal or thalamostriatal terminals forming asymmetric synapses (Doig et al. 2010). Although each MSN receives 5–10,000 excitatory inputs, usually, only a few come from the same cortical neurons, and simultaneous firing of several inputs is necessary for their activation. Therefore, MSNs are wired to integrate many convergent excitatory inputs, while they are highly sensitive to inhibition by interneurons. DA boutons are found on the neck of some spines, and glutamatergic synapses are never far from DA synapses, within range of a "spillover" concentration of synaptically released dopamine sufficient to stimulate receptors (Moss and Bolam 2008). MSNs also receive inhibitory inputs from striatal GABAergic interneurons, strategically located on their perikarya, and from cholinergic interneurons (Tepper et al. 2010). MSNs are GABAergic efferent neurons which project to the globus pallidus (GP) and substantia nigra *pars reticulata* (SNR) and, in the case of NAc MSNs, to the ventral pallidum. MSNs are inhibitory neurons, a rather unusual feature for long projection neurons in the nervous system. Their target neurons in the SNR or GP are also GABAergic neurons with a high basal firing rate. Thus, the functional result of MSNs firing is to "disinhibit" the targets of SNR or GP neurons (Chevalier and Deniau 1990). In fact, MSNs have fundamentally different functional properties depending on their targets (Fig. 18.1d). About half of them are part of the direct pathway which projects "directly" to the SNR or internal GP, which are output stations of the basal ganglia (Gerfen 1992). Direct pathway neurons, thus, disinhibit circuits negatively controlled by the basal ganglia. Conversely, MSNs in the "indirect" pathway project to the external GP that projects to the subthalamic nucleus, which, in turn, sends excitatory projections to the SNR and GPi. The net result of the action of MSNs in the indirect pathway is exactly the opposite of those in the direct pathway, namely, they reinforce the inhibition exerted by the basal ganglia on their targets, by increasing the activity of the SNR and GPi output neurons (Fig. 18.1d). This organization is essential to understand the function of the basal ganglia; the two striatal efferent pathways have opposite functional effects: the direct pathway disinhibits the targets (e.g., thalamocortical or superior colliculus neurons), whereas the indirect pathway reinforces this inhibition. The balance between these two pathways is central for the action selection function of the basal ganglia: a specific set of neurons involved in a complex motor behavior would be disinhibited through selective activation of a limited number of striatonigral MSNs, while the rest of the relevant circuits would be inhibited through striatopallidal MSNs. Recent studies using gene targeting technologies, in which either the striatopallidal or the striatonigral MSNs were selectively impaired (Hikida et al. 2010) or destroyed (Durieux et al. 2009), or stimulated through channelrhodopsin (Kravitz et al. 2010), have elegantly confirmed the opposite actions of the direct and indirect pathways. Importantly, DA receptors are unevenly distributed among these two populations of neurons (Gerfen 2000). D1 receptors are enriched on MSNs of the direct pathway, whereas D2 receptors are mostly located on indirect pathway MSNs. Since D1 receptors have an overall excitatory effect and the D2 receptors an inhibitory effect, DA regulates the balance between the two efferent pathways. In the absence of DA, the predominant inhibition of basal ganglia targets accounts for the akinesia and rigidity of parkinsonian patients. Striatopallidal neurons selectively express enkephalin and A2a adenosine receptors, whereas striatonigral neurons express substance P, dynorphin, and muscarinic m4 acetylcholine receptors. Recently, the two populations of striatal MSNs have been labeled in transgenic mouse using bacterial artificial chromosomes driving the expression of a variety of marker proteins (EGFP, tomato, fusion proteins, Cre recombinase, and others...) under the control of D1, D2, m4 acetylcholine, or A2a adenosine receptors (see Valjent et al. 2009). These studies have further supported the differences between striatonigral and striatopallidal MSNs in terms of gene expression, signaling pathways, morphology, and electrophysiological properties.

Another level of heterogeneity of striatal neurons corresponds to the striosomes (or patch) and matrix compartments (Graybiel and Hickey 1982; Gerfen 1984). In most adult mammals, acetylcholinesterase activity and somatostatin-immunoreactive fibers are primarily localized to the matrix, as are GABAergic neurons that coexpress calbindin. In contrast, μ-opioid receptors, cholinergic muscarinic receptors, and substance P fibers are concentrated in the striosomes. The afferences and targets of MSNs in the striosomes and in the matrix are in part different, and they have been proposed to be organized as two imbedded loop systems, with the striosome-DA neurons controlling the matrix effector system (Gerfen 1992).

18.3 Major Neurotransmitters Regulating MSNs and Their Signaling Pathways

MSNs receive a massive glutamate excitatory input which stimulates ionotropic and metabotropic receptors. They are also innervated by GABA interneurons and collaterals from other MSNs which can act on GABA-A and GABA-B receptors, as well as by cholinergic interneurons stimulating muscarinic and cholinergic acetylcholine receptors. MSNs abundantly express DA receptors, which all belong to the G-protein-coupled receptors (GPCRs) family, in contrast to receptors of glutamate, GABA, and acetylcholine, which include both ionotropic and metabotropic receptors. The D1 class of DA receptors includes the D1 and D5 (also called D1a and D1b) receptors, and the D2 class includes D2, D3, and D4 receptors. The D1 and D5 receptors are able to activate adenylyl cyclase, whereas D2-type receptors have the opposite effect (Fig. 18.2). MSNs predominantly express D1 and D2 receptors, which are located in neurons of the direct and indirect pathway, respectively (Gerfen 2000; Valjent et al. 2009). In the ventral striatum, MSNs also express D3 receptors (Sokoloff et al. 1990). The MSNs express a large variety of other GPCRs for serotonin, neuropeptides, and orphan receptors. GPCRs abundant in MSNs include adenosine A2a receptors restricted to striatopallidal neurons

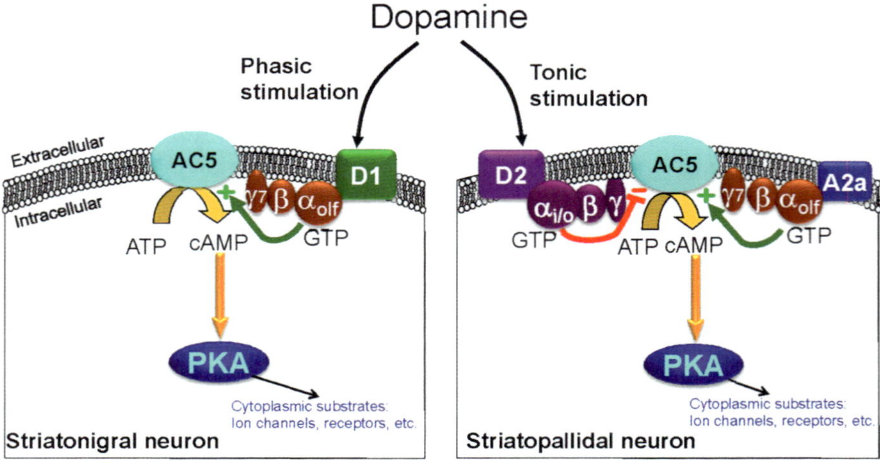

Fig. 18.2 *Coupling of striatal dopamine (DA) receptors.* For the most part, D1 and D2 receptors are located in striatonigral and striatopallidal neurons, respectively. D1 receptors are coupled to a heterotrimeric G protein comprising a $G\alpha_{olf}$, a β, and a γ_7 subunit. When bound to GTP and dissociated from the complex, $G\alpha_{olf}$ activates adenylyl cyclase, mostly AC5. D2 receptors inhibit AC5 through a Gi/o protein. In striatopallidal neurons, AC5 is activated by A2a adenosine receptors through the $G\alpha_{olf}$-β-γ_7 complex. D1 receptors are recruited by a phasic increase in DA release, while D2 receptors, which have a higher affinity for DA, are tonically activated by ambient levels of extracellular DA. Blockade of D2 receptors unmasks the action of A2a receptors

(Schiffmann and Vanderhaeghen 1993) and CB1 cannabinoid receptors which are highly enriched in both MSNs populations and are mostly located at their terminals (van der Stelt and Di Marzo 2003). Interestingly, GPCRs can form homo- and heterodimers, with possibly distinct coupling properties, whose physiological relevance is the topic of ongoing studies (Ferre et al. 2007). Finally, direct associations between DA and ionotropic receptors, with functional consequences on both types of receptors, have been identified, including D1 receptor binding to the intracellular loops of NR1 and NR2A subunits of the glutamate NMDA (*N*-methyl-D-aspartate) receptors (Lee et al. 2002a). In addition, DA and glutamate receptors cross regulate their expression, membrane localization, and function (Scott and Aperia 2009).

Beyond the receptor level, striatal neurons are endowed with a particular set of signaling proteins that distinguish them from other neurons. Adult MSNs contain very little, if any, $G\alpha_s$ subunit, and positive coupling of GPCRs to adenylyl cyclase is mediated by the $G\alpha_{olf}$ subunit, initially identified in the olfactory epithelium (Herve et al. 1993; Zhuang et al. 2000; Corvol et al. 2001). The associated gamma subunit is mostly γ_7 (Schwindinger et al. 2003, 2010). The functional consequences of the expression of these particular isoforms are not known, as their biochemical properties appear similar to the others. It is quite possible that an important aspect of the use of these specific genes is to allow high levels of expression. Indeed, $G\alpha_{olf}$ appears to be a limiting factor for D1 signaling since decreased levels of $G\alpha_{olf}$ subunit but not of D1 receptor profoundly alter cAMP signaling (Corvol et al. 2007).

The main adenylyl cyclase isoform in MSNs is the Ca^{2+}-insensitive AC5 isoform (Lee et al. 2002b). This is a major difference as compared to cortical and hippocampal neurons, for example, in which the presence of Ca^{2+}-activated cyclase provides a direct coupling between Ca^{2+} influx and cAMP production. In contrast, in MSNs cAMP production requires a second signal, such as DA in striatonigral or adenosine in striatopallidal neurons. cAMP is degraded by several phosphodiesterase (PDE) families expressed in striatal neurons (Menniti et al. 2006), including PDE1B (Polli and Kincaid 1994; Reed et al. 2002), PDE4B (Siuciak et al. 2008), and PDE10A (Fujishige et al. 1999; Siuciak et al. 2006).

The major target of cAMP is the cAMP-dependent protein kinase (PKA). The phosphorylation pathways activated by PKA have been extensively studied in the MSNs, in which it regulates a number of voltage-gated and ligand-gated ion channels (Surmeier et al. 2007). In addition, several protein substrates specifically enriched in the striatum have been identified, which are regulatory signaling proteins themselves regulated by phosphorylation. The best studied is dopamine- and cAMP-regulated phosphoprotein, Mr ~ 32,000 (DARPP-32), which is an inhibitor of protein phosphatase 1 (Walaas et al. 1983; Hemmings et al. 1984). Others include cAMP-regulated phosphoprotein, Mr ~ 21,000 (ARPP-21) (Hemmings and Greengard 1989; Ouimet et al. 1989), also termed regulator of calcium signaling (RCS), which is an inhibitor of Ca^{2+}/calmodulin targets when phosphorylated by PKA (Rakhilin et al. 2004), and cAMP-regulated phosphoprotein, Mr ~ 16,000 (ARPP-16) (Girault et al. 1990), recently shown to be an inhibitor of PP2A-55δ when phosphorylated by a Greatwall family kinase (Gharbi-Ayachi et al. 2010; Mochida et al. 2010). DARPP-32 inhibits PP1 when it is phosphorylated on Thr-34 by PKA (Hemmings et al. 1984). Phosphorylation on Thr-34 is enhanced by protein kinases CK2 and CK1, which increases its phosphorylation by PKA and prevents its dephosphorylation by calcineurin, respectively (Girault et al. 1989; Desdouits et al. 1995). Thus, DARPP-32 is a signaling hub that plays a critical role in striatal neurons and is involved in numerous physiological and pharmacological responses (Svenningsson et al. 2004; Le Novere et al. 2008). In the absence of DARPP-32, some responses are blunted while others are absent (Fienberg et al. 1998). For example, it is important for plasticity at corticostriatal synapses (Calabresi et al. 2000). In striatonigral neurons, DARPP-32 is phosphorylated on Thr-34 in response to stimulation of D1 receptors, whereas in striatopallidal neurons, the same response is triggered by blocking D2 receptors (Bateup et al. 2008). The effect of D2 antagonists unmasks a tonic D2 inhibition of adenylyl cyclase, which is also continuously activated by A2a (Svenningsson et al. 2000) and possibly other receptors stimulating adenylyl cyclase. The loss of DARPP-32 in striatonigral neurons decreases basal and cocaine-induced locomotion and abolishes dyskinetic behaviors in response to the Parkinson's disease drug L-dopa (Bateup et al. 2010). Conversely, the loss of DARPP-32 in striatopallidal neurons increases locomotor activity and strongly reduces the cataleptic response to an antipsychotic D2 antagonist (Bateup et al. 2010). Remarkably, when DARPP-32 is phosphorylated on Thr-75 by Cdk5, it becomes a potent inhibitor of PKA (Bibb et al. 1999). Regulation of Thr-75 phosphorylation appears to be to some extent a mirror image of regulation of Thr-34 phosphorylation (Bateup et al. 2008). Thus, DARPP-32 is a switch that can act as a feed-forward amplifier of PKA effects when phosphorylated on Thr-34 and dephosphorylated on Thr-75, or, in dramatic contrast, as a PKA inhibitor when phosphorylated on Thr-75 by Cdk5. Perhaps, this critical role in DA signaling and exquisite regulation accounts for the apparent correlation of DARPP-32 with cognitive abilities in humans (Meyer-Lindenberg et al. 2007) and mice (Kolata et al. 2010).

18.4 Cytonuclear Signaling in Striatal Neurons

Plasticity of corticostriatal synapses, and presumably also of the much less studied thalamostriatal synapses, is thought to play a central role in long-lasting behavioral adaptations that depend on striatal function. These behavioral modifications include habit learning (Barnes et al. 2005) and incentive learning (Dayan and Balleine 2002; Belin et al. 2009), as well as pathological modifications of striatal functions as exemplified by drug conditioning (Hyman et al. 2006) or L-dopa-induced dyskinesia in Parkinson's disease or its animal models (Santini et al. 2008). Dopamine plays a critical role in the control of corticostriatal plasticity, but the regulation of synaptic plasticity in the striatum is complex, and results are highly dependent on experimental conditions (Wickens 2009). Importantly, long-lasting behavioral adaptations such as locomotor sensitization induced by psychostimulants or opiates in rodents, as well as conditioned place preference (CPP), require gene transcription and protein synthesis. Therefore, much attention has been devoted to the regulation of gene expression in striatal neurons.

Early studies showed that the administration of nonspecific dopamine agonists such as cocaine, which prevents dopamine and other monoamine reuptake, or amphetamine, which promotes the release of dopamine and other monoamines, induced expression of immediate-early genes (IEG) in the striatum, including cFos, Zif268, ARC, and others (Gerfen 2000). Similar results were obtained with selective dopamine D1 agonists, but not D2 agonists. In contrast, D2 antagonists, not agonists, were able to induce IEG expression. Repeated administration of cocaine induced different patterns of gene expression, such as the slow-building levels of ΔFos-B, which was therefore proposed to play a selective role in the effects of drugs of abuse (Hope et al. 1994). The targets for such long-term regulations are not yet fully characterized but include negative feedback loops (Bibb et al. 2001). Recent work has started to identify the changes in gene expression involved in the regulation of spine density. For example, the downregulation of myocyte enhancer factor 2 (MEF2) is involved in cocaine-induced spine increase (Pulipparacharuvil et al. 2008), whereas its activation mediates the depolarization-induced spine pruning in a model of Parkinson's disease (Tian et al. 2010).

How are messages carried from the plasma membrane receptors to the nucleus? Although as in other cells, many pathways are probably used in MSNs, relatively few of them have been characterized, and this characterization revealed interesting specificities.

18.4.1 Extracellular Signal-Regulated Kinase (ERK) Pathway

Initial observations showed that neurotransmitters activating cAMP production, especially DA, induced transcriptional effects in striatal neurons. It was generally assumed that these effects were directly mediated by PKA since in other systems free PKA catalytic subunit is known to translocate to the nucleus and to phosphorylate nuclear proteins including the cAMP-response element-binding (CREB) protein (Montminy 1997). However, little evidence supported a direct role of PKA in the nucleus of MSNs. Recent work suggests that although this role remains likely, other pathways are very important in striatal neurons. The first line of evidence came from the demonstration that the ERK pathway is activated in striatal neurons following electrical stimulation of the cerebral cortex and that it mediates the phosphorylation of two transcription factors, CREB and Elk-1, and the induction of IEGs (cFos, Zif268, and MKP1) (Sgambato et al. 1998). ERK1 and ERK2 are two closely related mitogen-activated protein kinases (MAP-kinases) which are activated by phosphorylation of their activation loop by MAP-kinase and ERK-kinase (MEK 1 and 2). Glutamate was found capable to increase Elk-1 and CREB phosphorylation through activation of ERK1/2 and MEK1/2 in striatal slices (Vanhoutte et al. 1999). The same group then showed that injection of cocaine to mice induces the nuclear accumulation of active di-phospho ERK1/2 and that the activation of ERK is necessary for cocaine-induced IEG expression and CPP (Valjent et al. 2000). Further work showed that phosphorylation of ERK1/2 in the ventral striatum is a common effect of all drugs of abuse including cocaine, amphetamine, morphine, nicotine, Δ9-tetrahydrocannabinol, methamphetamine, 3,4-methylenedioxy-methamphetamine (MDMA, also known as ecstasy), and ethanol (Valjent et al. 2001; 2004; Salzmann et al. 2003; Ibba et al. 2009). This activation appeared to be functionally important since pharmacological blockade of MEK, usually with SL327, an inhibitor of MEK that crosses the blood-brain barrier, prevented the long-term behavioral effects of these drugs. ERK activation was also involved in the reconsolidation of drug-associated memories since it was reactivated by exposure to the drug-associated context and because MEK inhibition was able to erase previously acquired CPP (Miller and Marshall 2005; Valjent et al. 2006a). In animals, in which the DA neurons have been destroyed by a 6-OH-DA lesion, the DA precursor L-dopa, which is the standard treatment of Parkinson's disease, induces a strong activation of ERK in striatal neurons (Santini et al. 2007). L-dopa has no effect in the absence of lesion. Interestingly, pharmacological inhibition of MEK prevents the appearance of L-dopa-induced dyskinesia, which is thought to result at least in part from abnormal plasticity in the DA-depleted striatum subjected to strong intermittent DA stimulation (Santini et al. 2007).

The mechanism of activation of ERK involves both D1 DA receptors and NMDA glutamate receptors (Valjent et al. 2000, 2005) (Fig. 18.3). ERK activation is observed only in a subset of D1-expressing striatonigral neurons and is prevented by a D1 antagonist or in D1 knockout mice, or by an NMDA antagonist, indicating that it requires the concomitant stimulation of both D1 and NMDA receptors.

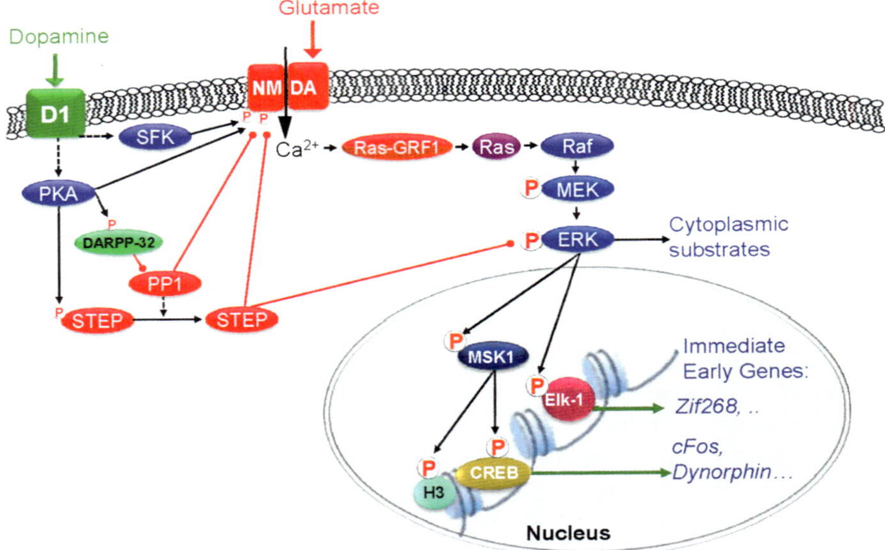

Fig. 18.3 *The extracellular signal-regulated kinase (ERK) pathway in striatal MSNs.* The ERK pathway is triggered by the coincident activation of dopamine D1 and glutamate NMDA receptors. D1 receptors potentiate the effects of glutamate through serine phosphorylation of NR1 subunit by PKA and tyrosine phosphorylation of NR2 subunits by a Src-family kinase (SFK). This may increase the conductance and membrane expression of the receptor. D1 also potentiates ERK activation through PKA phosphorylation of DARPP-32 which inhibits PP1. PP1 normally activates the striatal-enriched tyrosine phosphatase STEP which dephosphorylates the Tyr residues on ERK activation loop and NMDA receptors NR2 subunits. PP1 also dephosphorylates NMDA NR1 subunit on Ser. Activation of ERK is carried out in part by Ca^{2+}-activated Ras-GRF1. The Ras, Raf, MEK1/2 pathway activates ERK1/2 which has cytoplasmic and nuclear substrates. Among the latter, the transcription factor Elk-1 is a component of the ternary complex factor which binds to serum response elements (SRE). ERK1/2 also phosphorylates the nuclear kinase MSK1 which appears to play a prominent role in phosphorylation of histone H3 and cAMP-response element-binding (CREB) protein. This leads to the expression of immediate-early genes which are particularly sensitive to CREB (e.g., cFos) or ternary complex factor (e.g., Zif268 a.k.a. Egr-1). These two nuclear pathways downstream from ERK have distinct roles in long-term behavioral adaptations (see text). *Arrows* with *round ends* indicate inhibitory effects

Thus, the ERK pathway behaves as a coincidence detector or logical AND gate which detects the simultaneous activation of the contextual information coded by corticostriatal and thalamostriatal glutamate inputs, and the reward prediction error coded by DA neurons (Girault et al. 2007). This important property distinguishes the ERK pathway from the stimulation of the PKA pathway which results solely from the stimulation of D1 receptors and is more widespread in D1-expressing neurons (Valjent et al. 2005). The mechanism of activation of ERK upstream from MEK involves the Ca^{2+}-activated guanine nucleotide exchange factor Ras-GRF1 (Fasano et al. 2010). Other pathways must exist since ERK activation was decreased but not completely blocked in Ras-GRF1 null mice. They may include

CalDAG-GEF-1/2, two guanine nucleotide exchange factors activated by Ca^{2+} and diacylglycerol, present in striatal neurons, respectively enriched in the matrix and the striosomes (Toki et al. 2001; Crittenden et al. 2009). Interestingly, Ras-GRF1 and possibly CalDAG-GEF-1/2 dysregulation appears to be involved in the occurrence of L-dopa-induced dyskinesia in animal models of Parkinson's disease (Crittenden et al. 2009; Fasano et al. 2010). The cross talk between D1 and NMDA receptors takes place at multiple levels. Recent work showed that stimulation of D1 receptors increases responsiveness of NMDA receptors to glutamate through a cAMP-independent pathway which involves phosphorylation of NMDA receptor NR2B subunit by a Src-family tyrosine kinase (Pascoli et al. 2011). This cross talk increases Ca^{2+} influx through NMDA receptors and, consequently, ERK phosphorylation. In addition, PKA-mediated phosphorylation of DARPP-32 plays a critical role at several levels (Valjent et al. 2005). Phospho-Thr-34-DARPP-32 inhibits the PP1-induced activation of the striatal-enriched tyrosine phosphatase STEP, which plays a critical role in the dephosphorylation of ERK (Paul et al. 2003). DARPP-32 also acts upstream from MEK (Valjent et al. 2005), possibly by enhancing the phosphorylation of NMDA receptors on serine residues via PP1 inhibition (Snyder et al. 1998) and/or on tyrosine residues via STEP inhibition (Paul et al. 2003).

A critical question concerns the relevant targets of ERK1/2. Cytoplasmic substrates regulated by ERK may play an important role in its functional effects. For example, in hippocampal neurons, ERK phosphorylates and inhibits Kv4.2, a K^+ channel which controls the retropropagation of action potentials in the dendritic tree and the degree of depolarization sensed by NMDA receptors, a regulation critical for spike-timing-dependent plasticity (Sweatt 2004). However, this aspect has not been studied in striatal neurons. A recent publication has shown that ERK activation is necessary for corticostriatal LTP induced by electrical stimulation in slices or by cocaine injection in vivo (Pascoli et al. 2012). This work provided direct evidence for a role of this synaptic plasticity in cocaine-induced locomotor sensitization. On the other hand, phosphorylated active ERK accumulates rapidly in the nucleus, and its nuclear effects have been extensively studied. ERK can activate a family of related kinases, ribosomal S6 kinases 1 and 2 (RSK1/2), and mitogen- and stress-activated kinases 1 and 2 (MSK1/2) (Pearce et al. 2010). MSK1 is a nuclear kinase that is particularly enriched in striatal neurons, where it is slightly more abundant in striatonigral neurons (Bertran-Gonzalez et al. 2009). In MSK1 knockout mice, a number of responses to cocaine are blocked or blunted, including phosphorylation of CREB Ser-133 and histone H3 Ser-10 (H3S10), and expression of several IEGs including cFos and prodynorphin (Brami-Cherrier et al. 2005). In contrast, other IEGs such as Zif-268 are normally induced. Interestingly, locomotor sensitization to cocaine is diminished in MSK1 KO mice, whereas CPP is not altered (ibid.). This suggested that genes essential for CPP are MSK1 independent. Interestingly, CPP is blocked in Zif-268 KO mice (Valjent et al. 2006b), suggesting that this IEG, which is itself a transcription factor, specifically controls genes essential for the long-term behavioral consequences of the "rewarding" effects of cocaine.

Interestingly, although the core of the ERK pathway activation is controlled by D1 and NMDA receptors, other pathways and modulators play an important role. For example, stimulation of mGluR1/5 receptors is involved in amphetamine-induced ERK activation (Choe et al. 2002) and is necessary for drug-induced behavioral plasticity (Chiamulera et al. 2001). mGluR5 receptors can activate ERK through Ca^{2+} release from intracellular stores in synergy with D1 receptors and through interaction with Homer1b/c (Mao et al. 2005; Voulalas et al. 2005). In addition, the CB1 cannabinoid receptors in MSNs are necessary for cocaine-induced activation of ERK (Corbille et al. 2007). However, their precise level of action on the ERK pathway in this system is not known. Stimulation of CB1 receptors can directly activate the ERK pathway in hippocampus (Derkinderen et al. 2003), but in the striatum, the activation of ERK by Δ9-tetrahydrocannabinol (the main active compound in cannabis) is mediated by dopamine release and stimulation of D1 receptors (Valjent et al. 2001). Interestingly, stimulation of D2 receptors increases anandamide production in the striatum providing a mechanism by which these receptors may contribute to ERK regulation (Giuffrida et al. 1999). Thus, it appears that a cannabinoid-mediated cross talk between D2- and D1-expressing neurons may also contribute to the fine tuning of ERK activation.

Studies in BAC transgenic mice in which the striatonigral or striatopallidal neurons are labeled with EGFP expressed under the control of *drd1a* or *drd2* promoters, respectively, have shown that cocaine-induced phosphorylation of ERK, MSK1, and histone H3 occurs selectively in D1-positive neurons (Bertran-Gonzalez et al. 2008). The same was true for 6-OH-DA-lesioned mice treated with L-dopa (Santini et al. 2009). Surprisingly, these phosphoproteins were not detected in D2-positive neurons, even though a subset of neurons expresses the two markers (Bertran-Gonzalez et al. 2008). This raised questions about the ERK pathway in striatopallidal MSNs and the effects of D2 receptors. It turns out that blockade of D2 receptors activates selectively the phosphorylation of ERK in some striatopallidal neurons (Bertran-Gonzalez et al. 2008, 2009). The same pharmacological treatment increases MSK1 and H3 phosphorylation, as well as IEG expression in the same neurons (ibid.). These latter responses seem to be decoupled from ERK since they were insensitive to the MEK antagonist SL327. In contrast, they were shown to depend on the cAMP pathway since they were decreased by an A2a antagonist or in DARPP-32 Thr-34-Ala knockin mice (Bertran-Gonzalez et al. 2009). The strong effects of D2 antagonists on signaling responses show that the D2 receptors, which have a high affinity for DA, are activated in basal conditions by ambient DA in the striatum, as it was proposed for many years due to the spillover of DA from synapses. Basal activation of these D2 receptors prevents responses mediated by A2a receptors, and possibly others, which have the capacity to turn on the PKA pathway and to some extent the ERK pathway. Interestingly, the A2a receptor effect on ERK activation is strongly potentiated by its interaction with fibroblast growth factor (FGF) receptor, a receptor tyrosine kinase (Flajolet et al. 2008). The combined stimulation of FGF and A2a receptors increased an ERK-dependent corticostriatal LTP, selectively in striatopallidal neuron, an effect blocked by a D2 agonist. Thus, the blockade of D2 receptors by antipsychotic agents unleashes powerful signaling responses which alter synaptic plasticity and result in IEG expression. The functional consequences of these responses are unclear but may contribute to the acute and tardive side effects of antipsychotics attributed to their effects in the dorsal striatum, such as Parkinson-like symptoms and dyskinesia. It remains to explore whether these pathways can also be activated in physiological conditions, for example, following transient decreases in DA release linked to a negative reward prediction error. In addition, their contribution to the still mysterious delayed therapeutic effects of antipsychotics will be an interesting question for future work.

18.4.2 Nuclear Function of DARPP-32

Although it was apparent that some MSNs express nuclear DARPP-32 immunoreactivity (Ouimet and Greengard 1990), the significance of this observation was not known, and as mentioned above, DARPP-32 has been extensively characterized as a cytoplasmic signaling hub in both D1 and D2 neurons (Svenningsson et al. 2004). Recent work revealed a novel aspect of its functions, in the nucleus of MSNs, showing that treatment of mice with cocaine, amphetamine, or morphine, or a simple food-conditioned operant learning protocol, increased nuclear DARPP-32 and P-Thr-34-DARPP-32 immunoreactivity (Stipanovich et al. 2008). This response was absent in D1 knockout mice. Studies in non-neuronal cells and striatal neurons in culture showed that DARPP-32 undergoes a continuous cytonuclear shuttling and that its nuclear export depends on an incomplete nuclear export sequence (residues 103-111 in mouse), which is active when Ser-97 is phosphorylated by CK2 (Stipanovich et al. 2008) (Fig. 18.4). Dephosphorylation of Ser-97 induces the nuclear accumulation of DARPP-32. This dephosphorylation can be carried out by PP2A, and interestingly, several PP2A regulatory B subunits are enriched in MSNs. The B56δ subunit provides regulation by the cAMP pathway since it is phosphorylated by PKA phosphorylation (Ahn et al. 2007a), while PR72 has a Ca^{2+} binding motif (Ahn et al. 2007a). PP2A containing either of these two isoforms can dephosphorylate Thr-75 in response to cAMP and Ca^{2+} signals, respectively (Ahn et al. 2007a, b). Dephosphorylation of Ser-97 by B56δ is likely to contribute to the DA-induced nuclear accumulation of DARPP-32 (Stipanovich et al. 2008), although other factors are certainly involved. All PP1 catalytic subunit isoforms are found in cell nuclei, as well as their specific nuclear inhibitors, and PP1 is thought to play an important role, including in the cell cycle (Moorhead et al. 2007). In neurons, which do not divide, the potential functions of phosphatases in the nucleus are just starting to be explored. Nuclear DARPP-32 enhances phosphorylation of histone H3 on Ser-10 in response to D1 stimulation, and phosphorylation of both Thr-34 and Ser-97 is necessary for H3 phosphorylation in vivo (Stipanovich et al. 2008). Other effects of PP1 inhibition are likely to occur since expression in forebrain neurons of an inhibitor of PP1 targeted to the nucleus increased multiple posttranslational modifications of histones, including

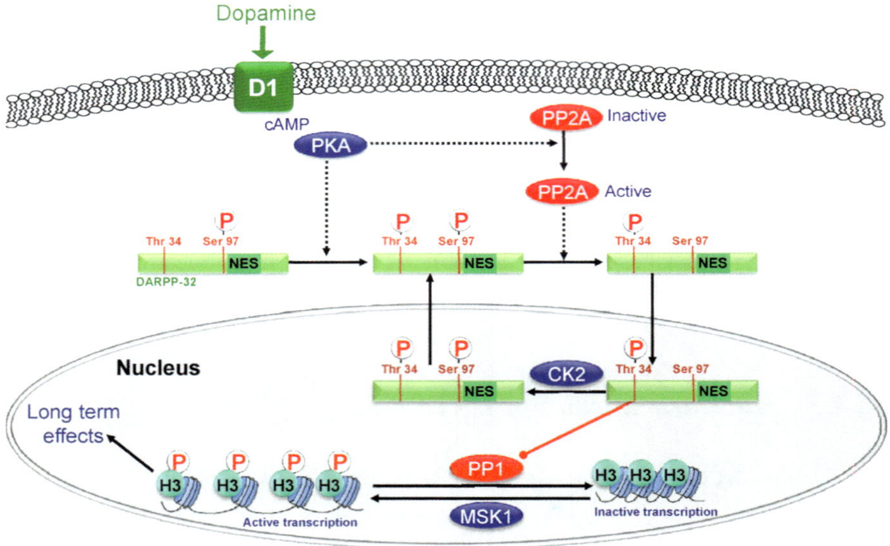

Fig. 18.4 *Nuclear role of DARPP-32.* DARPP-32 cycles continuously between cytoplasm and nucleus. Its nuclear export requires phosphorylation of Ser-97 by CK2. Following activation of D1 receptors, DARPP-32 is phosphorylated by PKA on Thr-34 and dephosphorylated by PP2A on Ser-97. Activation of PP2A results in part from the phosphorylation of its B56δ by PKA. DARPP-32 phosphorylated on Thr-34 can accumulate in the nucleus where it inhibits PP1. This potentiates the phosphorylation of H3 Ser-10 by MSK1 by preventing its dephosphorylation. Phosphorylation of H3 contributes to put the chromatin in an open conformation

not only phosphorylation of H3S10, but also acetylation of H2B, H3K14, and H4K5, as well as trimethylation of H3K36 (Koshibu et al. 2009). Additional potential substrates include the splicing machinery, with which not only PP1 but also DARPP-32 can be associated through its binding to tra2-beta1 (Benderska et al. 2010). Thus, neurotransmitters can trigger long-lasting changes in MSNs through a complex network of protein kinases and phosphatases converging on the nucleus and contributing to epigenetic regulations.

18.5   Epigenetics of MSNs

Epigenetic modifications are heritable traits that do not involve changes in DNA sequence. Strictly speaking, this would refer only to transgenerational stable changes, which may occur in some instances (Franklin et al. 2010). In a wider sense, epigenetic modifications are now emerging as fundamental mechanisms by which neurons adapt their transcriptional response to developmental and environmental cues (Riccio 2010; Zhang and Meaney 2010). Epigenetic alterations include covalent modifications of DNA, such as methylation of cytidine, but also hydroxylation (Kriaucionis and Heintz 2009), as well as modifications of histones, including changes in histone isoforms and covalent chemical modifications of their tails. It is important to emphasize that the apparent stability of epigenetic changes is not contradictory with the existence of active turnover mechanisms, as long as their net result is the maintenance of a steady-state level of the alteration. Additional mechanisms such as organization of nuclear chromatin territories are likely to play important roles (Schneider and Grosschedl 2007). In general, neurons have highly distinctive chromatin organization with characteristic patterns of euchromatin and heterochromatin. MSNs, for example, can be distinguished from other striatal cells only on the basis of their chromatin organization (Matamales et al. 2009). However, the functional meaning of nuclear architecture is only starting to be explored (Wittmann et al. 2009). Differentiation is to a large extent synonymous to specific patterns of gene expression, which depend on both the dynamic interplay of multiple transcription factors and other DNA binding proteins, and on epigenetic alterations. Dramatic differences in gene expression between D1-expresssing striatonigral and D2-expressing striatopallidal MSNs (Heiman et al. 2008) suggest that they carry distinct epigenetic modifications, although such modifications have still to be identified. Alterations in histone modifications in response to acute challenges by drugs of abuse or antipsychotics have been well documented in MSNs. As mentioned above, cocaine increases the global levels of H3S10 phosphorylation and H4K5 acetylation (Brami-Cherrier et al. 2005; Bertran-Gonzalez et al. 2008), whereas D2 antagonists increase H3S10 phosphorylation (Bertran-Gonzalez et al. 2008, 2009). These effects take place in D1 and D2 neurons, respectively. Study of specific genes has shown more precise patterns of histone modifications including H4 hyperacetylation of the cFos promoter within 30 min of a single cocaine and H3 hyperacetylation at the BDNF and Cdk5 promoters, two genes that are induced by chronic, but not acute, cocaine (Kumar et al. 2005). Histone H3K9 dimethylation through transcriptional regulation of the lysine dimethyltransferase G9a is involved in cocaine-induced structural and behavioral plasticity (Maze et al. 2010). DNA methylation is also modified by drugs in the striatum, through regulation of DNA methyltransferases (Dnmts) and possibly demethylases. For example, Dnmt3a expression is regulated in mouse nucleus accumbens (NAc) by chronic cocaine use and chronic social defeat stress (LaPlant et al. 2010). Methylation of CpG islands on the DNA provides patterns for recognition for a variety of protein with specific binding domains. Although methylation was initially thought to exert only a repressive effect on gene transcription, it appears that the situation is more complex and that the consequences depend on the pattern and location of methylation. Among the proteins, whose binding is increased by DNA methylation, MeCP2 has received particular attention since its mutation is responsible for the Rett syndrome which associates autistic behavior and mental retardation with some alterations of motricity. Recent work suggests that MeCP2 regulates the behavioral responses to cocaine through regulation of microRNA and BDNF expression (Deng et al. 2010; Im et al. 2010). Regulation of microRNAs and other small regulatory RNAs is another aspect of signaling that is likely to play an important role in striatal neurons as in other cells and that is only starting to be explored.

18.6 Conclusions and Perspectives

The striatum is a brain region that plays a specific and central role in information processing, which may be directed at action selection. Depending on the cortical regions to which it is applied, it may have various but always critical functional consequences. This processing depends on the existence of intermingled populations of efferent neurons with different properties and connections. These neurons have highly specialized signaling pathways that distinguishes them from other populations of brain principal neurons. We are only beginning to discover the specific properties of these signaling pathways and the functional consequences of their alterations using a wide variety of models. Most of this work is descriptive, and improvement of the modeling approaches may allow better understanding how the characteristics of these signaling pathways contribute to the striatal-information-processing capacities. One aspect that has seen much progress is the interplay between synaptic activity and nuclear functions. Changes in inputs activate multiple signaling pathways that alter gene transcription through posttranslational modifications of transcription factors, histones, and other proteins, as well as DNA methylation. As in other brain neurons, the stability of synaptic changes requires these changes in gene expression. They contribute to long-lasting neuronal alterations through induction of morphological changes, such as dendritic spines formation, but also through changes in DNA or histone-modifying enzymes which maintain epigenetic modifications. Thus, there is a constant dialogue between the pattern of synaptic inputs and weights and the pattern of gene expression which underlie the long-lasting alterations of striatal information processing and behavioral responses. It should be emphasized that there is an apparent contradiction between the existence of global alteration in gene expression and the idea of procedural learning and memory being borne by selective changes in synaptic weights. Solving this issue will be a major challenge for the future, important for understanding the mechanism of the basal ganglia function at the systems level.

References

Ahn, J. H., McAvoy, T., Rakhilin, S. V., Nishi, A., Greengard, P., & Nairn, A. C. (2007a). Protein kinase A activates protein phosphatase 2A by phosphorylation of the B56delta subunit. *Proceedings of the National Academy of Sciences of the United States of America, 104*, 2979–2984.

Ahn, J. H., Sung, J. Y., McAvoy, T., Nishi, A., Janssens, V., Goris, J., Greengard, P., & Nairn, A. C. (2007b). The B''/PR72 subunit mediates Ca^{2+}-dependent dephosphorylation of DARPP-32 by protein phosphatase 2A. *Proceedings of the National Academy of Sciences of the United States of America, 104*, 9876–9881.

Alheid, G. F., & Heimer, L. (1988). New perspectives in basal forebrain organization of special relevance for neuropsychiatric disorders: The striatopallidal, amygdaloid, and corticopetal components of substantia innominata. *Neuroscience, 27*, 1–39.

Barnes, T. D., Kubota, Y., Hu, D., Jin, D. Z., & Graybiel, A. M. (2005). Activity of striatal neurons reflects dynamic encoding and recoding of procedural memories. *Nature, 437*, 1158–1161.

Bateup, H. S., Svenningsson, P., Kuroiwa, M., Gong, S., Nishi, A., Heintz, N., & Greengard, P. (2008). Cell type-specific regulation of DARPP-32 phosphorylation by psychostimulant and antipsychotic drugs. *Nature Neuroscience, 11*, 932–939.

Bateup, H. S., Santini, E., Shen, W., Birnbaum, S., Valjent, E., Surmeier, D. J., Fisone, G., Nestler, E. J., & Greengard, P. (2010). Distinct subclasses of medium spiny neurons differentially regulate striatal motor behaviors. *Proceedings of the National Academy of Sciences of the United States of America, 107*, 14845–14850.

Belin, D., Jonkman, S., Dickinson, A., Robbins, T. W., & Everitt, B. J. (2009). Parallel and interactive learning processes within the basal ganglia: Relevance for the understanding of addiction. *Behavioural Brain Research, 199*, 89–102.

Benderska, N., Becker, K., Girault, J. A., Becker, C. M., Andreadis, A., & Stamm, S. (2010). DARPP-32 binds to tra2-beta1 and influences alternative splicing. *Biochimica et Biophysica Acta, 1799*, 448.

Bertran-Gonzalez, J., Bosch, C., Maroteaux, M., Matamales, M., Herve, D., Valjent, E., & Girault, J. A. (2008). Opposing patterns of signaling activation in dopamine D1 and D2 receptor-expressing striatal neurons in response to cocaine and haloperidol. *Journal of Neuroscience, 28*, 5671–5685.

Bertran-Gonzalez, J., Hakansson, K., Borgkvist, A., Irinopoulou, T., Brami-Cherrier, K., Usiello, A., Greengard, P., Herve, D., Girault, J. A., Valjent, E., & Fisone, G. (2009). Histone H3 phosphorylation is under the opposite tonic control of dopamine D2 and adenosine A2A receptors in striatopallidal neurons. *Neuropsychopharmacology, 34*, 1710–1720.

Bibb, J. A., Snyder, G. L., Nishi, A., Yan, Z., Meijer, L., Fienberg, A. A., Tsai, L. H., Kwon, Y. T., Girault, J. A., Czernik, A. J., Huganir, R. L., Hemmings, H. C., Jr., Nairn, A. C., & Greengard, P. (1999). Phosphorylation of DARPP-32 by Cdk5 modulates dopamine signalling in neurons. *Nature, 402*, 669–671.

Bibb, J. A., Chen, J., Taylor, J. R., Svenningsson, P., Nishi, A., Snyder, G. L., Yan, Z., Sagawa, Z. K., Ouimet, C. C., Nairn, A. C., Nestler, E. J., & Greengard, P. (2001). Effects of chronic exposure to cocaine are regulated by the neuronal protein Cdk5. *Nature, 410*, 376–380.

Brami-Cherrier, K., Valjent, E., Herve, D., Darragh, J., Corvol, J. C., Pages, C., Arthur, S. J., Girault, J. A., & Caboche, J. (2005). Parsing molecular and behavioral effects of cocaine in mitogen- and stress-activated protein kinase-1-deficient mice. *Journal of Neuroscience, 25*, 11444–11454.

Bromberg-Martin, E. S., Matsumoto, M., & Hikosaka, O. (2010). Dopamine in motivational control: Rewarding, aversive, and alerting. *Neuron, 68*, 815–834.

Calabresi, P., Gubellini, P., Centonze, D., Picconi, B., Bernardi, G., Chergui, K., Svenningsson, P., Fienberg, A. A., & Greengard, P. (2000). Dopamine and cAMP-regulated phosphoprotein 32 kDa controls both striatal long-term depression and long-term potentiation, opposing forms of synaptic plasticity. *Journal of Neuroscience, 20*, 8443–8451.

Chevalier, G., & Deniau, J. M. (1990). Disinhibition as a basic process in the expression of striatal functions. *Trends in Neurosciences, 13*, 277–280.

Chiamulera, C., Epping-Jordan, M. P., Zocchi, A., Marcon, C., Cottiny, C., Tacconi, S., Corsi, M., Orzi, F., & Conquet, F. (2001). Reinforcing and locomotor stimulant effects of cocaine are absent in mGluR5 null mutant mice. *Nature Neuroscience, 4*, 873–874.

Choe, E. S., Chung, K. T., Mao, L., & Wang, J. Q. (2002). Amphetamine increases phosphorylation of extracellular signal-regulated kinase and transcription factors in the rat striatum via group I metabotropic glutamate receptors. *Neuropsychopharmacology, 27*, 565–575.

Corbille, A. G., Valjent, E., Marsicano, G., Ledent, C., Lutz, B., Herve, D., & Girault, J. A. (2007). Role of cannabinoid type 1 receptors in locomotor activity and striatal signaling in response to psychostimulants. *Journal of Neuroscience, 27*, 6937–6947.

Corvol, J. C., Studler, J. M., Schonn, J. S., Girault, J. A., & Herve, D. (2001). Galpha (olf) is necessary for coupling D1 and A2a receptors to adenylyl cyclase in the striatum. *Journal of Neurochemistry, 76*, 1585–1588.

Corvol, J. C., Valjent, E., Pascoli, V., Robin, A., Stipanovich, A., Luedtke, R. R., Belluscio, L., Girault, J. A., & Herve, D. (2007). Quantitative changes in Galphaolf protein levels, but not D1 receptor, alter specifically acute responses to psychostimulants. *Neuropsychopharmacology, 32*, 1109–1121.

Crittenden, J. R., Cantuti-Castelvetri, I., Saka, E., Keller-McGandy, C. E., Hernandez, L. F., Kett, L. R., Young, A. B., Standaert, D. G., & Graybiel, A. M. (2009). Dysregulation of CalDAG-GEFI and CalDAG-GEFII predicts the severity of motor side-effects induced by antiparkinsonian therapy. *Proceedings of the National Academy of Sciences of the United States of America, 106*, 2892–2896.

Dayan, P., & Balleine, B. W. (2002). Reward, motivation, and reinforcement learning. *Neuron, 36*, 285–298.

Deng, J. V., Rodriguiz, R. M., Hutchinson, A. N., Kim, I. H., Wetsel, W. C., & West, A. E. (2010). MeCP2 in the nucleus accumbens contributes to neural and behavioral responses to psychostimulants. *Nature Neuroscience, 13*, 1128–1136.

Derkinderen, P., Valjent, E., Toutant, M., Corvol, J. C., Enslen, H., Ledent, C., Trzaskos, J., Caboche, J., & Girault, J. A. (2003). Regulation of extracellular signal-regulated kinase by cannabinoids in hippocampus. *Journal of Neuroscience, 23*, 2371–2382.

Desdouits, F., Siciliano, J. C., Greengard, P., & Girault, J. A. (1995). Dopamine- and cAMP-regulated phosphoprotein DARPP-32: Phosphorylation of Ser-137 by casein kinase I inhibits dephosphorylation of Thr-34 by calcineurin. *Proceedings of the National Academy of Sciences of the United States of America, 92*, 2682–2685.

Doig, N. M., Moss, J., & Bolam, J. P. (2010). Cortical and thalamic innervation of direct and indirect pathway medium-sized spiny neurons in mouse striatum. *Journal of Neuroscience, 30*, 14610–14618.

Durieux, P. F., Bearzatto, B., Guiducci, S., Buch, T., Waisman, A., Zoli, M., Schiffmann, S. N., & de Kerchove d'Exaerde, A. (2009). D2R striatopallidal neurons inhibit both locomotor and drug reward processes. *Nature Neuroscience, 12*, 393–395.

Fasano, S., Bezard, E., D'Antoni, A., Francardo, V., Indrigo, M., Qin, L., Dovero, S., Cerovic, M., Cenci, M. A., & Brambilla, R. (2010). Inhibition of Ras-guanine nucleotide-releasing factor 1 (Ras-GRF1) signaling in the striatum reverts motor symptoms associated with L-dopa-induced dyskinesia. *Proceedings of the National Academy of Sciences of the United States of America, 107*, 21824–21829.

Ferre, S., Ciruela, F., Woods, A. S., Lluis, C., & Franco, R. (2007). Functional relevance of neurotransmitter receptor heteromers in the central nervous system. *Trends in Neurosciences, 30*, 440–446.

Fienberg, A. A., et al. (1998). DARPP-32: Regulator of the efficacy of dopaminergic neurotransmission. *Science, 281*, 838–839.

Flajolet, M., Wang, Z., Futter, M., Shen, W., Nuangchamnong, N., Bendor, J., Wallach, I., Nairn, A. C., Surmeier, D. J., & Greengard, P. (2008). FGF acts as a co-transmitter through adenosine A(2A) receptor to regulate synaptic plasticity. *Nature Neuroscience, 11*, 1402–1409.

Franklin, T. B., Russig, H., Weiss, I. C., Graff, J., Linder, N., Michalon, A., Vizi, S., & Mansuy, I. M. (2010). Epigenetic transmission of the impact of early stress across generations. *Biological Psychiatry, 68*, 408–415.

Fujishige, K., Kotera, J., Michibata, H., Yuasa, K., Takebayashi, S., Okumura, K., & Omori, K. (1999). Cloning and characterization of a novel human phosphodiesterase that hydrolyzes both cAMP and cGMP (PDE10A). *Journal of Biological Chemistry, 274*, 18438–18445.

Gerfen, C. R. (1984). The neostriatal mosaic: Compartmentalization of corticostriatal input and striatonigral output systems. *Nature, 311*, 461–464.

Gerfen, C. R. (1992). The neostriatal mosaic: Multiple levels of compartmental organization. *Trends in Neurosciences, 15*, 133–139.

Gerfen, C. R. (2000). Molecular effects of dopamine on striatal-projection pathways. *Trends in Neurosciences, 23*, S64–S70.

Gharbi-Ayachi, A., Labbe, J. C., Burgess, A., Vigneron, S., Strub, J. M., Brioudes, E., Van-Dorsselaer, A., Castro, A., & Lorca, T. (2010). The substrate of Greatwall kinase, Arpp 19, controls mitosis by inhibiting protein phosphatase 2A. *Science, 330*, 1673–1677.

Girault, J. A., Hemmings, H. C., Jr., Williams, K. R., Nairn, A. C., & Greengard, P. (1989). Phosphorylation of DARPP-32, a dopamine- and cAMP-regulated phosphoprotein, by casein kinase II. *Journal of Biological Chemistry, 264*, 21748–21759.

Girault, J. A., Horiuchi, A., Gustafson, E. L., Rosen, N. L., & Greengard, P. (1990). Differential expression of ARPP-16 and ARPP-19, two highly related cAMP-regulated phosphoproteins, one of which is specifically associated with dopamine-innervated brain regions. *Journal of Neuroscience, 10*, 1124–1133.

Girault, J. A., Valjent, E., Caboche, J., & Herve, D. (2007). ERK2: A logical AND gate critical for drug-induced plasticity? *Current Opinion in Pharmacology, 7*, 77–85.

Giuffrida, A., Parsona, L. H., Kerr, T. M., Rodriguez de Fonseca, F., Navarro, M., & Piomelli, D. (1999). Dopamine activation of endogenous cannabinoid signaling in dorsal striatum. *Nature Neuroscience, 2*, 358–363.

Graybiel, A. M., & Hickey, T. L. (1982). Chemospecificity of ontogenetic units in the striatum: Demonstration by combining [3H]thymidine neuronography and histochemical staining. *Proceedings of the National Academy of Sciences of the United States of America, 79*, 198–202.

Heiman, M., Schaefer, A., Gong, S., Peterson, J. D., Day, M., Ramsey, K. E., Suarez-Farinas, M., Schwarz, C., Stephan, D. A., Surmeier, D. J., Greengard, P., & Heintz, N. (2008). A translational profiling approach for the molecular characterization of CNS cell types. *Cell, 135*, 738–748.

Hemmings, H. C., Jr., & Greengard, P. (1989). ARPP-21, a cAMP-regulated phosphoprotein Mr = 21,000 enriched in dopamine-innervated brain regions. I. Purification and characterization of the protein from bovine caudate nucleus. *Journal of Neuroscience, 9*, 851–864.

Hemmings, H. C., Jr., Greengard, P., Tung, H. Y. L., & Cohen, P. (1984). DARPP-32, a dopamine-regulated neuronal phosphoprotein, is a potent inhibitor of protein phosphatase-1. *Nature, 310*, 503–505.

Herve, D., Levi-Strauss, M., Marey-Semper, I., Verney, C., Tassin, J. P., Glowinski, J., & Girault, J. A. (1993). G(olf) and Gs in rat basal ganglia: Possible involvement of G(olf) in the coupling of dopamine D1 receptor with adenylyl cyclase. *Journal of Neuroscience, 13*, 2237–2248.

Hikida, T., Kimura, K., Wada, N., Funabiki, K., & Nakanishi, S. (2010). Distinct roles of synaptic transmission in direct and indirect striatal pathways to reward and aversive behavior. *Neuron, 66*, 896–907.

Hope, B. T., Nye, H. E., Kelz, M. B., Self, D. W., Iadarola, M. J., & Nestler, E. J. (1994). Induction of long-lasting AP1 complex composed of altered Fos-like proteins inbrain by chronic cocaine and other chronic treatments. *Neuron, 13*, 1235–1244.

Hyman, S. E., Malenka, R. C., & Nestler, E. J. (2006). Neural mechanisms of addiction: The role of reward-related learning and memory. *Annual Review of Neuroscience, 29*, 565–598.

Ibba, F., Vinci, S., Spiga, S., Peana, A. T., Assaretti, A. R., Spina, L., Longoni, R., & Acquas, E. (2009). Ethanol-induced extracellular signal regulated kinase: Role of dopamine D1 receptors. *Alcoholism, Clinical and Experimental Research, 33*, 858–867.

Im, H. I., Hollander, J. A., Bali, P., & Kenny, P. J. (2010). MeCP2 controls BDNF expression and cocaine intake through homeostatic interactions with microRNA-212. *Nature Neuroscience, 13*, 1120–1127.

Kolata, S., Light, K., Wass, C. D., Colas-Zelin, D., Roy, D., & Matzel, L. D. (2010). A dopaminergic gene cluster in the prefrontal cortex predicts performance indicative of general intelligence in genetically heterogeneous mice. *PLoS One, 5*, e14036.

Koshibu, K., Graff, J., Beullens, M., Heitz, F. D., Berchtold, D., Russig, H., Farinelli, M., Bollen, M., & Mansuy, I. M. (2009). Protein phosphatase 1 regulates the histone code for long-term memory. *Journal of Neuroscience, 29*, 13079–13089.

Kravitz, A. V., Freeze, B. S., Parker, P. R., Kay, K., Thwin, M. T., Deisseroth, K., & Kreitzer, A. C. (2010). Regulation of parkinsonian motor behaviours by optogenetic control of basal ganglia circuitry. *Nature, 466*, 622–626.

Kreitzer, A. C. (2009). Physiology and pharmacology of striatal neurons. *Annual Review of Neuroscience, 32*, 127–147.

Kriaucionis, S., & Heintz, N. (2009). The nuclear DNA base 5-hydroxymethylcytosine is present in Purkinje neurons and the brain. *Science, 324*, 929–930.

Kumar, A., Choi, K. H., Renthal, W., Tsankova, N. M., Theobald, D. E., Truong, H. T., Russo, S. J., Laplant, Q., Sasaki, T. S., Whistler, K. N., Neve, R. L., Self, D. W., & Nestler, E. J. (2005). Chromatin remodeling is a key mechanism underlying cocaine-induced plasticity in striatum. *Neuron, 48*, 303–314.

LaPlant, Q., et al. (2010). Dnmt3a regulates emotional behavior and spine plasticity in the nucleus accumbens. *Nature Neuroscience, 13*, 1137–1143.

Le Novere, N., Li, L., & Girault, J. A. (2008). DARPP-32: Molecular integration of phosphorylation potential. *Cellular and Molecular Life Sciences, 65*, 2125–2127.

Lee, F. J., Xue, S., Pei, L., Vukusic, B., Chery, N., Wang, Y., Wang, Y. T., Niznik, H. B., Yu, X. M., & Liu, F. (2002a). Dual regulation of NMDA receptor functions by direct protein-protein interactions with the dopamine D1 receptor. *Cell, 111*, 219–230.

Lee, K. W., Hong, J. H., Choi, I. Y., Che, Y., Lee, J. K., Yang, S. D., Song, C. W., Kang, H. S., Lee, J. H., Noh, J. S., Shin, H. S., & Han, P. L. (2002b). Impaired D2 dopamine receptor function in mice lacking type 5 adenylyl cyclase. *Journal of Neuroscience, 22*, 7931–7940.

Mao, L., Yang, L., Tang, Q., Samdani, S., Zhang, G., & Wang, J. Q. (2005). The scaffold protein Homer1b/c links metabotropic glutamate receptor 5 to extracellular signal-regulated protein kinase cascades in neurons. *Journal of Neuroscience, 25*, 2741–2752.

Matamales, M., Bertran-Gonzalez, J., Salomon, L., Degos, B., Deniau, J. M., Valjent, E., Herve, D., & Girault, J. A. (2009). Striatal medium-sized spiny neurons: Identification by nuclear staining and study of neuronal subpopulations in BAC transgenic mice. *PLoS One, 4*, e4770.

Maze, I., Covington, H. E., 3rd, Dietz, D. M., LaPlant, Q., Renthal, W., Russo, S. J., Mechanic, M., Mouzon, E., Neve, R. L., Haggarty, S. J., Ren, Y., Sampath, S. C., Hurd, Y. L., Greengard, P., Tarakhovsky, A., Schaefer, A., & Nestler, E. J. (2010). Essential role of the histone methyltransferase G9a in cocaine-induced plasticity. *Science, 327*, 213–216.

Menniti, F. S., Faraci, W. S., & Schmidt, C. J. (2006). Phosphodiesterases in the CNS: Targets for drug development. *Nature Reviews. Drug Discovery, 5*, 660–670.

Meyer-Lindenberg, A., Straub, R. E., Lipska, B. K., Verchinski, B. A., Goldberg, T., Callicott, J. H., Egan, M. F., Huffaker, S. S., Mattay, V. S., Kolachana, B., Kleinman, J. E., & Weinberger, D. R. (2007). Genetic evidence implicating DARPP-32 in human frontostriatal structure, function, and cognition. *The Journal of Clinical Investigation, 117*, 672–682.

Miller, C. A., & Marshall, J. F. (2005). Molecular substrates for retrieval and reconsolidation of cocaine-associated contextual memory. *Neuron, 47*, 873–884.

Mink, J. W. (1996). The basal ganglia: Focused selection and inhibition of competing motor programs. *Progress in Neurobiology, 50*, 381–425.

Mochida, S., Maslen, S. L., Skehel, M., & Hunt, T. (2010). Greatwall phosphorylates an inhibitor of protein phosphatase 2A that is essential for mitosis. *Science, 330*, 1670–1673.

Montminy, M. (1997). Transcriptional regulation by cyclic AMP. *Annual Review of Biochemistry, 66*, 807–822.

Moorhead, G. B. G., Trinkle-Mulcahy, L., & Ulke-Lemée, A. (2007). Emerging roles of nuclear protein phosphatases. *Nature Reviews Molecular Cell Biology, 7*, 235–244.

Moss, J., & Bolam, J. P. (2008). A dopaminergic axon lattice in the striatum and its relationship with cortical and thalamic terminals. *Journal of Neuroscience, 28*, 11221–11230.

Ouimet, C. C., & Greengard, P. (1990). Distribution of DARPP-32 in the basal ganglia: An electron microscopic study. *Journal of Neurocytology, 19*, 39–52.

Ouimet, C. C., Hemmings, H. C., Jr., & Greengard, P. (1989). ARPP-21, a cyclic AMP-regulated phosphoprotein enriched in dopamine-innervated brain regions. II. Immunocytochemical localization in rat brain. *Journal of Neuroscience, 9*, 865–875.

Pascoli, V., Besnard, A., Herve, D., Pages, C., Heck, N., Girault, J. A., Caboche, J., & Vanhoutte, P. (2011). Cyclic adenosine monophosphate-independent tyrosine phosphorylation of NR2B mediates cocaine-induced extracellular signal-regulated kinase activation. *Biological Psychiatry, 69*, 218–227.

Pascoli, V., Turiault, M., & Lüscher, C. (2012). Reversal of cocaine-evoked synaptic potentiation resets drug-induced adaptive behaviour. *Nature,* in press.

Paul, S., Nairn, A. C., Wang, P., & Lombroso, P. J. (2003). NMDA-mediated activation of the tyrosine phosphatase STEP regulates the duration of ERK signaling. *Nature Neuroscience, 6,* 34–42.

Pearce, L. R., Komander, D., & Alessi, D. R. (2010). The nuts and bolts of AGC protein kinases. *Nature Reviews Molecular Cell Biology, 11,* 9–22.

Polli, J. W., & Kincaid, R. L. (1994). Expression of a calmodulin-dependent phosphodiesterase isoform (PDE1B1) correlates with brain regions having extensive dopaminergic innervation. *Journal of Neuroscience, 14,* 1251–1261.

Pulipparacharuvil, S., Renthal, W., Hale, C. F., Taniguchi, M., Xiao, G., Kumar, A., Russo, S. J., Sikder, D., Dewey, C. M., Davis, M. M., Greengard, P., Nairn, A. C., Nestler, E. J., & Cowan, C. W. (2008). Cocaine regulates MEF2 to control synaptic and behavioral plasticity. *Neuron, 59,* 621–633.

Rakhilin, S. V., Olson, P. A., Nishi, A., Starkova, N. N., Fienberg, A. A., Nairn, A. C., Surmeier, D. J., & Greengard, P. (2004). A network of control mediated by regulator of calcium/calmodulin-dependent signaling. *Science, 306,* 698–701.

Redgrave, P., Prescott, T. J., & Gurney, K. (1999). The basal ganglia: A vertebrate solution to the selection problem? *Neuroscience, 89,* 1009–1023.

Reed, T. M., Repaske, D. R., Snyder, G. L., Greengard, P., & Vorhees, C. V. (2002). Phosphodiesterase 1B knock-out mice exhibit exaggerated locomotor hyperactivity and DARPP-32 phosphorylation in response to dopamine agonists and display impaired spatial learning. *Journal of Neuroscience, 22,* 5188–5197.

Riccio, A. (2010). Dynamic epigenetic regulation in neurons: Enzymes, stimuli and signaling pathways. *Nature Neuroscience, 13,* 1330–1337.

Salzmann, J., Marie-Claire, C., Le Guen, S., Roques, B. P., & Noble, F. (2003). Importance of ERK activation in behavioral and biochemical effects induced by MDMA in mice. *British Journal of Pharmacology, 140,* 831–838.

Santini, E., Valjent, E., Usiello, A., Carta, M., Borgkvist, A., Girault, J. A., Herve, D., Greengard, P., & Fisone, G. (2007). Critical involvement of cAMP/DARPP-32 and extracellular signal-regulated protein kinase signaling in L-DOPA-induced dyskinesia. *Journal of Neuroscience, 27,* 6995–7005.

Santini, E., Valjent, E., & Fisone, G. (2008). Parkinson's disease: Levodopa-induced dyskinesia and signal transduction. *FEBS Journal, 275,* 1392–1399.

Santini, E., Alcacer, C., Cacciatore, S., Heiman, M., Herve, D., Greengard, P., Girault, J. A., Valjent, E., & Fisone, G. (2009). L-DOPA activates ERK signaling and phosphorylates histone H3 in the striatonigral medium spiny neurons of hemiparkinsonian mice. *Journal of Neurochemistry, 108,* 621–633.

Schiffmann, S. N., & Vanderhaeghen, J.-J. (1993). Adenosine A₂ receptors regulate the gene expression of striatopallidal and striatonigral neurons. *Journal of Neuroscience, 13,* 1080–1087.

Schneider, R., & Grosschedl, R. (2007). Dynamics and interplay of nuclear architecture, genome organization, and gene expression. *Genes & Development, 21,* 3027–3043.

Schultz, W. (2010). Dopamine signals for reward value and risk: Basic and recent data. *Behavioral and Brain Functions, 6,* 24.

Schwindinger, W. F., Betz, K. S., Giger, K. E., Sabol, A., Bronson, S. K., & Robishaw, J. D. (2003). Loss of G protein gamma 7 alters behavior and reduces striatal alpha(olf) level and cAMP production. *Journal of Biological Chemistry, 278,* 6575–6579.

Schwindinger, W. F., Mihalcik, L. J., Giger, K. E., Betz, K. S., Stauffer, A. M., Linden, J., Herve, D., & Robishaw, J. D. (2010). Adenosine A2A receptor signaling and golf assembly show a specific requirement for the gamma7 subtype in the striatum. *Journal of Biological Chemistry, 285,* 29787–29796.

Scott, L., & Aperia, A. (2009). Interaction between N-methyl-D-aspartic acid receptors and D1 dopamine receptors: An important mechanism for brain plasticity. *Neuroscience, 158,* 62–66.

Sgambato, V., Pagès, C., Rogard, M., Besson, M. J., & Caboche, J. (1998). Extracellular signal-regulated kinase (ERK) controls immediate early gene induction on corticostriatal stimulation. *Journal of Neuroscience, 18*, 8814–8825.

Siuciak, J. A., McCarthy, S. A., Chapin, D. S., Fujiwara, R. A., James, L. C., Williams, R. D., Stock, J. L., McNeish, J. D., Strick, C. A., Menniti, F. S., & Schmidt, C. J. (2006). Genetic deletion of the striatum-enriched phosphodiesterase PDE10A: Evidence for altered striatal function. *Neuropharmacology, 51*, 374–385.

Siuciak, J. A., McCarthy, S. A., Chapin, D. S., & Martin, A. N. (2008). Behavioral and neurochemical characterization of mice deficient in the phosphodiesterase-4B (PDE4B) enzyme. *Psychopharmacology, 197*, 115–126.

Snyder, G. L., Fienberg, A. A., Huganir, R. L., & Greengard, P. (1998). A dopamine D1 receptor protein kinase A dopamine- and cAMP-regulated phosphoprotein (M_r 32 kDa) protein phosphatase-1 pathway regulates dephosphorylation of the NMDA receptor. *Journal of Neuroscience, 18*, 10297–10303.

Sokoloff, P., Giros, B., Martres, M.-P., Bouthenet, M.-L., & Schwartz, J.-C. (1990). Molecular cloning and characterization of a novel dopamine receptor (D_3) as a target for neuroleptics. *Nature, 347*, 146–151.

Stipanovich, A., Valjent, E., Matamales, M., Nishi, A., Ahn, J. H., Maroteaux, M., Bertran-Gonzalez, J., Brami-Cherrier, K., Enslen, H., Corbille, A. G., Filhol, O., Nairn, A. C., Greengard, P., Herve, D., & Girault, J. A. (2008). A phosphatase cascade by which rewarding stimuli control nucleosomal response. *Nature, 453*, 879–884.

Surmeier, D. J., Ding, J., Day, M., Wang, Z., & Shen, W. (2007). D1 and D2 dopamine-receptor modulation of striatal glutamatergic signaling in striatal medium spiny neurons. *Trends in Neurosciences, 30*, 228–235.

Svenningsson, P., Lindskog, M., Ledent, C., Parmentier, M., Greengard, P., Fredholm, B. B., & Fisone, G. (2000). Regulation of the phosphorylation of the dopamine- and cAMP-regulated phosphoprotein of 32 kDa in vivo by dopamine D1, dopamine D2, and adenosine A2A receptors. *Proceedings of the National Academy of Sciences of the United States of America, 97*, 1856–1860.

Svenningsson, P., Nishi, A., Fisone, G., Girault, J. A., Nairn, A. C., & Greengard, P. (2004). DARPP-32: An integrator of neurotransmission. *Annual Review of Pharmacology and Toxicology, 44*, 269–296.

Sweatt, J. D. (2004). Mitogen-activated protein kinases in synaptic plasticity and memory. *Current Opinion in Neurobiology, 14*, 311–317.

Tepper, J. M., Tecuapetla, F., Koos, T., & Ibáñez-Sandoval, O. (2010). Heterogeneity and diversity of striatal GABAergic interneurons. *Frontiers in Neuroanatomy, 4*, 150.

Tian, X., Kai, L., Hockberger, P. E., Wokosin, D. L., & Surmeier, D. J. (2010). MEF-2 regulates activity-dependent spine loss in striatopallidal medium spiny neurons. *Molecular and Cellular Neuroscience, 44*, 94–108.

Toki, S., Kawasaki, H., Tashiro, N., Housman, D. E., & Graybiel, A. M. (2001). Guanine nucleotide exchange factors CalDAG-GEFI and CalDAG-GEFII are colocalized in striatal projection neurons. *The Journal of Comparative Neurology, 437*, 398–407.

Valjent, E., Corvol, J. C., Pages, C., Besson, M. J., Maldonado, R., & Caboche, J. (2000). Involvement of the extracellular signal-regulated kinase cascade for cocaine-rewarding properties. *Journal of Neuroscience, 20*, 8701–8709.

Valjent, E., Pages, C., Rogard, M., Besson, M. J., Maldonado, R., & Caboche, J. (2001). Delta 9-tetrahydrocannabinol-induced MAPK/ERK and Elk-1 activation in vivo depends on dopaminergic transmission. *European Journal of Neuroscience, 14*, 342–352.

Valjent, E., Pages, C., Herve, D., Girault, J. A., & Caboche, J. (2004). Addictive and non-addictive drugs induce distinct and specific patterns of ERK activation in mouse brain. *European Journal of Neuroscience, 19*, 1826–1836.

Valjent, E., Pascoli, V., Svenningsson, P., Paul, S., Enslen, H., Corvol, J. C., Stipanovich, A., Caboche, J., Lombroso, P. J., Nairn, A. C., Greengard, P., Herve, D., & Girault, J. A. (2005). Regulation of a protein phosphatase cascade allows convergent dopamine and glutamate signals to activate ERK in the striatum. *Proceedings of the National Academy of Sciences of the United States of America, 102*, 491–496.

Valjent, E., Corbille, A. G., Bertran-Gonzalez, J., Herve, D., & Girault, J. A. (2006a). Inhibition of ERK pathway or protein synthesis during reexposure to drugs of abuse erases previously learned place preference. *Proceedings of the National Academy of Sciences of the United States of America, 103*, 2932–2937.

Valjent, E., Aubier, B., Corbille, A. G., Brami-Cherrier, K., Caboche, J., Topilko, P., Girault, J. A., & Herve, D. (2006b). Plasticity-associated gene Krox24/Zif268 Is required for long-lasting behavioral effects of cocaine. *Journal of Neuroscience, 26*, 4956–4960.

Valjent, E., Bertran-Gonzalez, J., Herve, D., Fisone, G., & Girault, J. A. (2009). Looking BAC at striatal signaling: Cell-specific analysis in new transgenic mice. *Trends in Neurosciences, 32*, 538–547.

van der Stelt, M., & Di Marzo, V. (2003). The endocannabinoid system in the basal ganglia and in the mesolimbic reward system: Implications for neurological and psychiatric disorders. *European Journal of Pharmacology, 480*, 133–150.

Vanhoutte, P., Barnier, J. V., Guibert, B., Pagès, C., Besson, M. J., Hipskind, R. A., & Caboche, J. (1999). Glutamate induces phosphorylation of Elk-1 and CREB, along with c-fos activation, via an extracellular signal-regulated kinase-dependent pathway in brain slices. *Molecular and Cellular Biology, 19*, 136–146.

Voorn, P., Vanderschuren, L. J., Groenewegen, H. J., Robbins, T. W., & Pennartz, C. M. (2004). Putting a spin on the dorsal-ventral divide of the striatum. *Trends in Neurosciences, 27*, 468–474.

Voulalas, P. J., Holtzclaw, L., Wolstenholme, J., Russell, J. T., & Hyman, S. E. (2005). Metabotropic glutamate receptors and dopamine receptors cooperate to enhance extracellular signal-regulated kinase phosphorylation in striatal neurons. *Journal of Neuroscience, 25*, 3763–3773.

Walaas, S. I., Aswad, D. W., & Greengard, P. (1983). A dopamine- and cyclic AMP-regulated phosphoprotein enriched in dopamine-innervated brain regions. *Nature, 301*, 69–71.

Wickens, J. R. (2009). Synaptic plasticity in the basal ganglia. *Behavioural Brain Research, 199*, 119–128.

Wickens, J. R., Horvitz, J. C., Costa, R. M., & Killcross, S. (2007). Dopaminergic mechanisms in actions and habits. *Journal of Neuroscience, 27*, 8181–8183.

Wittmann, M., Queisser, G., Eder, A., Wiegert, J. S., Bengtson, C. P., Hellwig, A., Wittum, G., & Bading, H. (2009). Synaptic activity induces dramatic changes in the geometry of the cell nucleus: Interplay between nuclear structure, histone H3 phosphorylation, and nuclear calcium signaling. *Journal of Neuroscience, 29*, 14687–14700.

Zhang, T. Y., & Meaney, M. J. (2010). Epigenetics and the environmental regulation of the genome and its function. *Annual Review of Psychology, 61*(439–466), C431–C433.

Zhuang, X., Belluscio, L., & Hen, R. (2000). GOLFalpha mediates dopamine D1 receptor signaling. *Journal of Neuroscience, 20*, RC91. 1–5.

Part IV
Synaptic Dysfunction and Synaptopathies

Chapter 19
Synaptic Dysfunction and Intellectual Disability

Pamela Valnegri, Carlo Sala, and Maria Passafaro

Abstract Intellectual disability (ID) is a common and highly heterogeneous paediatric disorder with a very severe social impact. Intellectual disability can be caused by environmental and/or genetic factors. Although in the last two decades a number of genes have been discovered whose mutations cause mental retardation, we are still far from identifying the impact of these mutations on brain functions. Many of the genes mutated in ID code for several proteins with a variety of functions: chromatin remodelling, pre-/post-synaptic activity, and intracellular trafficking. The prevailing hypothesis suggests that the ID phenotype could emerge from abnormal cellular processing leading to pre- and/or post-synaptic dysfunction. In this chapter, we focus on the role of small GTPases and adhesion molecules, and we discuss the mechanisms through which they lead to synaptic network dysfunction.

Keywords Dendritic spines • Intellectual disability • Mental retardation • Small GTPases • Synaptic cell adhesion molecules • Synaptopathies

19.1 Classification of ID

The central nervous system consists of more than 100 billion neurons, which process and transmit information in the form of electrical signals through specialised junctions called synapses. Precise control of synaptic development is critical for accurate neuronal network activity and normal brain function. It is clear that inappropriate loss of synaptic stability may lead to neurological diseases and the disruption of neuronal circuits.

P. Valnegri • C. Sala • M. Passafaro (✉)
CNR Institute of Neuroscience, Department of Pharmacology, University of Milan, Via Vanvitelli 32, 20129 Milan, Italy
e-mail: m.passafaro@cnr.it

M.R. Kreutz and C. Sala (eds.), *Synaptic Plasticity*,
Advances in Experimental Medicine and Biology 970,
DOI 10.1007/978-3-7091-0932-8_19, © Springer-Verlag/Wien 2012

Many neuropsychiatric diseases are characterised by an alteration in the morphology of dendritic spines and aberrant synaptic signalling and plasticity (Blanpied and Ehlers 2004). One of the most common neurodevelopmental disorders is mental retardation or intellectual disability (ID), characterised by an intelligence quotient (IQ) of 70 or below and deficits in at least two behaviours related to adaptive functioning diagnosed by the age of 18 years. Intellectual disability is a term used when a person, usually a child, has certain limitations in mental functioning and skills, such as communicating and caring for himself or herself.

The prevalence of ID is between 1% and 3% (Roeleveld et al. 1997; Leonard and Wen 2002), and it is the leading socioeconomic health-care problem. Moreover, approximately 30% more males are diagnosed with ID than females (McLaren and Bryson 1987); however, severe ID is more prevalent among females (Bradley et al. 2002).

ID is divided into five categories based on IQ tests: mild, moderate, severe, profound, and unable to classify. Epidemiological studies use a simplified classification, grouping subjects into mild ID (IQ = 50–70) and severe ID (IQ < 50) (Ropers 2010). While the prevalence of severe ID is relatively stable, the prevalence of mild ID is variable and often depends heavily on external environmental factors, such as level of maternal education and access to education and access to health care (Drews et al. 1995; Roeleveld et al. 1997; Leonard and Wen 2002). In addition to categorisation based on severity/IQ level, ID can also be grouped into syndromic intellectual disability (S-ID) and non-syndromic intellectual disability (NS-ID). In S-ID, patients present with one or more clinical features or co-morbidities in addition to ID. While S-ID has a clear definition, there is debate over the classification of NS-ID. NS-ID has been defined by the presence of intellectual disability as the sole clinical feature. The distinction between S-ID and NS-ID is not clear because it is has been difficult to exclude the presence of more subtle neurological anomalies and psychiatric disorders in these patients, as they may be less apparent or difficult to diagnose due to the cognitive impairment. Moreover, the symptoms of some syndromes may be subtle and difficult to diagnose unless they are specifically evaluated in the context of a known genetic defect associated with these features (Kaufman et al. 2010; Ropers 2010).

19.2 Causes of Intellectual Disability

Intellectual disability can be caused by environmental and/or genetic factors; however, in up to 60% of cases, there is no identifiable cause (Rauch et al. 2006). It is notable that the prevalence of ID is inversely correlated with socioeconomic standards, both within and between countries (Ropers 2010). In poor regions of the world, the prevalence of ID is two- to threefold higher than in western countries, and this is ascribed to low birth weight as well as pre- and perinatal complications, such as malnutrition, cultural deprivation, poor health care, and parental consanguinity (Seidman et al. 2000; Boulet et al. 2009). Other specific risk factors are exposure to certain genes, viruses or radiation, and severe head trauma or brain injury. In the western world, the most common preventable cause of ID is foetal alcohol syndrome (FAS) (May and Gossage 2001; Niccols 2007). In patients with FAS, average IQ scores range between 40 and 80 with a mean of 60–65, and they remain constant from infancy to adulthood (Spohr et al. 2007). However, only a minority of the children born to women classified as heavy drinkers have FAS, which highlights differences in foetal susceptibility to the neurotoxic effects of alcohol (Ropers 2010). Similarly, although it is notable that a very low birth weight is a major risk factor for ID, most individuals with very low weights function well as adults (Gäddlin et al. 2009).

Genetic causes of ID are thought to be responsible for 25–50% of cases, although this number has been identified (Rauch et al. 2006). Additionally, pathogenic copy number variants (CNV) have been associated with ID in a large number of studies and have contributed to the discovery of many genes that cause ID (Pinto et al. 2010). Over the past 15 years, many single-gene causes of NS-ID have been identified. Many of these NS-ID genes may also cause S-ID, autism, or other neurodevelopmental phenotypes, suggesting that other genetic modifiers or environmental factors may be involved in disease aetiology.

Among the genetic conditions associated with ID, the most frequent are X-linked intellectual disability (XLID) forms caused by single gene mutations on chromosome X, while autosomal gene mutations are mostly caused by subtle chromosomal rearrangements. The apparent excess of X-linked genes involved in ID disorders led to the hypothesis that a disproportionately high density of genes influencing cognitive abilities may reside on the human X chromosome.

Interestingly, with the exception of transcription and chromatin-remodelling factors, it is worth noting that more than 50% of ID-related proteins are enriched in the pre- and/or post-synaptic compartments and are probably involved in actin cytoskeleton rearrangement, synaptic plasticity, and synapse formation (Ropers and Hamel 2005). Because learning deficits are a constant feature of patients with ID, it is tempting to attribute some traits of ID to alterations in synaptic functions (Humeau et al. 2009). This hypothesis is supported by histological data. Post-mortem morphological analyses of neurons in patients with various forms of ID often show dendritic spines with altered shapes and densities (Purpura 1974; Kaufmann and Moser 2000; Fiala et al. 2002). The degree of these defects is correlated with the severity of ID. The multiplicity of genes and proteins involved also implicates specific signalling pathways, among which small GTPases and adhesion molecules appear to play a central role. The signalling pathways and mechanisms through which alteration of these proteins could contribute to dysfunction are analysed in this chapter.

19.3 RhoGTPase Proteins

As mentioned above, small GTPase signalling appears to play a predominant role in ID through the regulation of actin dynamics and receptor trafficking. A significant percentage of the genes implicated in ID code for synaptic proteins associated with GTPase signalling and function either as regulators or effectors of GTPases.

GTPases form a large family constituting almost 200 proteins characterised by their ability to bind and hydrolyse GTP. These proteins act as molecular switches, cycling between inactive GDP-bound and active GTP-bound states, and they are tightly regulated by a variety of modulators; guanine nucleotide exchange factors (GEFs) activate the switch by catalysing the exchange of GDP for GTP, whereas GTPase-activating proteins (GAPs) increase the intrinsic GTPase activity to inactivate the RhoGTPases switch and guanine nucleotide dissociation inhibitors (GDI) (Jaffe and Hall 2005). There are three primary interrelated functional systems regulated by GTPases at neuronal synapses: the actin cytoskeleton, local translation machinery, and receptor trafficking (Boda et al. 2010).

One group of proteins, the Rho subfamily of small GTP-binding proteins, and in particular Rac1, Cdc42, and RhoA GTPases, plays important roles in synaptic functions and various aspects of neuronal development (Moon and Zheng 2003), including dendritic branching (Threadgill et al. 1997), dendritic spine formation and maintenance (Govek et al. 2004), and neurite outgrowth and differentiation (Nasu-Nishimura et al. 2006) (Fig. 19.1). Rho proteins have been implicated in different aspects of neuronal morphogenesis, including dendritic arbor development and spine morphogenesis

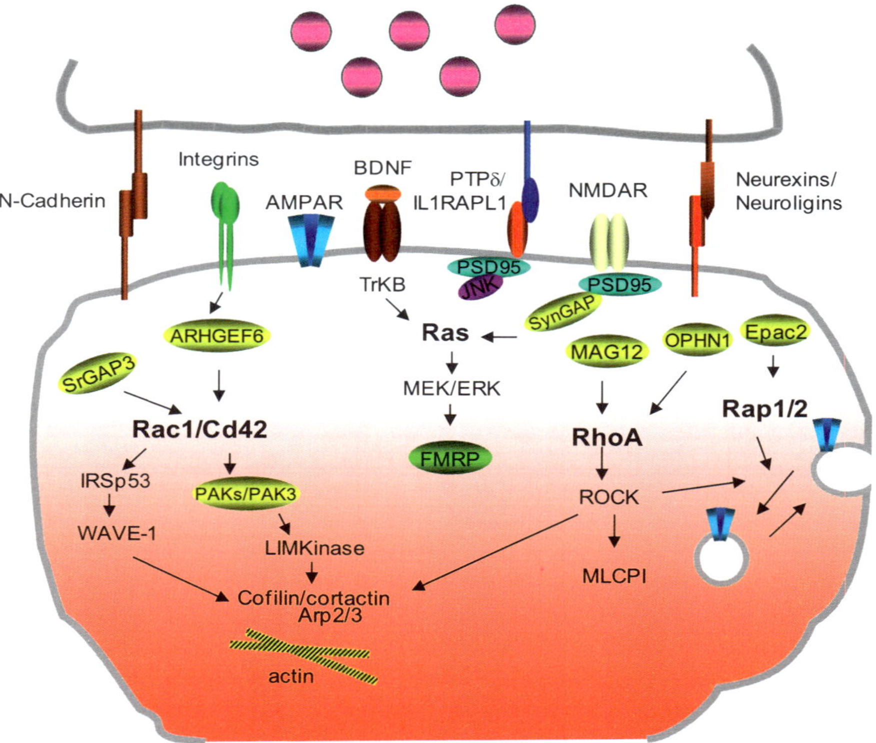

Fig. 19.1 Schematic diagram illustrating the gene products implicated in ID and associated at synapses. Signalling pathways associated with small GTPases are written in *yellow*

(Govek et al. 2005). In addition, a number of regulators and effectors that mediate the effects of Rho GTPases on the actin cytoskeleton and spine morphogenesis have been identified (Govek et al. 2005; Newey et al. 2005). The importance of proper RhoGTPase signalling in neuronal development and function has been highlighted by the identification of ID genes that encode regulators and effectors of RhoGTPases such as OPHN1, MAGI2, ARHGEF6, srGAP3, FGDI, PAK3, LIMK, and FMR1 (Fig. 19.1).

19.4 RhoA Signalling

The main evidence implicating RhoA signalling in ID was obtained in studies of the synaptic function of oligophrenin-1, a Rho-GAP for which several mutations or deletions have been identified in patients suffering from ID (Nadif Kasri and Van Aelst 2008). Oligophrenin-1 was found to negatively regulate RhoA (Ras homologous member A) and interact with the post-synaptic adaptor protein Homer (Govek et al. 2004). Importantly, Govek et al. (2004) showed that knock-down of oligophrenin-1 expression in CA1 pyramidal neurons resulted in a significant decrease in dendritic spine length. This effect could be mimicked by a constitutively active form of RhoA and could rescue an inhibitor of the RhoA effector Rho-kinase or ROCK, suggesting a regulation by RhoA of the spine actin cytoskeleton through action on LIM kinase (LIMK), myosin light chain (MLC), and/or MLC phosphatase (Govek et al. 2004; Nadif Kasri and Van Aelst 2008). Further studies of oligophrenin-1-deficient mice also revealed behavioural deficits in spatial memory, social recognition, and presynaptic facilitation (Khelfaoui et al. 2007). Moreover, it has been demonstrated that synaptic activity through NMDA receptor activation localises oligophrenin-1 to dendritic spines, where it forms a complex with AMPA receptors and selectively enhances AMPA-receptor-mediated synaptic transmission and spine size by stabilising synaptic AMPA receptors (Nakano-Kobayashi et al. 2009) (Fig. 19.1). In support of this idea, the authors showed that interference with AMPA receptor endocytosis could prevent the decrease in synaptic transmission and synapse density induced by oligophrenin-1 knock-down, possibly linking oligophrenin-1/RhoA signalling to AMPA receptor endocytosis. Additional evidence has shown that oligophrenin-1 is concentrated at endocytic sites and regulates AMPA receptor endocytosis at excitatory synapses by RhoA/ROCK signalling (Khelfaoui et al. 2009).

Another mechanism could involve the ID gene MAGI2, which codes for a scaffolding protein (S-SCAM) that has multiple functions at the synapse. One scaffolding protein could activate RhoA in response to NMDA receptor stimulation (Iida et al. 2007) (Fig. 19.1).

19.5 Rac1/Cdc42/PAKs Pathway

There are several proteins linked to ID implicated in Rac1/Cdc42/PAK signalling. These include upstream Rac/Cdc42 regulators such as ARHGEF6, a Rac/Cdc42 GEF; srGAP3 (MEGAP) for Slit Robo Rho GTPase-activating protein 3; and FGDI, a

Cdc42 GEF. These molecules are involved in regulating the active/inactive state of Rho proteins (Rac/Cdc42). There are also downstream effectors such as PAK3 and LIMK that regulate the actin cytoskeleton through phosphorylation of cofilin, one of the central functional regulators of actin dynamics (Bernstein and Bamburg 2010). Cofilin can be phosphorylated through two GTPase cascades mediated by LIMK: Rac/Cdc42/PAK or RhoA/ROCK (Bernstein and Bamburg 2010; Rex et al. 2009).

Three of these genes (ARHGEF6, PAK3, and LIMK) seem to be involved in the same signalling pathway; all have been shown to be expressed at the synapse and result in alterations of spine morphology or function upon knock-down or overexpression (Meng et al. 2002; Boda et al. 2004; Nodé-Langlois et al. 2006) (Fig. 19.1). ARHGEF6 (a Rho guanine nucleotide exchange factor 6) codes for a Cdc42/Rac exchange factor and is involved in integrin-mediated signalling leading to activation of the GTPases Rac1 (Ras-related C3 botulinum toxin substrate 1) and/or Cdc42 (cell division cycle 42) (Rosenberger et al. 2003). Mutations in ARHGEF6 have been demonstrated in patients with X-linked ID (Kutsche et al. 2000). Knock-down of ARHGEF6 using siRNA resulted in alterations of spine morphology, characterised by a decrease of large-mushroom-type spines and an increase of elongated spines and filopodia-like protrusions (Nodé-Langlois et al. 2006). This phenotype closely resembles the phenotype reported for neurons with knock-down PAK3 levels or expressing dominant-negative mutant forms of PAK3. Moreover, the defects observed in cells expressing mutant ARHGEF6 could be rescued by PAK3 activation, confirming that the two molecules are involved in the same signalling pathway (Nodé-Langlois et al. 2006).

PAK3 encodes a member of the large family of p21-activating kinases (PAK) and acts as a molecular effector on Rac1/Cdc42, mediating their effects on the cytoskeleton through LIMK activation and influencing downstream gene expression (Jaffer and Chernoff 2002; Bokoch 2003). Several mutations in PAK3 have been associated with X-linked ID (Allen et al. 1998; Bienvenu et al. 2000); these mutations have been associated with either loss of PAK3 protein or loss of its kinase activity. Several lines of evidence have demonstrated a role of PAK3 in regulating spine morphogenesis, synapse formation, and synaptic plasticity. Studies have shown that downregulation of PAK3 in rat hippocampal cultures results in the formation of abnormally elongated dendritic spines and filopodia-like protrusions and a decrease in mature spine synapses (Boda et al. 2004, 2008). The researchers noticed that these defects were associated with reduced expression of AMPARs at the synapse and LTP. Moreover, mice deficient in PAK3 expression exhibit deficits in hippocampal late-phase LTP, a distinct form of long-term synaptic plasticity involving de novo gene expression, and some deficiencies in learning and memory (Meng et al. 2005). PAK3 knockout mice exhibit no alterations in neuronal structure; however, a dramatic decrease was observed in the levels of the phosphorylated/active form of cAMP-responsive element-binding protein (CREB) in the hippocampus, a protein important for synaptic plasticity and memory formation in mice (Kandel 2001; Lonze and Ginty 2002). Therefore, reduced CREB function may be responsible for the impairment in late-phase hippocampal LTP in these mice. These morphologies and/or functions are a result of altered actin dynamics and/or transcriptional regulation (Nadif Kasri and Van Aelst 2008).

As mentioned previously, PAK and Rho-kinases both stimulate LIMK, a serine-only protein kinase that phosphorylates the actin depolymerisation factor cofilin, which is then unable to bind and depolymerise F-actin. LIMK1 is therefore a key component of the signal transduction network connecting extracellular stimuli to changes in cytoskeletal structures (Stanyon and Bernard 1999). The LIM-kinase 1 (LIMK1) gene is located on chromosome 7q11 and was found to be one of the genes heterozygously deleted in Williams syndrome (Tassabehji et al. 1996), a rare (1 in 25,000) mental disorder with profound deficits in visuospatial cognition. LIMK1 knockout mice show abnormal spine morphology, abnormal synaptic plasticity (including enhanced hippocampal LTP), and impaired spatial learning (Meng et al. 2002).

Regulation of the Rac/Cdc42/PAK signalling cascade is likely to be quite complex and involves several potential partners. For example, Rho GAP srGAP3, a gene located on chromosome 3p25, was found to be disrupted and functionally inactivated by a translocation breakpoint in a patient displaying severe ID (Endris et al. 2002). SrGAP3 associates with the scaffolding protein WAVE1 and inhibits Rac function (Soderling et al. 2007). However, no data are yet available to determine whether loss of srGAP3 impacts spine morphology.

The Rac/Cdc42/PAK signalling cascade integrates activity and trans-synaptic signals to fine-tune the spine actin cytoskeleton, controlling the growth, size, and morphology of spines and possibly also exocytosis and receptor trafficking. For example, Cdc42 has been directly implicated in the control of exocytosis and trafficking membrane proteins in other systems (Wu et al. 2008) and synaptic terminals (Doussau et al. 2000).

19.6 Regulation of Protein Synthesis

Recent studies have provided evidence that Rho GTPase signalling also plays a role in fragile X syndrome (FRAXA). Fragile X syndrome is caused by mutations in the Fmr1 gene that result in transcriptional silencing of the protein FMRP. The function of FMRP is still not completely understood, but it seems to act as an RNA-binding protein (Nimchinsky et al. 2002). Thus, it is thought that FMRP plays a key role in synaptic plasticity through the regulation of mRNA transport and translational inhibition of local protein synthesis at the synapse (D'Hulst and Kooy 2009). Among the target proteins that have exhibited an increased synthesis rate in Fmr1 knockout mice are notably PSD-95, Arc, and GluR1 (Muddashetty et al. 2007; Zalfa et al. 2007). Fmr1 knockout mice exhibit important defects affecting various aspects of synapse morphology, function, and plasticity (Pfeiffer and Huber 2009). Additionally, two pieces of evidence link FMRP to GTPase signalling. First, work in drosophila has provided evidence that the FMRP homologue in drosophila affects dendritic development by regulating the actin cytoskeleton through a translational suppression of Rac1 and profilin (Reeve et al. 2005). Second, Fmr1 knockout mice also have shown general impairment in LTP that can be rescued by BDNF or activation of the Ras/PI3K cascade, thereby suggesting a possible role of Ras signalling in this process (Hu et al. 2008) (Fig. 19.1). Therefore, signalling alterations involving GTPases might occur in Fmr1 knockout mice (see Chap. 25).

19.7 Ras/RAP Proteins

The Ras family of small GTPases (Ras, Rap1, and Rap2) and their downstream mitogen-activated protein kinases (ERK, JNK, and p38MAPK) and PIK3 signalling cascades control various physiological processes. In neuronal cells, Ras, Rap1, and Rap2 are differentially stimulated by different forms of synaptic activity via the activation of NMDA-Rs and calcium influx to independently control activity-dependent AMPAR trafficking events (Tada and Sheng 2006; Gu and Stornetta 2007; Thomas et al. 2008) (Fig. 19.1). Particularly, Ras promotes long-term potentiation (LTP) and the surface delivery of AMPA receptors (AMPARs), whereas Rap mediates long-term depression (LTD) or depotentiation and AMPAR internalisation (Zhu et al. 2002, 2005). Ras also stimulates the overproduction of dendritic protrusion or spines (Wu et al. 2001; Arendt et al. 2004), while Rap promotes spine loss (Pak et al. 2001; Fu et al. 2007; Ryu et al. 2008). Recent work, particularly in genetic screening, has linked genetic defects of various molecules causing aberrant Ras and Rap signalling with a number of mental disorders involving deficits in cognitive functioning and adaptive behaviours (Eng 2003; Zhu et al. 2004; Roberts et al. 2006; Schubbert et al. 2007; Ehninger et al. 2008; Orloff and Eng 2008; Levitt and Campbell 2009). These findings underscore the essential role of Ras and Rap signalling in controlling synaptic AMPAR trafficking during synaptic plasticity (Thomas and Huganir 2004; Gu and Stornetta 2007), a basic cellular mechanism of learning and memory (Kessels and Malinow 2009).

Their implication in intellectual disability has also been linked to one regulatory protein: SYNGAP, a Ras/Rap GTPase-activating protein recently found to be mutated in patients with mental retardation (Hamdan et al. 2009). The autosomal gene SYNGAP1 encodes a Ras GTPase-activating protein critical for cognition and synapse function, and mutations in this gene were found in 94 patients with non-syndromic intellectual disability (Hamdan et al. 2009). SYNGAP1 is selectively expressed in the brain and is a component of the NMDA-receptor complex, acting downstream of the receptor and blocking the insertion of the AMPA receptor at the post-synaptic membrane (Kim et al. 1998; Krapivinsky et al. 2004; Kim et al. 2005; Rumbaugh et al. 2006) by inhibiting the Ras-ERK pathway (Kim et al. 2005). A role for defective Ras signalling was recently proposed in the analyses of Fmr1 knockout mice: a selective deficit in the synaptic delivery of GluR1-containing AMPARs results in a loss of GluR1-dependent LTP (Hu et al. 2008). This effect was associated with a defect in the signalling between Ras, phosphoinositide 3-kinase (PI3K), and protein kinase B (PKB) consistent with inefficient signalling. More interestingly, enhancing Ras/PI3K/PKB signalling rescued synaptic delivery of GluR1-containing receptors and LTP, strengthening the important link existing between Ras signalling and AMPAR trafficking in Fmr1 knockout mice. This study makes a strong case for the hypothesis that Ras has an effect on the regulation of AMPAR delivery to synapses and that this mechanism could represent a primary cause for the alterations of spine morphology, function, and plasticity associated with alteration in this pathway (Boda et al. 2010).

19.8 Synaptic Adhesion Molecules Gene Mutations

It has been well defined that the correct development of synaptic specialisations and establishment of appropriate connectivity patterns are crucial for the assembly of functional neuronal circuits. Neuroligin proteins were first identified as neurexin binding partners; however, the neuroligin-neurexin protein complex was shown to be relevant for synapse formation when two proteins were found to associate in synapses and that the complex plays a major role in synapse formation and function (Fig. 19.1) (reviewed in Dean and Dresbach 2006; Dalva et al. 2007; Südhof 2008) (see Chap. 6).

This was further supported by the finding that deletions or mutations in neuroligins and neurexins in humans were associated with several cognitive disorders, including autism spectrum disorders (ASD), mental retardation, and schizophrenia. However, the exact correlation between the genetic and biochemical mechanisms by which these mutations contribute to these diseases remains unknown.

A deep DNA analysis in patients affected by autism revealed a strong association with mutations in the genes encoding Nrxn1, Nlgn1, Nlgn3, and Nlgn4. In autistic patients, seven point mutations, two distinct translocations, and four different large-scale deletions in the Nrxn1 gene have been identified (Feng et al. 2006; Szatmari et al. 2007; Kim et al. 2008; Marshall et al. 2008; Yan et al. 2008b; Zahir et al. 2008). For the Nlgn4 gene, at least ten different mutations have been observed (two frameshifts, five missense mutations, and three internal deletions), while for Nlgn3, a single point mutation has been identified (the R451C substitution) (Jamain et al. 2003; Laumonnier et al. 2004; Yan et al. 2005; Talebizadeh et al. 2006; Yan et al. 2008a). In addition to these mutations, deletions of X-chromosomal DNA including the Nlgn4 locus have been detected in autism patients (Chocholska et al. 2006; Macarov et al. 2007; Lawson-Yuen et al. 2008; Marshall et al. 2008).

These genetic findings appear to provide strong evidence for a role of the Nrxn/Nlgn complex in the pathogenesis of ASDs. However, it is important to note that the clinical manifestations in patients do not always correlate with the genetic mutations. For example, identical mutations are found in symptomatic patients and non-symptomatic relatives. The same mutations can also cause a completely different phenotype in different individuals. Lawson-Yuen et al. described a family where in which a microdeletion in Nlgn4 caused severe autism in one patient and Tourette's syndrome in the patient's brother (Lawson-Yuen et al. 2008). Mutations in Nrxn1α gene have also been found in individuals affected by schizophrenia (Kirov et al. 2008; Walsh et al. 2008), suggesting that dysfunctions in synaptic cell adhesion are characterised by the manifestation of a continuum of intellectual disabilities that includes autism and mental retardation. This is not surprising considering that completely different gene mutations can cause similar clinical manifestations, such as the ASDs (Morrow et al. 2008).

Although stronger genetic findings will better define the role of Nrxn/Nlgn mutations in humans, it seems clear that the mutations associated with ASD are not simply polymorphisms but always mutations that strongly affect the expression, structures, or functions of proteins. Some autism-like phenotypes have also been observed in Nlgn1, Nlgn3, and Nlgn4 mutant mice (Comoletti et al. 2004; Tabuchi et al. 2007; Jamain et al. 2008; Blundell et al. 2010).

However, there is not a linear relationship between the genetic alteration and phenotype manifestation in most human cases. It is still unclear why some non-symptomatic individuals have the same mutations as affected people. This suggests the existence of compensatory mechanisms or the concomitance of other unknown genetic or non-genetic co-factors.

Interestingly, Nrxn1α mutations have been linked to schizophrenia (Kirov et al. 2008; Walsh et al. 2008), while Nrxn3 alterations have been connected to different types of addiction (Hishimoto et al. 2007; Lachman et al. 2007). These data suggest that mutations in genes encoding Nrxns and Nlgns definitively alter a number of cognitive properties in humans.

The best method for ascertaining whether Nrxns and Nlgns mutations found in humans are directly correlated with ADS manifestation is to develop animal models for the same mutations. This has recently been done for three Nlgn mutations, the Nlgn3 R451C substitution and the Nlgn1 and Nlgn4 loss-of-function mutations (Tabuchi et al. 2007; Jamain et al. 2008; Blundell et al. 2010). Tabuchi et al. reported the characterisation of R451C knockin mice. They showed that the mice present normal motor and anxiety behaviours, but the social interaction was partially impaired, and the spatial learning capability was largely increased.

However, this behavioural phenotype does not completely recapitulate the human phenotype because the R415C substitution did not impair cognitive function in the mice but causes severe intellectual disability in humans (Jamain et al. 2003). Consistent with the idea that autism might be due to the alteration between the excitatory/ inhibitory balance, R451C mutant mice displayed an increase in inhibitory synaptic transmission in the somatosensory cortex. However, another study reported that the humanised R451C mutation in mice did not result in apparent autism-like phenotypes but produced detectable functional consequences that may be interpreted in terms of physical development and/or reduced sensitivity to stimuli (Chadman et al. 2008). Further research should aim to clarify the discrepancy between these two studies.

Nlgn3 knockout mice have shown a different phenotype compared with R451C mutant mice, suggesting that the point mutation confers a gain function (Jamain et al. 2003). The R451C mutation caused a local folding defect of the protein, which is partially retained in the ER and degraded (De Jaco et al. 2010). Thus, these data support the proposal that the R451C protein might sequester from synapse intracellular partners like PSD-95, supporting the hypothesis that behavioural changes can be linked to a subtle perturbation of synaptic functions.

Finally, both Nlgn4 and Nlgn1 deletions caused an autism-like phenotype in animal models (Jamain et al. 2008; Blundell et al. 2010). It is interesting to note that in general single mutation or the deletion of Nlgns or Nrxns does not perturb the overall synapse structure and formation, suggesting that these proteins are not implicated simply in building the synapses, because small changes in their functions can induce important changes in the neural network, causing cognitive impairments. Thus, it would not be surprising if other members of the synaptic adhesion molecules family are found to be mutated in ADS and intellectually disabled patients or if other proteins mutated in ASD are synaptogenic proteins.

This could be the case of *Interleukin-1 receptor accessory protein-like 1* (*IL1RAPL1*), in which mutations have been associated with cognitive impairment ranging from non-syndromic X-linked (the gene is on the X chromosome) mental retardation to autism (Carrie et al. 1999). IL1RAPL1 belongs to a novel family of Toll/ IL-1 receptors and shares 52% homology with the IL-1 receptor accessory protein (IL-1RacP). As with the other members of IL-1 receptor family, it is characterised by three extracellular Ig-like domains, a transmembrane domain, and an intracellular Toll/IL-1R homology domain (TIR domain). Additionally, unlike other members in the IL-1 receptor family, 150 additional amino acids (aa) occur at the C-terminal end. It has been previously shown that IL1RAPL1 interacts with NCS-1 through this intracellular region (Bahi et al. 2003) and that this interaction mediates the regulatory effect of IL1RAPL1 overexpression on N-type voltage-gated calcium channel (VGCC) activity in PC12 cells (Gambino et al. 2007).

More recently, we have shown that IL1RAPL1 can bind to PSD-95 and regulate its phosphorylation and synaptic association by activating c-Jun terminal kinase (JNK) (Pavlowsky et al. 2010) (Fig. 19.1). Interestingly, we also found that the extracellular domains of IL1RAPL1 and IL1RAPL2 (the paralogue) are able to induce presynaptic differentiation by binding to PTPδ (unpublished results). Thus, our data suggest that the IL1RAPL family of proteins have the function of a classical synaptic adhesion molecule.

In conclusion, all the genetic and functional data suggest that mutations in one of the multiple adhesion synaptic proteins were always associated with intellectual disabilities, suggesting that all of these molecules play an essential role in regulating synapses functions.

References

Allen, K. M., Gleeson, J. G., Bagrodia, S., Partington, M. W., MacMillan, J. C., Cerione, R. A., Mulley, J. C., & Walsh, C. A. (1998). PAK3 mutation in nonsyndromic X-linked mental retardation. *Nature Genetics, 20*, 25–30.

Arendt, T., Gärtner, U., Seeger, G., Barmashenko, G., Palm, K., Mittmann, T., Yan, L., Hümmeke, M., Behrbohm, J., Brückner, M. K., Holzer, M., Wahle, P., & Heumann, R. (2004). Neuronal activation of Ras regulates synaptic connectivity. *European Journal of Neuroscience, 19*, 2953–2966.

Bahi, N., Friocourt, G., Carrie, A., Graham, M. E., Weiss, J. L., Chafey, P., Fauchereau, F., Burgoyne, R. D., & Chelly, J. (2003). IL1 receptor accessory protein like, a protein involved in

X-linked mental retardation, interacts with Neuronal Calcium Sensor-1 and regulates exocytosis. *Human Molecular Genetics, 12,* 1415–1425.

Bernstein, B. W., & Bamburg, J. R. (2010). ADF/cofilin: A functional node in cell biology. *Trends in Cell Biology, 20,* 187–195.

Bienvenu, T., des Portes, V., McDonell, N., Carrié, A., Zemni, R., Couvert, P., Ropers, H. H., Moraine, C., van Bokhoven, H., Fryns, J. P., Allen, K., Walsh, C. A., Boué, J., Kahn, A., Chelly, J., & Beldjord, C. (2000). Missense mutation in PAK3, R67C, causes X-linked nonspecific mental retardation. *American Journal of Medical Genetics, 93,* 294–298.

Blanpied, T. A., & Ehlers, M. D. (2004). Microanatomy of dendritic spines: Emerging principles of synaptic pathology in psychiatric and neurological disease. *Biological Psychiatry, 55,* 1121–1127.

Blundell, J., Blaiss, C. A., Etherton, M. R., Espinosa, F., Tabuchi, K., Walz, C., Bolliger, M. F., Südhof, T. C., & Powell, C. M. (2010). Neuroligin-1 deletion results in impaired spatial memory and increased repetitive behavior. *Journal of Neuroscience, 30,* 2115–2129.

Boda, B., Alberi, S., Nikonenko, I., Node-Langlois, R., Jourdain, P., Moosmayer, M., Parisi-Jourdain, L., & Muller, D. (2004). The mental retardation protein PAK3 contributes to synapse formation and plasticity in hippocampus. *Journal of Neuroscience, 24,* 10816–10825.

Boda, B., Dubos, A., & Muller, D. (2010). Signaling mechanisms regulating synapse formation and function in mental retardation. *Current Opinion in Neurobiology, 20,* 519–527.

Boda, B., Jourdain, L., & Muller, D. (2008). Distinct, but compensatory roles of PAK1 and PAK3 in spine morphogenesis. *Hippocampus, 18,* 857–861.

Bokoch, G. M. (2003). Biology of the p21-activated kinases. *Annual Review of Biochemistry, 72,* 743–781.

Boulet, S. L., Schieve, L. A., & Boyle, C. A. (2009). Birth weight and health and developmental outcomes in US children, 1997–2005. *Matern Child Health Journal.* 2011. Oct:15(7):836–44. doi:10.1007/s10995-009-0538-2.

Bradley, E. A., Thompson, A., & Bryson, S. E. (2002). Mental retardation in teenagers: Prevalence data from the Niagara region, Ontario. *Canadian Journal of Psychiatry, 47,* 652–659.

Carrie, A., et al. (1999). A new member of the IL-1 receptor family highly expressed in hippocampus and involved in X-linked mental retardation. *Nature Genetics, 23,* 25–31.

Chadman, K. K., Gong, S., Scattoni, M. L., Boltuck, S. E., Gandhy, S. U., Heintz, N., & Crawley, J. N. (2008). Minimal aberrant behavioral phenotypes of neuroligin-3 R451C knockin mice. *Autism Research, 1,* 147–158.

Chocholska, S., Rossier, E., Barbi, G., & Kehrer-Sawatzki, H. (2006). Molecular cytogenetic analysis of a familial interstitial deletion Xp22.2-22.3 with a highly variable phenotype in female carriers. *American Journal of Medical Genetics A, 140,* 604–610.

Comoletti, D., De Jaco, A., Jennings, L. L., Flynn, R. E., Gaietta, G., Tsigelny, I., Ellisman, M. H., & Taylor, P. (2004). The Arg451Cys-neuroligin-3 mutation associated with autism reveals a defect in protein processing. *Journal of Neuroscience, 24,* 4889–4893.

D'Hulst, C., & Kooy, R. F. (2009). Fragile X syndrome: From molecular genetics to therapy. *Journal of Medical Genetics, 46,* 577–584.

Dalva, M. B., McClelland, A. C., & Kayser, M. S. (2007). Cell adhesion molecules: Signalling functions at the synapse. *Nature Reviews Neuroscience, 8,* 206–220.

De Jaco, A., Lin, M. Z., Dubi, N., Comoletti, D., Miller, M. T., Camp, S., Ellisman, M., Butko, M. T., Tsien, R. Y., & Taylor, P. (2010). Neuroligin trafficking deficiencies arising from mutations in the alpha/beta-hydrolase fold protein family. *The Journal of Biological Chemistry, 285,* 28674–28682.

Dean, C., & Dresbach, T. (2006). Neuroligins and neurexins: Linking cell adhesion, synapse formation and cognitive function. *Trends in Neurosciences, 29,* 21–29.

Doussau, F., Gasman, S., Humeau, Y., Vitiello, F., Popoff, M., Boquet, P., Bader, M. F., & Poulain, B. (2000). A Rho-related GTPase is involved in Ca(2+)-dependent neurotransmitter exocytosis. *The Journal of Biological Chemistry, 275,* 7764–7770.

Drews, C. D., Yeargin-Allsopp, M., Decouflé, P., & Murphy, C. C. (1995). Variation in the influence of selected sociodemographic risk factors for mental retardation. *American Journal of Public Health, 85,* 329–334.

Ehninger, D., Li, W., Fox, K., Stryker, M. P., & Silva, A. J. (2008). Reversing neurodevelopmental disorders in adults. *Neuron, 60*, 950–960.

Endris, V., Wogatzky, B., Leimer, U., Bartsch, D., Zatyka, M., Latif, F., Maher, E. R., Tariverdian, G., Kirsch, S., Karch, D., & Rappold, G. A. (2002). The novel Rho-GTPase activating gene MEGAP/srGAP3 has a putative role in severe mental retardation. *Proceedings of the National Academy of Sciences of the United States of America, 99*, 11754–11759.

Eng, C. (2003). PTEN: One gene, many syndromes. *Human Mutation, 22*, 183–198.

Feng, J., Schroer, R., Yan, J., Song, W., Yang, C., Bockholt, A., Cook, E. H., Skinner, C., Schwartz, C. E., & Sommer, S. S. (2006). High frequency of neurexin 1beta signal peptide structural variants in patients with autism. *Neuroscience Letters, 409*, 10–13.

Fiala, J. C., Spacek, J., & Harris, K. M. (2002). Dendritic spine pathology: Cause or consequence of neurological disorders? *Brain Research: Brain Research Reviews, 39*, 29–54.

Fu, Z., Lee, S. H., Simonetta, A., Hansen, J., Sheng, M., & Pak, D. T. (2007). Differential roles of Rap1 and Rap2 small GTPases in neurite retraction and synapse elimination in hippocampal spiny neurons. *Journal of Neurochemistry, 100*, 118–131.

Gäddlin, P. O., Finnström, O., Sydsjö, G., & Leijon, I. (2009). Most very low birth weight subjects do well as adults. *Acta Paediatrica, 98*, 1513–1520.

Gambino, F., Pavlowsky, A., Béglé, A., Dupont, J. L., Bahi, N., Courjaret, R., Gardette, R., Hadjkacem, H., Skala, H., Poulain, B., Chelly, J., Vitale, N., & Humeau, Y. (2007). IL1-receptor accessory protein-like 1 (IL1RAPL1), a protein involved in cognitive functions, regulates N-type Ca2+-channel and neurite elongation. *Proceedings of the National Academy of Sciences of the United States of America, 104*, 9063–9068.

Govek, E. E., Newey, S. E., Akerman, C. J., Cross, J. R., Van der Veken, L., & Van Aelst, L. (2004). The X-linked mental retardation protein oligophrenin-1 is required for dendritic spine morphogenesis. *Nature Neuroscience, 7*, 364–372.

Govek, E. E., Newey, S. E., & Van Aelst, L. (2005). The role of the Rho GTPases in neuronal development. *Genes and Development, 19*, 1–49.

Gu, Y., & Stornetta, R. L. (2007). Synaptic plasticity, AMPA-R trafficking, and Ras-MAPK signaling. *Acta Pharmacologica Sinica, 28*, 928–936.

Hamdan, F. F., et al. (2009). Mutations in SYNGAP1 in autosomal nonsyndromic mental retardation. *The New England Journal of Medicine, 360*, 599–605.

Hishimoto, A., Liu, Q. R., Drgon, T., Pletnikova, O., Walther, D., Zhu, X. G., Troncoso, J. C., & Uhl, G. R. (2007). Neurexin 3 polymorphisms are associated with alcohol dependence and altered expression of specific isoforms. *Human Molecular Genetics, 16*, 2880–2891.

Hu, H., Qin, Y., Bochorishvili, G., Zhu, Y., van Aelst, L., & Zhu, J. J. (2008). Ras signaling mechanisms underlying impaired GluR1-dependent plasticity associated with fragile X syndrome. *Journal of Neuroscience, 28*, 7847–7862.

Humeau, Y., Gambino, F., Chelly, J., & Vitale, N. (2009). X-linked mental retardation: Focus on synaptic function and plasticity. *Journal of Neurochemistry, 109*, 1–14.

Iida, J., Ishizaki, H., Okamoto-Tanaka, M., Kawata, A., Sumita, K., Ohgake, S., Sato, Y., Yorifuji, H., Nukina, N., Ohashi, K., Mizuno, K., Tsutsumi, T., Mizoguchi, A., Miyoshi, J., Takai, Y., & Hata, Y. (2007). Synaptic scaffolding molecule alpha is a scaffold to mediate N-methyl-D-aspartate receptor-dependent RhoA activation in dendrites. *Molecular and Cellular Biology, 27*, 4388–4405.

Jaffe, A. B., & Hall, A. (2005). Rho GTPases: Biochemistry and biology. *Annual Review of Cell and Developmental Biology, 21*, 247–269.

Jaffer, Z. M., & Chernoff, J. (2002). p21-activated kinases: Three more join the Pak. *The International Journal of Biochemistry and Cell Biology, 34*, 713–717.

Jamain, S., Quach, H., Betancur, C., Rastam, M., Colineaux, C., Gillberg, I. C., Soderstrom, H., Giros, B., Leboyer, M., Gillberg, C., & Bourgeron, T. (2003). Mutations of the X-linked genes encoding neuroligins NLGN3 and NLGN4 are associated with autism. *Nature Genetics, 34*, 27–29.

Jamain, S., Radyushkin, K., Hammerschmidt, K., Granon, S., Boretius, S., Varoqueaux, F., Ramanantsoa, N., Gallego, J., Ronnenberg, A., Winter, D., Frahm, J., Fischer, J., Bourgeron, T., Ehrenreich, H., & Brose, N. (2008). Reduced social interaction and ultrasonic communication in a mouse model of monogenic heritable autism. *Proceedings of the National Academy of Sciences of the United States of America, 105*, 1710–1715.

Kandel, E. R. (2001). The molecular biology of memory storage: A dialogue between genes and synapses. *Science, 294*, 1030–1038.

Kaufman, L., Ayub, M., & Vincent, J. B. (2010). The genetic basis of non-syndromic intellectual disability: A review. *Journal of Neurodevelopmental Disorders, 2*, 182–209.

Kaufmann, W. E., & Moser, H. W. (2000). Dendritic anomalies in disorders associated with mental retardation. *Cerebral Cortex, 10*, 981–991.

Kessels, H. W., & Malinow, R. (2009). Synaptic AMPA receptor plasticity and behavior. *Neuron, 61*, 340–350.

Khelfaoui, M., Denis, C., van Galen, E., de Bock, F., Schmitt, A., Houbron, C., Morice, E., Giros, B., Ramakers, G., Fagni, L., Chelly, J., Nosten-Bertrand, M., & Billuart, P. (2007). Loss of X-linked mental retardation gene oligophrenin1 in mice impairs spatial memory and leads to ventricular enlargement and dendritic spine immaturity. *Journal of Neuroscience, 27*, 9439–9450.

Khelfaoui, M., Pavlowsky, A., Powell, A. D., Valnegri, P., Cheong, K. W., Blandin, Y., Passafaro, M., Jefferys, J. G., Chelly, J., & Billuart, P. (2009). Inhibition of RhoA pathway rescues the endocytosis defects in Oligophrenin1 mouse model of mental retardation. *Human Molecular Genetics, 18*, 2575–2583.

Kim, M. J., Dunah, A. W., Wang, Y. T., & Sheng, M. (2005). Differential roles of NR2A- and NR2B-containing NMDA receptors in Ras-ERK signaling and AMPA receptor trafficking. *Neuron, 46*, 745–760.

Kim, J. H., Liao, D., Lau, L. F., & Huganir, R. L. (1998). SynGAP: A synaptic RasGAP that associates with the PSD-95/SAP90 protein family. *Neuron, 20*, 683–691.

Kim, H. G., et al. (2008). Disruption of neurexin 1 associated with autism spectrum disorder. *American Journal of Human Genetics, 82*, 199–207.

Kirov, G., Gumus, D., Chen, W., Norton, N., Georgieva, L., Sari, M., O'Donovan, M. C., Erdogan, F., Owen, M. J., Ropers, H. H., & Ullmann, R. (2008). Comparative genome hybridization suggests a role for NRXN1 and APBA2 in schizophrenia. *Human Molecular Genetics, 17*, 458–465.

Krapivinsky, G., Medina, I., Krapivinsky, L., Gapon, S., & Clapham, D. E. (2004). SynGAP-MUPP1-CaMKII synaptic complexes regulate p38 MAP kinase activity and NMDA receptor-dependent synaptic AMPA receptor potentiation. *Neuron, 43*, 563–574.

Kutsche, K., Yntema, H., Brandt, A., Jantke, I., Nothwang, H. G., Orth, U., Boavida, M. G., David, D., Chelly, J., Fryns, J. P., Moraine, C., Ropers, H. H., Hamel, B. C., van Bokhoven, H., & Gal, A. (2000). Mutations in ARHGEF6, encoding a guanine nucleotide exchange factor for Rho GTPases, in patients with X-linked mental retardation. *Nature Genetics, 26*, 247–250.

Lachman, H. M., Fann, C. S., Bartzis, M., Evgrafov, O. V., Rosenthal, R. N., Nunes, E. V., Miner, C., Santana, M., Gaffney, J., Riddick, A., Hsu, C. L., & Knowles, J. A. (2007). Genomewide suggestive linkage of opioid dependence to chromosome 14q. *Human Molecular Genetics, 16*, 1327–1334.

Laumonnier, F., Bonnet-Brilhault, F., Gomot, M., Blanc, R., David, A., Moizard, M. P., Raynaud, M., Ronce, N., Lemonnier, E., Calvas, P., Laudier, B., Chelly, J., Fryns, J. P., Ropers, H. H., Hamel, B. C., Andres, C., Barthélémy, C., Moraine, C., & Briault, S. (2004). X-linked mental retardation and autism are associated with a mutation in the NLGN4 gene, a member of the neuroligin family. *American Journal of Human Genetics, 74*, 552–557.

Lawson-Yuen, A., Saldivar, J. S., Sommer, S., & Picker, J. (2008). Familial deletion within NLGN4 associated with autism and Tourette syndrome. *European Journal of Human Genetics, 16*, 614–618.

Leonard, H., & Wen, X. (2002). The epidemiology of mental retardation: Challenges and opportunities in the new millennium. *Mental Retardation and Developmental Disabilities Research Reviews, 8*, 117–134.

Levitt, P., & Campbell, D. B. (2009). The genetic and neurobiologic compass points toward common signaling dysfunctions in autism spectrum disorders. *The Journal of Clinical Investigation, 119*, 747–754.

Lonze, B. E., & Ginty, D. D. (2002). Function and regulation of CREB family transcription factors in the nervous system. *Neuron, 35*, 605–623.

Macarov, M., Zeigler, M., Newman, J. P., Strich, D., Sury, V., Tennenbaum, A., & Meiner, V. (2007). Deletions of VCX-A and NLGN4: A variable phenotype including normal intellect. *Journal of Intellectual Disability Research, 51*, 329–333.

Marshall, C. R., et al. (2008). Structural variation of chromosomes in autism spectrum disorder. *American Journal of Human Genetics, 82*, 477–488.

May, P. A., & Gossage, J. P. (2001). Estimating the prevalence of fetal alcohol syndrome: A summary. *Alcohol Research and Health, 25*, 159–167.

McLaren, J., & Bryson, S. E. (1987). Review of recent epidemiological studies of mental retardation: Prevalence, associated disorders, and etiology. *American Journal of Mental Retardation, 92*, 243–254.

Meng, J., Meng, Y., Hanna, A., Janus, C., & Jia, Z. (2005). Abnormal long-lasting synaptic plasticity and cognition in mice lacking the mental retardation gene Pak3. *Journal of Neuroscience, 25*, 6641–6650.

Meng, Y., Zhang, Y., Tregoubov, V., Janus, C., Cruz, L., Jackson, M., Lu, W. Y., MacDonald, J. F., Wang, J. Y., Falls, D. L., & Jia, Z. (2002). Abnormal spine morphology and enhanced LTP in LIMK-1 knockout mice. *Neuron, 35*, 121–133.

Moon, S. Y., & Zheng, Y. (2003). Rho GTPase-activating proteins in cell regulation. *Trends in Cell Biology, 13*, 13–22.

Morrow, E. M., et al. (2008). Identifying autism loci and genes by tracing recent shared ancestry. *Science, 321*, 218–223.

Muddashetty, R. S., Kelić, S., Gross, C., Xu, M., & Bassell, G. J. (2007). Dysregulated metabotropic glutamate receptor-dependent translation of AMPA receptor and postsynaptic density-95 mRNAs at synapses in a mouse model of fragile X syndrome. *Journal of Neuroscience, 27*, 5338–5348.

Nadif Kasri, N., & Van Aelst, L. (2008). Rho-linked genes and neurological disorders. *Pflügers Archiv, 455*, 787–797.

Nakano-Kobayashi, A., Kasri, N. N., Newey, S. E., & Van Aelst, L. (2009). The Rho-linked mental retardation protein OPHN1 controls synaptic vesicle endocytosis via endophilin A1. *Current Biology, 19*, 1133–1139.

Nasu-Nishimura, Y., Hayashi, T., Ohishi, T., Okabe, T., Ohwada, S., Hasegawa, Y., Senda, T., Toyoshima, C., Nakamura, T., & Akiyama, T. (2006). Role of the Rho GTPase-activating protein RICS in neurite outgrowth. *Genes to Cells, 11*, 607–614.

Newey, S. E., Velamoor, V., Govek, E. E., & Van Aelst, L. (2005). Rho GTPases, dendritic structure, and mental retardation. *Journal of Neurobiology, 64*, 58–74.

Niccols, A. (2007). Fetal alcohol syndrome and the developing socio-emotional brain. *Brain and Cognition, 65*, 135–142.

Nimchinsky, E. A., Sabatini, B. L., & Svoboda, K. (2002). Structure and function of dendritic spines. *Annual Review of Physiology, 64*, 313–353.

Nodé-Langlois, R., Muller, D., & Boda, B. (2006). Sequential implication of the mental retardation proteins ARHGEF6 and PAK3 in spine morphogenesis. *Journal of Cell Science, 119*, 4986–4993.

Orloff, M. S., & Eng, C. (2008). Genetic and phenotypic heterogeneity in the PTEN hamartoma tumour syndrome. *Oncogene, 27*, 5387–5397.

Pak, D. T., Yang, S., Rudolph-Correia, S., Kim, E., & Sheng, M. (2001). Regulation of dendritic spine morphology by SPAR, a PSD-95-associated RapGAP. *Neuron, 31*, 289–303.

Pavlowsky, A., Gianfelice, A., Pallotto, M., Zanchi, A., Vara, H., Khelfaoui, M., Valnegri, P., Rezai, X., Bassani, S., Brambilla, D., Kumpost, J., Blahos, J., Roux, M. J., Humeau, Y., Chelly, J., Passafaro, M., Giustetto, M., Billuart, P., & Sala, C. (2010). A postsynaptic signaling pathway that may account for the cognitive defect due to IL1RAPL1 mutation. *Current Biology, 20*, 103–115.

Pfeiffer, B. E., & Huber, K. M. (2009). The state of synapses in fragile X syndrome. *The Neuroscientist, 15*, 549–567.

Pinto, D., et al. (2010). Functional impact of global rare copy number variation in autism spectrum disorders. *Nature, 466*, 368–372.

Purpura, D. P. (1974). Dendritic spine "dysgenesis" and mental retardation. *Science, 186*, 1126–1128.

Rauch, A., Hoyer, J., Guth, S., Zweier, C., Kraus, C., Becker, C., Zenker, M., Hüffmeier, U., Thiel, C., Rüschendorf, F., Nürnberg, P., Reis, A., & Trautmann, U. (2006). Diagnostic yield of various genetic approaches in patients with unexplained developmental delay or mental retardation. *American Journal of Medical Genetics A, 140*, 2063–2074.

Reeve, S. P., Bassetto, L., Genova, G. K., Kleyner, Y., Leyssen, M., Jackson, F. R., & Hassan, B. A. (2005). The Drosophila fragile X mental retardation protein controls actin dynamics by directly regulating profilin in the brain. *Current Biology, 15*, 1156–1163.

Rex, C. S., Chen, L. Y., Sharma, A., Liu, J., Babayan, A. H., Gall, C. M., & Lynch, G. (2009). Different Rho GTPase-dependent signaling pathways initiate sequential steps in the consolidation of long-term potentiation. *The Journal of Cell Biology, 186*, 85–97.

Roberts, A., Allanson, J., Jadico, S. K., Kavamura, M. I., Noonan, J., Opitz, J. M., Young, T., & Neri, G. (2006). The cardiofaciocutaneous syndrome. *Journal of Medical Genetics, 43*, 833–842.

Roeleveld, N., Zielhuis, G. A., & Gabreëls, F. (1997). The prevalence of mental retardation: A critical review of recent literature. *Developmental Medicine and Child Neurology, 39*, 125–132.

Ropers, H. H. (2010). Genetics of early onset cognitive impairment. *Annual Review of Genomics and Human Genetics, 11*, 161–187.

Ropers, H. H., & Hamel, B. C. (2005). X-linked mental retardation. *Nature Reviews Genetics, 6*, 46–57.

Rosenberger, G., Jantke, I., Gal, A., & Kutsche, K. (2003). Interaction of alphaPIX (ARHGEF6) with beta-parvin (PARVB) suggests an involvement of alphaPIX in integrin-mediated signaling. *Human Molecular Genetics, 12*, 155–167.

Rumbaugh, G., Adams, J. P., Kim, J. H., & Huganir, R. L. (2006). SynGAP regulates synaptic strength and mitogen-activated protein kinases in cultured neurons. *Proceedings of the National Academy of Sciences of the United States of America, 103*, 4344–4351.

Ryu, J., Futai, K., Feliu, M., Weinberg, R., & Sheng, M. (2008). Constitutively active Rap2 transgenic mice display fewer dendritic spines, reduced extracellular signal-regulated kinase signaling, enhanced long-term depression, and impaired spatial learning and fear extinction. *Journal of Neuroscience, 28*, 8178–8188.

Schubbert, S., Bollag, G., & Shannon, K. (2007). Deregulated Ras signaling in developmental disorders: New tricks for an old dog. *Current Opinion in Genetics and Development, 17*, 15–22.

Seidman, L. J., Buka, S. L., Goldstein, J. M., Horton, N. J., Rieder, R. O., & Tsuang, M. T. (2000). The relationship of prenatal and perinatal complications to cognitive functioning at age 7 in the New England Cohorts of the National Collaborative Perinatal Project. *Schizophrenia Bulletin, 26*, 309–321.

Soderling, S. H., Guire, E. S., Kaech, S., White, J., Zhang, F., Schutz, K., Langeberg, L. K., Banker, G., Raber, J., & Scott, J. D. (2007). A WAVE-1 and WRP signaling complex regulates spine density, synaptic plasticity, and memory. *Journal of Neuroscience, 27*, 355–365.

Spohr, H. L., Willms, J., & Steinhausen, H. C. (2007). Fetal alcohol spectrum disorders in young adulthood. *The Journal of Pediatrics, 150*, 175–179, 179.e171.

Stanyon, C. A., & Bernard, O. (1999). LIM-kinase1. *The International Journal of Biochemistry and Cell Biology, 31*, 389–394.

Südhof, T. C. (2008). Neuroligins and neurexins link synaptic function to cognitive disease. *Nature, 455*, 903–911.

Szatmari, P., et al. (2007). Mapping autism risk loci using genetic linkage and chromosomal rearrangements. *Nature Genetics, 39*, 319–328.

Tabuchi, K., Blundell, J., Etherton, M. R., Hammer, R. E., Liu, X., Powell, C. M., & Südhof, T. C. (2007). A neuroligin-3 mutation implicated in autism increases inhibitory synaptic transmission in mice. *Science, 318*, 71–76.

Tada, T., & Sheng, M. (2006). Molecular mechanisms of dendritic spine morphogenesis. *Current Opinion in Neurobiology, 16*, 95–101.

Talebizadeh, Z., Lam, D. Y., Theodoro, M. F., Bittel, D. C., Lushington, G. H., & Butler, M. G. (2006). Novel splice isoforms for NLGN3 and NLGN4 with possible implications in autism. *Journal of Medical Genetics, 43*, e21.

Tassabehji, M., Metcalfe, K., Fergusson, W. D., Carette, M. J., Dore, J. K., Donnai, D., Read, A. P., Pröschel, C., Gutowski, N. J., Mao, X., & Sheer, D. (1996). LIM-kinase deleted in Williams syndrome. *Nature Genetics, 13*, 272–273.

Thomas, G. M., & Huganir, R. L. (2004). MAPK cascade signalling and synaptic plasticity. *Nature Reviews Neuroscience, 5*, 173–183.

Thomas, G. M., Lin, D. T., Nuriya, M., & Huganir, R. L. (2008). Rapid and bi-directional regulation of AMPA receptor phosphorylation and trafficking by JNK. *EMBO Journal, 27*, 361–372.

Threadgill, R., Bobb, K., & Ghosh, A. (1997). Regulation of dendritic growth and remodeling by Rho, Rac, and Cdc42. *Neuron, 19*, 625–634.

Walsh, T., et al. (2008). Rare structural variants disrupt multiple genes in neurodevelopmental pathways in schizophrenia. *Science, 320*, 539–543.

Wu, F., Chen, Y., Li, Y., Ju, J., Wang, Z., & Yan, D. (2008). RNA-interference-mediated Cdc42 silencing down-regulates phosphorylation of STAT3 and suppresses growth in human 100. bladder-cancer cells. *Biotechnology and Applied Biochemistry, 49*, 121–128.

Wu, G. Y., Deisseroth, K., & Tsien, R. W. (2001). Spaced stimuli stabilize MAPK pathway activation and its effects on dendritic morphology. *Nature Neuroscience, 4*, 151–158.

Yan, J., Feng, J., Schroer, R., Li, W., Skinner, C., Schwartz, C. E., Cook, E. H., & Sommer, S. S. (2008a). Analysis of the neuroligin 4Y gene in patients with autism. *Psychiatric Genetics, 18*, 204–207.

Yan, J., Noltner, K., Feng, J., Li, W., Schroer, R., Skinner, C., Zeng, W., Schwartz, C. E., & Sommer, S. S. (2008b). Neurexin 1alpha structural variants associated with autism. *Neuroscience Letters, 438*, 368–370.

Yan, Y., Yang, D., Zarnowska, E. D., Du, Z., Werbel, B., Valliere, C., Pearce, R. A., Thomson, J. A., & Zhang, S. C. (2005). Directed differentiation of dopaminergic neuronal subtypes from human embryonic stem cells. *Stem Cells, 23*, 781–790.

Zahir, F. R., Baross, A., Delaney, A. D., Eydoux, P., Fernandes, N. D., Pugh, T., Marra, M. A., & Friedman, J. M. (2008). A patient with vertebral, cognitive and behavioural abnormalities and a de novo deletion of NRXN1alpha. *Journal of Medical Genetics, 45*, 239–243.

Zalfa, F., Eleuteri, B., Dickson, K. S., Mercaldo, V., De Rubeis, S., di Penta, A., Tabolacci, E., Chiurazzi, P., Neri, G., Grant, S. G., & Bagni, C. (2007). A new function for the fragile X mental retardation protein in regulation of PSD-95 mRNA stability. *Nature Neuroscience, 10*, 578–587.

Zhu, Y., Pak, D., Qin, Y., McCormack, S. G., Kim, M. J., Baumgart, J. P., Velamoor, V., Auberson, Y. P., Osten, P., van Aelst, L., Sheng, M., & Zhu, J. J. (2005). Rap2-JNK removes synaptic AMPA receptors during depotentiation. *Neuron, 46*, 905–916.

Zhu, J. J., Qin, Y., Zhao, M., Van Aelst, L., & Malinow, R. (2002). Ras and Rap control AMPA receptor trafficking during synaptic plasticity. *Cell, 110*, 443–455.

Zhu, X., Raina, A. K., Perry, G., & Smith, M. A. (2004). Alzheimer's disease: The two-hit hypothesis. *Lancet Neurology, 3*, 219–226.

Chapter 20
Synaptic Pathology of Down Syndrome

Craig C. Garner and Daniel Z. Wetmore

Abstract Down syndrome is characterized by mild to moderate cognitive impairments that are caused by trisomy of chromosome 21. Several anatomical, behavioral, electrophysiological, and developmental abnormalities have been associated with Down syndrome. In this review, the current knowledge about the neurobiology of this disease and future perspectives of pharmacological treatments for this condition will be discussed.

Keywords Developement • Pharmacotherapy • Spinogenesis • Trisomie 21 • Ts65Dn mice

20.1 Introduction

Down syndrome (DS) is caused by the triplication of chromosome 21 (Hsa21) and occurs in about 1/700 live birth, with approximately 5% arising from the partial triplication of Hsa21 genes (Egan et al. 2004; Morris and Alberman 2009). Individuals with DS experience mild to severe intellectual disability with IQs generally between 20 and 80 (Nadel 2003). Approximately 30% of individuals with DS have congenital heart disease that can be treated surgically in the first year of life (Roizen and Patterson 2003). Individuals with DS also exhibit craniofacial abnormalities, muscle hypotonia, hypothyroidism, and leukemia as well as the histopathology of Alzheimer's disease (plaques and tangles) beginning as early as the fourth decade of life (Wisniewski et al. 1985; Antonarakis et al. 2004). Infantile seizures occur more commonly in DS than the typically developing population, and seizure risk is also elevated for individuals over 40 concomitant with the emergence

C.C. Garner (✉) • D.Z. Wetmore
Department of Psychiatry and Behavioral Sciences, Nancy Pritzker Laboratory, Stanford University School of Medicine, 1201 Welch Rd., Palo Alto, CA, USA
e-mail: cgarner@stanford.edu

M.R. Kreutz and C. Sala (eds.), *Synaptic Plasticity*,
Advances in Experimental Medicine and Biology 970,
DOI 10.1007/978-3-7091-0932-8_20, © Springer-Verlag/Wien 2012

of Alzheimer's disease (Roizen and Patterson 2003). Children with DS exhibit developmental delays beginning in the first year of life including specific deficits in speech, language (Roizen and Patterson 2003), and cognitive tasks that depend on cerebellar, prefrontal, and hippocampal function (Nadel 2003; Pennington et al. 2003). Explicit long-term memory and verbal short-te m memory are more impaired in DS individuals, while associative learning nd implicit long-term memory are less affected (Lott and Dierssen 2010). Prior to 1960, the standard of care for individuals with DS was on average very poor with mean life expectancy of about 25 years. With improved medical care, individuals with DS are living longer and are able to integrate more fully within their families, social networks, and educational environments (Yang et al. 2002).

Since the seminal work by Lejeune demonstrating that the genetic cause of DS is the triplication of chromosome 21 (Lejeune et al. 1959), physicians and scientists have sought to understand how this trisomy gives rise to the varied phenotypes that define DS (Patterson 2009). Based on the human genome project and careful annotation of each genetic element, we now know that there are ~400 genes present on Hsa21 as well as several microRNAs (Antonarakis et al. 2004; Nikolaienko et al. 2005; Patterson 2009). Most, but not all, genes are expressed at 1.5-fold normal levels leading to the "gene-dosage hypothesis" that phenotypes associated with DS are caused by the increased expression of at least a subset of these genes (Rachidi and Lopes 2007). However, given the number of genes triplicated on Hsa21 as well as variability in expression of both Hsa21 and non-Hsa21 genes in individuals with DS, unraveling the genetics of individual phenotypes is a daunting task (Gardiner 2004). Particular insights have come from the analysis of human patients with partial Hsa21 trisomy (Korenberg 1990; Korenberg et al. 1994) as well as mouse models created since 1990 that contain a third copy of different sets of genes homologous to those on Hsa21 (Davisson et al. 1990; Reeves et al. 1995; Olson et al. 2004; Korbel et al. 2009). These studies have allowed investigators to identify genes linked to reduced cellularity in the brain, leukemia, and heart defects (Korenberg et al. 1992; Baek et al. 2009; Roper et al. 2009).

Insight concerning which genes are causally linked to cognitive impairment remains incomplete. Early studies of partial trisomy of Hsa21 proposed that a small chromosomal region, called the "Down syndrome critical region" (DSCR), accounted for hallmark features of DS (Delabar et al. 1993), but subsequent research disputes this finding (Korenberg et al. 1994). Mouse models of DS with trisomy for various portions of regions syntenic to Hsa21 have also reached contradictory and difficult to interpret conclusions about the DSCR (Patterson 2009). Olson and colleagues found that triplicating only the DSCR is necessary but not sufficient to cause brain phenotypes in mice (Olson et al. 2007), while another study found the DSCR to be sufficient for behavioral and physiological phenotypes in mouse models of DS (Belichenko et al. 2009a). Recent studies point to cognitive impairment being the consequence of complex alterations in the expression of both Hsa21 and non-Hsa21 genes (Gardiner 2004), suggesting that the assignment of specific genes to specific cognitive or behavioral phenotypes in DS may not be possible.

Classical translational strategies develop therapies based on knowledge of specific target genes (Wetmore and Garner 2010). In DS, this approach appears to be untenable due to the complexity of genetic interactions, leading many to conclude that DS is too complex to understand on a genetic basis much less to develop rational pharmacotherapies. However, in the last decade, neurobiological studies in mouse models of DS have described anatomical, behavioral, electrophysiological, and developmental abnormalities associated with DS. These studies, examining the output properties of synapses and neuronal circuits within a systems neurobiology framework, have ushered in an era of optimism for pharmacotherapies based on manipulating synaptic, neuronal, and circuit function. In mouse models of DS, a number of groups have used drugs to restore neural circuits and behavior to wild-type levels. Based on this research framework, several potential pharmacotherapies have or will soon begin testing in clinical trials, providing new hope for parents and individuals with DS. To better understand the mechanistic basis for these pharmacotherapeutic strategies, we present a brief review of basic research insight concerning brain function in DS.

20.2 Synaptic and Anatomical Abnormalities in DS

There is increasing evidence that disrupted cognitive function in DS is a consequence of altered synaptic function and a reduced capacity of neuronal circuits to acquire, store, and share information. Initial evidence that synaptopathies underlie cognitive dysfunction in DS came from anatomical studies by Marin-Padilla and Becker. These studies revealed dendritic dysgenesis, including reduced dendritic ramification and diminished synapse formation in infants with DS (Becker et al. 1991; Becker 1991). Abnormal spinogenesis within the first 2 years of life is also present in DS. Dendritic spines fail to achieve a normal mature morphology and/or exhibit enlarged atrophic spine head structures (Marin-Padilla 1976). Given the relationship between synaptogenesis and dendritic growth (Meyer and Smith 2006), these observations imply either direct synaptic dysfunction or shifts in the functional connectivity of neuronal populations within these circuits.

Magnetic resonance imaging of individuals with DS has shown that some brain regions are more severely affected in DS, including the hippocampus, cortex, and cerebellum (Kesslak et al. 1994). These anatomical findings are consistent with neuropsychological assessments demonstrating that learning and memory dysfunction in DS is not equally affected across all brain areas but disproportionately affects the hippocampus and prefrontal cortex (Nadel 2003; Pennington et al. 2003). As discussed further below, the reduced size of these brain regions is caused by reductions in the ramification of dendritic arbors, synaptogenesis, and neuronal cell numbers (Ross 1994; Weitzdoerfer et al. 2001). For example, there are specific reductions in the number of interneuron pools in the developed neocortex, granule cells in the cerebellum, and principle neurons in the hippocampus (Schmidt-Sidor et al. 1990; Golden and Hyman 1994; Weitzdoerfer et al. 2001).

20.3   Mouse Models of DS

Further insights into the contribution of synaptic dysfunction to intellectual disability in DS have come from the analysis of a growing number of mouse models that recapitulate many of the phenotypes observed in individuals with DS. Predictably, creating DS mouse models involves the triplication of mouse genes that are syntenic with Hsa21 genes (Moore and Roper 2007; Gardiner et al. 2010). In the mouse, Hsa21 genes are segregated into three large sections of the mouse genome, with mouse chromosomes 10 (Mmu10) and 17 (Mmu17) containing ~50 genes each and the distal end of Mmu16 an additional ~150 genes (Pletcher et al. 2001; Gardiner 2009; Patterson 2009). Two of the most well-studied mouse models of DS—the Ts65Dn and Ts1Cje lines—were created by Robertsonian translocations of the distal end of Mmu16 onto other chromosomes (Davisson et al. 1990; Villar et al. 2005). More recently, chromosomal engineering, which allows the duplication or deletion of large chromosomal segments, has permitted the creation of a set of partially trisomic mouse models that include subsets of genes from Mmu10, 16, and/or 17 (Olson et al. 2004; Patterson 2009; Pereira et al. 2009; Yu et al. 2010). Another mouse model of DS, though a mosaic, was generated by introducing nearly all of Hsa21 into the mouse genome (O'Doherty et al. 2005). As reviewed elsewhere, nearly all of these models exhibit reduced hippocampal learning and memory function, though the expression of other DS-related phenotypes are more restricted, allowing investigators to identify specific set of genes that appear to be causal for leukemias, decreased cellularity in the cerebellum, and heart disease (Liu et al. 2011; Patterson 2009; Gardiner et al. 2010; Das and Reeves 2011).

20.4   Neurobiological Studies in Mouse Models of DS

To date, the Ts65Dn mouse has been most thoroughly studied. Similar to individuals with DS, Ts65Dn mice exhibit craniofacial abnormalities, decreased brain size and cell counts, and reduced hippocampal-dependent long-term memory (Reeves et al. 1995; Fernandez et al. 2007; Fernandez and Garner 2008; Gardiner et al. 2010). Ts65Dn mice exhibit deficits navigating the Morris water maze (Reeves et al. 1995; Holtzman et al. 1996), reduced performance in the radial arm maze task (Demas et al. 1996, 1998), a lower percentage of alternation in a T-maze (Fernandez et al. 2007), and learning deficits during contextual fear conditioning (Hyde et al. 2001). In the novel object recognition (NOR) task, Ts65Dn mice fail to discriminate between familiar and novel items with a 24-hour delay between training and testing (Fernandez et al. 2007). Taken together, these observations suggest that Ts65Dn mice effectively recapitulate key features of DS despite not being trisomic for all Hsa21 genes. Neurobiological findings in Ts65Dn mice form the foundation for potential pharmacotherapies in DS.

Investigators have carefully examined the anatomical and physiological characteristics of neuronal circuits in Ts65Dn mice and identified abnormalities that parallel observations from individuals with DS (Aldridge et al. 2007). Ts65Dn mice exhibit hypocellularity and reduced synaptogenesis in neocortex (Aldridge et al. 2007; Chakrabarti et al. 2007), including less-complex dendritic arbors of layer 3 neurons in the frontal cortex (Kurt et al. 2000; Dierssen et al. 2003). Similarly, reductions in neuronal and excitatory synapse density have been observed in different subregions of the hippocampus, including CA1, CA3, and the dentate gyrus (DG) (Kurt et al. 2004). Conceptually, the lower number of excitatory synapses implies that trisomy may alter the rates of formation, stability, or plasticity of these synapses. Alternatively, disuse caused by increased inhibitory tone or altered neuromodulatory function (cholinergic, noradrenergic, or serotonergic) could account for these changes in dendritic complexity and excitatory synapse number. Initial electrophysiological analysis of synaptic plasticity in the CA1 region of the hippocampus of Ts65Dn mice revealed reduced NMDA receptor–dependent long-term potentiation (LTP) of excitatory synaptic transmission and enhanced long-term depression (LTD) (Siarey et al. 1999; Kleschevnikov et al. 2004; Siarey et al. 2006; Fernandez et al. 2007; Belichenko et al. 2009b). These findings suggest that excitatory synapses maintain the capacity for plasticity but have an altered set point biased toward long-term depression.

The net effect on neural circuit function of these various anatomical and electrophysiological findings requires further study to determine their combined effect on cognitive function. One recent study investigated synaptic connectivity and function into and within the CA3 region of hippocampus in Ts65Dn mice. Hanson and colleagues found decreased excitatory and inhibitory input to CA3 but excess connectivity and normal LTP among associational connections between CA3 pyramidal neurons (Hanson et al. 2007). But what is the cause of dysfunction in hippocampal circuits? One hypothesis posits that increased inhibition in the hippocampus in DS impairs synaptic plasticity. A number of anatomical, physiological, and genetic studies support this hypothesis (see below).

20.5 Genetic Causes of Synaptic Dysfunction that Lead to Over-Inhibition in Ts65Dn Mice

Functional, anatomical, and behavioral studies clearly suggest that one cause for impaired cognitive function in DS is excessive inhibitory tone in at least some neuronal circuits. What are the physiological, developmental, and genetic causes of this enhanced inhibitory drive in Ts65Dn mice? Mechanistically, changes in inhibitory tone and cognitive function could arise as a homeostatic response to periods of hyperexcitability as is thought to occur in Alzheimer's disease (Palop et al. 2007; Palop and Mucke 2009). Alternatively, over-inhibition could be a direct consequence of the triplication of Hsa21 genes that control inhibitory interneuron number or function. The analysis of partial trisomies in mouse models have failed to identify specific genes linked to reduced cognition. This suggests that imbalances in neuronal circuit function involve incremental contributions from many genes. Consistent with this concept, the analysis of genes triplicated on Hsa21 revealed that many encoded proteins regulate the transcription, translation, and activity of synaptic proteins, neuronal cell number, or the electrophysiological properties of neurons (see Gardiner and Costa 2006; Patterson 2009; Gardiner et al. 2010). As there are several excellent reviews on this topic, we will only touch on a few such genes present on Hsa21 with known links to synaptic function.

Two Hsa21 genes (ITSN1 and SYNJ1) encode proteins (intersectin and synaptojanin, respectively) involved in vesicle endocytosis at synapses. Intersectin is a multi-domain adaptor protein that functions in concert with dynamin and synaptojanin, among others, to orchestrate the retrieval of synaptic vesicle proteins in a clathrin-dependent manner (De Camilli 2004; Dittman and Ryan 2009). Synaptojanin 1, a phosphatidylinositol-4,5-bisphosphate [PtdIns(4,5)P2] phosphatase, dephosphorylates phospholipids at the site of synaptic vesicle endocytosis (Voronov et al. 2008). Altering the expression level of these proteins has been shown to impair the efficient retrieval of synaptic proteins at low stimulus frequency and cause deficits in spatial learning of mice in the Morris water maze (Voronov et al. 2008).

Intriguingly, the activity of synaptojanin 1, and other endocytic proteins, is regulated by a pair of Hsa21 genes, DYRK1a and DSCR1. DYRK1a encodes a protein kinase that phosphorylates synaptojanin 1, dynamin 1, and amphiphysin, regulating their interaction and synaptic vesicle endocytosis (Adayev et al. 2006; Murakami et al. 2006). Conversely, the Down syndrome critical region gene 1 (DSCR1) encodes a protein that regulates calcineurin, a calcium-sensitive protein phosphatase that dephosphorylates synaptojanin1 during nerve terminal depolarization (Rothermel et al. 2003; Lee et al. 2004) and genetically interacts with Dyrk1a (Arron et al. 2006). Functionally, the activity of Dscr1, also known as regulator of calcineurin (RCAN1), controls both synaptic vesicle fusion pore kinetics and endocytosis (Keating et al. 2008; Kim et al. 2010).

A simple interpretation of these data is that this collection of proteins alters presynaptic function in DS by changing synaptic vesicle release probability and/or the size of the vesicle pool during periods of neuronal excitability. How this might affect synaptic transmission and the properties of neuronal circuits is as yet unclear. Several issues contribute to this confusion. First, most published data on these molecules are from transgenic and knockout mice, and researchers have not examined how modulating the expression levels of these genes within trisomic neurons affects synaptic function. Second, studies have not explored whether these molecules have specific effects on inhibitory, excitatory, or neuromodulatory neurons—or more restricted subsets of cells. Finally, none of these molecules function solely within presynaptic boutons; each also participates in vesicle recycling and transport in other compartments. For example, Dyrk1a and calcineurin regulate the activity of NMDA receptors. Accordingly, these genes affect NMDA-dependent mechanisms of synaptic plasticity (Altafaj et al. 2008; Sanderson and Dell'acqua 2010). Moreover, they regulate neurogenesis (Dierssen et al. 2006), dopaminergic dysfunction in motor disorders (Martinez de Lagran et al. 2007), and the phosphorylation of the microtubule associated protein tau (Woods et al. 2001a).

20.6 Support for the Over-Inhibition Hypothesis for Cognitive Dysfunction in DS

The over-inhibition hypothesis of cognitive dysfunction in DS is based on studies investigating anatomy, development, and physiology in mouse models of DS. LTP is reduced in the DG (Kleschevnikov et al. 2004; Fernandez et al. 2007), but incubating slices from Ts65Dn mice with the noncompetitive $GABA_A$ receptor antagonist picrotoxin restored LTP in both the DG and CA1 (Kleschevnikov et al. 2004; Costa and Grybko 2005). Consistent with this finding, Belichenko and colleagues found that there were more GAD65 immunoreactive GABAergic synapses in the hippocampus and a shift in the location of inhibitory synapses onto dendritic spines, a connectivity pattern that occurs infrequently in wild-type mice (Belichenko et al. 2004). Intriguingly, these changes are associated with altered spine morphology in these circuits as seen in DS individuals. These inhibitory synapses have slightly larger apposition length, implying that these synapses may be stronger (Belichenko et al. 2004, 2009b). Functionally, inhibitory synapses onto DG cells exhibit higher synaptic vesicle release probability as well as increased mini-IPSC frequency with no change in amplitude (Kleschevnikov et al. 2004; Chakrabarti et al. 2010). Consistent with increased inhibitory drive suppressing otherwise normal plasticity at excitatory synapses, the reduced NMDA/AMPA ratio at excitatory Ts65Dn synapses is normalized by reducing extracellular Mg^{2+} to unblock NMDA receptors or by the addition of picrotoxin to block $GABA_A$ receptors (Kleschevnikov et al. 2004).

These studies raise a number of fundamental questions: Is there enhanced inhibitory tone within these circuits in vivo? Can increased inhibitory drive account for impaired hippocampal-based learning in Ts65Dn mice? What genes underlie enhanced inhibitory drive within these hippocampal circuits? Are changes in GABAergic synaptic function the sole cause of impaired learning and memory in Ts65Dn mice and individuals with DS? And is reduced learning in other neurodevelopmental or neurodegenerative mouse models caused by enhanced inhibitory tone?

Three Hsa21 genes—GIRK2, Olig1, and Olig2—have recently emerged as potential modulators of excessive inhibitory tone within neuronal circuits in mouse models of DS (Gardiner and Costa 2006; Patterson 2009; Chakrabarti et al. 2010; Cramer et al. 2010). GIRK2 is a subunit of a G protein coupled inwardly rectifying potassium channel whose gene, Kcnj6, is present on Hsa21 (Hattori et al. 2000). These channels localize within dendrites of CA1 pyramidal cells and other neurons (Drake et al. 1997; Koyrakh et al. 2005). GIRK2-containing channels reduce membrane potential and increase shunting, thus reducing neuronal excitability (Ehrengruber et al. 1997).

Lower excitability is thought to impede NMDA-dependent plasticity and lead to learning and memory dysfunction. This motivated Galdzicki and colleagues to examine its potential role in excessive inhibitory tone and impaired synaptic learning mechanisms in Ts65Dn mice (Best et al. 2007; Cramer et al. 2010). Baclofen, a $GABA_B$ agonist, caused a dramatic activation of GIRK channels in Ts65Dn compared to wild-type mice (Harashima et al. 2006; Best et al. 2007). This suggests that coupling between $GABA_B$ receptors and GIRK2 channels is enhanced in CA1 neurons and may potentially account for the impaired expression of LTP in the hippocampus (Siarey et al. 1999, 2006; Kleschevnikov et al. 2008).

An alternative/complementary mechanism of increased inhibitory tone within the DS hippocampus is due to increased output of inhibitory neurons. Several electrophysiological studies suggest that inhibitory GABAergic drive onto principle neurons in the cortex and hippocampus is elevated as measured by an increase in mini-IPSC frequency (Kleschevnikov et al. 2004; Chakrabarti et al. 2010). Kleschevnikov and colleagues observed paired pulse depression at inhibitory synapses in the hippocampus, suggesting that higher synaptic vesicle release probability may be one cause of over-inhibition (Kleschevnikov et al. 2004). A recent study by Chakrabarti et al. proposes an alternative or complementary mechanism of increased inhibition mediated by an increased number of parvalbumin, somatostatin, and calbindin interneurons in the neocortex and parvalbumin interneurons in the hippocampus (Chakrabarti et al. 2010). Specifically, these authors considered whether two basic helix-loop-helix (bHLH) transcription factors, Olig1 and Olig2, implicated in interneuron neurogenesis (Ma 2006; Miyoshi et al. 2007) and triplicated in DS, regulate interneuron number and enhance mini-IPSC frequency in Ts65Dn mice. Intriguingly, they found that genetically expressing Olig1 and Olig2 in two copies in Ts65Dn mice restored interneuron number and normalized spontaneous IPSCs onto CA1 pyramidal cells (Chakrabarti et al. 2010).

To examine the question of the effect of inhibition in vivo, Fabian Fernandez in my laboratory investigated whether low doses of one of three different noncompetitive $GABA_A$ receptor antagonists—picrotoxin, bilobalide, and pentylenetetrazol—administered once daily for ~2 weeks to young adult Ts65Dn mice (2–4 months of age) improved long-term memory (Fernandez et al. 2007). Intriguingly, all three drugs led to a long-lasting normalization of cognitive function (1 week to 2 months posttreatment), leading to improved performance in hippocampal-dependent tasks such as novel object recognition and T-maze alteration to levels identical to their wild-type littermates (Fernandez et al. 2007). The long-lasting nature of the improvement implies that stable neuroadaptive changes occur within these circuits due to drug therapy, a concept consistent with the capacity of these circuits to undergo LTP in slice studies without the addition of picrotoxin after $GABA_A$ antagonist therapy (Fernandez et al. 2007).

Together these data make a strong case that enhanced inhibitory tone is at least one cause of reduced cognitive function in Ts65Dn mice and possibly individuals with DS. Importantly, these studies provide the first clues that the triplication of Hsa21 genes does not permanently impair brain function, providing a potential path for developing effective therapies for treating cognitive impairment in DS (see below). Moreover, studies in mouse models of Alzheimer's disease or neurofibromatosis type 1 reveal that over-inhibition of neuronal circuits is a common cause of reduced cognitive function (Fernandez and Garner 2007; Cui et al. 2008; Yoshiike et al. 2008).

20.7 Emerging Pharmacotherapies in Down Syndrome

With an increased understanding of the cellular and molecular mechanisms underlying synaptic and circuit dysfunction in DS, it has become possible to consider potential drug therapies to normalize function. Although our discussion has focused primarily on synaptic dysfunction in DS, restoring cognitive function across brain areas can be achieved with therapies that normalize synaptic plasticity and/or address neuronal cell loss. As discussed above, there is reduced cellularity in the cerebellum, neocortex, hippocampus as well as neuronal atrophy in several brain nuclei. Studies by Roger Reeves and colleagues have recently shown that reduced cellularity in the cerebellum is caused by deficiencies in sonic hedge hog (Shh) signaling (Roper et al. 2006, 2009). Remarkably, administering a single dose of Shh pathway agonist SAG1.1 at postnatal day 1 returned cerebellar granule cell numbers to wild-type levels (Roper et al. 2006). However, developing such a treatment for DS could be quite challenging, as it would likely require an in utero intervention during human embryogenesis.

In the hippocampus, reduced cellularity in the adult has been linked to a slowing of cell progression through the cell cycle in neuronal stem cell populations (Chakrabarti et al. 2007; Contestabile et al. 2007). In wild-type mice, enriched environment, exercise, and approved selective serotonin reuptake inhibitor fluoxetine (Prozac™, Eli Lilly, Indianapolis, IN) have been found to increase hippocampal neurogenesis. In Ts65Dn mice, enriched environment and increased exercise have no effect on learning hippocampal neurogenesis in Ts65Dn mice (Martinez-Cue et al. 2002, 2005; Llorens-Martin et al. 2010). Intriguingly, several groups have observed that prolonged administration (~2 weeks) of low doses of fluoxetine normalized neurogenesis and cognition in adult Ts65Dn mice (Clark et al. 2006; Bianchi et al. 2010). These results indicate that restoring this neuronal population, thought to be critical for learning and memory, is feasible via an FDA-approved drug in individuals with DS.

Studies by Bill Mobley and colleagues have examined the cause of neuronal atrophy of basal forebrain cholinergic neurons (BFCN) as well as norepinephrinergic neurons in the locus coeruleus (LC) (Holtzman et al. 1996; Salehi et al. 2009). The health of both of these neuronal populations was found to be compromised by reduced nerve growth factor (NGF) signaling (Cooper et al. 2001). Importantly, this phenotype can be rescued by delivering exogenous NGF to BFCN and LC (Cooper et al. 2001; Salehi et al. 2009). Mechanistically, three copies of APP, a gene associated with Alzheimer's disease (Wisniewski et al. 1985; Kamenetz et al. 2003), were shown to be responsible for impaired NGF signaling by disrupting the retrograde transport of NGF signaling endosomes from axon terminals to the cell soma (Salehi et al. 2006). Conceptually, the administration of small molecule orthologs of NGF (Longo et al. 2007) could restore the health of these cells and ultimately the synaptic release of acetylcholine or norepinephrine, respectively.

Alternatively, one could imagine a therapy based on the exogenous delivery of drugs that enhance neurotransmitter function as currently used to treat Parkinson's disease (e.g., L-dopa) (Merims and Giladi 2008) and Alzheimer's disease (e.g., donepezil) (Prasher 2004; Birks and Harvey 2006). Due to the high prevalence of early onset Alzheimer's disease in DS, pharmacotherapies for DS based on drugs approved for Alzheimer's disease have been proposed and tested in both mice and humans. Strategies to boost acetylcholine levels include administering acetylcholinesterase inhibitors such as donepezil (Aricept™, Pfizer, New York, NY). Unfortunately, donepezil is not efficacious in Ts65Dn mice (Rueda et al. 2008), and efficacy in individuals with DS is inconclusive (Kishnani et al. 2009). Initial studies by Ahmad Salehi and colleagues to compensate for reduced norepinephrine levels in the hippocampus have met with some success (Salehi et al. 2009). His group administered a norepinephrine prodrug, L-DOPS (Droxidopa™, Sumitomo Pharmaceuticals, Tokyo, Japan and Chelsea Therapeutics, Charlotte, NC) or xamoterol, to Ts65Dn mice. These treatments improved contextual (but not cue) fear conditioning, a behavior linked to both hippocampal and amygdalar function, as well as nest building, an ADHD-related behavior (Salehi et al. 2009). However, these therapies are complicated as metabolites stimulate adrenergic receptors in the heart (Salehi et al. 2009). Alternatives include three FDA-approved drugs, guanfacine, an α2-adrenergic agonist, and the ADHD drugs Focalin and risperidone (Kolar et al. 2008). However, each requires formal testing in animal models and subjects with DS.

By many criteria, a root cause of delayed speech, language development, and higher order cognitive function in DS is the inability of individuals with DS to convert experiences and sensory information into long-term memory. Anatomical, physiological, and neuropsychological assessments point to impaired hippocampal-dependent learning and memory consolidation in the cortex. In the hippocampus, increased inhibitory tone has emerged as a likely cause of disrupted learning within these circuits (Fernandez and Garner 2007; Fernandez et al. 2007). As discussed above, increased inhibitory tone has been linked to the enhanced release of GABA from a larger number of inhibitory interneurons (Kleschevnikov et al. 2004; Chakrabarti et al. 2010), the hyperpolarization of dendritic arbors through the increased activity of G protein coupled inwardly rectifying potassium channels (Harashima et al. 2006; Best et al. 2007), and the inability to activate NMDA receptors due to Mg^{2+} block (Kleschevnikov et al. 2004). Since 2007, strategies designed to modulate each of these have been explored for their therapeutic potential. The first directly evaluated the concept of excessive inhibitory tone by administering low daily doses (3 mg/kg) of a noncompetitive $GABA_A$ receptor antagonist such as pentylenetetrazol (PTZ) (Fernandez and Garner 2007; Fernandez et al. 2007). This 2–3-week regimen was extremely robust generating improvements in learning that lasted for months (Fernandez et al. 2007; Rueda et al. 2008). It was also associated with restoration of hippocampal LTP, supporting a concept that this therapy elicits a neuroadaptive change in inhibitory tone that leads to long-lasting changes in neural circuit function that support improved synaptic plasticity and learning (Fernandez et al. 2007). Importantly, these studies were performed in adult mice indicating that trisomy does not permanently damage these circuits. Our recent studies showing that PTZ is efficacious in mice from 2 to 18 months support this concept (Colas et al. 2012). A principle concern of this strategy is that at high doses PTZ can induce seizure or reduce the threshold for seizure via "kindling" (Mason and Cooper 1972). Moreover, approximately 10% of young DS children have increased susceptibility to seizures, so clinical development of GABA$_A$ receptor drugs requires careful design and safety controls (Fernandez and Garner 2007). Preclinical studies on PTZ are helping to minimize these concerns. Specifically, we have found that PTZ is efficacious at doses more than 500-fold below doses that cause kindling and seizure. This treatment strategy does not increase the excitability of brain circuits in these animals nor reduce their threshold for seizure (Colas, Heller, Personal communication). A development path for PTZ is now in place allowing clinical studies to begin by spring 2012. Excitingly, drugs that target specific GABA$_A$ receptor subtypes, such as those containing the α5 subunit, could provide a larger therapeutic window for treatment targeting GABA$_A$ receptors (Delatour et al. 2009). However, more work is required to develop safe α5-specific compounds (Atack 2010).

Provocative alternatives to drugs targeting GABA$_A$ receptor are those designed to antagonize GABA$_B$ receptors, NMDA receptors, or Dyrk1a. As discussed above, GABA$_B$ receptors are G protein–coupled receptors capable of activating inwardly rectifying potassium channels. Initial studies with the GABA$_B$ antagonist CGP53432 improved LTP in Ts65Dn hippocampal slices (Kleschevnikov et al. 2008), indicating that this strategy may offer additional therapeutic potential. Given the important role of NMDA receptors in synaptic plasticity, the Costa group has initiated studies to examine whether the noncompetitive NMDA receptor antagonist memantine (Namenda™, Forest Laboratories, New York, NY), currently used to treat cognitive impairment in AD, can facilitate learning in DS. Their rationale is that three copies of DSCR1, an inhibitor of calcineurin, is predicted to increase NMDA receptor open time and open probability in DS (Costa et al. 2008). Together with GABA$_B$- and GIRK-induced hyperpolarization of dendritic membranes, this altered NMDA activation could lead to enhanced Mg^{2+} block of NMDA receptors and impaired synaptic plasticity. Although somewhat counterintuitive, this therapeutic strategy is based on the hypothesis that using memantine to reduce NMDA receptor activation would lead to fewer receptors stuck in an inactive state due to Mg^{2+} block, thus enabling synaptic plasticity to take place normally. Using acute dosing and fear conditioning in DS mouse models, Costa and colleagues showed that memantine improved contextual fear conditioning, a form of hippocampal-dependent learning (Costa et al. 2008; Siddiqui et al. 2008). Clinical studies with this FDA-approved drug are underway.

A final pharmacotherapy relevant to DS is based on reducing the activity of Dyrk1a. As discussed above, this kinase phosphorylates a large number of proteins associated with neurogenesis, neurotransmission, and microtubule assembly (Woods et al. 2001a, b; Murakami et al. 2006). Relevant targets in the synapse include the endocytic proteins synaptojanin, amphiphysin, and dynamin which are known to control both synaptic vesicle recycling and glutamate receptor function (Adayev et al. 2006; Altafaj et al. 2008; Kim et al. 2010). Epigallocatechin-3-gallate (EGCG), the major polyphenolic compound present in green tea, inhibits Dyrk1a activity and improves learning in transgenic Dyrk1a mice (Guedj et al. 2009). In hippocampal slices from Ts65Dn mice, EGCG enhances the induction of LTP in CA1 pyramidal cells (Xie et al. 2008). However, the mechanism of action of EGCG is also unclear as doses greater than 20 μM inhibit targets other than Dyrk1a (Xie et al. 2008). Nonetheless, these results are promising, and clinical trials with EGCG in DS are underway in Spain. Thus, while drinking green tea may have some cognitive benefits, employing EGCG as a pharmacotherapy in DS may require the development of compounds with yet higher specificity for Dyrk1a to reduce unpredictable off-target affects.

20.8 Closing Remarks and Future Directions

In the last decade, remarkable advances have been made in our understanding of the causes of cognitive dysfunction in DS and potential pharmacotherapeutic strategies to address them. For individuals with DS and their families, there is increasing hope that basic and clinical research will lead to approved drug therapies that allow individuals with DS to reach their full potential. However, much work remains for researchers and clinicians. Additional physiological studies, including in vivo recordings, in mouse models of DS are needed to better understand how trisomy of many hundreds of genes affects neuronal circuit function. In parallel, studies focused on the effect of trisomy on molecular pathways and genetic modules can contribute new insight about the causes of dysfunction and identify new drug targets that offer greater efficacy and safety. In addition to these basic research efforts, resources must be committed to translating these findings from bench to bedside through close partnership with clinicians, clinical development experts, and the biotech and pharmaceutical industries.

References

Adayev, T., Chen-Hwang, M. C., et al. (2006). MNB/DYRK1A phosphorylation regulates the interactions of synaptojanin 1 with endocytic accessory proteins. *Biochemical and Biophysical Research Communications, 351*(4), 1060–1065.

Aldridge, K., Reeves, R. H., et al. (2007). Differential effects of trisomy on brain shape and volume in related aneuploid mouse models. *American Journal of Medical Genetics A, 143A*(10), 1060–1070.

Altafaj, X., Ortiz-Abalia, J., et al. (2008). Increased NR2A expression and prolonged decay of NMDA-induced calcium transient in cerebellum of TgDyrk1A mice, a mouse model of Down syndrome. *Neurobiology of Disease, 32*(3), 377–384.

Antonarakis, S. E., Lyle, R., et al. (2004). Chromosome 21 and Down syndrome: From genomics to pathophysiology. *Nature Reviews Genetics, 5*(10), 725–738.

Arron, J. R., Winslow, M. M., et al. (2006). NFAT dysregulation by increased dosage of DSCR1 and DYRK1A on chromosome 21. *Nature, 441*(7093), 595–600.

Atack, J. R. (2010). Preclinical and clinical pharmacology of the GABA(A) receptor alpha5 subtype-selective inverse agonist alpha5IA. *Pharmacology and Therapeutics, 125*, 11–26.

Baek, K. H., Zaslavsky, A., et al. (2009). Down's syndrome suppression of tumour growth and the role of the calcineurin inhibitor DSCR1. *Nature, 459*(7250), 1126–1130.

Becker, L. E. (1991). Synaptic dysgenesis. *Canadian Journal of Neurological Sciences, 18*(2), 170–180.

Becker, L., Mito, T., et al. (1991). Growth and development of the brain in Down syndrome. *Progress in Clinical and Biological Research, 373*, 133–152.

Belichenko, N. P., Belichenko, P. V., et al. (2009a). The "Down syndrome critical region" is sufficient in the mouse model to confer behavioral, neurophysiological, and synaptic phenotypes characteristic of Down syndrome. *Journal of Neuroscience, 29*(18), 5938–5948.

Belichenko, P. V., Kleschevnikov, A. M., et al. (2009b). Excitatory-inhibitory relationship in the fascia dentata in the Ts65Dn mouse model of Down syndrome. *The Journal of Comparative Neurology, 512*(4), 453–466.

Belichenko, P. V., Masliah, E., et al. (2004). Synaptic structural abnormalities in the Ts65Dn mouse model of Down syndrome. *The Journal of Comparative Neurology, 480*(3), 281–298.

Best, T. K., Siarey, R. J., et al. (2007). Ts65Dn, a mouse model of Down syndrome, exhibits increased GABAB-induced potassium current. *Journal of Neurophysiology, 97*(1), 892–900.

Bianchi, P., Ciani, E., et al. (2010). Early pharmacotherapy restores neurogenesis and cognitive performance in the Ts65Dn mouse model for Down syndrome. *Journal of Neuroscience, 30* (26), 8769–8779.

Birks, J., & Harvey, R. J. (2006). Donepezil for dementia due to Alzheimer's disease. *Cochrane Database System Review, 1*, CD001190. DOI: 10.1002/14651858.CD001190.pub2.

Chakrabarti, L., Best, T. K., et al. (2010). Olig1 and Olig2 triplication causes developmental brain defects in Down syndrome. *Nature Neuroscience, 13*(8), 927–934.

Chakrabarti, L., Galdzicki, Z., et al. (2007). Defects in embryonic neurogenesis and initial synapse formation in the forebrain of the Ts65Dn mouse model of Down syndrome. *Journal of Neuroscience, 27*(43), 11483–11495.

Clark, S., Schwalbe, J., et al. (2006). Fluoxetine rescues deficient neurogenesis in hippocampus of the Ts65Dn mouse model for Down syndrome. *Experimental Neurology, 200*(1), 256–261.

Colas, D., Chuluun, B., et al. (2012). Assessment of the GABA$_A$ Antagonist Pentylenetetrazole as a Procognitive Therapy in a Mouse Model of Down Syndrome. *J Neuropsychopharmacology* (submitted)

Contestabile, A., Fila, T., et al. (2007). Cell cycle alteration and decreased cell proliferation in the hippocampal dentate gyrus and in the neocortical germinal matrix of fetuses with Down syndrome and in Ts65Dn mice. *Hippocampus, 17*(8), 665–678.

Cooper, J. D., Salehi, A., et al. (2001). Failed retrograde transport of NGF in a mouse model of Down's syndrome: Reversal of cholinergic neurodegenerative phenotypes following NGF infusion. *Proceedings of the National Academy of Sciences of the United States of America, 98*(18), 10439–10444.

Costa, A. C. S., & Grybko, M. J. (2005). Deficits in hippocampal CA1 LTP induced by TBS but not HFS in the Ts65Dn mouse: A model of Down syndrome. *Neuroscience Letters, 382*(3), 317–322.

Costa, A. C. S., Scott-McKean, J. J., et al. (2008). Acute injections of the NMDA receptor antagonist memantine rescue performance deficits of the Ts65Dn mouse model of Down syndrome on a fear conditioning test. *Neuropsychopharmacology, 33*(7), 1624–1632.

Cramer, N. P., Best, T. K., et al. (2010). GABAB-GIRK2-mediated signaling in Down syndrome. *Advances in Pharmacology, 58,* 397–426.

Cui, Y., Costa, R. M., et al. (2008). Neurofibromin regulation of ERK signaling modulates GABA release and learning. *Cell, 135*(3), 549–560.

Das, I., & Reeves, R. H. (2011). The use of mouse models to understand and improve cognitive deficits in Down syndrome. *Disease Models and Mechanisms, 4*(5), 596–606.

Davisson, M. T., Schmidt, C., et al. (1990). Segmental trisomy of murine chromosome 16: A new model system for studying Down syndrome. *Progress in Clinical and Biological Research, 360,* 263–280.

De Camilli, P. (2004). Molecular mechanisms in membrane traffic at the neuronal synapse: Role of protein-lipid interactions. *Harvey Lectures, 100,* 1–28.

Delabar, J. M., Theophile, D., et al. (1993). Molecular mapping of twenty-four features of Down syndrome on chromosome 21. *European Journal of Human Genetics, 1*(2), 114–124.

Delatour, B., Braudeau, J., et al. (2009). Alleviation of cognitive deficits in Ts65Dn mice modeling Down syndrome by pharmacological inhibition of GABAergic transmission. Program No. 829.16. 2009 Neuroscience Meeting Planner. Chicago, IL: Society for Neuroscience. Online.

Demas, G. E., Nelson, R. J., et al. (1996). Spatial memory deficits in segmental trisomic Ts65Dn mice. *Behavioural Brain Research, 82*(1), 85–92.

Demas, G. E., Nelson, R. J., et al. (1998). Impaired spatial working and reference memory in segmental trisomy (Ts65Dn) mice. *Behavioural Brain Research, 90*(2), 199–201.

Dierssen, M., Benavides-Piccione, R., et al. (2003). Alterations of neocortical pyramidal cell phenotype in the Ts65Dn mouse model of Down syndrome: Effects of environmental enrichment. *Cerebral Cortex, 13*(7), 758–764.

Dierssen, M., Ortiz-Abalia, J., et al. (2006). Pitfalls and hopes in Down syndrome therapeutic approaches: In the search for evidence-based treatments. *Behavior Genetics, 36*(3), 454–468.

Dittman, J., & Ryan, T. A. (2009). Molecular circuitry of endocytosis at nerve terminals. *Annual Review of Cell and Developmental Biology, 25,* 133–160.

Drake, C. T., Bausch, S. B., et al. (1997). GIRK1 immunoreactivity is present predominantly in dendrites, dendritic spines, and somata in the CA1 region of the hippocampus. *Proceedings of the National Academy of Sciences of the United States of America, 94*(3), 1007–1012.

Egan, J. F. X., Benn, P. A., et al. (2004). Down syndrome births in the United States from 1989 to 2001. *American Journal of Obstetrics and Gynecology, 191*(3), 1044–1048.

Ehrengruber, M. U., Doupnik, C. A., et al. (1997). Activation of heteromeric G protein-gated inward rectifier K+ channels overexpressed by adenovirus gene transfer inhibits the excitability of hippocampal neurons. *Proceedings of the National Academy of Sciences of the United States of America, 94*(13), 7070–7075.

Fernandez, F., & Garner, C. C. (2007). Over-inhibition: A model for developmental intellectual disability. *Trends in Neurosciences, 30*(10), 497–503.

Fernandez, F., & Garner, C. C. (2008). Episodic-like memory in Ts65Dn, a mouse model of Down syndrome. *Behavioural Brain Research, 188*(1), 233–237.

Fernandez, F., Morishita, W., et al. (2007). Pharmacotherapy for cognitive impairment in a mouse model of Down syndrome. *Nature Neuroscience, 10*(4), 411–413.

Gardiner, K. (2004). Gene-dosage effects in Down syndrome and trisomic mouse models. *Genome Biology, 5*(10), 244.

Gardiner, K. J. (2010). Molecular basis of pharmacotherapies for cognition in Down syndrome. *Trends Pharmacol Sci, 31*(2), 66–73.

Gardiner, K., & Costa, A. C. S. (2006). The proteins of human chromosome 21. *American Journal of Medical Genetics C, Seminars in Medical Genetics, 142C*(3), 196–205.

Gardiner, K., Herault, Y., et al. (2010). Down syndrome: From understanding the neurobiology to therapy. *Journal of Neuroscience, 30*(45), 14943–14945.

Golden, J. A., & Hyman, B. T. (1994). Development of the superior temporal neocortex is anomalous in trisomy 21. *Journal of Neuropathology and Experimental Neurology, 53*(5), 513–520.

Guedj, F., Sébrié, C., et al. (2009). Green tea polyphenols rescue of brain defects induced by overexpression of DYRK1A. *PLoS One, 4*(2), e4606.

Hanson, J. E., Blank, M., et al. (2007). The functional nature of synaptic circuitry is altered in area CA3 of the hippocampus in a mouse model of Down's syndrome. *Journal of Physiology (London), 579*(Pt 1), 53–67.

Harashima, C., Jacobowitz, D. M., et al. (2006). Elevated expression of the G-protein-activated inwardly rectifying potassium channel 2 (GIRK2) in cerebellar unipolar brush cells of a Down syndrome mouse model. *Cellular and Molecular Neurobiology, 26*(4–6), 719–734.

Hattori, M., Fujiyama, A., et al. (2000). The DNA sequence of human chromosome 21. *Nature, 405*(6784), 311–319.

Holtzman, D. M., Santucci, D., et al. (1996). Developmental abnormalities and age-related neurodegeneration in a mouse model of Down syndrome. *Proceedings of the National Academy of Sciences of the United States of America, 93*(23), 13333–13338.

Hyde, L. A., Frisone, D. F., et al. (2001). Ts65Dn mice, a model for Down syndrome, have deficits in context discrimination learning suggesting impaired hippocampal function. *Behavioural Brain Research, 118*(1), 53–60.

Kamenetz, F., Tomita, T., et al. (2003). APP processing and synaptic function. *Neuron, 37*(6), 925–937.

Keating, D. J., Dubach, D., et al. (2008). DSCR1/RCAN1 regulates vesicle exocytosis and fusion pore kinetics: Implications for Down syndrome and Alzheimer's disease. *Human Molecular Genetics, 17*(7), 1020–1030.

Kesslak, J. P., Nagata, S. F., et al. (1994). Magnetic resonance imaging analysis of age-related changes in the brains of individuals with Down's syndrome. *Neurology, 44*(6), 1039–1045.

Kim, Y., Park, J., et al. (2010). Overexpression of Dyrk1A causes the defects in synaptic vesicle endocytosis. *Neuro-Signals, 18*(3), 164–172.

Kishnani, P. S., Sommer, B. R., et al. (2009). The efficacy, safety, and tolerability of donepezil for the treatment of young adults with Down syndrome. *American Journal of Medical Genetics A, 149A*(8), 1641–1654.

Kleschevnikov, A. M., Belichenko, P. V., et al. (2008). Antagonists of the GABAB receptors enhance LTP and reduce pro-epileptiform activity in Ts65Dn mouse model of Down syndrome. Program No. 348.4. 2008 Neuroscience Meeting Planner. Washington, DC: Society for Neuroscience, 2008. Online.

Kleschevnikov, A. M., Belichenko, P. V., et al. (2004). Hippocampal long-term potentiation suppressed by increased inhibition in the Ts65Dn mouse, a genetic model of Down syndrome. *Journal of Neuroscience, 24*(37), 8153–8160.

Kolar, D., Keller, A., et al. (2008). Treatment of adults with attention-deficit/hyperactivity disorder. *Neuropsychiatric Disease and Treatment, 4*(1), 107–121.

Korbel, J. O., Tirosh-Wagner, T., et al. (2009). The genetic architecture of Down syndrome phenotypes revealed by high-resolution analysis of human segmental trisomies. *Proceedings of the National Academy of Sciences of the United States of America, 106*(29), 12031–12036.

Korenberg, J. R. (1990). Molecular mapping of the Down syndrome phenotype. *Progress in Clinical and Biological Research, 360*, 105–115.

Korenberg, J. R., Bradley, C., et al. (1992). Down syndrome: Molecular mapping of the congenital heart disease and duodenal stenosis. *American Journal of Human Genetics, 50*(2), 294–302.

Korenberg, J. R., Chen, X. N., et al. (1994). Down syndrome phenotypes: The consequences of chromosomal imbalance. *Proceedings of the National Academy of Sciences of the United States of America, 91*(11), 4997–5001.

Koyrakh, L., Lujan, R., et al. (2005). Molecular and cellular diversity of neuronal G-protein-gated potassium channels. *Journal of Neuroscience, 25*(49), 11468–11478.

Kurt, M. A., Davies, D. C., et al. (2000). Synaptic deficit in the temporal cortex of partial trisomy 16 (Ts65Dn) mice. *Brain Research, 858*(1), 191–197.

Kurt, M. A., Kafa, M. I., et al. (2004). Deficits of neuronal density in CA1 and synaptic density in the dentate gyrus, CA3 and CA1, in a mouse model of Down syndrome. *Brain Research, 1022* (1–2), 101–109.

Lee, S. Y., Wenk, M. R., et al. (2004). Regulation of synaptojanin 1 by cyclin-dependent kinase 5 at synapses. *Proceedings of the National Academy of Sciences of the United States of America, 101*(2), 546–551.

Lejeune, J., Turpin, R., et al. (1959). Mongolism; a chromosomal disease (trisomy). *Bulletin de l'Académie Nationale de Médecine, 143*(11–12), 256–265.

Liu, C., Morishima, M., et al. (2011). Genetic analysis of Down syndrome-associated heart defects in mice. *Human Genetics, 130*(5), 623–632.

Llorens-Martin, M. V., Rueda, N., et al. (2010). Effects of voluntary physical exercise on adult hippocampal neurogenesis and behavior of Ts65Dn mice, a model of Down syndrome. *Neuroscience, 171*(4), 1228–1240.

Longo, F. M., Yang, T., et al. (2007). Small molecule neurotrophin receptor ligands: Novel strategies for targeting Alzheimer's disease mechanisms. *Current Alzheimer Research, 4*(5), 503–506.

Lott, I. T., & Dierssen, M. (2010). Cognitive deficits and associated neurological complications in individuals with Down's syndrome. *Lancet Neurology, 9*(6), 623–633.

Ma, Q. (2006). Transcriptional regulation of neuronal phenotype in mammals. *Journal de Physiologie, 575*(Pt 2), 379–387.

Marin-Padilla, M. (1976). Pyramidal cell abnormalities in the motor cortex of a child with Down's syndrome. A Golgi study. *The Journal of Comparative Neurology, 167*(1), 63–81.

Martinez de Lagran, M., Bortolozzi, A., et al. (2007). Dopaminergic deficiency in mice with reduced levels of the dual-specificity tyrosine-phosphorylated and regulated kinase 1A, Dyrk1A(+/−). *Genes, Brain, and Behavior, 6*(6), 569–578.

Martinez-Cue, C., Baamonde, C., et al. (2002). Differential effects of environmental enrichment on behavior and learning of male and female Ts65Dn mice, a model for Down syndrome. *Behavioural Brain Research, 134*(1–2), 185–200.

Martinez-Cue, C., Rueda, N., et al. (2005). Behavioral, cognitive and biochemical responses to different environmental conditions in male Ts65Dn mice, a model of Down syndrome. *Behavioural Brain Research, 163*(2), 174–185.

Mason, C. R., & Cooper, R. M. (1972). A permanent change in convulsive threshold in normal and brain-damaged rats with repeated small doses of pentylenetetrazol. *Epilepsia, 13*(5), 663–674.

Merims, D., & Giladi, N. (2008). Dopamine dysregulation syndrome, addiction and behavioral changes in Parkinson's disease. *Parkinsonism & Related Disorders, 14*(4), 273–280.

Meyer, M. P., & Smith, S. J. (2006). Evidence from in vivo imaging that synaptogenesis guides the growth and branching of axonal arbors by two distinct mechanisms. *Journal of Neuroscience, 26*(13), 3604–3614.

Miyoshi, G., Butt, S. J., et al. (2007). Physiologically distinct temporal cohorts of cortical interneurons arise from telencephalic Olig2-expressing precursors. *Journal of Neuroscience, 27*(29), 7786–7798.

Moore, C. S., & Roper, R. J. (2007). The power of comparative and developmental studies for mouse models of Down syndrome. *Mammalian Genome, 18*(6–7), 431–443.

Morris, J. K., & Alberman, E. (2009). Trends in Down's syndrome live births and antenatal diagnoses in England and Wales from 1989 to 2008: Analysis of data from the National Down Syndrome Cytogenetic Register. *British Medical Journal, 339*, b3794.

Murakami, N., Xie, W., et al. (2006). Phosphorylation of amphiphysin I by minibrain kinase/dual-specificity tyrosine phosphorylation-regulated kinase, a kinase implicated in Down syndrome. *Journal of Biological Chemistry, 281*(33), 23712–23724.

Nadel, L. (2003). Down's syndrome: A genetic disorder in biobehavioral perspective. *Genes, Brain, and Behavior, 2*(3), 156–166.

Nikolaienko, O., Nguyen, C., et al. (2005). Human chromosome 21/Down syndrome gene function and pathway database. *Gene, 364*, 90–98.

O'Doherty, A., Ruf, S., et al. (2005). An aneuploid mouse strain carrying human chromosome 21 with Down syndrome phenotypes. *Science, 309*(5743), 2033–2037.

Olson, L. E., Roper, R. J., et al. (2004). Down syndrome mouse models Ts65Dn, Ts1Cje, and Ms1Cje/Ts65Dn exhibit variable severity of cerebellar phenotypes. *Developmental Dynamics, 230*(3), 581–589.

Olson, L. E., Roper, R. J., et al. (2007). Trisomy for the Down syndrome 'critical region' is necessary but not sufficient for brain phenotypes of trisomic mice. *Human Molecular Genetics, 16*(7), 774–782.

Palop, J. J., Chin, J., et al. (2007). Aberrant excitatory neuronal activity and compensatory remodeling of inhibitory hippocampal circuits in mouse models of Alzheimer's disease. *Neuron, 55*(5), 697–711.

Palop, J. J., & Mucke, L. (2009). Epilepsy and cognitive impairments in Alzheimer disease. *Archives of Neurology, 66*(4), 435–440.

Patterson, D. (2009). Molecular genetic analysis of Down syndrome. *Human Genetics, 126*(1), 195–214.

Pennington, B. F., Moon, J., et al. (2003). The neuropsychology of Down syndrome: Evidence for hippocampal dysfunction. *Child Development, 74*(1), 75–93.

Pereira, P. L., Magnol, L., et al. (2009). A new mouse model for the trisomy of the Abcg1-U2af1 region reveals the complexity of the combinatorial genetic code of down syndrome. *Human Molecular Genetics, 18*(24), 4756–4769.

Pletcher, M. T., Wiltshire, T., et al. (2001). Use of comparative physical and sequence mapping to annotate mouse chromosome 16 and human chromosome 21. *Genomics, 74*(1), 45–54.

Prasher, V. P. (2004). Review of donepezil, rivastigmine, galantamine and memantine for the treatment of dementia in Alzheimer's disease in adults with Down syndrome: Implications for the intellectual disability population. *International Journal of Geriatric Psychiatry, 19*(6), 509–515.

Rachidi, M., & Lopes, C. (2007). Mental retardation in Down syndrome: From gene dosage imbalance to molecular and cellular mechanisms. *Neuroscience Research, 59*(4), 349–369.

Reeves, R. H., Irving, N. G., et al. (1995). A mouse model for Down syndrome exhibits learning and behaviour deficits. *Nature Genetics, 11*(2), 177–184.

Roizen, N. J., & Patterson, D. (2003). Down's syndrome. *Lancet, 361*(9365), 1281–1289.

Roper, R. J., Baxter, L. L., et al. (2006). Defective cerebellar response to mitogenic Hedgehog signaling in Down [corrected] syndrome mice. *Proceedings of the National Academy of Sciences of the United States of America, 103*(5), 1452–1456.

Roper, R. J., Vanhorn, J. F., et al. (2009). A neural crest deficit in Down syndrome mice is associated with deficient mitotic response to Sonic hedgehog. *Mechanisms of Development, 126*(3–4), 212–219.

Ross, L. J. (1994). Developmental disabilities: Genetic implications. *Journal of Obstetric, Gynecologic, and Neonatal Nursing, 23*(6), 502–505.

Rothermel, B. A., Vega, R. B., et al. (2003). The role of modulatory calcineurin-interacting proteins in calcineurin signaling. *Trends in Cardiovascular Medicine, 13*(1), 15–21.

Rueda, N., Flórez, J., et al. (2008). Chronic pentylenetetrazole but not donepezil treatment rescues spatial cognition in Ts65Dn mice, a model for Down syndrome. *Neuroscience Letters, 433*(1), 22–27.

Salehi, A., Delcroix, J.-D., et al. (2006). Increased App expression in a mouse model of Down's syndrome disrupts NGF transport and causes cholinergic neuron degeneration. *Neuron, 51*(1), 29–42.

Salehi, A., Faizi, M., et al. (2009). Restoration of norepinephrine-modulated contextual memory in a mouse model of Down syndrome. *Science Translational Medicine, 1*(7), 7ra17.

Sanderson, J. L., & Dell'acqua, M. L. (2010). AKAP signaling complexes in regulation of excitatory synaptic plasticity. *The Neuroscientist, 17*(3), 321–336.

Schmidt-Sidor, B., Wisniewski, K. E., et al. (1990). Brain growth in Down syndrome subjects 15 to 22 weeks of gestational age and birth to 60 months. *Clinical Neuropathology, 9*(4), 181–190.

Siarey, R. J., Carlson, E. J., et al. (1999). Increased synaptic depression in the Ts65Dn mouse, a model for mental retardation in Down syndrome. *Neuropharmacology, 38*(12), 1917–1920.

Siarey, R. J., Kline-Burgess, A., et al. (2006). Altered signaling pathways underlying abnormal hippocampal synaptic plasticity in the Ts65Dn mouse model of Down syndrome. *Journal of Neurochemistry, 98*(4), 1266–1277.

Siddiqui, A., Lacroix, T., et al. (2008). Molecular responses of the Ts65Dn and Ts1Cje mouse models of Down syndrome to MK-801. *Genes, Brain, and Behavior, 7*(7), 810–820.

Villar, A. J., Belichenko, P. V., et al. (2005). Identification and characterization of a new Down syndrome model, Ts[Rb(12.1716)]2Cje, resulting from a spontaneous Robertsonian fusion between T(171)65Dn and mouse chromosome 12. *Mammalian Genome, 16*(2), 79–90.

Voronov, S. V., Frere, S. G., et al. (2008). Synaptojanin 1-linked phosphoinositide dyshomeostasis and cognitive deficits in mouse models of Down's syndrome. *Proceedings of the National Academy of Sciences of the United States of America, 105*(27), 9415–9420.

Weitzdoerfer, R., Dierssen, M., et al. (2001). Fetal life in Down syndrome starts with normal neuronal density but impaired dendritic spines and synaptosomal structure. *Journal of Neural Transmission Supplementum, 61*, 59–70.

Wetmore, D. Z., & Garner, C. C. (2010). Emerging pharmacotherapies for neurodevelopmental disorders. *Journal of Developmental and Behavioral Pediatrics, 31*(7), 564–581.

Wisniewski, K. E., Dalton, A. J., et al. (1985). Alzheimer's disease in Down's syndrome: Clinicopathologic studies. *Neurology, 35*(7), 957–961.

Woods, Y. L., Cohen, P., et al. (2001a). The kinase DYRK phosphorylates protein-synthesis initiation factor eIF2Bepsilon at Ser539 and the microtubule-associated protein tau at Thr212: Potential role for DYRK as a glycogen synthase kinase 3-priming kinase. *The Biochemical Journal, 355*(Pt 3), 609–615.

Woods, Y. L., Rena, G., et al. (2001b). The kinase DYRK1A phosphorylates the transcription factor FKHR at Ser329 in vitro, a novel in vivo phosphorylation site. *The Biochemical Journal, 355*(Pt 3), 597–607.

Xie, W., Ramakrishna, N., et al. (2008). Promotion of neuronal plasticity by (−)-epigallocatechin-3-gallate. *Neurochemical Research, 33*(5), 776–783.

Yang, Q., Rasmussen, S. A., et al. (2002). Mortality associated with Down's syndrome in the USA from 1983 to 1997: A population-based study. *Lancet, 359*(9311), 1019–1025.

Yoshiike, Y., Kimura, T., et al. (2008). GABA(A) receptor-mediated acceleration of aging-associated memory decline in APP/PS1 mice and its pharmacological treatment by picrotoxin. *PLoS One, 3*(8), e3029.

Yu, T., Li, Z., Jia, Z., Clapcote, S. J., Liu, C., Li, S., Asrar, S., Pao, A., Chen, R., Fan, N., Carattini-Rivera, S., Bechard, A. R., Spring, S., Henkelman, R. M., Stoica, G., Matsui, S., Nowak, N. J., Roder, J. C., Chen, C., Bradley, A., Yu, Y. E. (2010). A mouse model of Down syndrome trisomic for all human chromosome 21 syntenic regions. *Human Molecular Genetics, 19*(14), 2780–2791.

Chapter 21
The Synaptic Pathology of Drug Addiction

Michel C. Van den Oever, Sabine Spijker, and August B. Smit

Abstract A hallmark of drug addiction is the uncontrollable desire to consume drugs at the expense of severe negative consequences. Moreover, addicts that successfully refrain from drug use have a high vulnerability to relapse even after months or years of abstinence. In this chapter, we will discuss the current understanding of drug-induced neuroplasticity within the mesocorticolimbic brain system that contributes to the development of addiction and the persistence of relapse to drug seeking. I particular, we will focus at animal models that can be translated to human addiction. Although dopaminergic transmission is important for the acute effects of drug intake, the long-lived behavioral abnormalities associated with addiction are thought to arise from pathological plasticity in glutamatergic neurotransmission. The nature of changes in excitatory synaptic plasticity depends on several factors, including the type of drug, the brain area, and the time-point studied in the transition of drug exposure to withdrawal and relapse to drug seeking. Identification of drug-induced neuroplasticity is crucial to understand how molecular and cellular adaptations contribute to the end stage of addiction, which from a clinical perspective, is a time-point where pharmacotherapy may be most effectively employed.

Keywords Animal models • Dopamine • Drug induced neuroplasticity • Nucleus accumbens • Ventral tegmental area

M.C. Van den Oever • S. Spijker • A.B. Smit (✉)
Department of Molecular and Cellular Neurobiology, Center for Neurogenomics and Cognitive Research, Neuroscience Campus Amsterdam, VU University, De Boelelaan 1085, 1081 HV Amsterdam, The Netherlands
e-mail: guus.smit@cncr.vu.nl

M.R. Kreutz and C. Sala (eds.), *Synaptic Plasticity*,
Advances in Experimental Medicine and Biology 970,
DOI 10.1007/978-3-7091-0932-8_21, © Springer-Verlag/Wien 2012

21.1 Introduction

Drug addiction is characterized by compulsive drug-taking behavior and high rates of relapse. This addiction behavior cycle continues despite severe negative consequences for the affected individual. In addicts, the behavioral repertoire is narrowed down to drug seeking, drug taking, and recovering from drug use. Unfortunately, existing treatment of drug addiction is still relatively ineffective because it is compromised by high relapse rates. To develop better pharmacotherapy or other means of intervention, such as deep brain stimulation, research in the years to come must reveal the neurobiological underpinnings of addiction. In particular, compulsive drug-taking, long-term vulnerability to relapse to drug seeking and the mechanisms by which drug-associated stimuli (cues) control addictive behaviors need to be addressed. Accumulating evidence indicates that the development and persistence of addiction involves mechanisms of synaptic plasticity similar to traditional neuronal models of learning and memory, such as long-term potentiation (LTP) and long-term depression (LTD). Therefore, identification of the effects of drug exposure on cellular and molecular mechanisms of synaptic plasticity is thought to hold strong promise for the development of more effective pharmacotherapy. Also, understanding the neuronal circuitry of addiction in more detail will aid in designing intervention strategies using novel deep brain stimulation and optogenetics technologies.

From an evolutionary perspective, brain reward systems have evolved to mediate appropriate responses to natural rewards (e.g., food, sex), which are crucially important for survival of an organism. An integral part of the brain's reward circuitry is the mesocorticolimbic dopamine system, comprising dopaminergic neurons in the ventral tegmental area (VTA) and the brain regions to which these dopaminergic neurons project, including the ventral striatum (nucleus accumbens (NAc)), the dorsal striatum, the prefrontal cortex (PFC), and the amygdala (Hyman et al. 2006). Dopaminergic transmission within these brain regions is important for reinforcement learning, motivation, and goal-directed behavior (Schultz 1998; Tzschentke 2001), processes that drive the pursuit of rewards and the responding to reward-predictive stimuli. This neural circuitry that is essential for survival can be acted upon by powerful reinforcers, e.g., drugs of abuse, to take control over normal behavior and producing a state referred to as addiction.

To date, for most drugs of abuse, the molecular site(s) of action and the sequence of cellular events that follow acute drug administration have been identified. Although the direct pharmacological actions differ between these drugs, a similar direct or indirect effect has been observed for most drugs on dopamine release. For instance, opiates bind directly to μ-opioid receptors located on presynaptic terminals of GABAergic interneurons within the VTA (Gysling and Wang 1983). Stimulation of these receptors reduces tonic and stimulated release of GABA, thereby increasing the firing rate of VTA dopaminergic neurons (Di Chiara and North 1992). Amphetamine-like psychostimulants elevate extracellular levels of dopamine by limiting reuptake of dopamine by the dopamine transporter (Williams and Galli 2006), whereas nicotine increases firing frequency of dopaminergic neurons by directly binding to nicotinic acetylcholine receptors on dopaminergic cells (Mereu et al. 1987). The common action of most drugs of abuse to stimulate dopaminergic transmission suggests that the VTA and its dopaminergic target areas are likely substrates for drug-induced adaptations in synaptic plasticity that in turn sustain addictive behavior. Therefore, the majority of studies in the addiction field have focused on plasticity changes that occur in the mesocorticolimbic system as a consequence of drug exposure. In this chapter, we will discuss drug-induced effects on physiological plasticity in the mesocorticolimbic system and the potential molecular mechanisms that may support the addiction state.

21.2   Animal Models of Addiction

Behavioral symptoms characteristic of drug addiction can be successfully modeled in animals. This holds for a broad spectrum of scientific criteria, face, predictive, construct, and external validity (Epstein et al. 2006; Van der Staay et al. 2009).

Face validity: Both rat and mouse models offer face validity in the sense that homologous neuroanatomic structures are affected in these rodent models and in human addicts, namely the mesocorticolimbic system (Hyman 2005). Imaging studies and rodent models implicate frontal cortical areas in conditioned drug-seeking responses (Childress et al. 1999; Goldstein and Volkow 2002; Kalivas and Volkow 2005). However, the resolution and detailing of neuronal substrates and their spatiotemporal involvement need further investigation.

Predictive validity: Concerning the use of therapy, predictive validity is certainly true for the rat and mouse models (Crombag et al. 2008; Epstein et al. 2006). However, due to lack of detail in the mechanistic underpinning of addiction, this still has not led to systematic treatment in patients.

Construct validity: Because the nature of the addiction is quite well understood, construct validity seems adequately addressed. Animal models, in particular those involving self-administration (see below), mimic the pathology, the symptomatology, and the etiology of the disease to a large extent.

External validity: Rat and mouse self-administration models, including cue-induced relapse, are frequently used and are well translatable to the human condition (Epstein et al. 2006).

Taken together, animal models in addiction research are adequate at various levels of translation to the human condition. Further work into the spatiotemporal adaptation of circuitry, the mechanisms of relapse, and the (epi-)genetic differences in vulnerability needs to be addressed.

When interpreting the effects of drugs of abuse in animal models, it is important to consider the method of drug administration: noncontingent administration (i.e., passive exposure, administered by the experimenter) versus contingent administration (i.e., self-administration). Human drug addiction is usually the result of active control over drug intake and involves associative learning between the action (drug intake) and outcome (the experience of reward) as well as associative learning between the reinforcing properties of the drug and drug-related stimuli (e.g., locations were the drug is consumed and paraphernalia used to consume the drug). These pathological associations (maladaptive memories) are thought to have a critical role in the persistence of compulsive drug seeking and relapse to drug seeking during periods of abstinence. In this respect, the operant self-administration model has most resemblance to human drug addiction, as animals have voluntary control over drug intake. In the self-administration model, an operant response (e.g., pressing a lever or nose poking) results in the activation of a syringe pump that delivers the drug via a catheter that is implanted in the jugular vein (or orally in case of ethanol). Drugs that are self-administered by animals (psychostimulants, opiates, nicotine, ethanol) yield effects that correspond well with the high abuse potential in humans. Persistent associative memories (i.e., drug-associated environmental cues) have a pivotal role in relapse to drug taking. To study the involvement of these memories in compulsive drug seeking and relapse, drug delivery in animal models is often paired with the brief presentation of a visual (light stimulus) and/or auditory (tone) cue such that the animal learns to associate the audiovisual cues with the rewarding effects of the drug. This model has been used to study the mechanisms underlying the acquisition of drug self-administration, including the animal's motivation to obtain a drug reward (Gardner 2000). In addition, self-administration can be used to study the neurobiological underpinnings of long-lasting vulnerability to relapse by extending the model with a drug-free period after self-administration. Animals are confined to their home cage during this period (forced abstinence) or can undergo extinction training during which operant responding does not result in the infusion of the drug and presentation of drug-paired cues. Relapse to drug seeking (measured as a resumption of operant responding) is triggered by reexposing the animals to the previously administered drug or drug-associated cues or by exposure to stressors, stimuli that also precipitate relapse in human addicts (Epstein et al. 2006; Goeders 2003; O'Brien et al. 1986).

With noncontingent drug administration, animals receive a single drug injection or repeated drug injections, often delivered by the experimenter. Noncontingent drug exposure in animals is typically associated with an enhancement of the locomotor-activating effects of the drug, also known as behavioral sensitization (Robinson and Berridge 1993; Stewart and Badiani 1993; Vanderschuren et al. 1997). Although humans usually do not develop an addiction after involuntary drug administration, behavioral sensitization in animals is long lasting and can augment subsequent drug self-administration (Vezina 2004). Therefore, the neurobiological adaptations that mediate long-term behavioral sensitization are thought to at least partially overlap with the mechanisms that facilitate relapse to drug seeking after prolonged abstinence in the self-administration model, such as increased sensitivity to drug-priming injections and drug-associated cues. Conditioned place preference (CPP) is a well-established paradigm to study associative learning mechanisms of the rewarding effects of a drug and the environmental context (cue) in which the drug is administered (Bardo and Bevins 2000). During conditioning, animals (typically rats or mice) receive a drug and vehicle injection in two distinct separated contexts that can differ in visual, olfactory, and tactile cues. After conditioning, drug seeking is assessed by allowing the animals free access to the drug-paired and vehicle-paired context in the absence of drug reinforcement. Animals that remember the association between the drug-paired context and cocaine reward will spend more time in this context. Similar to the self-administration paradigm, CPP can be extended by an extinction phase, and drug seeking can subsequently be reinstated by drug priming or stressors (Aguilar et al. 2009).

Compared with self-administration, CPP is relatively easy to perform as animals do not require surgery and a preference for the drug-paired context can be established within one or two conditioning sessions (depending on the type of drug and dose) (Bardo and Bevins 2000). Therefore, CPP is often used to study the effects of drug reward and conditioning in mutant mouse lines. For example, the availability of the many mouse Cre driver lines (Cre recombinase expression driven by a neuronal subtype specific promoter) offers the possibility to dissect the neuronal circuitry of addiction with great precision using optogenetics technology (Lobo et al. 2010; Tsai et al. 2009; Witten et al. 2010). In the following sections, we provide an overview of neuroplasticity changes induced by drug exposure per brain region, and where data is available, we will discuss differential effects of noncontingent versus self-administration.

21.3 Drug-Induced Neuroplasticity in the VTA

Exposure to most drugs of abuse results in an increase in DA release in projection regions of VTA DA neurons. Moreover, an increase in DA release is also observed upon exposure to reward-associated stimuli (Schultz 1998). Increased DA transmission is thought to reflect a change in a tonic firing mode of VTA DA neurons to a phasic firing mode (Schultz 2007). This is supported by the recent observation that in the absence of drug reinforcement, phasic firing of DA neurons is sufficient to drive behavioral conditioning in a place preference paradigm (Tsai et al. 2009). The VTA receives extensive glutamatergic projections from corticolimbic brain structures (Geisler and Wise 2008), and glutamatergic input onto DA neurons may modulate the shift from a tonic to phasic firing mode (White 1996). Moreover, an NMDA receptor (NMDA-R)-dependent increase in glutamatergic synaptic strength has been reported in VTA DA neurons during the acquisition of a cue-reward association (Stuber et al. 2008). Hence, altered plasticity of glutamatergic synapses on dopaminergic neurons may serve as a neuronal substrate for drug-induced changes that support addictive behaviors.

The first evidence that in vivo cocaine administration induces changes in glutamatergic plasticity was provided by Ungless and coworkers. Animals that received a single noncontingent cocaine injection showed an increase in synaptic strength in DA neurons in the VTA (Ungless et al. 2001). Building on this finding, it was found that a single exposure to nearly all types of abused drugs induces a long-term potentiation of AMPA receptor (AMPA-R) responses in VTA DA neurons (Argilli et al. 2008; Saal et al. 2003; Ungless et al. 2001) (Table 21.1 and Fig. 21.1). Similar to cue-reward learning, the increase in synaptic strength is dependent on NMDA-R activation (Ungless et al. 2001) through stimulation of DA receptors locally in the VTA and requires protein synthesis (Argilli et al. 2008); the latter is typically required for long-term memory formation. Hence, the increase in synaptic strength may support an associative learning process. The potentiation of AMPA-R currents is paralleled by an increase in the synaptic expression of GluA2-lacking AMPA-Rs

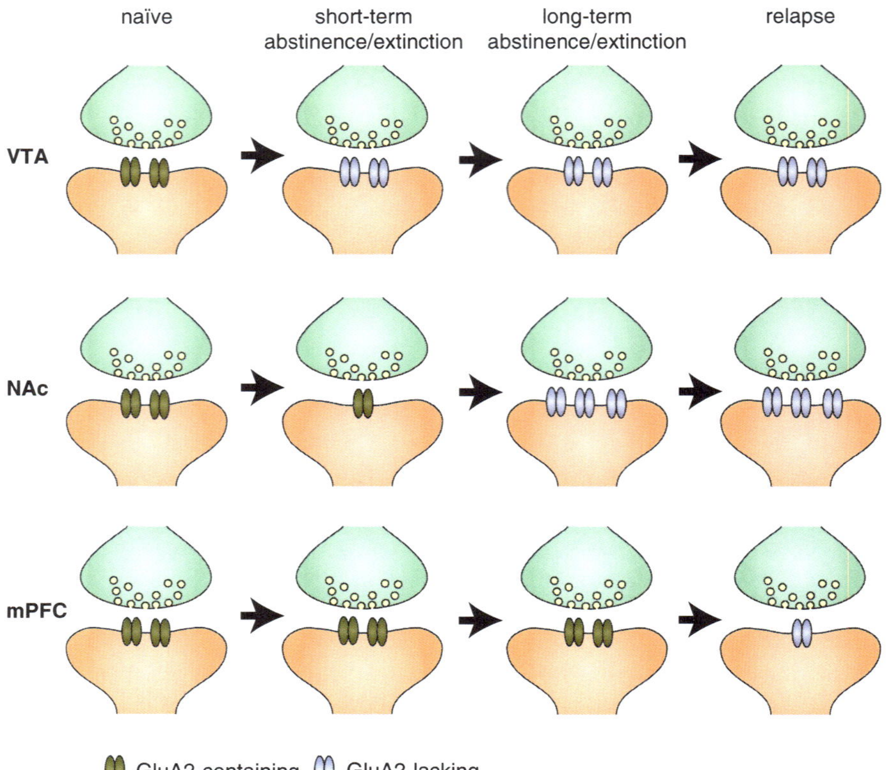

Fig. 21.1 Schematic overview of AMPA-R plasticity in the VTA, NAc, and mPFC at successive stages in the addiction cycle. Naïve animals (non-drug-exposed) typically express GluA2-containing AMPA-Rs. In the VTA, drug exposure results in a potentiation of AMPA-R currents and elevated expression of predominantly GluA2-lacking AMPA-Rs. This enhancement of synaptic strength is transient after noncontingent drug treatment but long lasting following drug self-administration. In the NAc, drug exposure (mainly observed with cocaine) induces a biphasic wave of AMPA-R plasticity. Whereas short-term withdrawal is associated with synaptic depression, an increase in GluA2-lacking AMPA-Rs is observed after long-term abstinence that remains enhanced after relapse. A different AMPA-R response was found in the mPFC (ventral subregion). Short- and long-term abstinence from heroin self-administration does not result in altered surface expression of AMPA-Rs; cue-induced relapse is mediated by the rapid endocytosis of GluA2 subunits

(Argilli et al. 2008; Bellone and Luscher 2006), in line with increased expression of GluA1, but not GluA2, in the VTA 24 h after cocaine treatment (Churchill et al. 1999; Fitzgerald et al. 1996). GluA2-lacking AMPA-Rs are calcium permeable, have greater channel conductance, and can thereby contribute to synaptic strengthening (Isaac et al. 2007). Interestingly, the drug-induced exchange of GluA2-containing with GluA2-lacking AMPA-Rs can be reversed by activation of mGluR1 receptors in the VTA (Bellone and Luscher 2006; Mameli et al. 2007) and disruption of mGluR1 function prolongs the cocaine-evoked increased expression of synaptic GluA2-lacking AMPA-Rs in VTA DA neurons (Mameli et al. 2009).

Table 21.1 Drug-induced glutamatergic plasticity in the mesocorticolimbic system

Brain area	Drug exposure	Short-term effect (<5 days abstinence)	Long-term effect (>5 days abstinence/extinction)	Challenge/Relapse (Up to 24 h after test)
VTA	Noncontingent	AMPA/NMDA ratio (1, 2, 3, 4)	AMPA/NMDA ratio unchanged (1, 4, 5)	N/A
		GluA2-lacking AMPA-Rs (3, 6)	N/A	N/A
		GluA1 (7, 8)	GluA1 unchanged (7)	N/A
		GluN1 (7, 8)	GluN1 unchanged (7)	N/A
	Self-administration	AMPA/NMDA ratio (5)	AMPA/NMDA ratio (5)	AMPA/NMDA ratio (5)
		N/A	LTP occluded (5)	N/A
		GluA2 (9)	GluA2 unchanged (9)	N/A
		GluN1 (9)	GluN1 (9)	N/A
NAc	Noncontingent	AMPA/NMDA ratio $\downarrow$ (10)	AMPA/NMDA ratio (10)	AMPA/NMDA ratio $\downarrow$ (10)
		N/A	mEPSC freq. + ampl. (10)	N/A
		GluA1 unchanged (7)	GluA1 (7)	N/A
		GluA1 surface unchanged (11, 12)	GluA1 surface (11, 12)	GluA1 surface $\downarrow$ (11, 12)
		GluA2/3 surface unchanged (11, 12)	GluA2 surface (11, 12)	GluA2 surface $\downarrow$ (11, 12)
	Self-administration	Synaptic depression (13)	N/A	N/A
		LTP and LTD occluded (14)	LTP and LTD occluded (14, 15, 16)	N/A
		GluA1 (9)	GluA1 (9, 18)	N/A
		GluA2 (9)	GluA2 up to 30d (9, 18)	N/A
		GluA1 surface unchanged (17)	GluA1 surface (17, 19)	GluA1 surface (17, 19)
		GluA2 surface unchanged (17)	GluA2 surface unchanged (17)	GluA2 surface $\downarrow$ (20)
		GluA1 surface unchanged (17)	GluA3 surface (17)	N/A
		Extracell. glutamate unchanged (21)	Extracell. glutamate $\downarrow$ (22, 21)	N/A
PFC	Noncontingent	LTP unchanged (24)	LTP facilitated (24)	N/A
		GABAergic inhibition unchanged (24)	GABAergic inhibition $\downarrow$ (24)	N/A
	Self-administration	N/A	AMPA/NMDA ratio unchanged (26)	AMPA/NMDA ratio $\downarrow$ (26)
		N/A	GABAergic inhibition unchanged (27)	GABAergic inhibition (27)
		GluA2/3 unchanged (8, 23)	GluA2/3 (23)	N/A
		GluA4 unchanged (23)	GluA4 (23)	N/A

(continued)

Table 21.1 (continued)

Brain area	Drug exposure	Short-term effect (<5 days abstinence)	Long-term effect (>5 days abstinence/extinction)	Challenge/Relapse (Up to 24 h after test)
		GluN2A unchanged (28)	GluN2A (28)	N/A
		GluN2B unchanged (23, 28)	GluN2B (23, 28)	N/A
		GluN3A unchanged (23)	GluN3A ↓ (23)	N/A
		N/A	GluA2 surface unchanged (26, 27)	GluA2 surface ↓ (26)
		N/A	GluA3 surface unchanged (27)	GluA3 surface ↓ (26)
		N/A	GluN2B surface unchanged (27)	GluN2B surface ↓ (26)
BLA	Self-administration	GluA1 (29)	GluA1 (29)	N/A
		GluN2A (29)	GluN2A unchanged (29)	N/A
		GluN2B unchanged (29)	GluN2B ↓ (29)	N/A
CeA	Noncontingent	LTP unchanged (30)	LTP facilitated (30)	N/A
		N/A	AMPA/NMDA ratio unchanged (30)	N/A
		N/A	GluN1 (30)	N/A
	Self-administration	GluA2 (29)	GluA2 (29)	N/A
		GluN1 unchanged (29)	GluN1 (29)	N/A

Summary of changes in AMPA-Rs and NMDA-Rs resulting from noncontingent drug exposure and drug self-administration. *N/A* data not available.
References: 1. Ungless et al. (2001); 2. Saal et al. (2003); 3. Argilli et al. (2008); 4. Borgland et al. (2004); 5. Chen et al. (2008); 6. Bellone and Luscher (2006); 7. Churchill et al. (1999); 8. Fitzgerald et al. (1996); 9. Lu et al. (2003); 10. Kourrich et al. (2007); 11. Boudreau and Wolf (2005); 12. Boudreau et al. (2007); 13. Schramm-Sapyta et al. (2006); 14. Martin et al. (2006); 15. Kasanetz et al. (2010); 16. Moussawi et al. (2009); 17. Conrad et al. (2008); 18. Sutton et al. (2003); 19. Anderson et al. (2008); 20. Famous et al. (2008); 21. Miguens et al. (2008); 22. Baker et al. (2003); 23. Tang et al. (2004); 24. Lu et al. (2010); 26. Van den Oever et al. (2008); 27. Van den Oever et al. (2010a); 28. Ben-Shahar et al. (2009); 29. Lu et al. (2005a); 30. Fu et al. (2007)

Synaptic potentiation of AMPA-R currents in VTA DA neurons is only observed up to 5 days after single or repeated noncontingent cocaine exposure (Borgland et al. 2004; Saal et al. 2003; Ungless et al. 2001), suggesting that it may contribute to early stages of the development of drug-seeking behaviors but not to the long-lived behavioral abnormalities (such as relapse) that characterize addiction. However, in contrast, mutant mice that lack expression of the NMDA-R GluN1 subunit and AMPA GluA1 subunit selectively in midbrain DA neurons show impaired potentiation of AMPA-R currents after cocaine treatment but normal acquisition of behavioral sensitization and CPP (Engblom et al. 2008). Even more surprising is the observation that extinction of cocaine CPP is absent in GluA1 mutants (Engblom et al. 2008), whereas in GluN1 mutant mice, reinstatement of cocaine CPP and behavioral sensitization to a cocaine challenge after forced abstinence is abolished (Engblom et al. 2008; Zweifel et al. 2008), indicating that synaptic strengthening in DA neurons may not mediate short-term behavioral effects but may contribute to the initiation of long-term neuroadaptations that maintain the persistence of drug-seeking behavior.

In line with noncontingent cocaine exposure, cocaine self-administration also results in a potentiation of AMPA receptor currents in VTA DA neurons (Chen et al. 2008). However, in contrast to the transient nature (up to ~5 days) of synaptic potentiation observed with repeated experimenter-administered cocaine (Borgland et al. 2004; Ungless et al. 2001), self-administration induced a potentiation that lasted for at least 3 months and that was resistant to extinction training (Chen et al. 2008) (Fig. 21.1). Moreover, synaptic potentiation was also observed in animals that self-administered natural rewarding substances (e.g., food and sucrose), but this effect was no longer observed after 21 days of forced abstinence (Chen et al. 2008). These findings suggest that associative learning mechanisms during cocaine self-administration strengthen and prolong the potentiation of AMPA-R responses in VTA DA neurons induced by cocaine and may therefore contribute to the enhanced storage and retrieval of a drug-related memory over a nondrug memory. At present, the origin of the glutamatergic projections that are strengthened remains to be elucidated. In fact, it cannot be excluded that the glutamatergic synapses that are transiently potentiated by noncontingent cocaine exposure and those that are persistently strengthened following self-administration receive input from different brain regions or different neuronal populations. This again argues for making a clear distinction in discussing noncontingent versus contingent paradigms. Nevertheless, these studies provide evidence that drugs of abuse can induce long-lasting plasticity at glutamatergic synapses and that the nature of these neuroadaptations depends on the method of drug administration.

21.4   Drug-Induced Neuroplasticity in the NAc

The NAc is a primary target of DA projections originating in the VTA. In addition to DA projections, the NAc receives dense glutamatergic input from cortical and subcortical areas, including the medial PFC (mPFC) and amygdala (Voorn et al. 2004).

The convergence of DA and glutamatergic afferents on to the same NAc neuron (Sesack et al. 2003) suggests that the NAc functions as a limbic-motor interface that processes the relevance of salient stimuli to initiate a behavioral response. Functionally and anatomically, the NAc can be divided into two regions, a core region that is thought to be important for conditioned responses and the attribution of salience to motivational relevant stimuli and a shell region that is thought to mediate the primary reinforcing properties of rewarding substances and novelty (Meredith et al. 2008; Zahm 1999), but these functional classifications may be an oversimplification. Substantial evidence indicates that glutamatergic transmission in the NAc is critically involved in the initiation of drug-seeking behavior. In particular, blockade of AMPA-R activation in the NAc prevents conditioned cocaine and heroin seeking (Cornish et al. 1999; Cornish and Kalivas 2000; Di Ciano and Everitt 2001; Ping et al. 2008). The glutamatergic input during reinstatement of drug seeking is thought to arise predominantly from mPFC projections to the NAc core (Kalivas et al. 2005; McFarland et al. 2003; Park et al. 2002).

In contrast to the VTA, changes in synaptic strength in NAc neurons have not been observed after a single drug exposure (Kourrich et al. 2007). However, repeated experimenter-administered cocaine injections result in a biphasic response in synaptic strength in NAc shell neurons (Fig. 21.1). After short-term withdrawal, namely 24 h following a repeated cocaine treatment regimen, AMPA-R currents are reduced in the NAc shell (Kourrich et al. 2007), and using in vivo single-cell recording, it was found that NAc neurons are less sensitive to glutamate after cocaine administration (White et al. 1995). Although the functional relevance of reduced AMPA-R currents remains to be determined, synaptic depression of NAc shell neurons during early withdrawal may impair the excitation of NAc neurons in response to natural rewarding stimuli and could thereby contribute to feelings of dysphoria and anhedonia. In contrast, prolonged abstinence (>7–10 days) from repeated noncontingent cocaine treatment results in a potentiation of AMPA-R function in NAc shell neurons (Boudreau et al. 2007; Boudreau and Wolf 2005; Kourrich et al. 2007). On the long-term, namely after 10–14 days of withdrawal, mice that received repeated cocaine treatment exhibit increased frequency and amplitude of AMPA-R miniature excitatory postsynaptic currents (mEPSCs), but not paired-pulse ratio in NAc shell neurons, pointing to an enhancement of postsynaptic AMPA-R responses (Kourrich et al. 2007). This is supported by the observation that long-term withdrawal from noncontingent cocaine treatment is associated with an increase in the surface expression of GluA1 and GluA2 subunits in the NAc (Boudreau et al. 2007, 2009; Boudreau and Wolf 2005). Moreover, the locomotor stimulating effect of AMPA infusion in the NAc is potentiated in animals that develop behavioral sensitization after repeated cocaine treatment (Bell and Kalivas 1996; Pierce et al. 1996).

Interestingly, potentiation of AMPA-R currents and surface expression is abruptly reversed by reexposing animals to a challenge cocaine injection (Boudreau et al. 2007; Kourrich et al. 2007; Thomas et al. 2001). This suggests that acute synaptic depression may mediate increased locomotor activity that is typically observed after a challenge injection of psychostimulants. Indeed, preventing the induction of synaptic depression by intra-NAc infusion of a synthetic peptide that disrupts clathrin-mediated endocytosis of GluA2 attenuates behavioral sensitization to an amphetamine challenge (Brebner et al. 2005). The precise molecular mechanisms by which AMPA-R surface expression are enhanced after prolonged withdrawal and reversed after a drug challenge are not known, but activation of the extracellular signal-regulated kinase (ERK; also known as mitogen-activated protein kinase) pathway may be involved. Phosphorylation of ERK can increase insertion of AMPA-Rs in the synaptic membrane during LTP (Zhu et al. 2002). In parallel with the biphasic changes in AMPA-R function, ERK phosphorylation is enhanced in the NAc after prolonged withdrawal (but not early withdrawal) from noncontingent cocaine treatment and normalizes following a cocaine challenge injection (Boudreau et al. 2007). Moreover, intra-accumbens injection of inhibitors of ERK phosphorylation blocks expression of cocaine-conditioned place preference (Gerdjikov et al. 2004; Miller and Marshall 2005), suggesting that ERK may have a role in conditioned drug seeking responses.

The effects of cocaine self-administration on glutamatergic synaptic plasticity in the NAc resemble those of repeated noncontingent cocaine exposure. Synaptic depression of NAc neurons is observed during early withdrawal (Schramm-Sapyta et al. 2006), whereas long-term abstinence is paralleled by a potentiation of synaptic transmission (Conrad et al. 2008). Interestingly, a history of cocaine self-administration impairs the ability to induce LTD in NAc shell and core neurons during early withdrawal (Martin et al. 2006), and both LTD and LTP are abolished in NAc core neurons after long-term cessation of self-administration (Martin et al. 2006; Moussawi et al. 2009). Notably, it was recently found that impaired LTD in the NAc occurs only in animals that develop the behavioral hallmarks of addiction and not in animals that maintain controlled cocaine intake (Kasanetz et al. 2010). The change in the ability to induce synaptic plasticity may reflect an already depressed or potentiated state of glutamatergic synapses in the NAc, a phenomenon known as metaplasticity (Abraham 2008). The enhanced surface expression of both GluA1 and GluA2 subunits following noncontingent cocaine exposure is also present acutely after contingent cocaine self-administration (Lu et al. 2003). However, in contrast to the long-term upregulation of both GluA1 and GluA2 in noncontingent exposure, increased synaptic strength during late withdrawal from cocaine self-administration is accompanied by an increase in surface expression of the GluA1, but not GluA2, subunit in NAc shell and core neurons (Anderson et al. 2008; Conrad et al. 2008; Lu et al. 2003). As mentioned above, GluA2-lacking AMPA-Rs have greater channel conductance and thereby contribute to LTP. The increase in GluA2-lacking AMPA-Rs in the NAc core is thought to underlie incubation of cocaine craving, i.e., the time-dependent augmentation of responding to drug-associated cues during abstinence (Conrad et al. 2008). An imbalance between synaptic and extrasynaptic extracellular glutamate levels in the NAc may support the persistent vulnerability to relapse to drug seeking (Kalivas 2009). Extracellular glutamate levels are reduced in the NAc core following both contingent and noncontingent drug exposure (Baker et al. 2003; Miguens et al. 2008), due to dysregulation of the glial cysteine-glutamate exchanger (Baker et al. 2003;

Knackstedt et al. 2009a) and diminished expression of glutamate transporter 1 (Glt1) (Knackstedt et al. 2010). Decreased basal extracellular glutamate levels contribute to increased synaptic glutamate release probability by reduced tonic activation of metabotropic glutamate receptors mGluR2 and mGluR3 on presynaptic glutamatergic afferents (Moran et al. 2005; Moussawi et al. 2011). Moreover, reduced activation of mGluR2/3 and mGluR5 underlies the impairment to induce LTP and LTD respectively following cocaine self-administration (Moussawi et al. 2009). Restoring glutamatergic tone on mGluR2/3 by N-acetylcysteine treatment attenuates reinstatement of drug seeking in animal models (Baker et al. 2003; Knackstedt et al. 2010; Moussawi et al. 2011; Peters and Kalivas 2006; Zhou and Kalivas 2008) and drug craving in humans (Knackstedt et al. 2009a; LaRowe et al. 2007; Mardikian et al. 2007). Enhanced synaptic glutamate transmission is thought to strengthen input from the dorsal mPFC to the NAc core upon exposure to drugs and drug-associated cues (Kalivas et al. 2005; LaLumiere and Kalivas 2008; McFarland et al. 2003).

Although potentiation of glutamatergic transmission in the NAc core drives drug-seeking responses, the functional consequences of enhanced expression of AMPA-R in the NAc shell are less clear. Numerous studies found that blockade of AMPA-Rs or inactivation of the NAc shell impairs reinstatement of drug seeking (Anderson et al. 2003, 2006, 2008; Bachtell et al. 2005; Bossert et al. 2006, 2007; Conrad et al. 2008; Famous et al. 2008; Fuchs et al. 2008; McFarland et al. 2004; Rogers et al. 2008; Schmidt and Pierce 2006; Xi et al. 2004), indicating that glutamatergic transmission supports relapse to drug seeking. Similar to NAc core, relapse may be facilitated by activation of GluA2-lacking AMPA-Rs in the NAc shell as it was found that GluA2 subunits internalize during cocaine-primed reinstatement and inhibition of GluA2 trafficking in the NAc core and shell attenuates subsequent cocaine seeking (Famous et al. 2008). In contrast to these findings, extinction training after cocaine self-administration increases the expression of GluA1 and GluA2/3 AMPA-Rs in the NAc shell, and viral-mediated overexpression of GluA1 facilitates extinction of responding (Sutton et al. 2003). This suggests that AMPA-R activation in the NAc shell can promote or suppress drug-seeking responses depending on the test conditions, emphasizing the need for future research to identify the population of neurons that exhibit increased synaptic strength after cocaine self-administration and the origin of glutamatergic input in the NAc shell during extinction and reinstatement testing.

21.5   Drug-Induced Neuroplasticity in the mPFC

Apart from the VTA and NAc, the mPFC comprises an integral part of the motivational circuit in the brain and is a major site of dopamine release and neuronal modulation from the VTA. There is a general agreement that the dopaminergic input to the mPFC is important in acquisition of drug self-administration (Schenk et al. 1991; Weissenborn et al. 1997) and that glutamatergic projections from the prefrontal cortex to the NAc are critical for reinstatement of drug-seeking behavior (Kalivas et al. 2005; LaLumiere and Kalivas 2008; McFarland et al. 2003). The prefrontal cortex can be divided into a dorsal part (including the prelimbic area) and a ventral part (including the infralimbic cortex). Anatomically, projections from the mPFC to the NAc are organized in a dorsal-ventral pattern, with the dorsal mPFC projecting predominantly to the NAc core and the ventral mPFC to the NAc shell (Heidbreder and Groenewegen 2003; Voorn et al. 2004; Sesack et al. 1989). A glutamatergic projection from the dorsal mPFC to the NAc core is thought to engage the motor circuitry, thereby driving drug-seeking responses (LaLumiere and Kalivas 2008; McFarland et al. 2003), whereas the ventral mPFC to the NAc shell connection is thought to be involved in extinction learning (Millan et al. 2011; Peters et al. 2009; Van den Oever et al. 2010b).

Relapse to cocaine and heroin seeking by (re-)exposure to drugs, cues, or stressors can be blocked by reversible pharmacological inactivation of the dorsal mPFC (Fuchs et al. 2005; McFarland and Kalivas 2001; McLaughlin and See 2003; Rogers et al. 2008) and ventral mPFC (Bossert et al. 2011; Koya et al. 2009), although the ventral mPFC may exert an opposite function after extinction of cocaine self-administration (Peters et al. 2008). Unfortunately, compared with the VTA and NAc, relatively little is known about short- and long-term changes in synaptic plasticity in the mPFC, and in most studies, dissociations between the dorsal and ventral mPFC have not been made. Withdrawal from repeated noncontingent cocaine administration is associated with a facilitation in the induction of LTP in excitatory synapses onto pyramidal cells in the mPFC (Lu et al. 2010), an effect that is apparent after 5 days of withdrawal (but not at earlier time-points) and mediated by brain-derived neurotrophic factor (BDNF)-induced suppression of GABAergic inhibition in the mPFC. Long-term (3 weeks) abstinence from repeated noncontingent cocaine administration is associated with an increase in mRNA levels of the GluA2 subunit but not in any of the other AMPA-R subunits (Ghasemzadeh et al. 1999). Similar to the VTA and NAc, contingent cocaine administration appears to have a more robust effect on glutamate receptor expression as an increase was observed in protein levels of GluA1, GluA2, and GluA4 in the mPFC following 2 weeks of abstinence from cocaine self-administration (Tang et al. 2004). Developments in the field of subcellular proteomics have enabled the analysis of changes in protein abundance in synaptic membrane fractions. Using such an approach, it was found that contingent heroin administration has no long-term effect on the synaptic membrane expression of AMPA-Rs in the mPFC (Van den Oever et al. 2010a); however, cue-induced relapse to heroin seeking is accompanied by an acute reduction in the synaptic membrane levels of GluA2 and GluA3 but not GluA1 (Van den Oever et al. 2008) (Fig. 21.1). The latter effect is paralleled by an acute synaptic depression of AMPA-R currents in mPFC pyramidal cells, and blocking this depression by injection of a peptide that prevents GluA2 endocytosis reduces cue-induced relapse to heroin seeking (Van den Oever et al. 2008). Moreover, exposure to heroin-conditioned cues results in an increase in the frequency of inhibitory postsynaptic currents received by mPFC pyramidal cells (Van den Oever et al. 2010a), which may further contribute to a depressed state of these neurons. These observations suggest that acutely reduced glutamatergic transmission in the ventral mPFC facilitates relapse to drug seeking. In support of this is the observation that activation of AMPA-Rs in the ventral mPFC reduces cocaine-primed reinstatement of drug seeking after cocaine seeking is extinguished (Peters et al. 2008). A glutamatergic projection from the infralimbic cortex to the NAc shell may mediate the expression of this extinction memory, as reversible unilateral inactivation of either region augments cocaine seeking under extinction conditions (Peters et al. 2008). In contrast, in nonextinguished animals, reversible inactivation of the ventral mPFC attenuates relapse to cocaine seeking (Koya et al. 2009), pointing to a complex role of this subregion of the mPFC in controlling drug-seeking responses.

In contrast to the effects observed on AMPA-R plasticity, NMDA-R plasticity in the mPFC is seemingly unaffected. Only few reports mention changes in NMDA-R subunits; long-term abstinence from contingent cocaine exposure is accompanied by increased expression of GluN2B (2 weeks abstinence) (Tang et al. 2004) and GluN2A (60 days abstinence), but the latter finding is specific for animals that experienced extended access to the drug (Ben-Shahar et al. 2009). In support of a role of NMDA-Rs in relapse, cue-induced heroin seeking is associated with a rapid downregulation in synaptic membrane expression of the GluN2B subunit (Van den Oever et al. 2008); however, changes in NMDA-R plasticity in mPFC pyramidal neurons were not observed. Taken together, several lines of evidence indicate that exposure to drugs of abuse results in adaptations in the functioning of synapses in the mPFC; however, more research is necessary to elucidate the long-term effects of drug self-administration and the formation of drug-cue associations on the mPFC neuronal circuitry and its consequences for glutamatergic output to target regions (i.e., NAc) and relapse vulnerability.

21.6 Drug-Induced Neuroplasticity in the Amygdala

The amygdala comprises several nuclei, including the basolateral nucleus (BLA) and central nucleus (CeA). The BLA is a key neuronal substrate that mediates the formation of drug-cue associations and controls stress- and cue-induced reinstatement of drug seeking (Buffalari and See 2010; Fuchs and See 2002; Gabriele and See 2010; Kruzich et al. 2001; McLaughlin and See 2003), whereas the CeA is thought to primarily mediate the expression of drug-seeking responses (Buffalari and See 2010; Kruzich and See 2001; Lu et al. 2005b). Despite this critical role of the BLA and CeA, data regarding the effect of drug exposure on synaptic plasticity in the amygdala is fairly limited.

Supporting a role for the lateral nucleus of the amygdala in the acquisition of drug-cue associations, an AMPA-R-mediated increase in synaptic strength was observed in thalamo-amygdala synapses during stimulus-reward learning for a natural reinforcer (sucrose) (Tye et al. 2008). In line with this observation, noncontingent cocaine treatment enhances glutamatergic transmission and occludes LTP in the lateral nucleus up to 3 days of withdrawal (Goussakov et al. 2006), but this effect is not present after 9 days of withdrawal. A longer lasting potentiation of glutamatergic transmission may occur after cocaine self-administration as short- (1 day) and long-term (30 days) abstinence from cocaine self-administration is associated with increased expression of GluA1, but not GluA2, subunits in the BLA (Lu et al. 2005a), potentially contributing to the presence of GluA2-lacking AMPA-Rs. With respect to NMDA-Rs, an increase in the expression of GluN2A was observed in the BLA after short-term (1 day) withdrawal and a downregulation of GluN2B after 30 days of abstinence (Lu et al. 2005a). In contrast, enhanced expression of GluA2 and GluN1, but not GluA1 nor GluN2a or GluN2B, was found in the CeA 30 days after cessation of contingent cocaine exposure (Lu et al. 2005a).

Following long-term abstinence from noncontingent cocaine, an enhanced cor-ticotrophin-releasing factor (CRF)-induced LTP has been observed in BLA to CeA synapses that is dependent on NMDA receptors and CRF1 receptor function (Fu et al. 2007; Pollandt et al. 2006). CRF-induced LTP after cocaine withdrawal is mediated through endogenous activation of both D1- and D2-like receptors (Krishnan et al. 2010). In support of a functional role for the CRF-induced LTP, cocaine-induced locomotor activity as well as stress-induced relapse is blocked by injection of a CRF antagonist (Erb et al. 1998; Sarnyai et al. 1992; Shaham et al. 1998).

Hence, although the amygdala is clearly implicated in acquisition and consoli-dation of cue-cocaine associations, as well as extinction learning and reinstatement of drug seeking, the functional role of specific subunits of glutamate receptors in the different subdivisions of the amygdala remains to be elucidated.

21.7 Conclusion

In animal models of drug addiction, exposure to drugs of abuse results in changes in the physiological properties of synapses in the mesocorticolimbic DA system. The nature of these adaptations depends on the type of drug, method of administration, brain area examined, and time-point in the development of addiction. In general, changes in synaptic plasticity resulting from drug self-administration appear to be more robust and longer lasting compared with changes induced by noncontingent drug administration. The time course of drug-induced changes in synaptic plasticity differs between the VTA and its target regions. In VTA DA neurons, an increase in synaptic strength is rapidly induced and already apparent after a single drug exposure, whereas potentiation of excitatory plasticity in the NAc and mPFC requires repeated drug administration and is only observed after long-term cessa-tion of drug administration. A recent study shows that the delayed cocaine-induced potentiation of synaptic strength in the NAc requires long-lasting enhancement of AMPA-R function in the VTA (Mameli et al. 2009), suggesting that changes in plasticity of synapses on DA neurons in the VTA may initiate the development of neuroplasticity changes in VTA target regions. Unfortunately, relatively little data is available on acute synaptic plasticity mechanisms that occur during tests for drug seeking and that may drive relapse. Investigating the spatiotemporal aspects of drug-induced plasticity changes that are specific for particular stages in the transition of initial consumption to compulsive drug taking and relapse is important to dissect which adaptations are crucially leading to and define the end-stage addictive behaviors.

Thus far, synaptic plasticity as a result of drug exposure has been predominantly studied in the VTA and NAc. Studies reporting drug-induced neuroadaptations in the mPFC and in amygdala need further efforts to gain better insight in the functional contribution of synaptic plasticity changes in these brain areas to addiction. With respect to the well-established involvement of the mPFC and amygdala in conditioned drug-seeking responses, identifying alterations in synaptic function and further detailing of the neuronal circuitry that contribute to the processing of drug-associated stimuli may yield new targets for pharmacotherapy and other intervention technologies aimed at reducing cue-evoked drug craving and relapse in addiction.

References

Abraham, W. C. (2008). Metaplasticity: Tuning synapses and networks for plasticity. *Nature Reviews Neuroscience, 9,* 387.

Aguilar, M. A., Rodriguez-Arias, M., & Minarro, J. (2009). Neurobiological mechanisms of the reinstatement of drug-conditioned place preference. *Brain Research Reviews, 59,* 253–277.

Anderson, S. M., Bari, A. A., & Pierce, R. C. (2003). Administration of the D1-like dopamine receptor antagonist SCH-23390 into the medial nucleus accumbens shell attenuates cocaine priming-induced reinstatement of drug-seeking behavior in rats. *Psychopharmacology (Berlin), 168,* 132–138.

Anderson, S. M., Famous, K. R., Sadri-Vakili, G., Kumaresan, V., Schmidt, H. D., Bass, C. E., Terwilliger, E. F., Cha, J. H., & Pierce, R. C. (2008). CaMKII: A biochemical bridge linking accumbens dopamine and glutamate systems in cocaine seeking. *Nature Neuroscience, 11,* 344–353.

Anderson, S. M., Schmidt, H. D., & Pierce, R. C. (2006). Administration of the D2 dopamine receptor antagonist sulpiride into the shell, but not the core, of the nucleus accumbens attenuates cocaine priming-induced reinstatement of drug seeking. *Neuropsychopharmacology, 31,* 1452–1461.

Argilli, E., Sibley, D. R., Malenka, R. C., England, P. M., & Bonci, A. (2008). Mechanism and time course of cocaine-induced long-term potentiation in the ventral tegmental area. *The Journal of Neuroscience, 28,* 9092–9100.

Bachtell, R. K., Whisler, K., Karanian, D., & Self, D. W. (2005). Effects of intra-nucleus accumbens shell administration of dopamine agonists and antagonists on cocaine-taking and cocaine-seeking behaviors in the rat. *Psychopharmacology (Berlin), 183,* 41–53.

Baker, D. A., McFarland, K., Lake, R. W., Shen, H., Tang, X. C., Toda, S., & Kalivas, P. W. (2003). Neuroadaptations in cystine-glutamate exchange underlie cocaine relapse. *Nature Neuroscience, 6,* 743–749.

Bardo, M. T., & Bevins, R. A. (2000). Conditioned place preference: What does it add to our preclinical understanding of drug reward? *Psychopharmacology (Berlin), 153,* 31–43.

Bell, K., & Kalivas, P. W. (1996). Context-specific cross-sensitization between systemic cocaine and intra-accumbens AMPA infusion in the rat. *Psychopharmacology (Berlin), 127*, 377–383.

Bellone, C., & Luscher, C. (2006). Cocaine triggered AMPA receptor redistribution is reversed in vivo by mGluR-dependent long-term depression. *Nature Neuroscience, 9*, 636–641.

Ben-Shahar, O., Obara, I., Ary, A. W., Ma, N., Mangiardi, M. A., Medina, R. L., & Szumlinski, K. K. (2009). Extended daily access to cocaine results in distinct alterations in Homer 1b/c and NMDA receptor subunit expression within the medial prefrontal cortex. *Synapse, 63*, 598–609.

Borgland, S. L., Malenka, R. C., & Bonci, A. (2004). Acute and chronic cocaine-induced potentiation of synaptic strength in the ventral tegmental area: Electrophysiological and behavioral correlates in individual rats. *The Journal of Neuroscience, 24*, 7482–7490.

Bossert, J. M., Gray, S. M., Lu, L., & Shaham, Y. (2006). Activation of group II metabotropic glutamate receptors in the nucleus accumbens shell attenuates context-induced relapse to heroin seeking. *Neuropsychopharmacology, 31*, 2197–2209.

Bossert, J. M., Poles, G. C., Wihbey, K. A., Koya, E., & Shaham, Y. (2007). Differential effects of blockade of dopamine D1-family receptors in nucleus accumbens core or shell on reinstatement of heroin seeking induced by contextual and discrete cues. *The Journal of Neuroscience, 27*, 12655–12663.

Bossert, J. M., Stern, A. L., Theberge, F. R., Cifani, C., Koya, E., Hope, B. T., & Shaham, Y. (2011). Ventral medial prefrontal cortex neuronal ensembles mediate context-induced relapse to heroin. *Nature Neuroscience, 14*, 420–422.

Boudreau, A. C., Ferrario, C. R., Glucksman, M. J., & Wolf, M. E. (2009). Signaling pathway adaptations and novel protein kinase A substrates related to behavioral sensitization to cocaine. *Journal of Neurochemistry, 110*, 363–377.

Boudreau, A. C., Reimers, J. M., Milovanovic, M., & Wolf, M. E. (2007). Cell surface AMPA receptors in the rat nucleus accumbens increase during cocaine withdrawal but internalize after cocaine challenge in association with altered activation of mitogen-activated protein kinases. *The Journal of Neuroscience, 27*, 10621–10635.

Boudreau, A. C., & Wolf, M. E. (2005). Behavioral sensitization to cocaine is associated with increased AMPA receptor surface expression in the nucleus accumbens. *The Journal of Neuroscience, 25*, 9144–9151.

Brebner, K., Wong, T. P., Liu, L., Liu, Y., Campsall, P., Gray, S., Phelps, L., Phillips, A. G., & Wang, Y. T. (2005). Nucleus accumbens long-term depression and the expression of behavioral sensitization. *Science, 310*, 1340–1343.

Buffalari, D. M., & See, R. E. (2010). Amygdala mechanisms of Pavlovian psychostimulant conditioning and relapse. *Current Topics in Behavioral Neurosciences, 3*, 73–99.

Chen, B. T., Bowers, M. S., Martin, M., Hopf, F. W., Guillory, A. M., Carelli, R. M., Chou, J. K., & Bonci, A. (2008). Cocaine but not natural reward self-administration nor passive cocaine infusion produces persistent LTP in the VTA. *Neuron, 59*, 288–297.

Childress, A. R., Mozley, P. D., McElgin, W., Fitzgerald, J., Reivich, M., & O'Brien, C. P. (1999). Limbic activation during cue-induced cocaine craving. *The American Journal of Psychiatry, 156*, 11–18.

Churchill, L., Swanson, C. J., Urbina, M., & Kalivas, P. W. (1999). Repeated cocaine alters glutamate receptor subunit levels in the nucleus accumbens and ventral tegmental area of rats that develop behavioral sensitization. *Journal of Neurochemistry, 72*, 2397–2403.

Conrad, K. L., Tseng, K. Y., Uejima, J. L., Reimers, J. M., Heng, L. J., Shaham, Y., Marinelli, M., & Wolf, M. E. (2008). Formation of accumbens GluR2-lacking AMPA receptors mediates incubation of cocaine craving. *Nature, 454*, 118–121.

Cornish, J. L., Duffy, P., & Kalivas, P. W. (1999). A role for nucleus accumbens glutamate transmission in the relapse to cocaine-seeking behavior. *Neuroscience, 93*, 1359–1367.

Cornish, J. L., & Kalivas, P. W. (2000). Glutamate transmission in the nucleus accumbens mediates relapse in cocaine addiction. *The Journal of Neuroscience, 20*, RC89.

Crombag, H. S., Bossert, J. M., Koya, E., & Shaham, Y. (2008). Review context-induced relapse to drug seeking: A review. *Philosophical Transactions of the Royal Society B: Biological Sciences, 363*, 3233–3243.

Di Chiara, G., & North, R. A. (1992). Neurobiology of opiate abuse. *Trends in Pharmacological Sciences, 13*, 185–193.

Di Ciano, P., & Everitt, B. J. (2001). Dissociable effects of antagonism of NMDA and AMPA/KA receptors in the nucleus accumbens core and shell on cocaine-seeking behavior. *Neuropsychopharmacology, 25*, 341–360.

Engblom, D., Bilbao, A., Sanchis-Segura, C., Dahan, L., Perreau-Lenz, S., Balland, B., Parkitna, J. R., Lujan, R., Halbout, B., Mameli, M., Parlato, R., Sprengel, R., Luscher, C., Schutz, G., & Spanagel, R. (2008). Glutamate receptors on dopamine neurons control the persistence of cocaine seeking. *Neuron, 59*, 497–508.

Epstein, D. H., Preston, K. L., Stewart, J., & Shaham, Y. (2006). Toward a model of drug relapse: An assessment of the validity of the reinstatement procedure. *Psychopharmacology (Berlin), 189*, 1–16.

Erb, S., Shaham, Y., & Stewart, J. (1998). The role of corticotropin-releasing factor and corticosterone in stress- and cocaine-induced relapse to cocaine seeking in rats. *The Journal of Neuroscience, 18*, 5529–5536.

Famous, K. R., Kumaresan, V., Sadri-Vakili, G., Schmidt, H. D., Mierke, D. F., Cha, J. H., & Pierce, R. C. (2008). Phosphorylation-dependent trafficking of GluR2-containing AMPA receptors in the nucleus accumbens plays a critical role in the reinstatement of cocaine seeking. *The Journal of Neuroscience, 28*, 11061–11070.

Fitzgerald, L. W., Ortiz, J., Hamedani, A. G., & Nestler, E. J. (1996). Drugs of abuse and stress increase the expression of GluR1 and NMDAR1 glutamate receptor subunits in the rat ventral tegmental area: Common adaptations among cross-sensitizing agents. *The Journal of Neuroscience, 16*, 274–282.

Fu, Y., Pollandt, S., Liu, J., Krishnan, B., Genzer, K., Orozco-Cabal, L., Gallagher, J. P., & Shinnick-Gallagher, P. (2007). Long-term potentiation (LTP) in the central amygdala (CeA) is enhanced after prolonged withdrawal from chronic cocaine and requires CRF1 receptors. *Journal of Neurophysiology, 97*, 937–941.

Fuchs, R. A., Evans, K. A., Ledford, C. C., Parker, M. P., Case, J. M., Mehta, R. H., & See, R. E. (2005). The role of the dorsomedial prefrontal cortex, basolateral amygdala, and dorsal hippocampus in contextual reinstatement of cocaine seeking in rats. *Neuropsychopharmacology, 30*, 296–309.

Fuchs, R. A., Ramirez, D. R., & Bell, G. H. (2008). Nucleus accumbens shell and core involvement in drug context-induced reinstatement of cocaine seeking in rats. *Psychopharmacology (Berlin), 200*, 545–556.

Fuchs, R. A., & See, R. E. (2002). Basolateral amygdala inactivation abolishes conditioned stimulus- and heroin-induced reinstatement of extinguished heroin-seeking behavior in rats. *Psychopharmacology (Berlin), 160*, 425–433.

Gabriele, A., & See, R. E. (2010). Reversible inactivation of the basolateral amygdala, but not the dorsolateral caudate putamen, attenuates consolidation of cocaine-cue associative learning in a reinstatement model of drug-seeking. *European Journal of Neuroscience, 32*, 1024–1029.

Gardner, E. L. (2000). What we have learned about addiction from animal models of drug self-administration. *The American Journal on Addictions, 9*, 285–313.

Geisler, S., & Wise, R. A. (2008). Functional implications of glutamatergic projections to the ventral tegmental area. *Reviews in the Neurosciences, 19*, 227–244.

Gerdjikov, T. V., Ross, G. M., & Beninger, R. J. (2004). Place preference induced by nucleus accumbens amphetamine is impaired by antagonists of ERK or p38 MAP kinases in rats. *Behavioral Neuroscience, 118*, 740–750.

Ghasemzadeh, M. B., Nelson, L. C., Lu, X. Y., & Kalivas, P. W. (1999). Neuroadaptations in ionotropic and metabotropic glutamate receptor mRNA produced by cocaine treatment. *Journal of Neurochemistry, 72*, 157–165.

Goeders, N. E. (2003). The impact of stress on addiction. *European Neuropsychopharmacology, 13*, 435–441.

Goldstein, R. Z., & Volkow, N. D. (2002). Drug addiction and its underlying neurobiological basis: Neuroimaging evidence for the involvement of the frontal cortex. *The American Journal of Psychiatry, 159*, 1642–1652.

Goussakov, I., Chartoff, E. H., Tsvetkov, E., Gerety, L. P., Meloni, E. G., Carlezon, W. A., Jr., & Bolshakov, V. Y. (2006). LTP in the lateral amygdala during cocaine withdrawal. *European Journal of Neuroscience, 23*, 239–250.

Gysling, K., & Wang, R. Y. (1983). Morphine-induced activation of A10 dopamine neurons in the rat. *Brain Research, 277*, 119–127.

Heidbreder, C. A., & Groenewegen, H. J. (2003). The medial prefrontal cortex in the rat: Evidence for a dorso-ventral distinction based upon functional and anatomical characteristics. *Neurosci Biobehav Rev, 27*, 555–579.

Hyman, S. E. (2005). Addiction: A disease of learning and memory. *The American Journal of Psychiatry, 162*, 1414–1422.

Hyman, S. E., Malenka, R. C., & Nestler, E. J. (2006). Neural mechanisms of addiction: The role of reward-related learning and memory. *Annual Review of Neuroscience, 29*, 565–598.

Isaac, J. T., Ashby, M., & McBain, C. J. (2007). The role of the GluR2 subunit in AMPA receptor function and synaptic plasticity. *Neuron, 54*, 859–871.

Kalivas, P. W. (2009). The glutamate homeostasis hypothesis of addiction. *Nature Reviews Neuroscience, 10*, 561–572.

Kalivas, P. W., & Volkow, N. D. (2005). The neural basis of addiction: A pathology of motivation and choice. *The American Journal of Psychiatry, 162*, 1403–1413.

Kalivas, P. W., Volkow, N., & Seamans, J. (2005). Unmanageable motivation in addiction: A pathology in prefrontal-accumbens glutamate transmission. *Neuron, 45*, 647–650.

Kasanetz, F., Deroche-Gamonet, V., Berson, N., Balado, E., Lafourcade, M., Manzoni, O., & Piazza, P. V. (2010). Transition to addiction is associated with a persistent impairment in synaptic plasticity. *Science, 328*, 1709–1712.

Knackstedt, L. A., LaRowe, S., Mardikian, P., Malcolm, R., Upadhyaya, H., Hedden, S., Markou, A., & Kalivas, P. W. (2009a). The role of cystine-glutamate exchange in nicotine dependence in rats and humans. *Biological Psychiatry, 65*, 841–845.

Knackstedt, L. A., Melendez, R. I., & Kalivas, P. W. (2010). Ceftriaxone restores glutamate homeostasis and prevents relapse to cocaine seeking. *Biological Psychiatry, 67*, 81–84.

Kourrich, S., Rothwell, P. E., Klug, J. R., & Thomas, M. J. (2007). Cocaine experience controls bidirectional synaptic plasticity in the nucleus accumbens. *The Journal of Neuroscience, 27*, 7921–7928.

Koya, E., Uejima, J. L., Wihbey, K. A., Bossert, J. M., Hope, B. T., & Shaham, Y. (2009). Role of ventral medial prefrontal cortex in incubation of cocaine craving. *Neuropharmacology, 56* (Suppl 1), 177–185.

Krishnan, B., Centeno, M., Pollandt, S., Fu, Y., Genzer, K., Liu, J., Gallagher, J. P., & Shinnick-Gallagher, P. (2010). Dopamine receptor mechanisms mediate corticotropin-releasing factor-induced long-term potentiation in the rat amygdala following cocaine withdrawal. *European Journal of Neuroscience, 31*, 1027–1042.

Kruzich, P. J., Congleton, K. M., & See, R. E. (2001). Conditioned reinstatement of drug-seeking behavior with a discrete compound stimulus classically conditioned with intravenous cocaine. *Behavioral Neuroscience, 115*, 1086–1092.

Kruzich, P. J., & See, R. E. (2001). Differential contributions of the basolateral and central amygdala in the acquisition and expression of conditioned relapse to cocaine-seeking behavior. *The Journal of Neuroscience, 21*, RC155.

LaLumiere, R. T., & Kalivas, P. W. (2008). Glutamate release in the nucleus accumbens core is necessary for heroin seeking. *The Journal of Neuroscience, 28*, 3170–3177.

LaRowe, S. D., Myrick, H., Hedden, S., Mardikian, P., Saladin, M., McRae, A., Brady, K., Kalivas, P. W., & Malcolm, R. (2007). Is cocaine desire reduced by N-acetylcysteine? *The American Journal of Psychiatry, 164*, 1115–1117.

Lobo, M. K., Covington, H. E., 3rd, Chaudhury, D., Friedman, A. K., Sun, H., Damez-Werno, D., Dietz, D. M., Zaman, S., Koo, J. W., Kennedy, P. J., Mouzon, E., Mogri, M., Neve, R. L., Deisseroth, K., Han, M. H., & Nestler, E. J. (2010). Cell type-specific loss of BDNF signaling mimics optogenetic control of cocaine reward. *Science, 330*, 385–390.

Lu, H., Cheng, P. L., Lim, B. K., Khoshnevisrad, N., & Poo, M. M. (2010). Elevated BDNF after cocaine withdrawal facilitates LTP in medial prefrontal cortex by suppressing GABA inhibition. *Neuron, 67*, 821–833.

Lu, L., Dempsey, J., Shaham, Y., & Hope, B. T. (2005a). Differential long-term neuroadaptations of glutamate receptors in the basolateral and central amygdala after withdrawal from cocaine self-administration in rats. *Journal of Neurochemistry, 94*, 161–168.

Lu, L., Grimm, J. W., Shaham, Y., & Hope, B. T. (2003). Molecular neuroadaptations in the accumbens and ventral tegmental area during the first 90 days of forced abstinence from cocaine self-administration in rats. *Journal of Neurochemistry, 85*, 1604–1613.

Lu, L., Hope, B. T., Dempsey, J., Liu, S. Y., Bossert, J. M., & Shaham, Y. (2005b). Central amygdala ERK signaling pathway is critical to incubation of cocaine craving. *Nature Neuroscience, 8*, 212–219.

Mameli, M., Balland, B., Lujan, R., & Luscher, C. (2007). Rapid synthesis and synaptic insertion of GluR2 for mGluR-LTD in the ventral tegmental area. *Science, 317*, 530–533.

Mameli, M., Halbout, B., Creton, C., Engblom, D., Parkitna, J. R., Spanagel, R., & Luscher, C. (2009). Cocaine-evoked synaptic plasticity: Persistence in the VTA triggers adaptations in the NAc. *Nature Neuroscience, 12*, 1036–1041.

Mardikian, P. N., LaRowe, S. D., Hedden, S., Kalivas, P. W., & Malcolm, R. J. (2007). An open-label trial of N-acetylcysteine for the treatment of cocaine dependence: A pilot study. *Progress in Neuro-Psychopharmacology & Biological Psychiatry, 31*, 389–394.

Martin, M., Chen, B. T., Hopf, F. W., Bowers, M. S., & Bonci, A. (2006). Cocaine self-administration selectively abolishes LTD in the core of the nucleus accumbens. *Nature Neuroscience, 9*, 868–869.

McFarland, K., Davidge, S. B., Lapish, C. C., & Kalivas, P. W. (2004). Limbic and motor circuitry underlying footshock-induced reinstatement of cocaine-seeking behavior. *The Journal of Neuroscience, 24*, 1551–1560.

McFarland, K., & Kalivas, P. W. (2001). The circuitry mediating cocaine-induced reinstatement of drug-seeking behavior. *The Journal of Neuroscience, 21*, 8655–8663.

McFarland, K., Lapish, C. C., & Kalivas, P. W. (2003). Prefrontal glutamate release into the core of the nucleus accumbens mediates cocaine-induced reinstatement of drug-seeking behavior. *The Journal of Neuroscience, 23*, 3531–3537.

McLaughlin, J., & See, R. E. (2003). Selective inactivation of the dorsomedial prefrontal cortex and the basolateral amygdala attenuates conditioned-cued reinstatement of extinguished cocaine-seeking behavior in rats. *Psychopharmacology (Berlin), 168*, 57–65.

Meredith, G. E., Baldo, B. A., Andrezjewski, M. E., & Kelley, A. E. (2008). The structural basis for mapping behavior onto the ventral striatum and its subdivisions. *Brain Structure & Function, 213*, 17–27.

Mereu, G., Yoon, K. W., Boi, V., Gessa, G. L., Naes, L., & Westfall, T. C. (1987). Preferential stimulation of ventral tegmental area dopaminergic neurons by nicotine. *European Journal of Pharmacology, 141*, 395–399.

Miguens, M., Del Olmo, N., Higuera-Matas, A., Torres, I., Garcia-Lecumberri, C., & Ambrosio, E. (2008). Glutamate and aspartate levels in the nucleus accumbens during cocaine self-administration and extinction: A time course microdialysis study. *Psychopharmacology (Berlin), 196*, 303–313.

Millan, E. Z., Marchant, N. J., & McNally, G. P. (2011). Extinction of drug seeking. *Behavioural Brain Research, 217*, 454–462.

Miller, C. A., & Marshall, J. F. (2005). Molecular substrates for retrieval and reconsolidation of cocaine-associated contextual memory. *Neuron, 47*, 873–884.

Moran, M. M., McFarland, K., Melendez, R. I., Kalivas, P. W., & Seamans, J. K. (2005). Cystine/glutamate exchange regulates metabotropic glutamate receptor presynaptic inhibition of excitatory transmission and vulnerability to cocaine seeking. *The Journal of Neuroscience, 25*, 6389–6393.

Moussawi, K., Pacchioni, A., Moran, M., Olive, M. F., Gass, J. T., Lavin, A., & Kalivas, P. W. (2009). N-Acetylcysteine reverses cocaine-induced metaplasticity. *Nature Neuroscience, 12*, 182–189.

Moussawi, K., Zhou, W., Shen, H., Reichel, C. M., See, R. E., Carr, D. B., & Kalivas, P. W. (2011). Reversing cocaine-induced synaptic potentiation provides enduring protection from relapse. *Proceedings of the National Academy of Sciences of the United States of America, 108*, 385–390.

O'Brien, C. P., Ehrman, R. N., & Ternes, J. W. (1986). *Classical conditioning in human opioid dependence*. Academic Press, Orlando.

Park, W. K., Jey, A. R., Anderson, S. M., Spealman, R. D., Rowlett, J. K., & Pierce, R. C. (2002). Cocaine administered into the medial prefrontal cortex reinstates cocaine-seeking behavior by increasing AMPA receptor-mediated glutamate transmission in the nucleus accumbens. *The Journal of Neuroscience, 22*, 2916–2925.

Peters, J., & Kalivas, P. W. (2006). The group II metabotropic glutamate receptor agonist, LY379268, inhibits both cocaine- and food-seeking behavior in rats. *Psychopharmacology (Berlin), 186*, 143–149.

Peters, J., Kalivas, P. W., & Quirk, G. J. (2009). Extinction circuits for fear and addiction overlap in prefrontal cortex. *Learning & Memory, 16*, 279–288.

Peters, J., LaLumiere, R. T., & Kalivas, P. W. (2008). Infralimbic prefrontal cortex is responsible for inhibiting cocaine seeking in extinguished rats. *The Journal of Neuroscience, 28*, 6046–6053.

Pierce, R. C., Bell, K., Duffy, P., & Kalivas, P. W. (1996). Repeated cocaine augments excitatory amino acid transmission in the nucleus accumbens only in rats having developed behavioral sensitization. *The Journal of Neuroscience, 16*, 1550–1560.

Ping, A., Xi, J., Prasad, B. M., Wang, M. H., & Kruzich, P. J. (2008). Contributions of nucleus accumbens core and shell GluR1 containing AMPA receptors in AMPA- and cocaine-primed reinstatement of cocaine-seeking behavior. *Brain Research, 1215*, 173–182.

Pollandt, S., Liu, J., Orozco-Cabal, L., Grigoriadis, D. E., Vale, W. W., Gallagher, J. P., & Shinnick-Gallagher, P. (2006). Cocaine withdrawal enhances long-term potentiation induced by corticotropin-releasing factor at central amygdala glutamatergic synapses via CRF, NMDA receptors and PKA. *European Journal of Neuroscience, 24*, 1733–1743.

Robinson, T. E., & Berridge, K. C. (1993). The neural basis of drug craving: An incentive-sensitization theory of addiction. *Brain Research: Brain Research Reviews, 18*, 247–291.

Rogers, J. L., Ghee, S., & See, R. E. (2008). The neural circuitry underlying reinstatement of heroin-seeking behavior in an animal model of relapse. *Neuroscience, 151*, 579–588.

Saal, D., Dong, Y., Bonci, A., & Malenka, R. C. (2003). Drugs of abuse and stress trigger a common synaptic adaptation in dopamine neurons. *Neuron, 37*, 577–582.

Sarnyai, Z., Hohn, J., Szabo, G., & Penke, B. (1992). Critical role of endogenous corticotropin-releasing factor (CRF) in the mediation of the behavioral action of cocaine in rats. *Life Sciences, 51*, 2019–2024.

Schenk, S., Horger, B. A., Peltier, R., & Shelton, K. (1991). Supersensitivity to the reinforcing effects of cocaine following 6-hydroxydopamine lesions to the medial prefrontal cortex in rats. *Brain Research, 543*, 227–235.

Schmidt, H. D., & Pierce, R. C. (2006). Cooperative activation of D1-like and D2-like dopamine receptors in the nucleus accumbens shell is required for the reinstatement of cocaine-seeking behavior in the rat. *Neuroscience, 142*, 451–461.

Schramm-Sapyta, N. L., Olsen, C. M., & Winder, D. G. (2006). Cocaine self-administration reduces excitatory responses in the mouse nucleus accumbens shell. *Neuropsychopharmacology, 31*, 1444–1451.

Schultz, W. (1998). Predictive reward signal of dopamine neurons. *Journal of Neurophysiology, 80*, 1–27.

Schultz, W. (2007). Multiple dopamine functions at different time courses. *Annual Review of Neuroscience, 30*, 259–288.

Sesack, S. R., Deutch, A. Y., Roth, R. H., & Bunney, B. S. (1989). Topographical organization of the efferent projections of the medial prefrontal cortex in the rat: An anterograde tract-tracing study with Phaseolus vulgaris leucoagglutinin. *J Comp Neurol, 290*, 213–242.

Sesack, S. R., Carr, D. B., Omelchenko, N., & Pinto, A. (2003). Anatomical substrates for glutamate-dopamine interactions: Evidence for specificity of connections and extrasynaptic actions. *Annals of the New York Academy of Sciences, 1003*, 36–52.

Shaham, Y., Erb, S., Leung, S., Buczek, Y., & Stewart, J. (1998). CP-154,526, a selective, non-peptide antagonist of the corticotropin-releasing factor1 receptor attenuates stress-induced relapse to drug seeking in cocaine- and heroin-trained rats. *Psychopharmacology (Berlin), 137*, 184–190.

Stewart, J., & Badiani, A. (1993). Tolerance and sensitization to the behavioral effects of drugs. *Behavioural Pharmacology, 4*, 289–312.

Stuber, G. D., Klanker, M., de Ridder, B., Bowers, M. S., Joosten, R. N., Feenstra, M. G., & Bonci, A. (2008). Reward-predictive cues enhance excitatory synaptic strength onto midbrain dopamine neurons. *Science, 321*, 1690–1692.

Sutton, M. A., Schmidt, E. F., Choi, K. H., Schad, C. A., Whisler, K., Simmons, D., Karanian, D. A., Monteggia, L. M., Neve, R. L., & Self, D. W. (2003). Extinction-induced upregulation in AMPA receptors reduces cocaine-seeking behaviour. *Nature, 421*, 70–75.

Tang, W., Wesley, M., Freeman, W. M., Liang, B., & Hemby, S. E. (2004). Alterations in ionotropic glutamate receptor subunits during binge cocaine self-administration and withdrawal in rats. *Journal of Neurochemistry, 89*, 1021–1033.

Thomas, M. J., Beurrier, C., Bonci, A., & Malenka, R. C. (2001). Long-term depression in the nucleus accumbens: A neural correlate of behavioral sensitization to cocaine. *Nature Neuroscience, 4*, 1217–1223.

Tsai, H. C., Zhang, F., Adamantidis, A., Stuber, G. D., Bonci, A., de Lecea, L., & Deisseroth, K. (2009). Phasic firing in dopaminergic neurons is sufficient for behavioral conditioning. *Science, 324*, 1080–1084.

Tye, K. M., Stuber, G. D., de Ridder, B., Bonci, A., & Janak, P. H. (2008). Rapid strengthening of thalamo-amygdala synapses mediates cue-reward learning. *Nature, 453*, 1253–1257.

Tzschentke, T. M. (2001). Pharmacology and behavioral pharmacology of the mesocortical dopamine system. *Progress in Neurobiology, 63*, 241–320.

Ungless, M. A., Whistler, J. L., Malenka, R. C., & Bonci, A. (2001). Single cocaine exposure in vivo induces long-term potentiation in dopamine neurons. *Nature, 411*, 583–587.

Van den Oever, M. C., Goriounova, N. A., Li, K. W., Van der Schors, R. C., Binnekade, R., Schoffelmeer, A. N., Mansvelder, H. D., Smit, A. B., Spijker, S., & De Vries, T. J. (2008). Prefrontal cortex AMPA receptor plasticity is crucial for cue-induced relapse to heroin-seeking. *Nature Neuroscience, 11*, 1053–1058.

Van den Oever, M. C., Lubbers, B. R., Goriounova, N. A., Li, K. W., Van der Schors, R. C., Loos, M., Riga, D., Wiskerke, J., Binnekade, R., Stegeman, M., Schoffelmeer, A. N., Mansvelder, H. D., Smit, A. B., De Vries, T. J., & Spijker, S. (2010a). Extracellular matrix plasticity and GABAergic inhibition of prefrontal cortex pyramidal cells facilitates relapse to heroin seeking. *Neuropsychopharmacology, 35*, 2120–2133.

Van den Oever, M. C., Spijker, S., Smit, A. B., & De Vries, T. J. (2010b). Prefrontal cortex plasticity mechanisms in drug seeking and relapse. *Neuroscience and Biobehavioral Reviews, 35*, 276–284.

Van der Staay, F. J., Arndt, S. S., & Nordquist, R. E. (2009). Evaluation of animal models of neurobehavioral disorders. *Behavioral and Brain Functions, 5*, 11.

Vanderschuren, L. J., Tjon, G. H., Nestby, P., Mulder, A. H., Schoffelmeer, A. N., & De Vries, T. J. (1997). Morphine-induced long-term sensitization to the locomotor effects of morphine and amphetamine depends on the temporal pattern of the pretreatment regimen. *Psychopharmacology (Berlin), 131*, 115–122.

Vezina, P. (2004). Sensitization of midbrain dopamine neuron reactivity and the self-administration of psychomotor stimulant drugs. *Neuroscience and Biobehavioral Reviews, 27*, 827–839.

Voorn, P., Vanderschuren, L. J., Groenewegen, H. J., Robbins, T. W., & Pennartz, C. M. (2004). Putting a spin on the dorsal-ventral divide of the striatum. *Trends in Neurosciences, 27*, 468–474.

Weissenborn, R., Robbins, T. W., & Everitt, B. J. (1997). Effects of medial prefrontal or anterior cingulate cortex lesions on responding for cocaine under fixed-ratio and second-order schedules of reinforcement in rats. *Psychopharmacology (Berlin), 134*, 242–257.

White, F. J. (1996). Synaptic regulation of mesocorticolimbic dopamine neurons. *Annual Review of Neuroscience, 19*, 405–436.

White, F. J., Hu, X. T., Zhang, X. F., & Wolf, M. E. (1995). Repeated administration of cocaine or amphetamine alters neuronal responses to glutamate in the mesoaccumbens dopamine system. *Journal of Pharmacology and Experimental Therapeutics, 273*, 445–454.

Williams, J. M., & Galli, A. (2006). The dopamine transporter: A vigilant border control for psychostimulant action. *Handbook of Experimental Pharmacology, 175*, 215–232.

Witten, I. B., Lin, S. C., Brodsky, M., Prakash, R., Diester, I., Anikeeva, P., Gradinaru, V., Ramakrishnan, C., & Deisseroth, K. (2010). Cholinergic interneurons control local circuit activity and cocaine conditioning. *Science, 330*, 1677–1681.

Xi, Z. X., Gilbert, J., Campos, A. C., Kline, N., Ashby, C. R., Jr., Hagan, J. J., Heidbreder, C. A., & Gardner, E. L. (2004). Blockade of mesolimbic dopamine D3 receptors inhibits stress-induced reinstatement of cocaine-seeking in rats. *Psychopharmacology (Berlin), 176*, 57–65.

Zahm, D. S. (1999). Functional-anatomical implications of the nucleus accumbens core and shell subterritories. *Annals of the New York Academy of Sciences, 877*, 113–128.

Zhou, W., & Kalivas, P. W. (2008). N-acetylcysteine reduces extinction responding and induces enduring reductions in cue- and heroin-induced drug-seeking. *Biological Psychiatry, 63*, 338–340.

Zhu, J. J., Qin, Y., Zhao, M., Van Aelst, L., & Malinow, R. (2002). Ras and Rap control AMPA receptor trafficking during synaptic plasticity. *Cell, 110*, 443–455.

Zweifel, L. S., Argilli, E., Bonci, A., & Palmiter, R. D. (2008). Role of NMDA receptors in dopamine neurons for plasticity and addictive behaviors. *Neuron, 59*, 486–496.

Chapter 22
Synaptic Dysfunction in Schizophrenia

Dong-Min Yin, Yong-Jun Chen, Anupama Sathyamurthy,
Wen-Cheng Xiong, and Lin Mei

Abstract Schizophrenia alters basic brain processes of perception, emotion, and judgment to cause hallucinations, delusions, thought disorder, and cognitive deficits. Unlike neurodegeneration diseases that have irreversible neuronal degeneration and death, schizophrenia lacks agreeable pathological hallmarks, which makes it one of the least understood psychiatric disorders. With identification of schizophrenia susceptibility genes, recent studies have begun to shed light on underlying pathological mechanisms. Schizophrenia is believed to result from problems during neural development that lead to improper function of synaptic transmission and plasticity, and in agreement, many of the susceptibility genes encode proteins critical for neural development. Some, however, are also expressed at high levels in adult brain. Here, we will review evidence for altered neurotransmission at glutamatergic, GABAergic, dopaminergic, and cholinergic synapses in schizophrenia and discuss roles of susceptibility genes in neural development as well as in synaptic plasticity and how their malfunction may contribute to pathogenic mechanisms of schizophrenia. We propose that mouse models with precise temporal and spatial control of mutation or overexpression would be useful to delineate schizophrenia pathogenic mechanisms.

Keywords Excitatory synaptic transmission • Inhibitory synaptic transmission • Neuromodulators • Schizophrenia • Schizophrenia susceptibility genes

D.-M. Yin • Y.-J. Chen • A. Sathyamurthy •
W.-C. Xiong • L. Mei (✉)
Department of Neurology, Institute of Molecular Medicine and Genetics, Georgia Health
Sciences University, 30912 Augusta, GA, USA
e-mail: lmei@georgiahealth.edu

M.R. Kreutz and C. Sala (eds.), *Synaptic Plasticity*,
Advances in Experimental Medicine and Biology 970,
DOI 10.1007/978-3-7091-0932-8_22, © Springer-Verlag/Wien 2012

22.1 Introduction

Schizophrenia alters basic brain processes of perception, emotion, and judgment to cause hallucinations, delusions, thought disorder, and cognitive deficits. It is a mental disorder that affects 0.5–1% of the population worldwide with devastating consequences for affected individuals and their families and is the seventh most costly illness in the USA. Unlike neurodegenerative diseases such as Alzheimer's disease, Parkinson's disease, and amyotrophic lateral sclerosis (ALS) that have irreversible neuronal degeneration and death, nerve cells in schizophrenia generally do not degenerate or die. Because of the lack of pathological hallmarks, schizophrenia remains to be one of the least understood psychiatric disorders. With identification of schizophrenia susceptibility genes, recent studies have begun to shed light on underlying pathological mechanisms. All brain functions depend on the function of synapses, connections between neurons. It is now widely believed that schizophrenia results from problems during neural development that lead to improper function of synaptic transmission and plasticity (Eastwood 2004; McCullumsmith et al. 2004; Mirnics et al. 2001; Nikolaus et al. 2009; Stephan et al. 2006). Intriguingly, many of the schizophrenia susceptibility genes encode proteins that have been implicated in synapse formation and/or function. This chapter focuses on the relationship between synaptic transmission and schizophrenia. We will first review evidence for altered neurotransmission at glutamatergic, GABAergic, dopaminergic, and cholinergic synapses in schizophrenia and discuss the roles of susceptibility genes in neural development and synaptic plasticity and how their malfunction may contribute to the pathogenic mechanisms of schizophrenia.

22.2 Altered Synaptic Transmission in Schizophrenia

22.2.1 The Glutamatergic Pathway

The interest in alterations of glutamatergic neurotransmission as potential pathological mechanisms in schizophrenia was raised when phencyclidine (PCP) was found to reduce noncompetitively excitation of neurons by NMDA (Anis et al. 1983). Earlier, PCP had been shown to produce transient psychotic symptoms in healthy individuals including thought disorder, blunted affect, and cognitive impairments that resemble those in schizophrenic patients (Fauman et al. 1976; Luby et al. 1959). Ketamine, a PCP derivative and a dissociative anesthetic drug, was also able to generate in healthy individuals transient schizophrenia-like (positive and negative) symptoms and impair cognitive functions that depend on the prefrontal cortex (PFC) (Adler et al. 1999; Krystal et al. 1994; Lahti et al. 2001; Malhotra et al. 1997). In schizophrenic patients, ketamine exacerbates preexisting symptoms (Lahti et al. 1995; Malhotra et al. 1997). Taken together, these results suggest a role of reduced glutamatergic function in schizophrenic pathology.

In agreement with this hypothesis were findings that glutamate levels, which inversely correlate with the severity of positive symptoms (Faustman et al. 1999), are significantly lower in the cerebrospinal fluid (CSF) and in brain tissues of schizophrenic patients (Kim et al. 1980; Tsai et al. 1995). Glutamate release from synaptosomes prepared from frozen brain samples of schizophrenics was reduced in response to NMDA or kainic acid (Sherman et al. 1991b). In addition, postmortem analysis shows reduced mRNA and enzymatic activity of glutamate carboxypeptidase II (GCP II), the enzyme that degrades the neuropeptide N-acetylaspartylglutamate (NAAG), which is a reversible antagonist of NMDA receptors (Hakak et al. 2001; Tsai et al. 1995). It is controversial whether levels of NMDA or AMPA receptors are reduced in schizophrenics. Increased mRNA levels were reported in some studies (Akbarian et al. 1996; Dracheva et al. 2001; Kristiansen et al. 2006) while other studies showed a decrease (Akbarian et al. 1995, 1996; Dracheva et al. 2001; Kristiansen et al. 2006; Mirnics et al. 2000). Morphologically, dendritic length and dendritic spine density are reduced in the cerebral cortex of schizophrenic patients (Garey et al. 1998; Glantz and Lewis 2000) although the density of pyramidal neurons was shown to be increased in the dorsal lateral PFC (DLPFC) in schizophrenics (Selemon and Goldman-Rakic 1999).

Adult rodents, when treated with NMDA antagonists, become hyperactive (Nabeshima et al. 1983; Sturgeon et al. 1979) and are impaired in prepulse inhibition (Bakshi and Geyer 1995; Bakshi et al. 1994), a behavioral deficit thought to model psychotic symptoms. They are also deficient in social interactions, a negative symptom (Sams-Dodd 1995, 1996) and cognition functions such as working memory (Jentsch et al. 1997). Mutant mice which expressed 5% of normal level of NR1 showed behavioral deficits relevant to schizophrenia including hyperactivity, impaired social interaction, and cognitive dysfunction, which can be ameliorated by antipsychotic treatments (Mohn et al. 1999).

Glutamatergic synapses are present on projection cells as well as interneurons. Both could be the target of "glutamatergic hypofunction." Interestingly, in acutely prepared hippocampal slices, GABAergic interneurons were tenfold more sensitive to NMDA receptor inhibitors than were pyramidal neurons (Grunze et al. 1996). Therefore, GABAergic interneurons should be more vulnerable than pyramidal cells to glutamatergic hypofunction. Hypoactivity of GABAergic neurons would result in impaired inhibition of projection cells and thus cognitive deficits. When the essential subunit of NMDA receptor NR1 was selectively eliminated in parvalbumin (PV)-positive interneurons, mutant mice are impaired in spatial working memory, but their spatial open field exploratory activity and their social activity are normal (Korotkova et al. 2010). Interestingly, when NR1 is ablated in about 50% of cortical interneurons during postnatal development, mutant mice exhibit novelty-induced hyperlocomotion and are impaired in mating and nest building (Belforte et al. 2010). These observations suggest that NMDA receptors in different types of interneurons could have distinct functions. Metabotropic glutamate receptors have also been implicated in schizophrenia. Pretreatment with LY354740, a selective agonist for metabotropicglutamate 2/3 (mGlu2/3) receptors, attenuated the disruptive effects of PCP on locomotion, stereotypy, working memory, and cortical glutamate efflux (Moghaddam and Adams 1998). These results suggest that mGlu2/3 receptor agonists have antipsychotic properties and may provide a new alternative for the treatment of schizophrenia.

22.2.2 The GABAergic Pathway

Dysfunctions of GABA transmission have also been implicated in the processes leading to psychosis (Keverne 1999; Lacroix et al. 2000). Psychotic symptoms in schizophrenia have been found to be correlated with reduced GABAergic inhibition in the medial temporal region (Busatto et al. 1997). GABAergic interneurons, representing about 20–30% of neocortical neurons, are a population that is extremely heterogeneous, varying in morphology, expression of markers, laminar distribution, and electrophysiological properties (Ascoli et al. 2008; Markram et al. 2004). Embedded in the network of principal cells, they innervate different domains of these neurons. For example, basket cells target the somata and proximal dendrites, chandelier cells form axoaxonic synapses on the axon initial segments. Somatostatin (SOM)-positive or Martinotti interneurons innervate distal dendrites and presumably regulate other inputs of principle cells. Thus, it is generally believed that GABAergic interneurons play a critical role in controlling cell excitability, spike timing, synchrony, and oscillatory activity in the mammalian central nervous system (McBain and Kauer 2009). Albeit fewer in number than principal cells, a single GABAergic neuron can innervate multiple principle cells and thus could potentially alter the activity of thousands of downstream neurons.

In situ hybridization studies demonstrated overall reduced levels of the 67-kDa isoform of glutamic acid decarboxylase (GAD67), the primary enzyme of GABA synthesis, in the PFC area 9 of the left hemisphere of schizophrenic brains (Akbarian et al. 1995). Similar results were obtained in a better controlled study of PFC area 9 of the right hemisphere (Volk et al. 2000). The reduction in GAD67 expression may not be due to antipsychotic medications because long-term treatment with haloperidol did not affect GAD67 mRNA expression in the PFC of monkeys (Volk et al. 2000). Moreover, the activity of GAD was significantly reduced in nucleus accumbens, amygdala, hippocampus, and putamen from schizophrenic postmortem brains (Bird et al. 1977). In agreement, GABA release from synaptosomes of schizophrenic brains was decreased (Sherman et al. 1991a, b). These results suggest that decreased GAD67 mRNA expression in the association regions of the neocortex may be a frequent feature of schizophrenia. Moreover, the binding of [^{3}H]nipecotic acid, a ligand for labeling GABA uptake sites, was reported to be reduced in schizophrenic brains (Reynolds et al. 1990; Simpson et al. 1989). In addition, also the mRNA and protein levels of GAT1 (GABA membrane transporter 1), a protein responsible for reuptake of released GABA into nerve terminals, are reduced in the DLPFC of subjects with schizophrenia (Lewis et al. 1999; Volk et al. 2001).

Early studies reported a loss of small neurons in cortical layer II (Benes et al. 1991). However, subsequent studies failed to see a significant reduction of GAD67-

positive neurons (Akbarian et al. 1995; Volk et al. 2000). Similarly, parvalbumin (PV)-positive interneurons were found to be reduced (Beasley and Reynolds 1997) or unchanged (Woo et al. 1997) in DLPFC in schizophrenia. Nevertheless, evidence appeared to be compelling that GABAergic function is reduced in the DLPFC of schizophrenic patients. Maybe as a compensatory mechanism, expression of $GABA_A$ receptor in superficial layers of the cortex of schizophrenic brains was increased (Benes et al. 1992; Hanada et al. 1987).

Intriguingly, GABAergic alternation in schizophrenia appears to be interneuron type specific. GAD67 expression is normal in 70% of GABAergic interneurons in the DLPFC but reduced or undetectable in the remaining 30% GABAergic neurons (Akbarian et al. 1995; Volk et al. 2000). The affected interneurons express PV, whereas those expressing calretinin appeared to be normal (Hashimoto et al. 2003). PV-positive neurons include basket cells that form perisomatic synapses onto pyramidal neurons and chandelier cells that form characteristic linear arrays of terminals (termed cartridges) on the axon initial segments of pyramidal neurons. GAT1 levels appear to be selectively reduced in chandelier axon cartridges in the DLFC of schizophrenic patients (Woo et al. 1998). On the other hand, $GABA_A$ receptors are upregulated on the postsynaptic membranes facing the axon initial segments, probably to compensate deficient GABAergic transmission (Volk et al. 2002).

Reduced GABA signaling from chandelier cells to pyramidal neurons could contribute to the pathophysiology of working memory dysfunction. Networks of PV-positive GABA neurons, formed by both chemical and electrical synapses, give rise to oscillatory activity in the gamma band range, the synchronized firing of a neuronal population at 30–80 Hz (Whittington et al. 2011). Thus, decreased inhibitory GABA transmission in schizophrenic patients might contribute to psychotic symptoms in schizophrenia. Consistent with this hypothesis, disinhibition of the ventral hippocampus by the $GABA_A$ antagonist picrotoxin would result in similar psychosis-related behavioral disturbances such as hyperactivity and decreased PPI (Bast et al. 2001).

22.2.3 The Cholinergic Pathway

The association of cholinergic pathways with schizophrenia was as ancient as the illness was diagnosed. Schizophrenic patients are often heavy smokers (Lohr and Flynn 1992), and acetylcholine-induced convulsion and atropine-induced coma were used to treat schizophrenia (Forrer and Miller 1958). Substantial evidence has accumulated over the years that suggests the involvement of dysfunction, mostly hypofunction, of cholinergic transmission in schizophrenia (Neubauer et al. 1975; Tandon et al. 1989). Acetylcholine modulates transmission of various neurotransmitters including glutamate, GABA, dopamine, and serotonin. Postmortem studies of brains of schizophrenic patients were ambiguous about protein levels and activity of choline acetyltransferase (ChAT), the enzyme crucially involved in the synthesis of acetylcholine, and AChE, the enzyme that degrades acetylcholine. Protein or activity levels were reported as increased, decreased, or unchanged.

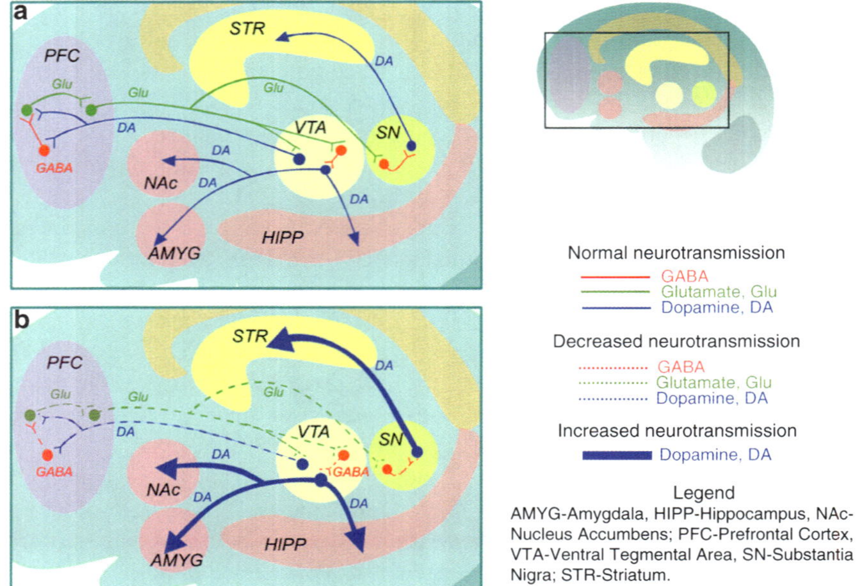

Fig. 22.1 Neurotransmitter pathways in schizophrenia

A more recent study suggested decreased levels of ChAT mRNA and a decreased number of ChAT-positive cells in striatum, particularly in the ventral striatum (Holt et al. 1999, 2005).

Acetylcholine acts by stimulating two types of receptors in the brain: nicotinic and muscarinic receptors. For neuronal nicotinic receptors, there are nine α and three β subunits; the predominant subtypes are the homomeric α7 and heteromeric α4 β2 subtypes (Paterson and Nordberg 2000). There are five types of muscarinic receptors (M1–5), each encoded by an individual gene. A region of chromosome 15, 15q13-14, that contains the α7 AChR subunit gene has been associated with schizophrenia, and SNPs have been described in the promoter region of the α7 subunit gene (Freedman et al. 1997). Studies using postmortem tissue suggest a decreased density of the α7 nicotinic subtype in the brains of schizophrenics (Freedman et al. 1995; Kucinski et al. 2010; Marutle et al. 2001). However, α7 AChR null mutant mice are normal in prepulse inhibition, water maze test, and fear conditioning except for increased anxiety in the open field test (Paylor et al. 1998). Animal studies demonstrate that α7-specific agonists can ameliorate positive and negative symptoms, improve learning and memory (water maze and Y maze), and attentional deficits (auditory gating) (Thomsen et al. 2010; Tregellas et al. 2011). In patients with schizophrenia, α7 agonists appeared to have procognitive effects (Thomsen et al. 2010). These observations suggest that this receptor subtype may be responsible for the inheritance of a pathophysiological aspect of the illness.

As mentioned above, many schizophrenic patients are extremely heavy nicotine users, even in comparison with other psychiatric patients (de Leon et al. 1995;

Hamera et al. 1995). α7 subunit mRNA and protein levels are lower in schizophrenic nonsmokers compared to control nonsmokers and are brought to control levels in schizophrenic smokers (Mexal et al. 2010). Intriguingly, several types of sensory processing deficits, including auditory sensory processing and eye-tracking abnormalities, could be normalized by nicotine, delivered as gum, or by smoking (Adler et al. 1993; Olincy et al. 1998). These observations suggest that schizophrenic patients may smoke to self-medicate endogenous behavioral deficits (Goff et al. 1992).

Initial investigations with quinuclidinyl benzilate (QNB), an antagonist that binds to all five subtypes of muscarinic receptors, were inconsistent on levels of muscarinic receptors in brains of schizophrenic patients. Ligand-binding studies with pirenzepine, an M1-specific antagonist, revealed consistently decreased levels in the DLPFC tissues from subjects with schizophrenia (Scarr et al. 2009). A reduction of pirenzepine binding may be schizophrenia-specific because it was not observed in patients with bipolar disorder or major depression (Zavitsanou et al. 2004). In primates, M1 muscarinic receptors are located postsynaptically in noncholinergic asymmetric and cholinergic symmetric synapses in cortical layers III and V/VI (Mrzljak et al. 1993). They may modulate the cholinergic input from the basal forebrain and intrinsic cortical cholinergic activity (Zhang et al. 2006). M1 mutant mice were normal in hippocampal learning and memory (Miyakawa et al. 2001; Shinoe et al. 2005) but were impaired in behavioral tasks requiring interactions between the hippocampus and cortex (Anagnostaras et al. 2003).

22.2.4 The Dopaminergic Pathway

The original dopamine hypothesis of schizophrenia, proposed over 40 years ago, associates hyperactivity of dopamine transmission with schizophrenia. It was based on effective antipsychotic drugs that appear to act by blocking dopamine D2 receptors and their antipsychotic potency as usually positively correlated with their D2 antagonistic activity (van Rossum 1966). Drugs which inhibit the reuptake of dopamine such as amphetamine can induce schizophrenia-like psychosis in nonpsychotic subjects (Angrist and Gershon 1970; Bell 1973; Gardner and Connell 1972) and exacerbate psychotic symptoms in schizophrenic patients (Laruelle et al. 1999; Lieberman et al. 1987). It was then believed that schizophrenia is associated with hyperactivity of subcortical mesolimbic D2 pathways in the brain. In support of this notion, positron emission tomography studies indicate that schizophrenia is associated with elevated amphetamine-induced synaptic dopamine concentrations (Breier et al. 1997; Laruelle et al. 1996). Striatal dopamine overactivity was observed in patients with "at risk mental states" (ARMS) that might eventually lead to the outbreak of psychosis (Howes et al. 2006).

D2-dependent antipsychotics are effective for positive symptoms but not negative symptoms and cognitive deficits in schizophrenic patients. These functions are mainly controlled by the neocortex where the density of D2 receptors is several times lower than that of D1 receptors (De Keyser et al. 1988; Hall et al. 1994). D1 receptor–mediated signaling regulates the critical patterns of sustained neuronal firing in the DLPFC during working memory tasks (Sawaguchi 2001; Williams and Goldman-Rakic 1995) and has been shown to be critical for cognitive functions subserved by the DLPFC, such as executive cognition and working memory (Sawaguchi and Goldman-Rakic 1991, 1994). Recent postmortem and imaging studies have suggested that the mesocortical dopaminergic projection to the PFC may be hypoactive (Toda and Abi-Dargham 2007). Dopaminergic axons from mesocortical regions were reduced in the DLPFC of schizophrenic patients (Akil et al. 1999). Probably to compensate for the reduced dopaminergic input, D1 receptor binding in the DLPFC was increased in in vivo imaging studies of drug-free and drug-naive schizophrenia subjects (Abi-Dargham et al. 2002). In some case, the D1 receptor binding was decreased in schizophrenic patients (Okubo et al. 1997). In summary, the D1 upregulation does not actually contribute to the impairment of working memory as D1 receptor antagonist worsen cognitive deficits in schizophrenia (Abi-Dargham and Moore 2003).

22.3 Functions of Schizophrenia Susceptibility Genes in Synapse Formation and Transmission

Many of the schizophrenia susceptibility genes have been implicated in neural development. In addition, recent evidence suggests that they may also regulate neurotransmission and synaptic plasticity. A comprehensive overview about the synaptic function of various schizophrenia susceptibility genes is given below.

22.3.1 *D2 DR*

Brain imaging studies have found an increase in the density and occupancy of D2 receptors in the striatum of schizophrenic patients (Abi-Dargham et al. 1998; Abi-Dargham et al. 2000; Wong et al. 1986). Also, several studies suggest that at least in a subpopulation of patients the observed increase in D2 receptor binding may be genetically determined (Hirvonen et al. 2004, 2005; Lawford et al. 2005; Zvara et al. 2005). D2 receptors are localized at the postsynaptic membrane of medium spiny neurons in the striatum (Gerfen 1992). In the PFC, where the expression levels of dopamine transporters are low (Sesack et al. 1998), the D2 receptor is localized at dopaminergic terminals to control the reuptake and the release of dopamine (Usiello et al. 2000) and at GABAergic terminals to control the release of GABA (Tseng and O'Donnell 2004). These D2 receptors are thought to fine-tune the firing of pyramidal neurons. Consistent with a major function of D2R as autoreceptors, the ability of dopamine to inhibit the firing of neurons in the midbrain or to inhibit the dopamine release in striatal projection areas is lost in D2R KO mice (Mercuri et al. 1997; Rouge-Pont et al. 2002). However, no in vivo genetic studies clarified the functions of D2 receptor in GABAergic interneurons. Overexpression of D2 receptor in medium spiny neurons in the striatum causes impairments in cognitive processes in the transgenic mice (Kellendonk et al. 2006). The transgenic mice are also impaired in incentive motivation that relates to negative symptoms. Interestingly, the cognitive, but not motivational, deficits persisted long after D2 receptor expression was switched off, suggesting that transient expression during prenatal development was sufficient to cause cognitive deficits in adulthood.

22.3.2   DISC1

The disrupted in schizophrenia (DISC) gene locus was first identified as a risk factor for major mental illness through study of a large Scottish family in which a balanced translocation between chromosomes 1 and 11 cosegregates with schizophrenia, bipolar disorder, and recurrent major depression (Millar et al. 2000; St Clair et al. 1990). This translocation directly disrupts the DISC1 protein and leads to a C-terminal truncated mutation of DISC1 (Millar et al. 2000). In addition to the translocation, several putative pathogenic mutations have been identified through sequencing DISC1 exons in patients (Song et al. 2008). DISC1 seems to serve as a scaffolding protein interacting with many proteins ranging from transcription factors, phosphodiesterases, and proteins implicated in cytoskeletal and centrosomal organization (Kamiya et al. 2008; Millar et al. 2003, 2005; Miyoshi et al. 2003; Morris et al. 2003; Ozeki et al. 2003). Consistent with this idea, studies in cell culture as well as in *Drosophila* and mice suggest that DISC1 may be involved in neuronal migration, positioning, differentiation, and neurite extension (Duan et al. 2007; Kamiya et al. 2005). DISC1 is expressed at the postsynaptic membrane of asymmetric synapses in human neocortex (Kirkpatrick et al. 2006). Mutant mice were generated to carry a 25-bp deletion in exon 6 of the Disc1 gene, which express a truncated DISC1 protein mimicking the mutant DISC1 found in the Scottish family (Kvajo et al. 2008). These mice exhibit fewer synaptic spines in the dentate gyrus, deficits in short-term plasticity at CA3/CA1 synapses, and impaired working memory (Kvajo et al. 2008). Depletion of DISC1 in newborn neurons in adult mice causes their mispositioning and accelerated formation of dendritic spines and synapses. DISC1-deficient newborn neurons also exhibit enhanced excitability (Duan et al. 2007).

22.3.3   DTNBP1/Dysbindin

Both linkage and association studies have implicated dystrobrevin-binding protein 1 (Dysbindin or DTNBP1) as a promising susceptibility gene for schizophrenia (Kirov et al. 2004; Schwab et al. 2003; Straub et al. 1995, 2002; Tang et al. 2003).

mRNA or protein levels of dysbindin were decreased in prefrontal cortex (PFC) and hippocampus (Talbot et al. 2004; Tang et al. 2009; Weickert et al. 2004, 2008) from schizophrenic patients. Dysbindin is a member of a protein complex, known as biogenesis of lysosome-related organelle complex 1 (BLOC-1). This complex is involved in vesicle trafficking and dendritic branching (Ghiani et al. 2010). In cultured neurons, increase and suppression of dysbindin expression can promote and inhibit glutamate release, respectively (Numakawa et al. 2004). The Sandy mice, which lack dysbindin protein owing to a deletion in the gene Dtnbp1 (encoding dysbindin) (Li et al. 2003), have a decreased rate of vesicle release, a correlated decrease in vesicle pool size, and an increased thickness of the postsynaptic density (Chen et al. 2008). In Sandy mice, deep-layer pyramidal neurons in the PFC showed reduced miniature and evoked EPSCs, and impaired paired-pulse facilitation, suggesting that dysbindin may regulate excitatory transmission in the PFC possibly by a presynaptic mechanism (Jentsch et al. 2009). Decreased levels of dysbindin are associated with reduction in NMDA-evoked currents in PFC pyramidal neurons and in NR1 expression (Karlsgodt et al. 2011). The Sandy mice showed mild deficit in spatial working memory (Jentsch et al. 2009), which appears to correlate with levels of NR1 expression (Karlsgodt et al. 2011).

22.3.4   *NRG1 and ErbB4*

Several linkage studies in independent populations have identified neuregulin 1 (NRG1) and its receptor ErbB4 as susceptibility genes of schizophrenia (Nicodemus et al. 2006; Norton et al. 2006; Stefansson et al. 2002, 2003; Yang et al. 2003). NRG1 isoforms (types I and IV) and the ErbB4 isoform (JMa, CYT1) are expressed at higher levels in the PFC and hippocampus of schizophrenic patients (Hashimoto et al. 2004; Law et al. 2007; Law et al. 2006; Silberberg et al. 2006). Another group reported a marked increase in NRG1-induced ErbB4 activation in the prefrontal cortex in schizophrenia, while the total level of NRG1 and ErbB4 did not alter (Hahn et al. 2006). NRG1 is a family of EGF domain–containing trophic factors that acts by activating ErbB tyrosine kinases (Mei and Xiong 2008). In vitro studies suggest that NRG1-ErbB4 signaling may regulate neuronal migration and gene expression of NMDA and GABA receptors (Mei and Xiong 2008). However, these notions were challenged by studies of mutant mice (Barros et al. 2009; Brinkmann et al. 2008; Chen et al. 2010a; Gajendran et al. 2009).

ErbB4 in rodents is enriched in GABAergic interneurons (Fazzari et al. 2010; Huang et al. 2000; Lai and Lemke 1991; Vullhorst et al. 2009; Yau et al. 2003). During development, NRG1-ErbB4 appears to play a role in the formation of excitatory synapses on GABAergic interneurons and inhibitory synapses on projection cells (Fazzari et al. 2010; Ting et al. 2011). Both NRG1 and ErbB4 are expressed in adult brain. Acute treatment of hippocampal slices with soluble NRG1 suppresses the induction of long-term potentiation (LTP) (Huang et al. 2000).

Evidence suggests that this effect is mediated by enhanced GABAergic transmission. We have recently demonstrated that NRG1 acts to promote GABA release and thus control the firing of pyramidal neurons and suppresses long-term potentiation (LTP) (Chen et al. 2010b; Huang et al. 2000; Wen et al. 2010; Woo et al. 2007). Ablation of ErbB4 in parvalbumin-positive interneurons causes schizophrenia-relevant phenotypes in mutant mice including hyperactivity, impaired prepulse inhibition, and working memory deficits (Wen et al. 2010).

In addition to inhibitory neurons, ErbB4 is highly expressed in midbrain dopaminergic neurons in rodents, monkeys, and humans (Abe et al. 2009; Steiner et al. 1999; Zheng et al. 2009). NRG1 has been shown to promote dopamine release in the striatum, hippocampus, and medial prefrontal cortex (Kato et al. 2010; Kwon et al. 2008; Yurek et al. 2004). In vitro studies suggest that NRG1 enhances the survival of dopaminergic neurons (Zhang et al. 2004). However, mutant mice where ErbB4 is ablated in the entire brain showed normal structure of the substantial nigra pars compacta and no deficits in motor performance, suggesting that ErbB4 is not required for the development or survival of dopaminergic neurons (Thuret et al. 2004). It will be interesting to generate dopaminergic neuron–specific ErbB4 mutant mice to determine whether NRG1-ErbB4 signaling is important for neurotransmission at dopaminergic synapses.

It is controversial whether NRG1 regulates excitatory synapse formation in pyramidal neurons and glutamatergic transmission. Overexpression of ErbB4 and suppression of its expression by ErbB4 shRNA promoted or inhibited the formation of glutamatergic synapses in pyramidal neurons of neonatal hippocampal slices (Li et al. 2007), suggesting a potential role in excitatory synapse formation. However, when ErbB4 is ablated specifically in CaMKII-positive neurons, it had no effect on basal glutamatergic transmission (Chen et al. 2010b). Acute treatment of soluble NRG1 did not alter paired-pulse facilitation (PPF) (Huang et al. 2000; Iyengar and Mott 2008), suggesting no effects of NRG1 on glutamate release. However, NRG1 mutant mice showed altered PPF and short-term plasticity (Bjarnadottir et al. 2007). Treatment with NRG1 decreased NMDAR-mediated excitatory postsynaptic currents in PFC slices and reduced whole-cell NMDAR currents in acutely isolated PFC pyramidal neurons by elevating intracellular Ca^{2+} and stimulating ERK activity (Gu et al. 2005). In hippocampal slices, however, NRG1 appeared to have little effect on NMDAR- or AMPAR-mediated basic transmission (Chen et al. 2010b). In human postmortem hippocampal tissues, NRG1 could attenuate ligand-induced phosphorylation of NMDA receptors and its association with signaling partners (Hahn et al. 2006).

NRG1 regulates the expression of the α7 nicotinic acetylcholine receptors (nAChRs) (Liu et al. 2001; Sandrock et al. 1997; Usdin and Fischbach 1986; Yang et al. 1998). Consistent with these reports, decreased α7 nAChR mRNA and protein in schizophrenic patients is associated with the genetic variation of NRG1 (Mathew et al. 2007). Recent studies of NRG1 mutant mice indicate that type III NRG1 regulates the axonal targeting of α7 nAChR and is required for the enhancement of hippocampal transmission by nicotine (Hancock et al. 2008; Zhong et al. 2008).

22.3.5 *Future Directions*

It is clear that synaptic transmission and plasticity are disrupted in schizophrenia. The disruption could be caused by problems that occurred during neural development and/or after brain wiring is complete. Interestingly, Rett syndrome–like neurological deficits of MeCP2 mutant mice can be reversed in adult stage (Guy et al. 2007). It would be important to determine whether this occurs to mutant mice of schizophrenia candidate genes, which would require the reversible transgenic or knockout strategies. Tet-Off system is commonly used to overexpress individual genes which can be reversed by doxycycline (Mayford et al. 1996). Tamoxifen-inducible Cre mice were generated to reactivate the genes by removing the loxP-STOP-loxP cassette (Guy et al. 2007; Hayashi and McMahon 2002). Another important question is to demonstrate the deficit in neural circuitry in schizophrenia. For example, recent studies showed impaired hippocampal-prefrontal synchrony in a genetic mouse model of schizophrenia which has the microdeletion on the human chromosome 22 (Sigurdsson et al. 2010). More recent paper reported that the efficacy of ventral hippocampus input to the nucleus accumbens is reduced in the type III NRG1 heterozygotes mutant mice (Nason et al. 2011). The third question to be addressed is how the dysfunction of different types of GABAergic interneurons contributes to the schizophrenia. Optogenetics, a new emerging technique which enables the activation or inactivation of different types of neurons with spatial and temporal control (Boyden et al. 2005; Gradinaru et al. 2009; Petreanu et al. 2009), is obviously of great advantage to address this question. Recent study demonstrated the critical roles of parvalbumin-positive interneurons in gamma-frequency synchronization in vivo using optogenetics (Sohal et al. 2009). Finally, how can we test the hypothesis that a synaptic defect is responsible for schizophrenia in humans? A direct way would be to study synaptic behavior in the brains of affected individuals, but this can not yet be done in the intact human brain. A possible alternative route involves the production of induced pluripotent stem cells (Takahashi et al. 2007; Yu et al. 2007) from adult cells derived from schizophrenic patients and then inducing these iPS cells to form neurons and synapses. The neuronal culture is also potentially useful in screening the individual antischizophrenia drugs.

References

Abe, Y., Namba, H., Zheng, Y., & Nawa, H. (2009). In situ hybridization reveals developmental regulation of ErbB1-4 mRNA expression in mouse midbrain: Implication of ErbB receptors for dopaminergic neurons. *Neuroscience, 161*, 95–110.

Abi-Dargham, A., Gil, R., Krystal, J., Baldwin, R. M., Seibyl, J. P., Bowers, M., van Dyck, C. H., Charney, D. S., Innis, R. B., & Laruelle, M. (1998). Increased striatal dopamine transmission in schizophrenia: Confirmation in a second cohort. *The American Journal of Psychiatry, 155*, 761–767.

Abi-Dargham, A., Mawlawi, O., Lombardo, I., Gil, R., Martinez, D., Huang, Y., Hwang, D. R., Keilp, J., Kochan, L., Van Heertum, R., et al. (2002). Prefrontal dopamine D1 receptors and working memory in schizophrenia. *Journal of Neuroscience, 22*, 3708–3719.

Abi-Dargham, A., & Moore, H. (2003). Prefrontal DA transmission at D1 receptors and the pathology of schizophrenia. *The Neuroscientist, 9*, 404–416.

Abi-Dargham, A., Rodenhiser, J., Printz, D., Zea-Ponce, Y., Gil, R., Kegeles, L. S., Weiss, R., Cooper, T. B., Mann, J. J., Van Heertum, R. L., et al. (2000). Increased baseline occupancy of D2 receptors by dopamine in schizophrenia. *Proceedings of the National Academy of Sciences of the United States of America, 97*, 8104–8109.

Adler, L. E., Hoffer, L. D., Wiser, A., & Freedman, R. (1993). Normalization of auditory physiology by cigarette smoking in schizophrenic patients. *The American Journal of Psychiatry, 150*, 1856–1861.

Adler, C. M., Malhotra, A. K., Elman, I., Goldberg, T., Egan, M., Pickar, D., & Breier, A. (1999). Comparison of ketamine-induced thought disorder in healthy volunteers and thought disorder in schizophrenia. *The American Journal of Psychiatry, 156*, 1646–1649.

Akbarian, S., Kim, J. J., Potkin, S. G., Hagman, J. O., Tafazzoli, A., Bunney, W. E., Jr., & Jones, E. G. (1995). Gene expression for glutamic acid decarboxylase is reduced without loss of neurons in prefrontal cortex of schizophrenics. *Archives of General Psychiatry, 52*, 258–266.

Akbarian, S., Sucher, N. J., Bradley, D., Tafazzoli, A., Trinh, D., Hetrick, W. P., Potkin, S. G., Sandman, C. A., Bunney, W. E., Jr., & Jones, E. G. (1996). Selective alterations in gene expression for NMDA receptor subunits in prefrontal cortex of schizophrenics. *Journal of Neuroscience, 16*, 19–30.

Akil, M., Pierri, J. N., Whitehead, R. E., Edgar, C. L., Mohila, C., Sampson, A. R., & Lewis, D. A. (1999). Lamina-specific alterations in the dopamine innervation of the prefrontal cortex in schizophrenic subjects. *The American Journal of Psychiatry, 156*, 1580–1589.

Anagnostaras, S. G., Murphy, G. G., Hamilton, S. E., Mitchell, S. L., Rahnama, N. P., Nathanson, N. M., & Silva, A. J. (2003). Selective cognitive dysfunction in acetylcholine M1 muscarinic receptor mutant mice. *Nature Neuroscience, 6*, 51–58.

Angrist, B. M., & Gershon, S. (1970). The phenomenology of experimentally induced amphetamine psychosis–preliminary observations. *Biological Psychiatry, 2*, 95–107.

Anis, N. A., Berry, S. C., Burton, N. R., & Lodge, D. (1983). The dissociative anaesthetics, ketamine and phencyclidine, selectively reduce excitation of central mammalian neurones by N-methyl-aspartate. *British Journal of Pharmacology, 79*, 565–575.

Ascoli, G. A., Alonso-Nanclares, L., Anderson, S. A., Barrionuevo, G., Benavides-Piccione, R., Burkhalter, A., Buzsaki, G., Cauli, B., Defelipe, J., Fairen, A., et al. (2008). Petilla terminology: Nomenclature of features of GABAergic interneurons of the cerebral cortex. *Nature Reviews Neuroscience, 9*, 557–568.

Bakshi, V. P., & Geyer, M. A. (1995). Antagonism of phencyclidine-induced deficits in prepulse inhibition by the putative atypical antipsychotic olanzapine. *Psychopharmacology, 122*, 198–201.

Bakshi, V. P., Swerdlow, N. R., & Geyer, M. A. (1994). Clozapine antagonizes phencyclidine-induced deficits in sensorimotor gating of the startle response. *Journal of Pharmacology and Experimental Therapeutics, 271*, 787–794.

Barros, C. S., Calabrese, B., Chamero, P., Roberts, A. J., Korzus, E., Lloyd, K., Stowers, L., Mayford, M., Halpain, S., & Muller, U. (2009). Impaired maturation of dendritic spines without disorganization of cortical cell layers in mice lacking NRG1/ErbB signaling in the central nervous system. *Proceedings of the National Academy of Sciences of the United States of America, 106*, 4507–4512.

Bast, T., Zhang, W. N., & Feldon, J. (2001). Hyperactivity, decreased startle reactivity, and disrupted prepulse inhibition following disinhibition of the rat ventral hippocampus by the GABA(A) receptor antagonist picrotoxin. *Psychopharmacology, 156*, 225–233.

Beasley, C. L., & Reynolds, G. P. (1997). Parvalbumin-immunoreactive neurons are reduced in the prefrontal cortex of schizophrenics. *Schizophrenia Research, 24*, 349–355.

Belforte, J. E., Zsiros, V., Sklar, E. R., Jiang, Z., Yu, G., Li, Y., Quinlan, E. M., & Nakazawa, K. (2010). Postnatal NMDA receptor ablation in corticolimbic interneurons confers schizophrenia-like phenotypes. *Nature Neuroscience, 13*, 76–83.

Bell, D. S. (1973). The experimental reproduction of amphetamine psychosis. *Archives of General Psychiatry, 29*, 35–40.

Benes, F. M., McSparren, J., Bird, E. D., SanGiovanni, J. P., & Vincent, S. L. (1991). Deficits in small interneurons in prefrontal and cingulate cortices of schizophrenic and schizoaffective patients. *Archives of General Psychiatry, 48*, 996–1001.

Benes, F. M., Vincent, S. L., Alsterberg, G., Bird, E. D., & SanGiovanni, J. P. (1992). Increased GABAA receptor binding in superficial layers of cingulate cortex in schizophrenics. *Journal of Neuroscience, 12*, 924–929.

Bird, E. D., Spokes, E. G., Barnes, J., MacKay, A. V., Iversen, L. L., & Shepherd, M. (1977). Increased brain dopamine and reduced glutamic acid decarboxylase and choline acetyl transferase activity in schizophrenia and related psychoses. *Lancet, 2*, 1157–1158.

Bjarnadottir, M., Misner, D. L., Haverfield-Gross, S., Bruun, S., Helgason, V. G., Stefansson, H., Sigmundsson, A., Firth, D. R., Nielsen, B., Stefansdottir, R., et al. (2007). Neuregulin1 (NRG1) signaling through Fyn modulates NMDA receptor phosphorylation: Differential synaptic function in NRG1+/– knock-outs compared with wild-type mice. *Journal of Neuroscience, 27*, 4519–4529.

Boyden, E. S., Zhang, F., Bamberg, E., Nagel, G., & Deisseroth, K. (2005). Millisecond-timescale, genetically targeted optical control of neural activity. *Nature Neuroscience, 8*, 1263–1268.

Breier, A., Su, T. P., Saunders, R., Carson, R. E., Kolachana, B. S., de Bartolomeis, A., Weinberger, D. R., Weisenfeld, N., Malhotra, A. K., Eckelman, W. C., et al. (1997). Schizophrenia is associated with elevated amphetamine-induced synaptic dopamine concentrations: Evidence from a novel positron emission tomography method. *Proceedings of the National Academy of Sciences of the United States of America, 94*, 2569–2574.

Brinkmann, B. G., Agarwal, A., Sereda, M. W., Garratt, A. N., Muller, T., Wende, H., Stassart, R. M., Nawaz, S., Humml, C., Velanac, V., et al. (2008). Neuregulin-1/ErbB signaling serves distinct functions in myelination of the peripheral and central nervous system. *Neuron, 59*, 581–595.

Busatto, G. F., Pilowsky, L. S., Costa, D. C., Ell, P. J., David, A. S., Lucey, J. V., & Kerwin, R. W. (1997). Correlation between reduced in vivo benzodiazepine receptor binding and severity of psychotic symptoms in schizophrenia. *The American Journal of Psychiatry, 154*, 56–63.

Chen, X. W., Feng, Y. Q., Hao, C. J., Guo, X. L., He, X., Zhou, Z. Y., Guo, N., Huang, H. P., Xiong, W., Zheng, H., et al. (2008). DTNBP1, a schizophrenia susceptibility gene, affects kinetics of transmitter release. *The Journal of Cell Biology, 181*, 791–801.

Chen, Y., Hancock, M. L., Role, L. W., & Talmage, D. A. (2010a). Intramembranous valine linked to schizophrenia is required for neuregulin 1 regulation of the morphological development of cortical neurons. *Journal of Neuroscience, 30*, 9199–9208.

Chen, Y. J., Zhang, M., Yin, D. M., Wen, L., Ting, A., Wang, P., Lu, Y. S., Zhu, X. H., Li, S. J., Wu, C. Y., et al. (2010b). ErbB4 in parvalbumin-positive interneurons is critical for neuregulin 1 regulation of long-term potentiation. *Proceedings of the National Academy of Sciences of the United States of America, 107*, 21818–21823.

De Keyser, J., Claeys, A., De Backer, J. P., Ebinger, G., Roels, F., & Vauquelin, G. (1988). Autoradiographic localization of D1 and D2 dopamine receptors in the human brain. *Neuroscience Letters, 91*, 142–147.

de Leon, J., Dadvand, M., Canuso, C., White, A. O., Stanilla, J. K., & Simpson, G. M. (1995). Schizophrenia and smoking: An epidemiological survey in a state hospital. *The American Journal of Psychiatry, 152*, 453–455.

Dracheva, S., Marras, S. A., Elhakem, S. L., Kramer, F. R., Davis, K. L., & Haroutunian, V. (2001). N-methyl-D-aspartic acid receptor expression in the dorsolateral prefrontal cortex of elderly patients with schizophrenia. *The American Journal of Psychiatry, 158*, 1400–1410.

Duan, X., Chang, J. H., Ge, S., Faulkner, R. L., Kim, J. Y., Kitabatake, Y., Liu, X. B., Yang, C. H., Jordan, J. D., Ma, D. K., et al. (2007). Disrupted-In-Schizophrenia 1 regulates integration of newly generated neurons in the adult brain. *Cell, 130*, 1146–1158.

Eastwood, S. L. (2004). The synaptic pathology of schizophrenia: Is aberrant neurodevelopment and plasticity to blame? *International Review of Neurobiology, 59*, 47–72.

Fauman, B., Aldinger, G., Fauman, M., & Rosen, P. (1976). Psychiatric sequelae of phencyclidine abuse. *Clinical Toxicology, 9*, 529–538.

Faustman, W. O., Bardgett, M., Faull, K. F., Pfefferbaum, A., & Csernansky, J. G. (1999). Cerebrospinal fluid glutamate inversely correlates with positive symptom severity in unmedicated male schizophrenic/schizoaffective patients. *Biological Psychiatry, 45*, 68–75.

Fazzari, P., Paternain, A. V., Valiente, M., Pla, R., Lujan, R., Lloyd, K., Lerma, J., Marin, O., & Rico, B. (2010). Control of cortical GABA circuitry development by Nrg1 and ErbB4 signalling. *Nature, 464*, 1376–1380.

Forrer, G. R., & Miller, J. J. (1958). Atropine coma: A somatic therapy in psychiatry. *The American Journal of Psychiatry, 115*, 455–458.

Freedman, R., Coon, H., Myles-Worsley, M., Orr-Urtreger, A., Olincy, A., Davis, A., Polymeropoulos, M., Holik, J., Hopkins, J., Hoff, M., et al. (1997). Linkage of a neurophysiological deficit in schizophrenia to a chromosome 15 locus. *Proceedings of the National Academy of Sciences of the United States of America, 94*, 587–592.

Freedman, R., Hall, M., Adler, L. E., & Leonard, S. (1995). Evidence in postmortem brain tissue for decreased numbers of hippocampal nicotinic receptors in schizophrenia. *Biological Psychiatry, 38*, 22–33.

Gajendran, N., Kapfhammer, J. P., Lain, E., Canepari, M., Vogt, K., Wisden, W., & Brenner, H. R. (2009). Neuregulin signaling is dispensable for NMDA- and GABA(A)-receptor expression in the cerebellum in vivo. *Journal of Neuroscience, 29*, 2404–2413.

Gardner, R., & Connell, P. H. (1972). Amphetamine and other non-opioid drug users attending a special drug dependence clinic. *British Medical Journal, 2*, 322–325.

Garey, L. J., Ong, W. Y., Patel, T. S., Kanani, M., Davis, A., Mortimer, A. M., Barnes, T. R., & Hirsch, S. R. (1998). Reduced dendritic spine density on cerebral cortical pyramidal neurons in schizophrenia. *Journal of Neurology, Neurosurgery & Psychiatry, 65*, 446–453.

Gerfen, C. R. (1992). The neostriatal mosaic: Multiple levels of compartmental organization. *Trends in Neurosciences, 15*, 133–139.

Ghiani, C. A., Starcevic, M., Rodriguez-Fernandez, I. A., Nazarian, R., Cheli, V. T., Chan, L. N., Malvar, J. S., de Vellis, J., Sabatti, C., & Dell'Angelica, E. C. (2010). The dysbindin-containing complex (BLOC-1) in brain: Developmental regulation, interaction with SNARE proteins and role in neurite outgrowth. *Molecular Psychiatry, 15*, 204–215.

Glantz, L. A., & Lewis, D. A. (2000). Decreased dendritic spine density on prefrontal cortical pyramidal neurons in schizophrenia. *Archives of General Psychiatry, 57*, 65–73.

Goff, D. C., Henderson, D. C., & Amico, E. (1992). Cigarette smoking in schizophrenia: Relationship to psychopathology and medication side effects. *The American Journal of Psychiatry, 149*, 1189–1194.

Gradinaru, V., Mogri, M., Thompson, K. R., Henderson, J. M., & Deisseroth, K. (2009). Optical deconstruction of parkinsonian neural circuitry. *Science, 324*, 354–359.

Grunze, H. C., Rainnie, D. G., Hasselmo, M. E., Barkai, E., Hearn, E. F., McCarley, R. W., & Greene, R. W. (1996). NMDA-dependent modulation of CA1 local circuit inhibition. *Journal of Neuroscience, 16*, 2034–2043.

Gu, Z., Jiang, Q., Fu, A. K., Ip, N. Y., & Yan, Z. (2005). Regulation of NMDA receptors by neuregulin signaling in prefrontal cortex. *Journal of Neuroscience, 25*, 4974–4984.

Guy, J., Gan, J., Selfridge, J., Cobb, S., & Bird, A. (2007). Reversal of neurological defects in a mouse model of Rett syndrome. *Science, 315*, 1143–1147.

Hahn, C. G., Wang, H. Y., Cho, D. S., Talbot, K., Gur, R. E., Berrettini, W. H., Bakshi, K., Kamins, J., Borgmann-Winter, K. E., Siegel, S. J., et al. (2006). Altered neuregulin 1-erbB4 signaling contributes to NMDA receptor hypofunction in schizophrenia. *Nature Medicine, 12*, 824–828.

Hakak, Y., Walker, J. R., Li, C., Wong, W. H., Davis, K. L., Buxbaum, J. D., Haroutunian, V., & Fienberg, A. A. (2001). Genome-wide expression analysis reveals dysregulation of myelination-related genes in chronic schizophrenia. *Proceedings of the National Academy of Sciences of the United States of America, 98*, 4746–4751.

Hall, H., Sedvall, G., Magnusson, O., Kopp, J., Halldin, C., & Farde, L. (1994). Distribution of D1- and D2-dopamine receptors, and dopamine and its metabolites in the human brain. *Neuropsychopharmacol, 11*, 245–256.

Hamera, E., Schneider, J. K., & Deviney, S. (1995). Alcohol, cannabis, nicotine, and caffeine use and symptom distress in schizophrenia. *The Journal of Nervous and Mental Disease, 183*, 559–565.

Hanada, S., Mita, T., Nishino, N., & Tanaka, C. (1987). [3H]muscimol binding sites increased in autopsied brains of chronic schizophrenics. *Life Sciences, 40*, 259–266.

Hancock, M. L., Canetta, S. E., Role, L. W., & Talmage, D. A. (2008). Presynaptic type III neuregulin1-ErbB signaling targets {alpha}7 nicotinic acetylcholine receptors to axons. *The Journal of Cell Biology, 181*, 511–521.

Hashimoto, R., Straub, R. E., Weickert, C. S., Hyde, T. M., Kleinman, J. E., & Weinberger, D. R. (2004). Expression analysis of neuregulin-1 in the dorsolateral prefrontal cortex in schizophrenia. *Molecular Psychiatry, 9*, 299–307.

Hashimoto, T., Volk, D. W., Eggan, S. M., Mirnics, K., Pierri, J. N., Sun, Z., Sampson, A. R., & Lewis, D. A. (2003). Gene expression deficits in a subclass of GABA neurons in the prefrontal cortex of subjects with schizophrenia. *Journal of Neuroscience, 23*, 6315–6326.

Hayashi, S., & McMahon, A. P. (2002). Efficient recombination in diverse tissues by a tamoxifen-inducible form of Cre: A tool for temporally regulated gene activation/inactivation in the mouse. *Developmental Biology, 244*, 305–318.

Hirvonen, M., Laakso, A., Nagren, K., Rinne, J. O., Pohjalainen, T., & Hietala, J. (2004). C957T polymorphism of the dopamine D2 receptor (DRD2) gene affects striatal DRD2 availability in vivo. *Molecular Psychiatry, 9*, 1060–1061.

Hirvonen, J., van Erp, T. G., Huttunen, J., Aalto, S., Nagren, K., Huttunen, M., Lonnqvist, J., Kaprio, J., Hietala, J., & Cannon, T. D. (2005). Increased caudate dopamine D2 receptor availability as a genetic marker for schizophrenia. *Archives of General Psychiatry, 62*, 371–378.

Holt, D. J., Bachus, S. E., Hyde, T. M., Wittie, M., Herman, M. M., Vangel, M., Saper, C. B., & Kleinman, J. E. (2005). Reduced density of cholinergic interneurons in the ventral striatum in schizophrenia: An in situ hybridization study. *Biological Psychiatry, 58*, 408–416.

Holt, D. J., Herman, M. M., Hyde, T. M., Kleinman, J. E., Sinton, C. M., German, D. C., Hersh, L. B., Graybiel, A. M., & Saper, C. B. (1999). Evidence for a deficit in cholinergic interneurons in the striatum in schizophrenia. *Neuroscience, 94*, 21–31.

Howes, O. D., Smith, S., Gaughran, F. P., Amiel, S. A., Murray, R. M., & Pilowsky, L. S. (2006). The relationship between prolactin levels and glucose homeostasis in antipsychotic-treated schizophrenic patients. *Journal of Clinical Psychopharmacology, 26*, 629–631.

Huang, Y. Z., Won, S., Ali, D. W., Wang, Q., Tanowitz, M., Du, Q. S., Pelkey, K. A., Yang, D. J., Xiong, W. C., Salter, M. W., et al. (2000). Regulation of neuregulin signaling by PSD-95 interacting with ErbB4 at CNS synapses. *Neuron, 26*, 443–455.

Iyengar, S. S., & Mott, D. D. (2008). Neuregulin blocks synaptic strengthening after epileptiform activity in the rat hippocampus. *Brain Research, 1208*, 67–73.

Jentsch, J. D., Tran, A., Le, D., Youngren, K. D., & Roth, R. H. (1997). Subchronic phencyclidine administration reduces mesoprefrontal dopamine utilization and impairs prefrontal cortical-dependent cognition in the rat. *Neuropsychopharmacol, 17*, 92–99.

Jentsch, J. D., Trantham-Davidson, H., Jairl, C., Tinsley, M., Cannon, T. D., & Lavin, A. (2009). Dysbindin modulates prefrontal cortical glutamatergic circuits and working memory function in mice. *Neuropsychopharmacol, 34*, 2601–2608.

Kamiya, A., Kubo, K., Tomoda, T., Takaki, M., Youn, R., Ozeki, Y., Sawamura, N., Park, U., Kudo, C., Okawa, M., et al. (2005). A schizophrenia-associated mutation of DISC1 perturbs cerebral cortex development. *Nature Cell Biology, 7*, 1167–1178.

Kamiya, A., Tan, P. L., Kubo, K., Engelhard, C., Ishizuka, K., Kubo, A., Tsukita, S., Pulver, A. E., Nakajima, K., Cascella, N. G., et al. (2008). Recruitment of PCM1 to the centrosome by the cooperative action of DISC1 and BBS4: A candidate for psychiatric illnesses. *Archives of General Psychiatry, 65*, 996–1006.

Karlsgodt, K. H., Robleto, K., Trantham-Davidson, H., Jairl, C., Cannon, T. D., Lavin, A., & Jentsch, J. D. (2011). Reduced dysbindin expression mediates N-methyl-D-aspartate receptor hypofunction and impaired working memory performance. *Biological Psychiatry, 69*, 28–34.

Kato, T., Abe, Y., Sotoyama, H., Kakita, A., Kominami, R., Hirokawa, S., Ozaki, M., Takahashi, H., & Nawa, H. (2010). Transient exposure of neonatal mice to neuregulin-1 results in hyperdopaminergic states in adulthood: Implication in neurodevelopmental hypothesis for schizophrenia. *Molecular Psychiatry, 16*, 307–320.

Kellendonk, C., Simpson, E. H., Polan, H. J., Malleret, G., Vronskaya, S., Winiger, V., Moore, H., & Kandel, E. R. (2006). Transient and selective overexpression of dopamine D2 receptors in the striatum causes persistent abnormalities in prefrontal cortex functioning. *Neuron, 49*, 603–615.

Keverne, E. B. (1999). GABA-ergic neurons and the neurobiology of schizophrenia and other psychoses. *Brain Research Bulletin, 48*, 467–473.

Kim, J. S., Kornhuber, H. H., Schmid-Burgk, W., & Holzmuller, B. (1980). Low cerebrospinal fluid glutamate in schizophrenic patients and a new hypothesis on schizophrenia. *Neuroscience Letters, 20*, 379–382.

Kirkpatrick, B., Xu, L., Cascella, N., Ozeki, Y., Sawa, A., & Roberts, R. C. (2006). DISC1 immunoreactivity at the light and ultrastructural level in the human neocortex. *The Journal of Comparative Neurology, 497*, 436–450.

Kirov, G., Ivanov, D., Williams, N. M., Preece, A., Nikolov, I., Milev, R., Koleva, S., Dimitrova, A., Toncheva, D., O'Donovan, M. C., et al. (2004). Strong evidence for association between the dystrobrevin binding protein 1 gene (DTNBP1) and schizophrenia in 488 parent-offspring trios from Bulgaria. *Biological Psychiatry, 55*, 971–975.

Korotkova, T., Fuchs, E. C., Ponomarenko, A., von Engelhardt, J., & Monyer, H. (2010). NMDA receptor ablation on parvalbumin-positive interneurons impairs hippocampal synchrony, spatial representations, and working memory. *Neuron, 68*, 557–569.

Kristiansen, L. V., Beneyto, M., Haroutunian, V., & Meador-Woodruff, J. H. (2006). Changes in NMDA receptor subunits and interacting PSD proteins in dorsolateral prefrontal and anterior cingulate cortex indicate abnormal regional expression in schizophrenia. *Molecular Psychiatry, 11*, 705.

Krystal, J. H., Karper, L. P., Seibyl, J. P., Freeman, G. K., Delaney, R., Bremner, J. D., Heninger, G. R., Bowers, M. B., Jr., & Charney, D. S. (1994). Subanesthetic effects of the noncompetitive NMDA antagonist, ketamine, in humans. Psychotomimetic, perceptual, cognitive, and neuroendocrine responses. *Archives of General Psychiatry, 51*, 199–214.

Kucinski, A. J., Stachowiak, M. K., Wersinger, S. R., Lippiello, P. M., & Bencherif, M. (2010). Alpha7 neuronal nicotinic receptors as targets for novel therapies to treat multiple domains of schizophrenia. *Current Pharmaceutical Biotechnology, 12*, 437–448.

Kvajo, M., McKellar, H., Arguello, P. A., Drew, L. J., Moore, H., MacDermott, A. B., Karayiorgou, M., & Gogos, J. A. (2008). A mutation in mouse Disc1 that models a schizophrenia risk allele leads to specific alterations in neuronal architecture and cognition. *Proceedings of the National Academy of Sciences of the United States of America, 105*, 7076–7081.

Kwon, O. B., Paredes, D., Gonzalez, C. M., Neddens, J., Hernandez, L., Vullhorst, D., & Buonanno, A. (2008). Neuregulin-1 regulates LTP at CA1 hippocampal synapses through activation of dopamine D4 receptors. *Proceedings of the National Academy of Sciences of the United States of America, 105*, 15587–15592.

Lacroix, L., Spinelli, S., Broersen, L. M., & Feldon, J. (2000). Blockade of latent inhibition following pharmacological increase or decrease of GABA(A) transmission. *Pharmacology, Biochemistry, and Behavior, 66*, 893–901.

Lahti, A. C., Koffel, B., LaPorte, D., & Tamminga, C. A. (1995). Subanesthetic doses of ketamine stimulate psychosis in schizophrenia. *Neuropsychopharmacol, 13*, 9–19.

Lahti, A. C., Weiler, M. A., Tamara Michaelidis, B. A., Parwani, A., & Tamminga, C. A. (2001). Effects of ketamine in normal and schizophrenic volunteers. *Neuropsychopharmacol, 25*, 455–467.

Lai, C., & Lemke, G. (1991). An extended family of protein-tyrosine kinase genes differentially expressed in the vertebrate nervous system. *Neuron, 6*, 691–704.

Laruelle, M., Abi-Dargham, A., Gil, R., Kegeles, L., & Innis, R. (1999). Increased dopamine transmission in schizophrenia: Relationship to illness phases. *Biological Psychiatry, 46*, 56–72.

Laruelle, M., Abi-Dargham, A., van Dyck, C. H., Gil, R., D'Souza, C. D., Erdos, J., McCance, E., Rosenblatt, W., Fingado, C., Zoghbi, S. S., et al. (1996). Single photon emission computerized tomography imaging of amphetamine-induced dopamine release in drug-free schizophrenic subjects. *Proceedings of the National Academy of Sciences of the United States of America, 93*, 9235–9240.

Law, A. J., Kleinman, J. E., Weinberger, D. R., & Weickert, C. S. (2007). Disease-associated intronic variants in the ErbB4 gene are related to altered ErbB4 splice-variant expression in the brain in schizophrenia. *Human Molecular Genetics, 16*, 129–141.

Law, A. J., Lipska, B. K., Weickert, C. S., Hyde, T. M., Straub, R. E., Hashimoto, R., Harrison, P. J., Kleinman, J. E., & Weinberger, D. R. (2006). Neuregulin 1 transcripts are differentially expressed in schizophrenia and regulated by 5' SNPs associated with the disease. *Proceedings of the National Academy of Sciences of the United States of America, 103*, 6747–6752.

Lawford, B. R., Young, R. M., Swagell, C. D., Barnes, M., Burton, S. C., Ward, W. K., Heslop, K. R., Shadforth, S., van Daal, A., & Morris, C. P. (2005). The C/C genotype of the C957T polymorphism of the dopamine D2 receptor is associated with schizophrenia. *Schizophrenia Research, 73*, 31–37.

Lewis, D. A., Pierri, J. N., Volk, D. W., Melchitzky, D. S., & Woo, T. U. (1999). Altered GABA neurotransmission and prefrontal cortical dysfunction in schizophrenia. *Biological Psychiatry, 46*, 616–626.

Li, B., Woo, R. S., Mei, L., & Malinow, R. (2007). The neuregulin-1 receptor erbB4 controls glutamatergic synapse maturation and plasticity. *Neuron, 54*, 583–597.

Li, W., Zhang, Q., Oiso, N., Novak, E. K., Gautam, R., O'Brien, E. P., Tinsley, C. L., Blake, D. J., Spritz, R. A., Copeland, N. G., et al. (2003). Hermansky-Pudlak syndrome type 7 (HPS-7) results from mutant dysbindin, a member of the biogenesis of lysosome-related organelles complex 1 (BLOC-1). *Nature Genetics, 35*, 84–89.

Lieberman, J. A., Kane, J. M., & Alvir, J. (1987). Provocative tests with psychostimulant drugs in schizophrenia. *Psychopharmacology, 91*, 415–433.

Liu, Y., Ford, B., Mann, M. A., & Fischbach, G. D. (2001). Neuregulins increase alpha7 nicotinic acetylcholine receptors and enhance excitatory synaptic transmission in GABAergic interneurons of the hippocampus. *Journal of Neuroscience, 21*, 5660–5669.

Lohr, J. B., & Flynn, K. (1992). Smoking and schizophrenia. *Schizophrenia Research, 8*, 93–102.

Luby, E. D., Cohen, B. D., Rosenbaum, G., Gottlieb, J. S., & Kelley, R. (1959). Study of a new schizophrenomimetic drug: Sernyl. *American Medical Association: Archives of Neurological Psychiatry, 81*, 363–369.

Malhotra, A. K., Pinals, D. A., Adler, C. M., Elman, I., Clifton, A., Pickar, D., & Breier, A. (1997). Ketamine-induced exacerbation of psychotic symptoms and cognitive impairment in neuroleptic-free schizophrenics. *Neuropsychopharmacology, 17*, 141–150.

Markram, H., Toledo-Rodriguez, M., Wang, Y., Gupta, A., Silberberg, G., & Wu, C. (2004). Interneurons of the neocortical inhibitory system. *Nature Reviews Neuroscience, 5*, 793–807.

Marutle, A., Zhang, X., Court, J., Piggott, M., Johnson, M., Perry, R., Perry, E., & Nordberg, A. (2001). Laminar distribution of nicotinic receptor subtypes in cortical regions in schizophrenia. *Journal of Chemical Neuroanatomy, 22*, 115–126.

Mathew, S. V., Law, A. J., Lipska, B. K., Davila-Garcia, M. I., Zamora, E. D., Mitkus, S. N., Vakkalanka, R., Straub, R. E., Weinberger, D. R., Kleinman, J. E., et al. (2007). Alpha7 nicotinic acetylcholine receptor mRNA expression and binding in postmortem human brain are associated with genetic variation in neuregulin 1. *Human Molecular Genetics, 16*, 2921–2932.

Mayford, M., Bach, M. E., Huang, Y. Y., Wang, L., Hawkins, R. D., & Kandel, E. R. (1996). Control of memory formation through regulated expression of a CaMKII transgene. *Science, 274*, 1678–1683.

McBain, C. J., & Kauer, J. A. (2009). Presynaptic plasticity: Targeted control of inhibitory networks. *Current Opinion in Neurobiology, 19*, 254–262.

McCullumsmith, R. E., Clinton, S. M., & Meador-Woodruff, J. H. (2004). Schizophrenia as a disorder of neuroplasticity. *International Review of Neurobiology, 59*, 19–45.

Mei, L., & Xiong, W. C. (2008). Neuregulin 1 in neural development, synaptic plasticity and schizophrenia. *Nature Reviews Neuroscience, 9*, 437–452.

Mercuri, N. B., Saiardi, A., Bonci, A., Picetti, R., Calabresi, P., Bernardi, G., & Borrelli, E. (1997). Loss of autoreceptor function in dopaminergic neurons from dopamine D2 receptor deficient mice. *Neuroscience, 79*, 323–327.

Mexal, S., Berger, R., Logel, J., Ross, R. G., Freedman, R., & Leonard, S. (2010). Differential regulation of alpha7 nicotinic receptor gene (CHRNA7) expression in schizophrenic smokers. *Journal of Molecular Neuroscience, 40*, 185–195.

Millar, J. K., Christie, S., & Porteous, D. J. (2003). Yeast two-hybrid screens implicate DISC1 in brain development and function. *Biochemical and Biophysical Research Communications, 311*, 1019–1025.

Millar, J. K., Pickard, B. S., Mackie, S., James, R., Christie, S., Buchanan, S. R., Malloy, M. P., Chubb, J. E., Huston, E., Baillie, G. S., et al. (2005). DISC1 and PDE4B are interacting genetic factors in schizophrenia that regulate cAMP signaling. *Science, 310*, 1187–1191.

Millar, J. K., Wilson-Annan, J. C., Anderson, S., Christie, S., Taylor, M. S., Semple, C. A., Devon, R. S., St Clair, D. M., Muir, W. J., Blackwood, D. H., et al. (2000). Disruption of two novel genes by a translocation co-segregating with schizophrenia. *Human Molecular Genetics, 9*, 1415–1423.

Mirnics, K., Middleton, F. A., Lewis, D. A., & Levitt, P. (2001). Analysis of complex brain disorders with gene expression microarrays: Schizophrenia as a disease of the synapse. *Trends in Neurosciences, 24*, 479–486.

Mirnics, K., Middleton, F. A., Marquez, A., Lewis, D. A., & Levitt, P. (2000). Molecular characterization of schizophrenia viewed by microarray analysis of gene expression in prefrontal cortex. *Neuron, 28*, 53–67.

Miyakawa, T., Yamada, M., Duttaroy, A., & Wess, J. (2001). Hyperactivity and intact hippocampus-dependent learning in mice lacking the M1 muscarinic acetylcholine receptor. *Journal of Neuroscience, 21*, 5239–5250.

Miyoshi, K., Honda, A., Baba, K., Taniguchi, M., Oono, K., Fujita, T., Kuroda, S., Katayama, T., & Tohyama, M. (2003). Disrupted-In-Schizophrenia 1, a candidate gene for schizophrenia, participates in neurite outgrowth. *Molecular Psychiatry, 8*, 685–694.

Moghaddam, B., & Adams, B. W. (1998). Reversal of phencyclidine effects by a group II metabotropic glutamate receptor agonist in rats. *Science, 281*, 1349–1352.

Mohn, A. R., Gainetdinov, R. R., Caron, M. G., & Koller, B. H. (1999). Mice with reduced NMDA receptor expression display behaviors related to schizophrenia. *Cell, 98*, 427–436.

Morris, J. A., Kandpal, G., Ma, L., & Austin, C. P. (2003). DISC1 (Disrupted-In-Schizophrenia 1) is a centrosome-associated protein that interacts with MAP1A, MIPT3, ATF4/5 and NUDEL: Regulation and loss of interaction with mutation. *Human Molecular Genetics, 12*, 1591–1608.

Mrzljak, L., Levey, A. I., & Goldman-Rakic, P. S. (1993). Association of m1 and m2 muscarinic receptor proteins with asymmetric synapses in the primate cerebral cortex: Morphological evidence for cholinergic modulation of excitatory neurotransmission. *Proceedings of the National Academy of Sciences of the United States of America, 90*, 5194–5198.

Nabeshima, T., Yamada, K., Yamaguchi, K., Hiramatsu, M., Furukawa, H., & Kameyama, T. (1983). Effect of lesions in the striatum, nucleus accumbens and medial raphe on phencyclidine-induced stereotyped behaviors and hyperactivity in rats. *European Journal of Pharmacology, 91*, 455–462.

Nason, M. W., Jr., Adhikari, A., Bozinoski, M., Gordon, J. A., & Role, L. W. (2011). Disrupted activity in the hippocampal-accumbens circuit of type III neuregulin 1 mutant mice. *Neuropsychopharmacology, 36*, 488–496.

Neubauer, H., Adams, M., & Redfern, P. (1975). The role of central cholinergic mechanisms in schizophrenia. *Medical Hypotheses, 1*, 32–34.

Nicodemus, K. K., Luna, A., Vakkalanka, R., Goldberg, T., Egan, M., Straub, R. E., & Weinberger, D. R. (2006). Further evidence for association between ErbB4 and schizophrenia and influence on cognitive intermediate phenotypes in healthy controls. *Molecular Psychiatry, 11*, 1062–1065.

Nikolaus, S., Antke, C., & Muller, H. W. (2009). In vivo imaging of synaptic function in the central nervous system: I. Movement disorders and dementia. *Behavioural Brain Research, 204*, 1–31.

Norton, N., Moskvina, V., Morris, D. W., Bray, N. J., Zammit, S., Williams, N. M., Williams, H. J., Preece, A. C., Dwyer, S., Wilkinson, J. C., et al. (2006). Evidence that interaction between neuregulin 1 and its receptor erbB4 increases susceptibility to schizophrenia. *American Journal of Medical Genetics B: Neuropsychiatric Genetics, 141B*, 96–101.

Numakawa, T., Yagasaki, Y., Ishimoto, T., Okada, T., Suzuki, T., Iwata, N., Ozaki, N., Taguchi, T., Tatsumi, M., Kamijima, K., et al. (2004). Evidence of novel neuronal functions of dysbindin, a susceptibility gene for schizophrenia. *Human Molecular Genetics, 13*, 2699–2708.

Okubo, Y., Suhara, T., Suzuki, K., Kobayashi, K., Inoue, O., Terasaki, O., Someya, Y., Sassa, T., Sudo, Y., Matsushima, E., et al. (1997). Decreased prefrontal dopamine D1 receptors in schizophrenia revealed by PET. *Nature, 385*, 634–636.

Olincy, A., Ross, R. G., Young, D. A., Roath, M., & Freedman, R. (1998). Improvement in smooth pursuit eye movements after cigarette smoking in schizophrenic patients. *Neuropsychopharmacol, 18*, 175–185.

Ozeki, Y., Tomoda, T., Kleiderlein, J., Kamiya, A., Bord, L., Fujii, K., Okawa, M., Yamada, N., Hatten, M. E., Snyder, S. H., et al. (2003). Disrupted-in-Schizophrenia-1 (DISC-1): Mutant truncation prevents binding to NudE-like (NUDEL) and inhibits neurite outgrowth. *Proceedings of the National Academy of Sciences of the United States of America, 100*, 289–294.

Paterson, D., & Nordberg, A. (2000). Neuronal nicotinic receptors in the human brain. *Progress in Neurobiology, 61*, 75–111.

Paylor, R., Nguyen, M., Crawley, J. N., Patrick, J., Beaudet, A., & Orr-Urtreger, A. (1998). Alpha7 nicotinic receptor subunits are not necessary for hippocampal-dependent learning or sensorimotor gating: A behavioral characterization of Acra7-deficient mice. *Learning and Memory, 5*, 302–316.

Petreanu, L., Mao, T., Sternson, S. M., & Svoboda, K. (2009). The subcellular organization of neocortical excitatory connections. *Nature, 457*, 1142–1145.

Reynolds, G. P., Czudek, C., & Andrews, H. B. (1990). Deficit and hemispheric asymmetry of GABA uptake sites in the hippocampus in schizophrenia. *Biological Psychiatry, 27*, 1038–1044.

Rouge-Pont, F., Usiello, A., Benoit-Marand, M., Gonon, F., Piazza, P. V., & Borrelli, E. (2002). Changes in extracellular dopamine induced by morphine and cocaine: Crucial control by D2 receptors. *Journal of Neuroscience, 22*, 3293–3301.

Sams-Dodd, F. (1995). Automation of the social interaction test by a video-tracking system: Behavioural effects of repeated phencyclidine treatment. *Journal of Neuroscience Methods, 59*, 157–167.

Sams-Dodd, F. (1996). Phencyclidine-induced stereotyped behaviour and social isolation in rats: A possible animal model of schizophrenia. *Behavioural Pharmacology, 7*, 3–23.

Sandrock, A. W., Jr., Dryer, S. E., Rosen, K. M., Gozani, S. N., Kramer, R., Theill, L. E., & Fischbach, G. D. (1997). Maintenance of acetylcholine receptor number by neuregulins at the neuromuscular junction in vivo. *Science, 276*, 599–603.

Sawaguchi, T. (2001). The effects of dopamine and its antagonists on directional delay-period activity of prefrontal neurons in monkeys during an oculomotor delayed-response task. *Neuroscience Research, 41*, 115–128.

Sawaguchi, T., & Goldman-Rakic, P. S. (1991). D1 dopamine receptors in prefrontal cortex: Involvement in working memory. *Science, 251*, 947–950.

Sawaguchi, T., & Goldman-Rakic, P. S. (1994). The role of D1-dopamine receptor in working memory: Local injections of dopamine antagonists into the prefrontal cortex of rhesus monkeys performing an oculomotor delayed-response task. *Journal of Neurophysiology, 71*, 515–528.

Scarr, E., Cowie, T. F., Kanellakis, S., Sundram, S., Pantelis, C., & Dean, B. (2009). Decreased cortical muscarinic receptors define a subgroup of subjects with schizophrenia. *Molecular Psychiatry, 14*, 1017–1023.

Schwab, S. G., Knapp, M., Mondabon, S., Hallmayer, J., Borrmann-Hassenbach, M., Albus, M., Lerer, B., Rietschel, M., Trixler, M., Maier, W., et al. (2003). Support for association of schizophrenia with genetic variation in the 6p22.3 gene, dysbindin, in sib-pair families with linkage and in an additional sample of triad families. *American Journal of Human Genetics, 72*, 185–190.

Selemon, L. D., & Goldman-Rakic, P. S. (1999). The reduced neuropil hypothesis: A circuit based model of schizophrenia. *Biological Psychiatry, 45*, 17–25.

Sesack, S. R., Hawrylak, V. A., Matus, C., Guido, M. A., & Levey, A. I. (1998). Dopamine axon varicosities in the prelimbic division of the rat prefrontal cortex exhibit sparse immunoreactivity for the dopamine transporter. *Journal of Neuroscience, 18*, 2697–2708.

Sherman, A. D., Davidson, A. T., Baruah, S., Hegwood, T. S., & Waziri, R. (1991a). Evidence of glutamatergic deficiency in schizophrenia. *Neuroscience Letters, 121*, 77–80.

Sherman, A. D., Hegwood, T. S., Baruah, S., & Waziri, R. (1991b). Deficient NMDA-mediated glutamate release from synaptosomes of schizophrenics. *Biological Psychiatry, 30*, 1191–1198.

Shinoe, T., Matsui, M., Taketo, M. M., & Manabe, T. (2005). Modulation of synaptic plasticity by physiological activation of M1 muscarinic acetylcholine receptors in the mouse hippocampus. *Journal of Neuroscience, 25*, 11194–11200.

Sigurdsson, T., Stark, K. L., Karayiorgou, M., Gogos, J. A., & Gordon, J. A. (2010). Impaired hippocampal-prefrontal synchrony in a genetic mouse model of schizophrenia. *Nature, 464*, 763–767.

Silberberg, G., Darvasi, A., Pinkas-Kramarski, R., & Navon, R. (2006). The involvement of ErbB4 with schizophrenia: Association and expression studies. *American Journal of Medical Genetics B: Neuropsychiatric Genetics, 141B*, 142–148.

Simpson, M. D., Slater, P., Deakin, J. F., Royston, M. C., & Skan, W. J. (1989). Reduced GABA uptake sites in the temporal lobe in schizophrenia. *Neuroscience Letters, 107*, 211–215.

Sohal, V. S., Zhang, F., Yizhar, O., & Deisseroth, K. (2009). Parvalbumin neurons and gamma rhythms enhance cortical circuit performance. *Nature, 459*, 698–702.

Song, W., Li, W., Feng, J., Heston, L. L., Scaringe, W. A., & Sommer, S. S. (2008). Identification of high risk DISC1 structural variants with a 2% attributable risk for schizophrenia. *Biochemical and Biophysical Research Communications, 367*, 700–706.

St Clair, D., Blackwood, D., Muir, W., Carothers, A., Walker, M., Spowart, G., Gosden, C., & Evans, H. J. (1990). Association within a family of a balanced autosomal translocation with major mental illness. *Lancet, 336*, 13–16.

Stefansson, H., Sarginson, J., Kong, A., Yates, P., Steinthorsdottir, V., Gudfinnsson, E., Gunnarsdottir, S., Walker, N., Petursson, H., Crombie, C., et al. (2003). Association of neuregulin 1 with schizophrenia confirmed in a Scottish population. *American Journal of Human Genetics, 72*, 83–87.

Stefansson, H., Sigurdsson, E., Steinthorsdottir, V., Bjornsdottir, S., Sigmundsson, T., Ghosh, S., Brynjolfsson, J., Gunnarsdottir, S., Ivarsson, O., Chou, T. T., et al. (2002). Neuregulin 1 and susceptibility to schizophrenia. *American Journal of Human Genetics, 71*, 877–892.

Steiner, H., Blum, M., Kitai, S. T., & Fedi, P. (1999). Differential expression of ErbB3 and ErbB4 neuregulin receptors in dopamine neurons and forebrain areas of the adult rat. *Experimental Neurology, 159*, 494–503.

Stephan, K. E., Baldeweg, T., & Friston, K. J. (2006). Synaptic plasticity and dysconnection in schizophrenia. *Biological Psychiatry, 59*, 929–939.

Straub, R. E., MacLean, C. J., Ma, Y., Webb, B. T., Myakishev, M. V., Harris-Kerr, C., Wormley, B., Sadek, H., Kadambi, B., O'Neill, F. A., et al. (2002). Genome-wide scans of three independent sets of 90 Irish multiplex schizophrenia families and follow-up of selected regions in all families provides evidence for multiple susceptibility genes. *Molecular Psychiatry, 7*, 542–559.

Straub, R. E., MacLean, C. J., O'Neill, F. A., Burke, J., Murphy, B., Duke, F., Shinkwin, R., Webb, B. T., Zhang, J., Walsh, D., et al. (1995). A potential vulnerability locus for schizophrenia on chromosome 6p24-22: Evidence for genetic heterogeneity. *Nature Genetics, 11*, 287–293.

Sturgeon, R. D., Fessler, R. G., & Meltzer, H. Y. (1979). Behavioral rating scales for assessing phencyclidine-induced locomotor activity, stereotyped behavior and ataxia in rats. *European Journal of Pharmacology, 59*, 169–179.

Takahashi, K., Tanabe, K., Ohnuki, M., Narita, M., Ichisaka, T., Tomoda, K., & Yamanaka, S. (2007). Induction of pluripotent stem cells from adult human fibroblasts by defined factors. *Cell, 131*, 861–872.

Talbot, K., Eidem, W. L., Tinsley, C. L., Benson, M. A., Thompson, E. W., Smith, R. J., Hahn, C. G., Siegel, S. J., Trojanowski, J. Q., Gur, R. E., et al. (2004). Dysbindin-1 is reduced in intrinsic, glutamatergic terminals of the hippocampal formation in schizophrenia. *The Journal of Clinical Investigation, 113*, 1353–1363.

Tandon, R., Dutchak, D., & Greden, J. F. (1989). Cholinergic syndrome following anticholinergic withdrawal in a schizophrenic patient abusing marijuana. *The British Journal of Psychiatry, 154*, 712–714.

Tang, J., LeGros, R. P., Louneva, N., Yeh, L., Cohen, J. W., Hahn, C. G., Blake, D. J., Arnold, S. E., & Talbot, K. (2009). Dysbindin-1 in dorsolateral prefrontal cortex of schizophrenia cases is reduced in an isoform-specific manner unrelated to dysbindin-1 mRNA expression. *Human Molecular Genetics, 18*, 3851–3863.

Tang, J. X., Zhou, J., Fan, J. B., Li, X. W., Shi, Y. Y., Gu, N. F., Feng, G. Y., Xing, Y. L., Shi, J. G., & He, L. (2003). Family-based association study of DTNBP1 in 6p22.3 and schizophrenia. *Molecular Psychiatry, 8*, 717–718.

Thomsen, M. S., Hansen, H. H., Timmerman, D. B., & Mikkelsen, J. D. (2010). Cognitive improvement by activation of alpha7 nicotinic acetylcholine receptors: From animal models to human pathophysiology. *Current Pharmaceutical Design, 16*, 323–343.

Thuret, S., Alavian, K. N., Gassmann, M., Lloyd, C. K., Smits, S. M., Smidt, M. P., Klein, R., Dyck, R. H., & Simon, H. H. (2004). The neuregulin receptor, ErbB4, is not required for normal development and adult maintenance of the substantia nigra pars compacta. *Journal of Neurochemistry, 91*, 1302–1311.

Ting, A. K., Chen, Y., Wen, L., Yin, D. M., Shen, C., Tao, Y., Liu, X., Xiong, W. C., & Mei, L. (2011). Neuregulin 1 promotes excitatory synapse development and function in GABAergic interneurons. *Journal of Neuroscience, 31*, 15–25.

Toda, M., & Abi-Dargham, A. (2007). Dopamine hypothesis of schizophrenia: Making sense of it all. *Current Psychiatry Reports, 9*, 329–336.

Tregellas, J. R., Tanabe, J., Rojas, D. C., Shatti, S., Olincy, A., Johnson, L., Martin, L. F., Soti, F., Kem, W. R., Leonard, S., et al. (2011). Effects of an alpha 7-nicotinic agonist on default network activity in schizophrenia. *Biological Psychiatry, 69*, 7–11.

Tsai, G., Passani, L. A., Slusher, B. S., Carter, R., Baer, L., Kleinman, J. E., & Coyle, J. T. (1995). Abnormal excitatory neurotransmitter metabolism in schizophrenic brains. *Archives of General Psychiatry, 52*, 829–836.

Tseng, K. Y., & O'Donnell, P. (2004). Dopamine-glutamate interactions controlling prefrontal cortical pyramidal cell excitability involve multiple signaling mechanisms. *Journal of Neuroscience, 24*, 5131–5139.

Usdin, T. B., & Fischbach, G. D. (1986). Purification and characterization of a polypeptide from chick brain that promotes the accumulation of acetylcholine receptors in chick myotubes. *The Journal of Cell Biology, 103*, 493–507.

Usiello, A., Baik, J. H., Rouge-Pont, F., Picetti, R., Dierich, A., LeMeur, M., Piazza, P. V., & Borrelli, E. (2000). Distinct functions of the two isoforms of dopamine D2 receptors. *Nature, 408*, 199–203.

van Rossum, J. M. (1966). The significance of dopamine-receptor blockade for the mechanism of action of neuroleptic drugs. *Archives Internationales de Pharmacodynamie et de Thérapie, 160*, 492–494.

Volk, D. W., Austin, M. C., Pierri, J. N., Sampson, A. R., & Lewis, D. A. (2000). Decreased glutamic acid decarboxylase67 messenger RNA expression in a subset of prefrontal cortical gamma-aminobutyric acid neurons in subjects with schizophrenia. *Archives of General Psychiatry, 57*, 237–245.

Volk, D., Austin, M., Pierri, J., Sampson, A., & Lewis, D. (2001). GABA transporter-1 mRNA in the prefrontal cortex in schizophrenia: Decreased expression in a subset of neurons. *The American Journal of Psychiatry, 158*, 256–265.

Volk, D. W., Pierri, J. N., Fritschy, J. M., Auh, S., Sampson, A. R., & Lewis, D. A. (2002). Reciprocal alterations in pre- and postsynaptic inhibitory markers at chandelier cell inputs to pyramidal neurons in schizophrenia. *Cerebral Cortex, 12*, 1063–1070.

Vullhorst, D., Neddens, J., Karavanova, I., Tricoire, L., Petralia, R. S., McBain, C. J., & Buonanno, A. (2009). Selective expression of ErbB4 in interneurons, but not pyramidal cells, of the rodent hippocampus. *Journal of Neuroscience, 29*, 12255–12264.

Weickert, C. S., Rothmond, D. A., Hyde, T. M., Kleinman, J. E., & Straub, R. E. (2008). Reduced DTNBP1 (dysbindin-1) mRNA in the hippocampal formation of schizophrenia patients. *Schizophrenia Research, 98*, 105–110.

Weickert, C. S., Straub, R. E., McClintock, B. W., Matsumoto, M., Hashimoto, R., Hyde, T. M., Herman, M. M., Weinberger, D. R., & Kleinman, J. E. (2004). Human dysbindin (DTNBP1) gene expression in normal brain and in schizophrenic prefrontal cortex and midbrain. *Archives of General Psychiatry, 61*, 544–555.

Wen, L., Lu, Y. S., Zhu, X. H., Li, X. M., Woo, R. S., Chen, Y. J., Yin, D. M., Lai, C., Terry, A. V., Jr., Vazdarjanova, A., et al. (2010). Neuregulin 1 regulates pyramidal neuron activity via ErbB4 in parvalbumin-positive interneurons. *Proceedings of the National Academy of Sciences of the United States of America, 107*, 1211–1216.

Whittington, M. A., Cunningham, M. O., LeBeau, F. E., Racca, C., & Traub, R. D. (2011). Multiple origins of the cortical gamma rhythm. *Developmental Neurobiology, 71*, 92–106.

Williams, G. V., & Goldman-Rakic, P. S. (1995). Modulation of memory fields by dopamine D1 receptors in prefrontal cortex. *Nature, 376*, 572–575.

Wong, D. F., Wagner, H. N., Jr., Tune, L. E., Dannals, R. F., Pearlson, G. D., Links, J. M., Tamminga, C. A., Broussolle, E. P., Ravert, H. T., Wilson, A. A., et al. (1986). Positron emission tomography reveals elevated D2 dopamine receptors in drug-naive schizophrenics. *Science, 234*, 1558–1563.

Woo, R. S., Li, X. M., Tao, Y., Carpenter-Hyland, E., Huang, Y. Z., Weber, J., Neiswender, H., Dong, X. P., Wu, J., Gassmann, M., et al. (2007). Neuregulin-1 enhances depolarization-induced GABA release. *Neuron, 54*, 599–610.

Woo, T. U., Miller, J. L., & Lewis, D. A. (1997). Schizophrenia and the parvalbumin-containing class of cortical local circuit neurons. *The American Journal of Psychiatry, 154*, 1013–1015.

Woo, T. U., Whitehead, R. E., Melchitzky, D. S., & Lewis, D. A. (1998). A subclass of prefrontal gamma-aminobutyric acid axon terminals are selectively altered in schizophrenia. *Proceedings of the National Academy of Sciences of the United States of America, 95*, 5341–5346.

Yang, X., Kuo, Y., Devay, P., Yu, C., & Role, L. (1998). A cysteine-rich isoform of neuregulin controls the level of expression of neuronal nicotinic receptor channels during synaptogenesis. *Neuron, 20*, 255–270.

Yang, J. Z., Si, T. M., Ruan, Y., Ling, Y. S., Han, Y. H., Wang, X. L., Zhou, M., Zhang, H. Y., Kong, Q. M., Liu, C., et al. (2003). Association study of neuregulin 1 gene with schizophrenia. *Molecular Psychiatry, 8*, 706–709.

Yau, H. J., Wang, H. F., Lai, C., & Liu, F. C. (2003). Neural development of the neuregulin receptor ErbB4 in the cerebral cortex and the hippocampus: Preferential expression by interneurons tangentially migrating from the ganglionic eminences. *Cerebral Cortex, 13*, 252–264.

Yu, J., Vodyanik, M. A., Smuga-Otto, K., Antosiewicz-Bourget, J., Frane, J. L., Tian, S., Nie, J., Jonsdottir, G. A., Ruotti, V., Stewart, R., et al. (2007). Induced pluripotent stem cell lines derived from human somatic cells. *Science, 318*, 1917–1920.

Yurek, D. M., Zhang, L., Fletcher-Turner, A., & Seroogy, K. B. (2004). Supranigral injection of neuregulin1-beta induces striatal dopamine overflow. *Brain Research, 1028*, 116–119.

Zavitsanou, K., Katsifis, A., Mattner, F., & Huang, X. F. (2004). Investigation of m1/m4 muscarinic receptors in the anterior cingulate cortex in schizophrenia, bipolar disorder, and major depression disorder. *Neuropsychopharmacol, 29*, 619–625.

Zhang, L., Fletcher-Turner, A., Marchionni, M. A., Apparsundaram, S., Lundgren, K. H., Yurek, D. M., & Seroogy, K. B. (2004). Neurotrophic and neuroprotective effects of the neuregulin glial growth factor-2 on dopaminergic neurons in rat primary midbrain cultures. *Journal of Neurochemistry, 91*, 1358–1368.

Zhang, Y., Hamilton, S. E., Nathanson, N. M., & Yan, J. (2006). Decreased input-specific plasticity of the auditory cortex in mice lacking M1 muscarinic acetylcholine receptors. *Cerebral Cortex, 16*, 1258–1265.

Zheng, Y., Watakabe, A., Takada, M., Kakita, A., Namba, H., Takahashi, H., Yamamori, T., & Nawa, H. (2009). Expression of ErbB4 in substantia nigra dopamine neurons of monkeys and humans. *Progress in Neuro-Psychopharmacology & Biological Psychiatry, 33*, 701–706.

Zhong, C., Du, C., Hancock, M., Mertz, M., Talmage, D. A., & Role, L. W. (2008). Presynaptic type III neuregulin 1 is required for sustained enhancement of hippocampal transmission by nicotine and for axonal targeting of alpha7 nicotinic acetylcholine receptors. *Journal of Neuroscience, 28*, 9111–9116.

Zvara, A., Szekeres, G., Janka, Z., Kelemen, J. Z., Cimmer, C., Santha, M., & Puskas, L. G. (2005). Over-expression of dopamine D2 receptor and inwardly rectifying potassium channel genes in drug-naive schizophrenic peripheral blood lymphocytes as potential diagnostic markers. *Disease Markers, 21*, 61–69.

Chapter 23
Molecular and Cellular Aspects of Mental Retardation in the Fragile X Syndrome: From Gene Mutation/s to Spine Dysmorphogenesis

Silvia De Rubeis, Esperanza Fernández, Andrea Buzzi, Daniele Di Marino, and Claudia Bagni

Abstract The Fragile X syndrome (FXS) is the most frequent form of inherited mental retardation and also considered a monogenic cause of Autism Spectrum Disorder. FXS symptoms include neurodevelopmental delay, anxiety, hyperactivity, and autistic-like behavior. The disease is due to mutations or loss of the Fragile X Mental Retardation Protein (FMRP), an RNA-binding protein abundant in the brain and gonads, the two organs mainly affected in FXS patients. FMRP has multiple functions in RNA metabolism, including mRNA decay, dendritic targeting of mRNAs, and protein synthesis. In neurons lacking FMRP, a wide array of mRNAs encoding proteins involved in synaptic structure and function are altered. As a result of this complex dysregulation, in the absence of FMRP, spine morphology and functioning is impaired. Consistently, model organisms for the study of the syndrome recapitulate the phenotype observed in FXS patients, such as dendritic spine anomalies and defects in learning.

Here, we review the fundamentals of genetic and clinical aspects of FXS, devoting a specific attention to ASD comorbidity and FXS-related diseases. We also review the current knowledge on FMRP functions through structural, molecular, and cellular findings. Finally, we discuss the neuroanatomical, electrophysiological, and

S. De Rubeis • E. Fernández • A. Buzzi • D. Di Marino
Center for Human Genetics, Katholieke Universiteit Leuven, 3000 Leuven, Belgium

Center for the Biology of Disease, Flanders Institute for Biotechnology (VIB), 3000 Leuven, Belgium

C. Bagni (✉)
Center for Human Genetics, Katholieke Universiteit Leuven, 3000 Leuven, Belgium

Center for the Biology of Disease, Flanders Institute for Biotechnology (VIB), 3000 Leuven, Belgium

Department of Experimental Medicine and Biochemical Sciences, University "Tor Vergata", 00133 Rome, Italy
e-mail: claudia.bagni@uniroma2.it; claudia.bagni@med.kuleuven.be

M.R. Kreutz and C. Sala (eds.), *Synaptic Plasticity*,
Advances in Experimental Medicine and Biology 970,
DOI 10.1007/978-3-7091-0932-8_23, © Springer-Verlag/Wien 2012

behavioral defects caused by FMRP loss, as well as the current treatments able to partially revert some of the FXS abnormalities.

Keywords FMR1 • FMRP • Fragile X • Messenger ribonucleoparticles • Spinogenesis

23.1 Genetics of Fragile X (FXS) and Fragile X Tremor Ataxia (FXTAS) Syndromes

23.1.1 Fragile Mental Retardation 1: A Gene Associated to Two Neurological Diseases

The Fragile X syndrome (FXS) is the most frequent form of inherited intellectual disability (Jacquemont et al. 2007). Patients with FXS show physical features, such as large ears, elongated face, and high-arched palate, which have been reported in 60% of prepubertal FXS boys. Other symptoms include connective tissue anomalies, which can lead to mitral valve prolapse, scoliosis, flat feet, and joint laxity. Recurrent otitis media and strabismus are also common. Macroorchidism due to a hypothalamic dysfunction affects about 90% of boys with FXS by the age of 14 (Jacquemont et al. 2007).

The neurological involvement displays a broad spectrum of cognitive and behavioral deficits. The developmental delay is the most consistent feature, with a mean IQ of 42 in boys and severe mental retardation in about 25% of cases. Since the disorder is X-linked and the penetrance is variable, females are usually in a low–normal range, with an IQ ranging from 70 to 90 (Jacquemont et al. 2007). Moreover, epilepsy has been described in 13–18% of boys and 4% in girls, but normally, the seizures and EEG alterations tend to resolve during childhood or early adulthood (Berry-Kravis 2002; Musumeci et al. 1999). Despite the severe neurobehavioral symptoms, the anatomical studies revealed minor abnormalities in postmortem brains from FXS patients (Hallahan et al. 2010; Reiss et al. 1995). The most prominent neuroanatomical feature is the dysgenesis of the dendritic spines, which appear longer and thinner than normal, likely due to a developmental delay in spine dynamics and transition from immature to mature spines (Cruz-Martin et al. 2010; Irwin et al. 2001). FXS is also the most common monogenic cause of Autism Spectrum Disorder, ASD (Hatton et al. 2006), a heterogenous group of neurodevelopmental pathologies affecting approximately 37 individuals in 10,000 (Fombonne 2005) and present in more than 40% of patients with intellectual disability (Moss and Howlin 2009). About 25% of FXS boys and 6% of girls meet criteria for ASD, while 1–2% of patients affected by ASD have FXS (Abrahams and Geschwind 2008; Hatton et al. 2006). In particular, recent reports estimated that about 30% of FXS subjects meet criteria for Autistic Disorder and 30% for Pervasive Developmental Disorder Not Otherwise Specified (Harris et al. 2008). However, up to 90% of children with Fragile X display behavioral alterations which resemble ASD, such as social anxiety, gaze avoidance, delayed speech development, echolalia, sensory hypersensitivity, tactile defensiveness, stereotypic movements, and poor motor coordination (Belmonte and Bourgeron 2006; Hernandez et al. 2009). The cognitive delay is more severe in FXS children with ASD, and additional neurological disorders, genetic problems, or seizures may increase the risk of autism (Garcia-Nonell et al. 2008).

FXS is due to triplet repeat expansion or point mutations in the Fragile X mental retardation 1 (FMR1) gene, located on chromosome Xq27-3 (Fig. 23.1). A severe FXS form has also been documented in a patient with a mutation in the coding region of the gene, leading to the substitution of isoleucine 304 for asparagine (Ile304Asn, see below) (De Boulle et al. 1993). Few cases with deletions in the coding regions have also been identified (Gedeon et al. 1992; Meijer et al. 1994; Mila et al. 2000; Wohrle et al. 1992). In over 90% of patients, a CGG triplet in the 5′ UTR of the gene is expanded over 200 copies, leading to hypermethylation of the CGG, transcriptional silencing, and abolished production of the Fragile X Mental Retardation Protein (FMRP) (Jacquemont et al. 2007). The CGG triplet region is highly polymorphic in the population. Normal alleles (5–44 CGG copies) are stably transmitted to the offspring; "gray-zone" alleles (45–54 copies) and "premutation" alleles (55–200 copies) are rather unstable and can evolve into a "full mutation" (>200 repeats) during the maternal transmission (Fig. 23.1). While the gray-zone alleles require at least two generations before expanding to a full mutation (Fernandez-Carvajal et al. 2009), the premutation is highly unstable, and the risk of transmitting an allele with the full mutation is function of the repeat length (Hagerman and Hagerman 2002). The carriers of premutation alleles were considered clinically unaffected since the discovery of a dominant late-onset neurodegenerative disorder: the Fragile X Tremor Ataxia Syndrome (FXTAS). In 2001, the Hagerman laboratory described for the first time action tremor associated with executive function impairments and brain atrophy in five elderly men with the premutation (Hagerman et al. 2001). The frequency of premutation carriers has

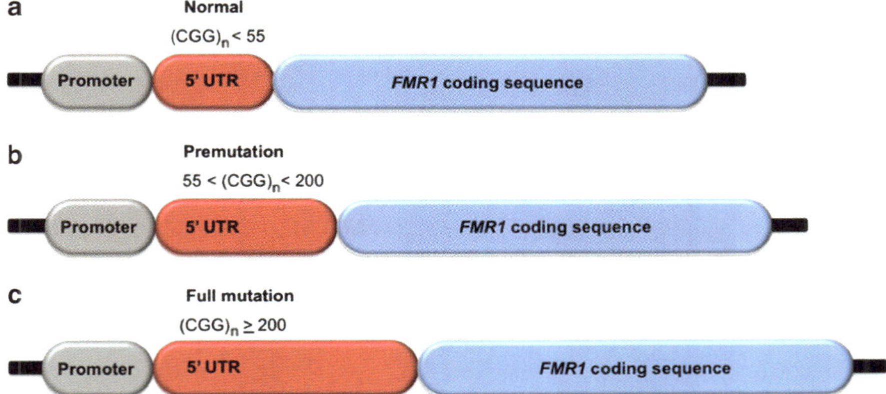

Fig. 23.1 Scheme of the *FMR1* gene which includes the promoter, the 5′ UTR, and the *FMR1-*coding sequence in a normal allele (5–44 CGG copies) (**a**), a premutated allele (55–200 copies) (**b**), and a full mutated allele (>200 repeats) (**c**)

been estimated in 1:800 in men and 1:300 in women, although FXTAS displays reduced penetrance (~33% in men, not yet determined in women). Major behavioral diagnostic criteria for FXTAS are gait ataxia and intention tremor, eventually associated with parkinsonism and cognitive decline, often progressing to dementia (Hagerman and Hagerman 2007). Psychiatric disturbances often observed in FXTAS patients include anxiety, depression, and hostility (Bacalman et al. 2006). In some premutation patients, a psychiatric phenotype with features resembling attention deficit hyperactivity disorder (ADHD) and ASD can also appear in childhood or adolescence (Farzin et al. 2006). Further studies on postmortem brains from premutation carriers revealed a generalized brain atrophy, white matter disease, and middle cerebellar peduncle lesions (Hagerman and Hagerman 2007). One of the key cytological hallmarks of FXTAS is the presence of intranuclear ubiquitin-positive inclusions in neurons and astrocytes throughout the brain (Greco et al. 2006). The intranuclear foci were consistently observed in model organisms for FXTAS, both mouse (Berman and Willemsen 2009) and *Drosophila* (Jin et al. 2007; Sofola et al. 2007). Remarkably, the number of inclusions correlates with the size of the CGG expansion (Greco et al. 2002, 2006).

FXS and FXTAS are both due to triplet expansions. While FXS is a loss-of-function disease, FXTAS is thought to be a consequence of RNA toxic gain-of-function mechanism. First, a consistent molecular feature in both FXTAS patients and mouse models is the elevation of aberrant CGG expanded *FMR1* mRNA levels (Allen et al. 2004; Kenneson et al. 2001; Tassone et al. 2000), due to increased transcription (Tassone et al. 2007). Nevertheless, carriers of premutation alleles show decreased levels of FMRP (Brouwer et al. 2008; Entezam et al. 2007) caused by the reduced translational efficiency of the *FMR1* mRNA carrying the CGG expansion (Primerano et al. 2002), as well as by a differential use of the *FMR1* mRNA 3' UTR (Tassone et al. 2011). It has been proposed that the CGG expansion in the 5' UTR would form a secondary structure inhibiting the ribosome scanning and thus leading to a scarce translational efficiency (Feng et al. 1995). The intranuclear foci contain in addition to the aberrant *FMR1* mRNA (Greco et al. 2002; Tassone et al. 2004) a variety of RNA-binding proteins (RBPs) interacting with the rCGG tract, such as PURalpha, hnRNP A2/B1, and CUG-BP1 (Iwahashi et al. 2006).

The *FMR1* gene encodes for FMRP (Bassell and Warren 2008; De Rubeis and Bagni 2010), an RNA-binding protein that contributes to the posttranscriptional control of gene expression (see below). In neurons, FMRP is part of messenger ribonucleoparticles (mRNPs) and regulates dendritic transport of associated mRNAs, their stability and local translation (Bassell and Warren 2008; De Rubeis and Bagni 2010). The roles of FMRP have been mainly addressed by using animal models that mimic FXS.

23.1.2 Model Organisms for the Study of the Fragile X Syndrome

The *FMR1* gene is conserved along evolution, and this allowed researchers to develop murine (Bakker 1994; Mientjes et al. 2006), *Drosophila* (Zhang et al. 2001), and zebrafish (den Broeder et al. 2009; Tucker et al. 2006) animal models to study the molecular, cellular, and behavioral phenotypes of the syndrome.

Mouse models. The first model available, the *Fmr1* knockout (KO) mouse, was generated by interrupting exon 5 of the *Fmr1* gene with a neomycin cassette (Bakker 1994). Although the insertional mutation does not mimic FXS in humans, it leads to the functional ablation of *Fmr1* gene since the interrupted *Fmr1* mRNA prevents the translation of a functional FMRP (Bakker 1994). This mouse model presents an array of anatomical, behavioral, and neurological similarities to FXS patients (see below). Recently, a conditional KO (*Fmr1* CKO) and a second generation *Fmr1* KO null for *Fmr1* mRNA (*Fmr1* KO 2) have been generated by flanking the murine promoter and the first exon with loxP sites (Mientjes et al. 2006). In *Fmr1* CKO, *Fmr1* expression can be suppressed at specific developmental stages or in specific cell types, as showed by crossing these mice with a line carrying the Cre recombinase driven by a Purkinje cell–specific promoter (Mientjes et al. 2006). Moreover, a mouse model mimicking the mutation Ile304Asn that leads to a severe FXS manifestation (De Boulle et al. 1993) has been recently generated; of interest, this model phenocopies the behavioral and electrophysiological defects observed in *Fmr1* KO mice (Zang et al. 2009) (see below).

Fruit fly Models. In *Drosophila melanogaster*, several loss-of-function mutations, ranging from hypomorphs to nulls, have been generated (Dockendorff et al. 2002; Inoue et al. 2002; Morales et al. 2002; Zhang et al. 2001). Such models display a variety of behavioral and developmental defects (Zarnescu et al. 2005).

Zebrafish Models. The first attempt to produce a model for FXS in *Danio rerio* was performed in 2006 using a knockdown approach by microinjecting morpholinos in early embryos (Tucker et al. 2006). Although the authors described defects in craniofacial development and neuronal branching in embryos, further studies failed in reproducing this phenotype in two *Fmr1* KO lines (den Broeder et al. 2009).

23.2 An Insight into the Structure of the Fragile X Mental Retardation Protein

The *FMR1* gene is composed of 17 exons and subjected to alternative splicing, occurring preferentially at the level of exons 12, 14, 15, and 17. This generates up to 12 different protein isoforms, with a molecular weight ranging between 70 and 80 kDa, with the longest isoform containing 632 amino acidic residues (Bassell and Warren 2008). The role of each isoform still needs to be clarified (Bassell and Warren 2008). Whereas FMRP isoforms are similarly expressed in many tissues and organs, the relative abundance of each isoform seems to be tissue specific (Kaufmann et al. 2002; Xie et al. 2009).

FMRP is a multidomain RNA-binding protein able to recognize several coding and noncoding RNAs, including the brain cytoplasmic RNA BC1/BC200 (Ashley et al. 1993; Johnson et al. 2006; Napoli et al. 2008; O'Donnell and Warren 2002;

Siomi et al. 1993; Zalfa et al. 2005) and microRNAs (Edbauer et al. 2010). Moreover, FMRP homodimerizes and interacts with several cytoplasmic and nuclear proteins, including the two paralogs Fragile-X-related proteins 1 and 2 (FXRP1 and FXRP2) (O'Donnell and Warren 2002; Tamanini et al. 1999; Zhang et al. 1995).

The protein can be structurally divided into three main regions: N-terminal region, central region, and C-terminal region (Fig. 23.2a). The N-terminal region is characterized by the presence of two Tudor domains (TD), a putative Helix-Loop-Helix domain (HLH), and a Nuclear Localization Signal (NLS) (Sjekloca et al. 2009) (Fig. 23.2a). The central region contains two K Homology domains (KH) that share a high degree of homology with the hnRNP K domain and a Nuclear Export Signal (NES) (Valverde et al. 2008) (Fig. 23.2a). The C-terminal region, which is the less conserved region among the different species, is characterized by the presence of an RGG box containing a conserved Arg-Gly-Gly triplet (Darnell et al. 2001; Menon et al. 2004; Sjekloca et al. 2009) (Fig. 23.2a).

A region modulated by phosphorylation is localized between the FMRP central portion and the RGG box, specifically located between the amino acids 483 and 521 and conserved along different species. Ceman and colleagues showed that the phosphorylation of serine 499 triggers hierarchical phosphorylation events of nearby serines (Ceman et al. 2003). This phosphorylation modulates the association of FMRP with Dicer involving FMRP in the miRNA pathway (Cheever and Ceman 2009).

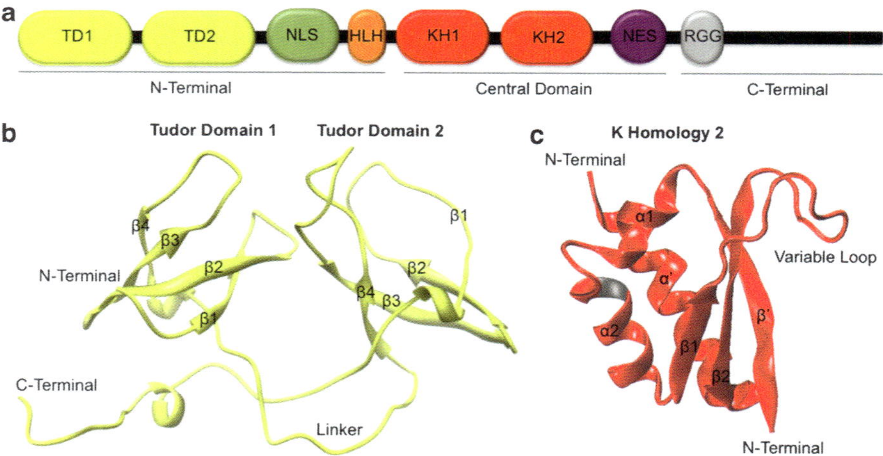

Fig. 23.2 Structural modules of FMRP. (a) The different domains that compose the protein are represented following the color code: *yellow* = Tudor domains, *blue* = nuclear localization signal, *green* = helix-loop-helix motif, *red* = K homology domains, *purple* = nuclear export signal, *gray* = RGG box. The division in N-terminal, central, and C-terminal regions is also represented. (b) Ribbon representation of the structure of the two Tudor domains represented in *yellow* and (c) of the KH2 domain represented in *red*. In both panels, the succession of the secondary structure elements is *underlined*. The position of residue Ile304 is highlighted in *gray*

The N-terminal region of FMRP contains two Tudor domains able to bind single-strand (ss) nucleic acids. The same activity is shared by the two KH domains in the central region (Darnell et al. 2001; Musco et al. 1996; Ramos et al. 2006; Valverde et al. 2007) and the RGG box in the C-terminal region. The structure of the entire protein has not been resolved so far, but NMR and X-ray structures of single domains are available (Ramos et al. 2006; Valverde et al. 2007). The structure of the first 134 residues of the N-terminal domain, resolved by NMR, reveals the three-dimensional organization of the two Tudor domains, each one formed by a barrel-like fold made of four-stranded antiparallel β sheet (Fig. 23.2b). The two Tudor domains are linked by an unstructured fragment (linker) (Ramos et al. 2006) (Fig. 23.2b). The structure of the first 134 residues reveals three structural motifs, the two Tudor domains repeats and one α helix (Ramos et al. 2006) (Fig. 23.2b). Extensive interactions are observed between these elements, strongly suggesting that all the elements are necessary for the stability of the overall N-terminal domain (Ramos et al. 2006). The structure also reveals hydrophobic pockets on the surface of the two Tudor domains, in analogy with other Tudor domains, such as the Survival of Motor Neuron (SMN) Tudor domain and the heterochromatin-associated protein 1 (HP1) chromo domains (Nielsen et al. 2002b; Sprangers et al. 2003). These hydrophobic pockets can bind methylated amino acids (Ramos et al. 2006).

The three-dimensional structure of the two KH domains located in the central region of FMRP has been solved by X-ray (Valverde et al. 2007). The KH domains, usually present in multiple copies in a protein, contain the consensus (ILV)-I-G-X_2-G-X_2-I sequence and are responsible for the interaction with ssDNA, mRNA, and rRNA. The KH domains consist of a $\beta^1\alpha^1\alpha^2\beta^2\beta'\alpha'$ fold, three-dimensionally oriented as a three-stranded β-sheet domain opposed to a three α-helices domain (Valverde et al. 2008), as evidenced by the X-ray structure (Valverde et al. 2007) (Fig. 23.2c). The KH domains are classified as Type I or Type II folds. Both contain the minimal KH motif but with different C- or N-terminal extensions giving $\beta^1\alpha^1\alpha^2\beta^1\beta'\alpha'$ and $\alpha'\beta'\beta^1\alpha^1\alpha^2\beta^2$ for Type I and Type II, respectively (Valverde et al. 2007) (Fig. 23.2c). KH domains in eukaryotic proteins are exclusively Type I, whereas in prokaryotic proteins are exclusively Type II (Grishin 2001; Siomi et al. 1993). β-Sheets β^2 and β' are connected by a variable loop, while α-helices α^2 and α' are connected by the so-called GXXG loop (Fig. 23.2c). This HLH (Helix-Loop-Helix) region of the domain forms the nucleic acid binding site (Valverde et al. 2008).

The best characterized missense mutation for FMRP is Ile304Asn, located on the KH2 domain (De Boulle et al. 1993). As previously mentioned, this mutation has been reported in an individual with a severe manifestation of FXS, including very low IQ, macroorchidism, and severe social and behavioral impairment (De Boulle et al. 1993). This underlines the importance of the KH2 domain for the neuronal functions of FMRP. The structure of the two KH domains revealed that the Ile304 residue is located on helix α2 of the KH2 domain (Fig. 23.2c). This residue is part of the hydrophobic core that stabilizes the three-dimensional folding of the domain. Indeed, the hydrophobic residues present at the interface of the α-helices and β-sheets domain establish a hydrophobic network of interactions that maintains the domain structure. This network of hydrophobic packing and van der Waals interactions is hypothesized to be conserved among all the Fragile-X-related proteins, such as FXRP1 and FXRP2, since all the hydrophobic residues abovementioned are conserved (Valverde et al. 2007). The substitution of residue Ile304 by an Asn, located in this hydrophobic core may disrupt this network, affecting the structure of the domain (Fig. 23.2c). Furthermore, being Ile304 buried in the domain core and not solvent accessible, it seems that the residue is not directly involved in the binding with nucleic acids, but the structural rearrangements caused by the mutation could affect the nucleic acid binding (Valverde et al. 2007).

FMRP belongs to a multiprotein complex and not only interacts with nucleic acids but also with a series of other proteins. Among all, the best characterized are the Cytoplasmic Fragile X Mental-Retardation-Interacting Protein 1 and 2 (CYFIP1 and 2), the Eukaryotic Translation Initiation Factor 4E (eIF4E), the Insulin-like Growth Factor 2 mRNA-Binding Protein (IGF2BP1), the Survival of Motor Neuron (SMN), the Tudor-Domain-Containing Protein 3 (TDRD3), the FXR1 and 2, the 58-kDa Microspherule Protein (MSP58), and the Nuclear Fragile X Mental-Retardation-Interacting Protein 1 and 2 (NUFIP1 and 2). FMRP also interacts with a series of other proteins involved in several diseases (Table 23.1). The extensive network of interactions explains the presence of FMRP in several multiprotein complexes.

Table 23.1 FMRP interactors found in mammalian cells and tissues. (*ND*) Indicates no diagnosed disease associated to the protein

Protein name	Molecular function/properties
FXR1P	RNA-binding protein
FXR2P	RNA-binding protein
CYFIP1	Rac1-binding protein translational repressor tumor suppressor
CYFIP2	Regulator of actin cytoskeleton
NUFIP1	DNA and RNA binding
82-FIP/NUFP2	ND
NUCLEOLIN	DNA and RNA binding
YB1/p50	DNA and RNA binding
STAUFEN1	Double-stranded RNA binding
PURα	DNA and RNA binding
PURβ	DNA and RNA binding
MYOSIN VA	mRNP and organelle transporter
RanBPM	Scaffolding protein: protein interaction and cytoskeletal-binding domain
elF2C2/AGO1	mRNA processing, translational control
DICER	RNase III endonuclease, RNA interference pathway
PABP1	RNA binding
Kinesin heavy chain 5A/C	Motor protein
Dynein intermediate chain	Microtubule and protein binding, motor activity
elF4E	RNA and protein binding
TDRD3	Nucleic acid binding
UBE21(UBC9)	Ubiquitin-ligase activity, SUMO-ligase activity, transcription factor binding
NXR2	Nuclear mRNA export
	Posttranscriptional mRNA metabolism
APC	Protein and microtubule binding

Most of the protein–protein and protein–RNA interactions occur at the N-terminal and central regions of FMRP. These two regions display a high degree of conservation among different family members (FMRP-FXR1P–FXR2P), while the C-terminal region, where the RGG box is located, is found to be the less conserved (Menon et al. 2004).

Among the best characterized interactions, the region encoded by exon 7 (residues 173–218) of FMRP is responsible for the interaction with CYFIP1 and 2. This interaction leads to the formation of the eIF4E-CYFIP1-FMRP complex that blocks mRNAs translation process (Napoli et al. 2008) (see below). Interaction of FMRP with NUFIP (Nuclear FMRP Interacting Protein) occurs through the N-terminal region (residues 1–217) (Bardoni et al. 2003). Residues 470–485 of the N terminus are also essential for the interaction with SMN, while residues 430–486 as well as the second KH domain are crucial for the binding with TDRD3 (Linder et al. 2008). Interestingly, the Ile304Asn mutation affects the FMRP–TDRD3 interaction (Linder et al. 2008). The C terminus is involved in the interaction with RanBPM, and this interaction also modulates the FMRP RNA binding activity (Menon et al. 2004).

23.3   Cellular and Molecular Functions of FMRP

FMRP is mainly expressed in the brain and gonads (Khandjian et al. 1995; Verheij et al. 1993) where it is mostly confined to the cytoplasm. FMRP has also been localized in nucleus (Willemsen et al. 1996). Despite existing a clear shuttling process of the protein from both compartments (Eberhart et al. 1996; Sittler et al. 1996), the neuronal cytoplasmic function of FMRP has been mainly addressed. FMRP forms large cytoplasmic ribonucleoparticles (RNPs) containing several other proteins and RNAs (Johnson et al. 2006; Zalfa et al. 2005; Zalfa and Bagni 2005; Zalfa et al. 2003). FMRP-RNPs have also been found to cosediment with both polyribosomes and mRNPs (Zalfa et al. 2006) consequently being involved not only in the traffic and stability of the transported mRNAs but also in their translation. FMRP has also been detected in P bodies (PB) and stress granules (SG) containing translationally silent preinitiation complexes (Anderson and Kedersha 2006). Several studies have shown that FMRP plays a critical role in regulating mRNA translation, transport, and stability (Bagni and Greenough 2005; Bassell and Warren 2008; De Rubeis and Bagni 2010) (see Table 23.2 for a list of validated mRNA targets). In addition, the expression of FMRP in dendrites increases after synaptic stimulation suggesting a direct link between FMRP function and synaptic plasticity activation (Antar et al. 2004, 2006; Ferrari et al. 2007).

23.3.1   *Regulation of Protein Synthesis*

The translational dysregulation of a subset of FMRP target mRNAs is probably the major contribution to FXS (Table 23.2) (Bassell and Warren 2008). In neurons, protein synthesis occurs not only in the soma but also along axons (Holt and

Table 23.2 Shortlist of FMRP mRNA targets whose association has been validated by applying in vivo or in vitro methods

mRNA	Dendritic localization	References
App		Westmark and Malter (2007)
Arc	+	Zalfa et al. (2003), Park et al. (2008)
CamKlla	+	Zalfa et al. (2003), Hou et al. (2006), Muddashetty et al. (2007)
eEF1A	+	Sung et al. (2003)
Fmr1	+	Weiler et al. (1997), Schaeffer et al. (2001)
GluR1/2	+	Muddashetty et al. (2007)
Map1b	+	Brown et al. (2001), Darnell et al. (2001), Zalfa et al. (2003)
NR1/NR2B		Schütt et al. (2009)
NR2A	+	Edbauer et al. 2010
PSD-95	+	Zalfa et al. (2007), Muddashetty et al. (2007)
SAPAP 1/2/3/4	+	Brown et al. (2001), Kindler et al. (2004), Narayanan et al. (2007), Dictenberg et al. (2008), Schütt et al. (2009)
Shank1/2	+	Schütt et al. (2009)
Rgs5	+	Miyashiro et al. (2003), Dictenberg et al. (2008)
GABA-Ad	+	Miyashiro et al. (2003), Dictenberg et al. (2008)

(+) Indicates the evidence for dendritic mRNA localization or synaptic synthesis, while no symbol reflects lack of any experimental conclusive result. This table has been updated from Table 1 of Bassell and Warren (2008)

Bullock 2009), dendrites, and postsynaptic sites (Steward and Schuman 2003). Local protein synthesis is required for long-lasting forms of synaptic plasticity that underlie consolidation of long-term memories (Flavell and Greenberg 2008).

In the brain, protein synthesis is a mechanism that follows different states of synaptic plasticity activation, and it is orchestrated by the action of glutamate receptors. The ionotropic receptors N-methyl-D-aspartate (NMDAR), alpha-amino-3-hydroxy-5-methyl-4-isoxazolepropionic acid (AMPAR), and the metabotrobic glutamate receptor (mGluR) play a key role in basic synaptic plasticity as well as in the activation of different synaptic plasticity states (Massey and Bashir 2007; Shi et al. 1999). In vitro models to study synaptic plasticity have been developed by depolarization of the postsynaptic membrane. Specifically, activation of mGluR by applying dihydroxyphenylglycine (DHPG) induces a synaptic plasticity state called long-term depression (LTD) (DHPG-induced LTD) (Massey and Bashir 2007) which involves several synaptic events including mRNA targeting and local protein synthesis and degradation (Gladding et al. 2009). However, in the *Fmr1* KO mice, DHPG-induced LTD plasticity is strongly increased (Huber et al. 2002), and it is also protein synthesis independent. This effect on LTD is likely due to deregulated local protein synthesis (Lu et al. 2004; Muddashetty et al. 2007; Zalfa et al. 2007; Huber et al. 2002; Nosyreva and Huber 2006; Ronesi and Huber 2008) and has settled the bases to describe the "mGluR theory" (Bear et al. 2004) (see below).

FMRP is involved in both basal and activity-dependent local protein synthesis by repressing in vivo and in vitro translation (Laggerbauer et al. 2001; Li et al. 2001; Lu et al. 2004; Muddashetty et al. 2007; Napoli et al. 2008; Zalfa et al. 2003). This has been proved in lymphoblastoid cells from individuals affected by FXS, in which 251 FMRP mRNA targets showed an abnormal polysomal distribution, explaining an increased translation (Brown et al. 2001). Moreover, protein synthesis of FMRP target mRNAs is increased in *Fmr1* KO mice, especially in purified synaptosomes, extending to synapses the function of FMRP as a translation repressor (Muddashetty et al. 2007; Schütt et al. 2009; Zalfa et al. 2003). Regarding proteins localized near the postsynaptic membrane, a single study performed by Schütt and colleagues detected an increase in the expression levels of the postsynaptic proteins SAPAP1, SAPAP2, SAPAP3, Shank1, Shank3, IRSp53 as well as the NMDA receptor subunits NR1 and NR2B and GluR1 (Schütt et al. 2009). These differences were either cortical or hippocampus specific. However, while FMRP was shown to bind the mRNAs encoding SAPAP1, SAPAP2, SAPAP3, Shank1, and the NMDA receptor subunits NR1 and NR2B, the loss of FMRP did not affect their total and synaptic mRNA levels indicating the role of FMRP in their translation control and not in mRNA stability (Schütt et al. 2009). In addition to this, Bassell and collaborators have recently found that FMRP binds and represses the translation of the mRNA encoding phosphatidy-linositol-4,5-bisphosphate 3-kinase 110 kDa catalytic subunit beta (p110β), the catalytic subunit of PI3K, a signaling molecule downstream activation of mGluRs (Gross et al. 2010). Further studies extending the proteomic analysis to the entire synaptosome are required to extend the role of FMRP and its mRNA targets at synapses.

FMRP expression and consequently its function can be regulated by posttranslational modifications such as ubiquitination (Hou et al. 2006) and/or phosphorylation. FMRP is rapidly translated at synapses in response to chemically induced LTD (Antar et al. 2004; Ferrari et al. 2007; Kao et al. 2010) and followed by a quick degradation (5 min) through the ubiquitin–proteasome system after LTD induction (Hou et al. 2006). In parallel, FMRP has also been found to be highly phosphorylated when it cosediments with polyribosomes whereas its dephosphorylation releases FMRP from polysomes allowing protein synthesis (Ceman et al. 2003). The activation of mTOR pathway, through protein phosphatase 2A (PP2A) and ribosomal protein S6 kinase (S6K) activation, seems to be involved in the dephosphorylation and phosphorylation of FMRP at different time points during LTD stimulation (Narayanan et al. 2007). These investigations highlight the fine-tuned mechanism that regulates translation and ultimately gene expression. However, whether FMRP represses translation during initiation and/or elongation step is still a controversy. While some laboratories have found FMRP mainly cosedimenting with polyribosomes (Ceman et al. 2003; Khandjian et al. 2004; Stefani et al. 2004), others have found FMRP cosedimenting with mRNPs (Ishizuka et al. 2002; Monzo et al. 2006; Napoli et al. 2008; Papoulas et al. 2010; Siomi et al. 1996; Zalfa et al. 2003), and one laboratory found FMRP equally distributed among the two fractions (Brown et al. 2001). The different distribution could be explained through the association of FMRP to a variety of mRNPs that may aggregate and form different neuronal granules such as P bodies, stress, and transport granules (Anderson and Kedersha 2006; Kanai et al. 2004; Zalfa et al. 2006). The shuttle of FMRP from polysomes to mRNPs might be influenced also by post-translational modifications (Ceman et al. 2003). However, recent findings suggested that FMRP is involved in the repression of translation initiation through CYFIP1, early identified as a partner of FMRP in neurons (Napoli et al. 2008; Schenck et al. 2003). In fact, CYFIP1 can act as a binding protein for the eukaryotic initiation factor 4E (eIF-4E), sequestering and repressing the assembly of the translation machinery. Specific mRNAs are tethered on CYFIP1-eIF4E by FMRP, and thus, only a subclass of mRNAs is repressed in a CYFIP1-dependent manner. Upon stimuli, CYFIP1-FMRP is released from eIF4E, and translation is activated (Napoli et al. 2008).

23.3.2 Regulation of mRNA Transport

It is worthwhile to mention that the levels of all FMRP-bound mRNAs are not necessarily translationally dysregulated in the absence of FMRP. Indeed, some experimental evidence indicate that FMRP is also involved in mRNA transport by delivering mRNAs which are thought to be in a dormant state from cell body, through dendrites, to spines where protein synthesis occurs (Fig. 23.3)

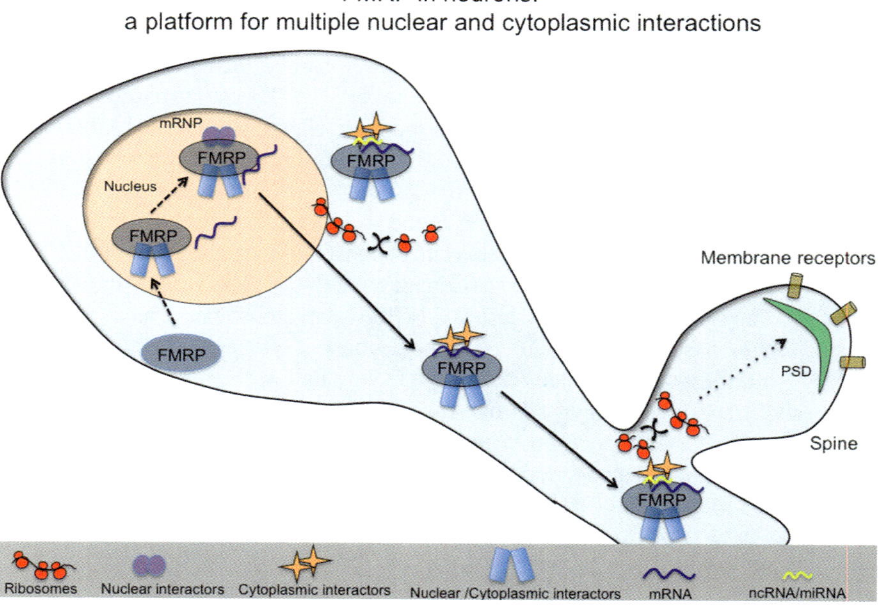

Fig. 23.3 FMRP forms part of a protein complex together with translationally arrested mRNAs. FMRP travels within an RNA–protein complex from the cell body to the synapses transporting dendritically localized mRNAs. After synaptic stimulation, FMRP liberates its mRNA targets allowing their local translation. The reversible translational repression and activation of the mRNA targets are regulated by a signaling pathway described in the text (see Sect. 23.3). Transported mRNAs are then locally translated in dendrites contributing to local protein synthesis and synaptic rearrangement that occurs after synaptic stimulation. FMRP can bind its mRNA targets through direct interaction or through noncoding RNAs such as BC1 RNA and microRNAs

(Bramham and Wells 2007). Upon synaptic LTD stimulation with DHPG, *Fmr1* mRNA is transported to dendrites and newly synthesized in proximity to metabotrobic receptor mGluR5 (Antar et al. 2004; Ferrari et al. 2007; Kao et al. 2010) where is further recruited along microtubules. Following DHPG stimulation, FMRP also interacts with motor proteins on microtubules, promoting the activity-dependent localization of bound mRNAs into synaptic spines (Antar et al. 2004, 2005; Davidovic et al. 2007; Dictenberg et al. 2008; Ferrari et al. 2007; Kanai et al. 2004). Therefore, certain FMRP targets such as *Map1b* and *Sapap4* have been found to be mislocalized in neurons of *Fmr1* KO mice (Table 23.2) (Dictenberg et al. 2008; Kao et al. 2010).

23.3.3   Regulation of mRNA Stability

For some FMRP-bound mRNAs, FMRP is a direct modulator of mRNA stability either by sustaining or preventing mRNA decay (De Rubeis and Bagni 2010). This has been revealed by the difference on the abundance of several mRNAs in *Fmr1* KO mice in comparison with wild type (WT) in three independent upscale screenings (Brown et al. 2001; Gantois et al. 2006; Miyashiro et al. 2003). In the first analysis, Brown et al. identified 144 mRNAs dysregulated in lymphoblastoid cells from FXS patients (Brown et al. 2001). Some of these mRNAs were also found to be dysregulated in a second analysis carried out in hippocampus of *Fmr1* KO mice (Miyashiro et al. 2003). Further analysis on the shared mRNAs showed that the dendritic localization of the mRNAs coding for the ribosomal component p40/LRP and the G-protein-coupled receptor kinase 4 (GRK4) was unaffected, while for the *dystroglycan-associated glycoprotein 1* (*DAG1*) mRNA, both localization and abundance were reduced (Miyashiro et al. 2003). The latter is an example of how FMRP may act on two different regulation mechanisms on the same mRNA.

Another example which shows how FMRP can regulate an mRNA at different levels is described by its action on *PSD-95* mRNA. Specifically, FMRP protects *PSD-95* mRNA from decay (Zalfa et al. 2007), and its stabilization is activity dependent. Downregulation of *PSD-95* mRNA occurs only in hippocampus and not in cortex, leading to decreased protein levels in hippocampus of *Fmr1* KO mice (Zalfa et al. 2007). However, PSD-95 protein levels are also downregulated in cortex of *Fmr1* KO mice indicating a defect on the synaptic translation of this *mRNA* in the cortex (Muddashetty et al. 2007). A different effect of FMRP on stability has been reported for the *Nxf1* mRNA. In this context, FMRP together with the nuclear export factor NXF2 facilitates the decay of *Nxf1* mRNA in a neuroblastoma cell line (Zhang et al. 2007). Following NXF2 overexpression, *Nxf1* is rapidly degraded. However, the degradation is impaired in absence of FMRP, suggesting that FMRP mediates *Nxf1* mRNA decay induced by NXF2 (Zhang et al. 2007).

In a genome-wide expression profiling study performed in hippocampus of *Fmr1* KO mice (Gantois et al. 2006), 224 mRNAs showed differences in the expression levels between *fmr1* KO and WT, being 143 mRNA underexpressed and 81 overexpressed. However, further analyses using microarrays and real-time PCR confirmed that only eight mRNAs were underexpressed in *Fmr1* KO mice. Among those eight mRNAs, it is worth to highlight that the $GABA_A$ receptor subunit δ shows almost two times of underexpression (Gantois et al. 2006), decrease that was already reported by El Idrissi et al. (2005). These results together with the finding that $GABA_A$ receptor subunit δ mRNA was found to be a FMRP target (Dictenberg et al. 2008; Miyashiro et al. 2003) further suggest the role of FMRP as a modulator of this mRNA stability. Noteworthy, eight GABA receptor subunits(α_1, α_3, α_4, β_1, β_2, γ_1, γ_2, and δ) were significantly reduced in cortex but not in cerebellum of *Fmr1* KO mice (D'Hulst et al. 2006).

All these results lead to the hypothesis that different FMRP protein complexes might play different roles in cortex and hippocampus and that FMRP regulates mRNAs through different mechanisms accordingly to cell type and subcellular localization.

23.3.4 RNA Sequence–Structure Recognition

Up to now, several different mechanisms through which FMRP binds to its mRNA targets have been described. One mechanism is mediated through the direct binding of FMRP to the mRNAs, as described for *PSD-95* mRNA (Zalfa et al. 2007). In this case, FMRP binds to G-rich sequences that can, in some cases (i.e., *Map1B* mRNA), also be folded as G-quartets (Darnell et al. 2001) Another mechanism is through the binding of FMRP to noncoding RNAs. The first example is via *BC1*, a small noncoding RNA that acts as a bridge between FMRP and its mRNA targets (Zalfa et al. 2003). A third mechanism of action involves the interaction of FMRP with its targets through microRNAs. Edbauer and colleagues have recently reported that FMRP is associated with at least 12 different miRNAs, and few of them have indeed a relevant effect on spine morphology (Edbauer et al. 2010). NMDA receptor subunit 2A (NR2A) mRNA is an FMRP target that indeed partially depends on miR-125b binding and whose expression has a direct impact on synaptic plasticity (Edbauer et al. 2010). Further studies are required to address which are the other molecular mechanisms mediated by FMRP which affect NR2A translation in neurons.

Considering the involvement of FMRP in the regulation of several genes, both in the cell body and at synapses, as a consequence, its absence causes several cellular phenotypic abnormalities including dendritic spine dysmorphogenesis and behavioral deficits that summarize the impaired molecular synaptic plasticity events. Figure 23.3 summarizes the current model on the multiple FMRP functions in neurons.

23.4 Learning, Memory, and Behavioral Phenotypes: Learning from the Mouse Model

Behavior impairment is one of the most compelling evidence in FXS. Since the animal model for FXS (*Fmr1* KO) has been developed (Bakker 1994), several behavioral analysis have been performed on this mouse model. *Fmr1* KO mice showed behavioral similarities to the syndrome affecting human individuals (see Table 23.3). This includes hyperactivity, abnormal anxiety-related responses, hyperreactivity to auditory stimuli, abnormal sensorimotor gating, and impaired motor coordination (Bakker 1994; Peier et al. 2000). Learning and memory tests performed with *Fmr1* KO mice have shown minor differences compared to their control WT. Morris water maze and radial arm maze tests showed only mild cognitive impairment in *Fmr1* KO mice (Bakker 1994; D'Hooge et al. 1997; Dobkin et al. 2000; Kooy et al. 1996; Mineur et al. 2002; Paradee et al. 1999; Peier et al. 2000; Van Dam et al. 2000; Yan et al. 2004), indicating that *Fmr1* KO mice have only a slight impairment in spatial learning. Only recently, *Fmr1* KO mice generated in the C57 albino genetic background showed a consistent impairment in spatial learning (Baker et al. 2010). Moreover, no impairment in associative aversive learning or memory has been reported for the *Fmr1* KO, since they successfully expressed conditioned taste aversion (Nielsen et al. 2009). Transfer of learning, or reversal learning, based on measures of learning rate is not impaired either (Moon et al. 2008). The major impairment revealing a learning and memory deficiency is reported for tests such as object recognition, eyeblink conditioning, and lever press avoidance (Brennan et al. 2006; Koekkoek et al. 2005; Ventura et al. 2004; Yan et al. 2004). In the learning paradigm of conditioned and contextual fear tests, which involve hippocampal and amygdaloid tasks, in two separate studies, *Fmr1* KO mice show no significant differences compared to their WT (Peier et al. 2000; Van Dam et al. 2000).

Genetic background has also a major impact on behavioral phenotypes (Bucan and Abel 2002; Wolfer and Lipp 2000). For example, the recently generated *Fmr1* KO C57-albino mice showed impaired contextual fear but unaltered conditioned fear (Baker et al. 2010). However, these characteristics were not fully detected in other *Fmr1* KO mice generated in other genetic background (FVB-129 mice) (Zhao et al. 2005) (Table 23.3). In prepulse inhibition (PPI) test, measuring sensory gating, *Fmr1* KO mice perform even better, and these findings contrast the human phenotype in which there is a decrease in PPI (Frankland et al. 2004; Van Dam et al. 2000). Nonetheless, in the PPI test, *Fmr1* KO mice are more reactive than their controls (Chen and Toth 2001; Nielsen et al. 2002a) but only at a near-threshold level of the startle stimulus (Chen and Toth 2001; Nielsen et al. 2002a), indicating a general impairment in sensorimotor gating. This is somehow in agreement with the overall decrease in functioning of the neuronal network of Fragile X patients and also with the decreased ratio of excitatory to inhibitory amino acids observed in the brain of *Fmr1* KO mice (Gruss and Braun 2001; Kooy 2003). Importantly, impaired attention and inhibitory control, two features clearly impaired in humans with FXS (Cornish et al. 2008; Garber et al. 2008), have not been studied in the mouse model for FXS, although a heightened emotional reactivity has been described in *Fmr1* KO by Moon and colleagues (Moon et al. 2008).

Table 23.3 Comparison between human and mouse *FMR1* deletion phenotypes

	Human	Mouse	References
Physical phenotype	Facial features (large ears, elongated face, and high-arched palate in ~60% of prepubertal FXS boys)	No major body phenotype	Bakker et al. (1994)
	Macroorchidism (~40% of prepubertal FXS boys; ~90% adult males)	Macroorchidism	Bakker et al. (1994), Kooy et al. (1996), Peier et al. (2000), Nielsen et al. (2002a)
Neuroanatomy	Increased spine density, excessive immature spines	Increased spine density, excessive immature spines	Rudelli et al. (1985), Comery et al. (1997), Irwin et al. (2001), Nimchinsky et al. (2001), Irwin et al. (2002), Galvez and Greenough (2005), Restivo et al. (2005), Antar et al. (2006), Dolen et al. (2007), Bilousova et al. (2009), Cruz-Martin et al. (2010), Grossman et al. (2010), Pan et al. (2010)
Cognition	Developmental delay (mean IQ ~42 in boys)	Learning and memory impairment	Bakker (1994), Kooy et al. (1996), D'Hooge et al. (1997), Paradee et al. (1999), Dobkin et al. (2000), Peier et al. (2000), Van Dam et al. (2000), Mineur et al. (2002), Frankland et al. (2004), Ventura et al. (2004), Yan et al. (2004), Koekkoek et al. (2005), Brennan et al. (2006), Moon et al. (2008), Nielsen et al. (2009), Baker et al. (2010)
Behavior	ADHD(~80%)	Hyperactivity	Bakker (1994), Peier et al. (2000), Liu and Smith (2009)
	ASD (25% boys, 6% girls)	Social behavior alterations	Spencer et al. (2005), Mineur et al. (2006), McNaughton et al. (2008), Mines et al. (2010)
	Epilepsy (13–18% boys, 4% girls)	Higher susceptibility to audiogenic seizures	Musumeci et al. (2000), Todd and Mack (2000), Chen and Toth (2001), Kooy (2003)

23.4.1 Social Behavior Phenotypes

Fmr1 KO mice are impaired in social dominance with unfamiliar mice and, even though they show interest in social interaction, they spend a longer period of time than WT mice before approaching an unfamiliar mouse (Spencer et al. 2005). Anxiety is a psychological and physiological state characterized by somatic, emotional, cognitive, and behavioral components observed in patients with FXS (Cornish et al. 2008; Garber et al. 2008). When the *Fmr1* KO were studied for this specific state, Mines and colleagues reported that they displayed more anxiety-related behaviors during social interaction (grooming, rearing, and digging) than WT mice (Mines et al. 2010) and reduced social approach and response to social novelty (Liu and Smith 2009). *Fmr1* KO males have also a reduced interest in social interaction with novel females (Mineur et al. 2006). Moreover, KO mice seem to prefer social involvement as their WT controls, even though they show social anxiety in approaching the novel conspecific (McNaughton et al. 2008). *Fmr1* KO mice also show impairment in social discrimination between positive and negative social interactions (McNaughton et al. 2008). In the acoustic startle reflex test, a response of mind and body to a sudden unexpected stimulus, and a measure of anxiety, *Fmr1* KO mice react less that their WT controls (Nielsen et al. 2002a) in contrast with the human behavior, in which FXS patients show an excessive reaction to external stimuli (Jacquemont et al. 2007). Finally, *Fmr1* KO mice show hyperactivity, decreased spatial and environmental anxiety-related responses, and altered motor coordination (Bakker 1994; Liu and Smith 2009; Peier et al. 2000). Since sleep problems are common in children with FXS, it was interesting that Zhang and colleagues reported an altered expression of clock genes in the FXS mouse model (Zhang et al. 2008).

Finally, the recently generated mouse model carrying the Ile304Asn point mutation in the C57BL/6J background seems to recapitulate some of the above reported behavioral characteristics (Zang et al. 2009).

Recently, Spencer and colleague also demonstrate that almost all the abovementioned behavioral characteristics depend on the genetic background of the FXS mouse model; therefore, modifier genes may play a role in phenotype expression (Spencer et al. 2011). This implicates a strong correlation with the variable phenotype in patients and also the difficulty to have a reliable animal model for the study of this pathology.

23.4.2 Rescue of the FXS Behavioral Phenotypes

Genetic, behavioral, and pharmacological rescues have been developed over the last few years. Overexpression of FMRP in the *Fmr1* KO mouse model using a YAC vector reduces to some extent the described behavioral abnormalities (Paylor et al. 2008; Spencer et al. 2005, 2008). Improvement in social behavior, such as in chamber test of social interaction, direct social interaction test, and resident–intruder task, has been demonstrated after enhanced neuroligin1 expression in *Fmr1* KO, even though no rescue is seen in learning and memory tasks (novel object recognition and plus shaped water maze tests) (Dahlhaus and El-Husseini 2010). One of the FMRP-dysregulated mRNAs encodes for the RGS4 (regulator of G-protein signaling) (Tervonen et al. 2005). *Fmr1* × RGS4 double KO shows a rescue of some of the behavioral phenotypes observed in the *Fmr1* KO such as tube test for social dominance, conditioned place preference, and reduced susceptibility to audiogenic seizures (Pacey et al. 2009, 2011). Moreover, treatment with group I mGluRs antagonist (MPEP) and lithium, through intervention on glycogen synthase kinase-3 (GSK3), ameliorates several behavioral aspects (Choi et al. 2010; Dolen et al. 2010; Liu et al. 2010; Min et al. 2009; Mines et al. 2010; Yuskaitis et al. 2010). Finally, it has been proposed that the absence of FMRP leads to higher levels of matrix metalloproteinase-9 activity (MMP-9) in the brain. In agreement, minocycline inhibits MMP-9 activity and alleviates behavioral measurement of the Aberrant Behavior Checklist – Community Edition (ABC-C) (irritability subscale, clinical global improvement scale (CGI), and the visual analog scale for behavior (VAS)) and synapse abnormalities in *Fmr1* knockout mice (Bilousova et al. 2009). Recently, minocycline administration was shown to provide significant functional benefits to FXS patients. These findings are consistent with the *Fmr1* knockout mouse model results, suggesting that minocycline modifies underlying neural defects that account for behavioral abnormalities. As the authors correctly report, a placebo-controlled trial of minocycline in FXS is warranted (Bilousova et al. 2009; Paribello et al. 2010).

As described below, absence of FMRP leads to an impaired GABA pathway due to its control on the mRNAs encoding different $GABA_A$ receptor subunits; therefore, a GABAergic approach to treat FXS treatment has been considered through the use of taurine, a $GABA_A$ agonist (El Idrissi et al. 2009). The authors show that taurine supplementation to Fragile X mice resulted in a significant improvement in acquisition of a passive avoidance task. Since taurine is an agonist for $GABA_A$ receptor, they suggest that chronic activation of $GABA_A$ receptors may have beneficial effects in ameliorating the learning deficits characteristic of the Fragile X syndrome (El Idrissi et al. 2009). Additionally, environmental enrichment has been proved of some efficacy in ameliorating the outcome of the pathology (Restivo et al. 2005). In this study, the authors showed that some behavior phenotypes such as habituation to object and motor activity as well as spine morphology were rescued. In addition, an increase of glutamate receptor subunit 1 (GluR1) levels in both genotypes was observed suggesting that FMRP-independent pathways activating glutamatergic signaling are preserved in *Fmr1* KO mice and that they can be elicited by environmental stimulation (Restivo et al. 2005). These findings indicate that the environment is of extreme importance for the patients. A follow-up study showed that in *Fmr1* KO mice that were raised in enriched environments, LTP was restored to WT levels, indicating that mechanisms for synaptic plasticity are in place in the *Fmr1* KO mouse but require stronger neuronal activity to be triggered (Meredith et al. 2007). Patients with FXS show in about 20% of the cases epileptic seizures (Garber et al. 2008). Even if the mouse model for FXS does not display spontaneous seizures, it is susceptible to audiogenic-induced seizures (Bakker 1994; Chen and Toth 2001; Kooy 2003; Kooy et al. 1996; Musumeci et al. 2000). Although seizure age dependency is still debated, a trend in the impaired response to acoustic seizures at P17 and P21 (postnatal days 17 and 21) has been reported (Chen and Toth 2001; Kooy 2003; Musumeci et al. 2000). This effect may be related to an increased cortical excitability or due to a deficit in long-term plasticity (Kooy 2003). Chemically induced seizures, by kainic acid, bicuculline, and pentylenetatrazole injection, do not show difference between *Fmr1* KO and their controls, suggesting that *Fmr1* KO mice have specific susceptibility to audiogenic stimuli (Chen and Toth 2001; Todd and Mack 2000) This is in agreement with a possible developmental impairment of the auditory system in Fragile X mice (Brown et al. 2010) proposed by Chen and Toth (Chen and Toth 2001). Only recently, a study on amygdala showed that *Fmr1* KO mice have a more accelerated kindling development and longer electrographic seizure duration. Both NMDA antagonist, MK-801, and mGluR5 antagonist, MPEP, were able to repress accelerated rate of kindling development (Qiu et al. 2009).

23.4.3 Electrophysiological Phenotypes

Due to the learning and memory alterations, characteristics of FXS, synaptic plasticity events monitored through *ex vivo* LTP and LTD paradigms have been extensively investigated in Fmr1 KO mice.

Effects on LTP. In the *Fmr1* KO mice, LTP has been found altered in hippocampus only after theta burst stimulation (Lauterborn et al. 2007), a stimulation able to induce LTP in close resemblance to physiological hippocampal frequency of theta rhythm (5–10 Hz) (Capocchi et al. 1992). Only recently, a reduction of LTP in dentate gyrus has been observed in *Fmr1* KO mice possibly due to a reduction of NMDA excitatory postsynaptic currents (EPSCs), which is a consequence of a reduced ratio of NMDA/AMPA receptors (Eadie et al. 2010; Yun and Trommer 2011). On the other hand, high-frequency stimulation (100–400 Hz) does not affect LTP in hippocampus of *Fmr1* KO mice (Godfraind et al. 1996; Li et al. 2002; Paradee et al. 1999; Zhang et al. 2009). At the level of cortex, the LTP responses are different and indeed appear impaired in the *Fmr1* KO mice (Hayashi et al. 2007; Larson et al. 2005; Li et al. 2002; Meredith et al. 2007; Zhao et al. 2005). Additionally, induction of LTP has also been shown to be altered in lateral amygdala (Suvrathan et al. 2010; Zhao et al. 2005). FMRP is required for glycine-induced LTP (Gly-LTP) in the CA1 of hippocampus. This form of LTP requires activation of postsynaptic NMDA receptors and metabotropic glutamatergic receptors, as well as the subsequent activation of extracellular signal-regulated kinase (ERK) 1/2. Genetic deletion of FMRP interrupts the phosphorylation of ERK1/2, suggesting the possible role of FMRP in the regulation of the activity of ERK1/2 (Shang et al. 2009).

Effects on LTD. LTD is altered in hippocampus of *Fmr1* KO mice either applying low-frequency-paired-pulse stimulation (PP-LFS), which is rescued by FMRP replacement through viral strategy (Zeier et al. 2009), or with application of the group I mGluR agonist DHPG (Huber et al. 2000). mGluR5-LTD and FMRP are also connected: DHPG induces FMRP expression in synaptosomes and cultured neurons (Antar et al. 2004; Ferrari et al. 2007; Kao et al. 2010; Weiler et al. 1997), and DHPG-induced LTD is increased in *Fmr1* KO mice (Hou et al. 2006; Huber et al. 2002). Therefore, mGluR5-induced LTD in the context of Fragile X has received a large attention. CA1 LTD is reported to be translation dependent and transcription independent when explored in WT mice (Hou and Klann 2004; Huber et al. 2000, 2002), in line with the signaling pathways described above and activated by mGluR5 which involve protein synthesis (Dolen et al. 2010). mGluR LTD occurs in a protein-synthesis-dependent manner and correlates glutamatergic receptors (NMDA and AMPA) regulation to morphologic changes in spine number, shape, and size (Dolen et al. 2010).

mGluR-induced LTD (via DHPG) is increased in *Fmr1* KO mice (Huber et al. 2002) and is protein synthesis independent, possibly due to deregulated local protein synthesis (Lu et al. 2004; Muddashetty et al. 2007; Zalfa et al. 2007), which in turn enhances a constant AMPAR internalization (Huber et al. 2002; Nosyreva and Huber 2006; Ronesi and Huber 2008). This mechanism has been described as the base of the so-called mGluR theory (Bear et al. 2004) which hypothesize that an excessive group 1 mGluRs activation, upstream to the lack of FMRP, increases the protein synthesis and induces excessive AMPAR internalization, responsible for the consequent increase in LTD. The theory has been validated by a genetic rescue of the *Fmr1* KO mouse model crossing it with the heterozygous mouse for mGluR5. The crossed *Fmr1* KO/mGluR5 (+/−) mice demonstrated amelioration of several mouse phenotypes. Specifically, 50% reduction of mGluR5 level of expression in the *Fmr1* KO background restored the altered protein expression, which in turn reduced the abnormal response to mGluR5-induced LTD, rescued the abnormal spine morphology, and ameliorated some behavioral phenotypes including a reduced incidence of the audiogenic seizures (AGS) (Dolen et al. 2007).

Nevertheless, the excessive activity of mGluR5 seems not to be the only possible cause of seizures in the *Fmr1* KO mouse model. As previously mentioned, another possible cause to consider is the deficit of GABA neurotransmission. Indeed, several lines of research have proposed that absence of FMRP leads to a dysfunction in the GABAergic system (Olmos-Serrano et al. 2010). Recent studies demonstrate that GABAergic inhibition is impaired at cellular (Selby et al. 2007), physiological (Centonze et al. 2008; Curia et al. 2009), and molecular level (Curia et al. 2009; D'Hulst et al. 2006; El Idrissi et al. 2005). Therefore, decrease of interneuron number, altered GABAergic transmission, and/or altered GABA$_A$ subunit expression may be the cause for epileptic seizures and/or EEG abnormalities associated to FXS. Moreover, an involvement of G-protein-coupled GABA$_B$ receptors demonstrated to be effective in attenuating the AGS phenotype in *Fmr1* KO mice (Pacey et al. 2009). The mechanism, possibly involved in efficacy for GABA$_B$ receptors, seems to be the coupling of GABA$_B$ receptors to regulator of RGS4 (Fowler et al. 2007) which in turn results as an inhibitor of Gp1 mGluRs (Saugstad et al. 1998). In agreement with this hypothesis, a double KO for *Fmr1* and RGS4 mouse model has reduced susceptibility to seizures (Pacey et al. 2009).

23.4.4 Spine Dysgenesis

Alterations of dendritic spines represent a common hallmark of mental retardation diseases and other synaptopathies (Purpura 1974). Although FXS is not characterized by gross brain defects, a consistent microanatomical phenotype is an increased spine density and an altered ratio of mature and immature spines (Comery et al. 1997; Irwin et al. 2001). During development, spines are stabilized and change their dynamic properties and morphology. Motility and turnover decrease, and thin, filopodia-like protrusions mature in stubby and mushroom-like shapes (Portera-Cailliau et al. 2003). A variety of intermediate shapes can exist, giving rise to a filopodia–spine continuum (Irwin et al. 2001).

The first evidence of spine dysgenesis in individuals with FXS was obtained by Golgi impregnation of autopsy material from a 62-year-old FXS patient (Rudelli et al. 1985). This analysis revealed long, tortuous, and thin dendritic spines on apical dendrites of pyramidal neurons from layers III and V of the parieto-occipital cortex (Rudelli et al. 1985). Further studies of three FXS adult individuals estimated higher spine density and length along the entire dendritic tree of cortical layer V pyramidal neurons (Irwin et al. 2001).

Mouse models of FXS (*Fmr1* KO) recapitulate the spine dysgenesis observed in patients. Mutant mice present increase density of long, immature spines in visual cortex, somatosensory cortex, and hippocampal dentate gyrus (Comery et al. 1997; Dolen et al. 2007; Galvez and Greenough 2005; Grossman et al. 2010; Irwin et al. 2002; Nimchinsky et al. 2001; Restivo et al. 2005). Spine defects in cortical neurons were detected during early postnatal development (1–3 weeks) and adulthood but not in 4-week-old mice (Galvez and Greenough 2005; Nimchinsky et al. 2001). However, in contrast with neocortex, spine abnormalities in the dentate gyrus remain constant during development (Grossman et al. 2010). Data obtained on *ex vivo* and *in vitro* systems do not consistently corroborate in vivo observations. Spine defects were not detected in neocortical and hippocampal organotypic cultures (Nimchinsky et al. 2001; Pfeiffer et al. 2010). Increased spine length and excessive filopodia protrusions were reported in primary *Fmr1* KO hippocampal neurons (Antar et al. 2006; Bilousova et al. 2009), although earlier reports produced divergent results (Braun and Segal 2000).

Excess of spines with immature morphology may be due to defects in spine dynamics and maturation, especially altered pruning (Galvez et al. 2003). Two recent studies examined spine plasticity in living *Fmr1* KO animals using transcranial two-photon imaging of somatosensory cortex. Both reports highlighted that the major abnormality is the augmented spine turnover, a process that includes formation of new spines and elimination of existing spines (Cruz-Martin et al. 2010; Pan et al. 2010).

In *Fmr1* KO mice, spine turnover fails to rapidly decrease during the first 2 postnatal weeks, leading to a delay in spine stabilization and transition from immature to mature spine types (Cruz-Martin et al. 2010). Increased spine turnover is maintained in late development (4 postnatal weeks) and in adulthood and may be due to a larger population of short-lived spines observed in KO mice (Pan et al. 2010). Since these transient spines display smaller head and longer neck, they could contribute to the immature spine morphology in KO animals (Pan et al. 2010).

Spine dynamics are known to change in response to experience-dependent modulation of specific circuits in the somatosensory and visual cortex (Holtmaat and Svoboda 2009). Although *Fmr1* KO animals display hypersensitivity to sensory stimuli (Chen and Toth 2001), this effect does not cause the enhanced spine dynamics, which in fact was not hampered by somatosensory deprivation (whisker trimming) (Pan et al. 2010). Furthermore, KO mice lack spine plasticity to somatosensory modulation (Pan et al. 2010). However, other circuits of experience-dependent neuronal plasticity seem to be preserved, since environmental enrichment rescues the spine abnormalities in visual cortex (Restivo et al. 2005).

Evidence of defects in spine maturation are compatible with the overall decrease of functional synapses, measured as dendritic protrusions juxtaposed with presynaptic markers in cultured hippocampal neurons (Antar et al. 2006; Braun and Segal 2000). Of interest, loss of synapses in *Fmr1* KO mice is also corroborated by electrophysiological data (Pfeiffer and Huber 2007; Pfeiffer et al. 2010).

The mechanisms leading to spine dysgenesis are not fully understood. Despite recent reports suggesting the involvement of the transcription regulator MEF2 (Pfeiffer et al. 2010) and neuronal microRNAs (Edbauer et al. 2010), the spine phenotype is likely multifactorial. As mentioned above, FMRP regulates the synthesis of a variety of proteins crucial for proper synaptic morphology and functionality (Bassell and Warren 2008). However, interfering with some of the signaling pathways altered in *Fmr1* KO mice – group I mGluRs, PI3K, and PAK1 RGS4 – is effective to partially rescue the spine anomalies.

As for the mGluR cascade, DHPG administration in cultured hippocampal neurons increases spine length in a protein-synthesis-dependent manner (Vanderklish and Edelman 2002). Hampering the excessive mGluR signaling by administration of the antagonist MPEP (de Vrij et al. 2008; Su et al. 2010) or genetic reduction of mGluR5 (Dolen et al. 2007) ameliorates the spine phenotype in *Fmr1* KO neurons. Furthermore, inhibition of PI3K, a signaling molecule excessively translated in the absence of FMRP, reduces spine density in cultured neurons (Gross et al. 2010). Similarly, inhibition of PAK1, a signaling cascade controlling actin cytoskeleton, partially rescues the defects in spine density and length (Hayashi et al. 2007). Finally, pharmacological treatments with lithium and minocycline, whose mechanisms are not fully elucidated, may also alleviate the spine phenotype in FXS mice (Bilousova et al. 2009; Liu et al. 2010).

Acknowledgments This study was supported by Telethon, Compagnia di San Paolo, COFIN, FWO, and FP7 (SynSys). We are thankful to the Associazione Italiana Sindrome X Fragile (http://www.xfragile.net/).

References

Abrahams, B. S., & Geschwind, D. H. (2008). Advances in autism genetics: On the threshold of a new neurobiology. *Nature Reviews Genetics, 9*, 341–355.

Allen, E. G., He, W., Yadav-Shah, M., & Sherman, S. L. (2004). A study of the distributional characteristics of FMR1 transcript levels in 238 individuals. *Human Genetics, 114*, 439–447.

Anderson, P., & Kedersha, N. (2006). RNA granules. *The Journal of Cell Biology, 172*, 803–808.

Antar, L. N., Afroz, R., Dictenberg, J. B., Carroll, R. C., & Bassell, G. J. (2004). Metabotropic glutamate receptor activation regulates fragile X mental retardation protein and FMR1 mRNA localization differentially in dendrites and at synapses. *Journal of Neuroscience, 24*, 2648–2655.

Antar, L. N., Dictenberg, J. B., Plociniak, M., Afroz, R., & Bassell, G. J. (2005). Localization of FMRP-associated mRNA granules and requirement of microtubules for activity-dependent trafficking in hippocampal neurons. *Genes, Brain, and Behavior, 4*, 350–359.

Antar, L. N., Li, C., Zhang, H., Carroll, R. C., & Bassell, G. J. (2006). Local functions for FMRP in axon growth cone motility and activity-dependent regulation of filopodia and spine synapses. *Molecular and Cellular Neurosciences, 32*, 37–48.

Ashley, C. T., Jr., Wilkinson, K. D., Reines, D., & Warren, S. T. (1993). FMR1 protein: Conserved RNP family domains and selective RNA binding. *Science, 262*, 563–566.

Bacalman, S., Farzin, F., Bourgeois, J. A., Cogswell, J., Goodlin-Jones, B. L., Gane, L. W., Grigsby, J., Leehey, M. A., Tassone, F., & Hagerman, R. J. (2006). Psychiatric phenotype of the fragile X-associated tremor/ataxia syndrome (FXTAS) in males: Newly described fronto-subcortical dementia. *The Journal of Clinical Psychiatry, 67*, 87–94.

Bagni, C., & Greenough, W. T. (2005). From mRNP trafficking to spine dysmorphogenesis: The roots of fragile X syndrome. *Nature Reviews Neuroscience, 6*, 376–387.

Baker, K. B., Wray, S. P., Ritter, R., Mason, S., Lanthorn, T. H., & Savelieva, K. V. (2010). Male and female Fmr1 knockout mice on C57 albino background exhibit spatial learning and memory impairments. *Genes, Brain, and Behavior, 9*, 562–574.

Bardoni, B., Willemsen, R., Weiler, I. J., Schenck, A., Severijnen, L. A., Hindelang, C., Lalli, E., & Mandel, J. L. (2003). NUFIP1 (nuclear FMRP interacting protein 1) is a nucleocytoplasmic shuttling protein associated with active synaptoneurosomes. *Experimental Cell Research, 289*, 95–107.

Bassell, G. J., & Warren, S. T. (2008). Fragile X syndrome: Loss of local mRNA regulation alters synaptic development and function. *Neuron, 60*, 201–214.

Bear, M. F., Huber, K. M., & Warren, S. T. (2004). The mGluR theory of fragile X mental retardation. *Trends in Neurosciences, 27*, 370–377.

Belmonte, M. K., & Bourgeron, T. (2006). Fragile X syndrome and autism at the intersection of genetic and neural networks. *Nature Neuroscience, 9*, 1221–1225.

Berman, R. F., & Willemsen, R. (2009). Mouse models of fragile x-associated tremor ataxia. *Journal of Investigative Medicine, 57*, 837–841.

Berry-Kravis, E. (2002). Epilepsy in fragile X syndrome. *Developmental Medicine and Child Neurology, 44*, 724–728.

Bilousova, T. V., Dansie, L., Ngo, M., Aye, J., Charles, J. R., Ethell, D. W., & Ethell, I. M. (2009). Minocycline promotes dendritic spine maturation and improves behavioural performance in the fragile X mouse model. *Journal of Medical Genetics, 46*, 94–102.

Bramham, C. R., & Wells, D. G. (2007). Dendritic mRNA: Transport, translation and function. *Nature Reviews Neuroscience, 8*, 776–789.

Braun, K., & Segal, M. (2000). FMRP involvement in formation of synapses among cultured hippocampal neurons. *Cerebral Cortex, 10*, 1045–1052.

Brennan, F. X., Albeck, D. S., & Paylor, R. (2006). Fmr1 knockout mice are impaired in a leverpress escape/avoidance task. *Genes, Brain, and Behavior, 5*, 467–471.

Brouwer, J. R., Huizer, K., Severijnen, L. A., Hukema, R. K., Berman, R. F., Oostra, B. A., & Willemsen, R. (2008). CGG-repeat length and neuropathological and molecular correlates in a mouse model for fragile X-associated tremor/ataxia syndrome. *Journal of Neurochemistry, 107*, 1671–1682.

Brown, V., Jin, P., Ceman, S., Darnell, J. C., O'Donnell, W. T., Tenenbaum, S. A., Jin, X., Feng, Y., Wilkinson, K. D., Keene, J. D., Darnell, R. B., & Warren, S. T. (2001). Microarray identification of FMRP-associated brain mRNAs and altered mRNA translational profiles in fragile X syndrome. *Cell, 107*, 477–487.

Bucan, M., & Abel, T. (2002). The mouse: Genetics meets behaviour. *Nature Reviews Genetics, 3*, 114–123.

Brown, M. R., Kronengold, J., Gazula, V.-R., Chen, Y., Strumbos, J. G., Sigworth, F. J., Navaratnam, D., & Kaczmarek. L. K., (2010). Fragile X mental retardation protein controls gating of the sodium-activated potassium channel Slack. *Nature Neuroscience Vol 13 (7)* Jul 2010.

Bakker, C. (1994). Fmr1 knockout mice: A model to study fragile X mental retardation. The Dutch-Belgian fragile X consortium. *Cell, 78*, 23–33.

Capocchi, G., Zampolini, M., & Larson, J. (1992). Theta burst stimulation is optimal for induction of LTP at both apical and basal dendritic synapses on hippocampal CA1 neurons. *Brain Research, 591*, 332–336.

Ceman, S., O'Donnell, W. T., Reed, M., Patton, S., Pohl, J., & Warren, S. T. (2003). Phosphorylation influences the translation state of FMRP-associated polyribosomes. *Human Molecular Genetics, 12*, 3295–3305.

Centonze, D., Rossi, S., Mercaldo, V., Napoli, I., Ciotti, M. T., De Chiara, V., Musella, A., Prosperetti, C., Calabresi, P., Bernardi, G., & Bagni, C. (2008). Abnormal striatal GABA transmission in the mouse model for the fragile X syndrome. *Biological Psychiatry, 63*, 963–973.

Chen, L., & Toth, M. (2001). Fragile X mice develop sensory hyperreactivity to auditory stimuli. *Neuroscience, 103*, 1043–1050.

Cheever, A., Ceman, S., (2009). Phosphorylation of FMRP inhibits association with Dicer. RNA. 15(3):362–6.

Choi, C. H., Schoenfeld, B. P., Bell, A. J., Hinchey, P., Kollaros, M., Gertner, M. J., Woo, N. H., Tranfaglia, M. R., Bear, M. F., Zukin, R. S., McDonald, T. V., Jongens, T. A., & McBride, S. M. (2010). Pharmacological reversal of synaptic plasticity deficits in the mouse model of fragile X syndrome by group II mGluR antagonist or lithium treatment. *Brain Research, 1380*, 106–119.

Comery, T. A., Harris, J. B., Willems, P. J., Oostra, B. A., Irwin, S. A., Weiler, I. J., & Greenough, W. T. (1997). Abnormal dendritic spines in fragile X knockout mice: Maturation and pruning deficits. *Proceedings of the National Academy of Sciences of the United States of America, 94*, 5401–5404.

Cornish, K. M., Li, L., Kogan, C. S., Jacquemont, S., Turk, J., Dalton, A., Hagerman, R. J., & Hagerman, P. J. (2008). Age-dependent cognitive changes in carriers of the fragile X syndrome. *Cortex, 44*, 628–636.

Cruz-Martin, A., Crespo, M., & Portera-Cailliau, C. (2010). Delayed stabilization of dendritic spines in fragile X mice. *Journal of Neuroscience, 30*, 7793–7803.

Curia, G., Papouin, T., Seguela, P., & Avoli, M. (2009). Downregulation of tonic GABAergic inhibition in a mouse model of fragile X syndrome. *Cerebral Cortex, 19*, 1515–1520.

D'Hooge, R., Nagels, G., Franck, F., Bakker, C. E., Reyniers, E., Storm, K., Kooy, R. F., Oostra, B. A., Willems, P. J., & De Deyn, P. P. (1997). Mildly impaired water maze performance in male Fmr1 knockout mice. *Neuroscience, 76*, 367–376.

D'Hulst, C., De Geest, N., Reeve, S. P., Van Dam, D., De Deyn, P. P., Hassan, B. A., & Kooy, R. F. (2006). Decreased expression of the GABAA receptor in fragile X syndrome. *Brain Research, 1121*, 238–245.

Dahlhaus, R., & El-Husseini, A. (2010). Altered neuroligin expression is involved in social deficits in a mouse model of the fragile X syndrome. *Behavioural Brain Research, 208*, 96–105.

Darnell, J. C., Jensen, K. B., Jin, P., Brown, V., Warren, S. T., & Darnell, R. B. (2001). Fragile X mental retardation protein targets G quartet mRNAs important for neuronal function. *Cell, 107*, 489–499.

Davidovic, L., Jaglin, X. H., Lepagnol-Bestel, A. M., Tremblay, S., Simonneau, M., Bardoni, B., & Khandjian, E. W. (2007). The fragile X mental retardation protein is a molecular adaptor between the neurospecific KIF3C kinesin and dendritic RNA granules. *Human Molecular Genetics, 16*, 3047–3058.

De Boulle, K., Verkerk, A. J., Reyniers, E., Vits, L., Hendrickx, J., Van Roy, B., Van den Bos, F., de Graaff, E., Oostra, B. A., & Willems, P. J. (1993). A point mutation in the FMR-1 gene associated with fragile X mental retardation. *Nature Genetics, 3*, 31–35.

De Rubeis, S., & Bagni, C. (2010). Fragile X mental retardation protein control of neuronal mRNA metabolism: Insights into mRNA stability. *Molecular and Cellular Neurosciences, 43*, 43–50.

de Vrij, F. M., Levenga, J., van der Linde, H. C., Koekkoek, S. K., De Zeeuw, C. I., Nelson, D. L., Oostra, B. A., & Willemsen, R. (2008). Rescue of behavioral phenotype and neuronal protrusion morphology in Fmr1 KO mice. *Neurobiology of Disease, 31*, 127–132.

den Broeder, M. J., van der Linde, H., Brouwer, J. R., Oostra, B. A., Willemsen, R., & Ketting, R. F. (2009). Generation and characterization of FMR1 knockout zebrafish. *PLoS One, 4*, e7910.

Dictenberg, J. B., Swanger, S. A., Antar, L. N., Singer, R. H., & Bassell, G. J. (2008). A direct role for FMRP in activity-dependent dendritic mRNA transport links filopodial-spine morphogenesis to fragile X syndrome. *Developmental Cell, 14*, 926–939.

Dobkin, C., Rabe, A., Dumas, R., El Idrissi, A., Haubenstock, H., & Brown, W. T. (2000). Fmr1 knockout mouse has a distinctive strain-specific learning impairment. *Neuroscience, 100*, 423–429.

Dockendorff, T. C., Su, H. S., McBride, S. M., Yang, Z., Choi, C. H., Siwicki, K. K., Sehgal, A., & Jongens, T. A. (2002). Drosophila lacking dfmr1 activity show defects in circadian output and fail to maintain courtship interest. *Neuron, 34*, 973–984.

Dolen, G., Carpenter, R. L., Ocain, T. D., & Bear, M. F. (2010). Mechanism-based approaches to treating fragile X. *Pharmacology and Therapeutics, 127*, 78–93.

Dolen, G., Osterweil, E., Rao, B. S., Smith, G. B., Auerbach, B. D., Chattarji, S., & Bear, M. F. (2007). Correction of fragile X syndrome in mice. *Neuron, 56*, 955–962.

Eadie, B. D., Cushman, J., Kannangara, T. S., Fanselow, M. S., & Christie, B. R. (2010). NMDA receptor hypofunction in the dentate gyrus and impaired context discrimination in adult Fmr1 knockout mice. *Hippocampus.* 3 NOV 2010, DOI: 10.1002/hipo.20890.

Eberhart, D. E., Malter, H. E., Feng, Y., & Warren, S. T. (1996). The fragile X mental retardation protein is a ribonucleoprotein containing both nuclear localization and nuclear export signals. *Human Molecular Genetics, 5*, 1083–1091.

Edbauer, D., Neilson, J. R., Foster, K. A., Wang, C. F., Seeburg, D. P., Batterton, M. N., Tada, T., Dolan, B. M., Sharp, P. A., & Sheng, M. (2010). Regulation of synaptic structure and function by FMRP-associated microRNAs miR-125b and miR-132. *Neuron, 65*, 373–384.

El Idrissi, A., Boukarrou, L., Dokin, C., & Brown, W. T. (2009). Taurine improves congestive functions in a mouse model of fragile X syndrome. *Advances in Experimental Medicine and Biology, 643*, 191–198.

El Idrissi, A., Ding, X. H., Scalia, J., Trenkner, E., Brown, W. T., & Dobkin, C. (2005). Decreased GABA(A) receptor expression in the seizure-prone fragile X mouse. *Neuroscience Letters, 377*, 141–146.

Entezam, A., Biacsi, R., Orrison, B., Saha, T., Hoffman, G. E., Grabczyk, E., Nussbaum, R. L., & Usdin, K. (2007). Regional FMRP deficits and large repeat expansions into the full mutation range in a new fragile X premutation mouse model. *Gene, 395*, 125–134.

Farzin, F., Perry, H., Hessl, D., Loesch, D., Cohen, J., Bacalman, S., Gane, L., Tassone, F., Hagerman, P., & Hagerman, R. (2006). Autism spectrum disorders and attention-deficit/hyperactivity disorder in boys with the fragile X premutation. *Journal of Developmental and Behavioral Pediatrics, 27*, S137–144.

Feng, Y., Zhang, F., Lokey, L. K., Chastain, J. L., Lakkis, L., Eberhart, D., & Warren, S. T. (1995). Translational suppression by trinucleotide repeat expansion at FMR1. *Science, 268*, 731–734.

Fernandez-Carvajal, I., Lopez Posadas, B., Pan, R., Raske, C., Hagerman, P. J., & Tassone, F. (2009). Expansion of an FMR1 grey-zone allele to a full mutation in two generations. *Journal of Molecular Diagnostics, 11*, 306–310.

Ferrari, F., Mercaldo, V., Piccoli, G., Sala, C., Cannata, S., Achsel, T., & Bagni, C. (2007). The fragile X mental retardation protein-RNP granules show an mGluR-dependent localization in the post-synaptic spines. *Molecular and Cellular Neurosciences, 34*, 343–354.

Flavell, S. W., & Greenberg, M. E. (2008). Signaling mechanisms linking neuronal activity to gene expression and plasticity of the nervous system. *Annual Review of Neuroscience, 31*, 563–590.

Fombonne, E. (2005). Epidemiology of autistic disorder and other pervasive developmental disorders. *The Journal of Clinical Psychiatry, 66*(Suppl 10), 3–8.

Fowler, C. E., Aryal, P., Suen, K. F., & Slesinger, P. A. (2007). Evidence for association of GABA (B) receptors with Kir3 channels and regulators of G protein signalling (RGS4) proteins. *The Journal of Physiology, 580*, 51–65.

Frankland, P. W., Wang, Y., Rosner, B., Shimizu, T., Balleine, B. W., Dykens, E. M., Ornitz, E. M., & Silva, A. J. (2004). Sensorimotor gating abnormalities in young males with fragile X syndrome and Fmr1-knockout mice. *Molecular Psychiatry, 9*, 417–425.

Galvez, R., Gopal, A. R., & Greenough, W. T. (2003). Somatosensory cortical barrel dendritic abnormalities in a mouse model of the fragile X mental retardation syndrome. *Brain Research, 971*, 83–89.

Galvez, R., & Greenough, W. T. (2005). Sequence of abnormal dendritic spine development in primary somatosensory cortex of a mouse model of the fragile X mental retardation syndrome. *American Journal of Medical Genetics A, 135*, 155–160.

Gantois, I., Vandesompele, J., Speleman, F., Reyniers, E., D'Hooge, R., Severijnen, L. A., Willemsen, R., Tassone, F., & Kooy, R. F. (2006). Expression profiling suggests underexpression of the GABA(A) receptor subunit delta in the fragile X knockout mouse model. *Neurobiology of Disease, 21*, 346–357.

Garber, K. B., Visootsak, J., & Warren, S. T. (2008). Fragile X syndrome. *European Journal of Human Genetics, 16*, 666–672.

Garcia-Nonell, C., Ratera, E. R., Harris, S., Hessl, D., Ono, M. Y., Tartaglia, N., Marvin, E., Tassone, F., & Hagerman, R. J. (2008). Secondary medical diagnosis in fragile X syndrome with and without autism spectrum disorder. *American Journal of Medical Genetics A, 146A*, 1911–1916.

Gedeon, A. K., Baker, E., Robinson, H., Partington, M. W., Gross, B., Manca, A., Korn, B., Poustka, A., Yu, S., Sutherland, G. R., et al. (1992). Fragile X syndrome without CCG amplification has an FMR1 deletion. *Nature Genetics, 1*, 341–344.

Gladding, C. M., Fitzjohn, S. M., & Molnar, E. (2009). Metabotropic glutamate receptor-mediated long-term depression: Molecular mechanisms. *Pharmacological Reviews, 61*, 395–412.

Godfraind, J. M., Reyniers, E., De Boulle, K., D'Hooge, R., De Deyn, P. P., Bakker, C. E., Oostra, B. A., Kooy, R. F., & Willems, P. J. (1996). Long-term potentiation in the hippocampus of fragile X knockout mice. *American Journal of Medical Genetics, 64*, 246–251.

Greco, C. M., Berman, R. F., Martin, R. M., Tassone, F., Schwartz, P. H., Chang, A., Trapp, B. D., Iwahashi, C., Brunberg, J., Grigsby, J., Hessl, D., Becker, E. J., Papazian, J., Leehey, M. A., Hagerman, R. J., & Hagerman, P. J. (2006). Neuropathology of fragile X-associated tremor/ataxia syndrome (FXTAS). *Brain, 129*, 243–255.

Greco, C. M., Hagerman, R. J., Tassone, F., Chudley, A. E., Del Bigio, M. R., Jacquemont, S., Leehey, M., & Hagerman, P. J. (2002). Neuronal intranuclear inclusions in a new cerebellar tremor/ataxia syndrome among fragile X carriers. *Brain, 125*, 1760–1771.

Grishin, N. V. (2001). KH domain: One motif, two folds. *Nucleic Acids Research, 29*, 638–643.

Gross, C., Nakamoto, M., Yao, X., Chan, C. B., Yim, S. Y., Ye, K., Warren, S. T., & Bassell, G. J. (2010). Excess phosphoinositide 3-kinase subunit synthesis and activity as a novel therapeutic target in fragile X syndrome. *Journal of Neuroscience, 30*, 10624–10638.

Grossman, A. W., Aldridge, G. M., Lee, K. J., Zeman, M. K., Jun, C. S., Azam, H. S., Arii, T., Imoto, K., Greenough, W. T., & Rhyu, I. J. (2010). Developmental characteristics of dendritic spines in the dentate gyrus of Fmr1 knockout mice. *Brain Research, 1355*, 221–227.

Gruss, M., & Braun, K. (2001). Alterations of amino acids and monoamine metabolism in male Fmr1 knockout mice: A putative animal model of the human fragile X mental retardation syndrome. *Neural Plasticity, 8,* 285–298.

Hagerman, R. J., & Hagerman, P. J. (2002). The fragile X premutation: Into the phenotypic fold. *Current Opinion in Genetics and Development, 12,* 278–283.

Hagerman, P. J., & Hagerman, R. J. (2007). Fragile X-associated tremor/ataxia syndrome – An older face of the fragile X gene. *Nature Clinical Practice Neurology, 3,* 107–112.

Hagerman, R. J., Leehey, M., Heinrichs, W., Tassone, F., Wilson, R., Hills, J., Grigsby, J., Gage, B., & Hagerman, P. J. (2001). Intention tremor, parkinsonism, and generalized brain atrophy in male carriers of fragile X. *Neurology, 57,* 127–130.

Hallahan, B. P., Craig, M. C., Toal, F., Daly, E. M., Moore, C. J., Ambikapathy, A., Robertson, D., Murphy, K. C., & Murphy, D. G. (2010). In vivo brain anatomy of adult males with fragile X syndrome: an MRI study. *NeuroImage,* 54(1):16–24.

Harris, S. W., Hessl, D., Goodlin-Jones, B., Ferranti, J., Bacalman, S., Barbato, I., Tassone, F., Hagerman, P. J., Herman, H., & Hagerman, R. J. (2008). Autism profiles of males with fragile X syndrome. *American Journal on Mental Retardation, 113,* 427–438.

Hatton, D. D., Sideris, J., Skinner, M., Mankowski, J., Bailey, D. B., Jr., Roberts, J., & Mirrett, P. (2006). Autistic behavior in children with fragile X syndrome: Prevalence, stability, and the impact of FMRP. *American Journal of Medical Genetics A, 140A,* 1804–1813.

Hayashi, M. L., Rao, B. S., Seo, J. S., Choi, H. S., Dolan, B. M., Choi, S. Y., Chattarji, S., & Tonegawa, S. (2007). Inhibition of p21-activated kinase rescues symptoms of fragile X syndrome in mice. *Proceedings of the National Academy of Sciences of the United States of America, 104,* 11489–11494.

Hernandez, R. N., Feinberg, R. L., Vaurio, R., Passanante, N. M., Thompson, R. E., & Kaufmann, W. E. (2009). Autism spectrum disorder in fragile X syndrome: A longitudinal evaluation. *American Journal of Medical Genetics A, 149A,* 1125–1137.

Holt, C. E., & Bullock, S. L. (2009). Subcellular mRNA localization in animal cells and why it matters. *Science, 326,* 1212–1216.

Holtmaat, A., & Svoboda, K. (2009). Experience-dependent structural synaptic plasticity in the mammalian brain. *Nature Reviews Neuroscience, 10,* 647–658.

Hou, L., Antion, M. D., Hu, D., Spencer, C. M., Paylor, R., & Klann, E. (2006). Dynamic translational and proteasomal regulation of fragile X mental retardation protein controls mGluR-dependent long-term depression. *Neuron, 51,* 441–454.

Hou, L., & Klann, E. (2004). Activation of the phosphoinositide 3-kinase-Akt-mammalian target of rapamycin signaling pathway is required for metabotropic glutamate receptor-dependent long-term depression. *Journal of Neuroscience, 24,* 6352–6361.

Huber, K. M., Gallagher, S. M., Warren, S. T., & Bear, M. F. (2002). Altered synaptic plasticity in a mouse model of fragile X mental retardation. *Proceedings of the National Academy of Sciences of the United States of America, 99,* 7746–7750.

Huber, K. M., Kayser, M. S., & Bear, M. F. (2000). Role for rapid dendritic protein synthesis in hippocampal mGluR-dependent long-term depression. *Science, 288,* 1254–1257.

Inoue, S., Shimoda, M., Nishinokubi, I., Siomi, M. C., Okamura, M., Nakamura, A., Kobayashi, S., Ishida, N., & Siomi, H. (2002). A role for the Drosophila fragile X-related gene in circadian output. *Current Biology, 12,* 1331–1335.

Irwin, S. A., Idupulapati, M., Gilbert, M. E., Harris, J. B., Chakravarti, A. B., Rogers, E. J., Crisostomo, R. A., Larsen, B. P., Mehta, A., Alcantara, C. J., Patel, B., Swain, R. A., Weiler, I. J., Oostra, B. A., & Greenough, W. T. (2002). Dendritic spine and dendritic field characteristics of layer V pyramidal neurons in the visual cortex of fragile-X knockout mice. *American Journal of Medical Genetics, 111,* 140–146.

Irwin, S. A., Patel, B., Idupulapati, M., Harris, J. B., Crisostomo, R. A., Larsen, B. P., Kooy, F., Willems, P. J., Cras, P., Kozlowski, P. B., Swain, R. A., Weiler, I. J., & Greenough, W. T. (2001). Abnormal dendritic spine characteristics in the temporal and visual cortices of patients with fragile-X syndrome: A quantitative examination. *American Journal of Medical Genetics, 98,* 161–167.

Ishizuka, A., Siomi, M. C., & Siomi, H. (2002). A Drosophila fragile X protein interacts with components of RNAi and ribosomal proteins. *Genes & Development, 16*, 2497–2508.

Iwahashi, C. K., Yasui, D. H., An, H. J., Greco, C. M., Tassone, F., Nannen, K., Babineau, B., Lebrilla, C. B., Hagerman, R. J., & Hagerman, P. J. (2006). Protein composition of the intranuclear inclusions of FXTAS. *Brain, 129*, 256–271.

Jacquemont, S., Hagerman, R. J., Hagerman, P. J., & Leehey, M. A. (2007). Fragile-X syndrome and fragile X-associated tremor/ataxia syndrome: Two faces of FMR1. *Lancet Neurology, 6*, 45–55.

Jin, P., Duan, R., Qurashi, A., Qin, Y., Tian, D., Rosser, T. C., Liu, H., Feng, Y., & Warren, S. T. (2007). Pur alpha binds to rCGG repeats and modulates repeat-mediated neurodegeneration in a Drosophila model of fragile X tremor/ataxia syndrome. *Neuron, 55*, 556–564.

Johnson, E. M., Kinoshita, Y., Weinreb, D. B., Wortman, M. J., Simon, R., Khalili, K., Winckler, B., & Gordon, J. (2006). Role of Pur alpha in targeting mRNA to sites of translation in hippocampal neuronal dendrites. *Journal of Neuroscience Research, 83*, 929–943.

Kanai, Y., Dohmae, N., & Hirokawa, N. (2004). Kinesin transports RNA: Isolation and characterization of an RNA-transporting granule. *Neuron, 43*, 513–525.

Kao, D. I., Aldridge, G. M., Weiler, I. J., & Greenough, W. T. (2010). Altered mRNA transport, docking, and protein translation in neurons lacking fragile X mental retardation protein. *Proceedings of the National Academy of Sciences of the United States of America, 107*, 15601–15606.

Kaufmann, W. E., Cohen, S., Sun, H. T., & Ho, G. (2002). Molecular phenotype of fragile X syndrome: FMRP, FXRPs, and protein targets. *Microscopy Research and Technique, 57*, 135–144.

Kenneson, A., Zhang, F., Hagedorn, C. H., & Warren, S. T. (2001). Reduced FMRP and increased FMR1 transcription is proportionally associated with CGG repeat number in intermediate-length and premutation carriers. *Human Molecular Genetics, 10*, 1449–1454.

Khandjian, E. W., Fortin, A., Thibodeau, A., Tremblay, S., Cote, F., Devys, D., Mandel, J. L., & Rousseau, F. (1995). A heterogeneous set of FMR1 proteins is widely distributed in mouse tissues and is modulated in cell culture. *Human Molecular Genetics, 4*, 783–789.

Khandjian, E. W., Huot, M. E., Tremblay, S., Davidovic, L., Mazroui, R., & Bardoni, B. (2004). Biochemical evidence for the association of fragile X mental retardation protein with brain polyribosomal ribonucleoparticles. *Proceedings of the National Academy of Sciences of the United States of America, 101*, 13357–13362.

Kindler, S., Rehbein, M., Classen, B., Richter, D., Böckers, TM., (2004). Distinct spatiotemporal expression of SAPAP transcripts in the developing rat brain: a novel dendritically localized mRNA. *Brain Res Mol Brain Res.* 126(1):14–21.

Koekkoek, S. K., Yamaguchi, K., Milojkovic, B. A., Dortland, B. R., Ruigrok, T. J., Maex, R., De Graaf, W., Smit, A. E., VanderWerf, F., Bakker, C. E., Willemsen, R., Ikeda, T., Kakizawa, S., Onodera, K., Nelson, D. L., Mientjes, E., Joosten, M., De Schutter, E., Oostra, B. A., Ito, M., & De Zeeuw, C. I. (2005). Deletion of FMR1 in Purkinje cells enhances parallel fiber LTD, enlarges spines, and attenuates cerebellar eyelid conditioning in fragile X syndrome. *Neuron, 47*, 339–352.

Kooy, R. F. (2003). Of mice and the fragile X syndrome. *Trends in Genetics, 19*, 148–154.

Kooy, R. F., D'Hooge, R., Reyniers, E., Bakker, C. E., Nagels, G., De Boulle, K., Storm, K., Clincke, G., De Deyn, P. P., Oostra, B. A., & Willems, P. J. (1996). Transgenic mouse model for the fragile X syndrome. *American Journal of Medical Genetics, 64*, 241–245.

Laggerbauer, B., Ostareck, D., Keidel, E. M., Ostareck-Lederer, A., & Fischer, U. (2001). Evidence that fragile X mental retardation protein is a negative regulator of translation. *Human Molecular Genetics, 10*, 329–338.

Larson, J., Jessen, R. E., Kim, D., Fine, A. K., & du Hoffmann, J. (2005). Age-dependent and selective impairment of long-term potentiation in the anterior piriform cortex of mice lacking the fragile X mental retardation protein. *Journal of Neuroscience, 25*, 9460–9469.

Lauterborn, J. C., Rex, C. S., Kramar, E., Chen, L. Y., Pandyarajan, V., Lynch, G., & Gall, C. M. (2007). Brain-derived neurotrophic factor rescues synaptic plasticity in a mouse model of fragile X syndrome. *Journal of Neuroscience, 27*, 10685–10694.

Li, J., Pelletier, M. R., Perez Velazquez, J. L., & Carlen, P. L. (2002). Reduced cortical synaptic plasticity and GluR1 expression associated with fragile X mental retardation protein deficiency. *Molecular and Cellular Neurosciences, 19*, 138–151.

Li, Z., Zhang, Y., Ku, L., Wilkinson, K. D., Warren, S. T., & Feng, Y. (2001). The fragile X mental retardation protein inhibits translation via interacting with mRNA. *Nucleic Acids Research, 29*, 2276–2283.

Linder, B., Plottner, O., Kroiss, M., Hartmann, E., Laggerbauer, B., Meister, G., Keidel, E., & Fischer, U. (2008). Tdrd3 is a novel stress granule-associated protein interacting with the fragile-X syndrome protein FMRP. *Human Molecular Genetics, 17*, 3236–3246.

Liu, Z. H., Chuang, D. M., Smith, C. B., Lithium ameliorates phenotypic deficits in a mouse model of fragile X syndrome. *Int J Neuropsychopharmacol.* 2011 Jun;14(5):618-30. Epub 2010 May 25.

Liu, Z. H., & Smith, C. B. (2009). Dissociation of social and nonsocial anxiety in a mouse model of fragile X syndrome. *Neuroscience Letters, 454*, 62–66.

Lu, R., Wang, H., Liang, Z., Ku, L., O'Donnell, W. T., Li, W., Warren, S. T., & Feng, Y. (2004). The fragile X protein controls microtubule-associated protein 1B translation and microtubule stability in brain neuron development. *Proceedings of the National Academy of Sciences of the United States of America, 101*, 15201–15206.

Massey, P. V., & Bashir, Z. I. (2007). Long-term depression: Multiple forms and implications for brain function. *Trends in Neurosciences, 30*, 176–184.

McNaughton, C. H., Moon, J., Strawderman, M. S., Maclean, K. N., Evans, J., & Strupp, B. J. (2008). Evidence for social anxiety and impaired social cognition in a mouse model of fragile X syndrome. *Behavioral Neuroscience, 122*, 293–300.

Meijer, H., de Graaff, E., Merckx, D. M., Jongbloed, R. J., de Die-Smulders, C. E., Engelen, J. J., Fryns, J. P., Curfs, P. M., & Oostra, B. A. (1994). A deletion of 1.6 kb proximal to the CGG repeat of the FMR1 gene causes the clinical phenotype of the fragile X syndrome. *Human Molecular Genetics, 3*, 615–620.

Menon, R. P., Gibson, T. J., & Pastore, A. (2004). The C terminus of fragile X mental retardation protein interacts with the multi-domain Ran-binding protein in the microtubule-organising centre. *Journal of Molecular Biology, 343*, 43–53.

Meredith, R. M., Holmgren, C. D., Weidum, M., Burnashev, N., & Mansvelder, H. D. (2007). Increased threshold for spike-timing-dependent plasticity is caused by unreliable calcium signaling in mice lacking fragile X gene FMR1. *Neuron, 54*, 627–638.

Mientjes, E. J., Nieuwenhuizen, I., Kirkpatrick, L., Zu, T., Hoogeveen-Westerveld, M., Severijnen, L., Rife, M., Willemsen, R., Nelson, D. L., & Oostra, B. A. (2006). The generation of a conditional Fmr1 knock out mouse model to study Fmrp function in vivo. *Neurobiology of Disease, 21*, 549–555.

Mila, M., Castellvi-Bel, S., Sanchez, A., Barcelo, A., Badenas, C., Mallolas, J., & Estivill, X. (2000). Rare variants in the promoter of the fragile X syndrome gene (FMR1). *Molecular and Cellular Probes, 14*, 115–119.

Min, W. W., Yuskaitis, C. J., Yan, Q., Sikorski, C., Chen, S., Jope, R. S., & Bauchwitz, R. P. (2009). Elevated glycogen synthase kinase-3 activity in fragile X mice: Key metabolic regulator with evidence for treatment potential. *Neuropharmacology, 56*, 463–472.

Mines, M. A., Yuskaitis, C. J., King, M. K., Beurel, E., & Jope, R. S. (2010). GSK3 influences social preference and anxiety-related behaviors during social interaction in a mouse model of fragile X syndrome and autism. *PLoS One, 5*, e9706.

Mineur, Y. S., Huynh, L. X., & Crusio, W. E. (2006). Social behavior deficits in the Fmr1 mutant mouse. *Behavioural Brain Research, 168*, 172–175.

Mineur, Y. S., Sluyter, F., de Wit, S., Oostra, B. A., & Crusio, W. E. (2002). Behavioral and neuroanatomical characterization of the Fmr1 knockout mouse. *Hippocampus, 12*, 39–46.

Miyashiro, K. Y., Beckel-Mitchener, A., Purk, T. P., Becker, K. G., Barret, T., Liu, L., Carbonetto, S., Weiler, I. J., Greenough, W. T., & Eberwine, J. (2003). RNA cargoes associating with FMRP reveal deficits in cellular functioning in Fmr1 null mice. *Neuron, 37*, 417–431.

Monzo, K., Papoulas, O., Cantin, G. T., Wang, Y., Yates, J. R., 3rd, & Sisson, J. C. (2006). Fragile X mental retardation protein controls trailer hitch expression and cleavage furrow formation in Drosophila embryos. *Proceedings of the National Academy of Sciences of the United States of America, 103*, 18160–18165.

Moon, J., Ota, K. T., Driscoll, L. L., Levitsky, D. A., & Strupp, B. J. (2008). A mouse model of fragile X syndrome exhibits heightened arousal and/or emotion following errors or reversal of contingencies. *Developmental Psychobiology, 50*, 473–485.

Morales, J., Hiesinger, P. R., Schroeder, A. J., Kume, K., Verstreken, P., Jackson, F. R., Nelson, D. L., & Hassan, B. A. (2002). Drosophila fragile X protein, DFXR, regulates neuronal morphology and function in the brain. *Neuron, 34*, 961–972.

Moss, J., & Howlin, P. (2009). Autism spectrum disorders in genetic syndromes: Implications for diagnosis, intervention and understanding the wider autism spectrum disorder population. *Journal of Intellectual Disability Research, 53*, 852–873.

Muddashetty, R. S., Kelic, S., Gross, C., Xu, M., & Bassell, G. J. (2007). Dysregulated metabotropic glutamate receptor-dependent translation of AMPA receptor and postsynaptic density-95 mRNAs at synapses in a mouse model of fragile X syndrome. *Journal of Neuroscience, 27*, 5338–5348.

Musco, G., Stier, G., Joseph, C., Castiglione Morelli, M. A., Nilges, M., Gibson, T. J., & Pastore, A. (1996). Three-dimensional structure and stability of the KH domain: Molecular insights into the fragile X syndrome. *Cell, 85*, 237–245.

Musumeci, S. A., Bosco, P., Calabrese, G., Bakker, C., De Sarro, G. B., Elia, M., Ferri, R., & Oostra, B. A. (2000). Audiogenic seizures susceptibility in transgenic mice with fragile X syndrome. *Epilepsia, 41*, 19–23.

Musumeci, S. A., Hagerman, R. J., Ferri, R., Bosco, P., Dalla Bernardina, B., Tassinari, C. A., De Sarro, G. B., & Elia, M. (1999). Epilepsy and EEG findings in males with fragile X syndrome. *Epilepsia, 40*, 1092–1099.

Napoli, I., Mercaldo, V., Boyl, P. P., Eleuteri, B., Zalfa, F., De Rubeis, S., Di Marino, D., Mohr, E., Massimi, M., Falconi, M., Witke, W., Costa-Mattioli, M., Sonenberg, N., Achsel, T., & Bagni, C. (2008). The fragile X syndrome protein represses activity-dependent translation through CYFIP1, a new 4E-BP. *Cell, 134*, 1042–1054.

Narayanan, U., Nalavadi, V., Nakamoto, M., Pallas, D. C., Ceman, S., Bassell, G. J., & Warren, S. T. (2007). FMRP phosphorylation reveals an immediate-early signaling pathway triggered by group I mGluR and mediated by PP2A. *Journal of Neuroscience, 27*, 14349–14357.

Nielsen, D. M., Derber, W. J., McClellan, D. A., & Crnic, L. S. (2002a). Alterations in the auditory startle response in Fmr1 targeted mutant mouse models of fragile X syndrome. *Brain Research, 927*, 8–17.

Nielsen, D. M., Evans, J. J., Derber, W. J., Johnston, K. A., Laudenslager, M. L., Crnic, L. S., & Maclean, K. N. (2009). Mouse model of fragile X syndrome: Behavioral and hormonal response to stressors. *Behavioral Neuroscience, 123*, 677–686.

Nielsen, P. R., Nietlispach, D., Mott, H. R., Callaghan, J., Bannister, A., Kouzarides, T., Murzin, A. G., Murzina, N. V., & Laue, E. D. (2002b). Structure of the HP1 chromodomain bound to histone H3 methylated at lysine 9. *Nature, 416*, 103–107.

Nimchinsky, E. A., Oberlander, A. M., & Svoboda, K. (2001). Abnormal development of dendritic spines in FMR1 knock-out mice. *Journal of Neuroscience, 21*, 5139–5146.

Nosyreva, E. D., & Huber, K. M. (2006). Metabotropic receptor-dependent long-term depression persists in the absence of protein synthesis in the mouse model of fragile X syndrome. *Journal of Neurophysiology, 95*, 3291–3295.

O'Donnell, W. T., & Warren, S. T. (2002). A decade of molecular studies of fragile X syndrome. *Annual Review of Neuroscience, 25*, 315–338.

Olmos-Serrano, J. L., Paluszkiewicz, S. M., Martin, B. S., Kaufmann, W. E., Corbin, J. G., & Huntsman, M. M. (2010). Defective GABAergic neurotransmission and pharmacological rescue of neuronal hyperexcitability in the amygdala in a mouse model of fragile X syndrome. *Journal of Neuroscience, 30*, 9929–9938.

Pacey, L. K., Doss, L., Cifelli, C., der Kooy, D. V., Heximer, S. P., & Hampson, D. R. (2011). Genetic deletion of regulator of G-protein signaling 4 (RGS4) rescues a subset of fragile X related phenotypes in the FMR1 knockout mouse. *Molecular and Cellular Neurosciences, 46*(3), 563–72.

Pacey, L. K., Heximer, S. P., & Hampson, D. R. (2009). Increased GABA(B) receptor-mediated signaling reduces the susceptibility of fragile X knockout mice to audiogenic seizures. *Molecular Pharmacology, 76*, 18–24.

Pan, F., Aldridge, G. M., Greenough, W. T., & Gan, W. B. (2010). Dendritic spine instability and insensitivity to modulation by sensory experience in a mouse model of fragile X syndrome. *Proceedings of the National Academy of Sciences of the United States of America, 107*, 17768–17773.

Papoulas, O., Monzo, K. F., Cantin, G. T., Ruse, C., Yates, J. R., 3rd, Ryu, Y. H., & Sisson, J. C. (2010). dFMRP and Caprin, translational regulators of synaptic plasticity, control the cell cycle at the Drosophila mid-blastula transition. *Development, 137*, 4201–4209.

Paradee, W., Melikian, H. E., Rasmussen, D. L., Kenneson, A., Conn, P. J., & Warren, S. T. (1999). Fragile X mouse: Strain effects of knockout phenotype and evidence suggesting deficient amygdala function. *Neuroscience, 94*, 185–192.

Paribello, C., Tao, L., Folino, A., Berry-Kravis, E., Tranfaglia, M., Ethell, I. M., & Ethell, D. W. (2010). Open-label add-on treatment trial of minocycline in fragile X syndrome. *BMC Neurology, 10*, 91.

Park, S., Park, J. M., Kim, S., Kim, J. A., Shepherd, J. D., Smith-Hicks, C. L., Chowdhury, S., Kaufmann, W., Kuhl, D., Ryazanov, A. G., Huganir, R. L., Linden, D. J., Worley, P. F., (2008). Elongation factor 2 and fragile X mental retardation protein control the dynamic translation of Arc/Arg3.1 essential for mGluR-LTD. *Neuron.* 59(1):70–83.

Paylor, R., Yuva-Paylor, L. A., Nelson, D. L., & Spencer, C. M. (2008). Reversal of sensorimotor gating abnormalities in Fmr1 knockout mice carrying a human Fmr1 transgene. *Behavioral Neuroscience, 122*, 1371–1377.

Peier, A. M., McIlwain, K. L., Kenneson, A., Warren, S. T., Paylor, R., & Nelson, D. L. (2000). (Over)correction of FMR1 deficiency with YAC transgenics: Behavioral and physical features. *Human Molecular Genetics, 9*, 1145–1159.

Pfeiffer, B. E., & Huber, K. M. (2007). Fragile X mental retardation protein induces synapse loss through acute postsynaptic translational regulation. *Journal of Neuroscience, 27*, 3120–3130.

Pfeiffer, B. E., Zang, T., Wilkerson, J. R., Taniguchi, M., Maksimova, M. A., Smith, L. N., Cowan, C. W., & Huber, K. M. (2010). Fragile X mental retardation protein is required for synapse elimination by the activity-dependent transcription factor MEF2. *Neuron, 66*, 191–197.

Portera-Cailliau, C., Pan, D. T., & Yuste, R. (2003). Activity-regulated dynamic behavior of early dendritic protrusions: Evidence for different types of dendritic filopodia. *Journal of Neuroscience, 23*, 7129–7142.

Primerano, B., Tassone, F., Hagerman, R. J., Hagerman, P., Amaldi, F., & Bagni, C. (2002). Reduced FMR1 mRNA translation efficiency in fragile X patients with premutations. *RNA, 8*, 1482–1488.

Purpura, D. P. (1974). Dendritic spine "dysgenesis" and mental retardation. *Science, 186*, 1126–1128.

Qiu, L. F., Lu, T. J., Hu, X. L., Yi, Y. H., Liao, W. P., & Xiong, Z. Q. (2009). Limbic epileptogenesis in a mouse model of fragile X syndrome. *Cerebral Cortex, 19*, 1504–1514.

Ramos, A., Hollingworth, D., Adinolfi, S., Castets, M., Kelly, G., Frenkiel, T. A., Bardoni, B., & Pastore, A. (2006). The structure of the N-terminal domain of the fragile X mental retardation protein: A platform for protein–protein interaction. *Structure, 14*, 21–31.

Reiss, A. L., Abrams, M. T., Greenlaw, R., Freund, L., & Denckla, M. B. (1995). Neurodevelopmental effects of the FMR-1 full mutation in humans. *Nature Medicine, 1*, 159–167.

Restivo, L., Ferrari, F., Passino, E., Sgobio, C., Bock, J., Oostra, B. A., Bagni, C., & Ammassari-Teule, M. (2005). Enriched environment promotes behavioral and morphological recovery in a mouse model for the fragile X syndrome. *Proceedings of the National Academy of Sciences of the United States of America, 102*, 11557–11562.

Ronesi, J. A., & Huber, K. M. (2008). Metabotropic glutamate receptors and fragile X mental retardation protein: Partners in translational regulation at the synapse. *Science Signaling, 1*, pe6.

Rudelli, R. D., Brown, W. T., Wisniewski, K., Jenkins, E. C., Laure-Kamionowska, M., Connell, F., & Wisniewski, H. M. (1985). Adult fragile X syndrome. Clinico-neuropathologic findings. *Acta Neuropathologica, 67*, 289–295.

Saugstad, J. A., Marino, M. J., Folk, J. A., Hepler, J. R., & Conn, P. J. (1998). RGS4 inhibits signaling by group I metabotropic glutamate receptors. *Journal of Neuroscience, 18*, 905–913.

Schenck, A., Bardoni, B., Langmann, C., Harden, N., Mandel, J. L., & Giangrande, A. (2003). CYFIP/Sra-1 controls neuronal connectivity in Drosophila and links the Rac1 GTPase pathway to the fragile X protein. *Neuron, 38*, 887–898.

Schütt, J., Falley, K., Richter, D., Kreienkamp, H. J., & Kindler, S. (2009). Fragile X mental retardation protein regulates the levels of scaffold proteins and glutamate receptors in postsynaptic densities. *Journal of Biological Chemistry, 284*, 25479–25487.

Selby, L., Zhang, C., & Sun, Q. Q. (2007). Major defects in neocortical GABAergic inhibitory circuits in mice lacking the fragile X mental retardation protein. *Neuroscience Letters, 412*, 227–232.

Schaeffer, C., Bardoni, B., Mandel, J. L., Ehresmann, B., Ehresmann, C., Moine, H,. (2001). The fragile X mental retardation protein binds specifically to its mRNA via a purine quartet motif. *EMBO J.* 20(17):4803–13.

Shang, Y., Wang, H., Mercaldo, V., Li, X., Chen, T., & Zhuo, M. (2009). Fragile X mental retardation protein is required for chemically-induced long-term potentiation of the hippocampus in adult mice. *Journal of Neurochemistry, 111*, 635–646.

Shi, S. H., Hayashi, Y., Petralia, R. S., Zaman, S. H., Wenthold, R. J., Svoboda, K., & Malinow, R. (1999). Rapid spine delivery and redistribution of AMPA receptors after synaptic NMDA receptor activation. *Science, 284*, 1811–1816.

Siomi, H., Siomi, M. C., Nussbaum, R. L., & Dreyfuss, G. (1993). The protein product of the fragile X gene, FMR1, has characteristics of an RNA-binding protein. *Cell, 74*, 291–298.

Siomi, M. C., Zhang, Y., Siomi, H., & Dreyfuss, G. (1996). Specific sequences in the fragile X syndrome protein FMR1 and the FXR proteins mediate their binding to 60S ribosomal subunits and the interactions among them. *Molecular and Cellular Biology, 16*, 3825–3832.

Sittler, A., Devys, D., Weber, C., & Mandel, J. L. (1996). Alternative splicing of exon 14 determines nuclear or cytoplasmic localisation of fmr1 protein isoforms. *Human Molecular Genetics, 5*, 95–102.

Sjekloca, L., Konarev, P. V., Eccleston, J., Taylor, I. A., Svergun, D. I., & Pastore, A. (2009). A study of the ultrastructure of fragile-X-related proteins. *Biochemical Journal, 419*, 347–357.

Sofola, O. A., Jin, P., Qin, Y., Duan, R., Liu, H., de Haro, M., Nelson, D. L., & Botas, J. (2007). RNA-binding proteins hnRNP A2/B1 and CUGBP1 suppress fragile X CGG premutation repeat-induced neurodegeneration in a Drosophila model of FXTAS. *Neuron, 55*, 565–571.

Spencer, C. M., Alekseyenko, O., Hamilton, S. M., Thomas, A. M., Serysheva, E., Yuva-Paylor, L. A., & Paylor, R. (2011). Modifying behavioral phenotypes in Fmr1KO mice: Genetic background differences reveal autistic-like responses. *Autism Research, 4*, 40–56.

Spencer, C. M., Alekseyenko, O., Serysheva, E., Yuva-Paylor, L. A., & Paylor, R. (2005). Altered anxiety-related and social behaviors in the Fmr1 knockout mouse model of fragile X syndrome. *Genes, Brain, and Behavior, 4*, 420–430.

Spencer, C. M., Graham, D. F., Yuva-Paylor, L. A., Nelson, D. L., & Paylor, R. (2008). Social behavior in Fmr1 knockout mice carrying a human FMR1 transgene. *Behavioral Neuroscience, 122*, 710–715.

Sprangers, R., Groves, M. R., Sinning, I., & Sattler, M. (2003). High-resolution X-ray and NMR structures of the SMN Tudor domain: Conformational variation in the binding site for symmetrically dimethylated arginine residues. *Journal of Molecular Biology, 327*, 507–520.

Stefani, G., Fraser, C. E., Darnell, J. C., & Darnell, R. B. (2004). Fragile X mental retardation protein is associated with translating polyribosomes in neuronal cells. *Journal of Neuroscience, 24*, 7272–7276.

Steward, O., & Schuman, E. M. (2003). Compartmentalized synthesis and degradation of proteins in neurons. *Neuron, 40*, 347–359.

Su, T., Fan, H. X., Jiang, T., Sun, W. W., Den, W. Y., Gao, M. M., Chen, S. Q., Zhao, Q. H., & Yi, Y. H. (2010). Early continuous inhibition of group 1 mGlu signaling partially rescues dendritic spine abnormalities in the Fmr1 knockout mouse model for fragile X syndrome. *Psychopharmacology* (Berlin), 215(2):291–300.

Sung, Y. J., Dolzhanskaya, N., Nolin, S. L., Brown, T., Currie, J. R., Denman, R. B,. (2003). The fragile X mental retardation protein FMRP binds elongation factor 1A mRNA and negatively regulates its translation in vivo. *J Biol Chem.* 278(18):15669–78. Epub 2003 Feb 19.

Suvrathan, A., Hoeffer, C. A., Wong, H., Klann, E., & Chattarji, S. (2010). Characterization and reversal of synaptic defects in the amygdala in a mouse model of fragile X syndrome. *Proceedings of the National Academy of Sciences of the United States of America, 107*, 11591–11596.

Tamanini, F., Van Unen, L., Bakker, C., Sacchi, N., Galjaard, H., Oostra, B. A., & Hoogeveen, A. T. (1999). Oligomerization properties of fragile-X mental-retardation protein (FMRP) and the fragile-X-related proteins FXR1P and FXR2P. *Biochemical Journal, 343*(Pt 3), 517–523.

Tassone, F., Beilina, A., Carosi, C., Albertosi, S., Bagni, C., Li, L., Glover, K., Bentley, D., & Hagerman, P. J. (2007). Elevated FMR1 mRNA in premutation carriers is due to increased transcription. *RNA, 13*, 555–562.

Tassone, F., De Rubeis, S., Carosi, C., La Fata, G., Serpa, G., Raske, C., Willemsen, R., Hagerman, P. J., Bagni, C., (2011). Differential usage of transcriptional start sites and polyadenylation sites in FMR1 premutation alleles. *Nucleic Acids Res*, 2011. Aug: 39(14):6172–85. Epub 2011 Apr 7.

Tassone, F., Hagerman, R. J., Taylor, A. K., Gane, L. W., Godfrey, T. E., & Hagerman, P. J. (2000). Elevated levels of FMR1 mRNA in carrier males: A new mechanism of involvement in the fragile-X syndrome. *American Journal of Human Genetics, 66*, 6–15.

Tassone, F., Iwahashi, C., & Hagerman, P. J. (2004). FMR1 RNA within the intranuclear inclusions of fragile X-associated tremor/ataxia syndrome (FXTAS). *RNA Biology, 1*, 103–105.

Tervonen, T., Akerman, K., Oostra, B. A., & Castren, M. (2005). Rgs4 mRNA expression is decreased in the brain of Fmr1 knockout mouse. *Brain Research. Molecular Brain Research, 133*, 162–165.

Todd, P. K., & Mack, K. J. (2000). Sensory stimulation increases cortical expression of the fragile X mental retardation protein in vivo. *Brain Research: Molecular Brain Research, 80*, 17–25.

Tucker, B., Richards, R. I., & Lardelli, M. (2006). Contribution of mGluR and Fmr1 functional pathways to neurite morphogenesis, craniofacial development and fragile X syndrome. *Human Molecular Genetics, 15*, 3446–3458.

Valverde, R., Edwards, L., & Regan, L. (2008). Structure and function of KH domains. *FEBS Journal, 275*, 2712–2726.

Valverde, R., Pozdnyakova, I., Kajander, T., Venkatraman, J., & Regan, L. (2007). Fragile X mental retardation syndrome: Structure of the KH1-KH2 domains of fragile X mental retardation protein. *Structure, 15*, 1090–1098.

Van Dam, D., D'Hooge, R., Hauben, E., Reyniers, E., Gantois, I., Bakker, C. E., Oostra, B. A., Kooy, R. F., & De Deyn, P. P. (2000). Spatial learning, contextual fear conditioning and conditioned emotional response in Fmr1 knockout mice. *Behavioural Brain Research, 117*, 127–136.

Vanderklish, P. W., & Edelman, G. M. (2002). Dendritic spines elongate after stimulation of group 1 metabotropic glutamate receptors in cultured hippocampal neurons. *Proceedings of the National Academy of Sciences of the United States of America, 99*, 1639–1644.

Ventura, R., Pascucci, T., Catania, M. V., Musumeci, S. A., & Puglisi-Allegra, S. (2004). Object recognition impairment in Fmr1 knockout mice is reversed by amphetamine: Involvement of dopamine in the medial prefrontal cortex. *Behavioural Pharmacology, 15*, 433–442.

Verheij, C., Bakker, C. E., de Graaff, E., Keulemans, J., Willemsen, R., Verkerk, A. J., Galjaard, H., Reuser, A. J., Hoogeveen, A. T., & Oostra, B. A. (1993). Characterization and localization of the FMR-1 gene product associated with fragile X syndrome. *Nature, 363*, 722–724.

Weiler, I. J., Irwin, S. A., Klintsova, A. Y., Spencer, C. M., Brazelton, A. D., Miyashiro, K., Comery, T. A., Patel, B., Eberwine, J., & Greenough, W. T. (1997). Fragile X mental retardation protein is translated near synapses in response to neurotransmitter activation. *Proceedings of the National Academy of Sciences of the United States of America, 94*, 5395–5400.

Willemsen, R., Bontekoe, C., Tamanini, F., Galjaard, H., Hoogeveen, A., & Oostra, B. (1996). Association of FMRP with ribosomal precursor particles in the nucleolus. *Biochemical and Biophysical Research Communications, 225*, 27–33.

Westmark, C. J., Malter, J. S., (2007). FMRP mediates mGluR5-dependent translation of amyloid precursor protein. PLoS Biol. 5(3):e52.

Wohrle, D., Kotzot, D., Hirst, M. C., Manca, A., Korn, B., Schmidt, A., Barbi, G., Rott, H. D., Poustka, A., Davies, K. E., et al. (1992). A microdeletion of less than 250 kb, including the proximal part of the FMR-I gene and the fragile-X site, in a male with the clinical phenotype of fragile-X syndrome. *American Journal of Human Genetics, 51*, 299–306.

Wolfer, D. P., & Lipp, H. P. (2000). Dissecting the behaviour of transgenic mice: Is it the mutation, the genetic background, or the environment? *Experimental Physiology, 85*, 627–634.

Xie, W., Dolzhanskaya, N., LaFauci, G., Dobkin, C., & Denman, R. B. (2009). Tissue and developmental regulation of fragile X mental retardation 1 exon 12 and 15 isoforms. *Neurobiology of Disease, 35*, 52–62.

Yan, Q. J., Asafo-Adjei, P. K., Arnold, H. M., Brown, R. E., & Bauchwitz, R. P. (2004). A phenotypic and molecular characterization of the fmr1-tm1Cgr fragile X mouse. *Genes, Brain, and Behavior, 3*, 337–359.

Yun, S. H., & Trommer, B. L. (2011). Fragile X mice: Reduced long-term potentiation and N-methyl-D-aspartate receptor-mediated neurotransmission in dentate gyrus. *Journal of Neuroscience Research, 89*, 176–182.

Yuskaitis, C. J., Mines, M. A., King, M. K., Sweatt, J. D., Miller, C. A., & Jope, R. S. (2010). Lithium ameliorates altered glycogen synthase kinase-3 and behavior in a mouse model of fragile X syndrome. *Biochemical Pharmacology, 79*, 632–646.

Zalfa, F., Achsel, T., & Bagni, C. (2006). mRNPs, polysomes or granules: FMRP in neuronal protein synthesis. *Current Opinion in Neurobiology, 16*, 265–269.

Zalfa, F., Adinolfi, S., Napoli, I., Kuhn-Holsken, E., Urlaub, H., Achsel, T., Pastore, A., & Bagni, C. (2005). Fragile X mental retardation protein (FMRP) binds specifically to the brain cytoplasmic RNAs BC1/BC200 via a novel RNA-binding motif. *Journal of Biological Chemistry, 280*, 33403–33410.

Zalfa, F., & Bagni, C. (2005). Another view of the role of FMRP in translational regulation. *Cellular and Molecular Life Sciences, 62*, 251–252.

Zalfa, F., Eleuteri, B., Dickson, K. S., Mercaldo, V., De Rubeis, S., di Penta, A., Tabolacci, E., Chiurazzi, P., Neri, G., Grant, S. G., & Bagni, C. (2007). A new function for the fragile X mental retardation protein in regulation of PSD-95 mRNA stability. *Nature Neuroscience, 10*, 578–587.

Zalfa, F., Giorgi, M., Primerano, B., Moro, A., Di Penta, A., Reis, S., Oostra, B., & Bagni, C. (2003). The fragile X syndrome protein FMRP associates with BC1 RNA and regulates the translation of specific mRNAs at synapses. *Cell, 112*, 317–327.

Zang, J. B., Nosyreva, E. D., Spencer, C. M., Volk, L. J., Musunuru, K., Zhong, R., Stone, E. F., Yuva-Paylor, L. A., Huber, K. M., Paylor, R., Darnell, J. C., & Darnell, R. B. (2009). A mouse model of the human fragile X syndrome I304N mutation. *PLoS Genetics, 5*, e1000758.

Zarnescu, D. C., Shan, G., Warren, S. T., & Jin, P. (2005). Come FLY with us: Toward understanding fragile X syndrome. *Genes, Brain, and Behavior, 4*, 385–392.

Zeier, Z., Kumar, A., Bodhinathan, K., Feller, J. A., Foster, T. C., & Bloom, D. C. (2009). Fragile X mental retardation protein replacement restores hippocampal synaptic function in a mouse model of fragile X syndrome. *Gene Therapy, 16*, 1122–1129.

Zhang, Y. Q., Bailey, A. M., Matthies, H. J., Renden, R. B., Smith, M. A., Speese, S. D., Rubin, G. M., & Broadie, K. (2001). Drosophila fragile X-related gene regulates the MAP1B homolog Futsch to control synaptic structure and function. *Cell, 107*, 591–603.

Zhang, J., Fang, Z., Jud, C., Vansteensel, M. J., Kaasik, K., Lee, C. C., Albrecht, U., Tamanini, F., Meijer, J. H., Oostra, B. A., & Nelson, D. L. (2008). Fragile X-related proteins regulate mammalian circadian behavioral rhythms. *American Journal of Human Genetics, 83*, 43–52.

Zhang, J., Hou, L., Klann, E., & Nelson, D. L. (2009). Altered hippocampal synaptic plasticity in the FMR1 gene family knockout mouse models. *Journal of Neurophysiology, 101*, 2572–2580.

Zhang, Y., O'Connor, J. P., Siomi, M. C., Srinivasan, S., Dutra, A., Nussbaum, R. L., & Dreyfuss, G. (1995). The fragile X mental retardation syndrome protein interacts with novel homologs FXR1 and FXR2. *EMBO Journal, 14*, 5358–5366.

Zhang, M., Wang, Q., & Huang, Y. (2007). Fragile X mental retardation protein FMRP and the RNA export factor NXF2 associate with and destabilize Nxf1 mRNA in neuronal cells. *Proceedings of the National Academy of Sciences of the United States of America, 104*, 10057–10062.

Zhao, M. G., Toyoda, H., Ko, S. W., Ding, H. K., Wu, L. J., & Zhuo, M. (2005). Deficits in trace fear memory and long-term potentiation in a mouse model for fragile X syndrome. *Journal of Neuroscience, 25*, 7385–7392.

Chapter 24
Synaptic Dysfunction in Parkinson's Disease

Barbara Picconi, Giovanni Piccoli, and Paolo Calabresi

Abstract Activity-dependent modifications in synaptic efficacy, such as long-term depression (LTD) and long-term potentiation (LTP), represent key cellular substrates for adaptive motor control and procedural memory. The impairment of these two forms of synaptic plasticity in the nucleus striatum could account for the onset and the progression of motor and cognitive symptoms of Parkinson's disease (PD), characterized by the massive degeneration of dopaminergic neurons. In fact, both LTD and LTP are peculiarly controlled and modulated by dopaminergic transmission coming from nigrostriatal terminals.

Changes in corticostriatal and nigrostriatal neuronal excitability may influence profoundly the threshold for the induction of synaptic plasticity, and changes in striatal synaptic transmission efficacy are supposed to play a role in the occurrence of PD symptoms. Understanding of these maladaptive forms of synaptic plasticity has mostly come from the analysis of experimental animal models of PD. A series of cellular and synaptic alterations occur in the striatum of experimental parkinsonism in response to the massive dopaminergic loss. In particular, dysfunctions in trafficking and subunit composition of glutamatergic NMDA receptors on striatal efferent neurons contribute to the clinical features of the experimental parkinsonism.

Interestingly, it has become increasingly evident that in striatal spiny neurons, the correct assembly of NMDA receptor complex at the postsynaptic site is a major

B. Picconi
Laboratorio di Neurofisiologia, Fondazione Santa Lucia I.R.C.C.S., 00143 Rome, Italy

G. Piccoli
CNR Institute of Neuroscience, Milan, Italy

P. Calabresi (✉)
Laboratorio di Neurofisiologia, Fondazione Santa Lucia I.R.C.C.S., 00143 Rome, Italy

Clinica Neurologica, Facoltà di Medicina e Chirurgia, Università degli Studi di Perugia, Ospedale S. Maria della Misericordia, 06156 Perugia, Italy
e-mail: calabre@unipg.it

M.R. Kreutz and C. Sala (eds.), *Synaptic Plasticity*,
Advances in Experimental Medicine and Biology 970,
DOI 10.1007/978-3-7091-0932-8_24, © Springer-Verlag/Wien 2012

player in early phases of PD, and it is sensitive to distinct degrees of DA denervation. The molecular defects at the basis of PD progression may be not confined just at the postsynaptic neuron: accumulating evidences have recently shown that the genes linked to PD play a critical role at the presynaptic site. DA release into the synaptic cleft relies on a proper presynaptic vesicular transport; impairment of SV trafficking, modification of DA flow, and altered presynaptic plasticity have been described in several PD animal models. Furthermore, an impaired DA turnover has been described in presymptomatic PD patients. Thus, given the pathological events occurring precociously at the synapses of PD patients, post- and presynaptic sites may represent an adequate target for early therapeutic intervention.

Keywords α-synuclein • LRRK2 • NMDA receptors • Postsynaptic density

24.1 Introduction

Parkinson's disease (PD) is one of the most frequent human neurodegenerative disorders associated with the process of cerebral aging. PD physiopathology is linked to a widespread process of degeneration of dopamine (DA)-secreting neurons in the substantia nigra *pars compacta* (SNc), with the consequent loss of the neurons projecting to the striatum (Lang and Lozano 1998a, b). The parkinsonian symptoms appear when brain levels of DA reach the 70–80% of the normal levels. The main clinical features of PD are the direct consequences of a dysfunction occurring within both the striatum and the entire basal ganglia system (Calabresi et al. 2007). Bradykinesia, rigidity, tremor at rest, postural instability, micrographia, and shuffling gait represent the principal motor symptoms that allow the diagnosis of PD (Jankovic 2008). The clinical detection of these motor symptoms is often accompanied by autonomic, cognitive, and psychiatric problems (Calabresi et al. 2006; Kehagia et al. 2010). Rare forms of PD resulted from missense mutations of α-synuclein as well as increased expression of normal α-synuclein are characterized by early onset and autosomal-dominant inheritance (Polymeropoulos et al. 1997; Singleton et al. 2003). Intracytoplasmic inclusions called Lewy bodies and the progressive loss of DA-containing neurons in the SNc represent the main neuropathological features of the genetic forms of PD (Spillantini et al. 1998; Dickson et al. 2009; Schulz-Schaeffer 2010).

Mutations in seven genes have been implicated in various forms of familial parkinsonism. Two autosomal-dominant genes (α-synuclein and LRRK2) and three autosomal-recessive genes (Parkin, DJ-1, and PINK1) have been definitively associated with inherited PD (Nussbaum and Polymeropoulos 1997; Polymeropoulos et al. 1997; Healy et al. 2004; Paisan-Ruiz et al. 2004; Valente et al. 2004). As well as these, other mutations have been reported in UCHL-1, synphilin-1, and NR4A2 that may or may not be biologically significant (Leroy et al. 1998; Le et al. 2003; Marx et al. 2003). Synaptic loss is one of the major neurobiological dysfunction occurring in several neurological diseases (Wishart et al. 2006), for example, synaptic failure happens in a very early phase in both patients and animal models during the progression of Alzheimer's disease (Selkoe 2002).

24.2 Parkinson's Disease and Presynaptic Dysfunction

Accumulating evidence has convincingly demonstrated that the genes linked to PD play a critical role at the presynaptic site. α-synuclein is a 140–amino acid protein present in almost all subcellular compartments but particularly enriched in the presynaptic terminals where it is loosely associated with the distal reserve pool of synaptic vesicles (Lavedan 1998; Yu et al. 2007). Structural and functional studies have shown that α-synuclein is involved in the trafficking of synaptic vesicles. In fact, the presynaptic boutons of cultures lacking α-synuclein presented a marked reduction in the number of vesicles present in the distal pool although the number of vesicles docked at the synaptic plasma remained unaltered (Murphy et al. 2000). Accordingly, α-synuclein knockout (KO) mice showed a marked decrease in the pool of undocked synaptic vesicles and significantly impaired hippocampal response to long-lasting low-frequency stimulation (Cabin et al. 2002). Furthermore, α-synuclein KO mice are characterized by an increased evoked DA release: these observations might imply that α-synuclein normally acts as a negative regulator of DA neurotransmission in an activity-dependent fashion (Abeliovich et al. 2000). Strikingly, overexpression of α-synuclein inhibits neurotransmitter release affecting specifically the size of the synaptic vesicle recycling pool (Nemani et al. 2010). Thus, α-synuclein seems to be deeply implicated in the synaptic vesicle trafficking required for a proper presynaptic DA release by keeping low the amount of DA within the presynaptic bouton (Sidhu et al. 2004; Yu et al. 2005). Given that cytosolic DA might be converted into highly reactive oxidative molecules, it can be speculated that pathological mutations or aggregation of α-synuclein might prejudice normal α-synuclein functions. This impairment may bring to accumulation of DA and thus to the generation of toxic moieties. Interestingly, also DJ-1 and PINK1 KO mice exhibit presynaptic defects. DJ-1 is a redox-sensitive molecular chaperone, and it has been proposed that it inhibits protein aggregation, including α-synuclein formations (Shendelman et al. 2004; Wilson et al. 2004; Moore et al. 2006; Gasser 2009). DJ-1 is expressed widely throughout the tissues, and it is subcellularly localized to the cytosol, mitochondrial matrix, and intermembrane space (Zhang et al. 2005). Acute slice preparation from DJ-1 KO mice showed a reduce DA overflow and impaired LTD. Furthermore, the mice had a poor performance in terms of spontaneous activities and generalized hypokinesia in open field (Goldberg et al. 2003, 2005). DJ-1 has been reported to sustain also hippocampal LTD consolidation, suggesting a potential involvement for this protein in modulating hippocampal dependent cognitive dysfunctions reported in PD (Wang et al. 2008). PINK1 instead is a serine/threonine kinase localized in the mitochondria (Silvestri et al. 2005; Zhou et al. 2008). If PINK1 KO mice failed to exhibit any major abnormality, they showed clear deficits in nigrostriatal DA neurotransmission. Robust evidence supports the conclusion that loss of PINK1 function causes a selective impairment in exocytotic DA release (Kitada et al. 2007; Gispert et al. 2009). Actual knowledge about PINK1 suggests that it may reside in the mitochondria but, given that its kinase domain faces the cytosol, it may have extramitochondrial phosphotargets (Silvestri et al. 2005). Therefore, it might be argued that PINK1 can modify via phosphorylation the activity of proteins involved in DA release. Noteworthy, it has been demonstrated that Parkin, PINK1, and DJ-1 interact physically and functionally. In fact, these three proteins form a ternary complex that promotes ubiquitination and degradation of aberrantly expressed and heat shock–induced Parkin substrates, as Parkin itself and synphilin-1. Pathogenic mutants might reduce the activity of the degradative complex (Xiong et al. 2009).

Mutations in LRRK2 gene account for up to 13% of familial PD cases compatible with dominant inheritance (Paisan-Ruiz et al. 2004, 2008; Zimprich et al. 2004) and 1–2% of sporadic PD patients, thus suggesting this protein as the most significant player in PD pathogenesis identified to date (Aasly et al. 2005; Berg et al. 2005; Taylor et al. 2006). Clinically and pathologically, the features of LRRK2-associated parkinsonism are often indistinguishable from idiopathic PD, although pathologic variability exists even within PARK8-linked kindred, ranging from nigral neuronal loss only to general neuronal loss with α-synuclein, ubiquitin, or tau inclusions [reviewed in (Whaley et al. 2006)]. Furthermore, the neuropathology demonstrated in postmortem brain examinations of patients with LRRK2 mutations most often involves synucleinopathy, but occasionally tauopathy, suggesting a role for LRRK2 that is upstream of protein inclusion pathology (Zimprich et al. 2004; Taymans and Cookson 2010; Wider et al. 2010). The LRRK2 protein has a molecular weight of approximately 280 kDa and contains several domains including a Ras/GTPase like (Roc), a C-terminal of Roc (COR), a kinase (similar to mitogen-activated protein kinase), and a WD40 domain (Bosgraaf and Van Haastert 2003; Guo et al. 2006). Phylogenetically, the LRRK2 kinase domain belongs to the TKL (tyrosine like kinases) and shows high similarity to mixed lineage kinases (MLKs) (Manning et al. 2002; Marin 2006). Few LRRK2 substrates, including moesin, 4E-BP, MKKs, tubulin beta, and α-synuclein, have been found so far in in vitro assays (Jaleel et al. 2007; Imai et al. 2008; Gillardon 2009; Gloeckner et al. 2009; Qing et al. 2009). Several single nucleotide alterations have been identified in LRRK2 (Lesage et al. 2005; Mata et al. 2005), covering all functional domains, but only five missense mutations clearly segregate with PD in large family studies (Goldwurm et al. 2005; Bonifati 2006a, b). Disease-segregating mutations in LRRK2 have been reported in the kinase domain (G2019S, I2020T), in the Roc domain (R1441C/G), and in the COR domain (Y1699C) [reviewed in Mata et al. (2005)].

The most common mutation found in western countries kindred, G2019S, falls in the kinase domain and increases LRRK2 kinase activity while mutations in the Roc domain appear to decrease the GTPase activity of LRRK2 to affect protein dimerization and to slightly increase kinase activity [reviewed in more detail in

(Moore 2008)]. The G2019S mutation has been identified also in parkinsonian patients with no family history of disease (Gilks et al. 2005; Healy et al. 2008); other LRRK2 variants affecting kinase activity appear to be important risk factors in two genome-wide association studies of sporadic PD (Simon-Sanchez et al. 2009). Although studies show little concordance regarding the level of LRRK2 mRNA/protein expression in the SN, LRRK2 protein expression has been demonstrated in tyrosine-hydroxylase positive neurons of the SNc and in medium-sized spiny neurons of the striatum (Galter et al. 2006; Melrose et al. 2006; Higashi et al. 2007a, b). Cortical regions that are affected in dementia associated with PD, including pyramidal neurons of the cerebral cortex and of Ammon's horn, also demonstrate relatively high levels of LRRK2 (Biskup et al. 2006; Higashi et al. 2007b). At the subcellular level, precedent studies showed LRRK2 is mainly associated with mitochondria but also with multiple vesicles structure, including synaptic vesicles (Biskup et al. 2006). Despite its predominance in PD, the physiological function of LRRK2 is not known, and therefore, its precise role in the etiology of PD is far from being understood.

Neurotransmission defects have been repeatedly observed in different LRRK2 models (Li et al. 2009; Tong et al. 2009; Xiong et al. 2009; Li et al. 2010). Functional impairments in nigrostriatal dopaminergic innervation and degeneration of the nigrostriatal projections have been demonstrated in R1441C-LRRK2 homozygous knock-in mice (Tong et al. 2009) and in R1441C-LRRK2 BAC transgenic mice (Li et al. 2009), respectively. G2019S BAC transgenic mice show deficiencies in striatal dopamine release and enhanced striatal tau immunoreactivity without dopaminergic neuron loss in the substantia nigra (Li et al. 2010). Recent studies have enlightened that LRRK2 acts directly at the secretory and endocytic molecular machinery (Shin et al. 2008; Xiong et al. 2010). Finally, it has been shown that electrophysiological properties as well as proper vesicular trafficking and spatial distribution in the presynaptic pool depend on the presence of LRRK2 as an integral part of presynaptic protein complex (Piccoli et al. 2011). Presynaptic proteins – NSF, AP-2 complex subunits, SV2A, synapsin, syntaxin 1 (Piccoli et al. 2011), and Rab5b (Shin et al. 2008) – as well as actin (Meixner et al. 2010) have been found to interact, at least in vitro, with LRRK2 (Fig 24.1). These proteins have been previously described as key elements of synaptic vesicle trafficking. NSF catalyzes the release of the SNARE complex (SNAP 25, syntaxin 1, and VAMP) and allows the first step of the endocytic cycle where also Rab5 proteins are called in action. The clathrin complex [clathrin, AP-2 adaptor complex, and accessory proteins as dynamin and AP180] constitutes one of the major pathways for SV recycling from the membrane to the resting pool (RP). The control of storage and mobilization of SV in the RRP depends instead on the synaptic vesicle glycoproteins SV2A and B while synapsins are thought to immobilize SV in the RP by cross-linking vesicles to the actin cytoskeleton. Strikingly, an increased DA turnover has been noticed in presymptomatic LRRK2 mutation carriers (Sossi et al. 2010). Increased turnover might arise as a compensatory mechanism to counteract DA-neurons loss (Adams et al. 2005), but it has also been suggested that increased DA turnover might by itself

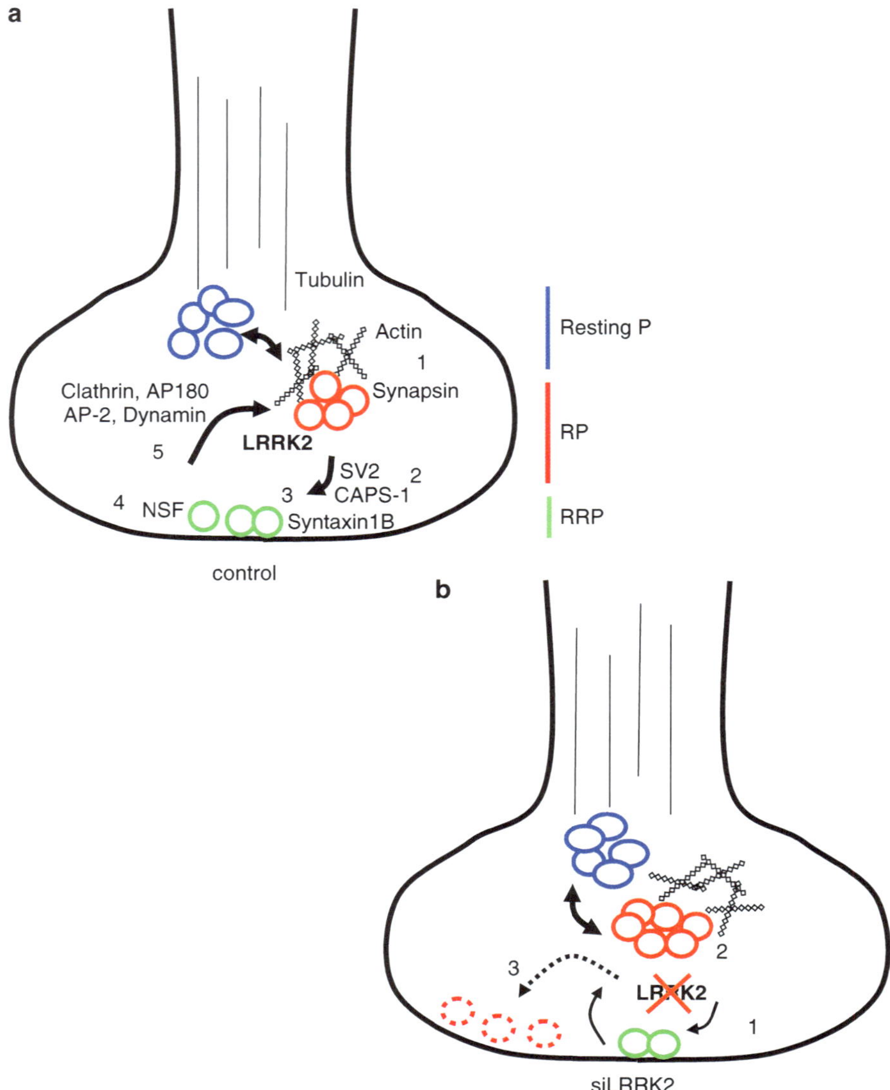

Fig. 24.1 Model of LRRK2 function at the presynaptic site. Given the interaction between LRRK2, cytoskeletal elements, and presynaptic proteins (Shin et al. 2008; Meixner et al. 2010; Piccoli et al. 2011), it has been proposed LRRK2 is part of the molecular complex that controls SV fusion rate. It might modulate SV storage in the RP and SV trafficking between the RP and the membrane. (**a**) SV actively cycles between the RRP and the RP, even if the major part of SV belongs to an apparently inactive resting pool. (1) SV is maintained in the RP by synapsin-actin cytoskeleton interaction. (2) SV2A and calcium-dependent secretion activator 1 (CASP1) convert the vesicles into fusion-responsive state. (3) SNAREs dock SV to the presynaptic membrane in preparation for fusion. (4) After fusion-pore opening, vesicle-fusing ATPase (NSF) disrupts the SNARE complex releasing SV. (5) SV recycles to the RP mainly through clathrin-coated pits

contribute to disease progression secondary to DA-associated toxicity (Smith et al. 2002; Zigmond et al. 2002). Therefore, accumulating evidences suggest that synaptic dysfunction is a primary effect of LRRK2 gene mutations and that synaptic failure is intimately involved in LRRK2 due PD pathogenesis.

24.3 Postsynaptic Dysfunction in Parkinson's Disease

The natural history of PD is complex and involves differential mechanisms during its various clinical phases. Most of the evidence on pathogenic pathways in PD has been obtained using experimental models of complete striatal DA depletion mimicking advanced PD such as rats lesioned with 6-hydroxydopamine (6-OHDA) (Schwarting and Huston 1996) and macaques lesioned with 1-methyl-4-phenyl-1,2,3,6-tetrahydropyridine (MPTP) (Jenner and Marsden 1986).

The massive denervation of the dopaminergic nigrostriatal terminals, as observed in advanced PD, is associated to maladaptive plasticity (Calabresi et al. 2007), alteration of striatal dendritic spines (Anglade et al. 1996; Day et al. 2006), and changes of glutamatergic signaling (Betarbet et al. 2000; Picconi et al. 2004). In advanced PD, spontaneous excitatory glutamatergic synaptic activity can be dramatically altered. These pathological events may also alter the amplitude and the direction of long-term changes of excitatory transmission induced by repetitive synaptic activation. Moreover, changes in neuronal phasic and/or tonic firing discharge may occur. Even slight changes in corticostriatal and nigrostriatal neuronal excitability may influence profoundly the threshold for the induction of synaptic plasticity. Changes in striatal synaptic transmission efficacy are supposed to be the cellular basis for such complex integrative functions (Calabresi et al. 2006, 2007), and experimental findings show that short- and long-term changes in corticostriatal synaptic plasticity may play a role in PD (Gubellini et al. 2002; Picconi et al. 2003).

Two cardinal features of PD pathophysiology are represented by the alteration of glutamatergic synapses paradoxically accompanied by the described increase of glutamatergic transmission within the striatum. The real mechanisms underlying this increased excitatory drive remains unknown. Recently, the synaptic changes in both corticostriatal and thalamostriatal afferents have been studied in MPTP-treated monkeys taking as main markers the vesicular glutamate transporters (vGluTs) 1 and 2 (Raju et al. 2008). This study demonstrates the increased presence of vGluT1 in the striatum of MPTP monkeys without any significant change in the pattern of

Fig. 24.1 (continued) endocytosis (5). (**b**) Impairment of LRRK2 levels/function might impair the functionality of the exo-endo machinery. In absence of LRRK2, SV might not properly cycle between the (1) RRP and (2) the RP. (3) The reduction of the molecular constrain represented by LRRK2 and LRRK2-associated protein might increase SV probability to reach the membrane and fuse (In the cartoon are depicted only presynaptic proteins putatively interacting with LRRK2)

synaptic connectivity. However, a clear degree of synaptic reorganization of the thalamostriatal system has been found. These findings suggest a differential degree of plasticity between the two systems in parkinsonian primates.

In the last decades have been described and extensively studied two forms of striatal synaptic plasticity (long-term depression (LTD) and long-term potentiation (LTP)) thought to underlie cognitive performance both in vitro (Calabresi et al. 1992b, c; Lovinger et al. 1993; Walsh 1993; Walsh and Dunia 1993; Partridge et al. 2000) and in vivo (Charpier and Deniau 1997; Reynolds and Wickens 2000; Mahon et al. 2004).

A high-frequency stimulation (HFS) protocol of the corticostriatal fibers (Calabresi et al. 1992b, c; Lovinger et al. 1993) allows to induce both forms of synaptic plasticity, the type of the long-lasting changes being critically dependent upon the level of membrane depolarization and on the ionotropic glutamate receptor subtype activated during the HFS. A third form of synaptic plasticity (depotentiation) results from the reversal of an established LTP by the application of a low-frequency stimulation of corticostriatal fibers (O'Dell and Kandel 1994; Picconi et al. 2003).

Compared to other brain areas, in which synaptic plasticity has been extensively studied, the striatum has the peculiar feature of receiving a massive dopaminergic input arising from SNc. Accordingly, a unique characteristic of striatal LTD is the requirement of DA receptor activation by endogenous DA (Calabresi et al. 2007). In fact, this form of synaptic plasticity is lost after massive DA denervation both in 6-OHDA rats (Calabresi et al. 1992c) and MPTP-treated monkeys (Quik et al. 2006).

The absence of LTD in the striatum of parkinsonian animals can be attributed to the failed activation of DA receptors during the induction phase of this form of synaptic plasticity. LTD, in fact, can be restored after DA denervation by ensuring DA receptor activation through the application of exogenous DA or by the coactivation of both D1 and D2 receptors (Calabresi et al. 1992a, 2007). Similarly, massive nigrostriatal denervation blocks corticostriatal LTP (Picconi et al. 2003; Calabresi et al. 2007). Interestingly, a "balanced" DA/DARPP-32 pathway is required for the corticostriatal system to be able to express both LTD and LTP (Calabresi et al. 2000).

It is of interest to note that distinct degrees of DA denervation may differentially affect the induction and maintenance of these two distinct and opposite forms of corticostriatal synaptic changes (Paille et al. 2010). An incomplete DA denervation does not affect corticostriatal LTD which is, however, abolished by a complete lesion suggesting that a low, although critical, level of DA is required for this form of synaptic plasticity. Conversely, an incomplete DA denervation dramatically alters the maintenance of LTP confirming a critical role of this form of synaptic plasticity in the early motor parkinsonian symptoms (Paille et al. 2010).

Recently, to understand the early synaptic mechanisms occurring in PD, the striatal dysfunctions have been studied in mice overexpressing human A53T-alfa-synuclein (Kurz et al. 2010). A53T-alfa-synuclein overexpressing mice, in their advanced stage, present dysfunctional DA neurotransmission and consequently an impaired striatal LTD, confirming, once more, the relevant role of an intact and correct balance in the dopaminergic nigrostriatal transmission for a physiological synaptic activity.

The pathophysiological picture emerging from the last years of experimental approach shows that the strength of glutamatergic signals from the cortex to the striatum might be dynamically regulated during the progression of the disease. In fact, bidirectional changes in corticostriatal synaptic plasticity are critically controlled by the different degree of nigral denervation which influences the endogenous DA levels and the assembly of striatal N-methyl-D-aspartate (NMDA)–type glutamate receptor subunits.

NMDA receptors are glutamate ion channels and represent the key elements in the regulation of synaptic function in the central nervous system. They resulted from the coassembly of three different receptor subunit families: NMDA receptor 1 (NR1), NR2A-NR2D, and NR3A-NR3B (Dingledine et al. 1999; Nishi et al. 2001). NMDA receptors are highly permeable to Ca^{2+}, and its influx through the receptor channel is essential for the synaptogenesis, the synaptic remodeling, and the long-lasting changes in synaptic efficacy such as synaptic plasticity (Collingridge et al. 2004).

In the neuronal synapses, NMDA receptors are clustered in the postsynaptic density (PSD) that consists of numerous scaffolding cytoskeletal and signaling proteins, some of which are in close contact with the cytoplasmic domain of glutamate ionotropic receptors in the postsynaptic membrane (Kennedy 2000; Gardoni et al. 2001). This accumulation of NMDA receptors at the postsynaptic compartment ensures a rapid response to neurotransmitter release and provides a molecular mechanism for linking the transmembrane ion flux to the signaling machinery responsible for specific second messenger pathways. Among the protein complex governing the response of the signaling cascade, the α-calcium-calmodulin-dependent protein kinase II (α-CaMKII) is directly linked to the NR2A/NR2B subunits (Gardoni et al. 1998; Strack et al. 2000) and competes in NR2A binding with PSD-95 (Gardoni et al. 2001). Interestingly, CaMKII- and tyrosine-dependent phosphorylation of NMDA receptors is altered in experimental model of PD (Oh et al. 1999).

In the striatum as well as in other brain areas, LTP requires activation of NMDA receptors (Calabresi et al. 1992b, 2007; Collingridge and Bliss 1995; Malenka and Bear 2004). Interestingly, it has become increasingly evident that in striatal spiny neurons, NMDA receptor complex is also profoundly altered in experimental PD (Ulas and Cotman 1996; Dunah and Standaert 2001).

Early studies evaluated NMDA receptor abundance, composition, and phosphorylation in advanced model of PD. In the DA, denervated striatum has been found a decreased level of NR1 and NR2B subunits in striatal membranes, while the abundance of NR2A was unchanged (Ulas and Cotman 1996; Dunah and Standaert 2001). Further studies in the 6-OHDA model showed similar results and associated to alterations in synaptic plasticity (Picconi et al. 2003, 2004; Gardoni et al. 2006). In particular, NR2B subunit was specifically reduced in the synaptic density from advanced parkinsonian rats when compared with sham-lesioned rats in the absence of parallel alterations of NR1 and NR2A (Picconi et al. 2003, 2004; Gardoni et al. 2006). Interestingly, these molecular alterations have been further confirmed in parkinsonian macaques (Hallett et al. 2005). Hallett's group shows that in the striatum of MPTP-lesioned macaques the DA depletion induces massive changes in the levels of striatal NMDA receptor proteins, such as a reduction in the abundance of NR1 and NR2B but not NR2A subunit. Moreover, in the denervated striatum of parkinsonian animals, the alteration of NMDA receptor subunit localization at synaptic sites is accompanied by a decreased recruitment of PSD-95 to NR2A–NR2B subunits; these events are paralleled by an increased activation of the pool of α-CaMKII associated to the NMDA receptor complex (Picconi et al. 2004). Further, other studies reported that experimental Parkinsonism in rats appears to be associated with decreased synaptic membrane localization and increased vesicular localization of PSD-95 and SAP97 members of the PSD-MAGUK family (Nash et al. 2005) that could account for dysregulation of NMDA receptors at synapses.

While in advanced parkinsonism LTP is completely lost and this synaptic alteration is coupled to specifically reduced levels of NR2B subunits in the PSD compartment (Gardoni et al. 2006), the picture found in the early parkinsonian rats is quite different.

As mentioned above, the incomplete DA denervation dramatically alters the maintenance of LTP. This synaptic alteration recorded in striatal spiny neurons is also accompanied by a dramatic increase in the NR2A NMDA receptor subunits in the striatal synapses, suggesting the presence of a profound rearrangement of the receptor complex composition (Paille et al. 2010). These profound differences in NMDA receptors in the postsynaptic compartment of partially versus fully lesioned rats suggest that NR2-type regulatory subunits are sensitive to plastic changes induced by the differential degree of DA denervation.

Moreover, NMDA receptor subunits NR2A and NR2B interact with membrane-associated guanylate kinases (MAGUK); this interaction governs their trafficking and clustering at synaptic sites (Kim and Sheng 2004). The analysis of PSD-95, SAP97, and SAP102 in the postsynaptic compartment reveals a significant reduction of the three proteins in advanced parkinsonian rats compared with sham-operated rats (Gardoni et al. 2006). In contrast, in early parkinsonian animals, the level of these proteins is the same as in the sham-operated animals, suggesting that in this model of "early" PD, no alteration of MAGUK protein distribution at the synapse is present. These data suggest that the NR2A subunit level at the synaptic site is a major player in early phases of PD, and it is sensitive to distinct degrees of DA denervation; thus, it may represent an adequate target for early therapeutic intervention.

In the PSD, other important receptors included in the glutamatergic ionotropic receptors class, and mediating the functions of glutamate, are represented by alpha-amino-3-hydroxy-5-methyl-4-isoxazolepropionic acid (AMPA) receptors, tetrameric proteins composed of subunits GluR1-4. Upon binding with glutamate, synaptic AMPA receptors induce membrane depolarization and after removing magnesium (Mg^{2+}) block from NMDA allow to reduce the threshold to induce long-term increases of the synaptic responses. AMPA receptor–dependent depolarization also opens L-type calcium (Ca^{2+}) channels and leads to activation of CRE elements that are responsible for gene transcription.

Recently, a critical role of AMPA receptors in PD has been shown (Lee et al. 2008). Lee and colleagues found that paraquat, a putative causative agent for PD, inhibits postsynaptic AMPA receptors on dopaminergic neurons in the SNc. However, there is still no general consensus on the mechanism underlying dysregulation of AMPA receptor distribution or composition changes in PD. GluR1 subunit of AMPA receptor has not found changed in the striatum of parkinsonian rats (Bernard et al. 1996; Betarbet et al. 2000), while GluR1 immunoreactivity is increased in the caudate and putamen of MPTP monkeys (Betarbet et al. 2000). Evidence has been provided that GluR1 immunoreactivity is decreased in striatal spiny neurons (Lai et al. 2003) and in striatal membrane fractions of parkinsonian rats (Gardoni et al. 2006); on the contrary, no alteration of GluR1 levels in the postsynaptic density has been found in 6-OHDA-lesioned rats (Picconi et al. 2004).

24.4 Conclusions

Given the correlation recently described between LRRK2 and α-synuclein (Lin et al. 2009; Carballo-Carbajal et al. 2010), the impact of α-synuclein on synaptic vesicle recycling (Fortin et al. 2010; Nemani et al. 2010), and the functional links among DJ-1, Parkin, PINK1, and α-synuclein (Shendelman et al. 2004; Xiong et al. 2009), the regulation of DA release might arise as one the main biological pathway compromised during PD onset. The molecular mechanisms underlying these synaptic transmission defects, however, remain largely elusive. Although little is known about the precise mechanisms of exocytotic DA release, it likely uses a similar mechanism as glutamatergic synapses, in which release is energy-dependent, is mediated by the SNARE-dependent fusion of synaptic vesicles and is triggered by Ca^{2+} binding to synaptotagmins. Synaptic vesicles undergo in the nerve terminal to high-frequency trafficking cycles thanks to the presence of extremely specialized machinery, allowing very rapid triggering and switching off of synaptic vesicle exocytosis in response to depolarization-evoked Ca^{2+} influx. A major goal in neurobiology in recent years has been to gain insight into the molecular machinery that mediates neurotransmitter release. More than 1,000 proteins function in the presynaptic nerve terminal, and hundreds are thought to participate in exo-endocytosis. The processes are finely tuned and depend on the interaction between protein expressed on SV membranes and protein expressed on the presynaptic membranes (Rizo and Rosenmund 2008; Sudhof and Rothman 2009).

This complex network of interaction is plastically shaped by posttranslational modifications: the presynaptic modulation of neurotransmitter release is in fact altered by protein kinases and protein phosphatases (Turner et al. 1999; Fdez and Hilfiker 2006) and by protein degradation (Ehlers 2003; Yao et al. 2007). One possibility worth to be explored is that PD-related proteins alter SV trafficking via modification of presynaptic proteins.

Cellular and postsynaptic alterations occurring in the striatum of experimental parkinsonism in response to the massive dopaminergic loss may lead to synaptic dysfunction and corticostriatal transmission instabilities. In particular, maladaptive forms of synaptic plasticity consequently to the alteration in the subunit composition of glutamatergic ionotropic receptors, that is, NMDA receptors, contribute to the clinical features of PD. Interestingly, it has become increasingly evident that the correct assembly of NMDA receptor complex at the synaptic site is a major player in early phases of PD and it is sensitive to distinct degrees of DA denervation; thus, it may represent an adequate target for early therapeutic intervention.

References

Aasly, J. O., Toft, M., Fernandez-Mata, I., Kachergus, J., Hulihan, M., White, L. R., & Farrer, M. (2005). Clinical features of LRRK2-associated Parkinson's disease in central Norway. *Annals of Neurology, 57*, 762–765.

Abeliovich, A., Schmitz, Y., Farinas, I., Choi-Lundberg, D., Ho, W. H., Castillo, P. E., Shinsky, N., Verdugo, J. M., Armanini, M., Ryan, A., Hynes, M., Phillips, H., Sulzer, D., & Rosenthal, A. (2000). Mice lacking alpha-synuclein display functional deficits in the nigrostriatal dopamine system. *Neuron, 25*, 239–252.

Adams, J. R., van Netten, H., Schulzer, M., Mak, E., McKenzie, J., Strongosky, A., Sossi, V., Ruth, T. J., Lee, C. S., Farrer, M., Gasser, T., Uitti, R. J., Calne, D. B., Wszolek, Z. K., & Stoessl, A. J. (2005). PET in LRRK2 mutations: Comparison to sporadic Parkinson's disease and evidence for presymptomatic compensation. *Brain, 128*, 2777–2785.

Anglade, P., Mouatt-Prigent, A., Agid, Y., & Hirsch, E. (1996). Synaptic plasticity in the caudate nucleus of patients with Parkinson's disease. *Neurodegeneration, 5*, 121–128.

Berg, D., Schweitzer, K., Leitner, P., Zimprich, A., Lichtner, P., Belcredi, P., Brussel, T., Schulte, C., Maass, S., & Nagele, T. (2005). Type and frequency of mutations in the LRRK2 gene in familial and sporadic Parkinson's disease*. *Brain, 128*, 3000–3011.

Bernard, V., Gardiol, A., Faucheux, B., Bloch, B., Agid, Y., & Hirsch, E. C. (1996). Expression of glutamate receptors in the human and rat basal ganglia: Effect of the dopaminergic denervation on AMPA receptor gene expression in the striatopallidal complex in Parkinson's disease and rat with 6-OHDA lesion. *The Journal of Comparative Neurology, 368*, 553–568.

Betarbet, R., Porter, R. H., & Greenamyre, J. T. (2000). GluR1 glutamate receptor subunit is regulated differentially in the primate basal ganglia following nigrostriatal dopamine denervation. *Journal of Neurochemistry, 74*, 1166–1174.

Biskup, S., Moore, D. J., Celsi, F., Higashi, S., West, A. B., Andrabi, S. A., Kurkinen, K., Yu, S. W., Savitt, J. M., Waldvogel, H. J., Faull, R. L., Emson, P. C., Torp, R., Ottersen, O. P., Dawson, T. M., & Dawson, V. L. (2006). Localization of LRRK2 to membranous and vesicular structures in mammalian brain. *Annals of Neurology, 60*, 557–569.

Bonifati, V. (2006a). Parkinson's disease: The LRRK2-G2019S mutation: Opening a novel era in Parkinson's disease genetics. *European Journal of Human Genetics, 14*, 1061–1062.

Bonifati, V. (2006b). The pleomorphic pathology of inherited Parkinson's disease: Lessons from LRRK2. *Current Neurology and Neuroscience Reports, 6*, 355–357.

Bosgraaf, L., & Van Haastert, P. J. (2003). Roc, a Ras/GTPase domain in complex proteins. *Biochimica et Biophysica Acta, 1643*, 5–10.

Cabin, D. E., Shimazu, K., Murphy, D., Cole, N. B., Gottschalk, W., McIlwain, K. L., Orrison, B., Chen, A., Ellis, C. E., Paylor, R., Lu, B., & Nussbaum, R. L. (2002). Synaptic vesicle depletion correlates with attenuated synaptic responses to prolonged repetitive stimulation in mice lacking alpha-synuclein. *The Journal of Neuroscience, 22*, 8797–8807.

Calabresi, P., Maj, R., Mercuri, N. B., & Bernardi, G. (1992a). Coactivation of D1 and D2 dopamine receptors is required for long-term synaptic depression in the striatum. *Neuroscience Letters, 142*, 95–99.

Calabresi, P., Pisani, A., Mercuri, N. B., & Bernardi, G. (1992b). Long-term potentiation in the striatum is unmasked by removing the voltage-dependent magnesium block of NMDA receptor channels. *European Journal of Neuroscience, 4*, 929–935.

Calabresi, P., Picconi, B., Parnetti, L., & Di Filippo, M. (2006). A convergent model for cognitive dysfunctions in Parkinson's disease: The critical dopamine-acetylcholine synaptic balance. *Lancet Neurology, 5*, 974–983.

Calabresi, P., Picconi, B., Tozzi, A., & Di Filippo, M. (2007). Dopamine-mediated regulation of corticostriatal synaptic plasticity. *Trends in Neurosciences, 30*, 211–219.

Calabresi, P., Maj, R., Pisani, A., Mercuri, N. B., & Bernardi, G. (1992c). Long-term synaptic depression in the striatum: Physiological and pharmacological characterization. *The Journal of Neuroscience, 12*, 4224–4233.

Calabresi, P., Gubellini, P., Centonze, D., Picconi, B., Bernardi, G., Chergui, K., Svenningsson, P., Fienberg, A. A., & Greengard, P. (2000). Dopamine and cAMP-regulated phosphoprotein 32 kDa controls both striatal long-term depression and long-term potentiation, opposing forms of synaptic plasticity. *The Journal of Neuroscience, 20*, 8443–8451.

Carballo-Carbajal, I., Weber-Endress, S., Rovelli, G., Chan, D., Wolozin, B., Klein, C. L., Patenge, N., Gasser, T., & Kahle, P. J. (2010). Leucine-rich repeat kinase 2 induces alpha-synuclein expression via the extracellular signal-regulated kinase pathway. *Cellular Signalling, 22*, 821–827.

Charpier, S., & Deniau, J. M. (1997). In vivo activity-dependent plasticity at cortico-striatal connections: Evidence for physiological long-term potentiation. *Proceedings of the National Academy of Sciences of the United States of America, 94*, 7036–7040.

Collingridge, G. L., & Bliss, T. V. (1995). Memories of NMDA receptors and LTP. *Trends in Neurosciences, 18*, 54–56.

Collingridge, G. L., Isaac, J. T., & Wang, Y. T. (2004). Receptor trafficking and synaptic plasticity. *Nature Reviews: Neuroscience, 5*, 952–962.

Day, M., Wang, Z., Ding, J., An, X., Ingham, C. A., Shering, A. F., Wokosin, D., Ilijic, E., Sun, Z., Sampson, A. R., Mugnaini, E., Deutch, A. Y., Sesack, S. R., Arbuthnott, G. W., & Surmeier, D. J. (2006). Selective elimination of glutamatergic synapses on striatopallidal neurons in Parkinson disease models. *Nature Neuroscience, 9*, 251–259.

Dickson, D. W., Braak, H., Duda, J. E., Duyckaerts, C., Gasser, T., Halliday, G. M., Hardy, J., Leverenz, J. B., Del Tredici, K., Wszolek, Z. K., & Litvan, I. (2009). Neuropathological assessment of Parkinson's disease: Refining the diagnostic criteria. *Lancet Neurology, 8*, 1150–1157.

Dingledine, R., Borges, K., Bowie, D., & Traynelis, S. F. (1999). The glutamate receptor ion channels. *Pharmacological Reviews, 51*, 7–61.

Dunah, A. W., & Standaert, D. G. (2001). Dopamine D1 receptor-dependent trafficking of striatal NMDA glutamate receptors to the postsynaptic membrane. *The Journal of Neuroscience, 21*, 5546–5558.

Ehlers, M. D. (2003). Activity level controls postsynaptic composition and signaling via the ubiquitin-proteasome system. *Nature Neuroscience, 6*, 231–242.

Fdez, E., & Hilfiker, S. (2006). Vesicle pools and synapsins: New insights into old enigmas. *Brain Cell Biology, 35*, 107–115.

Fortin, D. L., Nemani, V. M., Nakamura, K., & Edwards, R. H. (2010). The behavior of alpha-synuclein in neurons. *Movement Disorders, 25*(Suppl 1), S21–26.

Galter, D., Westerlund, M., Carmine, A., Lindqvist, E., Sydow, O., & Olson, L. (2006). LRRK2 expression linked to dopamine-innervated areas. *Annals of Neurology, 59*, 714–719.

Gardoni, F., Caputi, A., Cimino, M., Pastorino, L., Cattabeni, F., & Di Luca, M. (1998). Calcium/calmodulin-dependent protein kinase II is associated with NR2A/B subunits of NMDA receptor in postsynaptic densities. *Journal of Neurochemistry, 71*, 1733–1741.

Gardoni, F., Schrama, L. H., Kamal, A., Gispen, W. H., Cattabeni, F., & Di Luca, M. (2001). Hippocampal synaptic plasticity involves competition between Ca^{2+}/calmodulin-dependent protein kinase II and postsynaptic density 95 for binding to the NR2A subunit of the NMDA receptor. *The Journal of Neuroscience, 21*, 1501–1509.

Gardoni, F., Picconi, B., Ghiglieri, V., Polli, F., Bagetta, V., Bernardi, G., Cattabeni, F., Di Luca, M., & Calabresi, P. (2006). A critical interaction between NR2B and MAGUK in L-DOPA induced dyskinesia. *The Journal of Neuroscience, 26*, 2914–2922.

Gasser, T. (2009). Molecular pathogenesis of Parkinson disease: Insights from genetic studies. *Expert Reviews in Molecular Medicine, 11*, e22.

Gilks, W. P., Abou-Sleiman, P. M., Gandhi, S., Jain, S., Singleton, A., Lees, A. J., Shaw, K., Bhatia, K. P., Bonifati, V., Quinn, N. P., Lynch, J., Healy, D. G., Holton, J. L., Revesz, T., & Wood, N. W. (2005). A common LRRK2 mutation in idiopathic Parkinson's disease. *Lancet, 365*, 415–416.

Gillardon, F. (2009). Leucine-rich repeat kinase 2 phosphorylates brain tubulin-beta isoforms and modulates microtubule stability – a point of convergence in parkinsonian neurodegeneration? *Journal of Neurochemistry, 110*, 1514–1522.

Gispert, S., Ricciardi, F., Kurz, A., Azizov, M., Hoepken, H. H., Becker, D., Voos, W., Leuner, K., Muller, W. E., Kudin, A. P., Kunz, W. S., Zimmermann, A., Roeper, J., Wenzel, D., Jendrach, M., Garcia-Arencibia, M., Fernandez-Ruiz, J., Huber, L., Rohrer, H., Barrera, M., Reichert, A. S., Rub, U., Chen, A., Nussbaum, R. L., & Auburger, G. (2009). Parkinson phenotype in aged PINK1-deficient mice is accompanied by progressive mitochondrial dysfunction in absence of neurodegeneration. *PLoS One, 4*, e5777.

Gloeckner, C. J., Schumacher, A., Boldt, K., & Ueffing, M. (2009). The Parkinson disease-associated protein kinase LRRK2 exhibits MAPKKK activity and phosphorylates MKK3/6 and MKK4/7, in vitro. *Journal of Neurochemistry, 109*, 959–968.

Goldberg, M. S., Pisani, A., Haburcak, M., Vortherms, T. A., Kitada, T., Costa, C., Tong, Y., Martella, G., Tscherter, A., Martins, A., Bernardi, G., Roth, B. L., Pothos, E. N., Calabresi, P., & Shen, J. (2005). Nigrostriatal dopaminergic deficits and hypokinesia caused by inactivation of the familial Parkinsonism-linked gene DJ-1. *Neuron, 45*, 489–496.

Goldberg, M. S., Fleming, S. M., Palacino, J. J., Cepeda, C., Lam, H. A., Bhatnagar, A., Meloni, E. G., Wu, N., Ackerson, L. C., Klapstein, G. J., Gajendiran, M., Roth, B. L., Chesselet, M. F., Maidment, N. T., Levine, M. S., & Shen, J. (2003). Parkin-deficient mice exhibit nigrostriatal deficits but not loss of dopaminergic neurons. *The Journal of Biological Chemistry, 278*, 43628–43635.

Goldwurm, S., Di Fonzo, A., Simons, E. J., Rohe, C. F., Zini, M., Canesi, M., Tesei, S., Zecchinelli, A., Antonini, A., Mariani, C., Meucci, N., Sacilotto, G., Sironi, F., Salani, G., Ferreira, J., Chien, H. F., Fabrizio, E., Vanacore, N., Dalla Libera, A., Stocchi, F., Diroma, C., Lamberti, P., Sampaio, C., Meco, G., Barbosa, E., Bertoli-Avella, A. M., Breedveld, G. J., Oostra, B. A., Pezzoli, G., & Bonifati, V. (2005). The G6055A (G2019S) mutation in LRRK2 is frequent in both early and late onset Parkinson's disease and originates from a common ancestor. *Journal of Medical Genetics, 42*, e65.

Gubellini, P., Picconi, B., Bari, M., Battista, N., Calabresi, P., Centonze, D., Bernardi, G., Finazzi-Agro, A., & Maccarrone, M. (2002). Experimental Parkinsonism alters endocannabinoid degradation: Implications for striatal glutamatergic transmission. *The Journal of Neuroscience, 22*, 6900–6907.

Guo, L., Wang, W., & Chen, S. G. (2006). Leucine-rich repeat kinase 2: Relevance to Parkinson's disease. *The International Journal of Biochemistry & Cell Biology, 38*, 1469–1475.

Hallett, P. J., Dunah, A. W., Ravenscroft, P., Zhou, S., Bezard, E., Crossman, A. R., Brotchie, J. M., & Standaert, D. G. (2005). Alterations of striatal NMDA receptor subunits associated with the development of dyskinesia in the MPTP-lesioned primate model of Parkinson's disease. *Neuropharmacology, 48*, 503–516.

Healy, D. G., Wood, N. W., & Schapira, A. H. (2008). Test for LRRK2 mutations in patients with Parkinson's disease. *Practical Neurology, 8*, 381–385.

Healy, D. G., Abou-Sleiman, P. M., Valente, E. M., Gilks, W. P., Bhatia, K., Quinn, N., Lees, A. J., & Wood, N. W. (2004). DJ-1 mutations in Parkinson's disease. *Journal of Neurology, Neurosurgery, and Psychiatry, 75*, 144–145.

Higashi, S., Moore, D. J., Colebrooke, R. E., Biskup, S., Dawson, V. L., Arai, H., Dawson, T. M., & Emson, P. C. (2007a). Expression and localization of Parkinson's disease-associated leucine-rich repeat kinase 2 in the mouse brain. *Journal of Neurochemistry, 100*, 368–381.

Higashi, S., Biskup, S., West, A. B., Trinkaus, D., Dawson, V. L., Faull, R. L., Waldvogel, H. J., Arai, H., Dawson, T. M., Moore, D. J., & Emson, P. C. (2007b). Localization of Parkinson's disease-associated LRRK2 in normal and pathological human brain. *Brain Research, 1155*, 208–219.

Imai, Y., Gehrke, S., Wang, H. Q., Takahashi, R., Hasegawa, K., Oota, E., & Lu, B. (2008). Phosphorylation of 4E-BP by LRRK2 affects the maintenance of dopaminergic neurons in Drosophila. *The EMBO Journal, 27*, 2432–2443.

Jaleel, M., Nichols, R. J., Deak, M., Campbell, D. G., Gillardon, F., Knebel, A., & Alessi, D. R. (2007). LRRK2 phosphorylates moesin at threonine-558: Characterization of how Parkinson's disease mutants affect kinase activity. *The Biochemical Journal, 405*, 307–317.

Jankovic, J. (2008). Parkinson's disease: Clinical features and diagnosis. *Journal of Neurology, Neurosurgery, and Psychiatry, 79*, 368–376.

Jenner, P., & Marsden, C. D. (1986). The actions of 1-methyl-4-phenyl-1,2,3,6-tetrahydropyridine in animals as a model of Parkinson's disease. *Journal of Neural Transmission: Supplementum, 20*, 11–39.

Kehagia, A. A., Barker, R. A., & Robbins, T. W. (2010). Neuropsychological and clinical heterogeneity of cognitive impairment and dementia in patients with Parkinson's disease. *Lancet Neurology, 9*, 1200–1213.

Kennedy, M. B. (2000). Signal-processing machines at the postsynaptic density. *Science, 290*, 750–754.

Kim, E., & Sheng, M. (2004). PDZ domain proteins of synapses. *Nature Reviews: Neuroscience, 5*, 771–781.

Kitada, T., Pisani, A., Porter, D. R., Yamaguchi, H., Tscherter, A., Martella, G., Bonsi, P., Zhang, C., Pothos, E. N., & Shen, J. (2007). Impaired dopamine release and synaptic plasticity in the striatum of PINK1-deficient mice. *Proceedings of the National Academy of Sciences of the United States of America, 104*, 11441–11446.

Kurz, A., Double, K. L., Lastres-Becker, I., Tozzi, A., Tantucci, M., Bockhart, V., Bonin, M., Garcia-Arencibia, M., Nuber, S., Schlaudraff, F., Liss, B., Fernandez-Ruiz, J., Gerlach, M., Wullner, U., Luddens, H., Calabresi, P., Auburger, G., & Gispert, S. (2010). A53T-alpha-synuclein overexpression impairs dopamine signaling and striatal synaptic plasticity in old mice. *PLoS One, 5*, e11464.

Lai, S. K., Tse, Y. C., Yang, M. S., Wong, C. K., Chan, Y. S., & Yung, K. K. (2003). Gene expression of glutamate receptors GluR1 and NR1 is differentially modulated in striatal neurons in rats after 6-hydroxydopamine lesion. *Neurochemistry International, 43*, 639–653.

Lang, A. E., & Lozano, A. M. (1998a). Parkinson's disease. Second of two parts. *The New England Journal of Medicine, 339*, 1130–1143.

Lang, A. E., & Lozano, A. M. (1998b). Parkinson's disease. First of two parts. *The New England Journal of Medicine, 339*, 1044–1053.

Lavedan, C. (1998). The synuclein family. *Genome Research, 8*, 871–880.

Le, W. D., Xu, P., Jankovic, J., Jiang, H., Appel, S. H., Smith, R. G., & Vassilatis, D. K. (2003). Mutations in NR4A2 associated with familial Parkinson disease. *Nature Genetics, 33*, 85–89.

Lee, C. Y., Lee, C. H., Shih, C. C., & Liou, H. H. (2008). Paraquat inhibits postsynaptic AMPA receptors on dopaminergic neurons in the substantia nigra pars compacta. *Biochemical Pharmacology, 76,* 1155–1164.

Leroy, E., Boyer, R., Auburger, G., Leube, B., Ulm, G., Mezey, E., Harta, G., Brownstein, M. J., Jonnalagada, S., Chernova, T., Dehejia, A., Lavedan, C., Gasser, T., Steinbach, P. J., Wilkinson, K. D., & Polymeropoulos, M. H. (1998). The ubiquitin pathway in Parkinson's disease. *Nature, 395,* 451–452.

Lesage, S., Leutenegger, A. L., Ibanez, P., Janin, S., Lohmann, E., Durr, A., & Brice, A. (2005). LRRK2 haplotype analyses in European and North African families with Parkinson disease: A common founder for the G2019S mutation dating from the 13th century. *The American Society of Human Genetics, 77,* 330–332.

Li, X., Patel, J. C., Wang, J., Avshalumov, M. V., Nicholson, C., Buxbaum, J. D., Elder, G. A., Rice, M. E., & Yue, Z. (2010). Enhanced striatal dopamine transmission and motor performance with LRRK2 overexpression in mice is eliminated by familial Parkinson's disease mutation G2019S. *The Journal of Neuroscience, 30,* 1788–1797.

Li, Y., Liu, W., Oo, T. F., Wang, L., Tang, Y., Jackson-Lewis, V., Zhou, C., Geghman, K., Bogdanov, M., Przedborski, S., Beal, M. F., Burke, R. E., & Li, C. (2009). Mutant LRRK2 (R1441G) BAC transgenic mice recapitulate cardinal features of Parkinson's disease. *Nature Neuroscience, 12,* 826–828.

Lin, X., Parisiadou, L., Gu, X. L., Wang, L., Shim, H., Sun, L., Xie, C., Long, C. X., Yang, W. J., Ding, J., Chen, Z. Z., Gallant, P. E., Tao-Cheng, J. H., Rudow, G., Troncoso, J. C., Liu, Z., Li, Z., & Cai, H. (2009). Leucine-rich repeat kinase 2 regulates the progression of neuropathology induced by Parkinson's-disease-related mutant alpha-synuclein. *Neuron, 64,* 807–827.

Lovinger, D. M., Tyler, E. C., & Merritt, A. (1993). Short- and long-term synaptic depression in rat neostriatum. *Journal of Neurophysiology, 70,* 1937–1949.

Mahon, S., Deniau, J. M., & Charpier, S. (2004). Corticostriatal plasticity: Life after the depression. *Trends in Neurosciences, 27,* 460–467.

Malenka, R. C., & Bear, M. F. (2004). LTP and LTD: An embarrassment of riches. *Neuron, 44,* 5–21.

Manning, G., Whyte, D. B., Martinez, R., Hunter, T., & Sudarsanam, S. (2002). The protein kinase complement of the human genome. *Science, 298,* 1912–1934.

Marin, I. (2006). The Parkinson disease gene LRRK2: Evolutionary and structural insights. *Molecular Biology and Evolution, 23,* 2423–2433.

Marx, F. P., Holzmann, C., Strauss, K. M., Li, L., Eberhardt, O., Gerhardt, E., Cookson, M. R., Hernandez, D., Farrer, M. J., Kachergus, J., Engelender, S., Ross, C. A., Berger, K., Schols, L., Schulz, J. B., Riess, O., & Kruger, R. (2003). Identification and functional characterization of a novel R621C mutation in the synphilin-1 gene in Parkinson's disease. *Human Molecular Genetics, 12,* 1223–1231.

Mata, I. F., Kachergus, J. M., Taylor, J. P., Lincoln, S., Aasly, J., Lynch, T., Hulihan, M. M., Cobb, S. A., Wu, R. M., Lu, C. S., Lahoz, C., Wszolek, Z. K., & Farrer, M. J. (2005). Lrrk2 pathogenic substitutions in Parkinson's disease. *Neurogenetics, 6,* 171–177.

Meixner, A., Boldt, K., Van Troys, M., Askenazi, M., Gloeckner, C. J., Bauer, M., Marto, J. A., Ampe, C., Kinkl, N., & Ueffing, M. (2010). A QUICK screen for Lrrk2 interaction partners–leucine-rich repeat kinase 2 is involved in actin cytoskeleton dynamics. *Molecular and Cellular Proteomics, 10,* M110 001172.

Melrose, H., Lincoln, S., Tyndall, G., Dickson, D., & Farrer, M. (2006). Anatomical localization of leucine-rich repeat kinase 2 in mouse brain. *Neuroscience, 139,* 791–794.

Moore, D. J. (2008). The biology and pathobiology of LRRK2: Implications for Parkinson's disease. *Parkinsonism & Related Disorders, 14*(Suppl 2), S92–98.

Moore, D. J., Dawson, V. L., & Dawson, T. M. (2006). Lessons from Drosophila models of DJ-1 deficiency. *Science of Aging Knowledge Environment, 2006,* pe2.

Murphy, D. D., Rueter, S. M., Trojanowski, J. Q., & Lee, V. M. (2000). Synucleins are developmentally expressed, and alpha-synuclein regulates the size of the presynaptic vesicular pool in primary hippocampal neurons. *The Journal of Neuroscience, 20*, 3214–3220.

Nash, J. E., Johnston, T. H., Collingridge, G. L., Garner, C. C., & Brotchie, J. M. (2005). Subcellular redistribution of the synapse-associated proteins PSD-95 and SAP97 in animal models of Parkinson's disease and L-DOPA-induced dyskinesia. *The FASEB Journal, 19*, 583–585.

Nemani, V. M., Lu, W., Berge, V., Nakamura, K., Onoa, B., Lee, M. K., Chaudhry, F. A., Nicoll, R. A., & Edwards, R. H. (2010). Increased expression of alpha-synuclein reduces neurotransmitter release by inhibiting synaptic vesicle reclustering after endocytosis. *Neuron, 65*, 66–79.

Nishi, M., Hinds, H., Lu, H. P., Kawata, M., & Hayashi, Y. (2001). Motoneuron-specific expression of NR3B, a novel NMDA-type glutamate receptor subunit that works in a dominant-negative manner. *The Journal of Neuroscience, 21*, RC185.

Nussbaum, R. L., & Polymeropoulos, M. H. (1997). Genetics of Parkinson's disease. *Human Molecular Genetics, 6*, 1687–1691.

O'Dell, T. J., & Kandel, E. R. (1994). Low-frequency stimulation erases LTP through an NMDA receptor-mediated activation of protein phosphatases. *Learning & Memory, 1*, 129–139.

Oh, J. D., Vaughan, C. L., & Chase, T. N. (1999). Effect of dopamine denervation and dopamine agonist administration on serine phosphorylation of striatal NMDA receptor subunits. *Brain Research, 821*, 433–442.

Paille, V., Picconi, B., Bagetta, V., Ghiglieri, V., Sgobio, C., Di Filippo, M., Viscomi, M. T., Giampa, C., Fusco, F. R., Gardoni, F., Bernardi, G., Greengard, P., Di Luca, M., & Calabresi, P. (2010). Distinct levels of dopamine denervation differentially alter striatal synaptic plasticity and NMDA receptor subunit composition. *The Journal of Neuroscience, 30*, 14182–14193.

Paisan-Ruiz, C., Nath, P., Washecka, N., Gibbs, J. R., & Singleton, A. B. (2008). Comprehensive analysis of LRRK2 in publicly available Parkinson's disease cases and neurologically normal controls. *Human Mutation, 29*, 485–490.

Paisan-Ruiz, C., Jain, S., Evans, E. W., Gilks, W. P., Simon, J., van der Brug, M., Lopez de Munain, A., Aparicio, S., Gil, A. M., Khan, N., Johnson, J., Martinez, J. R., Nicholl, D., Carrera, I. M., Pena, A. S., de Silva, R., Lees, A., Marti-Masso, J. F., Perez-Tur, J., Wood, N. W., & Singleton, A. B. (2004). Cloning of the gene containing mutations that cause PARK8-linked Parkinson's disease. *Neuron, 44*, 595–600.

Partridge, J. G., Tang, K. C., & Lovinger, D. M. (2000). Regional and postnatal heterogeneity of activity-dependent long-term changes in synaptic efficacy in the dorsal striatum. *Journal of Neurophysiology, 84*, 1422–1429.

Piccoli, G., Condliffe, S. B., Bauer, M., Giesert, F., Boldt, K., De Astis, S., Meixner, A., Sarioglu, H., Vogt-Weisenhorn, D. M., Wurst, W., Gloeckner, C. J., Matteoli, M., Sala, C., & Ueffing, M. (2011). LRRK2 controls synaptic vesicle storage and mobilization within the recycling pool. *The Journal of Neuroscience, 31*, 2225–2237.

Picconi, B., Centonze, D., Hakansson, K., Bernardi, G., Greengard, P., Fisone, G., Cenci, M. A., & Calabresi, P. (2003). Loss of bidirectional striatal synaptic plasticity in L-DOPA-induced dyskinesia. *Nature Neuroscience, 6*, 501–506.

Picconi, B., Gardoni, F., Centonze, D., Mauceri, D., Cenci, M. A., Bernardi, G., Calabresi, P., & Di Luca, M. (2004). Abnormal Ca2 + −calmodulin-dependent protein kinase II function mediates synaptic and motor deficits in experimental parkinsonism. *The Journal of Neuroscience, 24*, 5283–5291.

Polymeropoulos, M. H., Lavedan, C., Leroy, E., Ide, S. E., Dehejia, A., Dutra, A., Pike, B., Root, H., Rubenstein, J., Boyer, R., Stenroos, E. S., Chandrasekharappa, S., Athanassiadou, A., Papapetropoulos, T., Johnson, W. G., Lazzarini, A. M., Duvoisin, R. C., Di Iorio, G., Golbe, L. I., & Nussbaum, R. L. (1997). Mutation in the alpha-synuclein gene identified in families with Parkinson's disease. *Science, 276*, 2045–2047.

Qing, H., Wong, W., McGeer, E. G., & McGeer, P. L. (2009). Lrrk2 phosphorylates alpha synuclein at serine 129: Parkinson disease implications. *Biochemical and Biophysical Research Communications, 387*, 149–152.

Quik, M., Chen, L., Parameswaran, N., Xie, X., Langston, J. W., & McCallum, S. E. (2006). Chronic oral nicotine normalizes dopaminergic function and synaptic plasticity in 1-methyl-4-phenyl-1,2,3,6-tetrahydropyridine-lesioned primates. *The Journal of Neuroscience, 26*, 4681–4689.

Raju, D. V., Ahern, T. H., Shah, D. J., Wright, T. M., Standaert, D. G., Hall, R. A., & Smith, Y. (2008). Differential synaptic plasticity of the corticostriatal and thalamostriatal systems in an MPTP-treated monkey model of parkinsonism. *European Journal of Neuroscience, 27*, 1647–1658.

Reynolds, J. N., & Wickens, J. R. (2000). Substantia nigra dopamine regulates synaptic plasticity and membrane potential fluctuations in the rat neostriatum, in vivo. *Neuroscience, 99*, 199–203.

Rizo, J., & Rosenmund, C. (2008). Synaptic vesicle fusion. *Nature Structural and Molecular Biology, 15*, 665–674.

Schulz-Schaeffer, W. J. (2010). The synaptic pathology of alpha-synuclein aggregation in dementia with Lewy bodies, Parkinson's disease and Parkinson's disease dementia. *Acta Neuropathologica, 120*, 131–143.

Schwarting, R. K., & Huston, J. P. (1996). The unilateral 6-hydroxydopamine lesion model in behavioral brain research. Analysis of functional deficits, recovery and treatments. *Progress in Neurobiology, 50*, 275–331.

Selkoe, D. J. (2002). Alzheimer's disease is a synaptic failure. *Science, 298*, 789–791.

Shendelman, S., Jonason, A., Martinat, C., Leete, T., & Abeliovich, A. (2004). DJ-1 is a redox-dependent molecular chaperone that inhibits alpha-synuclein aggregate formation. *PLoS Biology, 2*, e362.

Shin, N., Jeong, H., Kwon, J., Heo, H. Y., Kwon, J. J., Yun, H. J., Kim, C. H., Han, B. S., Tong, Y., Shen, J., Hatano, T., Hattori, N., Kim, K. S., Chang, S., & Seol, W. (2008). LRRK2 regulates synaptic vesicle endocytosis. *Experimental Cell Research, 314*, 2055–2065.

Sidhu, A., Wersinger, C., & Vernier, P. (2004). Does alpha-synuclein modulate dopaminergic synaptic content and tone at the synapse? *The FASEB Journal, 18*, 637–647.

Silvestri, L., Caputo, V., Bellacchio, E., Atorino, L., Dallapiccola, B., Valente, E. M., & Casari, G. (2005). Mitochondrial import and enzymatic activity of PINK1 mutants associated to recessive parkinsonism. *Human Molecular Genetics, 14*, 3477–3492.

Simon-Sanchez, J., Schulte, C., Bras, J. M., Sharma, M., Gibbs, J. R., Berg, D., Paisan-Ruiz, C., Lichtner, P., Scholz, S. W., Hernandez, D. G., Kruger, R., Federoff, M., Klein, C., Goate, A., Perlmutter, J., Bonin, M., Nalls, M. A., Illig, T., Gieger, C., Houlden, H., Steffens, M., Okun, M. S., Racette, B. A., Cookson, M. R., Foote, K. D., Fernandez, H. H., Traynor, B. J., Schreiber, S., Arepalli, S., Zonozi, R., Gwinn, K., van der Brug, M., Lopez, G., Chanock, S. J., Schatzkin, A., Park, Y., Hollenbeck, A., Gao, J., Huang, X., Wood, N. W., Lorenz, D., Deuschl, G., Chen, H., Riess, O., Hardy, J. A., Singleton, A. B., & Gasser, T. (2009). Genome-wide association study reveals genetic risk underlying Parkinson's disease. *Nature Genetics, 41*, 1308–1312.

Singleton, A. B., Farrer, M., Johnson, J., Singleton, A., Hague, S., Kachergus, J., Hulihan, M., Peuralinna, T., Dutra, A., Nussbaum, R., Lincoln, S., Crawley, A., Hanson, M., Maraganore, D., Adler, C., Cookson, M. R., Muenter, M., Baptista, M., Miller, D., Blancato, J., Hardy, J., & Gwinn-Hardy, K. (2003). Alpha-Synuclein locus triplication causes Parkinson's disease. *Science, 302*, 841.

Smith, A. D., Castro, S. L., & Zigmond, M. J. (2002). Stress-induced Parkinson's disease: A working hypothesis. *Physiology and Behavior, 77*, 527–531.

Sossi, V., de la Fuente-Fernandez, R., Nandhagopal, R., Schulzer, M., McKenzie, J., Ruth, T. J., Aasly, J. O., Farrer, M. J., Wszolek, Z. K., & Stoessl, J. A. (2010). Dopamine turnover increases in asymptomatic LRRK2 mutations carriers. *Movement Disorders, 25*, 2717–2723.

Spillantini, M. G., Crowther, R. A., Jakes, R., Hasegawa, M., & Goedert, M. (1998). Alpha-Synuclein in filamentous inclusions of Lewy bodies from Parkinson's disease and dementia with lewy bodies. *Proceedings of the National Academy of Sciences of the United States of America, 95*, 6469–6473.

Strack, S., McNeill, R. B., & Colbran, R. J. (2000). Mechanism and regulation of calcium/calmodulin-dependent protein kinase II targeting to the NR2B subunit of the N-methyl-D-aspartate receptor. *The Journal of Biological Chemistry, 275*, 23798–23806.

Sudhof, T. C., & Rothman, J. E. (2009). Membrane fusion: Grappling with SNARE and SM proteins. *Science, 323*, 474–477.

Taylor, J. P., Mata, I. F., & Farrer, M. J. (2006). LRRK2: A common pathway for parkinsonism, pathogenesis and prevention? *Trends in Molecular Medicine, 12*, 76–82.

Taymans, J. M., & Cookson, M. R. (2010). Mechanisms in dominant parkinsonism: The toxic triangle of LRRK2, alpha-synuclein, and tau. *Bioessays, 32*, 227–235.

Tong, Y., Pisani, A., Martella, G., Karouani, M., Yamaguchi, H., Pothos, E. N., & Shen, J. (2009). R1441C mutation in LRRK2 impairs dopaminergic neurotransmission in mice. *Proceedings of the National Academy of Sciences of the United States of America, 106*, 14622–14627.

Turner, K. M., Burgoyne, R. D., & Morgan, A. (1999). Protein phosphorylation and the regulation of synaptic membrane traffic. *Trends in Neurosciences, 22*, 459–464.

Ulas, J., & Cotman, C. W. (1996). Dopaminergic denervation of striatum results in elevated expression of NR2A subunit. *Neuroreport, 7*, 1789–1793.

Valente, E. M., Abou-Sleiman, P. M., Caputo, V., Muqit, M. M., Harvey, K., Gispert, S., Ali, Z., Del Turco, D., Bentivoglio, A. R., Healy, D. G., Albanese, A., Nussbaum, R., Gonzalez-Maldonado, R., Deller, T., Salvi, S., Cortelli, P., Gilks, W. P., Latchman, D. S., Harvey, R. J., Dallapiccola, B., Auburger, G., & Wood, N. W. (2004). Hereditary early-onset Parkinson's disease caused by mutations in PINK1. *Science, 304*, 1158–1160.

Walsh, J. P. (1993). Depression of excitatory synaptic input in rat striatal neurons. *Brain Research, 608*, 123–128.

Walsh, J. P., & Dunia, R. (1993). Synaptic activation of N-methyl-D-aspartate receptors induces short-term potentiation at excitatory synapses in the striatum of the rat. *Neuroscience, 57*, 241–248.

Wang, Y., Chandran, J. S., Cai, H., & Mattson, M. P. (2008). DJ-1 is essential for long-term depression at hippocampal CA1 synapses. *Neuromolecular Medicine, 10*, 40–45.

Whaley, N. R., Uitti, R. J., Dickson, D. W., Farrer, M. J., & Wszolek, Z. K. (2006). Clinical and pathologic features of families with LRRK2-associated Parkinson's disease. *Journal of Neural Transmission. Supplementum, 70*, 221–229.

Wider, C., Dickson, D. W., & Wszolek, Z. K. (2010). Leucine-rich repeat kinase 2 gene-associated disease: redefining genotype-phenotype correlation. *Neurodegenerative Diseases, 7*, 175–179.

Wilson, M. A., St Amour, C. V., Collins, J. L., Ringe, D., & Petsko, G. A. (2004). The 1.8-A resolution crystal structure of YDR533Cp from Saccharomyces cerevisiae: a member of the DJ-1/ThiJ/PfpI superfamily. *Proceedings of the National Academy of Sciences of the United States of America, 101*, 1531–1536.

Wishart TM, Parson SH, Gillingwater TH. (2006). Synaptic vulnerability in neurodegenerative disease. *Journal of Neuropathology and Experimental Neurology, 65*, 733–739.

Xiong, H., Wang, D., Chen, L., Choo, Y. S., Ma, H., Tang, C., Xia, K., Jiang, W., Ronai, Z., Zhuang, X., & Zhang, Z. (2009). Parkin, PINK1, and DJ-1 form a ubiquitin E3 ligase complex promoting unfolded protein degradation. *The Journal of Clinical Investigation, 119*, 650–660.

Xiong, Y., Coombes, C. E., Kilaru, A., Li, X., Gitler, A. D., Bowers, W. J., Dawson, V. L., Dawson, T. M., & Moore, D. J. (2010). GTPase activity plays a key role in the pathobiology of LRRK2. *PLoS Genetics, 6*, e1000902.

Yao, I., Takagi, H., Ageta, H., Kahyo, T., Sato, S., Hatanaka, K., Fukuda, Y., Chiba, T., Morone, N., Yuasa, S., Inokuchi, K., Ohtsuka, T., Macgregor, G. R., Tanaka, K., & Setou, M. (2007). SCRAPPER-dependent ubiquitination of active zone protein RIM1 regulates synaptic vesicle release. *Cell, 130*, 943–957.

Yu, S., Ueda, K., & Chan, P. (2005). Alpha-synuclein and dopamine metabolism. *Molecular Neurobiology, 31*, 243–254.
Yu, S., Li, X., Liu, G., Han, J., Zhang, C., Li, Y., Xu, S., Liu, C., Gao, Y., Yang, H., Ueda, K., & Chan, P. (2007). Extensive nuclear localization of alpha-synuclein in normal rat brain neurons revealed by a novel monoclonal antibody. *Neuroscience, 145*, 539–555.
Zhang, L., Shimoji, M., Thomas, B., Moore, D. J., Yu, S. W., Marupudi, N. I., Torp, R., Torgner, I. A., Ottersen, O. P., Dawson, T. M., & Dawson, V. L. (2005). Mitochondrial localization of the Parkinson's disease related protein DJ-1: Implications for pathogenesis. *Human Molecular Genetics, 14*, 2063–2073.
Zhou, C., Huang, Y., Shao, Y., May, J., Prou, D., Perier, C., Dauer, W., Schon, E. A., & Przedborski, S. (2008). The kinase domain of mitochondrial PINK1 faces the cytoplasm. *Proceedings of the National Academy of Sciences of the United States of America, 105*, 12022–12027.
Zigmond, M. J., Hastings, T. G., & Perez, R. G. (2002). Increased dopamine turnover after partial loss of dopaminergic neurons: Compensation or toxicity? *Parkinsonism & Related Disorders, 8*, 389–393.
Zimprich, A., Biskup, S., Leitner, P., Lichtner, P., Farrer, M., Lincoln, S., Kachergus, J., Hulihan, M., Uitti, R. J., Calne, D. B., Stoessl, A. J., Pfeiffer, R. F., Patenge, N., Carbajal, I. C., Vieregge, P., Asmus, F., Muller-Myhsok, B., Dickson, D. W., Meitinger, T., Strom, T. M., Wszolek, Z. K., & Gasser, T. (2004). Mutations in LRRK2 cause autosomal-dominant parkinsonism with pleomorphic pathology. *Neuron, 44*, 601–607.

Chapter 25
Synaptic Dysfunction in Alzheimer's Disease

Elena Marcello, Roberta Epis, Claudia Saraceno, and Monica Di Luca

Abstract Generation of amyloid peptide (Aβ) is at the beginning of a cascade that leads to Alzheimer's disease (AD). Amyloid precursor protein (APP), as well as β- and γ-secretases, is the principal player involved in Aβ production, while α-secretase cleavage on APP prevents Aβ deposition. Recent studies suggested that soluble assembly states of Aβ peptides can cause cognitive problems by disrupting synaptic function in the absence of significant neurodegeneration. Therefore, current research investigates the relative importance of these various soluble Aβ assemblies in causing synaptic dysfunction and cognitive deficits. Several Aβ oligomers targets and cellular mechanisms responsible of Aβ-induced synaptic failure have been identified. The first and most important mechanism impugns a toxic gain of function for Aβ which results due to self-association and attainment of new structures capable of novel interactions that lead to impaired plasticity. Other scenarios predicate that Aβ has a normal physiological role. On the one hand, insufficient Aβ could lead to a loss of normal function, whereas excess Aβ may precipitate dysfunction. How this occurs and which the main target/s is/are for the synaptic action of Aβ remains to be fully understood and would certainly represent one of the main challenges to future AD research.

Keywords Alzheimer's disease • Amyloid ß • Amyloid Precursor Protein • Glutamate receptors • Secretases

E. Marcello (✉) • R. Epis • C. Saraceno • M. Di Luca
Department of Pharmacological Sciences and Centre of Excellence on Neurodegenerative Diseases, University of Milan, Via Balzaretti 9, 20133 Milan, Italy
e-mail: elena.marcello@unimi.it; monica.diluca@unimi.it

M.R. Kreutz and C. Sala (eds.), *Synaptic Plasticity*,
Advances in Experimental Medicine and Biology 970,
DOI 10.1007/978-3-7091-0932-8_25, © Springer-Verlag/Wien 2012

25.1 Introduction

Dementia may affect adults of all ages, but the risk increases with age. According to European epidemiological studies, dementia affects 6–7% of the population over 65.

In Europe (the European Union, Iceland, Norway and Switzerland), the estimated number of patients aged 65 and over who have dementia is 4.9 million, with an estimated annual incidence approaching one million. As Europe's population ages, these numbers are expected to increase dramatically. At present, more than half of these patients have Alzheimer's disease (AD) (Di Luca et al. 2011).

However, when Alois Alzheimer, a Bavarian psychiatrist, first defined the clinical–pathological syndrome that bears his name at a meeting in Munich in 1906, neither he nor his audience recognized that the disorder he described in a woman in her early 50s might ultimately turn out to be indistinguishable from common senile dementia. Indeed, it was in the late 1960s that AD became generally accepted as the most common basis for senile dementia.

Alzheimer's original patient, a woman referred to as Auguste D. in his report, exemplified several cardinal features of the disorder that we still observe in most patients nowadays: progressive memory impairment, disordered cognitive function, altered behaviour including paranoia, delusions, loss of social appropriateness and a progressive decline in language function.

Indeed, AD is a slowly progressive disorder, with an insidious onset and a progressive impairment of episodic memory; instrumental signs include aphasia, apraxia and agnosia, together with general cognitive symptoms, such as impaired judgement, decision-making and orientation.

Past and current criteria for a diagnosis of AD rely upon the presence at autopsy of characteristic neuropathological lesions: senile plaques formed from aggregated amyloid protein ($A\beta$) and neurofibrillary tangles (NFTs), which are intraneuronal accumulation of aberrant forms of hyperphosphorylated tau (Glenner and Wong 1984).

The various neurochemical, neurological, psychological and also physical changes observed in AD patients suggest that AD is a multifactorial disease. This has been widely discussed in the recent decades. Although many theories on the cause of AD have surfaced over the past quarter of a century, only some of them have survived the test of time, with the most widely accepted theory being the 'amyloid hypothesis'.

A new understanding of the amyloid cascade hypothesis proposes an alternative mechanism for memory loss based on the impact of small, soluble amyloid β oligomers (Hardy and Selkoe 2002; Klein et al. 2001) on synaptic function. Indeed, early memory loss originates from synapse failure before neuron death, and synapse failure derives from actions of amyloid β oligomers rather than fibrils. In support of this hypothesis, many studies have demonstrated that the best statistical correlation occurs between measures of synaptic density and degree of dementia (DeKosky and Scheff 1990; Terry et al. 1991) and have documented a significant decrease in synaptic density in the association cortices and hippocampus of AD brain (Bertoni-Freddari et al. 1996; DeKosky and Scheff 1990; Terry et al. 1991).

25.2 The Amyloid Hypothesis

The 'amyloid hypothesis' was first proposed from research conducted in the middle of the 1980s showing that senile plaques found in AD brain tissue were composed mainly of a sticky Aβ peptide (Masters et al. 1985). This hypothesis was formalized by Hardy and Higgins (1992) who stated that Aβ 'precipitates to form amyloid and, in turn, causes NFTs and cell death' (Hardy and Higgins 1992). Up to now, most investigators believe that the production and cerebral deposition of amyloid plaques composed of the 38–42 amino acids (aa) Aβ peptide are central to the development of AD (Selkoe 2000). According to the amyloid hypothesis, deposition and accumulation of Aβ in the brain is the primary factor driving AD pathogenesis (Selkoe 1991; Hardy and Selkoe 2002). In animal models, Aβ deposition has also been observed to develop prior to the tangle pathology (Oddo et al. 2003).

Therefore, the basic biochemical formula for Aβ production was investigated in minute details to determine the aetiology of the disease.

25.2.1 The Characters of the Amyloid Cascade

In the late 1980s, it was first recognized that Aβ peptide derives from its large precursor protein, amyloid precursor protein (APP), by sequential proteolytic cleavages (Kang et al. 1987). Aβ domain is located within the type I transmembrane protein APP at the junction between the intraluminal and transmembrane domains.

Aβ production turned out to be one paradigmatic example of a more general biological process called regulated intramembrane proteolysis. Membrane proteins, as APP or Notch, firstly undergo a shedding process leading to the release of ectodomains in extracellular fluids. Secondly, the membrane-retained stubs can be cleaved within their intracellular domains, giving rise to small hydrophobic peptides released into extracellular fluids as well as to intracellular domains into the cytoplasm. These small intracytoplasmic peptides may possess different functions including activation of nuclear signalling (Haass 2004).

As regards APP, the shedding process is mediated by α– or β-secretases, and the cleavage of the membrane-retained stubs is due to γ-secretase (Haass and Selkoe 1993).

The production of Aβ is mediated by the concerted action of β-secretase (β-site APP cleaving enzyme, BACE1) (Vassar et al. 1999) and γ-secretase, a multimeric complex thought to be made up of an essential quartet of transmembrane proteins— presenilin 1 (PS1) or presenilin 2 (PS2), nicastrin, APH1 and PEN2 (Edbauer et al. 2003).

BACE1 cleaves APP at the N-terminus of Aβ sequence, leaving a 99-aa-long C-terminal fragment (CTF99) attached to the extracellular membrane and releasing

 E. Marcello et al.

a soluble fragment, sAPPβ, into the extracellular space. CTF99 can then be cleaved by γ-secretase at the C-terminus of Aβ sequence; this processing allows the release of the amyloidogenic Aβ fragment and the amyloid intracellular domain (AICD) (Fig. 25.1).

The cleavage of γ-secretase releases Aβ peptides of varying length from the plasma membrane, depending on the site of cleavage. Of these, Aβ42 has an

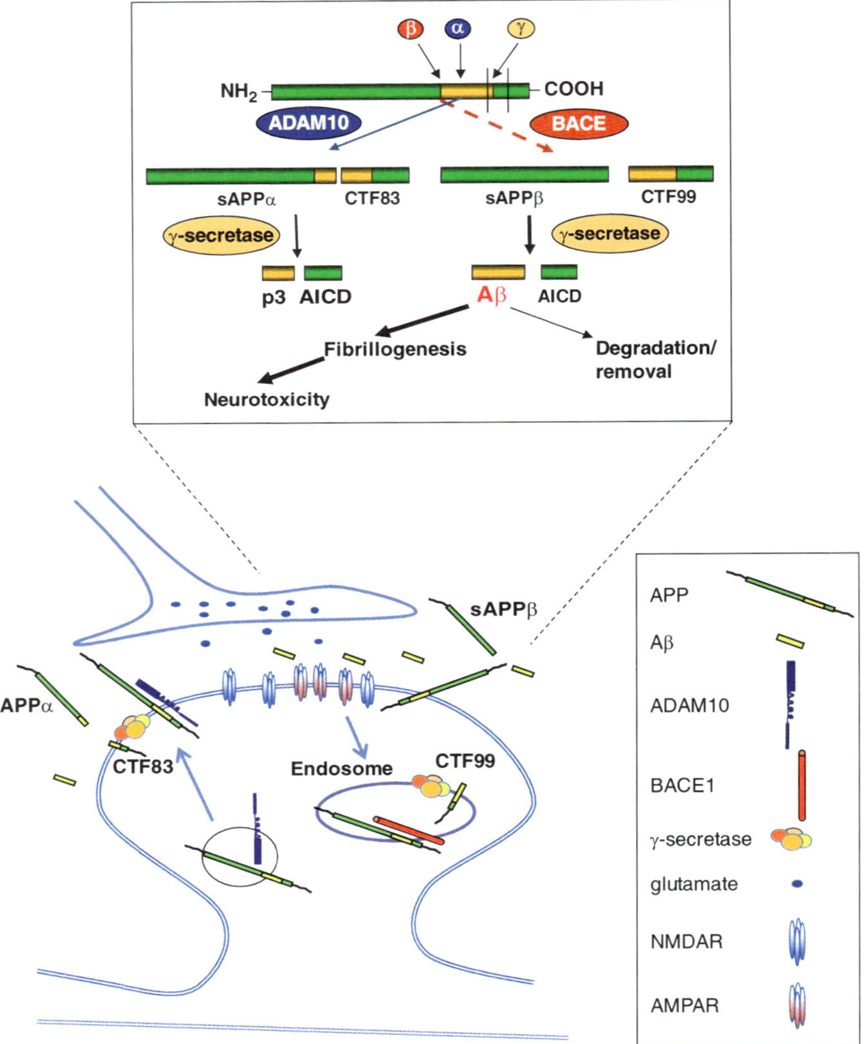

Fig. 25.1 Scheme of the proteolytic events and cleavage products that are generated during the processing of APP. APP is delivered to the surface membrane, where it is cleaved by α-secretase within the sequence of Aβ, thus precluding the formation of the amyloidogenic fragment. APP molecules that fail to be cleaved by α-secretase can be internalized into endocytic compartments and subsequently cleaved by β-secretase (BACE1) and γ-secretase to generate Aβ

increased propensity to form the fibrillar amyloid aggregates that are found in the brains of AD patients. Aβ42 is widely regarded as the main pathogenic species causing AD, unlike its more common but less fibrillogenic relative, Aβ40 (Jarrett et al. 1993).

One of the most important characteristics of Aβ metabolism is that Aβ is synthesized under regular physiological conditions. The abnormal accumulation of Aβ in AD is the result of an imbalance between the levels of Aβ production, aggregation and clearance.

Due to the heterogeneous cleavage of γ-secretase, also the AICD length varies generally from less than 57–59 amino acids. The biological functions of AICD are mediated by interactions with specific binding factors which might regulate its stability and cellular localization, but its role in the pathogenesis of AD is still under investigation (Buoso et al. 2010).

On the other hand, the main protagonist of the physiological APP metabolic pathway is α-secretase, which cleaves APP within the sequence corresponding to Aβ, thus preventing its formation. Two recent studies finally demonstrated that the constitutively cleaving α-secretase activity in neurons is selectively mediated by ADAM10 (Jorissen et al. 2010; Kuhn et al. 2010), a member of 'a disintegrin and metalloprotease' (ADAMs) proteins, which are key components in protein ectodomain shedding. ADAM10-mediated non-amyloidogenic pathway on APP releases one soluble, neurotrophic fragment called sAPPα and one membrane-associated stub, called CTF83, which can then be cleaved by the γ-secretase complex, liberating extracellular p3 and the AICD (Fig. 25.1).

Since APP and the secretases are all integral transmembrane proteins, the formation of Aβ could be modulated by sorting mechanisms. Moreover, the amyloidogenic and non-amyloidogenic pathways are differentially segregated within the cells, being α-secretase activity localized in the trans-Golgi network or at the plasma membrane (Lammich et al. 1999), whereas BACE1 activity is mainly confined to the endoplasmic reticulum and the endosomal/lysosomal system (Kinoshita et al. 2003) (Fig. 25.1).

Regarding Aβ clearance, this process is mediated by proteolytic enzymes such as neprilysin (Iwata et al. 2001) and insulin-degrading enzyme (IDE) (Qiu et al. 1998), chaperone molecules such as apolipoprotein E (ApoE) (Kim et al. 2009), lysosomal [e.g. autophagy (Bendiske and Bahr 2003)] and non-lysosomal pathways [e.g. proteasome (Marambaud et al. 2005)]. While in familial forms of AD, mutations result in an increased Aβ production or aggregation, in sporadic AD, failure of the clearance mechanisms might play a central role.

25.2.2 The Amyloid Cascade as the Primary Event

The cloning of the gene encoding APP and its localization to chromosome 21 (Goldgaber et al. 1987; Kang et al. 1987; Robakis et al. 1987; Tanzi et al. 1987), coupled with the earlier recognition that trisomy 21 (Down syndrome) leads invariably to the neuropathology of AD (Olson and Shaw 1969), set the stage for the proposal that Aβ accumulation is the primary event in AD pathogenesis. In addition, the identification of mutations in the APP gene that cause hereditary cerebral haemorrhage with amyloidosis (Dutch type) showed that APP mutations could cause Aβ deposition, albeit largely outside the brain parenchyma (Levy et al. 1990; Van Broeckhoven et al. 1990).

Soon, the first genetic mutations causing AD were discovered in the APP gene (Goate et al. 1991; Hardy 1992; Hendriks et al. 1992; Mullan et al. 1992). The contemporaneous discovery that Aβ was a normal product of APP metabolism throughout life and could be measured in culture medium, cerebrospinal fluid and plasma (Haass et al. 1992; Seubert et al. 1992; Shoji et al. 1992) allowed scientists to quickly establish the biochemical abnormalities caused by APP mutations. The majority of the mutations cluster at or very near the sites within APP that are normally cleaved by secretases. In accordance with this, these mutations promote generation of Aβ by favouring proteolytic processing of APP by β- or γ-secretase or increase the relative production of Aβ42 compared to Aβ40 (Citron et al. 1992; Cai et al. 1993; Suzuki et al. 1994). Furthermore, APP mutations internal to the Aβ sequence heighten the self-aggregation of Aβ into amyloid fibrils (Wisniewski et al. 1991).

These exciting developments provided the genetic framework for the emerging amyloid hypothesis (Selkoe 1991; Hardy and Higgins 1992).

In the past years, bolstered particularly by the cloning of PSs proteins (Levy-Lahad et al. 1995; Sherrington et al. 1995) and the demonstration that AD-causing mutations in PS1 and PS2 also enhance the processing of APP to form Aβ (Scheuner et al. 1996), the amyloid hypothesis has become the focus of AD research.

In addition to the cloning of PS1 and PS2 and the discovery that they alter APP metabolism (Borchelt et al. 1996; Duff et al. 1996; Citron et al. 1997) through a direct effect on the γ-secretase protease (De Strooper et al. 1998; Wolfe et al. 1999), there have been four conceptually important observations that strongly support the amyloid hypothesis.

First, mutations in the gene encoding the tau protein, the main component of NFTs, cause frontotemporal dementia with parkinsonism (Hutton et al. 1998; Poorkaj et al. 1998; Spillantini et al. 1998).

This neurodegenerative disorder is characterized by severe deposition of tau in NFTs in the brain but no deposition of amyloid. The clear implication is that even the most severe consequences of tau alteration—profound NFTs formation leading to fatal neurodegeneration—are not sufficient to induce the amyloid plaques characteristic of AD. Thus, the NFTs of wild-type tau seen in AD brains are likely to have been deposited after changes in Aβ metabolism and initial plaque formation, rather than before (Hardy et al. 1998).

Second, transgenic mice overexpressing both mutant human APP and mutant human tau undergo increased formation of tau-positive tangles (as compared with mice overexpressing tau alone), whereas the structure and number of their amyloid plaques are essentially unaltered (Lewis et al. 2001). This finding suggests that altered APP processing occurs before tau alterations in the pathogenic cascade of

AD, a notion bolstered by the recent observation that in mouse hippocampal primary neuronal cultures, Aβ toxicity is tau dependent (Rapoport et al. 2002).

Third, crossing APP transgenic mice with ApoE-deficient mice markedly reduces cerebral Aβ deposition in the offspring (Bales et al. 1997), providing strong evidence that the pathogenic role of genetic variability at the human ApoE locus (Corder et al. 1993) is very likely to involve Aβ metabolism. And fourth, growing evidence indicates that genetic variability in Aβ catabolism and clearance may contribute to the risk of late-onset AD (Wavrant-DeVrieze et al. 1999; Bertram et al. 2000; Ertekin-Taner et al. 2000; Myers et al. 2000; Olson et al. 2001).

Taken together, these findings are consistent with the notion that cerebral Aβ accumulation is the primary mechanism of AD pathogenesis and that the rest of the disease process, including tau tangle formation, results from an imbalance between Aβ production and Aβ clearance.

25.3 Amyloid Cascade and Synaptic Failure

25.3.1 Updating the Amyloid Hypothesis

As above described, the amyloid cascade hypothesis of AD, as initially formulated, proposed that the hallmark progressive deposition of insoluble fibrillar Aβ in plaques triggered neurodegeneration which, in turn, caused the insidious escalation of debilitating symptoms, including progression through the different stages of clinical dementia. Support for this proposal came from the discovery that application of fibril-containing Aβ to cultured neurons was highly toxic in vitro (Lorenzo and Yankner 1996) and that intracerebral injection of fibril-containing Aβ caused a neurodegeneration-associated disruption of performance of cognitive tasks in animals (McDonald et al. 1994; Nitta et al. 1994; Maurice et al. 1996; Stephan et al. 2001). However, the relatively poor correlation between the severity of clinical dementia at the time of death of patients with AD and either the magnitude of fibrillar Aβ load or the extent of neuron loss in the brain provided a major challenge for the original amyloid cascade hypothesis (Terry 1996).

In fact, many studies demonstrated that the best statistical correlation occurs between measures of synaptic density and degree of dementia (DeKosky and Scheff 1990; Terry et al. 1991). Data obtained by electron microscopy (Davies et al. 1987; Scheff et al. 1990; Seabrook et al. 1999; Scheff et al. 2006, 2007), immunocytochemical and biochemical analyses on synaptic marker proteins in AD biopsies and autopsies (Terry et al. 1991; Honer et al. 1992; Dickson et al. 1995) indicate that synaptic loss in the hippocampus and neocortex is an early event (Masliah et al. 1994) and the major structural correlate to cognitive dysfunction (Gibson 1983; Hamos et al. 1989; Bertoni-Freddari et al. 1990; DeKosky and Scheff 1990). Not NFTs, senile plaques, nor even neuronal loss show such a strong statistical correlation with dementia (Terry et al. 1991; Masliah and Terry 1993).

Moreover, the decrease in synapse number and density seems disproportionate to the loss of neuronal cell bodies (Davies et al. 1987; DeKosky and Scheff 1990; Bertoni-Freddari et al. 1996), suggesting that pruning of synaptic endings may precede the demise of the neuron in the disease process. Furthermore, some changes in the brains of AD patients and APP transgenic mice suggest that synaptic function is compromised prior to the physical deterioration of neuronal structures (Oddo et al. 2003; Palop et al. 2003; Westphalen et al. 2003; Yao et al. 2003).

This evidence, coupled with the fact that large fibrillar plaques present much less Aβ surface area to neuronal membranes than do a multitude of small oligomers that can diffuse into synaptic clefts, indicates that such soluble assembly forms are better candidates for inducing neuronal and/or synaptic dysfunction than plaques per se. Indeed, human Aβ can exist in diverse assembly states, including monomers, dimers, trimers, tetramers, dodecamers, higher-order oligomers and protofibrils, as well as mature fibrils, which can form microscopically visible amyloid plaques in brain tissues (Glabe 2008).

Therefore, a new understanding of the amyloid cascade hypothesis proposes an alternative mechanism for memory loss based on the impact of small, soluble Aβ oligomers (Klein et al. 2001; Hardy and Selkoe 2002). Indeed, different soluble molecular species that are generated at very early stages of the disease and that only at more advanced stages are deposited in an aggregated form could be involved in synaptic failure. It has thus been suggested that soluble assembly states of Aβ peptides can cause cognitive problems by disrupting synaptic function in the absence of significant neurodegeneration.

Therefore, current research investigates the relative importance of these various soluble Aβ assemblies in causing synaptic dysfunction and cognitive deficits.

25.3.2 *A Snapshot on Aβ Oligomers*

In light of the evidence that soluble oligomers of Aβ, rather than fibrils or plaques, can selectively impair the synaptic plasticity mechanisms necessary for memory processing, the research carried out in the recent years aimed at studying the conversion of normally Aβ non-toxic monomers to toxic oligomers and at defining which Aβ aggregate is responsible for synaptic failure.

Much evidence suggests that Aβ oligomers are more potent than Aβ fibrils and amyloid deposits in eliciting abnormalities in synaptic functions and neural network activity (Klein et al. 2001; Cleary et al. 2005; Lesne et al. 2006; Shankar et al. 2007; Walsh and Selkoe 2007; Selkoe 2008; Shankar et al. 2008). Therefore, many recent studies focusing on functional Aβ effects have used oligomers of human Aβ prepared from synthetic Aβ peptides (Wang et al. 2004), isolated from transfected cell lines (Walsh and Selkoe 2007) or purified from brains affected by AD (Shankar et al. 2008).

Nevertheless, a large and confusing body of literature describes many types of assembly forms of synthetic Aβ, including protofibrils (PFs), annular structures, paranuclei, Aβ-derived diffusible ligands (ADDLs), globulomers and amyloid fibrils (Teplow 1998; Caughey and Lansbury 2003). In general, soluble oligomers are defined as Aβ assemblies that are not pelleted from physiological fluids by high-speed centrifugation, and not all of the aforementioned synthetic assembly forms fulfil this definition. Moreover, soluble oligomers can bind to other macromolecules or to cell membranes and can therefore become insoluble.

PFs are intermediates that were observed in the course of studying the fibrillization of synthetic Aβ (Harper et al. 1997; Hartley et al. 1999; Walsh et al. 1999). They are flexible structures that can continue to polymerize in vitro to form amyloid fibrils or can depolymerize to lower-order species. PFs are narrower than *bona fide* amyloid fibrils (approx 5 nm versus approx 10 nm). Ultrastructural analyses of synthetic PF preparations by electron microscopy and atomic force microscopy have revealed both straight and curved assemblies of up to 150 nm in length. Synthetic Aβ PFs have been shown to contain substantial sheet structure, as they can bind to Congo red or thioflavin T in an ordered manner. Annular assemblies of synthetic Aβ are doughnut-like structures, with an outer diameter of 8–12 nm and an inner diameter of 2.0–2.5 nm, that can be distinguished from PFs by atomic force microscopy and electron microscopy (Lashuel et al. 2002; Bitan et al. 2003).

Some laboratories have observed smaller oligomeric species of synthetic Aβ than annuli and have designated these ADDLs (Lambert et al. 1998). Apparent ADDL-like oligomeric assemblies have been isolated from postmortem AD brains, and their presence correlated with memory loss (Gong et al. 2003). Chemical stabilization of synthetic Aβ assembly intermediates has revealed an apparent hexamer periodicity, with hexamer, dodecamer and octadecamer structures observed (Bitan et al. 2003). Whether the recently described Aβ*56, an apparent dodecamer of natural Aβ detected in the brains of an APP transgenic mouse line (Bitan et al. 2003), might represent an in vivo analogue of synthetic ADDLs remains unclear, as direct structural comparisons have not been possible.

Whereas most of the Aβ assembly intermediates described above have only been observed upon in vitro incubation of synthetic Aβ, small oligomeric Aβ forms occur in vivo and might therefore be relevant to disease pathogenesis. Intracellular and secreted soluble dimeric and trimeric oligomers have been described in cultured cells (Podlisny et al. 1995; Walsh et al. 2000), and SDS-stable oligomers of varying sizes have also been detected by Western blotting in APP transgenic mouse brain and human brain (Enya et al. 1999; Funato et al. 1999; Lesne et al. 2006). Such natural (i.e. non-synthetic) Aβ oligomers can be resistant not only to SDS but also to the Aβ-degrading protease IDE, which can only digest monomeric Aβ (Walsh et al. 2002). Naturally secreted monomeric and oligomeric Aβ species are being characterized in experiments in vivo to decipher their effects on synaptic structure and function (Walsh et al. 2002; Kamenetz et al. 2003). Aβ oligomers produced by cultured cells could be related to the aforementioned Aβ*56 (Lesne et al. 2006), which seems to represent a brain-derived soluble dodecamer that has amnestic activity. Like the Aβ oligomers produced from cultured cells (Walsh et al. 2002), Aβ*56 might disrupt synaptic function and therefore affect memory

(Lesne et al. 2006). Whether Aβ*56 and species that are similar to it are stable assemblies of only Aβ under native conditions, or whether smaller oligomeric assemblies can associate with another protein, is currently unknown. However, Aβ*56 and Aβ trimers secreted by cultured cells could turn out to share common synaptotoxic properties.

In light of the above, the complete characterization of the different assembly forms and conformations of Aβ is important to discern which Aβ aggregate is harmful and triggers synaptic failure. Indeed, it is quite possible that different Aβ assemblies or conformation have diverse targets, thus, depending on the form of Aβ used, research groups can report differing results, as described in the following section.

25.4　Targets of Aβ at the Synapse

The subcellular sites from which Aβ acts have been identified: overexpression of APP in either dendritic or axonal compartments led to a reduction in spine density and plasticity in nearby neurons (Wei et al. 2010). In particular, Aβ oligomers bind to synaptic sites (Lacor et al. 2004) and reduce the density of spines in organotypic hippocampal slice cultures (Hsieh et al. 2006; Shrestha et al. 2006; Shankar et al. 2007), dissociated cultured neurons (Calabrese et al. 2007; Lacor et al. 2007; Evans et al. 2008) and transgenic mouse models (Spires et al. 2005; Jacobsen et al. 2006).

Nevertheless, the molecular link between Aβ oligomers and the occurrence of early spine loss remains elusive.

Several studies, reported below, describe a number of targets and mechanisms responsible for Aβ-dependent effects on synaptic function.

25.4.1　Acetylcholine Receptors

Interactions between Aβ and various acetylcholine receptors have been shown through biochemical and pharmacologic techniques. Given the profound loss of cholinergic transmission in AD, nicotinic and muscarinic acetylcholine receptors have drawn considerable attention. Synthetic Aβ has been shown to bind the calcium-permeable α7 nicotinic acetylcholine receptors (nAChRs) with high affinity (Wang et al. 2000). Functionally, this interaction has been proposed to account for the internalization of N-methyl-D-aspartate (NMDA) receptors through a calcineurin-dependent pathway (Snyder et al. 2005; Dewachter et al. 2009). Because these studies focus on postsynaptic cholinergic transmission, it is unclear whether interactions with acetylcholine receptor signalling directly account for the disruption of presynaptic cholinergic projections in AD, such as those extending from the nucleus basalis of Meynert, one of the areas mainly affected in AD patients.

25.4.2 Glutamate Receptors

Glutamate receptors are central to synaptic functioning, and excitatory synaptic transmission is tightly regulated by the number of active NMDA receptors and α-amino-3-hydroxy-5-methyl-4-isoxazolepropionic acid (AMPA) receptors at the synapse. NMDA receptor activation has a central role, as it can induce either long-term potentiation (LTP) or long-term depression (LTD), depending on the extent of the resultant intracellular calcium ($[Ca^{2+}]i$) rise in the dendritic spines and on the downstream activation of specific intracellular cascades (Kasai et al. 2010). Activation of synaptic NMDA receptors and large increases in $[Ca^{2+}]i$ are required for LTP, whereas LTD implies a differential mobilization of NMDA receptors in different CA regions of the hippocampus; in fact, it has been reported that in CA1, lateral diffusion occurs during LTD induction, but in CA3, endocytosis of NMDA receptors is predominant (Lau and Zukin 2007). LTP induction promotes recruitment of AMPA receptors and growth of dendritic spines, whereas LTD induces spine shrinkage and synaptic loss (De Roo et al. 2008).

Their modulation seems implicated in deleterious synaptic Aβ action. Aβ number of studies have reported that the effects of Aβ on the viability, morphology and physiology of neurons are dependent on NMDA receptor activation (Ye et al. 2004; Lacor et al. 2007; Shankar et al. 2007). Memantine, an activity-dependent NMDA receptor antagonist, is used for the treatment of AD. Initially it was proposed to mitigate glutamate excitotoxicity, but it may also block more subtle effects of NMDA receptor activation that lead to synaptic depression and loss.

Pathologically elevated Aβ may indirectly cause a partial block of NMDA receptors and shift the activation of NMDA receptor–dependent signalling cascades towards pathways involved in the induction of LTD and synaptic loss (Kamenetz et al. 2003; Hsieh et al. 2006; Shankar et al. 2007). This model is consistent with the fact that Aβ impairs LTP (Walsh et al. 2002; Cleary et al. 2005) and enhances LTD (Kim et al. 2001; Hsieh et al. 2006; Li et al. 2009). Although the mechanisms underlying Aβ-induced LTD have not yet been fully elucidated, they may involve receptor internalization (Snyder et al. 2005; Hsieh et al. 2006) or desensitization (Liu et al. 2004) and subsequent collapse of dendritic spines (Snyder et al. 2005; Hsieh et al. 2006).

Moreover, recent findings suggest that pathologically elevated Aβ blocks neuronal glutamate uptake at synapses, leading to increased glutamate at the synaptic cleft (Li et al. 2009). A rise in glutamate would initially activate synaptic NMDA receptors, which might be followed by desensitization of the receptors and, ultimately, synaptic depression. A second effect of increased glutamate would be a spillover and activation of extra- or perisynaptic NMDA receptors enriched of the subunit 2B, which have a key role in LTD induction (Liu et al. 2004). Also, the activation of metabotropic glutamate receptors (mGluRs) may be involved in the facilitation of LTD by Aβ (Hsieh et al. 2006; Li et al. 2009). Activation of mGluR recruits a number of signalling pathways (such as p38 MAP kinase or ERK pathways), stimulates release of intracellular $[Ca^{2+}]i$ stores through generation of inositol triphosphate or modulates associated ionic channels. Various groups have reported that Aβ mediates synaptic depression and loss through activation of group I mGluRs with p38 MAP kinase and calcineurin as downstream effectors (Wang et al. 2004; Hsieh et al. 2006; Shankar et al. 2008; Dewachter et al. 2009; Li et al. 2009). These effects may result in postsynaptic AMPA receptor endocytosis (Snyder et al. 2005) and decreased presynaptic neurotransmitter release probability (Zakharenko et al. 2002), both of which decrease synaptic strength. Understanding the contribution of mGluR receptors to Aβ-mediated synaptic depression is difficult because of the high variability in the mechanisms associated to mGluRs across brain regions and in developmental time periods.

Thus, Aβ-induced synaptic depression may result from an initial increase in synaptic activation of NMDA receptors by glutamate, followed by synaptic NMDA receptor desensitization; NMDA receptor and AMPA receptor internalization; and activation of mGluRs (Renner et al. 2010). Aβ-induced LTD-like processes may underlie Aβ-induced LTP deficits, as blocking LTD-related signalling cascades, such as those mediated by mGluR or p38 MAPK, prevents Aβ-dependent inhibition of LTP (Wang et al. 2004).

25.4.3 Beyond Receptors: Other Aβ Targets

In addition to receptors well known for their involvement in synaptic plasticity, Aβ appears to affect other synapse proteins.

For example, Cisse et al. (2011) found out that Aβ binds to EphB2, a protein that interacts with NMDA receptors (Dalva et al. 2000), and whose deficiency reduces LTP (Grunwald et al. 2001; Henderson et al. 2001). Aβ binding decreases EphB2 levels. Reducing EphB2 levels in the dentate gyrus—the input region of the hippocampus—in normal mouse brain mimics the reduced NMDA receptor currents and impaired LTP that occur in an AD mouse model (Cisse et al. 2011). Remarkably, the authors reported that virus-mediated expression of EphB2 in the dentate gyrus of an AD mouse model 'cures' the mice, with both NMDA receptor–mediated synaptic responses and LTP returning to normal levels (Cisse et al. 2011).

Another study proposes that cellular prion protein (PrP^C) functions as a receptor to mediate the deleterious effects of Aβ oligomers, which bind with nanomolar affinity to PrP^C. This interaction does not require the infectious PrP^{Sc} conformation. This hypothesis is supported by isolation of PrP^C as an Aβ42-oligomer-binding site in an unbiased genome-wide screen, by the match between PrP^C expression and the properties of Aβ42-oligomer-binding sites and by the localization of Aβ binding to a neurodegeneration-associated domain of PrP^C (Lauren et al. 2009).

PrP^C has also been shown to interact with NMDA receptor subunit 2D and to modulate its function (Khosravani et al. 2008). The authors wondered whether Aβ interaction might regulate glutamate receptors directly through PrP^C. Synaptic responsiveness in hippocampal slices from young adult PrP null mice is normal, but the Aβ oligomer blockade of LTP is absent. Anti-PrP antibodies prevent Aβ-oligomer binding to PrP^C and rescue synaptic plasticity in hippocampal slices treated with oligomeric Aβ (Lauren et al. 2009).

A recent study put forward a new target of Aβ: D'Amelio et al. (2011) described the activation of caspase-3 in hippocampal dendritic spines, corresponding to the onset of memory decline in Tg2576 mice, a frequently used mouse model of AD. In spines, caspase-3 activates calcineurin, which in turn triggers dephosphorylation and removal of the GluR1 subunit of AMPA-type receptor from postsynaptic sites. These molecular modifications led to alterations of glutamatergic synaptic transmission and plasticity and correlated with spine degeneration and a deficit in hippocampal-dependent memory (D'Amelio et al. 2011). The caspase activation is clearly a result of the presence of soluble Aβ. It occurs in the mice months before plaques are deposited and can be blocked by blocking Aβ synthesis. The results are surprising, as activation of the 'executioner' caspase is not associated with neuronal cell death (D'Amelio et al. 2011).

These results, illustrating a non-apoptotic activation of caspase-3, are reminiscent of a recent study that found that caspase-3 is activated during normal physiological LTD (Li et al. 2010). The parallels between the caspase-3-triggered synaptic dysfunction in Tg2576 mice and those reported previously for LTD are striking, with both caspase-mediated cascades leading to removal of AMPA-type receptors from postsynaptic sites. The suggestion that there is a chronic, non-apoptotic activation of caspases in AD models presents a new picture of the pace of the final steps of neurodegeneration, in which Aβ-induced initiation of cell death–associated cascades leads not to acute cell death but to a slower loss with a prominent role for remodelling of dendrites, spines and synaptic connections.

25.4.4 Bridging Aβ and Tau at the Synapse

Recent studies shed lights on the mechanisms linking Aβ toxicity and tau. As described above, according to the amyloid cascade hypothesis, Aβ formation is the critical step in driving AD pathogenesis (Hardy and Selkoe 2002); thus, a crucial question is where tau is to be placed in the amyloid cascade. Ittner et al. (2010) found out a new role for tau in Aβ downstream toxicity mechanisms (Ittner et al. 2010). First of all, they show that tau, known as axonal protein, has a dendritic function in postsynaptic targeting of the Src kinase Fyn. At postsynaptic sites, Fyn phosphorylates the NMDA receptor subunit 2B (NR2B), thereby mediating complex formation of NMDA receptors with the postsynaptic density protein 95 (PSD95) (Salter and Kalia 2004). This NMDA receptor–PSD95 interaction is required for excitotoxic downstream signalling (Salter and Kalia 2004). As a consequence of the reduced NMDA receptors–PSD95 interaction (Ittner et al. 2010), tau−/− mice are less susceptible to both experimental seizures and Aβ toxicity (Ittner et al. 2010; Roberson et al. 2007). In summary, although tau is predominantly found in axons, its newly discovered dendritic functions are pivotal in healthy neurons and, when disturbed, seem to have a role in disease.

Based on these recent findings, a novel 'tau axis hypothesis' which links amyloid β and tau pathology in the dendritic compartment has been postulated (Ittner and Gotz 2011).

This hypothesis consists of two parts.

First, postsynaptic toxicity of Aβ is tau dependent. More precisely, tau interacts with Fyn and thereby increases targeting and/or scaffolding of Fyn to the postsynaptic compartment, where Fyn links NMDA receptors to downstream signalling pathways. This sensitizes NMDA receptors and makes them responsive to Aβ toxicity. This mode of tau-dependent Aβ toxicity in the dendritic compartment of neurons involves excitotoxic signalling.

Second, exposure of neurons to Aβ—and in particular prolonged exposure—has multiple toxic effects. Importantly, Aβ triggers progressively increased phosphorylation (hyperphosphorylation) of tau. As a consequence, tau binding to microtubules is compromised, causing tau to accumulate at an increasing pace in the somatodendritic compartment of diseased neurons. Moreover, phosphorylated tau has an increased affinity for Fyn (Bhaskar et al. 2005).

In conclusion, this results in high levels of postsynaptic Fyn and in sensitization of NMDA receptors, which renders dendrites even more susceptible to Aβ toxicity (Ittner and Gotz 2011).

25.5 From Pathology to Physiology

It can be argued that pathology never exists for pathology's sake—pathogenic mechanisms do not exist solely to induce disease. Instead, they are a reflection of aberrations in normal physiological processes.

APP processing constitutes a complex signalling centre that serves multiple physiological functions that could trigger pathological events when deregulated during disease.

Since the studies mentioned above led to the concept that the synapse, and in particular the postsynaptic compartment, is the subcellular *locus* where AD pathogenesis takes place, the comprehension of the physiological function of APP and its metabolites at the synaptic level becomes more and more important.

25.5.1 Physiological Role of APP and Its Metabolites

Since the discovery of APP, a number of physiological roles have been attributed to the molecule, some unique to certain isoforms, but its actual functions remain unclear. Suffice to say that a number of functional domains have been mapped to the extra- and intracellular regions of APP. These include metal (copper and zinc) and extracellular matrix components (heparin, collagen and laminin) binding; neurotrophic and adhesion domains; and protease inhibition (the Kunitz protease inhibitor domain present in APP751 and APP770 isoforms) domain.

The overall structure of the protein suggests that APP could be a receptor or growth factor (Rossjohn et al. 1999), but the functions of APP and its homologues in vivo remain poorly understood.

APP has been shown to stimulate neurite outgrowth from a variety of settings. This phenotype is compatible with the upregulation of APP expression during neuronal maturation (Hung et al. 1992).

The N-terminal heparin-binding domain of APP (residues 28–123) upstream from the pentapeptide domain RERMS sequence also stimulates neurite outgrowth and promotes synaptogenesis. Interestingly, the crystal structure of this domain shows similarities to known cysteine-rich growth factors (Rossjohn et al. 1999). Conversely, injection of anti-APP antibodies directly into the brain led to impairment in behavioural tasks in adult rats (Meziane et al. 1998).

Shedding new light on the matter, Nikolaev et al. (2009) revealed a physiological mechanism in which an APP product (N-APP) binds directly to a death receptor to trigger axonal pruning and neuronal culling during development. These processes are thought to be activated by a lack of trophic factors that induced cleavage of APP by β-secretase, resulting in formation of sAPPβ and subsequently N-APP. Surprisingly, N-APP acts as a necessary and sufficient ligand for DR6, inducing axonal and neuronal degeneration after trophic factor removal. However, neither inhibition of α-secretase nor antibody-mediated blocking of Aβ42 affected this pathway, although Aβ42 did trigger axonal degeneration in a DR6-independent manner. An oversprouting axon phenotype similar to that in mice lacking DR6 was seen in mice lacking both APP and a closely related protein, APLP2, further supporting a role in pruning (Nikolaev et al. 2009).

Moreover, in neuronal cells, full-length APP may play important roles in maintaining nerve cell structure and signal transduction. APP may have a range of physiological functions associated with developing and adult neurons that are modulated through its sequential processing pathways and mediated through specific interactions with cell-surface proteins (for secreted species) and intracellular proteins. In the case of APP-null mutations, mice show a variety of alterations in neural structure and function, including gliosis, decreased neocortical and hippocampal levels of synaptophysin, lowered dendritic lengths in hippocampal neurons, reduced survival of cultured neurons and impaired LTP (Perez et al. 1997; Dawson et al. 1999; Seabrook and Rosahl 1999). However, these effects could be due as much to the loss of APP neurotrophic derivative sAPPα as to the loss of activity by full-length APP.

In fact, among the proposed physiological functions for APP and its products (reviewed in Mattson (1997), Thinakaran and Koo (2008)), the best established one is the role of sAPPα in promoting neuronal survival. Indeed, secreted APP exerts proliferative actions in a variety of cell types as well as neurotrophic and neuroprotective effects (Mucke et al. 1996). In general, these effects are induced by sAPPα approximately 100 times more strongly than by sAPPβ. In one study examining neurite outgrowth, sAPPα actually lowered growth below control levels (Li et al. 1997).

sAPPα stimulates the proliferation of neural stem cells from embryonic rat neocortex and from adult mouse brain (Ohsawa et al. 1999; Caille et al. 2004). sAPPα has neurotrophic and neuroprotective properties, and recently, it was shown to increase LTP and spatial memory (Mattson 1997; Gralle and Ferreira 2007; Taylor et al. 2008). Specific domains of sAPPα have been identified that contribute to neuroprotection and others to the stimulation of neurite outgrowth in vitro (Mattson 1997). Two domains located between residues 96–110 and 319–335 in sAPPα were reported to contribute to neurite outgrowth. The former region is also a binding site for heparan sulfate proteoglycans (HSPG) (Ninomiya et al. 1994; Small et al. 1994). The signalling pathways involved in sAPPα neuroprotection have been characterized. Less well known are the signalling pathways involved in sAPPα neurotrophic properties. Recently, it has been shown that mitogen-activated protein kinase (MAPK)/ERK pathway is activated during neurite outgrowth of neural stem cell–derived neurons or primary neurons in response to sAPPα (Greenberg et al. 1995; Gakhar-Koppole et al. 2008; Rohe et al. 2008).

Moreover, sAPPα formation in neuronal cells is selectively mediated by ADAM10 (Jorissen et al. 2010; Kuhn et al. 2010), which represents a valuable target for AD therapy. In fact, ADAM10 overexpression in an AD animal model reverses impaired LTP and cognitive deficits early in life before plaque formation occurs. As reported, a neuron-specific knockout of PS1 prevented amyloid plaque formation but did not improve cognitive deficits of APP [V717I] mice (Dewachter et al. 2002), the mouse model used in the Postina investigation (Postina et al. 2004). In their study, the beneficial effect of increased ADAM10 activity, including cognitive improvements, can most likely be attributed to the combined effects of decreased levels of toxic Aβ peptides and endogenously increased amounts of neuroprotective sAPPα (Postina et al. 2004).

This enzyme is specifically localized in the postsynaptic density of excitatory synapses (Marcello et al. 2007). ADAM10 synaptic localization is relevant for neuronal APP processing and Aβ production (Marcello et al. 2007) because the mechanism that cause APP and the secretases to colocalize in the same membranous compartment plays important roles in the regulation of Aβ production. Indeed, the mechanisms underlying ADAM10 trafficking in neurons and responsible for its synaptic localization have been elucidated. Synapse-associated protein 97 (SAP97), a protein involved in dynamic trafficking of proteins to the excitatory synapse, is responsible for driving ADAM10 to the postsynaptic membrane by a direct interaction through its SH3 domain. NMDA receptor activation mediates this event and positively modulates α-secretase activity. Furthermore, perturbing ADAM10/SAP97 association in vivo by cell-permeable peptides impairs ADAM10 localization in postsynaptic membranes and consequently decreases the APP physiological metabolism (Marcello et al. 2007).

Moreover, ADAM10 trafficking mechanism and ADAM10/SAP97 association are involved in AD pathogenesis. Indeed, ADAM10 synaptic levels and ADAM10/SAP97 association are reduced in the hippocampus of AD patients at an early stage of disease (Marcello et al. 2010), and interfering with ADAM10/SAP97 complex for 2 weeks by means of a cell-permeable peptide strategy in mice is sufficient to increase amyloid levels and leads to the reproduction of initial phases of sporadic AD (Epis et al. 2010).

Thus, these studies put forward the importance of APP functions and ADAM10-mediated physiological metabolism, which could be considered a strategic target for the development of AD therapies.

25.5.2 *Physiological Role of Aβ*

A growing body of literature supports a physiological role for Aβ in normal synapse function. For instance, in organotypic hippocampal slices, β-secretase activity is increased by synaptic activity, and the resulting Aβ peptides depress excitatory transmission through AMPA and NMDA receptors, suggesting a role for Aβ in homeostatic plasticity (Kamenetz et al. 2003). Indeed, in APP transgenic mouse brain, there is a strong positive correlation between synaptic activity and the concentration of Aβ in the interstitial fluid (Cirrito et al. 2005), and in humans, cerebral Aβ concentration increases as neuronal function and mental status recover in patients with traumatic brain injury (Brody et al. 2008).

Notably, the production of Aβ and its secretion into the extracellular space are tightly regulated by neuronal activity in vitro (Kamenetz et al. 2003) and in vivo (Cirrito et al. 2005). Increased neuronal activity enhances Aβ production, and blocking neuronal activity has the opposite effect (Kamenetz et al. 2003). This synaptic regulation of Aβ production is mediated, at least in part, by clathrin-dependent endocytosis of surface APP at presynaptic terminals, endosomal proteo-lytic cleavage of APP and Aβ release at synaptic terminals (Cirrito et al. 2005). In addition, pathogenic Aβ species can also be released from dendrites (Wei et al. 2010). This tight neuronal activity–dependent regulation of Aβ secretion has been observed during pathological events, such as epileptiform activity induced by electrical stimulation (Cirrito et al. 2005), as well as during normal physiological processes, such as the sleep–wake cycle (Kang et al. 2009). It is also supported by the earlier development of amyloid plaques in patients with epilepsy (Mackenzie and Miller 1994). These findings support the notion that APP and Aβ are part of a feedback loop that controls neuronal excitability (Kamenetz et al. 2003). In this model, Aβ production is enhanced by action potential–dependent synaptic activity, leading to increased Aβ at synapses and reduction of excitatory transmission postsynaptically. Pathological elevation of Aβ would be expected to put this negative feedback regulator into overdrive, suppressing excitatory synaptic activity at the postsynaptic level.

Aβ may also have a role in regulating well-described forms of synaptic plastic-ity. A recent study suggests that Aβ also acts as a positive regulator at presynaptic level. In this study, relatively small increases in endogenous Aβ abundance (~1.5-fold), induced by inhibition of extracellular Aβ degradation in otherwise unmanipulated wild-type neurons, enhanced the release probability of synaptic vesicles and increased neuronal activity in neuronal culture (Abramov et al. 2009). Enhanced extracellular Aβ increased spontaneous excitatory postsynaptic currents without significantly altering inhibitory currents. All these effects were exclusively presynaptic and dependent on firing rates, with less facilitation seen in neurons with higher firing rates. Thus, small increases of Aβ may facilitate presynaptic glutamatergic release in neurons with low activity but not in neurons with high activity.

Consistent with this finding, application of low (picomolar range) concentrations of Aβ markedly potentiates synaptic transmission, whereas higher concentrations (low nanomolar range) of Aβ cause the expected synaptic depression (Puzzo et al. 2008). The potentiating effect of Aβ does not affect postsynaptic NMDA and AMPA receptors' currents but is dependent on nAChR activation, suggesting a presynaptic mechanism mediated by buildup of Ca^{2+} in presynaptic terminals. Thus, Aβ may directly act on presynaptic α7-nAChR (Dineley et al. 2002) and be part of a positive feedback loop that increases presynaptic Ca^{2+} and Aβ secretion. Indeed, blocking nAChRs or removing α7-nAChRs decreases Aβ secretion and blocks Aβ-induced facilitation (Wei et al. 2010). Of particular importance, Aβ-induced presynaptic facilitation depends on an optimal Aβ concentration, with higher or lower concentrations impairing synaptic transmission (Abramov et al. 2009). A positive modulatory effect of Aβ on synaptic transmission is further supported indirectly by the finding that an abnormally low Aβ level in mice deficient in APP (Seabrook et al. 1999), PS1 (Saura et al. 2004) or BACE1 (Laird et al. 2005) is associated with synaptic transmission deficits.

While a physiologic role for Aβ can be surmised from such studies, the Aβ assembly form responsible for these effects is not known. However, it seems likely that monomeric Aβ would mediate these effects, not least since monomer would be the predominant form of Aβ present in freshly reconstituted synthetic peptide or cortical Aβ. Consequently, pathologic effects on synapse physiology may not only arise from the appearance of higher-order Aβ assemblies, which assume a toxic gain of function, but also rising level of monomeric Aβ. Thus, increased concentrations of Aβ could lead to synaptic dysregulation mediate by abnormally high levels of monomer and the formation of toxic oligomers. For instance, it seems plausible that the increase in non-convulsive seizures observed in APP transgenic mice results from an Aβ-dependent imbalance of excitatory and inhibitory activities (Palop et al. 2007). In this scenario, Aβ promotes neuronal overexcitability, which results in GABAergic sprouting of inhibitory synapses as a compensatory mechanism. Similarly, using multiphoton imaging of intraneuronal calcium fluctuations, high focal levels of Aβ were shown to increase heterogeneity in the excitability of neurons within 60 μm of amyloid plaques (Busche et al. 2008). Moreover, dendritic spine loss observed within 20 μm of amyloid plaques in the Tg2576 APP transgenic mouse provides a structural correlate to the physiologic findings (Spires et al. 2005).

25.6 Conclusions

How Aβ mediates its effects on synaptic plasticity may take many years to fully understand, but already we know that it is likely to involve three different levels. The first and most important mechanism impugns a toxic gain of function for Aβ which results due to self-association and attainment of new structures capable of novel interactions that lead to impaired plasticity. The other two scenarios predicate that Aβ has a normal physiological role. On the one hand, insufficient Aβ could lead to a loss of normal function, whereas excess Aβ may precipitate dysfunction.

How this occurs and which the main target/s is/are for the synaptic action of Aβ remains to be fully understood and would certainly represent one of the main challenges to future AD research.

References

Abramov, E., Dolev, I., Fogel, H., Ciccotosto, G. D., Ruff, E., & Slutsky, I. (2009). Amyloid-beta as a positive endogenous regulator of release probability at hippocampal synapses. *Nature Neuroscience, 12*, 1567–1576.

Bales, K. R., Verina, T., Dodel, R. C., Du, Y., Altstiel, L., Bender, M., Hyslop, P., Johnstone, E. M., Little, S. P., Cummins, D. J., Piccardo, P., Ghetti, B., & Paul, S. M. (1997). Lack of apolipoprotein E dramatically reduces amyloid beta-peptide deposition. *Nature Genetics, 17*, 263–264.

Bendiske, J., & Bahr, B. A. (2003). Lysosomal activation is a compensatory response against protein accumulation and associated synaptopathogenesis–an approach for slowing Alzheimer disease? *Journal of Neuropathology and Experimental Neurology, 62*, 451–463.

Bertoni-Freddari, C., Fattoretti, P., Casoli, T., Meier-Ruge, W., & Ulrich, J. (1990). Morphological adaptive response of the synaptic junctional zones in the human dentate gyrus during aging and Alzheimer's disease. *Brain Research, 517*, 69–75.

Bertoni-Freddari, C., Fattoretti, P., Casoli, T., Caselli, U., & Meier-Ruge, W. (1996). Deterioration threshold of synaptic morphology in aging and senile dementia of Alzheimer's type. *Analytical and Quantitative Cytology and Histology, 18*, 209–213.

Bertram, L., Blacker, D., Mullin, K., Keeney, D., Jones, J., Basu, S., Yhu, S., McInnis, M. G., Go, R. C., Vekrellis, K., Selkoe, D. J., Saunders, A. J., & Tanzi, R. E. (2000). Evidence for genetic linkage of Alzheimer's disease to chromosome 10q. *Science, 290*, 2302–2303.

Bhaskar, K., Yen, S. H., & Lee, G. (2005). Disease-related modifications in tau affect the interaction between Fyn and Tau. *The Journal of Biological Chemistry, 280*, 35119–35125.

Bitan, G., Kirkitadze, M. D., Lomakin, A., Vollers, S. S., Benedek, G. B., & Teplow, D. B. (2003). Amyloid beta -protein (Abeta) assembly: Abeta 40 and Abeta 42 oligomerize through distinct pathways. *Proceedings of the National Academy of Sciences of the United States of America, 100*, 330–335.

Borchelt, D. R., Thinakaran, G., Eckman, C. B., Lee, M. K., Davenport, F., Ratovitsky, T., Prada, C. M., Kim, G., Seekins, S., Yager, D., Slunt, H. H., Wang, R., Seeger, M., Levey, A. I., Gandy, S. E., Copeland, N. G., Jenkins, N. A., Price, D. L., Younkin, S. G., & Sisodia, S. S. (1996). Familial Alzheimer's disease-linked presenilin 1 variants elevate Abeta1-42/1-40 ratio in vitro and in vivo. *Neuron, 17*, 1005–1013.

Brody, D. L., Magnoni, S., Schwetye, K. E., Spinner, M. L., Esparza, T. J., Stocchetti, N., Zipfel, G. J., & Holtzman, D. M. (2008). Amyloid-beta dynamics correlate with neurological status in the injured human brain. *Science, 321*, 1221–1224.

Buoso, E., Lanni, C., Schettini, G., Govoni, S., & Racchi, M. (2010). Beta-Amyloid precursor protein metabolism: Focus on the functions and degradation of its intracellular domain. *Pharmacological Research, 62*(4), 308–317.

Busche, M. A., Eichhoff, G., Adelsberger, H., Abramowski, D., Wiederhold, K. H., Haass, C., Staufenbiel, M., Konnerth, A., & Garaschuk, O. (2008). Clusters of hyperactive neurons near amyloid plaques in a mouse model of Alzheimer's disease. *Science, 321*, 1686–1689.

Cai, X. D., Golde, T. E., & Younkin, S. G. (1993). Release of excess amyloid beta protein from a mutant amyloid beta protein precursor. *Science, 259*, 514–516.

Caille, I., Allinquant, B., Dupont, E., Bouillot, C., Langer, A., Muller, U., & Prochiantz, A. (2004). Soluble form of amyloid precursor protein regulates proliferation of progenitors in the adult subventricular zone. *Development, 131*, 2173–2181.

Calabrese, B., Shaked, G. M., Tabarean, I. V., Braga, J., Koo, E. H., & Halpain, S. (2007). Rapid, concurrent alterations in pre- and postsynaptic structure induced by naturally-secreted amyloid-beta protein. *Molecular and Cellular Neurosciences, 35*, 183–193.

Caughey, B., & Lansbury, P. T. (2003). Protofibrils, pores, fibrils, and neurodegeneration: Separating the responsible protein aggregates from the innocent bystanders. *Annual Review of Neuroscience, 26*, 267–298.

Cirrito, J. R., Yamada, K. A., Finn, M. B., Sloviter, R. S., Bales, K. R., May, P. C., Schoepp, D. D., Paul, S. M., Mennerick, S., & Holtzman, D. M. (2005). Synaptic activity regulates interstitial fluid amyloid-beta levels in vivo. *Neuron, 48*, 913–922.

Cisse, M., Halabisky, B., Harris, J., Devidze, N., Dubal, D. B., Sun, B., Orr, A., Lotz, G., Kim, D. H., Hamto, P., Ho, K., Yu, G. Q., & Mucke, L. (2011). Reversing EphB2 depletion rescues cognitive functions in Alzheimer model. *Nature, 469*, 47–52.

Citron, M., Oltersdorf, T., Haass, C., McConlogue, L., Hung, A. Y., Seubert, P., Vigo-Pelfrey, C., Lieberburg, I., & Selkoe, D. J. (1992). Mutation of the beta-amyloid precursor protein in familial Alzheimer's disease increases beta-protein production. *Nature, 360*, 672–674.

Citron, M., Westaway, D., Xia, W., Carlson, G., Diehl, T., Levesque, G., Johnson-Wood, K., Lee, M., Seubert, P., Davis, A., Kholodenko, D., Motter, R., Sherrington, R., Perry, B., Yao, H., Strome, R., Lieberburg, I., Rommens, J., Kim, S., Schenk, D., Fraser, P., St George Hyslop, P., & Selkoe, D. J. (1997). Mutant presenilins of Alzheimer's disease increase production of 42-residue amyloid beta-protein in both transfected cells and transgenic mice. *Nature Medicine, 3*, 67–72.

Cleary, J. P., Walsh, D. M., Hofmeister, J. J., Shankar, G. M., Kuskowski, M. A., Selkoe, D. J., & Ashe, K. H. (2005). Natural oligomers of the amyloid-beta protein specifically disrupt cognitive function. *Nature Neuroscience, 8*, 79–84.

Corder, E. H., Saunders, A. M., Strittmatter, W. J., Schmechel, D. E., Gaskell, P. C., Small, G. W., Roses, A. D., Haines, J. L., & Pericak-Vance, M. A. (1993). Gene dose of apolipoprotein E type 4 allele and the risk of Alzheimer's disease in late onset families. *Science, 261*, 921–923.

D'Amelio, M., Cavallucci, V., Middei, S., Marchetti, C., Pacioni, S., Ferri, A., Diamantini, A., De Zio, D., Carrara, P., Battistini, L., Moreno, S., Bacci, A., Ammassari-Teule, M., Marie, H., & Cecconi, F. (2011). Caspase-3 triggers early synaptic dysfunction in a mouse model of Alzheimer's disease. *Nature Neuroscience, 14*, 69–76.

Dalva, M. B., Takasu, M. A., Lin, M. Z., Shamah, S. M., Hu, L., Gale, N. W., & Greenberg, M. E. (2000). EphB receptors interact with NMDA receptors and regulate excitatory synapse formation. *Cell, 103*, 945–956.

Davies, C. A., Mann, D. M., Sumpter, P. Q., & Yates, P. O. (1987). A quantitative morphometric analysis of the neuronal and synaptic content of the frontal and temporal cortex in patients with Alzheimer's disease. *Journal of Neurological Sciences, 78*, 151–164.

Dawson, G. R., Seabrook, G. R., Zheng, H., Smith, D. W., Graham, S., O'Dowd, G., Bowery, B. J., Boyce, S., Trumbauer, M. E., Chen, H. Y., Van der Ploeg, L. H., & Sirinathsinghji, D. J. (1999). Age-related cognitive deficits, impaired long-term potentiation and reduction in synaptic marker density in mice lacking the beta-amyloid precursor protein. *Neuroscience, 90*, 1–13.

De Roo, M., Klauser, P., Garcia, P. M., Poglia, L., & Muller, D. (2008). Spine dynamics and synapse remodeling during LTP and memory processes. *Progress in Brain Research, 169*, 199–207.

De Strooper, B., Saftig, P., Craessaerts, K., Vanderstichele, H., Guhde, G., Annaert, W., Von Figura, K., & Van Leuven, F. (1998). Deficiency of presenilin-1 inhibits the normal cleavage of amyloid precursor protein. *Nature, 391*, 387–390.

DeKosky, S. T., & Scheff, S. W. (1990). Synapse loss in frontal cortex biopsies in Alzheimer's disease: Correlation with cognitive severity. *Annals of Neurology, 27*, 457–464.

Dewachter, I., Reverse, D., Caluwaerts, N., Ris, L., Kuiperi, C., Van den Haute, C., Spittaels, K., Umans, L., Serneels, L., Thiry, E., Moechars, D., Mercken, M., Godaux, E., & Van Leuven, F. (2002). Neuronal deficiency of presenilin 1 inhibits amyloid plaque formation and corrects hippocampal long-term potentiation but not a cognitive defect of amyloid precursor protein [V717I] transgenic mice. *The Journal of Neuroscience, 22*, 3445–3453.

Dewachter, I., Filipkowski, R. K., Priller, C., Ris, L., Neyton, J., Croes, S., Terwel, D., Gysemans, M., Devijver, H., Borghgraef, P., Godaux, E., Kaczmarek, L., Herms, J., & Van Leuven, F. (2009). Deregulation of NMDA-receptor function and down-stream signaling in APP[V717I] transgenic mice. *Neurobiology of Aging, 30*, 241–256.

Di Luca, M., Baker, M., Corradetti, R., Kettenmann, H., Mendlewicz, J., Olesen, J., Ragan, I., & Westphal, M. (2011). Consensus document on European brain research. *The European Journal of Neuroscience, 33*, 768–818.

Dickson, D. W., Crystal, H. A., Bevona, C., Honer, W., Vincent, I., & Davies, P. (1995). Correlations of synaptic and pathological markers with cognition of the elderly. *Neurobiology of Aging, 16*, 285–298. discussion 298–304.

Dineley, K. T., Bell, K. A., Bui, D., & Sweatt, J. D. (2002). Beta -Amyloid peptide activates alpha 7 nicotinic acetylcholine receptors expressed in Xenopus oocytes. *The Journal of Biological Chemistry, 277*, 25056–25061.

Duff, K., Eckman, C., Zehr, C., Yu, X., Prada, C. M., Perez-tur, J., Hutton, M., Buee, L., Harigaya, Y., Yager, D., Morgan, D., Gordon, M. N., Holcomb, L., Refolo, L., Zenk, B., Hardy, J., & Younkin, S. (1996). Increased amyloid-beta42(43) in brains of mice expressing mutant presenilin 1. *Nature, 383*, 710–713.

Edbauer, D., Winkler, E., Regula, J. T., Pesold, B., Steiner, H., & Haass, C. (2003). Reconstitution of gamma-secretase activity. *Nature Cell Biology, 5*, 486–488.

Enya, M., Morishima-Kawashima, M., Yoshimura, M., Shinkai, Y., Kusui, K., Khan, K., Games, D., Schenk, D., Sugihara, S., Yamaguchi, H., & Ihara, Y. (1999). Appearance of sodium dodecyl sulfate-stable amyloid beta-protein (Abeta) dimer in the cortex during aging. *The American Journal of Pathology, 154*, 271–279.

Epis, R., Marcello, E., Gardoni, F., Vastagh, C., Malinverno, M., Balducci, C., Colombo, A., Borroni, B., Vara, H., Dell'Agli, M., Cattabeni, F., Giustetto, M., Borsello, T., Forloni, G., Padovani, A., & Di Luca, M. (2010). Blocking ADAM10 synaptic trafficking generates a model of sporadic Alzheimer's disease. *Brain, 133*, 3323–3335.

Ertekin-Taner, N., Graff-Radford, N., Younkin, L. H., Eckman, C., Baker, M., Adamson, J., Ronald, J., Blangero, J., Hutton, M., & Younkin, S. G. (2000). Linkage of plasma Abeta42 to a quantitative locus on chromosome 10 in late-onset Alzheimer's disease pedigrees. *Science, 290*, 2303–2304.

Evans, N. A., Facci, L., Owen, D. E., Soden, P. E., Burbidge, S. A., Prinjha, R. K., Richardson, J. C., & Skaper, S. D. (2008). Abeta(1–42) reduces synapse number and inhibits neurite outgrowth in primary cortical and hippocampal neurons: A quantitative analysis. *Journal of Neuroscience Methods, 175*, 96–103.

Funato, H., Enya, M., Yoshimura, M., Morishima-Kawashima, M., & Ihara, Y. (1999). Presence of sodium dodecyl sulfate-stable amyloid beta-protein dimers in the hippocampus CA1 not exhibiting neurofibrillary tangle formation. *The American Journal of Pathology, 155*, 23–28.

Gakhar-Koppole, N., Hundeshagen, P., Mandl, C., Weyer, S. W., Allinquant, B., Muller, U., & Ciccolini, F. (2008). Activity requires soluble amyloid precursor protein alpha to promote neurite outgrowth in neural stem cell-derived neurons via activation of the MAPK pathway. *The European Journal of Neuroscience, 28*, 871–882.

Gibson, P. H. (1983). EM study of the numbers of cortical synapses in the brains of ageing people and people with Alzheimer-type dementia. *Acta Neuropathologica, 62*, 127–133.

Glabe, C. G. (2008). Structural classification of toxic amyloid oligomers. *The Journal of Biological Chemistry, 283*, 29639–29643.

Glenner, G. G., & Wong, C. W. (1984). Alzheimer's disease and Down's syndrome: Sharing of a unique cerebrovascular amyloid fibril protein. *Biochemical and Biophysical Research Communications, 122*, 1131–1135.

Goate, A., Chartier-Harlin, M. C., Mullan, M., Brown, J., Crawford, F., Fidani, L., Giuffra, L., Haynes, A., Irving, N., James, L., et al. (1991). Segregation of a missense mutation in the amyloid precursor protein gene with familial Alzheimer's disease. *Nature, 349*, 704–706.

Goldgaber, D., Lerman, M. I., McBride, O. W., Saffiotti, U., & Gajdusek, D. C. (1987). Characterization and chromosomal localization of a cDNA encoding brain amyloid of Alzheimer's disease. *Science, 235*, 877–880.

Gong, Y., Chang, L., Viola, K. L., Lacor, P. N., Lambert, M. P., Finch, C. E., Krafft, G. A., & Klein, W. L. (2003). Alzheimer's disease-affected brain: Presence of oligomeric A beta ligands (ADDLs) suggests a molecular basis for reversible memory loss. *Proceedings of the National Academy of Sciences of the United States of America, 100*, 10417–10422.

Gralle, M., & Ferreira, S. T. (2007). Structure and functions of the human amyloid precursor protein: The whole is more than the sum of its parts. *Progress in Neurobiology, 82*, 11–32.

Greenberg, S. M., Qiu, W. Q., Selkoe, D. J., Ben-Itzhak, A., & Kosik, K. S. (1995). Aminoterminal region of the beta-amyloid precursor protein activates mitogen-activated protein kinase. *Neuroscience Letters, 198*, 52–56.

Grunwald, I. C., Korte, M., Wolfer, D., Wilkinson, G. A., Unsicker, K., Lipp, H. P., Bonhoeffer, T., & Klein, R. (2001). Kinase-independent requirement of EphB2 receptors in hippocampal synaptic plasticity. *Neuron, 32*, 1027–1040.

Haass, C. (2004). Take five–BACE and the gamma-secretase quartet conduct Alzheimer's amyloid beta-peptide generation. *The EMBO Journal, 23*, 483–488.

Haass, C., & Selkoe, D. J. (1993). Cellular processing of beta-amyloid precursor protein and the genesis of amyloid beta-peptide. *Cell, 75*, 1039–1042.

Haass, C., Schlossmacher, M. G., Hung, A. Y., Vigo-Pelfrey, C., Mellon, A., Ostaszewski, B. L., Lieberburg, I., Koo, E. H., Schenk, D., Teplow, D. B., et al. (1992). Amyloid beta-peptide is produced by cultured cells during normal metabolism. *Nature, 359*, 322–325.

Hamos, J. E., DeGennaro, L. J., & Drachman, D. A. (1989). Synaptic loss in Alzheimer's disease and other dementias. *Neurology, 39*, 355–361.

Hardy, J. (1992). Framing beta-amyloid. *Nature Genetics, 1*, 233–234.

Hardy, J., & Selkoe, D. J. (2002). The amyloid hypothesis of Alzheimer's disease: Progress and problems on the road to therapeutics. *Science, 297*, 353–356.

Hardy, J., Duff, K., Hardy, K. G., Perez-Tur, J., & Hutton, M. (1998). Genetic dissection of Alzheimer's disease and related dementias: Amyloid and its relationship to tau. *Nature Neuroscience, 1*, 355–358.

Hardy, J. A., & Higgins, G. A. (1992). Alzheimer's disease: The amyloid cascade hypothesis. *Science, 256*, 184–185.

Harper, J. D., Wong, S. S., Lieber, C. M., & Lansbury, P. T. (1997). Observation of metastable Abeta amyloid protofibrils by atomic force microscopy. *Chemistry & Biology, 4*, 119–125.

Hartley, D. M., Walsh, D. M., Ye, C. P., Diehl, T., Vasquez, S., Vassilev, P. M., Teplow, D. B., & Selkoe, D. J. (1999). Protofibrillar intermediates of amyloid beta-protein induce acute electrophysiological changes and progressive neurotoxicity in cortical neurons. *The Journal of Neuroscience, 19*, 8876–8884.

Henderson, J. T., Georgiou, J., Jia, Z., Robertson, J., Elowe, S., Roder, J. C., & Pawson, T. (2001). The receptor tyrosine kinase EphB2 regulates NMDA-dependent synaptic function. *Neuron, 32*, 1041–1056.

Hendriks, L., van Duijn, C. M., Cras, P., Cruts, M., Van Hul, W., van Harskamp, F., Warren, A., McInnis, M. G., Antonarakis, S. E., Martin, J. J., et al. (1992). Presenile dementia and cerebral haemorrhage linked to a mutation at codon 692 of the beta-amyloid precursor protein gene. *Nature Genetics, 1*, 218–221.

Honer, W. G., Dickson, D. W., Gleeson, J., & Davies, P. (1992). Regional synaptic pathology in Alzheimer's disease. *Neurobiology of Aging, 13*, 375–382.

Hsieh, H., Boehm, J., Sato, C., Iwatsubo, T., Tomita, T., Sisodia, S., & Malinow, R. (2006). AMPAR removal underlies Abeta-induced synaptic depression and dendritic spine loss. *Neuron, 52*, 831–843.

Hung, A. Y., Koo, E. H., Haass, C., & Selkoe, D. J. (1992). Increased expression of beta-amyloid precursor protein during neuronal differentiation is not accompanied by secretory cleavage. *Proceedings of the National Academy of Sciences of the United States of America, 89*, 9439–9443.

Hutton, M., Lendon, C. L., Rizzu, P., Baker, M., Froelich, S., Houlden, H., Pickering-Brown, S., Chakraverty, S., Isaacs, A., Grover, A., Hackett, J., Adamson, J., Lincoln, S., Dickson, D., Davies, P., Petersen, R. C., Stevens, M., de Graaff, E., Wauters, E., van Baren, J., Hillebrand, M., Joosse, M., Kwon, J. M., Nowotny, P., Che, L. K., Norton, J., Morris, J. C., Reed, L. A., Trojanowski, J., Basun, H., Lannfelt, L., Neystat, M., Fahn, S., Dark, F., Tannenberg, T., Dodd, P. R., Hayward, N., Kwok, J. B., Schofield, P. R., Andreadis, A., Snowden, J., Craufurd, D., Neary, D., Owen, F., Oostra, B. A., Hardy, J., Goate, A., van Swieten, J., Mann, D., Lynch, T., & Heutink, P. (1998). Association of missense and 5'-splice-site mutations in tau with the inherited dementia FTDP-17. *Nature, 393*, 702–705.

Ittner, L. M., & Gotz, J. (2011). Amyloid-beta and tau–a toxic pas de deux in Alzheimer's disease. *Nature Reviews Neuroscience, 12*, 65–72.

Ittner, L. M., Ke, Y. D., Delerue, F., Bi, M., Gladbach, A., van Eersel, J., Wolfing, H., Chieng, B. C., Christie, M. J., Napier, I. A., Eckert, A., Staufenbiel, M., Hardeman, E., & Gotz, J. (2010). Dendritic function of tau mediates amyloid-beta toxicity in Alzheimer's disease mouse models. *Cell, 142*, 387–397.

Iwata, N., Tsubuki, S., Takaki, Y., Shirotani, K., Lu, B., Gerard, N. P., Gerard, C., Hama, E., Lee, H. J., & Saido, T. C. (2001). Metabolic regulation of brain Abeta by neprilysin. *Science, 292*, 1550–1552.

Jacobsen, J. S., Wu, C. C., Redwine, J. M., Comery, T. A., Arias, R., Bowlby, M., Martone, R., Morrison, J. H., Pangalos, M. N., Reinhart, P. H., & Bloom, F. E. (2006). Early-onset behavioral and synaptic deficits in a mouse model of Alzheimer's disease. *Proceedings of the National Academy of Sciences of the United States of America, 103*, 5161–5166.

Jarrett, J. T., Berger, E. P., & Lansbury, P. T. (1993). The carboxy terminus of the β amyloid protein is critical for the seeding of amyloid formation: implications for the pathogenesis of Alzheimer's disease. *Biochemistry, 32*, 4693–4697.

Jorissen, E., Prox, J., Bernreuther, C., Weber, S., Schwanbeck, R., Serneels, L., Snellinx, A., Craessaerts, K., Thathiah, A., Tesseur, I., Bartsch, U., Weskamp, G., Blobel, C. P., Glatzel, M., De Strooper, B., & Saftig, P. (2010). The disintegrin/metalloproteinase ADAM10 is essential for the establishment of the brain cortex. *The Journal of Neuroscience, 30*, 4833–4844.

Kamenetz, F., Tomita, T., Hsieh, H., Seabrook, G., Borchelt, D., Iwatsubo, T., Sisodia, S., & Malinow, R. (2003). APP processing and synaptic function. *Neuron, 37*, 925–937.

Kang, J., Lemaire, H. G., Unterbeck, A., Salbaum, J. M., Masters, C. L., Grzeschik, K. H., Multhaup, G., Beyreuther, K., & Muller-Hill, B. (1987). The precursor of Alzheimer's disease amyloid A4 protein resembles a cell-surface receptor. *Nature, 325*, 733–736.

Kang, J. E., Lim, M. M., Bateman, R. J., Lee, J. J., Smyth, L. P., Cirrito, J. R., Fujiki, N., Nishino, S., & Holtzman, D. M. (2009). Amyloid-beta dynamics are regulated by orexin and the sleep-wake cycle. *Science, 326*, 1005–1007.

Kasai, H., Fukuda, M., Watanabe, S., Hayashi-Takagi, A., & Noguchi, J. (2010). Structural dynamics of dendritic spines in memory and cognition. *Trends in Neurosciences, 33*(3), 121–9.

Khosravani, H., Zhang, Y., Tsutsui, S., Hameed, S., Altier, C., Hamid, J., Chen, L., Villemaire, M., Ali, Z., Jirik, F. R., & Zamponi, G. W. (2008). Prion protein attenuates excitotoxicity by inhibiting NMDA receptors. *The Journal of Cell Biology, 181*, 551–565.

Kim, J., Basak, J. M., & Holtzman, D. M. (2009). The role of apolipoprotein E in Alzheimer's disease. *Neuron, 63*, 287–303.

Kim, J. H., Anwyl, R., Suh, Y. H., Djamgoz, M. B., & Rowan, M. J. (2001). Use-dependent effects of amyloidogenic fragments of (beta)-amyloid precursor protein on synaptic plasticity in rat hippocampus in vivo. *The Journal of Neuroscience, 21*, 1327–1333.

Kinoshita, A., Fukumoto, H., Shah, T., Whelan, C. M., Irizarry, M. C., & Hyman, B. T. (2003). Demonstration by FRET of BACE interaction with the amyloid precursor protein at the cell surface and in early endosomes. *Journal of Cell Science, 116*, 3339–3346.

Klein, W. L., Krafft, G. A., & Finch, C. E. (2001). Targeting small Abeta oligomers: The solution to an Alzheimer's disease conundrum? *Trends in Neurosciences, 24*, 219–224.

Kuhn, P. H., Wang, H., Dislich, B., Colombo, A., Zeitschel, U., Ellwart, J. W., Kremmer, E., Rossner, S., & Lichtenthaler, S. F. (2010). ADAM10 is the physiologically relevant, constitutive alpha-secretase of the amyloid precursor protein in primary neurons. *The EMBO Journal, 29*, 3020–3032.

Lacor, P. N., Buniel, M. C., Furlow, P. W., Clemente, A. S., Velasco, P. T., Wood, M., Viola, K. L., & Klein, W. L. (2007). Abeta oligomer-induced aberrations in synapse composition, shape, and density provide a molecular basis for loss of connectivity in Alzheimer's disease. *The Journal of Neuroscience, 27*, 796–807.

Lacor, P. N., Buniel, M. C., Chang, L., Fernandez, S. J., Gong, Y., Viola, K. L., Lambert, M. P., Velasco, P. T., Bigio, E. H., Finch, C. E., Krafft, G. A., & Klein, W. L. (2004). Synaptic targeting by Alzheimer's-related amyloid beta oligomers. *The Journal of Neuroscience, 24*, 10191–10200.

Laird, F. M., Cai, H., Savonenko, A. V., Farah, M. H., He, K., Melnikova, T., Wen, H., Chiang, H. C., Xu, G., Koliatsos, V. E., Borchelt, D. R., Price, D. L., Lee, H. K., & Wong, P. C. (2005). BACE1, a major determinant of selective vulnerability of the brain to amyloid-beta amyloidogenesis, is essential for cognitive, emotional, and synaptic functions. *The Journal of Neuroscience, 25*, 11693–11709.

Lambert, M. P., Barlow, A. K., Chromy, B. A., Edwards, C., Freed, R., Liosatos, M., Morgan, T. E., Rozovsky, I., Trommer, B., Viola, K. L., Wals, P., Zhang, C., Finch, C. E., Krafft, G. A., & Klein, W. L. (1998). Diffusible, nonfibrillar ligands derived from Abeta1-42 are potent central nervous system neurotoxins. *Proceedings of the National Academy of Sciences of the United States of America, 95*, 6448–6453.

Lammich, S., Kojro, E., Postina, R., Gilbert, S., Pfeiffer, R., Jasionowski, M., Haass, C., & Fahrenholz, F. (1999). Constitutive and regulated alpha-secretase cleavage of Alzheimer's amyloid precursor protein by a disintegrin metalloprotease. *Proceedings of the National Academy of Sciences of the United States of America, 96*, 3922–3927.

Lashuel, H. A., Hartley, D., Petre, B. M., Walz, T., & Lansbury, P. T., Jr. (2002). Neurodegenerative disease: Amyloid pores from pathogenic mutations. *Nature, 418*, 291.

Lau, C. G., & Zukin, R. S. (2007). NMDA receptor trafficking in synaptic plasticity and neuropsychiatric disorders. *Nature Reviews Neuroscience, 8*(6), 413–26.

Lauren, J., Gimbel, D. A., Nygaard, H. B., Gilbert, J. W., & Strittmatter, S. M. (2009). Cellular prion protein mediates impairment of synaptic plasticity by amyloid-beta oligomers. *Nature, 457*, 1128–1132.

Lesne, S., Koh, M. T., Kotilinek, L., Kayed, R., Glabe, C. G., Yang, A., Gallagher, M., & Ashe, K. H. (2006). A specific amyloid-beta protein assembly in the brain impairs memory. *Nature, 440*, 352–357.

Levy-Lahad, E., Wasco, W., Poorkaj, P., Romano, D. M., Oshima, J., Pettingell, W. H., Yu, C. E., Jondro, P. D., Schmidt, S. D., Wang, K., et al. (1995). Candidate gene for the chromosome 1 familial Alzheimer's disease locus. *Science, 269*, 973–977.

Levy, E., Carman, M. D., Fernandez-Madrid, I. J., Power, M. D., Lieberburg, I., van Duinen, S. G., Bots, G. T., Luyendijk, W., & Frangione, B. (1990). Mutation of the Alzheimer's disease amyloid gene in hereditary cerebral hemorrhage, Dutch type. *Science, 248*, 1124–1126.

Lewis, J., Dickson, D. W., Lin, W. L., Chisholm, L., Corral, A., Jones, G., Yen, S. H., Sahara, N., Skipper, L., Yager, D., Eckman, C., Hardy, J., Hutton, M., & McGowan, E. (2001). Enhanced neurofibrillary degeneration in transgenic mice expressing mutant tau and APP. *Science, 293*, 1487–1491.

Li, H. L., Roch, J. M., Sundsmo, M., Otero, D., Sisodia, S., Thomas, R., & Saitoh, T. (1997). Defective neurite extension is caused by a mutation in amyloid beta/A4 (Aβ) protein precursor found in familial Alzheimer's disease. *Journal Neurobiology, 32*, 469–480.

Li, S., Hong, S., Shepardson, N. E., Walsh, D. M., Shankar, G. M., & Selkoe, D. (2009). Soluble oligomers of amyloid Beta protein facilitate hippocampal long-term depression by disrupting neuronal glutamate uptake. *Neuron, 62*, 788–801.

Li, Z., Jo, J., Jia, J. M., Lo, S. C., Whitcomb, D. J., Jiao, S., Cho, K., & Sheng, M. (2010). Caspase-3 activation via mitochondria is required for long-term depression and AMPA receptor internalization. *Cell, 141*, 859–871.

Liu, L., Wong, T. P., Pozza, M. F., Lingenhoehl, K., Wang, Y., Sheng, M., Auberson, Y. P., & Wang, Y. T. (2004). Role of NMDA receptor subtypes in governing the direction of hippocampal synaptic plasticity. *Science, 304*, 1021–1024.

Lorenzo, A., & Yankner, B. A. (1996). Amyloid fibril toxicity in Alzheimer's disease and diabetes. *Annals of the New York Academy of Sciences, 777*, 89–95.

Mackenzie, I. R., & Miller, L. A. (1994). Senile plaques in temporal lobe epilepsy. *Acta Neuropathologica, 87*, 504–510.

Marambaud, P., Zhao, H., & Davies, P. (2005). Resveratrol promotes clearance of Alzheimer's disease amyloid-beta peptides. *The Journal of Biological Chemistry, 280*, 37377–37382.

Marcello, E., Epis, R., Saraceno, C., Gardoni, F., Borroni, B., Cattabeni, F., Padovani, A., & Di Luca, M. (2010). SAP97-mediated local trafficking is altered in Alzheimer disease patients' hippocampus. *Neurobiology of Aging, 33(2)*, 422.e1-422.e10v.

Marcello, E., Gardoni, F., Mauceri, D., Romorini, S., Jeromin, A., Epis, R., Borroni, B., Cattabeni, F., Sala, C., Padovani, A., & Di Luca, M. (2007). Synapse-associated protein-97 mediates alpha-secretase ADAM10 trafficking and promotes its activity. *The Journal of Neuroscience, 27*, 1682–1691.

Masliah, E., & Terry, R. (1993). The role of synaptic proteins in the pathogenesis of disorders of the central nervous system. *Brain Pathology, 3*, 77–85.

Masliah, E., Mallory, M., Hansen, L., DeTeresa, R., Alford, M., & Terry, R. (1994). Synaptic and neuritic alterations during the progression of Alzheimer's disease. *Neuroscience Letters, 174*, 67–72.

Masters, C. L., Simms, G., Weinman, N. A., Multhaup, G., McDonald, B. L., & Beyreuther, K. (1985). Amyloid plaque core protein in Alzheimer disease and Down syndrome. *Proceedings of the National Academy of Sciences of the United States of America, 82*, 4245–4249.

Mattson, M. P. (1997). Cellular actions of beta-amyloid precursor protein and its soluble and fibrillogenic derivatives. *Physiological Reviews, 77*, 1081–1132.

Maurice, T., Lockhart, B. P., & Privat, A. (1996). Amnesia induced in mice by centrally administered beta-amyloid peptides involves cholinergic dysfunction. *Brain Research, 706*, 181–193.

McDonald, M. P., Dahl, E. E., Overmier, J. B., Mantyh, P., & Cleary, J. (1994). Effects of an exogenous beta-amyloid peptide on retention for spatial learning. *Behavioral and Neural Biology, 62*, 60–67.

Meziane, H., Dodart, J. C., Mathis, C., Little, S., Clemens, J., Paul, S. M., & Ungerer, A. (1998). Memory-enhancing effects of secreted forms of the beta-amyloid precursor protein in normal and amnestic mice. *Proceedings of the National Academy of Sciences of the United States of America, 95*, 12683–12688.

Mucke, L., Abraham, C. R., & Masliah, E. (1996). Neurotrophic and neuroprotective effects of hAPP in transgenic mice. *Annals of the New York Academy of Sciences, 777*, 82–88.

Mullan, M., Crawford, F., Axelman, K., Houlden, H., Lilius, L., Winblad, B., & Lannfelt, L. (1992). A pathogenic mutation for probable Alzheimer's disease in the APP gene at the N-terminus of beta-amyloid. *Nature Genetics, 1*, 345–347.

Myers, A., Holmans, P., Marshall, H., Kwon, J., Meyer, D., Ramic, D., Shears, S., Booth, J., DeVrieze, F. W., Crook, R., Hamshere, M., Abraham, R., Tunstall, N., Rice, F., Carty, S., Lillystone, S., Kehoe, P., Rudrasingham, V., Jones, L., Lovestone, S., Perez-Tur, J., Williams, J., Owen, M. J., Hardy, J., & Goate, A. M. (2000). Susceptibility locus for Alzheimer's disease on chromosome 10. *Science, 290*, 2304–2305.

Nikolaev, A., McLaughlin, T., O'Leary, D. D., & Tessier-Lavigne, M. (2009). APP binds DR6 to trigger axon pruning and neuron death via distinct caspases. *Nature, 457*, 981–989.

Ninomiya, H., Roch, J. M., Jin, L. W., & Saitoh, T. (1994). Secreted form of amyloid beta/A4 protein precursor (APP) binds to two distinct APP binding sites on rat B103 neuron-like cells through two different domains, but only one site is involved in neuritotropic activity. *Journal of Neurochemistry, 63*, 495–500.

Nitta, A., Itoh, A., Hasegawa, T., & Nabeshima, T. (1994). Beta-Amyloid protein-induced Alzheimer's disease animal model. *Neuroscience Letters, 170*, 63–66.

Oddo, S., Caccamo, A., Shepherd, J. D., Murphy, M. P., Golde, T. E., Kayed, R., Metherate, R., Mattson, M. P., Akbari, Y., & LaFerla, F. M. (2003). Triple-transgenic model of Alzheimer's disease with plaques and tangles: Intracellular Abeta and synaptic dysfunction. *Neuron, 39*, 409–421.

Ohsawa, I., Takamura, C., Morimoto, T., Ishiguro, M., & Kohsaka, S. (1999). Amino-terminal region of secreted form of amyloid precursor protein stimulates proliferation of neural stem cells. *The European Journal of Neuroscience, 11*, 1907–1913.

Olson, J. M., Goddard, K. A., & Dudek, D. M. (2001). The amyloid precursor protein locus and very-late-onset Alzheimer disease. *The American Journal of Human Genetics, 69*, 895–899.

Olson, M. I., & Shaw, C. M. (1969). Presenile dementia and Alzheimer's disease in mongolism. *Brain, 92*, 147–156.

Palop, J. J., Jones, B., Kekonius, L., Chin, J., Yu, G. Q., Raber, J., Masliah, E., & Mucke, L. (2003). Neuronal depletion of calcium-dependent proteins in the dentate gyrus is tightly linked to Alzheimer's disease-related cognitive deficits. *Proceedings of the National Academy of Sciences of the United States of America, 100*, 9572–9577.

Palop, J. J., Chin, J., Roberson, E. D., Wang, J., Thwin, M. T., Bien-Ly, N., Yoo, J., Ho, K. O., Yu, G. Q., Kreitzer, A., Finkbeiner, S., Noebels, J. L., & Mucke, L. (2007). Aberrant excitatory neuronal activity and compensatory remodeling of inhibitory hippocampal circuits in mouse models of Alzheimer's disease. *Neuron, 55*, 697–711.

Perez, R. G., Zheng, H., Van der Ploeg, L. H., & Koo, E. H. (1997). The beta-amyloid precursor protein of Alzheimer's disease enhances neuron viability and modulates neuronal polarity. *The Journal of Neuroscience, 17*, 9407–9414.

Podlisny, M. B., Ostaszewski, B. L., Squazzo, S. L., Koo, E. H., Rydell, R. E., Teplow, D. B., & Selkoe, D. J. (1995). Aggregation of secreted amyloid beta-protein into sodium dodecyl sulfate-stable oligomers in cell culture. *The Journal of Biological Chemistry, 270*, 9564–9570.

Poorkaj, P., Bird, T. D., Wijsman, E., Nemens, E., Garruto, R. M., Anderson, L., Andreadis, A., Wiederholt, W. C., Raskind, M., & Schellenberg, G. D. (1998). Tau is a candidate gene for chromosome 17 frontotemporal dementia. *Annals of Neurology, 43*, 815–825.

Postina, R., Schroeder, A., Dewachter, I., Bohl, J., Schmitt, U., Kojro, E., Prinzen, C., Endres, K., Hiemke, C., Blessing, M., Flamez, P., Dequenne, A., Godaux, E., van Leuven, F., & Fahrenholz, F. (2004). A disintegrin-metalloproteinase prevents amyloid plaque formation and hippocampal defects in an Alzheimer disease mouse model. *The Journal of Clinical Investigation, 113*, 1456–1464.

Puzzo, D., Privitera, L., Leznik, E., Fa, M., Staniszewski, A., Palmeri, A., & Arancio, O. (2008). Picomolar amyloid-beta positively modulates synaptic plasticity and memory in hippocampus. *The Journal of Neuroscience, 28*, 14537–14545.

Qiu, W. Q., Walsh, D. M., Ye, Z., Vekrellis, K., Zhang, J., Podlisny, M. B., Rosner, M. R., Safavi, A., Hersh, L. B., & Selkoe, D. J. (1998). Insulin-degrading enzyme regulates extracellular levels of amyloid beta-protein by degradation. *The Journal of Biological Chemistry, 273*, 32730–32738.

Rapoport, M., Dawson, H. N., Binder, L. I., Vitek, M. P., & Ferreira, A. (2002). Tau is essential to beta -amyloid-induced neurotoxicity. *Proceedings of the National Academy of Sciences of the United States of America, 99*, 6364–6369.

Renner, M., Lacor, P. N., Velasco, P. T., Xu, J., Contractor, A., Klein, W. L., & Triller, A. (2010). Deleterious effects of amyloid beta oligomers acting as an extracellular scaffold for mGluR5. *Neuron, 66*(5), 739–54.

Robakis, N. K., Ramakrishna, N., Wolfe, G., & Wisniewski, H. M. (1987). Molecular cloning and characterization of a cDNA encoding the cerebrovascular and the neuritic plaque amyloid peptides. *Proceedings of the National Academy of Sciences of the United States of America, 84*, 4190–4194.

Roberson, E. D., Scearce-Levie, K., Palop, J. J., Yan, F., Cheng, I. H., Wu, T., Gerstein, H., Yu, G. Q., & Mucke, L. (2007). Reducing endogenous tau ameliorates amyloid beta-induced deficits in an Alzheimer's disease mouse model. *Science, 316*, 750–754.

Rohe, M., Carlo, A. S., Breyhan, H., Sporbert, A., Militz, D., Schmidt, V., Wozny, C., Harmeier, A., Erdmann, B., Bales, K. R., Wolf, S., Kempermann, G., Paul, S. M., Schmitz, D., Bayer, T. A., Willnow, T. E., & Andersen, O. M. (2008). Sortilin-related receptor with A-type repeats (SORLA) affects the amyloid precursor protein-dependent stimulation of ERK signaling and adult neurogenesis. *The Journal of Biological Chemistry, 283*, 14826–14834.

Rossjohn, J., Cappai, R., Feil, S. C., Henry, A., McKinstry, W. J., Galatis, D., Hesse, L., Multhaup, G., Beyreuther, K., Masters, C. L., & Parker, M. W. (1999). Crystal structure of the N-terminal, growth factor-like domain of Alzheimer amyloid precursor protein. *Nature Structural Biology, 6*, 327–331.

Salter, M. W., & Kalia, L. V. (2004). Src kinases: A hub for NMDA receptor regulation. *Nature Reviews Neuroscience, 5*, 317–328.

Saura, C. A., Choi, S. Y., Beglopoulos, V., Malkani, S., Zhang, D., Shankaranarayana Rao, B. S., Chattarji, S., Kelleher, R. J., 3rd, Kandel, E. R., Duff, K., Kirkwood, A., & Shen, J. (2004). Loss of presenilin function causes impairments of memory and synaptic plasticity followed by age-dependent neurodegeneration. *Neuron, 42*, 23–36.

Scheff, S. W., DeKosky, S. T., & Price, D. A. (1990). Quantitative assessment of cortical synaptic density in Alzheimer's disease. *Neurobiology of Aging, 11*, 29–37.

Scheff, S. W., Price, D. A., Schmitt, F. A., & Mufson, E. J. (2006). Hippocampal synaptic loss in early Alzheimer's disease and mild cognitive impairment. *Neurobiology of Aging, 27*, 1372–1384.

Scheff, S. W., Price, D. A., Schmitt, F. A., DeKosky, S. T., & Mufson, E. J. (2007). Synaptic alterations in CA1 in mild Alzheimer disease and mild cognitive impairment. *Neurology, 68*, 1501–1508.

Scheuner, D., Eckman, C., Jensen, M., Song, X., Citron, M., Suzuki, N., Bird, T. D., Hardy, J., Hutton, M., Kukull, W., Larson, E., Levy-Lahad, E., Viitanen, M., Peskind, E., Poorkaj, P., Schellenberg, G., Tanzi, R., Wasco, W., Lannfelt, L., Selkoe, D., & Younkin, S. (1996). Secreted amyloid beta-protein similar to that in the senile plaques of Alzheimer's disease is increased in vivo by the presenilin 1 and 2 and APP mutations linked to familial Alzheimer's disease. *Nature Medicine, 2*, 864–870.

Seabrook, G. R., & Rosahl, T. W. (1999). Transgenic animals relevant to Alzheimer's disease. *Neuropharmacology, 38*, 1–17.

Seabrook, G. R., Smith, D. W., Bowery, B. J., Easter, A., Reynolds, T., Fitzjohn, S. M., Morton, R. A., Zheng, H., Dawson, G. R., Sirinathsinghji, D. J., Davies, C. H., Collingridge, G. L., & Hill, R. G. (1999). Mechanisms contributing to the deficits in hippocampal synaptic plasticity in mice lacking amyloid precursor protein. *Neuropharmacology, 38*, 349–359.

Selkoe, D. J. (1991). The molecular pathology of Alzheimer's disease. *Neuron, 6*, 487–498.

Selkoe, D. J. (2000). Toward a comprehensive theory for Alzheimer's disease. Hypothesis: Alzheimer's disease is caused by the cerebral accumulation and cytotoxicity of amyloid beta-protein. *The Annals of the New York Academy of Sciences, 924*, 17–25.

Selkoe, D. J. (2008). Soluble oligomers of the amyloid beta-protein impair synaptic plasticity and behavior. *Behavioural Brain Research, 192*, 106–113.

Seubert, P., Vigo-Pelfrey, C., Esch, F., Lee, M., Dovey, H., Davis, D., Sinha, S., Schlossmacher, M., Whaley, J., Swindlehurst, C., et al. (1992). Isolation and quantification of soluble Alzheimer's beta-peptide from biological fluids. *Nature, 359*, 325–327.

Shankar, G. M., Bloodgood, B. L., Townsend, M., Walsh, D. M., Selkoe, D. J., & Sabatini, B. L. (2007). Natural oligomers of the Alzheimer amyloid-beta protein induce reversible synapse loss by modulating an NMDA-type glutamate receptor-dependent signaling pathway. *The Journal of Neuroscience, 27*, 2866–2875.

Shankar, G. M., Li, S., Mehta, T. H., Garcia-Munoz, A., Shepardson, N. E., Smith, I., Brett, F. M., Farrell, M. A., Rowan, M. J., Lemere, C. A., Regan, C. M., Walsh, D. M., Sabatini, B. L., & Selkoe, D. J. (2008). Amyloid-beta protein dimers isolated directly from Alzheimer's brains impair synaptic plasticity and memory. *Nature Medicine, 14*, 837–842.

Sherrington, R., Rogaev, E. I., Liang, Y., Rogaeva, E. A., Levesque, G., Ikeda, M., Chi, H., Lin, C., Li, G., Holman, K., et al. (1995). Cloning of a gene bearing missense mutations in early-onset familial Alzheimer's disease. *Nature, 375*, 754–760.

Shoji, M., Golde, T. E., Ghiso, J., Cheung, T. T., Estus, S., Shaffer, L. M., Cai, X. D., McKay, D. M., Tintner, R., Frangione, B., et al. (1992). Production of the Alzheimer amyloid beta protein by normal proteolytic processing. *Science, 258*, 126–129.

Shrestha, B. R., Vitolo, O. V., Joshi, P., Lordkipanidze, T., Shelanski, M., & Dunaevsky, A. (2006). Amyloid beta peptide adversely affects spine number and motility in hippocampal neurons. *Molecular and Cellular Neurosciences, 33*, 274–282.

Small, D. H., Nurcombe, V., Reed, G., Clarris, H., Moir, R., Beyreuther, K., & Masters, C. L. (1994). A heparin-binding domain in the amyloid protein precursor of Alzheimer's disease is involved in the regulation of neurite outgrowth. *The Journal of Neuroscience, 14*, 2117–2127.

Snyder, E. M., Nong, Y., Almeida, C. G., Paul, S., Moran, T., Choi, E. Y., Nairn, A. C., Salter, M. W., Lombroso, P. J., Gouras, G. K., & Greengard, P. (2005). Regulation of NMDA receptor trafficking by amyloid-beta. *Nature Neuroscience, 8*, 1051–1058.

Spillantini, M. G., Bird, T. D., & Ghetti, B. (1998). Frontotemporal dementia and Parkinsonism linked to chromosome 17: A new group of tauopathies. *Brain Pathology, 8*, 387–402.

Spires, T. L., Meyer-Luehmann, M., Stern, E. A., McLean, P. J., Skoch, J., Nguyen, P. T., Bacskai, B. J., & Hyman, B. T. (2005). Dendritic spine abnormalities in amyloid precursor protein transgenic mice demonstrated by gene transfer and intravital multiphoton microscopy. *The Journal of Neuroscience, 25*, 7278–7287.

Stephan, A., Laroche, S., & Davis, S. (2001). Generation of aggregated beta-amyloid in the rat hippocampus impairs synaptic transmission and plasticity and causes memory deficits. *The Journal of Neuroscience, 21*, 5703–5714.

Suzuki, N., Cheung, T. T., Cai, X. D., Odaka, A., Otvos, L., Jr., Eckman, C., Golde, T. E., & Younkin, S. G. (1994). An increased percentage of long amyloid beta protein secreted by familial amyloid beta protein precursor (beta APP717) mutants. *Science, 264*, 1336–1340.

Tanzi, R. E., Gusella, J. F., Watkins, P. C., Bruns, G. A., St George-Hyslop, P., Van Keuren, M. L., Patterson, D., Pagan, S., Kurnit, D. M., & Neve, R. L. (1987). Amyloid beta protein gene: cDNA, mRNA distribution, and genetic linkage near the Alzheimer locus. *Science, 235*, 880–884.

Taylor, C. J., Ireland, D. R., Ballagh, I., Bourne, K., Marechal, N. M., Turner, P. R., Bilkey, D. K., Tate, W. P., & Abraham, W. C. (2008). Endogenous secreted amyloid precursor protein-alpha regulates hippocampal NMDA receptor function, long-term potentiation and spatial memory. *Neurobiology of Disease, 31*, 250–260.

Teplow, D. B. (1998). Structural and kinetic features of amyloid beta-protein fibrillogenesis. *Amyloid, 5*, 121–142.

Terry, R. D. (1996). The pathogenesis of Alzheimer disease: An alternative to the amyloid hypothesis. *Journal of Neuropathology and Experimental Neurology, 55*, 1023–1025.

Terry, R. D., Masliah, E., Salmon, D. P., Butters, N., DeTeresa, R., Hill, R., Hansen, L. A., & Katzman, R. (1991). Physical basis of cognitive alterations in Alzheimer's disease: Synapse loss is the major correlate of cognitive impairment. *Annals of Neurology, 30*, 572–580.

Thinakaran, G., & Koo, E. H. (2008). Amyloid precursor protein trafficking, processing, and function. *The Journal of Biological Chemistry, 283*, 29615–29619.

Van Broeckhoven, C., Haan, J., Bakker, E., Hardy, J. A., Van Hul, W., Wehnert, A., Vegter-Van der Vlis, M., & Roos, R. A. (1990). Amyloid beta protein precursor gene and hereditary cerebral hemorrhage with amyloidosis (Dutch). *Science, 248*, 1120–1122.

Vassar, R., Bennett, B. D., Babu-Khan, S., Kahn, S., Mendiaz, E. A., Denis, P., Teplow, D. B., Ross, S., Amarante, P., Loeloff, R., Luo, Y., Fisher, S., Fuller, J., Edenson, S., Lile, J., Jarosinski, M. A., Biere, A. L., Curran, E., Burgess, T., Louis, J. C., Collins, F., Treanor, J., Rogers, G., & Citron, M. (1999). Beta-secretase cleavage of Alzheimer's amyloid precursor protein by the transmembrane aspartic protease BACE. *Science, 286*, 735–741.

Walsh, D. M., & Selkoe, D. J. (2007). A beta oligomers - a decade of discovery. *Journal of Neurochemistry, 101*, 1172–1184.

Walsh, D. M., Tseng, B. P., Rydel, R. E., Podlisny, M. B., & Selkoe, D. J. (2000). The oligomerization of amyloid beta-protein begins intracellularly in cells derived from human brain. *Biochemistry, 39*, 10831–10839.

Walsh, D. M., Klyubin, I., Fadeeva, J. V., Cullen, W. K., Anwyl, R., Wolfe, M. S., Rowan, M. J., & Selkoe, D. J. (2002). Naturally secreted oligomers of amyloid beta protein potently inhibit hippocampal long-term potentiation in vivo. *Nature, 416*, 535–539.

Walsh, D. M., Hartley, D. M., Kusumoto, Y., Fezoui, Y., Condron, M. M., Lomakin, A., Benedek, G. B., Selkoe, D. J., & Teplow, D. B. (1999). Amyloid beta-protein fibrillogenesis. Structure and biological activity of protofibrillar intermediates. *Journal of Biological Chemistry, 274*, 25945–25952.

Wang, H. Y., Lee, D. H., Davis, C. B., & Shank, R. P. (2000). Amyloid peptide Abeta(1–42) binds selectively and with picomolar affinity to alpha7 nicotinic acetylcholine receptors. *Journal of Neurochemistry, 75*, 1155–1161.

Wang, Q., Walsh, D. M., Rowan, M. J., Selkoe, D. J., & Anwyl, R. (2004). Block of long-term potentiation by naturally secreted and synthetic amyloid beta-peptide in hippocampal slices is mediated via activation of the kinases c-Jun N-terminal kinase, cyclin-dependent kinase 5, and p38 mitogen-activated protein kinase as well as metabotropic glutamate receptor type 5. *The Journal of Neuroscience, 24*, 3370–3378.

Wavrant-DeVrieze, F., Lambert, J. C., Stas, L., Crook, R., Cottel, D., Pasquier, F., Frigard, B., Lambrechts, M., Thiry, E., Amouyel, P., Tur, J. P., Chartier-Harlin, M. C., Hardy, J., & Van Leuven, F. (1999). Association between coding variability in the LRP gene and the risk of late-onset Alzheimer's disease. *Human Genetics, 104*, 432–434.

Wei, W., Nguyen, L. N., Kessels, H. W., Hagiwara, H., Sisodia, S., & Malinow, R. (2010). Amyloid beta from axons and dendrites reduces local spine number and plasticity. *Nature Neuroscience, 13*, 190–196.

Westphalen, R. I., Scott, H. L., & Dodd, P. R. (2003). Synaptic vesicle transport and synaptic membrane transporter sites in excitatory amino acid nerve terminals in Alzheimer disease. *Journal of Neural Transmission, 110*, 1013–1027.

Wisniewski, T., Ghiso, J., & Frangione, B. (1991). Peptides homologous to the amyloid protein of Alzheimer's disease containing a glutamine for glutamic acid substitution have accelerated amyloid fibril formation. *Biochemical and Biophysical Research Communications, 179*, 1247–1254.

Wolfe, M. S., Xia, W., Ostaszewski, B. L., Diehl, T. S., Kimberly, W. T., & Selkoe, D. J. (1999). Two transmembrane aspartates in presenilin-1 required for presenilin endoproteolysis and gamma-secretase activity. *Nature, 398*, 513–517.

Yao, P. J., Zhu, M., Pyun, E. I., Brooks, A. I., Therianos, S., Meyers, V. E., & Coleman, P. D. (2003). Defects in expression of genes related to synaptic vesicle trafficking in frontal cortex of Alzheimer's disease. *Neurobiology of Disease, 12*, 97–109.

Ye, C., Walsh, D. M., Selkoe, D. J., & Hartley, D. M. (2004). Amyloid beta-protein induced electrophysiological changes are dependent on aggregation state: N-methyl-D-aspartate (NMDA) versus non-NMDA receptor/channel activation. *Neuroscience Letters, 366*, 320–325.

Zakharenko, S. S., Zablow, L., & Siegelbaum, S. A. (2002). Altered presynaptic vesicle release and cycling during mGluR-dependent LTD. *Neuron, 35*, 1099–1110.

Erratum to: Gliotransmission and the Tripartite Synapse

Mirko Santello, Corrado Calì, and Paola Bezzi

Erratum to:
Kreutz, Michael R., Sala Carlo (Eds.), *Synaptic Plasticity* , Dynamics, Development, and Disease Series 'Advances in Experimental Medicine and Biology', Vol. 970, DOI 10.1007/978-3-7091-0932-8_14, Springer-Verlag/Wien 2012

The addresses for this chapter should read:

M. Santello, C. Calì, P. Bezzi (⊠)
DBCM - University of Lausanne, Rue du Bugnon 9, 1005, Lausanne, Switzerland
e-mail: paola.bezzi@unil.ch

M. Santello
Department of Physiology, University of Bern, Bühlplatz 5, 3012 Bern, Switzerland
(Current address)

The online version of the original chapter can be found under
DOI 10.1007/978-3-7091-0932-8

M.R. Kreutz and C. Sala (eds.), *Synaptic Plasticity*,
Advances in Experimental Medicine and Biology 970,
DOI 10.1007/978-3-7091-0932-8_26, © Springer-Verlag/Wien 2012

Contributors

Hilmar Bading Department of Neurobiology, Interdisciplinary Centre for Neurosciences (IZN), University of Heidelberg, Heidelberg, Germany, Hilmar. Bading@uni-hd.de

Claudia Bagni Center for Human Genetics, Katholieke Universiteit Leuven, Leuven, Belgium, claudia.bagni@uniroma2.it; claudia.bagni@med.kuleuven.be

Julia Bär PG Neuroplasticity, Leibniz Institute for Neurobiology, Magdeburg, Germany

Robert van den Berg Cell Biology, Utrecht University, Utrecht, The Netherlands

Paola Bezzi DBCM, Department of Physiology, University of Bern, Bern, Switzerland, paola.bezzi@unil.ch

Tobias M. Boeckers Institute of Anatomy and Cell Biology, Ulm University Faculty of Medicine, Ulm, Germany, tobias.boeckers@uni-ulm.de

Miquel Bosch Department of Brain and Cognitive Sciences, The Picower Institute for Learning and Memory, Massachusetts Institute of Technology, Cambridge, MA, USA

Olena Bukalo National Institute of Child Health and Human Development, National Institutes of Health, Bethesda, MD, USA, bukalool@mail.nih.gov

Andrea Buzzi Center for Human Genetics, Katholieke Universiteit Leuven, Leuven, Belgium

Paolo Calabresi Clinica Neurologica, Università degli Studi di Perugia, Ospedale S. Maria della Misericordia, Perugia, Italy, calabre@unipg.it

M.R. Kreutz and C. Sala (eds.), *Synaptic Plasticity*,
Advances in Experimental Medicine and Biology 970,
DOI 10.1007/978-3-7091-0932-8, © Springer-Verlag/Wien 2012

Corrado Calì DBCM, Department of Physiology, University of Bern, Bern, Switzerland

Yong-Jun Chen Department of Neurology, Institute of Molecular Medicine and Genetics, Georgia Health Sciences University, Augusta, GA, USA

Jun-Hyeok Choi National Creative Research Initiative Center for Memory, Department of Biological Sciences, College of Natural Sciences, Seoul National University, Seoul, South Korea

Alexander Dityatev Department of Neuroscience and Brain Technologies, Istituto Italiano di Tecnologia, Genoa, Italy, alexander.dityatev@iit.it

Roberta Epis Department of Pharmacological Sciences and Centre of Excellence on Neurodegenerative Diseases, University of Milan, Milan, Italy

Laurent Fagni Institute of Functional Genomics, CNRS-UMR5203, INSERM-U661, Université Montpellier 1, Université Montpellier 2, Montpellier, France, laurent.fagni@igf.cnrs.fr

Esperanza Fernández Center for Human Genetics, Katholieke Universiteit Leuven, Leuven, Belgium

Renato Frischknecht Leibniz Institute for Neurobiology, Magdeburg, Germany, rfrischk@ifn-magdeburg.de

Kensuke Futai Department of Psychiatry, Brudnick Neuropsychiatric Research Institute, University of Massachusetts Medical School, Worcester, MA, USA

Craig C. Garner Department of Psychiatry and Behavioral Sciences, Nancy Pritzker Laboratory, Stanford University School of Medicine, Palo Alto, CA, USA, cgarner@stanford.edu

Jean-Antoine Girault Institut du Fer à Moulin, UMR-S 839, Inserm and Université Pierre et Marie Curie (UPMC), Paris, France, jean-antoine.girault@inserm.fr

Ingo H. Greger Neurobiology Division, MRC Laboratory of Molecular Biology, Cambridge, UK, ig@mrc-lmb.cam.ac.uk

Eckart D. Gundelfinger Leibniz Institute for Neurobiology, Magdeburg, Germany

Yasunori Hayashi Brain Science Institute, RIKEN, Wako, Saitama, Japan, yhayashi@brain.riken.jp

Martin Heine Research group Molecular Physiology, Leibniz Institute for Neurobiology, Magdeburg, Germany, Martin.Heine@ifn-magdeburg.de

Casper C. Hoogenraad Cell Biology, Utrecht University, Utrecht, The Netherlands, c.hoogenraad@uu.nl

Jacek Jaworski International Institute of Molecular and Cell Biology, Warsaw, Poland, jaworski@iimcb.gov.pl

Bong-Kiun Kaang National Creative Research Initiative Center for Memory, Department of Biological Sciences, College of Natural Sciences, Seoul National University, Seoul, South Korea, kaang@snu.ac.kr

Anna Karpova PG Neuroplasticity, Leibniz Institute for Neurobiology, Magdeburg, Germany

Stefan Kindler University Medical Center Hamburg-Eppendorf, Institute for Human Genetics, Hamburg, Germany, kindler@uke.de

Hans-Jürgen Kreienkamp University Medical Center Hamburg-Eppendorf, Institute for Human Genetics, Hamburg, Germany, kreienkamp@uke.de

Michael R. Kreutz PG Neuroplasticity, Leibniz Institute for Neurobiology, Magdeburg, Germany, kreutz@ifn-magdeburg.de

Monica Di Luca Department of Pharmacological Sciences and Centre of Excellence on Neurodegenerative Diseases, University of Milan, Milan, Italy

Elena Marcello Department of Pharmacological Sciences, University of Milan, Milan, Italy, Elena.marcello@unimi.it

Daniele Di Marino Center for Human Genetics, Katholieke Universiteit Leuven, Leuven, Belgium

Lin Mei Department of Neurology, Institute of Molecular Medicine and Genetics, Georgia Health Sciences University, Augusta, GA, USA, lmei@mcg.edu

Michel C. Van den Oever Department of Molecular and Cellular Neurobiology, Center for Neurogenomics and Cognitive Research, Neuroscience Campus Amsterdam, VU University, Amsterdam, The Netherlands

Shigeo Okabe Department of Cellular Neurobiology, Graduate School of Medicine, The University of Tokyo, Tokyo, Japan, okabe@m.u-tokyo.ac.jp

Ken-ichi Okamoto Samuel Lunenfeld Research Institute, Mount Sinai Hospital, Toronto, ON, Canada

Maria Passafaro Department of Pharmacology, CNR Institute of Neuroscience, University of Milan, Milan, Italy, m.passafaro@cnr.it

Andrew C. Penn CNRS, Interdisciplinary Institute for Neuroscience, Bordeaux, France

C. Peter Bengtson Department of Neurobiology, Interdisciplinary Centre for Neurosciences (IZN), University of Heidelberg, Heidelberg, Germany, Bengtson@nbio.uni-heidelberg.de

Peter Penzes Department of Physiology, Northwestern University Feinberg School of Medicine, Chicago, USA, p-penzes@northwestern.edu

Giovanni Piccoli CNR Institute of Neuroscience, Milan, Italy

Barbara Picconi Laboratorio di Neurofisiologia, Fondazione Santa Lucia, I.R.C. C.S, Rome, Italy

Igor Rafalovich Department of Psychiatry and Behavioral Sciences, Northwestern University Feinberg School of Medicine, Chicago, USA

Silvia De Rubeis Center for Human Genetics, Katholieke Universiteit Leuven, Leuven, Belgium

Carlo Sala Department of Pharmacology, CNR Institute of Neuroscience, University of Milan, Milan, Italy, c.sala@cnr.it

Claudia Saraceno Department of Pharmacological Sciences and Centre of Excellence on Neurodegenerative Diseases, University of Milan, Milan, Italy

Mirko Santello DBCM, Department of Physiology, University of Bern, Bern, Switzerland

Anupama Sathyamurthy Department of Neurology, Institute of Molecular Medicine and Genetics, Georgia Health Sciences University, Augusta, GA, USA

Michael J. Schmeisser Institute of Anatomy and Cell Biology, Ulm University Faculty of Medicine, Ulm, Germany

Stephan J. Sigrist Genetics Institute of Biology, Freie Universität Berlin, Berlin, Germany, stephan.sigrist@fu-berlin.de

August B. Smit Department of Molecular and Cellular Neurobiology, Center for Neurogenomics and Cognitive Research, Neuroscience Campus Amsterdam, VU University, Amsterdam, The Netherlands, guus.smit@cncr.vu.nl

Sabine Spijker Department of Molecular and Cellular Neurobiology, Center for Neurogenomics and Cognitive Research, Neuroscience Campus Amsterdam, VU University, Amsterdam, The Netherlands

Madhav Sukumaran Laboratory of Cellular and Synaptic MRC LMB and Neurophysiology, National Institutes of Health, Eunice Kennedy Shriver National Institute of Child Health and Human Development, Bethesda, MD, USA

Lukasz Swiech International Institute of Molecular and Cell Biology, Warsaw, Poland

Ulrich Thomas Leibniz Insitute for Neurobiology, Magdeburg, Germany, thomas@ifn-magdeburg.de

Malgorzata Urbanska International Institute of Molecular and Cell Biology, Warsaw, Poland

Pamela Valnegri Department of Pharmacology, CNR Institute of Neuroscience, University of Milan, Milan, Italy

Chiara Verpelli Department of Pharmacology, CNR Institute of Neuroscience, University of Milan, Milan, Italy

Daniel Z. Wetmore Department of Psychiatry and Behavioral Sciences, Nancy Pritzker Laboratory, Stanford University School of Medicine, Palo Alto, CA, USA

Wen-Cheng Xiong Department of Neurology, Institute of Molecular Medicine and Genetics, Georgia Health Sciences University, Augusta, GA, USA

Dong-Min Yin Department of Neurology, Institute of Molecular Medicine and Genetics, Georgia Health Sciences University, Augusta, GA, USA

Index

A

A2a receptor, 418
Acetylcholine
  ChAT, 497
  receptors, 582
Actin, 82, 175–185, 188, 271, 339
  dynamics, 133, 139, 147
  filament, 134, 139, 144, 146
  nucleation, 276
Actin cytoskeleton
  actin, 98, 100, 104, 107, 108, 111, 113
  ankyrin, 106
  catenin, 100–104, 108, 109, 115–116
  cofilin, 111
  rearrangement, 435
Action potentials, 316–317
Active zone (AZ), 4–6, 11–14, 16
Activity-regulated genes, 337–338
AD. *See* Alzheimer's disease (AD)
ADAM10, 577, 588–589
Adaptor proteins, 183–186
ADAR1, 250
Addiction, 470
Adhesion single particle tracking (SPT), 200,
      201, 203, 205, 208, 211
Advanced parkinsonism, 562
Aggrecan, 154, 155, 164
ALS. *See* Amyotrophic lateral sclerosis (ALS)
Alternative splicing, 249
Alzheimer's disease (AD), 573–591
AMPA receptors (AMPARs), 32, 44, 88, 133,
      140, 144, 146, 157, 160–162, 164,
      204–209, 211, 437
  trafficking events, 187, 440
AMPARs. *See* AMPA receptors (AMPARs)
AMPA-type, 242
Amygdala, 470, 482–483

basolateral nucleus (BLA), 482
central nucleus (CeA), 482
Amyloid β, 574, 586
  cascade, 574–582, 585
  hypothesis, 574–580
  oligomers, 574, 580–582, 584–585
Amyloid precursor protein (APP), 575–582,
      586–590
Amyotrophic lateral sclerosis (ALS), 185,
      188, 256
Analytical ultracentrifugation, 247
Anisomycin, 228
AP-1, 9–10
ApoE. *See* Apolipoprotein E (ApoE)
Apolipoprotein E (ApoE), 577, 579
APP. *See* Amyloid precursor protein (APP)
Arcadin, 346
Arc/Arg3.1, 338, 341
ARHGEF6, PAK3, and LIMK, 438
Arp2/3, 275, 276
ARPP-16, 413
ARPP-21, 413
ASDs. *See* Autism spectrum disorders (ASDs)
Aspartate, 312
Assembly, 241
Associate, 256
Associative learning, 471
Astrocytes, 308–313, 315, 317, 319
ATP, 312, 313
Autism spectrum disorders (ASDs), 47, 518
Auxiliary factors, 252

B

BACE1, 575–577, 590
Bacterial artificial chromosomes, 411
Barrel cortex, 255

M.R. Kreutz and C. Sala (eds.), *Synaptic Plasticity*,
Advances in Experimental Medicine and Biology 970,
DOI 10.1007/978-3-7091-0932-8, © Springer-Verlag/Wien 2012

Printed by Printforce, the Netherlands